T0358966

Ultrashort Laser Pulse Phenomena

Ultrashort Laser Pulse Phenomena

Fundamentals, Techniques, and Applications on a Femtosecond Time Scale

Second Edition

JEAN-CLAUDE DIELS
Department of Physics and Astronomy
University of New Mexico
Albuquerque, NM

WOLFGANG RUDOLPH
Department of Physics and Astronomy
University of New Mexico
Albuquerque, NM

AMSTERDAM • BOSTON • HEIDELBERG • LONDON
NEW YORK • OXFORD • PARIS • SAN DIEGO
SAN FRANCISCO • SINGAPORE • SYDNEY • TOKYO

Academic Press is an imprint of Elsevier

Academic Press is an imprint of Elsevier
30 Corporate Drive, Suite 400, Burlington, MA 01803, USA
525 B Street, Suite 1900, San Diego, California 92101-4495, USA
84 Theobald's Road, London WC1X 8RR, UK

This book is printed on acid-free paper. 

Library of Congress Cataloging-in-Publication Data
Application submitted

British Library Cataloguing-in-Publication Data
A catalogue record for this book is available from the British Library.

ISBN 13: 978-0-12-215493-5
ISBN 10: 0-12-215493-2

For information on all Academic Press publications
visit our Web site at www.books.elsevier.com

Transferred to Digital Printing 2011

Contents

Preface

Almost 10 years have passed since the first edition of *Ultrashort Laser Pulse Phenomena*. The field of ultrafast optics and spectroscopy has evolved and matured tremendously; tools and techniques available only in research laboratories 10 years ago are now common in many laboratories outside physics and engineering and have been commercialized. During the same period the field has progressed at an astonishing speed, opening new directions, constantly challenging the frontiers of high field science and ultrafast spectroscopy. Our provocative statement from the first edition predicting attosecond pulses at the end of the 1990s materialized. To properly account for the developments of the past decade each chapter of the first edition would need to be expanded into a whole book.

Having said this it is clear that this second edition, like the first edition, cannot be an attempt to review and summarize the latest developments in the field. Periodic updates can be found in the proceedings of the conferences on ultrafast phenomena and on ultrafast optics held alternately every other year. However, as is typical for a mature scientific area, despite the dramatic progress a number of fundamental subjects have emerged. These topics, not much different from the material covered in the first edition, are what students and researchers entering the field need to learn.

In line with the scope of the first edition, the second edition is also intended to bridge the gap between a textbook and a monograph. Written at the level of senior undergraduate students from physics, chemistry, or engineering it represents a mix of tutorial sections and more advanced writings motivating further study of the original literature.

Compared to the first edition, changes have been made in particular in Chapters 1, 2, 3, 5, 9, and 13. The tutorial aspect was emphasized more, and material useful for the researcher was added. The original Chapter 5 on "Ultrashort Sources" has been expanded and split into two chapters, Chapter 5 on "Fundamentals" and Chapter 6 on "Examples." Some newer developments were added to Chapter 9 on "Diagnostic Techniques" and to Chapter 13 on "Selected Applications." Except for some updates and corrections, Chapters 7, 8, 10, 11, and 12 are essentially unchanged.

We would like to express our gratitude to all our colleagues and students who have supported us with numerous suggestions and corrections. In particular, we are indebted to current and former students L. Arissian, J. Biegert, M. Dennis, S. Diddams, P. Dorn, J. Jasapara, J. Jones, M. Kempe, A. Knorr, M. Mero,

J. Nicholson, P. Rambo, A. Schmidt-Sody, A. Velten, J. Zeller, X. Zhao, and to Professors D. Budker, M. Lenzner, G. Reali, M. Sheik-Bahae, B. Wilhelmi, K. Wodkiewicz, A. Zewail, and W. Zinth.

We are grateful also to the contributions of all the students who took courses in the development stage of the first and second editions of this book and proofread individual sections.

Last but not least, we are grateful to our wives, who watched the years go by as our lives became hostage to this endeavor.

Our apologies again to anyone whose work has not been adequately recognized, as we could not possibly cover completely the macrocosm of the temporal microcosm.

Albuquerque, December 2005

Preface to the First Edition

What do we understand about "ultrashort laser pulse phenomena"? It really takes a whole book to define the term. By ultrashort we mean femtosecond (fs), which is a unit of time equal to 10^{-15} s. This time scale becomes accessible because of progress in the generation, amplification, and measurement of ultrashort light pulses. Ultrashort phenomena involve more than just the study of ultrashort lived events. Because of the large energy concentration in a fs optical pulse, this topic encompasses the study of the interaction of intense laser light with matter, as well as the transient response of atoms and molecules and basic properties of the fs radiation itself.

This book is intended as an introduction to ultrashort phenomena to researchers and graduate and senior undergraduate students in optics, physics, chemistry, and engineering. A preliminary version of this book has been used at the University of New Mexico, Jena and Pavia, as a course for graduate and advanced undergraduate students. The femtosecond light gives a different illumination to some classical problems in electromagnetism, optics, quantum mechanics, and electrical engineering. We believe therefore that this book can provide useful illustrations for instructors in these fields.

It is *not* the goal of this book to represent a complete overview of *the latest* progress in the field. We wish to apologize in advance for all the important and pioneering fs work that we failed to cite. For space limitation, we have chosen to present only a few examples of application in the various fields. We are not offering different theoretical aspects of any particular phenomenon, but rather choose to select a description that is consistent throughout the book. Our aim is to cover the basic techniques and applications rather than enter into details of the most fashionable topic of the day. We have attempted to use simple notations and to remain within the MKS system of units.

Consistent with the instructional goal of this book, the first chapter is an extensive review of propagation properties of light in time and frequency domains. Classical optics is reviewed in the next chapter, in light of the particular propagation properties of fs pulses. Some aspects of white light optics—such as coherence and focusing—can be explained in the simplest manner by picturing incoherent radiation as a random sequence of fs pulses. Femtosecond pulses are generally meant to interact with matter. Therefore, a review of this aspect is given in Chapter 3. The latter serves as introduction to the most startling, unexpected, complex properties of transient interaction of coherent fs pulses with resonant physical and chemical systems (Chapter 4). This is a subfield of which the basic foundations are well understood, but it is still open to numerous experimental demonstrations and applications. Chapters 5 through 9 review practical aspects of femtosecond physics, such as sources, amplifiers, pulse shapers, diagnostic techniques, and measurement techniques.

The last three chapters are examples of application of ultrafast techniques. In Chapter 10, the frontier between quantum mechanics and classical mechanics is being probed with fs pulses. New techniques make it possible to "visualize" electrons in Rydberg orbits or the motion of atoms in molecules. The examples of ultrafast processes in matter are presented in this chapter by order of increasing system complexity (from the orbiting electron to the biological complex).

Femtosecond pulses of high peak powers lead to the generation of extremely short wavelength electron and X-ray pulses, as well as to extremely long wavelengths. Some of these techniques are reviewed in Chapter 11. On the long wavelength end of the spectrum, fs pulses are used as Dirac delta function on antennas for submillimeter radiation (frequencies in the THz range). This is a recent application of ultrafast solid-state photoconductive switches.

A few applications that exploit the short duration (range gating imaging), the high coherence, or the high intensity (solitons or filamentation in air) have been selected for the final Chapter 13.

Problems are given at the end of most chapters. Some are typical textbook problems with a straightforward solution. Other problems are designed to put the student in a realistic research situation.

Why Ultrashort Pulse Phenomena?

Yes, you are right! You can be happy without femtosecond pulses and, maybe, consider yourself lucky enough not to be involved with it too deeply. Nevertheless it is a fascinating as well as challenging task to observe and to control processes in nature on a time scale of several femtoseconds. Note, one femtosecond (1 fs) is the $1 : 10^{15}$th part of a second and corresponds to about half a period of red light. The ratio of one fs to one second is about the ratio of 5 minutes to the age of the earth. During one fs, visible light travels over a distance of several hundred nanometers, which is hardly of any concern to us in our daily routine. However, this pathlength corresponds to several thousand elementary cells in a solid which is quite a remarkable number of atomic distances. This suggests the importance the fs time scale might have in the microcosm. Indeed, various essential processes in atoms and molecules, as well as interactions among them, proceed faster than what can be resolved on a picosecond time scale ($1\ \text{ps} = 10^{-12}\text{s}$). Their relevance results simply from the fact that these events are the primary steps for most (macroscopic) reactions in physics, chemistry, and biology.

To illustrate the latter point, let us have a look at the simplest atom—the hydrogen atom—consisting of a positively charged nucleus and a negatively charged electron. Quantum mechanics tells us that an atomic system exists in discrete energy states described by a quantum number n. In the classical picture this corresponds to an electron (wave packet) circulating around the proton on paths with radius $R \propto n^2$. From simple textbook physics, the time T_R necessary for one round-trip can be estimated with $T_R = 4n^2h^3\epsilon_0^2/(e^4 m_e)$, where h is Planck's constant, ϵ_0 is the permittivity of free space, and m_e, e are the electron mass and charge [1, 2]. For $n = 26$, for instance, we obtain a period of about 100 fs. Consequently, an (hydrogen) atom excited to a high Rydberg state is expected to show some macroscopic properties changing periodically on a fs time scale.

Let us next consider atoms bound in a molecule. Apart from translation, the isolated molecule has various internal degrees of freedom for periodical motion—rotation and vibration as well as for conformation changes. Depending on the binding forces, potentials, and masses of the constituents, the corresponding periods may range from the ps to the fs scale. Another example of ultrafast dynamics in the molecular world is the chemical reaction, for instance, the simple dissociation $(AB)^* \rightarrow A + B$. Here the breaking of the bond is accompanied by a geometrical separation of the two components caused by a repulsive potential. Typical recoil velocities are of the order of 1 km/s, which implies that the transition from the bound state to the isolated complexes proceeds within 100 fs. Similar time intervals, of course, can be expected if separated particles undergo a chemical reaction.

Additional processes come into play if the particle we look at is not isolated but under the influence of surrounding atoms or molecules, which happens in a gas (mixture) or solution. Strong effects are expected as a result of collisions. Moreover, even a simple translation or rotation that alters the relative position of the molecule to the neighboring particles may lead to a variation of the molecular properties because of a changed local field. The characteristic time constants depend on the particle density and the translation–rotation velocity, which in turn is determined by the temperature and strength of interaction with the neighbors. The characteristic times can be comparatively long in diluted gases (ns to μs) and can be short in solutions at room temperature (fs).

Finally, let us have a look at a solid where the atomic particles are usually trapped at a relatively well-defined position in the lattice. Their motion is restricted usually to lattice vibrations (phonons) with possible periods in the order of 100 fs, which corresponds to phonon energies of several tens of millielectronvolts (for instance, the longitudinal optical or LO phonon in GaAs has an energy of about 35 meV).

The fundamental problem to be solved is to find tools and techniques that allow us to observe and manipulate on a fs time scale. At present, speaking about tools and techniques for fs physics means dealing with laser physics, in particular with ultrashort light pulses produced in lasers. Shortly after the invention of the laser in 1960, methods were developed to use them for the generation of light pulses. In the sixties, the microsecond (μs) and nanosecond (ns) range were extensively studied. In the seventies, progress in laser physics opened up the ps range, and the eighties were characterized by the broad introduction of fs techniques (extrapolating this dramatic development we may expect the attosecond physics in the late nineties). Optical methods have taken precedence over electronics in time resolving fast events ever since light pulses shorter than a few ps have become available. It should also be mentioned that the shortest electrical and X-ray pulses are now being produced by means of fs light pulses, which in turn enlarges the application field of ultrafast techniques.

Femtosecond technology opens up new fascinating possibilities based on the unique properties of femtosecond light pulses:

- Energy can be concentrated in a temporal interval as short as several 10^{-15} s, which corresponds to only a few optical cycles in the visible range.
- The pulse peak power can be extremely large even at moderate pulse energies. For instance, a 50-fs pulse with an energy of 1 mJ ($\approx 3 * 10^{15}$ "red" photons) exhibits an average power of 20 Gigawatt. Focusing this pulse to a 100-μm^2 spot yields an intensity of 20 Petawatt/cm^2 (20 10^{15} W/cm^2!), which means an electric field strength of about 3 GV/cm. This value is larger than a typical inner-atomic field of 1 GV/cm.
- The geometrical length of a fs pulse amounts only to several micrometers (10 fs corresponds to 3 μm in vacuum). Such a coherence length is usually associated with incoherent light. The essential difference is that incoherent light is generally spread over a much longer distance.

The attractiveness of fs light pulses not only lies in the possibility to trace processes in their ultrafast dynamics, but also in the fact that one simply can do things faster. Of course only a few, but essential, parts in modern technology can be accelerated by using ultrashort (fs) light pulses. Of primary importance are data transfer and data processing utilizing the high carrier frequency of light and the subsequent large possible bandwidths. In this respect one of the most spectacular goals is to create an optical computer. Moreover, techniques are being developed that allow distortionless propagation of ultrashort light pulses over long distances (several thousand kilometers) through optical fibers, a precondition for a future Terahertz information transfer.

A variety of nonlinear processes, reversible as well as irreversible ones, become accessible thanks to the large intensities of fs pulses. There are proposals to use such pulses for laser fusion. To reach TW intensities, tabletop devices are replacing the building size high energy facilities previously required. First attempts to generate short X-ray pulses by using fs pulse–induced plasmas have already proven successful.

The short geometrical lengths of fs light pulses suggest interesting applications for optical ranging with micrometer resolution, as well as for combinations of micrometer spatial resolution with femtosecond temporal resolution.

The ultrashort phenomena to which this book refers are created by *light pulses*, which are wave packets of electromagnetic waves oscillating at optical frequencies. The emphasis of this book is not on the optical frequency range but on physical phenomena associated with *ultrafast* electromagnetic pulses. The latter will be ephemeral when consisting of only a small number of optical periods and spatially confined when made up of a small number of wavelengths. Another criterion for short is that the length of the pulse be small compared with the distance over which it propagates, particularly when large changes of shape and modulation take place. In the particular area of light–matter interaction, a pulse is generally considered as a δ function excitation when its duration is small compared to that of all atomic or molecular relaxations.

BIBLIOGRAPHY

[1] J. Orear. *Physik.* Carl Hansen Verlag, Munchen, Vienna, Austria, 1982.

[2] R. A. Serway, C. J. Moses, and C. A. Moyer. *Modern Physics.* Saunders College Publishers, Philadelphia, PA, 1989.

1

Fundamentals

1.1. CHARACTERISTICS OF FEMTOSECOND LIGHT PULSES

Femtosecond (fs) light pulses are electromagnetic wave packets and as such are fully described by the time and space dependent electric field. In the frame of a semiclassical treatment the propagation of such fields and the interaction with matter are governed by Maxwell's equations with the material response given by a macroscopic polarization. In this first chapter we will summarize the essential notations and definitions used throughout the book. The pulse is characterized by measurable quantities that can be directly related to the electric field. A complex representation of the field amplitude is particularly convenient in dealing with propagation problems of electromagnetic pulses. The next section expands on the choice of field representation.

1.1.1. Complex Representation of the Electric Field

Let us consider first the temporal dependence of the electric field neglecting its spatial and polarization dependence, i.e., $\mathbf{E}(x, y, z, t) = E(t)$. A complete description can be given either in the time or the frequency domain. Even though the measured quantities are real, it is generally more convenient to use complex representation. For this reason, starting with the real $E(t)$, one defines the complex spectrum of the field strength $\tilde{E}(\Omega)$, through the complex Fourier transform ($\mathcal{F}$):

$$\tilde{E}(\Omega) = \mathcal{F}\{E(t)\} = \int_{-\infty}^{\infty} E(t)e^{-i\Omega t}dt = |\tilde{E}(\Omega)|e^{i\Phi(\Omega)} \tag{1.1}$$

In the definition (1.1), $|\tilde{E}(\Omega)|$ denotes the spectral amplitude, and $\Phi(\Omega)$ is the spectral phase. Here and in what follows, complex quantities related to the field are typically written with a tilde.

Because $E(t)$ is a real function, $\tilde{E}(\Omega) = \tilde{E}^*(-\Omega)$ holds. Given $\tilde{E}(\Omega)$, the time dependent electric field is obtained through the inverse Fourier transform ($\mathcal{F}^{-1}$):

$$E(t) = \mathcal{F}^{-1}\left\{\tilde{E}(\Omega)\right\} = \frac{1}{2\pi}\int_{-\infty}^{\infty} \tilde{E}(\Omega)e^{i\Omega t}d\Omega. \tag{1.2}$$

For practical reasons it may not be convenient to use functions that are nonzero for negative frequencies, as needed in the evaluation of Eq. (1.2). Frequently a complex representation of the electric field in the time domain is also desired. Both aspects can be satisfied by introducing a complex electric field as

$$\tilde{E}^+(t) = \frac{1}{2\pi}\int_{0}^{\infty} \tilde{E}(\Omega)e^{i\Omega t}d\Omega \tag{1.3}$$

and a corresponding spectral field strength that contains only positive frequencies:

$$\tilde{E}^+(\Omega) = |\tilde{E}(\Omega)|e^{i\Phi(\Omega)} = \begin{cases} \tilde{E}(\Omega) & \text{for} \quad \Omega \geq 0 \\ 0 & \text{for} \quad \Omega < 0. \end{cases} \tag{1.4}$$

$\tilde{E}^+(t)$ and $\tilde{E}^+(\Omega)$ are related to each other through the complex Fourier transform defined in Eq. (1.1) and Eq. (1.2), i.e.,

$$\tilde{E}^+(t) = \frac{1}{2\pi}\int_{-\infty}^{\infty} \tilde{E}^+(\Omega)e^{i\Omega t}d\Omega \tag{1.5}$$

and

$$\tilde{E}^+(\Omega) = \int_{-\infty}^{\infty} \tilde{E}^+(t)e^{-i\Omega t}dt. \tag{1.6}$$

The real physical electric field $E(t)$ and its complex Fourier transform can be expressed in terms of the quantities derived in Eq. (1.5) and Eq. (1.6) and the corresponding quantities $\tilde{E}^{-}(t)$, $\tilde{E}^{-}(\Omega)$ for the negative frequencies. These quantities relate to the real electric field:

$$E(t) = \tilde{E}^{+}(t) + \tilde{E}^{-}(t) \tag{1.7}$$

and its complex Fourier transform:

$$\tilde{E}(\Omega) = \tilde{E}^{+}(\Omega) + \tilde{E}^{-}(\Omega). \tag{1.8}$$

It can be shown that $\tilde{E}^{+}(t)$ can also be calculated through analytic continuation of $E(t)$

$$\tilde{E}^{+}(t) = E(t) + iE'(t) \tag{1.9}$$

where $E'(t)$ and $E(t)$ are Hilbert transforms of each other. In this sense $\tilde{E}^{+}(t)$ can be considered as the complex analytical correspondent of the real function $E(t)$.

The complex electric field $\tilde{E}^{+}(t)$ is usually represented by a product of an amplitude function and a phase term:

$$\tilde{E}^{+}(t) = \frac{1}{2}\mathcal{E}(t)e^{i\Gamma(t)} \tag{1.10}$$

In most practical cases of interest here the spectral amplitude will be centered around a mean frequency ω_ℓ and will have appreciable values only in a frequency interval $\Delta\omega$ small compared to ω_ℓ. In the time domain this suggests the convenience of introducing a carrier frequency ω_ℓ and of writing $\tilde{E}^{+}(t)$ as:

$$\boxed{\tilde{E}^{+}(t) = \frac{1}{2}\mathcal{E}(t)e^{i\varphi_0}e^{i\varphi(t)}e^{i\omega_\ell t} = \frac{1}{2}\tilde{\mathcal{E}}(t)e^{i\omega_\ell t}} \tag{1.11}$$

where $\varphi(t)$ is the time-dependent phase, $\tilde{\mathcal{E}}(t)$ is called the complex field envelope and $\mathcal{E}(t)$ the real field envelope, respectively. The constant phase term $e^{i\varphi_0}$ is most often of no relevance, and can be neglected. There are however particular circumstances pertaining to short pulses where the outcome of the pulse interaction with matter depends on φ_0, often referred to as "carrier to envelope phase." The measurement and control of φ_0 can therefore be quite important. Figure 1.1 shows the electric field of two pulses with identical $\mathcal{E}(t)$ but different φ_0. We will discuss the carrier to envelope phase in more detail in Chapters 5 and 13.

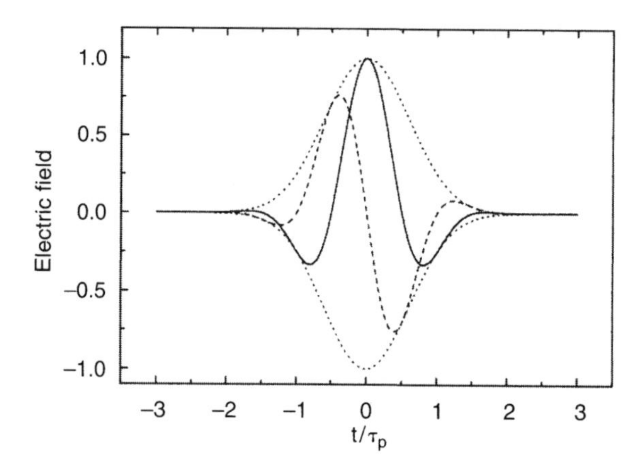

Figure 1.1 Electric field of two extremely short pulses, $E(t) = \exp[-2\ln 2(t/\tau_p)^2]\cos(\omega_\ell t + \varphi_0)$ with $\varphi_0 = 0$ (solid line) and $\varphi_0 = \pi/2$ (dashed line). Both pulses have the same envelope (dotted line). The full width of half maximum of the intensity envelope, τ_p, was chosen as $\tau_p = \pi/\omega_\ell$.

Because the laser pulse represents a propagating electromagnetic wave packet, the dc component of its spectrum vanishes. Hence the time integral over the electric field is zero.

$$\int_{-\infty}^{\infty} E(t)dt = \int_{-\infty}^{\infty} E(t)e^{-i(\Omega=0)t}dt = \mathcal{F}\{E(t)\}_{\Omega=0} = 0. \tag{1.12}$$

The description of the field given by Eqs. (1.9) through (1.11) is quite general. However, the usefulness of the concept of an envelope and carrier frequency as defined in Eq. (1.11) is limited to the cases where the bandwidth is only a small fraction of the carrier frequency:

$$\frac{\Delta\omega}{\omega_\ell} \ll 1 \tag{1.13}$$

For inequality (1.13) to be satisfied, the temporal variation of $\mathcal{E}(t)$ and $\varphi(t)$ within an optical cycle $T = 2\pi/\omega_\ell$ ($T \approx 2$ fs for visible radiation) has to be small. The corresponding requirement for the complex envelope $\tilde{\mathcal{E}}(t)$ is

$$\left|\frac{d}{dt}\tilde{\mathcal{E}}(t)\right| \ll \omega_\ell \left|\tilde{\mathcal{E}}(t)\right|. \tag{1.14}$$

Keeping in mind that today the shortest light pulses contain only a few optical cycles, one has to carefully check whether a slowly varying envelope and phase can describe the pulse behavior satisfactorily. If they do, the theoretical description of pulse propagation and interaction with matter can be greatly simplified by applying the slowly varying envelope approximation (SVEA), as will be evident later in this chapter. Note that even for pulses consisting of a few optical cycles the electric field can formally be represented in a form similar to Eq. (1.11), as illustrated by Fig. 1.1. Equation (1.11) represents only a mathematical identity and does not imply that the SVEA is appropriate.

Given the spectral description of a signal, $\tilde{E}^+(\Omega)$, the complex envelope $\tilde{\mathcal{E}}(t)$ is simply the inverse transform of the translated spectral field:

$$\tilde{\mathcal{E}}(t) = \mathcal{E}(t)e^{i\varphi(t)} = \frac{1}{2\pi}\int_{-\infty}^{\infty} 2\tilde{E}^+(\Omega + \omega_\ell)e^{i\Omega t}d\Omega, \tag{1.15}$$

where the modulus $\mathcal{E}(t)$ in Eq. (1.15) represents the real envelope. The optimum "translation" in the spectral domain ω_ℓ is the one that gives the envelope $\tilde{\mathcal{E}}(t)$ with the least amount of modulation. Spectral translation of Fourier transforms is a standard technique to reconstruct the envelope of interference patterns, and is used in Chapter 9 on diagnostic techniques. The Fourier transform of the complex envelope $\tilde{\mathcal{E}}(t)$ is the spectral envelope function:

$$\tilde{\mathcal{E}}(\Omega) = \int_{-\infty}^{\infty} \tilde{\mathcal{E}}(t)e^{-i\Omega t}dt = 2\int_{-\infty}^{\infty} \tilde{E}^+(t)e^{-i(\Omega+\omega_\ell)t}dt. \tag{1.16}$$

The choice of ω_ℓ is such that the spectral amplitude $\tilde{\mathcal{E}}(\Omega)$ is centered at the origin $\Omega = 0$.

Let us now discuss more carefully the physical meaning of the phase function $\varphi(t)$. The choice of carrier frequency in Eq. (1.11) should be such as to minimize the variation of phase $\varphi(t)$. The first derivative of the phase factor $\Gamma(t)$ in Eq. (1.10) establishes a time-dependent carrier frequency (instantaneous frequency):

$$\boxed{\omega(t) = \omega_\ell + \frac{d}{dt}\varphi(t).} \tag{1.17}$$

Although Eq. (1.17) can be seen as a straightforward definition of an instantaneous frequency based on the temporal variation of the phase factor $\Gamma(t)$, we will see in Section 1.1.4 that it can be rigorously derived from the Wigner distribution. For $d\varphi/dt = b =$ const., a nonzero value of b just means a correction of the carrier frequency which is now $\omega'_\ell = \omega_\ell + b$. For $d\varphi/dt = f(t)$, the carrier frequency varies with time, and the corresponding pulse is said to be frequency modulated or chirped. For $d^2\varphi/dt^2 < (>) 0$, the carrier frequency decreases (increases) along the pulse, which then is called down (up) chirped.

From Eq. (1.10) it is obvious that the decomposition of $\Gamma(t)$ into ω and $\varphi(t)$ is not unique. The most useful decomposition is one that ensures the smallest $d\varphi/dt$ during the intense portion of the pulse. A common practice is to identify ω_ℓ with the carrier frequency at the pulse peak. A better definition—which is consistent in the time and frequency domains—is to use the intensity weighted *average* frequency:

$$\boxed{\langle\omega\rangle = \frac{\int_{-\infty}^{\infty}|\tilde{\mathcal{E}}(t)|^2\omega(t)dt}{\int_{-\infty}^{\infty}|\tilde{\mathcal{E}}(t)|^2dt} = \frac{\int_{-\infty}^{\infty}|\tilde{E}^+(\Omega)|^2\Omega d\Omega}{\int_{-\infty}^{\infty}|\tilde{E}^+(\Omega)|^2d\Omega}} \tag{1.18}$$

The various notations are illustrated in Figure 1.2 where a linearly up-chirped pulse is taken as an example. The temporal dependence of the real electric field is sketched in the top part of Fig 1.2. A complex representation in the time domain is illustrated with the amplitude and instantaneous frequency of the field. The positive and negative frequency components of the Fourier transform are shown in amplitude and phase in the bottom part of the figure.

1.1.2. Power, Energy, and Related Quantities

Let us imagine the practical situation in which the pulse propagates as a beam with cross section A, and with $E(t)$ as the relevant component of the electric field. The (instantaneous) pulse power (in Watt) in a dispersionless material of refractive index n can be derived from the Poynting theorem of electrodynamics [1] and is given by

$$\mathcal{P}(t) = \epsilon_0 cn \int_A dS \frac{1}{T}\int_{t-T/2}^{t+T/2} E^2(t')dt' \tag{1.19}$$

where c is the velocity of light in vacuum, ϵ_0 is the dielectric permittivity and $\int_A dS$ stands for integration over the beam cross section. The power can be measured by a detector (photodiode, photomultiplier, etc.) which integrates over the beam cross section. The temporal response of this device must be short as compared to the speed of variations of the field envelope to be measured. The temporal averaging is performed over one optical period $T = 2\pi/\omega_\ell$. Note that the instantaneous power as introduced in Eq. (1.19) is then just a convenient theoretical quantity. In a practical measurement T has to be replaced by the actual response time τ_R of the detector. Therefore, even with the fastest detectors available today ($\tau_R \approx 10^{-13} - 10^{-12}$ s), details of the envelope of fs light pulses cannot be resolved directly.

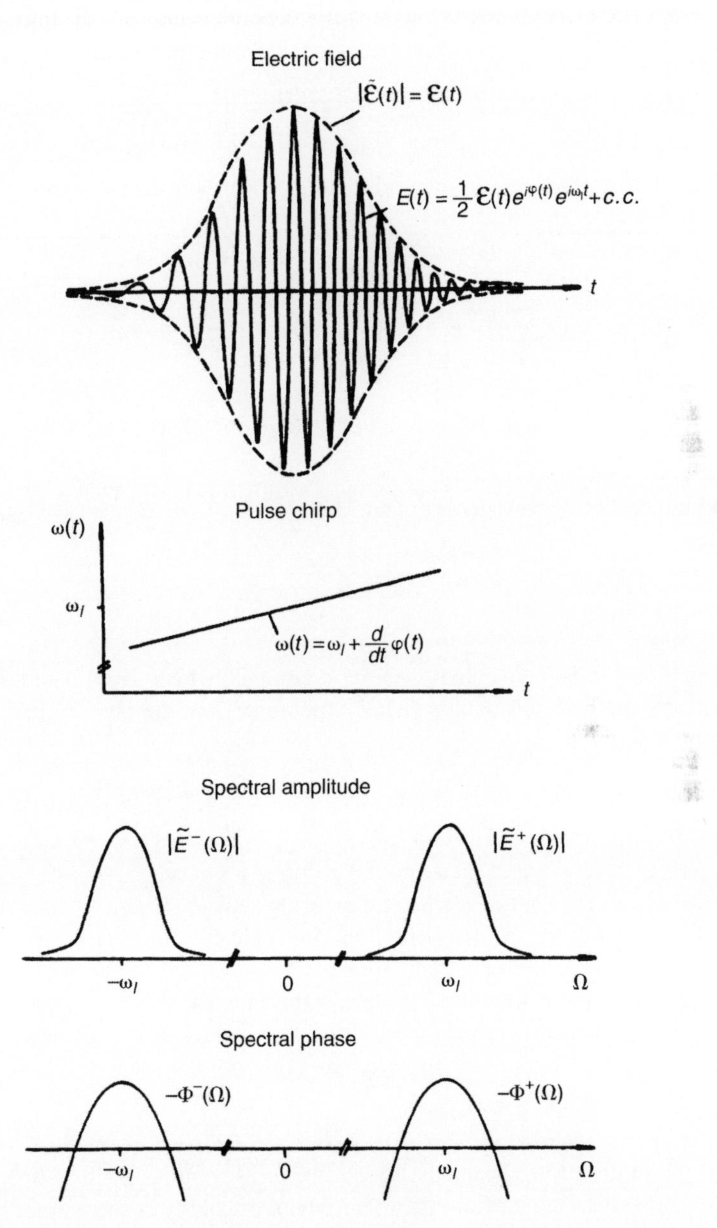

Figure 1.2 Electric field, time-dependent carrier frequency, and spectral amplitude of an upchirped pulse.

A temporal integration of the power yields the energy $\mathcal{W}$ (in Joules):

$$\mathcal{W} = \int_{-\infty}^{\infty} \mathcal{P}(t')dt' \tag{1.20}$$

where the upper and lower integration limits essentially mean "before" and "after" the pulse under investigation.

The corresponding quantity per unit area is the intensity (W/cm^2):

$$\begin{aligned} I(t) &= \epsilon_0 cn \frac{1}{T} \int_{t-T/2}^{t+T/2} E^2(t')dt' \\ &= \frac{1}{2}\epsilon_0 cn\mathcal{E}^2(t) = 2\epsilon_0 cn\tilde{E}^+(t)\tilde{E}^-(t) = \frac{1}{2}\epsilon_0 cn\tilde{\mathcal{E}}(t)\tilde{\mathcal{E}}^*(t) \end{aligned} \tag{1.21}$$

and the energy density per unit area (J/cm^2):

$$W = \int_{-\infty}^{\infty} I(t')dt'. \tag{1.22}$$

Sometimes it is convenient to use quantities which are related to photon numbers, such as the photon flux $\mathcal{F}$ (photons/s) or the photon flux density F (photons/s/cm^2):

$$\mathcal{F}(t) = \frac{\mathcal{P}(t)}{\hbar\omega_\ell} \quad \text{and} \quad F(t) = \frac{I(t)}{\hbar\omega_\ell} \tag{1.23}$$

where $\hbar\omega_\ell$ is the energy of one photon at the carrier frequency.

The spectral properties of the light are typically obtained by measuring the intensity of the field, without any time resolution, at the output of a spectrometer. The quantity that is measured is the spectral intensity:

$$S(\Omega) = |\eta(\Omega)\tilde{E}^+(\Omega)|^2 \tag{1.24}$$

where η is a scaling factor which accounts for losses, geometrical influences, and the finite resolution of the spectrometer. Assuming an ideal spectrometer, $|\eta|^2$ can be determined from the requirement of energy conservation:

$$|\eta|^2 \int_{-\infty}^{\infty} |\tilde{E}^+(\Omega)|^2 d\Omega = 2\epsilon_0 cn \int_{-\infty}^{\infty} \tilde{E}^+(t)\tilde{E}^-(t)dt \tag{1.25}$$

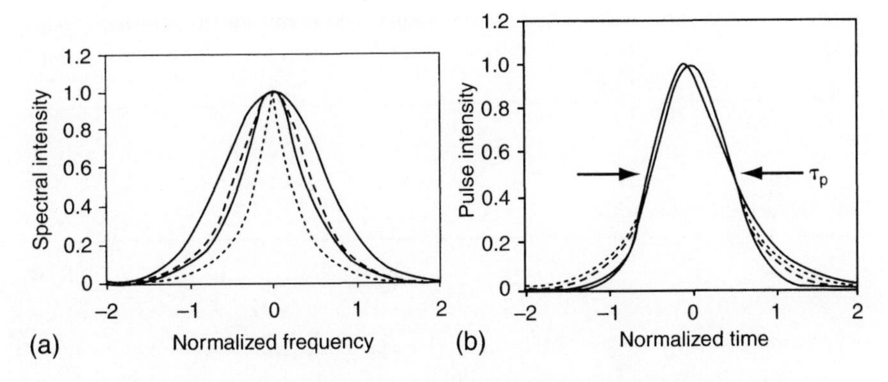

Figure 1.3 Temporal pulse profiles and the corresponding spectra (normalized).

——— Gaussian pulse	$\mathcal{E}(t) \propto \exp[-1.385(t/\tau_p)^2]$
- - - - - - - Sech pulse	$\mathcal{E}(t) \propto \mathrm{sech}[1.763(t/\tau_p)]$
········· Lorentzian pulse	$\mathcal{E}(t) \propto [1 + 1.656(t/\tau_p)^2]^{-1}$
——— Asymm. sech pulse	$\mathcal{E}(t) \propto [\exp(t/\tau_p) + \exp(-3t/\tau_p)]^{-1}$

and Parseval's theorem [2]:

$$\int_{-\infty}^{\infty} |\tilde{E}^+(t)|^2 dt = \frac{1}{2\pi} \int_0^{\infty} |\tilde{E}^+(\Omega)|^2 d\Omega \tag{1.26}$$

from which follows $|\eta|^2 = \epsilon_0 cn/\pi$. The complete expression for the spectral intensity [from Eq. (1.24)] is thus:

$$S(\Omega) = \frac{\epsilon_0 cn}{4\pi} \left|\tilde{\mathcal{E}}(\Omega - \omega_\ell)\right|^2. \tag{1.27}$$

Figure 1.3 gives examples of typical pulse shapes and the corresponding spectra.

The complex quantity $\tilde{E}^+$ will be used most often throughout the book to describe the electric field. Therefore, to simplify notations, we will omit the superscript + whenever this will not cause confusion.

1.1.3. Pulse Duration and Spectral Width

Unless specified otherwise, we define the pulse duration τ_p as the full width at half maximum (FWHM) of the intensity profile, $|\tilde{\mathcal{E}}(t)|^2$, and the spectral width $\Delta\omega_p$ as the FWHM of the spectral intensity $|\tilde{\mathcal{E}}(\Omega)|^2$. Making that statement is an obvious admission that other definitions exist. Precisely because of the difficulty of asserting the exact pulse shape, standard waveforms have been selected. The most commonly cited are the Gaussian, for which the temporal dependence of the field is:

$$\tilde{\mathcal{E}}(t) = \tilde{\mathcal{E}}_0 \exp\{-(t/\tau_G)^2\} \tag{1.28}$$

and the secant hyperbolic:

$$\tilde{\mathcal{E}}(t) = \tilde{\mathcal{E}}_0 \mathrm{sech}(t/\tau_s). \tag{1.29}$$

The parameters $\tau_G = \tau_p/\sqrt{2\ln 2}$ and $\tau_s = \tau_p/1.76$ are generally more convenient to use in theoretical calculations involving pulses with these assumed shapes than the FWHM of the intensity, τ_p.

Because the temporal and spectral characteristics of the field are related to each other through Fourier transforms, the bandwidth $\Delta\omega_p$ and pulse duration τ_p cannot vary independently of each other. There is a minimum duration–bandwidth product:

$$\Delta\omega_p \tau_p = 2\pi\Delta\nu_p\tau_p \geq 2\pi c_B. \tag{1.30}$$

c_B is a numerical constant on the order of 1, depending on the actual pulse shape. Some examples are shown in Table 1.1. The equality holds for pulses without frequency modulation (unchirped) which are called "bandwidth limited" or "Fourier limited." Such pulses exhibit the shortest possible duration at a given

Table 1.1

Examples of standard pulse profiles. The spectral values given are for unmodulated pulses. Note that the Gaussian is the shape with the minimum product of mean square deviation of the intensity and spectral intensity.

Shape	Intensity profile $I(t)$	τ_p FWHM	Spectral profile $S(\Omega)$	$\Delta\omega_p$ FWHM	c_B	$\langle\tau_p\rangle\langle\Delta\Omega_p\rangle$ MSQ
Gauss	$e^{-2(t/\tau_G)^2}$	$1.177\tau_G$	$e^{-\left(\frac{\Omega\tau_G}{2}\right)^2}$	$2.355/\tau_G$	0.441	0.5
Sech	$\mathrm{sech}^2(t/\tau_s)$	$1.763\tau_s$	$\mathrm{sech}^2\frac{\pi\Omega\tau_s}{2}$	$1.122/\tau_s$	0.315	0.525
Lorentz	$[1+(t/\tau_L)^2]^{-2}$	$1.287\tau_L$	$e^{-2\lvert\Omega\rvert\tau_L}$	$0.693/\tau_L$	0.142	0.7
Asym. sech	$\left[e^{t/\tau_a}+e^{-3t/\tau_a}\right]^{-2}$	$1.043\tau_a$	$\mathrm{sech}\frac{\pi\Omega\tau_a}{2}$	$1.677/\tau_a$	0.278	
Square	1 for $\lvert t/\tau_r\rvert \leq 1$, 0 elsewhere	τ_r	$\mathrm{sinc}^2(\Omega\tau_r)$	$2.78/\tau_r$	0.443	3.27

spectral width and pulse shape. We refer the reader to Section 1.1.4, for a more general discussion of the uncertainty relation between pulse and spectral width based on mean square deviations (MSQ).

The shorter the pulse duration, the more difficult it becomes to assert its detailed characteristics. In the femtosecond domain, even the simple concept of pulse duration seems to fade away in a cloud of mushrooming definitions. Part of the problem is that it is difficult to determine the exact pulse shape. For single pulses, the typical representative function that is readily accessible to the experimentalist is the intensity autocorrelation:

$$A_{\text{int}}(\tau) = \int_{-\infty}^{\infty} I(t)I(t-\tau)dt \tag{1.31}$$

The Fourier transform of the correlation (1.31) is the real function:

$$A_{\text{int}}(\Omega) = \tilde{\mathcal{I}}(\Omega)\tilde{\mathcal{I}}^*(\Omega) \tag{1.32}$$

where the notation $\tilde{\mathcal{I}}(\Omega)$ is the Fourier transform of the function $I(t)$ and should not be confused with the spectral intensity $S(\Omega)$. The fact that the autocorrelation function $A_{\text{int}}(\tau)$ is symmetric, hence its Fourier transform is real, [2] implies that little information about the pulse shape can be extracted from such a measurement. Furthermore, the intensity autocorrelation (1.31) contains no information about the pulse phase or coherence. This point is discussed in detail in Chapter 9.

Gaussian Pulses

Having introduced essential pulse characteristics, it seems convenient to discuss an example to which we can refer to in later chapters. We choose a Gaussian pulse with linear chirp. This choice is one of analytical convenience: the Gaussian shape is *not* the most commonly encountered temporal shape. The electric field is given by

$$\tilde{\mathcal{E}}(t) = \mathcal{E}_0 e^{-(1+ia)(t/\tau_G)^2} \tag{1.33}$$

with the pulse duration

$$\tau_p = \sqrt{2\ln 2}\,\tau_G. \tag{1.34}$$

Note that with the definition (1.33) the chirp parameter a is positive for a downchirp ($d\varphi/dt = -2at/\tau_G^2$). The Fourier transform of (1.33) yields

$$\tilde{\mathcal{E}}(\Omega) = \frac{\mathcal{E}_0\sqrt{\pi}\tau_G}{\sqrt[4]{1+a^2}} \exp\left\{i\Phi - \frac{\Omega^2\tau_G{}^2}{4(1+a^2)}\right\} \tag{1.35}$$

with the spectral phase given by:

$$\phi(\Omega) = -\frac{1}{2}\arctan(a) + \frac{a\tau_G{}^2}{4(1+a^2)}\Omega^2. \tag{1.36}$$

It can be seen from Eq. (1.35) that the spectral intensity is the Gaussian:

$$S(\omega_\ell + \Omega) = \frac{|\eta|^2\pi\mathcal{E}_0^2\tau_G^2}{\sqrt{1+a^2}}\exp\left\{-\frac{\Omega^2\tau_G{}^2}{2(1+a^2)}\right\} \tag{1.37}$$

with a FWHM given by:

$$\Delta\omega_p = 2\pi\Delta\nu_p = \frac{1}{\tau_G}\sqrt{8\ln 2(1+a^2)} \tag{1.38}$$

For the pulse duration–bandwidth product we find

$$\Delta\nu_p\tau_p = \frac{2\ln 2}{\pi}\sqrt{1+a^2} \tag{1.39}$$

Obviously, the occurrence of chirp ($a \neq 0$) results in additional spectral components which enlarge the spectral width and lead to a duration–bandwidth product exceeding the Fourier limit ($2\ln 2/\pi \approx 0.44$) by a factor $\sqrt{1+a^2}$, consistent with Eq. (1.30). We also want to point out that the spectral phase given by Eq. (1.36) changes quadratically with frequency if the input pulse is linearly chirped. Although this is exactly true for Gaussian pulses as can be seen from Eq. (1.36), it holds approximately for other pulse shapes. In the next section, we will develop a concept that allows one to discuss the pulse duration–bandwidth product from a more general point of view, which is independent of the actual pulse and spectral profile.

1.1.4. Wigner Distribution, Second-Order Moments, Uncertainty Relations

Wigner Distribution

The Fourier transform as defined in Section 1.1.1 is a widely used tool in beam and pulse propagation. In beam propagation, it leads directly to the far field pattern of a propagating beam (Fraunhofer approximation) of arbitrary transverse profile. Similarly, the Fourier transform leads directly to the pulse temporal profile, following propagation through a dispersive medium, as we will see at the end of this chapter. The Fourier transform gives a weighted average of the spectral components contained in a signal. Unfortunately, the exact spatial or temporal location of these spectral components is hidden in the phase of the spectral field. There has been therefore a need for new two-dimensional representation of the waves in either the plane of space–wave vector, or time–angular frequency. Such a function was introduced by Wigner [3] and applied to quantum mechanics. The same distribution was applied to the area of signal processing by Ville [4]. Properties and applications of the Wigner distribution in quantum mechanics and optics are reviewed in two recent books by Schleich [5] and Cohen [6]. A clear analysis of the close relationship between quantum mechanics and optics can be found in Praxmeir and Wókiewicz [7]. The Wigner distribution of a function $\tilde{E}(t)$ is defined by[1]:

$$\begin{aligned}\mathcal{W}_E(t,\Omega) &= \int_{-\infty}^{\infty} \tilde{E}\left(t+\frac{s}{2}\right)\tilde{E}^*\left(t-\frac{s}{2}\right)e^{-i\Omega s}ds \\ &= \frac{1}{2\pi}\int_{-\infty}^{\infty} \tilde{E}\left(\Omega+\frac{s}{2}\right)\tilde{E}^*\left(\Omega-\frac{s}{2}\right)e^{its}ds. \end{aligned} \tag{1.40}$$

One can see that the definition is a local representation of the spectrum of the signal, because:

$$\int_{-\infty}^{\infty} \mathcal{W}_E(t,\Omega)dt = \left|\tilde{E}(\Omega)\right|^2 \tag{1.41}$$

and

$$\int_{-\infty}^{\infty} \mathcal{W}_E(t,\Omega)d\Omega = 2\pi\left|\tilde{E}(t)\right|^2. \tag{1.42}$$

The subscript E refers to the use of the instantaneous complex electric field $\tilde{E}$ in the definition of the Wigner function, rather than the electric field envelope $\tilde{\mathcal{E}} = \mathcal{E}\exp[i\omega_\ell t + i\varphi(t)]$ defined at the beginning of this chapter. There is a simple relation between the Wigner distribution $\mathcal{W}_E$ of the instantaneous field $\tilde{E}$,

[1] t and Ω are conjugated variables as in Fourier transforms. The same definitions can be made in the space–wave vector domain, where the variables are then x and k.

and the Wigner distribution $\mathcal{W}_{\mathcal{E}}$ of the real envelope amplitude $\mathcal{E}$:

$$\begin{aligned}
\mathcal{W}_E(t,\Omega) &= \int_{-\infty}^{\infty} \mathcal{E}\left(t+\frac{s}{2}\right) e^{i[\omega_\ell(t+s/2)+\varphi(t+s/2)]} \\
&\quad \times\ \mathcal{E}^*\left(t-\frac{s}{2}\right) e^{-i[\omega_\ell(t-s/2)+\varphi(t-s/2)]} e^{-i\Omega s} ds \\
&= \int_{-\infty}^{\infty} \mathcal{E}\left(t+\frac{s}{2}\right) \mathcal{E}^*\left(t-\frac{s}{2}\right) e^{-i[\Omega-(\omega_\ell+\dot{\varphi}(t))]s} ds \\
&= \mathcal{W}_{\mathcal{E}}\{t,[\Omega-(\omega_\ell+\dot{\varphi})]\}. \qquad (1.43)
\end{aligned}$$

We will drop the subscript E and $\mathcal{E}$ for the Wigner function when the distinction is not essential.

The intensity and spectral intensities are directly proportional to frequency and time integrations of the Wigner function. In accordance with Eqs. (1.21) and (1.27):

$$\frac{1}{2\sqrt{\mu_0/\epsilon}} \int_{-\infty}^{\infty} \mathcal{W}_{\mathcal{E}}(t,\Omega) d\Omega = I(t) \qquad (1.44)$$

$$\frac{1}{2\sqrt{\mu_0/\epsilon}} \int_{-\infty}^{\infty} \mathcal{W}_{\mathcal{E}}(t,\Omega) dt = S(\Omega). \qquad (1.45)$$

Figure 1.4 shows the Wigner distribution of an unchirped Gaussian pulse [(a), left] versus a Gaussian pulse with a quadratic chirp [(b), right]. The introduction of a quadratic phase modulation leads to a tilt (rotation) and flattening of the distribution. This distortion of the Wigner function results directly from the relation (1.43) applied to a Gaussian pulse. We have defined in Eq. (1.33) the phase of the linearly chirped pulse as $\varphi(t) = -at^2/\tau_G^2$. If $\mathcal{W}_{\text{unchirp}}$ is the Wigner distribution of the unchirped pulse, the linear chirp transforms that function into:

$$\mathcal{W}_{\text{chirp}} = \mathcal{W}_{\text{unchirp}} \left(t, \Omega - \frac{2at}{\tau_G^2}\right), \qquad (1.46)$$

hence the tilt observed in Fig. 1.4. Mathematical tools have been developed to produce a pure rotation of the phase space (t, Ω). We refer the interested reader to the literature for details on the Wigner distribution and in particular on the fractional Fourier transform [8, 9]. It has been shown that such a rotation describes the propagation of a pulse through a medium with a quadratic dispersion (index of refraction being a quadratic function of frequency) [10].

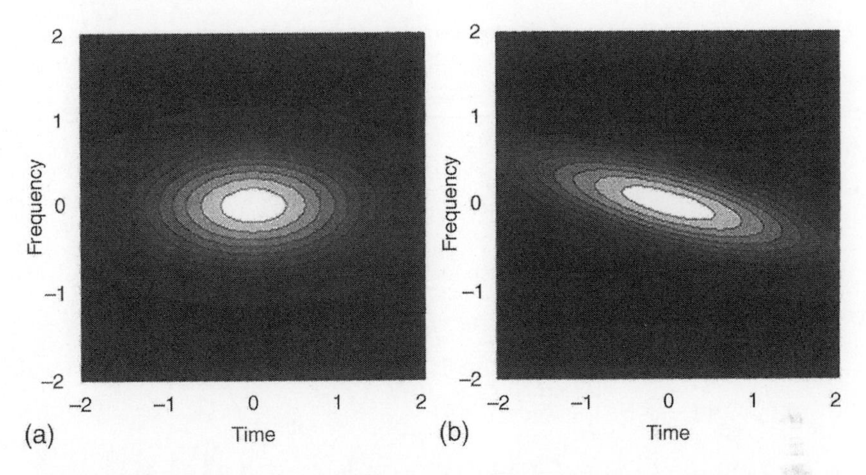

Figure 1.4 Wigner distribution for a Gaussian pulse. Left (a), the phase function $\varphi(t) = \varphi_0$ is a constant. On the right (b), Wigner distribution for a linearly chirped pulse, i.e., with a quadratic phase modulation $\varphi(t) = at^2$.

Moments of the Electric Field

It is mainly history and convenience that led to the adoption of the FWHM of the pulse intensity as the quantity representative of the pulse duration. Sometimes pulse duration and spectral width defined by the FWHM values are not suitable measures. This is, for instance, the case in pulses with substructure or broad wings causing a considerable part of the energy to lie outside the range given by the FWHM. In these cases it may be preferable to use averaged values derived from the appropriate second-order moments. It appears in fact, as will be shown in examples of propagation, that the second moment of the field distribution is a better choice.

For the sake of generality, let us designate by $f(x)$ the field as a function of the variable x (which can be the transverse coordinate, transverse wave vector, time or frequency). The moment of order n for the quantity x with respect to intensity is defined as:

$$\langle x^n \rangle = \frac{\int_{-\infty}^{\infty} x^n |f(x)|^2 dx}{\int_{-\infty}^{\infty} |f(x)|^2 dx} \tag{1.47}$$

The first-order moment, $\langle x \rangle$, is the "center of mass" of the intensity distribution and is most often chosen as reference, in such a way as to have a zero value.

For example, the center of the transverse distribution will be on axis, $x = 0$, or a Gaussian temporal intensity distribution $\mathcal{E}_0 \exp[-(t/\tau_G)^2]$ will be centered at $t = 0$. A good criterium for the width of a distribution is the mean square deviation (MSQ):

$$\langle \Delta x \rangle = \sqrt{\langle x^2 \rangle - \langle x \rangle^2}. \tag{1.48}$$

The explicit expressions in the time and frequency domains are:

$$\langle \tau_p \rangle = \langle \Delta t \rangle = \left[\frac{1}{W} \int_{-\infty}^{\infty} t^2 I(t) dt - \frac{1}{W^2} \left(\int_{-\infty}^{\infty} t I(t) dt \right)^2 \right]^{\frac{1}{2}} \tag{1.49}$$

$$\langle \Delta\omega_p \rangle = \langle \Delta\Omega \rangle = \left[\frac{1}{W} \int_{-\infty}^{\infty} \Omega^2 S(\Omega) d\Omega - \frac{1}{W^2} \left(\int_{-\infty}^{\infty} \Omega S(\Omega) d\Omega \right)^2 \right]^{\frac{1}{2}} \tag{1.50}$$

where $S(\Omega)$ is the spectral intensity defined in Eq. (1.24). Whenever appropriate we will assume that the first-order moments are zero, which yields $\langle \Delta x \rangle = \sqrt{\langle x^2 \rangle}$.

The second-order moments can also be defined using the Wigner distribution [Eq. (1.40)]:

$$\langle t^2 \rangle = \frac{\int\int_{-\infty}^{\infty} t^2 \mathcal{W}_E(t,\Omega) dt d\Omega}{\int\int_{-\infty}^{\infty} \mathcal{W}_E(t,\Omega) dt d\Omega} = \frac{\int_{-\infty}^{\infty} t^2 |\tilde{E}(t)|^2 dt}{\int_{-\infty}^{\infty} |\tilde{E}(t)|^2 dt} \tag{1.51}$$

$$\langle \Omega^2 \rangle = \frac{\int\int_{-\infty}^{\infty} \Omega^2 \mathcal{W}_E(t,\Omega) dt d\Omega}{\int\int_{-\infty}^{\infty} \mathcal{W}_E(t,\Omega) dt d\Omega} = \frac{\int_{-\infty}^{\infty} \Omega^2 |\tilde{E}(\Omega)|^2 d\Omega}{\int_{-\infty}^{\infty} |\tilde{E}(\Omega)|^2 d\Omega}. \tag{1.52}$$

Although the above equations do not bring anything new, the Wigner distribution lets us define another quantity, which describes the coupling between conjugated variables:

$$\langle t, \Omega \rangle = \frac{\int\int_{-\infty}^{\infty} (t - \langle t \rangle)(\Omega - \langle \Omega \rangle) \mathcal{W}_E(t,\Omega) dt d\Omega}{\int\int_{-\infty}^{\infty} \mathcal{W}_E(t,\Omega) dt d\Omega}. \tag{1.53}$$

A nonzero $\langle t, \Omega \rangle$ implies that the center of mass of the spectral intensity evolves with time, as in Fig. 1.4. One can thus define an instantaneous frequency:

$$\omega(t) = \frac{\int_{-\infty}^{\infty} \Omega \mathcal{W}_E(t,\Omega) d\Omega}{\int_{-\infty}^{\infty} \mathcal{W}_E(t,\Omega) d\Omega}. \tag{1.54}$$

By substituting the definition of the Wigner distribution Eq. (1.40) in Eq. (1.54), it is possible to demonstrate rigourously the relation (1.17). Indeed, substituting the definition (1.43) in Eq. (1.54) leads to:

$$\omega(t) = \frac{\int_{-\infty}^{\infty} \Omega \mathcal{W}_{\mathcal{E}}[t, \Omega - (\omega_\ell + \dot{\varphi})] d\Omega}{\int_{-\infty}^{\infty} \mathcal{W}_E(t, \Omega) d\Omega}$$

$$= \frac{\int_{-\infty}^{\infty} [\Omega' + \omega_\ell + \dot{\varphi}(t)] \mathcal{W}_{\mathcal{E}}[t, \Omega'] d\Omega'}{\int_{-\infty}^{\infty} \mathcal{W}_E(t, \Omega) d\Omega}$$

$$= \omega_\ell + \dot{\varphi}(t), \tag{1.55}$$

where we used the fact that $\int \Omega' \mathcal{W}_{\mathcal{E}}(t, \Omega') d\Omega' = 0$.

There is a well-known uncertainty principle between the second moment of conjugated variables. If k is the Fourier-conjugated variable of x, it is shown in Appendix A that:

$$\langle x^2 \rangle \langle k^2 \rangle = \frac{M^4}{4} \geq \frac{1}{4}, \tag{1.56}$$

where we have defined a shape factor M^2, which has been extensively used to describe the departure of beam profile from the "ideal Gaussian" [11]. This relation can be applied to time and frequency:

$$\langle t^2 \rangle \langle \Omega^2 \rangle = \frac{M^4}{4} \geq \frac{1}{4}. \tag{1.57}$$

Equality only holds for a Gaussian pulse (beam) shape free of any phase modulation, which implies that the Wigner distribution for a Gaussian shape occupies the smallest area in the time–frequency plane. It is also important to note that the uncertainty relations (1.56) and (1.57) only hold for the pulse widths defined as the MSQ. For Gaussian pulses, $\langle \tau_p \rangle \langle \Delta\omega_p \rangle = \sqrt{\langle t^2 \rangle \langle \Omega^2 \rangle} = 1/2$ is true, while for the products of the FWHM of the intensity and spectral intensity $c_B = \tau_p \Delta\nu_p = 0.441$. In fact, the pulse duration–bandwidth product *is not minimum* for a Gaussian pulse, as illustrated in Table 1.1, which gives the value of c_B for various pulse shapes without phase modulation. It remains that, for a given pulse shape, c_B is the smallest for pulses without frequency modulation (unchirped) which are called bandwidth limited or Fourier limited. Such pulses exhibit the shortest possible duration at a given spectral width and pulse shape.

If there is a frequency variation across a pulse, its spectrum will contain additional spectral components. Consequently, the modulated pulse possesses a spectral width that is larger than the Fourier limit given by column five in Table 1.1.

Chirped Pulses

A quadratic phase modulation plays an essential role in light propagation, be it in time or space. Because a spherical wavefront can be approximated by a quadratic phase ($\varphi(x) \propto x^2$, where x is the transverse dimension) near any propagation axis of interest, imparting a quadratic spatial phase modulation will lead to focusing or defocusing of a beam. The analogue is true in time: Imparting a quadratic phase modulation ($\varphi(t) \propto t^2$) will lead to pulse compression or broadening after propagation through a dispersive medium. These problems relating to pulse propagation will be discussed in several sections and chapters of this book. In this section we attempt to clarify quantitatively the relation between a quadratic chirp in the temporal or frequency space and the corresponding broadening of the spectrum or pulse duration, respectively. The results are interchangeable from frequency to temporal space.

Let us first assume that a laser pulse, initially unchirped, propagates through a dispersive material that leaves the pulse spectrum, $|\tilde{\mathcal{E}}(\Omega)|^2$, unchanged but produces a quadratic phase modulation in the frequency domain. The pulse spectrum is centered at the average frequency $\langle\Omega\rangle = \omega_\ell$. The average frequency does not change, hence the first nonzero term in the Taylor expansion of $\phi(\Omega)$ is

$$\phi(\Omega) = \frac{1}{2}\left.\frac{d^2\phi}{d\Omega^2}\right|_0 \langle\Omega^2\rangle, \tag{1.58}$$

where $\phi(\Omega)$ determines the phase factor of $\tilde{\mathcal{E}}(\Omega)$:

$$\tilde{\mathcal{E}}(\Omega) = \mathcal{E}(\Omega)e^{i\phi(\Omega)}. \tag{1.59}$$

The first- and second-order moments are, according to the definitions (1.47):

$$\langle t\rangle = \frac{\int_{-\infty}^{\infty} t\tilde{\mathcal{E}}(t)\tilde{\mathcal{E}}(t)^* dt}{\int_{-\infty}^{\infty} |\tilde{\mathcal{E}}(t)|^2 dt} = \frac{\int_{-\infty}^{\infty} \frac{d\tilde{\mathcal{E}}(\Omega)}{d\Omega}\tilde{\mathcal{E}}^*(\Omega)d\Omega}{\int_{-\infty}^{\infty} |\tilde{\mathcal{E}}(\Omega)|^2 d\Omega} = \left\langle\frac{d\phi}{d\Omega}\right\rangle \tag{1.60}$$

and

$$\langle t^2 \rangle = \frac{\int_{-\infty}^{\infty} t\tilde{\mathcal{E}}(t)t\tilde{\mathcal{E}}(t)^* dt}{\int_{-\infty}^{\infty} |\tilde{\mathcal{E}}(t)|^2 dt} = \frac{\int_{-\infty}^{\infty} \left| \frac{d\tilde{\mathcal{E}}(\Omega)}{d\Omega} \right|^2 d\Omega}{\int_{-\infty}^{\infty} |\tilde{\mathcal{E}}(t)|^2 dt}$$

$$= \frac{\int_{-\infty}^{\infty} \left[\frac{d\mathcal{E}(\Omega)}{d\Omega} \right]^2 d\Omega}{\int_{-\infty}^{\infty} |\tilde{\mathcal{E}}(\Omega)|^2 d\Omega} + \left\langle \left(\frac{d\phi}{d\Omega} \right)^2 \right\rangle. \tag{1.61}$$

It is left to a problem at the end of this chapter to derive these results. Because the initial pulse was unchirped and its spectral amplitude is not affected by propagation through the transparent medium, the first term in Eq. (1.61) represents the initial second-order moment $\langle t^2 \rangle_0$. Substituting the expression for the quadratic phase Eq. (1.58) into Eq. (1.47) for the first-order moment, we find from Eq. (1.61):

$$\langle t^2 \rangle = \langle t^2 \rangle_0 + \left[\left. \frac{d^2\phi}{d\Omega^2} \right|_0 \right]^2 \langle \Omega^2 \rangle. \tag{1.62}$$

The frequency chirp introduces a temporal broadening (of the second-order moment) directly proportional to the square of the chirp coefficient, $\left[\left. \frac{d^2\phi}{d\Omega^2} \right|_0 \right]^2$.

Likewise we can analyze the situation where a temporal phase modulation $\varphi(t) = \left. \frac{d\varphi}{dt} \right|_0 t^2$ is impressed on the pulse while the pulse envelope, $|\tilde{\mathcal{E}}(t)|^2$, remains unchanged. This temporal frequency modulation or chirp, characterized by the second derivative in the middle (center of mass) of the pulse, leads to a spectral broadening given by:

$$\langle \Omega^2 \rangle = \langle \Omega^2 \rangle_0 + \left[\left. \frac{d^2\varphi}{dt^2} \right|_0 \right]^2 \langle t^2 \rangle \tag{1.63}$$

where $\langle \Omega^2 \rangle_0$ refers to the spectrum of the input pulse and $\langle t^2 \rangle$ is the (constant) second-order moment of time.

Equations (1.62) and (1.63) demonstrate the advantage of using the MSQ to define the pulse duration and bandwidth, because it shows a simple relation between the broadening in the time or spectral domain, because of a chirp in the spectral or time domain, respectively independent of the pulse and spectral shape. For the two different situations described by Eqs. (1.62) and (1.63), we can apply the uncertainty relation, Eq. (1.57),

$$\langle t^2\rangle\langle\Omega^2\rangle = \frac{M^4}{4}\kappa_c \geq \frac{1}{4}. \tag{1.64}$$

We have introduced a factor of chirp κ_c, equal to

$$\kappa_c = 1 + \frac{M^4}{4\langle t^2\rangle_0^2}\left[\left.\frac{d^2\phi}{d\Omega^2}\right|_0\right]^2 \tag{1.65}$$

in case of a frequency chirp and constant spectrum, or

$$\kappa_c = 1 + \frac{M^4}{4\langle\Omega^2\rangle_0^2}\left[\left.\frac{d^2\varphi}{dt^2}\right|_0\right]^2 \tag{1.66}$$

in case of a temporal chirp and constant pulse envelope.

In summary, using the *mean square deviation (MSQ)* to define the pulse duration and bandwidth:

- The duration–bandwidth product $\sqrt{\langle t^2\rangle\langle\Omega^2\rangle}$ is minimum 0.5 for a Gaussian pulse shape, without phase modulation.
- For any pulse shape, one can define a shape factor M^2 equal to the minimum duration–bandwidth product for that particular shape.
- Any quadratic phase modulation—or linear chirp—whether in frequency or time, increases the bandwidth–duration product by a chirp factor κ_c. The latter increases proportionally to the second derivative of the phase modulation, whether in time or in frequency.

1.2. PULSE PROPAGATION

So far we have considered only temporal and spectral characteristics of light pulses. In this subsection we shall be interested in the propagation of such pulses through matter. This is the situation one always encounters when working with electromagnetic wave packets (at least until somebody succeeds in building a suitable trap). The electric field, now considered in its temporal and spatial dependence, is again a suitable quantity for the description of the propagating wave packet. In view of the optical materials that will be investigated, we can neglect external charges and currents and confine ourselves to nonmagnetic permeabilities and uniform media. A wave equation can be derived for the electric field vector **E** from Maxwell equations (see for instance [12]) which in Cartesian coordinates reads

$$\boxed{\left(\frac{\partial^2}{\partial x^2}+\frac{\partial^2}{\partial y^2}+\frac{\partial^2}{\partial z^2}-\frac{1}{c^2}\frac{\partial^2}{\partial t^2}\right)\mathbf{E}(x,y,z,t)=\mu_0\frac{\partial^2}{\partial t^2}\mathbf{P}(x,y,z,t)}, \tag{1.67}$$

where μ_0 is the magnetic permeability of free space. The source term of Eq. (1.67) contains the polarization **P** and describes the influence of the medium on the field as well as the response of the medium. Usually the polarization is decomposed into two parts:

$$\mathbf{P}=\mathbf{P}^L+\mathbf{P}^{NL}. \tag{1.68}$$

The decomposition of Eq. (1.68) is intended to distinguish a polarization that varies linearly ($\mathbf{P}^L$) from one that varies nonlinearly ($\mathbf{P}^{NL}$) with the field. Historically, $\mathbf{P}^L$ represents the medium response in the frame of "ordinary" optics, e.g., classical optics [13] and is responsible for effects such as diffraction, dispersion, refraction, linear losses and linear gain. Frequently, these processes can be attributed to the action of a host material which in turn may contain sources of a nonlinear polarization $\mathbf{P}^{NL}$. The latter is responsible for nonlinear optics [14–16] which includes, for instance, saturable absorption and gain, harmonic generation and Raman processes.

As will be seen in Chapters 3 and 4, both $\mathbf{P}^L$ and in particular $\mathbf{P}^{NL}$ are often related to the electric field by complicated differential equations. One reason is that no physical phenomenon can be truly instantaneous. In this chapter we will omit $\mathbf{P}^{NL}$. Depending on the actual problem under consideration, $\mathbf{P}^{NL}$ will have to be specified and added to the wave equation as a source term.

1.2.1. The Reduced Wave Equation

Equation (1.67) is of rather complicated structure and in general can solely be solved by numerical methods. However, by means of suitable approximations and simplifications, one can derive a "reduced wave equation" that will enable us to deal with many practical pulse propagation problems in a rather simple way. We assume the electric field to be linearly polarized and propagating in the z-direction as a plane wave, i.e., the field is uniform in the transverse x,y direction. The wave equation has now been simplified to:

$$\left(\frac{\partial^2}{\partial z^2}-\frac{1}{c^2}\frac{\partial^2}{\partial t^2}\right)E(z,t)=\mu_0\frac{\partial^2}{\partial t^2}P^L(z,t). \tag{1.69}$$

As known from classical electrodynamics [12] the linear polarization of a medium is related to the field through the dielectric susceptibility χ. In the frequency domain we have

$$\tilde{P}^L(\Omega, z) = \epsilon_0 \; \chi(\Omega)\tilde{E}(\Omega, z), \tag{1.70}$$

which is equivalent to a convolution integral in the time domain

$$P^L(t, z) = \epsilon_0 \int_{-\infty}^{t} dt' \; \chi(t')E(z, t - t'). \tag{1.71}$$

Here ϵ_0 is the permittivity of free space. The finite upper integration limit, t, expresses the fact that the response of the medium must be causal. For a nondispersive medium (which implies an infinite bandwidth for the susceptibility, $\chi(\Omega) = \text{const.}$) the medium response is instantaneous, i.e., memory free. In general, $\chi(t)$ describes a finite response time of the medium, which in the frequency domain, means nonzero dispersion. This simple fact has important implications for the propagation of short pulses and time varying radiation in general. We will refer to this point several times in later chapters—in particular when dealing with coherent interaction.

The Fourier transform of (1.69) together with (1.70) yields

$$\boxed{\left[\frac{\partial^2}{\partial z^2} + \Omega^2 \epsilon(\Omega)\mu_0\right] \tilde{E}(z, \Omega) = 0} \tag{1.72}$$

where we have introduced the dielectric constant

$$\epsilon(\Omega) = [1 + \chi(\Omega)]\epsilon_0. \tag{1.73}$$

For now we will assume a real susceptibility and dielectric constant. Later we will discuss effects associated with complex quantities. The general solution of (1.72) for the propagation in the $+z$ direction is

$$\tilde{E}(\Omega, z) = \tilde{E}(\Omega, 0)e^{-ik(\Omega)z}, \tag{1.74}$$

where the propagation constant $k(\Omega)$ is determined by the dispersion relation of linear optics

$$k^2(\Omega) = \Omega^2 \epsilon(\Omega)\mu_0 = \frac{\Omega^2}{c^2} n^2(\Omega), \tag{1.75}$$

and $n(\Omega)$ is the refractive index of the material. For further consideration we expand $k(\Omega)$ about the carrier frequency ω_ℓ

$$k(\Omega) = k(\omega_\ell) + \delta k, \tag{1.76}$$

where

$$\delta k = \left.\frac{dk}{d\Omega}\right|_{\omega_\ell} (\Omega - \omega_\ell) + \frac{1}{2} \left.\frac{d^2k}{d\Omega^2}\right|_{\omega_\ell} (\Omega - \omega_\ell)^2 + \cdots \tag{1.77}$$

and write Eq. (1.74) as

$$\tilde{E}(\Omega, z) = \tilde{E}(\Omega, 0)e^{-ik_\ell z}e^{-i\delta kz}, \tag{1.78}$$

where $k_\ell^2 = \omega_\ell^2 \epsilon(\omega_\ell)\mu_0 = \omega_\ell^2 n^2(\omega_\ell)/c^2$. In most practical cases of interest, the Fourier amplitude will be centered on a mean wave vector k_ℓ, and will have appreciable values only in an interval Δk small compared to k_ℓ. In analogy to the introduction of an envelope function slowly varying in time, after the separation of a rapidly oscillating term, cf. Eqs. (1.11)–(1.14), we can define now an amplitude which is slowly varying in the spatial coordinate

$$\tilde{\mathcal{E}}(\Omega, z) = \tilde{E}(\Omega + \omega_\ell, 0)e^{-i\delta kz}. \tag{1.79}$$

Again, for this concept to be useful we must require that

$$\left|\frac{d}{dz}\tilde{\mathcal{E}}(\Omega, z)\right| \ll k_\ell \left|\tilde{\mathcal{E}}(\Omega, z)\right| \tag{1.80}$$

which implies a sufficiently small wave number spectrum

$$\left|\frac{\Delta k}{k_\ell}\right| \ll 1. \tag{1.81}$$

In other words, the pulse envelope must not change significantly while travelling through a distance comparable with the wavelength $\lambda_\ell = 2\pi/\omega_\ell$. Fourier transforming of Eq. (1.78) into the time domain gives

$$\tilde{E}(t, z) = \frac{1}{2}\left\{\frac{1}{\pi}\int_{-\infty}^{\infty} d\Omega\, \tilde{E}(\Omega, 0)e^{-i\delta kz}e^{i(\Omega-\omega_\ell)t}\right\} e^{i(\omega_\ell t - k_\ell z)} \tag{1.82}$$

which can be written as

$$\boxed{\tilde{E}(t,z) = \frac{1}{2}\tilde{\mathcal{E}}(t,z)e^{i(\omega_\ell t - k_\ell z)}} \tag{1.83}$$

where $\tilde{\mathcal{E}}(t,z)$ is now the envelope varying slowly in space and time, defined by the term in the curled brackets in Eq. (1.82).

Further simplification of the wave equation requires a corresponding equation for $\tilde{\mathcal{E}}$ utilizing the envelope properties. Only a few terms in the expansion of $k(\Omega)$ and $\epsilon(\Omega)$, respectively, will be considered. To this effect we expand $\epsilon(\Omega)$ as series around ω_ℓ, leading to the following form for the linear polarization (1.70)

$$\tilde{P}^L(\Omega,z) = \left(\epsilon(\omega_\ell) - \epsilon_0 + \sum_{n=1}^{\infty} \frac{1}{n!}\frac{d^n\epsilon}{d\Omega^n}\bigg|_{\omega_\ell}(\Omega - \omega_\ell)^n\right)\tilde{E}(\Omega,z). \tag{1.84}$$

In terms of the pulse envelope, the above expression corresponds in the time domain to

$$\tilde{P}^L(t,z) = \frac{1}{2}\left\{[\epsilon(\omega_\ell) - \epsilon_0]\tilde{\mathcal{E}}(t,z) + \sum_{n=1}^{\infty}(-i)^n \frac{\epsilon^{(n)}(\omega_\ell)}{n!}\frac{\partial^n}{\partial t^n}\tilde{\mathcal{E}}(t,z)\right\}e^{i(\omega_\ell t - k_\ell z)}, \tag{1.85}$$

where $\epsilon^{(n)}(\omega_\ell) = \frac{\partial^n}{\partial\Omega^n}\epsilon\Big|_{\omega_\ell}$. The term in the curled brackets defines the slowly varying envelope of the polarization, $\tilde{\mathcal{P}}^L$. The next step is to replace the electric field and the polarization in the wave equation (1.69) by Eq. (1.82) and Eq. (1.85), respectively. We transfer thereafter to a coordinate system (η, ξ) moving with the group velocity $\nu_g = \left(\frac{dk}{d\Omega}\Big|_{\omega_\ell}\right)^{-1}$, which is the standard transformation to a "retarded" frame of reference:

$$\xi = z \qquad \eta = t - \frac{z}{\nu_g} \tag{1.86}$$

and

$$\frac{\partial}{\partial z} = \frac{\partial}{\partial \xi} - \frac{1}{\nu_g}\frac{\partial}{\partial \eta}; \qquad \frac{\partial}{\partial t} = \frac{\partial}{\partial \eta}. \tag{1.87}$$

A straightforward calculation leads to the final result:

$$\frac{\partial}{\partial \xi}\tilde{\mathcal{E}} - \frac{i}{2}k_\ell'' \frac{\partial^2}{\partial \eta^2}\tilde{\mathcal{E}} + \mathcal{D} = -\frac{i}{2k_\ell}\frac{\partial}{\partial \xi}\left(\frac{\partial}{\partial \xi} - \frac{2}{\nu_g}\frac{\partial}{\partial \eta}\right)\tilde{\mathcal{E}} \tag{1.88}$$

The quantity

$$\mathcal{D} = -\frac{i\mu_0}{2k_\ell}\sum_{n=3}^{\infty}\frac{(-i)^n}{n!}\left[\omega_\ell^2\epsilon^{(n)}(\omega_\ell) - 2n\omega_\ell\epsilon^{(n-1)}(\omega_\ell) + n(n-1)\epsilon^{(n-2)}(\omega_\ell)\right]\frac{\partial^n}{\partial \eta^n}\tilde{\mathcal{E}} \tag{1.89}$$

contains dispersion terms of higher order, and has been derived by taking directly the second order derivative of the polarization defined by the product of envelope and fast oscillating terms in Eq. (1.85). The indices of the three resulting terms have been redefined to factor out a single derivative of order (n) of the field envelope. The second derivative of k:

$$k_\ell'' = \left.\frac{\partial^2 k}{\partial \Omega^2}\right|_{\omega_\ell} = -\left.\frac{1}{\nu_g^2}\frac{d\nu_g}{d\Omega}\right|_{\omega_\ell} = \frac{1}{2k_\ell}\left[\frac{2}{\nu_g^2} - 2\mu_0\epsilon(\omega_\ell) - 4\omega_\ell\mu_0\epsilon^{(1)}(\omega_\ell) - \omega_\ell^2\mu_0\epsilon^{(2)}(\omega_\ell)\right] \tag{1.90}$$

is the group velocity dispersion (GVD) parameter. It should be mentioned that the GVD is usually defined as the derivative of ν_g with respect to λ, $d\nu_g/d\lambda$, related to k'' through

$$\frac{d\nu_g}{d\lambda} = \frac{\Omega^2\nu_g^2}{2\pi c}\frac{d^2k}{d\Omega^2}. \tag{1.91}$$

So far we have not made any approximations, and the structure of Eq. (1.88) is still rather complex. However, we can exploit at this point the envelope properties (1.14) and (1.80), which, in this particular situation, imply:

$$\left|\frac{1}{k_\ell}\left(\frac{\partial}{\partial \xi} - \frac{2}{\nu_g}\frac{\partial}{\partial \eta}\right)\tilde{\mathcal{E}}\right| = \left|\frac{1}{k_\ell}\left(\frac{\partial}{\partial z} - \frac{1}{\nu_g}\frac{\partial}{\partial t}\right)\tilde{\mathcal{E}}\right| \ll \left|\tilde{\mathcal{E}}\right|. \tag{1.92}$$

The right-hand side of (1.88) can thus be neglected if the prerequisites for introducing pulse envelopes are fulfilled. This procedure is SVEA and reduces the wave equation to first-order derivatives with respect to the spatial coordinate.

If the propagation of short pulses is computed over long distances, the cumulative error introduced by neglecting the right-hand side of Eq. (1.88) may be significant. In those cases, a direct numerical treatment of the second-order wave equation is required.

Further simplifications are possible for a broad class of problems of practical interest, where the dielectric constant changes slowly over frequencies within the pulse spectrum. In those cases, terms with $n \geq 3$ can be omitted too ($\mathcal{D} = 0$), leading to a greatly simplified reduced wave equation:

$$\boxed{\frac{\partial}{\partial \xi}\tilde{\mathcal{E}}(\eta, \xi) - \frac{i}{2}k_\ell'' \frac{\partial^2}{\partial \eta^2}\tilde{\mathcal{E}}(\eta, \xi) = 0,} \tag{1.93}$$

which describes the evolution of the complex pulse envelope as it propagates through a loss-free medium with GVD. The reader will recognize the structure of the one-dimensional Schrödinger equation.

1.2.2. Retarded Frame of Reference

In the case of zero GVD [$k_\ell'' = 0$ in Eq. (1.93)], the pulse envelope does not change at all in the system of local coordinates (η, ξ). This illustrates the usefulness of introducing a coordinate system moving at the group velocity. In the laboratory frame, the pulse travels at the group velocity without any distortion.

In dealing with short pulses as well as in dealing with white light (see Chapter 2) the appropriate retarded frame of reference is moving at the *group* rather than at the *wave* (*phase*) velocity. Indeed, while a monochromatic wave of frequency Ω travels at the phase velocity $v_p(\Omega) = c/n(\Omega)$, it is the superposition of many such waves with differing phase velocities that leads to a wave packet (pulse) propagating with the group velocity. The importance of the frame of reference moving at the group velocity is such that, in the following chapters, the notation z and t will be substituted for ξ and η, unless the laboratory frame is explicitly specified.

Some propagation problems—such as the propagation of coupled waves in nonlinear crystals discussed in Chapter 3—are more appropriately treated in the frequency domain. As a simple exercise, let us derive the group velocity directly from the solution of the wave equation in the form of Eq. (1.78)

$$\tilde{E}(\Omega, z) = \tilde{E}(\Omega, 0)e^{-ik_\ell z}e^{-i\delta k z}. \tag{1.94}$$

The Fourier transform amplitude $E(\Omega, 0)$ represented on the top left of Fig. 1.5 is not changed by propagation. On the top right, the time domain representation of the pulse, or the inverse transform of $E(\Omega, 0)$, is centered at $t = 0$ (solid line).

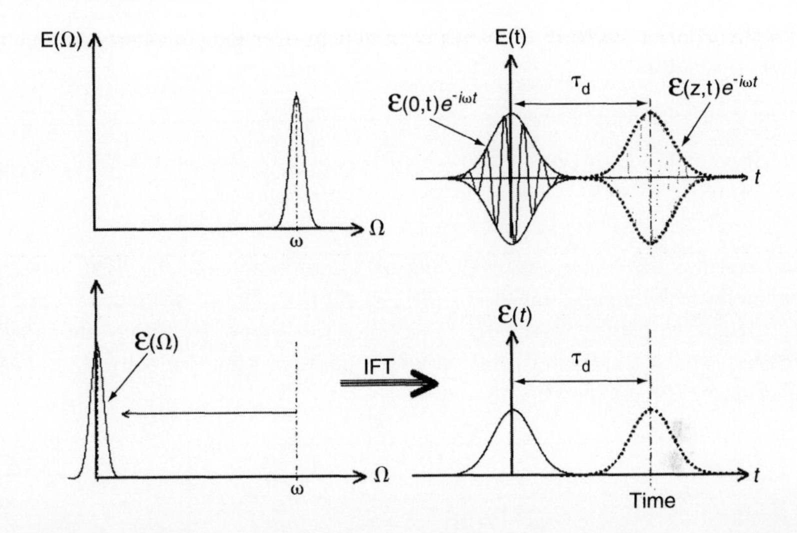

Figure 1.5 The Fourier transform amplitude $E(\Omega, 0)$ is sketched in the upper left, and the corresponding field in the time domain on the upper right (solid line). The lower part of the figure displays the field amplitudes, $\mathcal{E}(\Omega)$ on the left, centered at the origin of the frequency scale, and the corresponding inverse Fourier transform $\mathcal{E}(t)$. Propagation in the frequency domain is obtained by multiplying the field at $z = 0$ by the phase factor $\exp(-i\tau_d\Omega)$, where $\tau_d = z/v_g$ is the group delay. In the time domain, this corresponds to delaying the pulse by an amount τ_d (right). The delayed fields $|E(z,t)|$ and $\mathcal{E}(z,t)$ are shown in dotted lines on the right of the figure.

We assume that the expansion of the wave vector $k(\Omega)$, Eq. (1.76), can be terminated after the linear term, that is

$$\delta k = \left.\frac{dk}{d\Omega}\right|_{\omega_\ell} (\Omega - \omega_\ell). \tag{1.95}$$

The inverse Fourier transform of Eq. (1.94) now yields

$$\begin{aligned}\tilde{E}(t,z) &= e^{-ik_\ell z}\int_{-\infty}^{\infty}\tilde{E}(\Omega,0)\exp\left[-i\left.\frac{dk}{d\Omega}\right|_{\omega_\ell}(\Omega-\omega_\ell)z\right]e^{i\Omega t}d\Omega \qquad (1.96)\\ &= e^{i(\omega_\ell t - k_\ell z)}\int_{-\infty}^{\infty}\tilde{E}(\Omega'+\omega_\ell,0)\exp\left[i\left(t-\left.\frac{dk}{d\Omega}\right|_{\omega_\ell}z\right)\Omega'\right]d\Omega'\end{aligned}$$

where we substituted $\Omega = \Omega' + \omega_\ell$ to obtain the last equation. This equation is just the inverse Fourier transform of the field spectrum shifted to the origin (i.e., the spectrum of the envelope $\tilde{\mathcal{E}}(\Omega)$, represented on the lower left of Fig. 1.5) with the Fourier variable "time" now given by $t - \frac{dk}{d\Omega}|_{\omega_\ell} z$. Carrying out the transform yields

$$\tilde{E}(t,z) = \frac{1}{2}\tilde{\mathcal{E}}(t,z)e^{i(\omega_\ell t - k_\ell z)} = \frac{1}{2}\tilde{\mathcal{E}}\left(t - \left.\frac{dk}{d\Omega}\right|_{\omega_\ell} z, 0\right) e^{i(\omega_\ell t - k_\ell z)}. \quad (1.97)$$

We have thus the important result that, in the time domain, the light pulse has been delayed by an amount ($\tau_d = \frac{dk}{d\Omega}|_{\omega_\ell} z$) proportional to distance. Within the approximation that the wave vector is a linear function of frequency, the pulse is seen to propagate without distortion with a constant group velocity v_g given by either of the three expressions:

$$\frac{1}{v_g} = \left.\frac{dk}{d\Omega}\right|_{\omega_\ell} \quad (1.98)$$

$$\frac{1}{v_g} = \frac{n_0}{c} + \frac{\omega_\ell}{c}\left.\frac{dn}{d\Omega}\right|_{\omega_\ell} \quad (1.99)$$

$$\frac{1}{v_g} = \frac{n_0}{c} - \frac{\lambda}{c}\left.\frac{dn}{d\lambda}\right|_{\lambda}. \quad (1.100)$$

The first term in Eqs. (1.99) and (1.100) represent the phase delay per unit length, while the second term in these equations is the change in carrier to envelope phase per unit length. We note that the dispersion of the wave vector ($dk/d\Omega$) or of the index of refraction ($dn/d\lambda$) is responsible for a difference between the phase velocity $v_p = c/n_0$ and the group velocity v_g. In a frame of reference moving at the velocity v_g, $\tilde{\mathcal{E}}(z,t)$ remains identically unchanged. Pulse distortions thus only result from high order (higher than 1) terms in the Taylor series expansion of $k(\Omega)$. For this reason, most pulse propagation problems are treated in a retarded frame of reference, moving at the velocity v_g.

Forward–Backward Propagating Waves

We consider an ultrashort pulse plane wave propagating through a dielectric medium. Before the arrival of the pulse, there are no induced dipoles, and for the index of refraction we assume that of a vacuum ($n = 1$). As the dipoles are driven into motion by the first few cycles of the pulse, the index of refraction changes to the value n of the dielectric. One consequence of this causal phenomenon is the "precursor" predicted by Sommerfeld and Brillouin, see for example [12].

One might wonder if the discontinuity in index created by a short and intense pulse should not lead to a reflection for a portion of the pulse. This is an important question regarding the validity of the first-order approximation to Maxwell's propagation equations. If, at $t = 0$, a short wave packet is launched in the $+z$ direction in a homogeneous medium, is it legitimate to assume that there will be no pulse generated in the opposite direction?

The answer that we give in this section is, that in the framework of Maxwell's second-order equation and a linear polarization, there is no such "induced reflection." This property extends even to the nonlinear polarization created by the interaction of the light with a two-level system.

If we include the nonresonant part of the linear polarization in the index of refraction n (imaginary part of n), incorporate in the remainder polarization P all nonlinear and resonant interaction effects, and add a phenomenological scattering term σ we find to the following form for the second-order wave equation:

$$\left(\frac{\partial^2}{\partial z^2} - \frac{n^2}{c^2}\frac{\partial^2}{\partial t^2}\right)\tilde{E} = \mu_0 \frac{\partial^2}{\partial t^2}\tilde{P} + \frac{n\sigma}{c}\frac{\partial}{\partial t}\tilde{E} \tag{1.101}$$

The polarization appearing in the right-hand side can be instantaneous, or be the solution of a differential equation as in the case of most interactions with resonant atomic or molecular systems. Resonant light–matter interactions will be studied in detail in Chapters 3 and 4. The wave equation Eq. (1.101) can be written as a product of a forward and backward propagating operator. Instead of the variables t and z, it is more convenient to use the retarded time variable corresponding to the two possible wave velocities $\pm c/n$:

$$s = t - \frac{n}{c}z$$
$$r = t + \frac{n}{c}z. \tag{1.102}$$

In the new variables, Maxwell's equation (1.101) becomes:

$$\frac{\partial^2}{\partial s \partial r}\tilde{E} = \frac{c^2}{n^2}\left\{\frac{\mu_0}{4}\left(\frac{\partial}{\partial s} + \frac{\partial}{\partial r}\right)^2 \tilde{P} + \frac{n\sigma}{c}\left(\frac{\partial}{\partial s} + \frac{\partial}{\partial r}\right)\right\}\tilde{E}. \tag{1.103}$$

We seek a solution in the form of a forward and a backward propagating field of amplitude $\tilde{\mathcal{E}}_F$ and $\tilde{\mathcal{E}}_B$:

$$\tilde{E} = \frac{1}{2}\tilde{\mathcal{E}}_F e^{i\omega_\ell s} + \frac{1}{2}\tilde{\mathcal{E}}_B e^{i\omega_\ell r}. \tag{1.104}$$

Substitution into Maxwell's Eq. (1.101):

$$
\begin{aligned}
& e^{i\omega_\ell s}\left[2i\omega_\ell \frac{\partial}{\partial r} + \frac{\partial^2}{\partial s \partial r} + \frac{c\sigma}{2n}\left(\frac{\partial}{\partial s} + \frac{\partial}{\partial r} + 2i\omega_\ell\right)\right]\frac{1}{2}\tilde{\mathcal{E}}_F \\
& + e^{i\omega_\ell r}\left[2i\omega_\ell \frac{\partial}{\partial s} + \frac{\partial^2}{\partial s \partial r} + \frac{c\sigma}{2n}\left(\frac{\partial}{\partial s} + \frac{\partial}{\partial r} + 2i\omega_\ell\right)\right]\frac{1}{2}\tilde{\mathcal{E}}_B \\
& = -\frac{\mu_0 c^2}{4n^2}\left(\frac{\partial}{\partial s} + \frac{\partial}{\partial r}\right)^2 \tilde{P},
\end{aligned}
\tag{1.105}
$$

which we rewrite in an abbreviated way using the differential operators $\mathcal{L}$ and $\mathcal{M}$ for the forward and backward propagating waves, respectively:

$$
\mathcal{L}\tilde{\mathcal{E}}_F e^{i\omega_\ell s} + \mathcal{M}\tilde{\mathcal{E}}_B e^{i\omega_\ell r} = -\frac{\mu_0 c^2}{4n^2}\left(\frac{\partial}{\partial s} + \frac{\partial}{\partial r}\right)^2 \tilde{P}. \tag{1.106}
$$

In the case of a linear medium, the forward and backward wave travel independently. If, as initial condition, we choose $\tilde{\mathcal{E}}_B = 0$ along the line $r+s=0$ $(t=0)$, there will be no back scattered wave. If the polarization is written as a slowly varying amplitude:

$$
\tilde{P} = \frac{1}{2}\tilde{\mathcal{P}}_F e^{i\omega_\ell s} + \frac{1}{2}\tilde{\mathcal{P}}_B e^{i\omega_\ell r}, \tag{1.107}
$$

the equations for the forward and backward propagating wave also separate if $\tilde{\mathcal{P}}_F$ is only a function of $\tilde{\mathcal{E}}_F$, and $\tilde{\mathcal{P}}_B$ only a function of $\tilde{\mathcal{E}}_B$. This is because a source term for $\tilde{\mathcal{P}}_B$ can only be formed by a "grating" term, which involves a product of $\tilde{\mathcal{E}}_B\tilde{\mathcal{E}}_F$. It applies to a polarization created by near resonant interaction with a two-level system, using the semiclassical approximation, as will be considered in Chapters 3 and 4. The separation between forward and backward traveling waves has been demonstrated by Eilbeck [17, 18] outside of the slowly varying approximation. Within the slowly varying approximation, we generally write the second derivative with respect to time of the polarization as $-\omega_\ell^2\tilde{\mathcal{P}}$, and therefore, the forward and backward propagating waves are still uncoupled, even when $\tilde{\mathcal{P}} = \tilde{\mathcal{P}}(\tilde{\mathcal{E}}_F, \tilde{\mathcal{E}}_B)$, provided there is only a forward propagating beam as initial condition.

1.2.3. Dispersion

For nonzero GVD ($k_\ell'' \neq 0$) the propagation problem (1.93) can be solved either directly in the time or in the frequency domain. In the first case, the solution is given by a Poisson integral [19] which here reads

$$\tilde{\mathcal{E}}(t,z) = \frac{1}{\sqrt{2\pi i k_\ell'' z}} \int_{-\infty}^{t} \tilde{\mathcal{E}}(t', z=0) \exp\left(i\frac{(t-t')^2}{2k_\ell'' z}\right) dt'. \tag{1.108}$$

As we will see in subsequent chapters, it is generally more convenient to treat linear pulse propagation through transparent linear media in the frequency domain, because only the phase factor of the envelope $\tilde{\mathcal{E}}(\Omega)$ is affected by propagation.

It follows directly from the solution of Maxwell's equations in the frequency domain [for instance Eqs. (1.74) and (1.79)] that the spectral envelope after propagation through a thickness z of a linear transparent material is given by:

$$\tilde{\mathcal{E}}(\Omega, z) = \tilde{\mathcal{E}}(\Omega, 0) \exp\left(-\frac{i}{2}k_\ell''\Omega^2 z - \frac{i}{3!}k_\ell'''\Omega^3 z - \cdots\right). \tag{1.109}$$

Thus we have for the temporal envelope

$$\tilde{\mathcal{E}}(t, z) = \mathcal{F}^{-1}\left\{\tilde{\mathcal{E}}(\Omega, 0) \exp\left(-\frac{i}{2}k_\ell''\Omega^2 z - \frac{i}{3!}k_\ell'''\Omega^3 z - \cdots\right)\right\}. \tag{1.110}$$

If we limit the Taylor expansion of k to the GVD term k_ℓ'', we find that an initially bandwidth-limited pulse develops a spectral phase with a quadratic frequency dependence, resulting in chirp.

We had defined a "chirp coefficient"

$$\kappa_c = 1 + \frac{M^4}{4\langle t^2\rangle_0^2}\left[\left.\frac{d\phi}{d\Omega}\right|_{\omega_\ell}\right]^2$$

when considering in Section 1.1.4 the influence of quadratic chirp on the uncertainty relation Eq. (1.64) based on the successive moments of the field distribution. In the present case, we can identify the phase modulation:

$$\left.\frac{d\phi}{d\Omega}\right|_{\omega_\ell} = -k_\ell'' z. \tag{1.111}$$

Because the spectrum (in amplitude) of the pulse $|\tilde{\mathcal{E}}(\Omega, z)|^2$ remains constant [as shown for instance in Eq. (1.109)], the spectral components responsible for chirp must appear at the expense of the envelope shape, which has to become broader.

At this point we want to introduce some useful relations for the characterization of the dispersion. The dependence of a dispersive parameter can be given as a function of either the frequency Ω or the vacuum wavelength λ. The first-, second-, and third-order derivatives are related to each other by

$$\frac{d}{d\Omega} = -\frac{\lambda^2}{2\pi c}\frac{d}{d\lambda} \tag{1.112}$$

$$\frac{d^2}{d\Omega^2} = \frac{\lambda^2}{(2\pi c)^2}\left(\lambda^2\frac{d^2}{d\lambda^2} + 2\lambda\frac{d}{d\lambda}\right) \tag{1.113}$$

$$\frac{d^3}{d\Omega^3} = -\frac{\lambda^3}{(2\pi c)^3}\left(\lambda^3\frac{d^3}{d\lambda^3} + 6\lambda^2\frac{d^2}{d\lambda^2} + 6\lambda\frac{d}{d\lambda}\right). \tag{1.114}$$

The dispersion of the material is described by either the frequency dependence $n(\Omega)$ or the wavelength dependence $n(\lambda)$ of the index of refraction. The derivatives of the propagation constant used most often in pulse propagation problems, expressed in terms of the index n, are:

$$\frac{dk}{d\Omega} = \frac{n}{c} + \frac{\Omega}{c}\frac{dn}{d\Omega} = \frac{1}{c}\left(n - \lambda\frac{dn}{d\lambda}\right) \tag{1.115}$$

$$\frac{d^2k}{d\Omega^2} = \frac{2}{c}\frac{dn}{d\Omega} + \frac{\Omega}{c}\frac{d^2n}{d\Omega^2} = \left(\frac{\lambda}{2\pi c}\right)\frac{1}{c}\left(\lambda^2\frac{d^2n}{d\lambda^2}\right) \tag{1.116}$$

$$\frac{d^3k}{d\Omega^3} = \frac{3}{c}\frac{d^2n}{d\Omega^2} + \frac{\Omega}{c}\frac{d^3n}{d\Omega^3} = -\left(\frac{\lambda}{2\pi c}\right)^2\frac{1}{c}\left(3\lambda^2\frac{d^2n}{d\lambda^2} + \lambda^3\frac{d^3n}{d\lambda^3}\right). \tag{1.117}$$

The second equation, Eq. (1.116), defining the GVD is the frequency derivative of $1/v_g$. Multiplied by the propagation length L, it describes the frequency dependence of the group delay. It is sometimes expressed in $\text{fs}^2\ \mu\text{m}^{-1}$.

A positive GVD corresponds to

$$\frac{d^2k}{d\Omega^2} > 0. \tag{1.118}$$

1.2.4. Gaussian Pulse Propagation

For a more quantitative picture of the influence that GVD has on the pulse propagation we consider the linearly chirped Gaussian pulse of Eq. (1.33)

$$\tilde{\mathcal{E}}(t, z=0) = \mathcal{E}_0 e^{-(1+ia)(t/\tau_{G0})^2} = \mathcal{E}_0 e^{-(t/\tau_{G0})^2} e^{i\varphi(t,z=0)}$$

entering the sample. To find the pulse at an arbitrary position z, we multiply the field spectrum, Eq. (1.35), with the propagator $\exp\left(-i\frac{1}{2}k_\ell''\Omega^2 z\right)$ as done in Eq. (1.109), to obtain

$$\tilde{\mathcal{E}}(\Omega, z) = \tilde{A}_0 e^{-x\Omega^2} e^{iy\Omega^2} \tag{1.119}$$

where

$$x = \frac{\tau_{G0}^2}{4(1+a^2)} \tag{1.120}$$

and

$$y(z) = \frac{a\tau_{G0}^2}{4(1+a^2)} - \frac{k_\ell'' z}{2}. \tag{1.121}$$

$\tilde{A}_0$ is a complex amplitude factor which we will not consider in what follows and τ_{G0} describes the pulse duration at the sample input. The time dependent electric field that we obtain by Fourier transforming Eq. (1.119) can be written as

$$\tilde{\mathcal{E}}(t, z) = \tilde{A}_1 \exp\left\{-\left(1 + i\frac{y(z)}{x}\right)\left(\frac{t}{\sqrt{\frac{4}{x}[x^2 + y^2(z)]}}\right)^2\right\}. \tag{1.122}$$

Obviously, this describes again a linearly chirped Gaussian pulse. For the "pulse duration" (note $\tau_p = \sqrt{2\ln 2}\,\tau_G$) and phase at position z we find

$$\tau_G(z) = \sqrt{\frac{4}{x}[x^2 + y^2(z)]} \tag{1.123}$$

and

$$\varphi(t, z) = -\frac{y(z)}{4[x^2 + y^2(z)]} t^2. \tag{1.124}$$

Let us consider first an initially unchirped input pulse ($a = 0$). The pulse duration and chirp parameter develop as:

$$\tau_G(z) = \tau_{G0}\sqrt{1 + \left(\frac{z}{L_d}\right)^2} \tag{1.125}$$

$$\frac{\partial^2}{\partial t^2}\varphi(t, z) = \left(\frac{1}{\tau_{G0}^2}\right)\frac{2z/L_d}{1 + (z/L_d)^2}. \tag{1.126}$$

We have defined a characteristic length:

$$L_d = \frac{\tau_{G0}^2}{2\left|k_\ell''\right|}. \tag{1.127}$$

For later reference let us also introduce a so-called dispersive length defined as

$$L_D = \frac{\tau_{p0}^2}{|k_\ell''|} \tag{1.128}$$

where for Gaussian pulses $L_D \approx 2.77L_d$. Bandwidth-limited Gaussian pulses double their length after propagation of about $0.6L_D$. For propagation lengths $z \gg L_d$ the pulse broadening of an unchirped input pulse as described by Eq. (1.125) can be simplified to

$$\frac{\tau_G(z)}{\tau_{G0}} \approx \frac{z}{L_d} = \frac{2|k_\ell''|}{\tau_{G0}^2}z. \tag{1.129}$$

It is interesting to compare the result of Eq. (1.125) with that of Eq. (1.62), where we used the second moment as a measure for the pulse duration. Because the Gaussian is the shape for minimum uncertainty [Eq. (1.57)], and because $d^2\phi/d\Omega^2 = -k''z$, Eq. (1.125) is equivalent to

$$\langle t^2 \rangle = \langle t^2 \rangle_0 + 4\frac{(k'')^2 z^2}{\langle t^2 \rangle_0}.$$

If the input pulse is chirped ($a \neq 0$) two different behaviors can occur depending on the relative sign of a and k_ℓ''. In the case of opposite sign, $y^2(z)$ increases monotonously resulting in pulse broadening, cf. Eq. (1.123). If a and k''_ℓ have equal sign $y^2(z)$ decreases until it becomes zero after a propagation distance

$$z_c = \frac{\tau_{G0}^2 a}{2|k''_\ell|(1+a^2)}. \tag{1.130}$$

At this position the pulse reaches its shortest duration

$$\tau_G(z_c) = \tau_{Gmin} = \frac{\tau_{G0}}{\sqrt{1+a^2}} \tag{1.131}$$

and the time-dependent phase according to Eq. (1.124) vanishes. From here on the propagation behavior is that of an unchirped input pulse of duration τ_{Gmin}, that is, the pulse broadens and develops a time-dependent phase. The larger the input chirp ($|a|$), the shorter the minimum pulse duration that can be obtained [see Eq. (1.131)]. The underlying reason is that the excess bandwidth of a chirped pulse is converted into a narrowing of the envelope by chirp compensation, until the Fourier limit is reached. The whole procedure including the impression of chirp on a pulse will be treated in Chapter 8 in more detail.

There is a complete analogy between the propagation (diffraction) effects of a spatially Gaussian beam and the temporal evolution of a Gaussian pulse in a dispersive medium. For instance, the pulse duration and the slope of the chirp follow the same evolution with distance as the waist and curvature of a Gaussian beam, as detailed at the end of this chapter. A linearly chirped Gaussian pulse in a dispersive medium is completely characterized by the position and (minimum) duration of the unchirped pulse, just as a spatially Gaussian beam is uniquely defined by the position and size of its waist. To illustrate this point, let us consider a linearly chirped pulse whose "duration" τ_G and chirp parameter a are known at a certain position z_1. The position z_c of the minimum duration (unchirped pulse) is found again by setting $y = 0$ in Eq. (1.121):

$$z_c = z_1 + \frac{\tau_G^2}{2k''_\ell}\frac{a}{1+a^2} = z_1 + a\frac{\tau_{Gmin}^2}{2k''_\ell}. \tag{1.132}$$

The position z_c is after z_1 if a and k''_ℓ have the same sign[2]; before z_1 if they have opposite sign. All the temporal characteristics of the pulse are most conveniently defined in terms of the distance $L = z - z_c$ to the point of zero chirp, and the minimum duration τ_{Gmin}. This is similar to Gaussian beam propagation where

[2]For instance, an initially downchirped ($a > 0$) pulse at $z = z_c$ will be compressed in a medium with positive dispersion ($k'' > 0$).

the location of the beam waist often serves as reference. The chirp parameter a and the pulse duration τ_G at any point L are then simply given by

$$a(L) = L/L_d \tag{1.133}$$

$$\tau_G(L) = \tau_{Gmin}\sqrt{1 + [a(L)]^2} \tag{1.134}$$

where the dispersion parameter $L_d = \tau_{Gmin}^2/(2|k_\ell''|)$. The pulse duration-bandwidth product varies with distance L as

$$c_B(L) = \frac{2\ln 2}{\pi}\sqrt{1 + [a(L)]^2}. \tag{1.135}$$

To summarize, Figure 1.6 illustrates the behavior of a linearly chirped Gaussian pulse as it propagates through a dispersive sample.

Simple physical consideration can lead directly to a crude approximation for the maximum broadening that a bandwidth-limited pulse of duration τ_p and spectral width $\Delta\omega_p$ will experience. Each group of waves centered around a frequency Ω travels with its own group velocity $\nu_g(\Omega)$. The difference of group velocities over the pulse spectrum then becomes:

$$\Delta\nu_g = \left[\frac{d\nu_g}{d\Omega}\right]_{\omega_\ell} \Delta\omega_p. \tag{1.136}$$

Accordingly, after a travel distance L the pulse spread can be as large as

$$\Delta\tau_p = \left|\Delta\left(\frac{L}{\nu_g}\right)\right| \approx \frac{L}{\nu_g^2}|\Delta\nu_g| \tag{1.137}$$

which, by means of Eqs. (1.90) and (1.136), yields:

$$\Delta\tau_p = L|k_\ell''|\Delta\omega_p. \tag{1.138}$$

Approximating $\tau_p \approx \Delta\omega_p^{-1}$, a characteristic length after which a pulse has approximately doubled its duration can now be estimated as:

$$L_D' = \frac{1}{|k_\ell''|\Delta\omega_p^2}. \tag{1.139}$$

Measuring the length in meter and the spectral width in nm the GVD of materials is sometimes given in fs/(m nm) which pictorially describes the pulse broadening

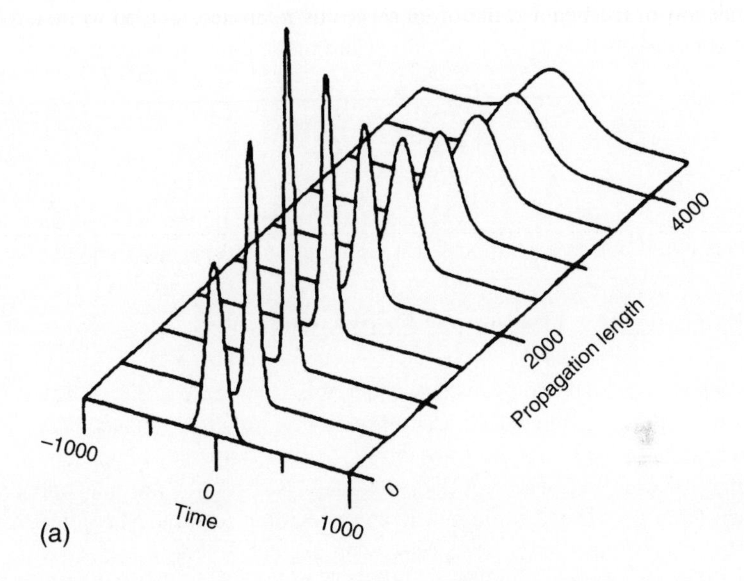

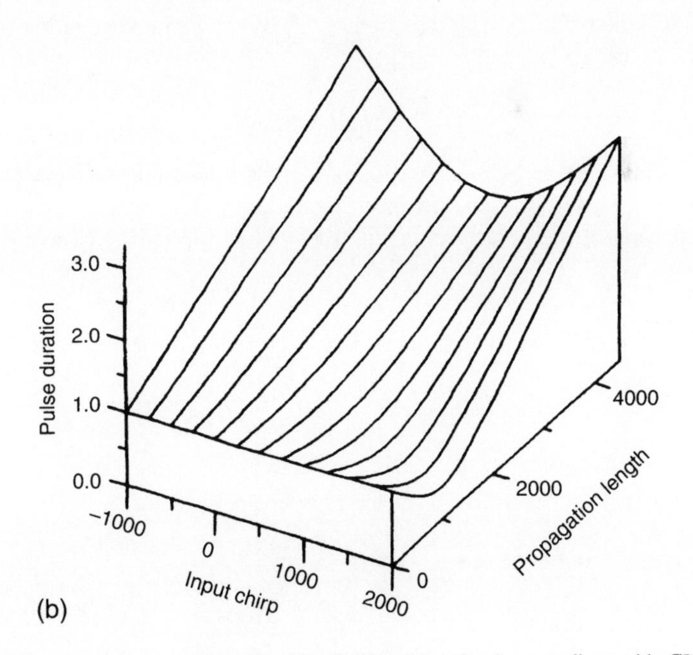

Figure 1.6 Propagation of a linearly chirped Gaussian pulse in a medium with GVD [pulse shape (a), pulse duration for different input chirp (b)].

per unit travel distance and unit spectral width. From Eq. (1.138) we find for the corresponding quantity

$$\boxed{\frac{\Delta\tau_p}{L\Delta\lambda} = 2\pi\frac{c}{\lambda_\ell^2}|k_\ell''|.} \tag{1.140}$$

For BK7 glass at 620 nm, $k_\ell'' \approx 1.02 \times 10^{-25}$ s^2/m, and the GVD as introduced above is about 500 fs per nm spectral width and m propagation length.

1.2.5. Complex Dielectric Constant

In general, the dielectric constant, which was introduced in Eq. (1.72) as a real quantity, is complex. Indeed a closer inspection of Eq. (1.71) shows that the finite memory time of matter requires not only ϵ, χ to be frequency dependent but also that they be complex. The real and imaginary part of $\tilde{\epsilon}$, $\tilde{\chi}$ are not independent of each other but related through a Kramers–Kronig relation. The consideration of a real $\epsilon(\Omega)$ is justified as long as we can neglect (linear) losses or gain. This is valid for transparent samples or propagation lengths which are too short for these processes to become essential for the pulse shaping. For completeness we will modify the reduced wave equation (1.93) by taking into account a complex dielectric constant $\tilde{\epsilon}(\Omega)$ represented as

$$\tilde{\epsilon}(\Omega) = \epsilon(\Omega) + i\epsilon_i(\Omega). \tag{1.141}$$

Let us assume $\tilde{\epsilon}(\Omega)$ to be weakly dispersive. The same procedure introduced to derive Eq. (1.93) can be used after inserting the complex dielectric constant $\tilde{\epsilon}$ into the expression of the polarization Eq. (1.84). Now the reduced wave equation becomes

$$\frac{\partial}{\partial z}\tilde{\mathcal{E}}(t,z) - \frac{i}{2}k''\frac{\partial^2}{\partial t^2}\tilde{\mathcal{E}}(t,z) = \kappa_1\tilde{\mathcal{E}}(t,z) + i\kappa_2\frac{\partial}{\partial t}\tilde{\mathcal{E}}(t,z) + \kappa_3\frac{\partial^2}{\partial t^2}\tilde{\mathcal{E}}(t,z) \tag{1.142}$$

where

$$\kappa_1 = \frac{\omega_\ell}{2}\eta_0\epsilon_i(\omega_\ell) \tag{1.143}$$

$$\kappa_2 = \frac{1}{2}\eta_0\left[2\epsilon_i(\omega_\ell) + \omega_\ell\frac{d}{d\Omega}\epsilon_i(\Omega)\bigg|_{\omega_\ell}\right] \tag{1.144}$$

$$\kappa_3 = \frac{1}{4\omega_\ell}\eta_0\left[2\epsilon_i(\omega_\ell) + 4\omega_\ell\frac{d}{d\Omega}\epsilon_i(\Omega)\bigg|_{\omega_\ell} + \omega_\ell^2\frac{d^2}{d\Omega^2}\epsilon_i(\Omega)\bigg|_{\omega_\ell}\right]. \tag{1.145}$$

In the preceding expressions, $\eta_0 = \sqrt{\mu_0/\epsilon_0} \approx 377\,\Omega\text{ms}$ is the characteristic impedance of vacuum. For zero GVD, and neglecting the two last terms in the right-hand side of Eq. (1.142), the pulse evolution with propagation distance z is described by

$$\frac{\partial}{\partial z}\tilde{\mathcal{E}}(t,z) - \kappa_1\tilde{\mathcal{E}}(t,z) = 0 \tag{1.146}$$

which has the solution

$$\tilde{\mathcal{E}}(t,z) = \tilde{\mathcal{E}}(t,0)e^{\kappa_1 z}. \tag{1.147}$$

The pulse experiences loss or gain depending on the sign of κ_1 and does not change its shape. Equation (1.147) states simply the Lambert-Beer law of linear optics.

An interesting situation is that in which there would be neither gain nor loss at the pulse carrier frequency, i.e., $\epsilon_i(\omega_\ell) = 0$ and $\frac{d}{d\Omega}\epsilon_i(\Omega)\Big|_{\omega_\ell} \neq 0$, which could occur between an absorption and amplification line. Neglecting the terms with the second temporal derivative of $\tilde{\mathcal{E}}$, the propagation problem is governed by the equation

$$\frac{\partial}{\partial z}\tilde{\mathcal{E}}(t,z) - i\kappa_2\frac{\partial}{\partial t}\tilde{\mathcal{E}}(t,z) = 0. \tag{1.148}$$

The solution of this equation is simply

$$\tilde{\mathcal{E}}(t,z) = \tilde{\mathcal{E}}(t + i\kappa_2 z, 0). \tag{1.149}$$

To get an intuitive picture on what happens with the pulse according to Eq. (1.149), let us choose an unchirped Gaussian pulse $\tilde{\mathcal{E}}(t,0)$ [see Eq. (1.33) for $a = 0$] entering the sample at $z = 0$. From Eq. (1.149) we find:

$$\tilde{\mathcal{E}}(t,z) = \tilde{\mathcal{E}}(t,0)\exp\left[\kappa_2^2(z/\tau_G)^2\right]\exp\left[-i2\kappa_2 tz/\tau_G^2\right]. \tag{1.150}$$

The pulse is amplified, and simultaneously its center frequency is shifted with propagation distance. The latter shift is because of the amplification of one part of the pulse spectrum (the high (low) frequency part if $\kappa_2 < (>)0$) while the other part is absorbed. The result is a continuous shift of the pulse spectrum in the corresponding direction and a net gain while the pulse shape is preserved.

In the beginning of this section we mentioned that there is always an imaginary contribution of the dielectric constant leading to gain or loss. The question arises whether a wave equation such as Eq. (1.93), where only the real part of $\tilde{\epsilon}$ was considered, is of any practical relevance for describing pulse propagation through matter. The answer is yes, because in (almost) transparent regions the pulse change owing to dispersion can be much larger than the change caused by losses. An impressive manifestation of this fact is pulse propagation through optical fibers. High-quality fibers made from fused silica can exhibit damping constants as low as 1 dB/km at wavelengths near 1 μm, where the GVD term is found to be $k'' \approx 75\ \mathrm{ps}^2/\mathrm{km}$, see for example [20]. Consequently, a 100 fs pulse launched into a 10 m fiber loses just about 2% of its energy while it broadens by about a factor of 150. To illustrate the physics underlying the striking difference between the action of damping and dispersion, let us consider a dielectric constant $\tilde{\epsilon}(\Omega)$ originating from a single absorption line.

We will use the simple model of a classical harmonic oscillator consisting of an electron bound to a nucleus to calculate the dispersion and absorption of that line. The equation of motion of the electron is:

$$\frac{d^2r}{dt^2} + \omega_0^2 r + \frac{1}{T_c}\frac{dr}{dt} = \frac{e}{m_e}E, \tag{1.151}$$

where $\omega_0 = \sqrt{C/m_e}$ (C being the "spring constant") is the resonance frequency, m_e the electron mass, e its charge, and $1/T_c$ the damping constant. Assuming an electric field of the form $E = (1/2)\tilde{\mathcal{E}}_0 \exp(i\Omega t)$, one finds the polarization $P = N_0 e r$ (N_0 being the number of oscillators (dipoles) per unit volume):

$$P(\Omega) = \frac{N_0 e^2}{m_e}\frac{E}{\omega_0^2 - \Omega^2 + i\Omega/T_c}. \tag{1.152}$$

Using the general relation between polarization and electric field $P = \epsilon_0 \chi E$ we obtain an expression for the complex susceptibility:

$$\chi(\Omega) = \frac{N_0^2 e^2}{\epsilon_0 m_e}\frac{1}{\omega_0^2 - \Omega^2 + i\Omega/T_c}. \tag{1.153}$$

The real and imaginary parts of the susceptibility χ can be calculated:

$$\chi_r = \frac{N_0 e^2}{\epsilon_0 m_e}\frac{(\omega_0^2 - \Omega^2)}{(\omega_0^2 - \Omega^2)^2 + \Omega^2/T_c^2} \approx \frac{N_0 e^2 T_2}{2 m_e \epsilon_0 \omega_0}\frac{\Delta\omega T_2}{1 + \Delta\omega^2 T_2^2} \tag{1.154}$$

$$\chi_i = -\frac{N_0 e^2}{\epsilon_0 m_e}\frac{(\Omega/T_c)}{(\omega_0^2 - \Omega^2)^2 + \Omega^2/T_c^2} \approx -\frac{N_0 e^2 T_2}{2 m_e \epsilon_0 \omega_0}\frac{1}{1 + \Delta\omega^2 T_2^2} \tag{1.155}$$

The second term of each preceding equation corresponds to the approximation of small detuning $\Delta\omega = \omega_0 - \Omega \ll \omega_0$. $1/T_2$ is the linewidth of the Lorentzian absorption line, and $T_2 = 2T_c$ will be assimilated in Chapters 3 and 4 to the phase relaxation time of the oscillators. The real and imaginary parts of the oscillator contribution to the susceptibility are responsible for a frequency dependence of the wave vector. One can write

$$k(\Omega) = \Omega\sqrt{\mu_0\epsilon_0\,[1 + \chi(\Omega)]} \approx \frac{\Omega}{c}\left[1 + \frac{1}{2}\chi(\Omega)\right] \tag{1.156}$$

For frequencies Ω being sufficiently far from resonance, i.e., $|(\omega_0 - \Omega)T_2)| = |\Delta\omega T_2| \gg 1$, but with $|\omega_\ell - \Omega| \ll \omega_\ell$ (narrow pulse spectrum), the real and imaginary parts of the propagation constant are given by:

$$k_r(\Omega) \simeq \frac{\Omega}{c} + B\frac{\Omega}{\Delta\omega T_2} \tag{1.157}$$

$$k_i(\Omega) \simeq -B\frac{\Omega}{(\Delta\omega T_2)^2}, \tag{1.158}$$

where $B = (N_0 e^2 T_2)/(4\epsilon_0\omega_0 c m_e)$. The GVD, responsible for pulse reshaping, is:

$$k''(\Omega) \simeq \frac{2BT_2^2\omega_0}{[\Delta\omega T_2]^3}. \tag{1.159}$$

For small travel distances L the relative change of pulse energy can be estimated from Eq. (1.74) and Eq. (1.20) to be:

$$\Delta\mathcal{W}_{rel} = 1 - \frac{\mathcal{W}(L)}{\mathcal{W}(0)} \approx -2k_i L. \tag{1.160}$$

The relative change of pulse duration because of GVD can be evaluated from Eq. (1.125) and we find:

$$\Delta\tau_{rel} = \frac{\tau_G(L)}{\tau_{G0}} - 1 \approx 2\left(\frac{k_\ell'' L}{\tau_{G0}^2}\right)^2. \tag{1.161}$$

To compare both relative pulse distortions we consider their ratio, using Eqs. (1.158), (1.159), (1.160), and (1.161):

$$\frac{\Delta\tau_{rel}}{\Delta\mathcal{W}_{rel}} = \Delta\mathcal{W}_{rel}\frac{2}{(\Delta\omega T_2)^2}\left(\frac{T_2}{\tau_{G0}}\right)^4. \tag{1.162}$$

At given material parameters and carrier frequency, shorter pulses always lead to a dominant pulse spreading. For $T_2 = 10^{-10}$ s (typical value for a single electronic resonance), and a detuning $\Delta\omega T_2 = 10^4$, we find for example:

$$\frac{\Delta\tau_{rel}}{\Delta\mathcal{W}_{rel}} \approx \Delta\mathcal{W}_{rel}\left(\frac{1200\,\text{fs}}{\tau_{G0}}\right)^4. \tag{1.163}$$

To summarize, a resonant transition of certain spectral width $1/T_2$ influences short pulse (pulse duration < 1ps) propagation outside resonance mainly because of dispersion. Therefore, the consideration of a transparent material ($\epsilon_i \approx 0$) with a frequency dependent, real dielectric constant $\epsilon(\Omega)$, which was necessary to derive Eq. (1.93), is justified in many practical cases involving ultrashort pulses.

1.3. INTERACTION OF LIGHT PULSES WITH LINEAR OPTICAL ELEMENTS

Even though this topic is treated in detail in Chapter 2, we want to discuss here some general aspects of pulse distortions induced by linear optical elements. These elements comprise typical optical components, such as mirrors, prisms, and gratings, which one usually finds in all optical setups. Here we shall restrict ourselves to the temporal and spectral changes the pulse experiences and shall neglect a possible change of the beam characteristics. A linear optical element of this type can be characterized by a complex optical transfer function

$$\tilde{H}(\Omega) = R(\Omega)e^{-i\Psi(\Omega)} \tag{1.164}$$

that relates the incident field spectrum $\tilde{E}_{in}(\Omega)$ to the field at the sample output $\tilde{E}(\Omega)$

$$\tilde{E}(\Omega) = R(\Omega)e^{-i\Psi(\Omega)}\tilde{E}_{in}(\Omega). \tag{1.165}$$

Here $R(\Omega)$ is the (real) amplitude response and $\Psi(\Omega)$ is the phase response. As can be seen from Eq. (1.165), the influence of $R(\Omega)$ is that of a frequency filter. The phase factor $\Psi(\Omega)$ can be interpreted as the phase delay which a spectral component of frequency Ω experiences. To get an insight of how the phase response affects the light pulse, we assume that $R(\Omega)$ does not change over the pulse spectrum whereas $\Psi(\Omega)$ does. Thus, we obtain for the output field from Eq. (1.165):

$$\tilde{E}(t) = \frac{1}{2\pi}R\int_{-\infty}^{+\infty}\tilde{E}_{in}(\Omega)e^{-i\Psi(\Omega)}e^{i\Omega t}\,d\Omega. \tag{1.166}$$

Replacing $\Psi(\Omega)$ by its Taylor expansion around the carrier frequency ω_ℓ of the incident pulse

$$\Psi(\Omega) = \sum_{n=0}^{\infty} b_n(\Omega - \omega_\ell)^n \tag{1.167}$$

with the expansion coefficients

$$b_n = \frac{1}{n!}\frac{d^n\Psi}{d\Omega^n}\bigg|_{\omega_\ell} \tag{1.168}$$

we obtain for the pulse

$$\begin{aligned}\tilde{E}(t) &= \frac{1}{2}\tilde{\mathcal{E}}(t)e^{i\omega_\ell t} \\ &= \frac{1}{2\pi}Re^{-ib_0}e^{i\omega_\ell t}\int_{-\infty}^{+\infty}\tilde{E}_{in}(\Omega) \\ &\quad\times \exp\left(-i\sum_{n=2}^{\infty} b_n(\Omega - \omega_\ell)^n\right)e^{i(\Omega-\omega_\ell)(t-b_1)}\,d\Omega. \end{aligned} \tag{1.169}$$

By means of Eq. (1.169) we can easily interpret the effect of the various expansion coefficients b_n. The term e^{-ib_0} is a constant phase shift (phase delay) having no effect on the pulse envelope. A nonvanishing b_1 leads solely to a shift of the pulse on the time axis t; the pulse would obviously keep its position on a time scale $t' = t - b_1$. The term b_1 determines a group delay in a similar manner as the first-order expansion coefficient of the propagation constant k defined a group velocity in Eq. (1.97). The higher-order expansion coefficients produce a nonlinear behavior of the spectral phase which changes the pulse envelope and chirp. The action of the term with $n = 2$, for example, producing a quadratic spectral phase, is analogous to that of GVD in transparent media.

If we decompose the input field spectrum into modulus and phase $\tilde{E}_{in}(\Omega) = |\tilde{E}_{in}(\Omega)|\exp(i\Phi_{in}(\Omega))$, we obtain from Eq. (1.165) for the spectral phase at the output

$$\Phi(\Omega) = \Phi_{in}(\Omega) - \sum_{n=0}^{\infty} b_n(\Omega - \omega_\ell)^n. \tag{1.170}$$

It is interesting to investigate what happens if the linear optical element is chosen to compensate for the phase of the input field. For Taylor coefficients with $n \geq 2$:

$$b_n = \frac{1}{n!} \frac{d^n}{d\Omega^n} \Phi_{in}(\Omega)\Big|_{\omega_\ell} . \tag{1.171}$$

A closer inspection of Eq. (1.169) shows that when Eq. (1.171) is satisfied, all spectral components are in phase for $t - b_1 = 0$, leading to a pulse with maximum peak intensity, as was discussed in previous sections. We will come back to this important point when discussing pulse compression. We want to point out the formal analogy between the solution of the linear wave equation (1.74) and Eq. (1.165) for $R(\Omega) = 1$ and $\Psi(\Omega) = k(\Omega)z$. This analogy expresses the fact that a dispersive transmission object is just one example of a linear element. In this case we obtain for the spectrum of the complex envelope

$$\tilde{\mathcal{E}}(\Omega, z) = \tilde{\mathcal{E}}_{in}(\Omega, 0) \exp\left[-i \sum_{n=0}^{\infty} \frac{1}{n!} k_\ell^{(n)} (\Omega - \omega_\ell)^n z\right] \tag{1.172}$$

where $k_\ell^{(n)} = (d^n/d\Omega^n) k(\Omega)|_{\omega_\ell}$.

Next let us consider a sequence of m optical elements. The resulting transfer function is given by the product of the individual contributions $\tilde{H}_j(\Omega)$

$$\tilde{H}(\Omega) = \prod_{j=1}^{m} \tilde{H}_j(\Omega) = \left(\prod_{j=1}^{m} R_j(\Omega)\right) \exp\left[-i \sum_{j=1}^{m} \Psi_j(\Omega)\right] \tag{1.173}$$

which means an addition of the phase responses in the exponent. Subsequently, by a suitable choice of elements, one can reach a zero-phase response so that the action of the device is through the amplitude response only. In particular, the quadratic phase response of an element (e.g., dispersive glass path) leading to pulse broadening can be compensated with an element having an equal phase response of opposite sign (e.g., grating pair) which automatically would recompress the pulse to its original duration. Such methods are of great importance for the handling of ultrashort light pulses. Corresponding elements will be discussed in Chapter 2.

1.4. GENERATION OF PHASE MODULATION

At this point let us briefly discuss essential physical mechanisms to produce a time-dependent phase of the pulse, i.e., a chirped light pulse. Processes resulting in a phase modulation can be divided into those that increase the pulse spectral width and those that leave the spectrum unchanged. The latter can be attributed to the action of linear optical processes. Any transparent linear medium, or spectrally "flat" reflector, can change the phase of a pulse, without affecting its spectral amplitude. The action of these elements is most easily analyzed in the frequency domain. As we have seen in the previous section, the phase modulation results from the different phase delays that different spectral components experience on interaction. The result for an initially bandwidth-limited pulse, in the time domain, is a temporally broadened pulse with a certain frequency distribution across the envelope, such that the spectral amplitude profile remains unchanged. For an element to act in this manner its phase response $\Psi(\Omega)$ must have nonzero derivatives of at least second order as explained in the previous section.

A phase modulation that leads to a spectral broadening is most easily discussed in the time domain. Let us assume that the action of a corresponding optical element on an unchirped input pulse can be formally written as:

$$\tilde{E}(t) = T(t)e^{i\Phi(t)}\tilde{E}_{in}(t) \tag{1.174}$$

where T and Φ define a time-dependent amplitude and phase response, respectively. For our simplified discussion here let us further assume that T = const., leaving the pulse envelope unaffected. Because the output pulse has an additional phase modulation $\Phi(t)$ its spectrum must have broadened during the interaction. If thc pulse under consideration is responsible for the time dependence of Φ, then we call the process self-phase modulation. If additional pulses cause the temporal change of the optical properties we will refer to it as cross-phase modulation. Often, phase modulation occurs through a temporal variation of the index of refraction n of a medium during the passage of the pulse. For a medium of length d the corresponding phase is:

$$\Phi(t) = -k(t)d = -\frac{2\pi}{\lambda}n(t)d. \tag{1.175}$$

In later chapters we will discuss in detail several nonlinear optical interaction schemes with short light pulses that can produce a time dependence of n.

A time dependence of n can also be achieved by applying a voltage pulse at an electro-optic material for example. However, with the view on phase shaping of femtosecond light pulses the requirements for the timing accuracy of the voltage pulse make this technique difficult.

1.5. BEAM PROPAGATION

1.5.1. General

So far we have considered light pulses propagating as plane waves, which allowed us to describe the time varying field with only one spatial coordinate. This simplification implies that the intensity across the beam is constant and, moreover, that the beam diameter is infinitely large. Both features hardly fit what we know from laser beams. Despite the fact that both features do not match the real world, such a description has been successfully applied for many practical applications and will be used in this book whenever possible. This simplified treatment is justified if the processes under consideration either do not influence the transverse beam profile (e.g., sufficiently short sample length) or allow one to discuss the change of beam profile and pulse envelope as if they occur independently from each other. The general case, where both dependencies mix, is often more complicated and, frequently, requires extensive numerical treatment. Here we will discuss solely the situation where the change of such pulse characteristics as duration, chirp, and bandwidth can be separated from the change of the beam profile. Again we restrict ourselves to a linearly polarized field which now has to be considered in its complete spatial dependence. Assuming a propagation in the z-direction, we can write the field in the form:

$$E = E(x, y, z, t) = \frac{1}{2}\tilde{u}(x, y, z)\tilde{\mathcal{E}}(t)e^{i(\omega t - k_\ell z)} + c.c. \tag{1.176}$$

In the definition (1.176) the scalar $\tilde{u}(x, y, z)$ is to describe the transverse beam profile and $\tilde{\mathcal{E}}(t)$ is the slowly varying complex envelope introduced in Eq. (1.82). Note that the rapid z-dependence of E is contained in the exponential function. Subsequently, $\tilde{u}$ is assumed to vary slowly with z. Under these conditions the insertion of Eq. (1.176) into the wave equation (1.67) yields after separation of the time dependent part in paraxial approximation [11]:

$$\boxed{\left(\frac{\partial^2}{\partial x^2} + \frac{\partial^2}{\partial y^2} - 2ik_\ell\frac{\partial}{\partial z}\right)\tilde{u}(x, y, z) = 0,} \tag{1.177}$$

which is usually solved by taking the Fourier transform along the space coordinates x and y, yielding:

$$\left[\frac{\partial}{\partial z} - \frac{i}{2k_\ell}\left(k_x^2 + k_y^2\right)\right]\tilde{u}(k_x, k_y, z) = 0, \tag{1.178}$$

where k_x and k_y are the Fourier variables (spatial frequencies, wave numbers). This equation can be integrated, to yield the integral form of Fresnel equation:

$$\tilde{u}(k_x, k_y, z) = \tilde{u}(k_x, k_y, 0)e^{\frac{i}{2k_\ell}\left(k_x^2+k_y^2\right)z}. \quad (1.179)$$

Paraxial approximation means that the transverse beam dimensions remain sufficiently small compared with typical travel distances of interest. An important particular solution of the wave equation within the paraxial approximation is the Gaussian beam (see [11]), which can be written in the form:

$$\tilde{u}(x, y, z) = \frac{u_0}{\sqrt{1 + z^2/\rho_0^2}} e^{-i\Theta(z)} e^{-ik_\ell(x^2+y^2)/2R(z)} e^{-(x^2+y^2)/w^2(z)}. \quad (1.180)$$

where

$$R(z) = z + \rho_0^2/z \quad (1.181)$$

$$w(z) = w_0\sqrt{1 + z^2/\rho_0^2} \quad (1.182)$$

$$\Theta(z) = \arctan(z/\rho_0) \quad (1.183)$$

$$\rho_0 = \frac{n\pi w_0^2}{\lambda}. \quad (1.184)$$

Sometimes it is convenient to write Eq. (1.180) as

$$\tilde{u}(x, y, z) = \frac{u_0}{\sqrt{1 + z^2/\rho_0^2}} e^{-i\Theta(z)} e^{-ik_\ell(x^2+y^2)/2\tilde{q}(z)} \quad (1.185)$$

where $\tilde{q}(z)$ is the complex beam parameter which is defined by:

$$\frac{1}{\tilde{q}(z)} = \frac{1}{R(z)} - \frac{i\lambda}{\pi w^2(z)} = \frac{1}{\tilde{q}(0) + z}. \quad (1.186)$$

Optical beams described by Eq. (1.180) exhibit a Gaussian intensity profile transverse to the propagation direction with $w(z)$ as a measure of the beam diameter, as sketched in Figure 1.7. The origin of the z-axis ($z = 0$) is chosen to be the position of the beam waist $w_0 = w(z = 0)$. The radius of curvature of planes of constant phase is $R(z)$. Its value is infinity at the beam waist

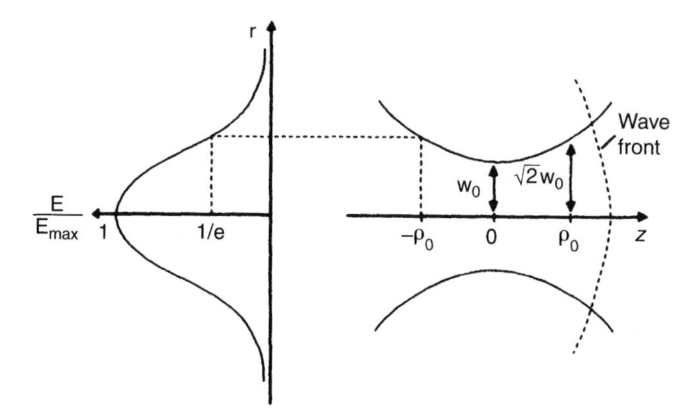

Figure 1.7 Parameters of Gaussian beams.

(plane phase front)[3] and at $z = \infty$. The length ρ_0 is called the Rayleigh range; $2\rho_0$ being the confocal parameter. For $-\rho_0 \leq z \leq \rho_0$, the beam size is within the limits $w_0 \leq w \leq \sqrt{2}w_0$. Given the amplitude u_0 at a given beam waist and wavelength λ, the field at an arbitrary position (x, y, z) is completely predictable by means of Eqs. (1.180) through (1.184).

Instead of using the differential equation (1.177), one can equivalently describe the field propagation by an integral equation. The basic approach is to start with Huygens' principle, and apply the Fresnel approximation assuming paraxial wave propagation [11]. Assuming that the field distribution (or beam profile) $\tilde{u}(x', y', z') = \tilde{u}_0(x', y')$ is known at a plane $z' = \text{const.}$; the field distribution $\tilde{u}(x, y, z)$ at a plane $z = z' + L$ is given by:

$$\tilde{u}(x, y, z) = \frac{ie^{ik_\ell L}}{\lambda L} \int_{-\infty}^{\infty} \int_{-\infty}^{\infty} \tilde{u}_0(x', y') e^{-ik_\ell[(x'-x)^2+(y'-y)^2]/(2L)} dx' dy'. \quad (1.187)$$

Note that both ways of describing the field variation because of diffraction are equivalent. One can easily show that the field (1.187) is a convolution of $\tilde{u}(x, y, 0)$ and $\exp[-ik(x^2 + y^2)/(2L)]$.

[3]The phase term $\Theta(z)$ in Eq. (1.180) takes on a constant value and need not be considered for $z \gg \rho_0$. A Gaussian beam at the position of its waist must not be confused with a plane wave.

1.5.2. Analogy between Pulse and Beam Propagation

Comparing the paraxial wave equation (1.177) and the reduced wave equation (1.93) describing pulse propagation through a GVD medium we notice an interesting correspondence. Both equations are of similar structure. In terms of the reduced wave equation the transverse space coordinates x, y in Eq. (1.177) seem to play the role of the time variable. This space–time analogy suggests the possibility of translating simply the effects related to dispersion into beam propagation properties. For instance, we may compare the temporal broadening of an unchirped pulse because of dispersion with the change of beam size because of diffraction. In this sense free-space propagation plays a similar role for the beam characteristics as a GVD medium does for the pulse envelope. To illustrate this in more detail let us start with Eq. (1.179), and, for simplicity, restrict ourselves to one dimension. The field spectrum at $z = L$

$$\tilde{u}(k_x, L) = \tilde{u}(k_x, z = 0)e^{ik_x^2 L/(2k_\ell)}, \tag{1.188}$$

which after inverse Fourier transform yields

$$\tilde{u}(x, L) \propto \mathcal{F}^{-1}\left\{\tilde{u}_0(k_x)e^{iLk_x^2/(2k_\ell)}\right\}. \tag{1.189}$$

Let us next recall Eq. (1.110) which described the temporal pulse envelope after a GVD medium of length L

$$\tilde{\mathcal{E}}(\eta, L) = \mathcal{F}^{-1}\left\{\tilde{\mathcal{E}}_0(\Omega)e^{-\frac{i}{2}k_\ell''\Omega^2 L}\right\}. \tag{1.190}$$

A comparison with Eq. (1.189) clearly shows the similarity between the diffraction and the dispersion problem. This is to be illustrated in more detail for Gaussian pulse profiles and Gaussian beams. As we have seen in the previous section the quadratic phase factor in Eq. (1.190) broadens an unchirped input pulse and leads to a (linear) frequency sweep across the pulse (chirp) although the pulse spectrum remains unchanged. In an analogous manner we can interpret Eq. (1.189) for the beam profile. A bandwidth-limited Gaussian beam means a beam without phase variation across the beam, which, in terms of Eq. (1.180), requires a radius of curvature of the phase front $R = \infty$. Thus, a Gaussian beam is bandwidth-limited at its waist where it takes on its minimum possible size (at a given spatial frequency spectrum). Multiplication with a quadratic phase factor to describe the beam propagation, cf. Eq. (1.189), leads to beam broadening and "chirp." The latter simply accounts for a finite phase front curvature. Roughly speaking, the spatial frequency components that are not needed to form the broadened beam profile are responsible for the beam divergence.

Table 1.2

Comparison of dispersion and diffraction.

Gaussian pulse	Gaussian beam
Bandwidth-limited pulse at $z = 0$ (unchirped pulse)	Beam waist at $z = 0$ (plane phase fronts)
$\tilde{\mathcal{E}}_0(t) \propto e^{-(t/\tau_{G0})^2}$	$\tilde{u}_0(x) \propto e^{-(x/w_0)^2}$
$\tilde{\mathcal{E}}_0(\Omega) \propto e^{-(\tau_{G0}\Omega/2)^2}$	$\tilde{u}_0(k_x) \propto e^{-(k_x w_0/2)^2}$
Propagation through a medium of length L (dispersion)	Free space propagation over distance L (diffraction)
$\tilde{\mathcal{E}}(\Omega, L) \propto \exp\left[-\left(\frac{\tau_{G0}\Omega}{2}\right)^2 - i\frac{k_\ell'' L\Omega^2}{2}\right]$	$\tilde{u}(k_x, L) \propto \exp\left[-\left(\frac{w_0 k_x}{2}\right)^2 + i\frac{L{k_x}^2}{2k_\ell}\right]$
$\tilde{\mathcal{E}}(t, L) \propto \exp\left[-(1+i\bar{a})\left(\frac{t}{\tau_G}\right)^2\right]$	$\tilde{u}(x, L) \propto \exp\left[-(1+i\bar{b})\left(\frac{x}{w}\right)^2\right]$
$\propto \exp\left[i\omega_\ell \frac{t^2}{2\tilde{p}(L)}\right]$	$\propto \exp\left[-ik_\ell \frac{x^2}{2\tilde{q}(L)}\right]$
$\bar{a} = L/L_d$	$\bar{b} = L/\rho_0$
$\tau_G(L) = \tau_{G0}\sqrt{1+\bar{a}^2}$	$w(L) = w_0\sqrt{1+\bar{b}^2}$
Chirp coefficient (slope)	Wavefront curvature
$\ddot{\varphi} = \frac{2\bar{a}}{1+\bar{a}^2}\frac{1}{\tau_{G0}^2}$	$\frac{1}{R} = \frac{\bar{b}}{1+\bar{b}^2}\frac{1}{\rho_0}$
Characteristic (dispersion) length	Characteristic (Rayleigh) length
$L_d = \frac{\tau_{G0}^2}{2\lvert k_\ell''\rvert}$	$\rho_0 = \frac{n\pi w_0^2}{\lambda_\ell} = \frac{k_\ell w_0^2}{2}$
Complex pulse parameter	Complex beam parameter
$\frac{1}{\tilde{p}(L)} = \frac{\ddot{\varphi}(L)}{\omega_\ell} + \frac{2i/\omega_\ell}{\tau_G^2(L)}$	$\frac{1}{\tilde{q}(L)} = \frac{1}{R(L)} + \frac{i\lambda_\ell}{\pi w^2(L)}$

Table 1.2 summarizes our discussion comparing the characteristics of Gaussian beam and pulse propagation.

1.5.3. Analogy between Spatial and Temporal Imaging

The analogy between pulse and beam propagation was applied to establish a time–domain analog of an optical imaging system by Kolner and Nazarathy [21].

Optical microscopy, for example, serves to magnify tiny structures so that they can be observed by a (relatively) low-resolution system such as our eyes. The idea of the "time lens" is to magnify ultrafast fs transients so that they can be resolved, for example, by a relatively slow oscilloscope. Of course, the opposite direction is also possible, which would lead to data compression in space or time. Although Table 1.2 illustrates the space–time duality for free-space propagation, we now need to look for devices resembling imaging elements such as lenses. From Fourier optics it is known that a lens introduces a quadratic phase factor, thus transforming a (Fourier-limited) input beam (parallel beam) into a spatially chirped (focused) beam. The "time equivalent" lens is a quadratic phase modulator. Quadratic dispersion through a medium with GVD is the temporal analogue of diffraction. Let us consider the lens arrangement of Figure 1.8, in which the light from an object—represented by the field envelope $\mathcal{E}(r)$—at a distance d_1 from the lens, is imaged on a screen at a distance d_2 from the lens. The real image is produced on the screen if the distance and focal distance of the lens satisfy the lens formula:

$$\frac{1}{d_1} + \frac{1}{d_2} = \frac{1}{f}. \tag{1.191}$$

With some approximations, one can derive the time–domain equivalent of the Gaussian lens formula, [21] for an optical system [Fig. 1.8(b)] in which the initial signal $\tilde{\mathcal{E}}(t)$ is propagated for a distance d_1 through a dispersive medium characterized by a wave vector k_2, is given a quadratic phase modulation by a time lens, and propagates for a distance d_2 through a medium of wave vector k_2:

$$\left(d_1 \frac{d^2 k_1}{d\Omega^2}\right)^{-1} + \left(d_2 \frac{d^2 k_2}{d\Omega^2}\right)^{-1} = (f_T/\omega_0)^{-1}. \tag{1.192}$$

In this temporal lens formula, $d_{1,2}(d^2 k_{1,2}/d\Omega^2)$ are the dispersion characteristics of the object and image side, respectively, and $\omega_0/f_T = \partial^2\phi/\partial t^2$ is the parameter of the quadratic phase modulation impressed by the modulator. As in optical imaging, to achieve large magnification with practical devices, short focal lengths are desired. For time imaging this translates into a short focal time f_T which in turn requires a suitably large phase modulation.

Note that the real image of an object can only be recognized on a screen located at a specific distance from the lens, i.e., in the image plane. At any other distance the intensity distribution visible on a screen usually does not resemble the object, because of diffraction. Likewise, the dispersive element broadens each individual pulse (if we assume zero input chirp). It is only after the time lens and a suitably designed second dispersive element that a "pulse train" with the same contrast as the input (but stretched or compressed) emerges.

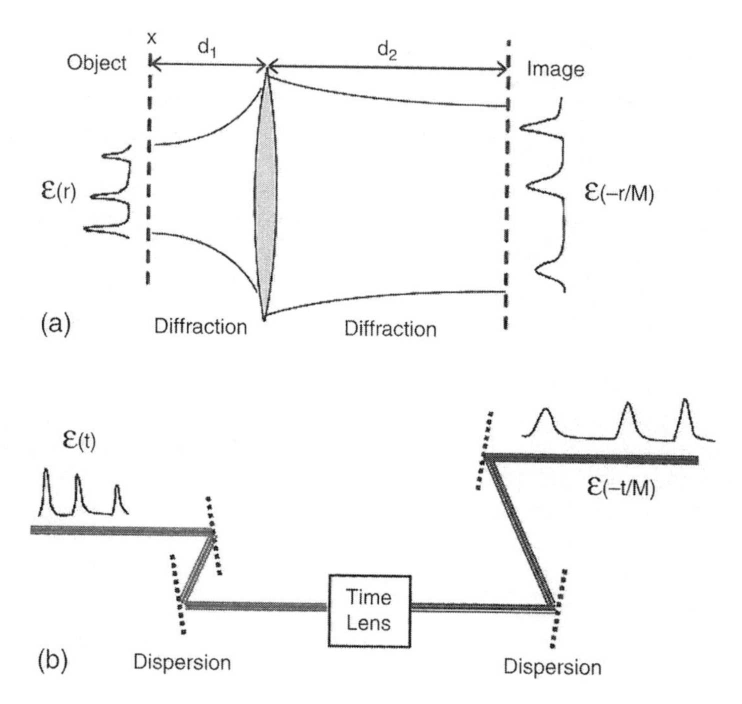

Figure 1.8 Space–time analogy of imaging. (a) Spatial imaging configuration. The "object" is a graphic representation of a three pulse sequence. The "real image" shows a magnified, inverted picture. (b) The temporal imaging configuration. A pair of gratings on either side of the time lens represents a dispersive length characterized by $d^2k/d\Omega^2$, see also Chapter 2. The object is a three pulse sequence. The "image" is a reversed, expanded three pulse sequence. Possible time lenses are explained in the text. (Adapted from [22].)

One possible approach to create a large phase modulation is cross-phase modulation, in which a properly shaped powerful "pump" pulse creates a large index sweep (quadratic with time) in the material of the time lens. Another approach is to use sum or difference frequency generation to impart the linear chirp of one pulse onto the pulse to be "imaged." The linear chirp can be obtained by propagating a strong pulse through a fiber. A detailed review of this "parametric temporal imaging" can be found in Bennett and Kolner [22, 23]. The time equivalent of a long propagation distance (large diffraction) is a large dispersion, which can be obtained with a pair of gratings, see Chapter 2. Note that in a standard magnifying optical system with a single lens, the real image is inverted with respect to the object. The same applies to the temporal imaging: The successive pulses appear in reverse order in the image.

1.6. NUMERICAL MODELING OF PULSE PROPAGATION

The generation and application of femtosecond light pulses requires one to study their propagation through linear and nonlinear optical media. Those studies have been undertaken not only to satisfy theorists. They are necessary to design and optimize experiments and to save time and money. Because of the complexity of interactions taking place numerical methods have to be used in many cases. From the mathematical point of view it is desirable to develop a numerical model optimized with respect to computer time and accuracy for each experimental situation to be described. In this section we will present a procedure that allows one to study pulse propagation through a variety of materials. This model is optimized neither with respect to computer time nor with respect to accuracy. However, it is universal and is directly associated with the physics of the problem. Moreover, it has been successfully applied to various situations. Among them are, for instance, pulse propagation through nonlinear optical fibers and amplifiers and pulse evolution in fs lasers. Without going into the numerical details, we will briefly describe the main features of this concept. In the course of the book we will then present various examples.

In the frame of approximations discussed in the section of beam propagation the electric field can be represented as

$$\begin{aligned} E(x,y,z,t) &= \frac{1}{2}\tilde{u}(x,y,z)\tilde{\mathcal{E}}(z,t)e^{i(\omega_\ell t - k_\ell z)} + c.c. \\ &= \frac{1}{2}\mathcal{U}(x,y,z,t)e^{i(\omega_\ell t - k_\ell z)} + c.c. \end{aligned} \tag{1.193}$$

where $\tilde{\mathcal{E}}$ is the complex pulse envelope, and $\tilde{u}$ describes the transverse beam profile. The medium through which the pulse travels is not to be specified. In general, it will respond linearly as well as nonlinearly to the electric field. For example, the pulse changes shape and chirp because of dispersion while it is amplified or absorbed nonlinearly because of a time-dependent gain coefficient. Therefore, the wave equation derived before, for the case of linear dispersive media, must be supplemented by certain nonlinear interaction terms. In following chapters we will discuss those nonlinear processes in detail. For the moment we will introduce them only formally. Let us first assume that a change in the beam profile can be neglected. Then the behavior of the field is fully described by its complex envelope $\tilde{\mathcal{E}}$. The propagation equation in local coordinates reads

$$\frac{\partial}{\partial z}\tilde{\mathcal{E}} = \frac{1}{2}ik_\ell''\frac{\partial^2}{\partial t^2}\tilde{\mathcal{E}} - \mathcal{D} + \mathcal{B}_1 + \mathcal{B}_2 + \cdots + \mathcal{B}_n \tag{1.194}$$

where the terms $\mathcal{B}_i$ stand for contributions from nonlinear light matter interaction. A direct numerical evaluation of Eq. (1.194) often requires solving a set of nonlinear, partial differential equations. Note that for the determination of the $\mathcal{B}_i$, additional (differential) equations describing the medium must be considered. As with partial differential equations in general, the numerical procedures are rather complicated. Moreover, they may differ largely from each other even when the problems seem to be similar from the physical point of view.

A more intuitive approach can be chosen, as outlined next. The sample of length L is divided into M slices of length $\Delta z = L/M$; each slice sufficiently thin as to induce only a small change in the pulse parameters. Assuming that the complex envelope at propagation distance $z = m\Delta z$ $(m = 1, 2, \ldots, M)$ is given by $\tilde{\mathcal{E}}(t, z)$, the envelope at the output of the next slice $(z + \Delta z)$ can be obtained from Eq. (1.194) as

$$\begin{aligned} \tilde{\mathcal{E}}(t, z + \Delta z) = \tilde{\mathcal{E}}(t, z) + \Bigg[\frac{1}{2} i k_\ell'' \frac{\partial^2}{\partial t^2} \tilde{\mathcal{E}}(t, z) - \mathcal{D} + \mathcal{B}_1(t, z, \tilde{\mathcal{E}}) \\ + \mathcal{B}_2(t, z, \tilde{\mathcal{E}}) + \cdots + \mathcal{B}_n(t, z, \tilde{\mathcal{E}}) \Bigg] \Delta z \end{aligned} \tag{1.195}$$

which can be written formally as

$$\begin{aligned} \tilde{\mathcal{E}}(t, z + \Delta z) = \tilde{\mathcal{E}}(t, z) + \delta_{k''} \tilde{\mathcal{E}}(t, z) + \delta_{\mathcal{D}} \tilde{\mathcal{E}}(t, z) + \delta_1 \tilde{\mathcal{E}}(t, z) \\ + \delta_2 \tilde{\mathcal{E}}(t, z) +, \cdots, + \delta_n \tilde{\mathcal{E}}(t, z). \end{aligned} \tag{1.196}$$

The quantities $\delta_i \tilde{\mathcal{E}}(z, t)$ represent the (small) envelope changes because of the various linear and nonlinear processes. For their calculation the envelope at z only is required. The action of the individual processes is treated as if they occur successively and independently in each slice. The pulse envelope at the end of each slice is then the sum of the input pulse plus the different contributions. The resulting envelope $\tilde{\mathcal{E}}(t, z + \Delta z)$ serves as input for the next slice, and so on until $z + \Delta z = L$.

The methods which can be applied to determine $\delta_i \tilde{\mathcal{E}}$ depend on the specific kind of interaction. For example, it may be necessary to solve a set of differential equations, but only with respect to the time coordinate. As mentioned before, the discussion of nonlinear optical processes will be the subject of following chapters.

This type of numerical calculation is critically dependent on the number of slices. It is the strongest interaction affecting the propagating pulse which determines the length of the slices. As a rule of thumb, the envelope distortion in each slice must not exceed a few percent, and doubling and halving of M must not change the results more than the required accuracy allows.

Many propagation problems have been investigated already with ps and ns light pulses, theoretically as well as experimentally. The severe problem when dealing with fs light pulses is dispersion, which enters Eq. (1.196) through $\delta_k''\tilde{\mathcal{E}}$ (GVD) and $\delta_{\mathcal{D}}\tilde{\mathcal{E}}$ (higher-order dispersion). From the discussion in the preceding sections we can easily derive expressions for these quantities. If only GVD needs to be considered, we can start from Eq. (1.109)

$$\tilde{\mathcal{E}}(\Omega, z + \Delta z) = \tilde{\mathcal{E}}(\Omega, z)e^{-ik_\ell''\Omega^2\Delta z/2} \tag{1.197}$$

which, for sufficiently small Δz, can be approximated as

$$\tilde{\mathcal{E}}(\Omega, z + \Delta z) \approx \tilde{\mathcal{E}}(\Omega, z) - \frac{1}{2}ik_\ell''\Omega^2\Delta z\tilde{\mathcal{E}}(\Omega, z). \tag{1.198}$$

Thus we have for $\delta_{k''}\tilde{\mathcal{E}}(t, z)$

$$\delta_{k''}\tilde{\mathcal{E}}(t, z) \approx \mathcal{F}^{-1}\left\{-\frac{1}{2}ik_\ell''\Omega^2\Delta z\tilde{\mathcal{E}}(\Omega, z)\right\}. \tag{1.199}$$

If additional dispersion terms matter, we can utilize Eq. (1.172) and obtain

$$\delta_{\mathcal{D}}\tilde{\mathcal{E}}(t, z) = \mathcal{F}^{-1}\left\{-i\sum_{n=3}^{\infty}\frac{1}{n!}k_\ell^{(n)}\Omega^n\Delta z\tilde{\mathcal{E}}(\Omega, z)\right\}. \tag{1.200}$$

Next, let us consider a change in the beam profile. This must be taken into account if the propagation length through the material is long as compared with the confocal length. In addition, beam propagation effects can play a role if the setup to be modeled consists of various individual elements separated from each other by air or vacuum. This is the situation that is, for instance, encountered in lasers. It is the evolution of $\tilde{\mathcal{U}} = \tilde{u}\tilde{\mathcal{E}}$ rather than only that of $\tilde{\mathcal{E}}$ that has to be modeled now. The change of $\tilde{\mathcal{U}}$ from z to $z + \Delta z$ is

$$\tilde{\mathcal{U}}(x, y, z + \Delta z, t) = \tilde{\mathcal{U}}(x, y, z, t) + \delta\tilde{\mathcal{U}} \tag{1.201}$$

where

$$\delta\tilde{\mathcal{U}} = \tilde{u}\delta\tilde{\mathcal{E}} + \tilde{\mathcal{E}}\delta\tilde{u}. \tag{1.202}$$

The change of the pulse envelope $\delta\tilde{\mathcal{E}}$ can be derived as described previously. For the determination of $\delta\tilde{u}$ we can evaluate the diffraction integral (1.187) over a propagation length or equivalently proceed to the Fourier space and use Eq. (1.179). For Gaussian beams we may simply use Eq. (1.185).

1.7. SPACE–TIME EFFECTS

For short pulses a coupling of spatial and temporal effects becomes important even for propagation in a nondispersive medium. The physical reason is that self-diffraction of a beam of finite transverse size (e.g., Gaussian beam) is wavelength dependent. A separation of time and frequency effects according to Eqs. (1.176) and (1.177) is clearly not feasible if such processes matter. One can construct a solution by solving the diffraction integral (1.187) for each spectral component. The superposition of these solutions and an inverse Fourier transform then yields the temporal field distribution. Starting with a field $\tilde{E}(x', y', \Omega) = \mathcal{F}\left\{\tilde{E}(x', y', t)\right\}$ in a plane $\Sigma'(x', y')$ at $z = 0$ we find for the field in a plane $\Sigma(x, y)$ at $z = L$:

$$\tilde{E}(x, y, L, t) = \mathcal{F}^{-1}\left\{\frac{i\Omega e^{-i\Omega L/c}}{2\pi c L}\int\int dx'dy'\,\tilde{E}(x', y', \Omega) \times \exp\left[-i\frac{\Omega}{2Lc}\left((x - x')^2 + (y - y')^2\right)\right]\right\} \quad (1.203)$$

where we have assumed a nondispersive medium with refractive index $n = 1$. Solutions can be found by solving numerically Eq. (1.203) starting with an arbitrary pulse and beam profile at a plane $z = 0$. Properties of these solutions were discussed by Christov [24]. They revealed that the pulse becomes phase modulated in space and time with a pulse duration that changes across the beam profile. Because of the stronger diffraction of long wavelength components the spectrum on axis shifts to shorter wavelengths.

For a Gaussian beam and pulse profile at $z = 0$, i.e., $\tilde{E}(x', y', 0, t) \propto \exp(-r'^2/w_0^2)\exp(-t^2/\tau_{G0}^2)\exp(i\omega_\ell t)$ with $r'^2 = x'^2 + y'^2$, the time–space distribution of the field at $z = L$ is of the form [24]:

$$\tilde{E}(r, z = L, t) \propto \exp\left(-\frac{\eta^2}{\tau_G^2}\right)\exp\left[\left(-\frac{w_0\omega_\ell\tau_{G0}}{2Lc\tau_G}r\right)^2\right]\exp\left(i\frac{\omega_\ell\tau_{G0}^2}{\tau_G^2}\eta\right) \quad (1.204)$$

where

$$\tau_G^2 = \tau_{G0}^2 + [w_0 r/(Lc)]^2 \quad (1.205)$$

and $\eta = \left[t - L/c - r^2/(2Lc)\right]$. This result shows a complex mixing of spatial and temporal pulse and beam characteristics. The first term in Eq. (1.204) indicates a pulse duration that increases with increasing distance r from the optical axis. For an order of magnitude estimation let us determine the input pulse duration τ_{G0} at which the pulse duration has increased to $2\tau_{G0}$ at a radial coordinate $r = w$ after the beam has propagated over a certain distance $L \gg \rho_0$. From Eq. (1.205) this is equivalent to $\tau_{G0} = w_0 r/(Lc)$. For $r = w$ with $w \approx L\lambda/(\pi w)$, cf. Eq. (1.182), the pulse duration becomes $\tau_{G0} \approx \lambda/(\pi c)$. Obviously, these effects become only important if the pulses approach the single-cycle regime.

1.8. PROBLEMS

1. Verify the c_B factors of the pulse duration–bandwidth product of a Gaussian and sech pulse as given in Table 1.1.
2. Calculate the pulse duration $\bar{\tau}_p$ defined as the second moment in Eq. (1.49) for a Gaussian pulse and compare with τ_p (FWHM).
3. Consider a medium consisting of particles that can be described by harmonic oscillators so that the linear susceptibility in the vicinity of a resonance is given by Eq. (1.153). Investigate the behavior of the phase and group velocity in the absorption region. You will find a region where $\nu_g > \nu_p$. Is the theory of relativity violated here?
4. Assume a Gaussian pulse which is linearly chirped in a phase modulator that leaves its envelope unchanged. The chirped pulse is then sent through a spectral amplitude only filter of spectral width (FWHM) $\Delta\omega_F$. Calculate the duration of the filtered pulse and determine an optimum spectral width of the filter for which the shortest pulses are obtained. (*Hint:* For simplification you may assume an amplitude only filter of Gaussian profile, i.e., $\tilde{H}(\omega - \omega) = \exp\left[-\ln 2\left(\frac{\omega-\omega}{\Delta\omega_F}\right)^2\right]$.)
5. Derive the general expression for $d^n/d\Omega^n$ in terms of derivatives with respect to λ.
6. Assume that both the temporal and spectral envelope functions $\mathcal{E}(t)$ and $\mathcal{E}(\Omega)$, respectively, are peaked at zero. Let us define a pulse duration τ_p^* and spectral width $\Delta\omega_p^*$ using the electric field and its Fourier transform by

$$\tau_p^* = \frac{1}{|\mathcal{E}(t=0)|}\int_{-\infty}^{\infty} |\mathcal{E}(t)|dt$$

and

$$\Delta\omega_p^* = \frac{1}{|\mathcal{E}(\Omega=0)|}\int_{-\infty}^{\infty} |\mathcal{E}(\Omega)|d\Omega.$$

Show that for this particular definition of pulse duration and spectral width the uncertainty relation reads

$$\tau_p^*\Delta\omega_p^* \geq 2\pi.$$

7. Derive Eqs. (1.60) and (1.61). *Hint:* Make use of Parsival's theorem

$$2\pi \int_{-\infty}^{\infty} |f(t)|^2 dt = \int_{-\infty}^{\infty} |f(\Omega)|^2 d\Omega$$

and the fact that $\mathcal{F}\{tf(t)\} = -i\frac{d}{d\Omega}\mathcal{F}\{f(t)\}$.

8. A polarization—to second-order in the electric field—is defined as $P^{(2)}(t) \propto \chi^{(2)}E^2(t)$. We have seen that the preferred representation for the field is the complex quantity $E^+(t) = \frac{1}{2}\mathcal{E}(t)\exp[i(\omega t + \varphi(t)]$. Give a convenient description of the nonlinear polarization in terms of $E^+(t)$, $\mathcal{E}(t)$ and $\varphi(t)$. Consider in particular second harmonic generation and optical rectification. Explain the physics associated with the various terms of $P^{(2)}$ (or $P^{+(2)}$, if you can define one).

BIBLIOGRAPHY

[1] E. Hecht and A. Zajac. *Optics*. Addison-Wesley, Menlo Park, CA, 1987.

[2] W. H. Press, B. P. Flannery, S. E. Teukolsky, and W. P. Vetterling. *Numerical Recipes*. Cambridge University Press, NY, 1986.

[3] E. Wigner. *Physical Review*, 40:749, 1932.

[4] J. Ville. *Cables et transmissions*, 2A:61, 1948.

[5] W. P. Schleich. *Quantum Optics in Phase Space*. Wiley-VCH, Weinheim, Germany, 2001.

[6] L. Cohen. *Time-frequency Analysis, Theory and Applications*. Prentice-Hall signal processing series, Saddle River, NJ, 1995.

[7] L. Praxmeier and K. Wókiewicz. Time and frequency description of optical pulses. *Laser Physics*, 15:1477–1485, 2005.

[8] A. W. Lohmann and B. H. Soffer. Relationships between the radon-wigner and fractional fourier transforms. *Journal of Optical Society of America A*, 11:1798, 1994.

[9] M. G. Raymer, D. F. McAlister, and M. Beck. Complex wave-field reconstruction using phase-space tomography. *Physical Review Letters*, 72:1137, 1994.

[10] H. M. Ozaktas and D. Mendlovic. Fourier transforms of fractional order and their optical interpretation. *Optics Communication*, 101:163, 1993.

[11] A. E. Siegman. *Lasers*. University Science Books, Mill Valley, CA, 1986.

[12] J. D. Jackson. *Classical Electrodynamics*. John Wiley & Sons, NY, 1975.

[13] M. Born and E. Wolf. *Principles of Optics—Electromagnetic Theory of Propagation, Interference and Diffraction of Light*. Pergamon Press, Oxford, England, 1980.

[14] Y. R. Shen. *The Principles of Nonlinear Optics*. John Wiley & Sons, NY, 1984.

[15] M. Schubert and B. Wilhelmi. *Nonlinear Optics and Quantum Electronics*. John Wiley & Sons, NY, 1978.

[16] R. Boyd. *Nonlinear Optics*. Academic Press, Boston, MA, 1977.

[17] J. C. Eilbeck and R. K. Bullough. The method of characteristics in the theory of resonant or nonresonant nonlinear optics. *Journal of Physics A: General Physics*, 5:820–830, 1972.

[18] J. C. Eilbeck, J. D. Gibbon, P. J. Caudrey, and R. K. Bullough. Solitons in nonlinear optics I: A more accurate description of the 2π pulse in self-induced transparency. *Journal of Physics A: Mathematical, Nuclear and General*, 6:1337–1345, 1973.

[19] E. Kamke. *Differentialgleichungen*. Geest and Portig, Leipzig, Germany, 1969.
[20] G. P. Agrawal. *Nonlinear Fiber Optics*. Academic Press, Boston, MA, 1995.
[21] B. H. Kolner and M. Nazarathy. Temporal imaging with a time lens. *Optics Letters*, 14:630–632, 1989.
[22] C. V. Bennett and B. H. Kolner. Principles of parametric temporal imaging, part II: System performance. *IEEE Journal of Quantum Electronics*, 36:649–655, 2000.
[23] C. V. Bennett and B. H. Kolner. Principles of parametric temporal imaging, part I: System configurations. *IEEE Journal of Quantum Electronics*, 36:430–437, 2000.
[24] I. P. Christov. Propagation of femtosecond light pulses. *Optics Communication*, 53:364–367, 1985.

TO CHIRP OR NOT TO CHIRP ...

... THAT IS THE CHALLENGE

2

Femtosecond Optics

2.1. INTRODUCTION

Whether short pulses or continuous radiation, light should follow the rules of "classical optics." There are, however, some properties related to the bending, propagation, and focalization of light that are specific to fs pulses. Ultrashort pulses are more "unforgiving" of some "defects" of optical systems, as compared to ordinary light of large spectral bandwidth, i.e., white light.[1] Studying optical systems with fs pulses helps in turn to improve the understanding and performances of these systems in white light. We will study properties of basic elements (coatings, lenses, prisms, gratings) and some simple combinations thereof. The dispersion of the index of refraction is the essential parameter for most of the effects to be discussed in this chapter. Some values are listed for selected optical materials in Table 2.1. As already noted in Chapter 1, the second derivative of the index of refraction is positive over the visible spectrum for most transparent materials, corresponding to a positive GVD. There is a sign reversal of the GVD in fused silica around 1.3 μm, which has led to zero dispersion or negative dispersion fibers.

Often a transparent material will be characterized by a fit of the index of refraction as a function of wavelength. Values for most nonlinear materials can

[1] Such light can be regarded as superposition of random fluctuations (short light pulses), the mean duration of which determines the spectral width. A measurement of the light intensity, however, averages over these fluctuations.

Table 2.1

Dispersion parameters for some optical materials. BK7 is the most commonly used optical glass. The SFS are dispersive heavy flint glasses. SQ1 is fused silica. The dispersion parameters for the glasses were calculated with Sellmeier's equations and data from various optical catalogs. The data for the UV wavelengths must be considered as order of magnitude approximations. The ZnSe data are taken from Duarte [2]. Using Eqs. (1.113)–(1.117), the dispersion values given in terms of $n(\Omega)$ can easily be transformed into the corresponding values for $k(\Omega)$.

Material	λ_ℓ (nm)	$n(\omega_\ell)$	$n'(\omega_\ell)$ 10^{-2} (fs)	$n'(\lambda_\ell)$ 10^{-2} (μm^{-1})	$n''(\omega_\ell)$ 10^{-3} (fs^2)	$n''(\lambda_\ell)$ (μm^{-2})	$n'''(\omega_\ell)$ 10^{-4} (fs^3)	$n'''(\lambda_\ell)$ (μm^{-3})
BK7	400	1.5307	1.13	−13	3	1.10	6.9	−12
	500	1.5213	0.88	−6.6	2.3	0.396	7.7	−3.5
	620	1.5154	0.75	−3.6	1.6	0.150	13	−1.1
	800	1.5106	0.67	−2	0.06	0.05	39	−0.29
	1000	1.5074	0.73	−1.4	−3.2	0.016	114	−0.09
SF6	400	1.8674	5.8	−67	30	7.40	214	−120
	500	1.8236	3.7	−28	16	2	86	−21
	620	1.8005	2.7	−13	12	0.70	50	−5.3
	800	1.7844	2	−5.9	8	0.22	56	−1.2
	1000	1.7757	1.71	−3.2	4	0.08	115	−0.36
SF10	400	1.7784	4.6	−54	24	5.9	183	−98
	500	1.7432	3	−22	13	1.6	69	−17
	620	1.7244	2.2	−11	9	0.56	42	−4.2
	800	1.7112	1.7	−5	6	0.17	58	−1
	1000	1.7038	1.5	−2.8	2	0.06	132	−0.3
SF14	400	1.8185	5.3	−62	27	6.8	187	−10.9
	500	1.7786	2.8	−25	15	1.9	85	−2
	620	1.7576	2.5	−12	10	0.63	50	−4.8
	800	1.7430	1.8	−5.5	7	0.20	54	−1.1
	1000	1.7349	1.6	−3.0	3.4	0.072	110	−0.33
SQ1	248	1.5121	2.36	−72	11	15	76	−520
	308	1.4858	1.35	−27	4.1	3.3	23	−66.0
	400	1.4701	0.93	−11	2.3	0.86	6	−9.80
	500	1.4623	0.73	−5.5	1.8	0.32	6	−2.80
	620	1.4574	0.62	−3	1.2	0.13	13	−0.89
	800	1.4533	0.58	−1.7	−0.4	0.04	41	−0.24
	1000	1.4504	0.67	−1.3	−3.8	0.012	121	−0.08
	1300	1.4469	1	−1.1	−14	−0.0003	446	−0.02
	1500	1.4446	1.4	−1.2	−27	−0.0031	915	−0.01
LaSF9	620	1.8463	2.28	−11.2	9.04	0.50		
	800	1.8326	1.76	−5.20	5.79	0.17		
ZnSe	620	2.586	14.24	−30	117.3	2		−15
	800	2.511	8.35	−15	63.3	0.69		−3

be found in Dmitriev *et al.* [1]. A common form is the Sellmeier equation:

$$n^2(\lambda_\ell) = 1 + \frac{B_1\lambda_\ell^2}{\lambda_\ell^2 - C_1} + \frac{B_2\lambda_\ell^2}{\lambda_\ell^2 - C_2} + \frac{B_3\lambda_\ell^2}{\lambda_\ell^2 - C_3}. \tag{2.1}$$

In the case of fused silica, the parameters are[2]:

B_1	$6.96166300 \cdot 10^{-1}$	μm^{-2}
B_2	$4.07942600 \cdot 10^{-1}$	μm^{-2}
B_3	$8.97479400 \cdot 10^{-1}$	μm^{-2}
C_1	$4.67914826 \cdot 10^{-3}$	μm^{2}
C_2	$1.35120631 \cdot 10^{-2}$	μm^{2}
C_3	$9.79340025 \cdot 10^{+1}$	μm^{2}

with the wavelength λ_ℓ expressed in microns. Another example of a possible fit function is the Laurent series formula:

$$n^2(\lambda_\ell) = A + B\lambda_\ell^2 + \frac{C}{\lambda_\ell^2} + \frac{D}{\lambda_\ell^4} + \frac{E}{\lambda_\ell^6} + \frac{F}{\lambda_\ell^8}. \tag{2.2}$$

For crystalline quartz with extraordinary and ordinary index n_e and n_o, respectively, the parameters are[2]:

Parameter	for n_e	for n_o	Unit
A	$2.38490000 \cdot 10^{+0}$	$2.35728000 \cdot 10^{+0}$	
B	$-1.25900000 \cdot 10^{-2}$	$-1.17000000 \cdot 10^{-2}$	μm^{-2}
C	$1.07900000 \cdot 10^{-2}$	$1.05400000 \cdot 10^{-2}$	μm^{2}
D	$1.65180000 \cdot 10^{-4}$	$1.34143000 \cdot 10^{-4}$	μm^{4}
E	$-1.94741000 \cdot 10^{-6}$	$-4.45368000 \cdot 10^{-7}$	μm^{6}
F	$9.36476000 \cdot 10^{-8}$	$5.92362000 \cdot 10^{-8}$	μm^{8}

With the wavelength λ_ℓ being expressed in microns.

An interesting material for its high index in the visible (VIS)–near infrared (NIR) spectral range is ZnS. The first- and second-order dispersion are plotted in Figure 2.1.

We shall start this chapter with an analysis of a simple Michelson interferometer.

---

[2]The values for fused silica and quartz are courtesy of CVI, Albuquerque, New Mexico. Available at: www.cvi.com.

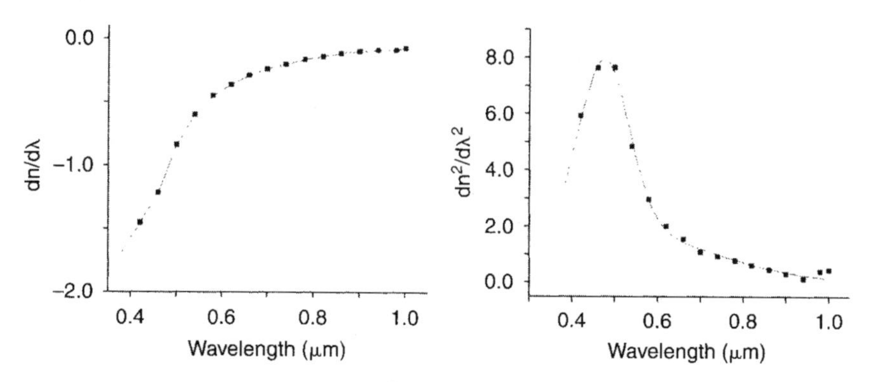

Figure 2.1 First-order dispersion in μm^{-1} (left) and second-order dispersion in μm^{-2} (right) of ZnS.

2.2. WHITE LIGHT AND SHORT PULSE INTERFEROMETRY

Incoherent radiation has received increasing attention as the poor man's fs source (even the wealthiest experimentalist will now treat bright incoherent sources with a certain amount of deference). The similarities between white light and femtosecond light pulses are most obvious when studying coherence properties, but definitely transcend the field of coherent interactions.

Let us consider the basic Michelson interferometer sketched in Figure 2.2. The *real* field on the detector, resulting from the interferences of E_1 and E_2,

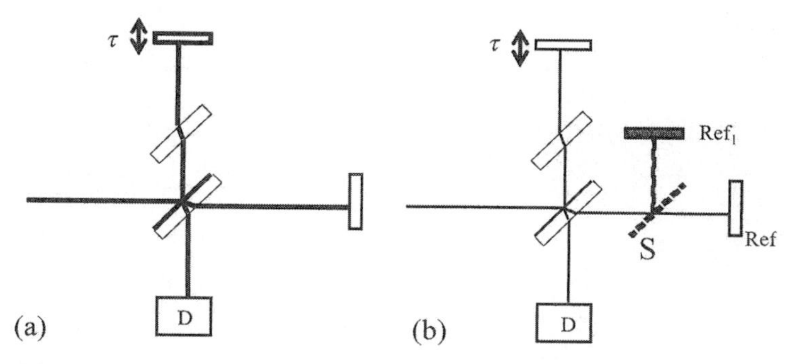

Figure 2.2 **Left:** Balanced Michelson interferometer. **Right:** For the measurement of mirror dispersion, a reflecting sample is inserted between the beam splitter and the reference mirror Ref. (dotted line). The deflected beam is shown as a dashed line orthogonal to the displaced reference mirror Ref_1.

is $E = E_1(t-\tau) + E_2(t)$ with τ being the delay parameter. The intensity at the output of the interferometer is given by the electric field squared averaged over one light period T [Eq. (1.21)]:

$$\begin{aligned} I(t,\tau) &= \epsilon_0 cn \frac{1}{T}\int_{t-T/2}^{t+T/2} [E_1(t'-\tau) + E_2(t')]^2 dt' \\ &= 2\epsilon_0 cn[\tilde{E}_1^+(t-\tau) + \tilde{E}_2^+(t)][\tilde{E}_1^-(t-\tau) + \tilde{E}_2^-(t)] \\ &= \frac{1}{2}\epsilon_0 cn \Big\{ \mathcal{E}_1^2(t-\tau) + \mathcal{E}_2^2(t) \\ &\quad + \tilde{\mathcal{E}}_1^*(t-\tau)\tilde{\mathcal{E}}_2(t)e^{i\omega_\ell \tau} + \tilde{\mathcal{E}}_1(t-\tau)\tilde{\mathcal{E}}_2^*(t)e^{-i\omega_\ell \tau} \Big\}. \end{aligned} \tag{2.3}$$

Here again, we have chosen to decompose the field in an amplitude function $\tilde{\mathcal{E}}$ and a phase function centered around a somewhat arbitrary average frequency of the radiation, ω_ℓ, as in Eqs. (1.10) and (1.11).

The actual signal recorded at the output of the interferometer is the intensity, $\bar{I}$, averaged over the response time τ_R of the detector. In the case of ultrashort pulses $\tau_R \gg \tau_p$ holds and what is being measured is the time integral $\int_{-\infty}^{+\infty} I(t',\tau)dt'$. We will use the notation $\langle\ \rangle$ for either integration or averaging, which results in a quantity that is time independent. Assuming thus that all fluctuations of the signal are averaged out by the detector's slow response, the measured signal reduces to the following expression:

$$\begin{aligned} \bar{I}(\tau) &= \frac{\epsilon_0 cn}{4}\left\{ \langle \tilde{\mathcal{E}}_1^2 \rangle + \langle \tilde{\mathcal{E}}_2^2 \rangle + \left\langle \tilde{\mathcal{E}}_1^*(t-\tau)\tilde{\mathcal{E}}_2(t)e^{i\omega_\ell \tau} + \tilde{\mathcal{E}}_1(t-\tau)\tilde{\mathcal{E}}_2^*(t)e^{-i\omega_\ell \tau} \right\rangle \right\} \\ &= \epsilon_0 cn \left\{ A_{11}(0) + A_{22}(0) + \tilde{A}_{12}^+(\tau) + \tilde{A}_{12}^-(\tau) \right\}. \end{aligned} \tag{2.4}$$

On the right hand side of the first line in Eq. (2.4) we recognize correlation functions similar to that in Eq. (1.31), except that they involve the electric fields rather than the intensities. In complete analogy with the definitions of the complex electric fields, the two complex functions correspond to positive and negative spectral components[3] of a correlation function $A_{12}(\tau) = \tilde{A}_{12}^+(\tau) + \tilde{A}_{12}^-(\tau)$, where,

[3] Spectrum is defined here with respect to the conjugate variable of the delay parameter τ.

e.g., the positive frequency component is defined as:

$$\tilde{A}_{12}^{+}(\tau) = \frac{1}{4}\langle \tilde{\mathcal{E}}_1^*(t-\tau)\tilde{\mathcal{E}}_2(t)e^{i\omega_\ell\tau}\rangle$$

$$= \frac{1}{2}\tilde{\mathcal{A}}_{12}(\tau)e^{i\omega_\ell\tau}. \tag{2.5}$$

The Fourier transform of the correlation of two functions is the product of the Fourier transforms [3]:

$$\tilde{A}_{12}^{+}(\Omega) = \int_{-\infty}^{\infty} \tilde{A}_{12}^{+}(\tau)e^{-i\Omega\tau}d\tau = \int_{-\infty}^{\infty} \tilde{\mathcal{A}}_{12}(\tau)e^{-i(\Omega\tau-\omega_\ell\tau)}d\tau$$

$$= \frac{1}{4}\tilde{\mathcal{E}}_1^*(\Omega-\omega_\ell)\tilde{\mathcal{E}}_2(\Omega-\omega_\ell)$$

$$= \tilde{E}_1^*(\Omega)\tilde{E}_2(\Omega). \tag{2.6}$$

In the ideal case of infinitely thin beam splitter, nondispersive broadband reflectors and beam splitters, $\tilde{E}_1 = \tilde{E}_2$, and the expression (2.6) is real. Correspondingly, the correlation defined by Eq. (2.5) is an electric field autocorrelation which is a symmetric function with respect to the delay origin $\tau = 0$. This fundamental property is of little practical importance when manipulating data from a real instrument, because, in the optical time domain, it is difficult to determine exactly the "zero delay" point, which requires measurement of the relative delays of the two arms with an accuracy better than 100 Å. It is therefore more convenient to use an arbitrary origin for the delay τ, and use the generally complex Fourier transformation of Eq. (2.6).

For an ideally balanced interferometer, the output from the two arms is identical, and the right-hand side of Eq. (2.6) is simply the spectral intensity of the light. This instrument is therefore referred to as a Fourier spectrometer.

Let us turn our attention to the slightly "unbalanced" Michelson interferometer. For instance, with a single beam splitter of finite thickness d', beam 2 will have traversed $L = d'/\cos(\theta_r) = d$ (θ_r being the angle of refraction) more glass than beam 1 (Figure 2.2). It is well-known that the "white light" interference fringes are particularly elusive, because of the short coherence length of the radiation, which translates into a restricted range of delays over which a fringe pattern can be observed. How will that fringe pattern be modified and shifted by having one beam traverse a path of length $2d$ in glass rather than in air (assumed here to be dispersionless)? Let $\tilde{E}_1(t)$ refer to the field amplitude at the detector, corresponding to the beam that has passed through the unmodified arm with the least amount of glass. Using Eq. (1.165) with $R = 1$, $\Psi(\Omega) = k(\Omega)L$ and considering only terms with $n \leq 2$ in the expansion of Ψ, cf. Eq. (1.167), we find the second beam through the simple transformation:

$$\tilde{E}_2(\Omega) = \tilde{E}_1(\Omega)\exp\{-iL\,[k(\Omega) - \Omega/c]\}$$
$$\approx \tilde{E}_1(\Omega)\exp\left\{-i\left[\left(k_\ell - \frac{\Omega}{c}\right)L + k'_\ell L(\Omega - \omega_\ell) + \frac{k''_\ell L}{2}(\Omega - \omega_\ell)^2\right]\right\} \tag{2.7}$$

where, as outlined earlier, $(k'_\ell)^{-1} = ([\frac{dk}{d\Omega}]_{\omega_\ell})^{-1}$ determines the group velocity of a wave packet centered around ω_ℓ and $k''_\ell = [\frac{d^2k}{d\Omega^2}]_{\omega_\ell}$ is responsible for GVD. The time-dependent electric field is given by the Fourier transform of Eq. (2.7). Neglecting GVD we find for the complex field envelope:

$$\tilde{\mathcal{E}}_2(t) = e^{-i(k_\ell + k'_\ell\omega_\ell)L}\tilde{\mathcal{E}}_1\left[t - (k'_\ell - 1/c)L\right]. \tag{2.8}$$

Apart from an unimportant phase factor the obvious change introduced by the glass path in one arm of the interferometer is a shift of time origin, i.e., a shift of the maximum of the correlation. This is a mere consequence of the longer time needed for light to traverse glass instead of air. The shift in "time origin" measured with the unbalanced versus "balanced" Michelson is

$$\Delta\tau = \left(k'_\ell - \frac{1}{c}\right)L$$
$$= \frac{L}{c}\left\{(n-1) + \omega_\ell\left[\frac{dn}{d\Omega}\right]_{\omega_\ell}\right\}$$
$$= \frac{L}{c}\left\{(n-1) - \lambda_\ell\left[\frac{dn}{d\lambda}\right]_{\lambda_\ell}\right\}, \tag{2.9}$$

where we replaced k'_ℓ by Eq. (1.116). The first term in the right-hand side of the second and third equation represents the temporal delay resulting from the difference of the optical pathlength in air ($n \approx 1$) and glass. The second term contains the contribution from the group velocity in glass. In the above derivation, we have not specifically assumed that the radiation consists of short pulses. It is also the case for white light continuous wave (cw) radiation that the group velocity contributes to the shift of zero delay introduced by an unbalance of dispersive media between the two arms of the interferometer.

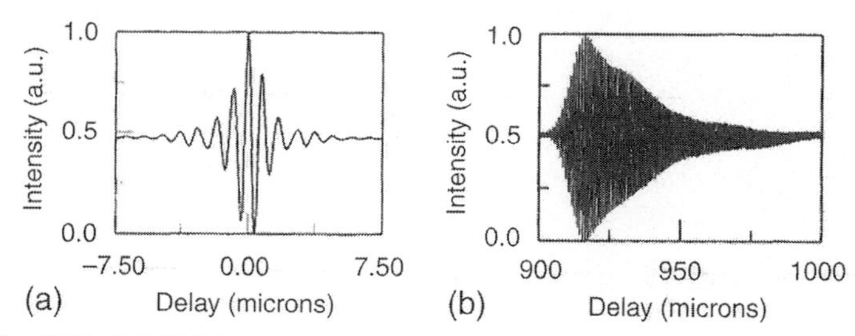

Figure 2.3 "White light" Michelson interferogram. The fringes of the balanced interferometer are shown on the left. The fringe pattern shifts to the right and is broadened by the insertion of a thin quartz plate in one arm of the interferometer.

The third (and following) terms of the expansion of $k(\Omega)$ account for the deformation of the fringe pattern observed in the recording of Figure 2.3. The propagation can be more easily visualized in the time domain for fs pulses. The group velocity delay is because of the pulse envelope "slipping" with respect to the waves. The GVD causes different parts of the pulse spectrum to travel at different velocity, resulting in pulse deformation. The result of the Michelson interferogram is a cross-correlation between the field amplitude of the "original" pulse and the signal propagated through glass.

The same considerations can be applied to white light, which can be viewed as a temporal random distribution of ultrashort pulses. The concept of incoherent radiation being constructed out of a statistical time sequence of ultrashort pulses is also useful for studying coherence in light–matter interactions, as will be studied in detail in Chapter 4 on coherent interactions. The correlation [Eqs. (2.4) and (2.5)] is maximum for exactly overlapping statistical phase and intensity fluctuations from both arms of the interferometer. These fluctuations have a duration of the order of the inverse bandwidth of the radiation, and so can be in the fs range for broad bandwidth light. Each of these individual fs spikes will travel at the group velocity. Dispersion in group velocity causes individual frequency components of these spikes to travel at different speeds, resulting most often in pulse stretching. If the source for the interferogram of Fig. 2.3 had been an fs pulse, the recording on the right of the figure would represent the cross-correlation between the field of the stretched-out pulse $\tilde{\mathcal{E}}_2(t)$ with the original (shorter) pulse $\tilde{\mathcal{E}}_1(t)$ (of which the autocorrelation is shown on the left of the figure). Such a measurement can be used to determine the shape of the field $\tilde{\mathcal{E}}_2(t)$. The limiting case of a cross-correlation between a δ function and an unknown function yields the function directly. Indeed, the unbalanced Michelson is a powerful tool that can be used for a complete determination of the shape of fs signals, in amplitude and phase, as will be seen in Chapter 9.

In the case of the incoherent radiation used for the recording of Fig. 2.3, the broadened signal on the right merely reflects the "stretching" of the statistical fluctuations of the white light. This measurement however provides important information on material properties essential in fs optics. To illustrate this point we will show how the displacement of the zero delay point in the interferogram of Fig. 2.3 can be used to determine the first terms of an expansion of the transfer function of linear optical elements.

According to Eq. (2.6), the Fourier transform of the correlation function $\tilde{A}_{11}^{+}(\tau)$ measured with the balanced interferometer[4] is simply the spectral field intensity of the source. It is difficult, and not essential, to determine exactly the zero point, and therefore the measurement generally provides $\tilde{A}_{11}^{+}(\tau + \tau_e)\exp(i\varphi_e)$, which is the function $\tilde{A}_{11}^{+}(\tau)$ with an unknown phase (φ_e) and delay (τ_e) error. Similarly, the cross-correlation measured after addition of a dielectric sample of thickness L_2 in one arm of the interferometer (right hand side of Fig. 2.3) is $\tilde{A}_{12}^{+}(\tau + \tau_f)\exp(i\varphi_f)$, which is the function $\tilde{A}_{12}^{+}(\tau)$ with an unknown phase (φ_f) and delay (τ_f) error. The ratio of the Fourier transforms of both measurements is:

$$
\begin{aligned}
\frac{\tilde{A}_{12}^{+}(\Omega)\, e^{-i(\Omega\tau_f - \varphi_f)}}{\tilde{A}_{11}^{+}(\Omega)\, e^{-i(\Omega\tau_e - \varphi_e)}} &= \frac{\tilde{E}_2(\Omega)}{\tilde{E}_1(\Omega)} e^{-i[\Omega(\tau_f - \tau_e) - (\varphi_f - \varphi_e)]} \\
&= e^{-i[k(\Omega)L + \Omega(\tau_f - \tau_e) - (\varphi_f - \varphi_e)]}, \qquad (2.10)
\end{aligned}
$$

where we have made use of Eqs. (2.6) and (2.7). Unless special instrumental provisions have been made to make ($\varphi_f = \varphi_e$), and ($\tau_f = \tau_e$), this measurement will not provide the first two terms of a Taylor expansion of the dispersion function $k(\Omega)$. This is generally not a serious limitation, because physically, the undetermined terms are only associated with a phase shift and delay of the fs pulses. The white light interferometer is an ideal instrument to determine the second- and higher-order dispersions of a sample. Writing the complex Fourier transforms of the interferograms in amplitude and phase:

$$
\begin{aligned}
\tilde{A}_{12}(\Omega) &= A_{12}(\Omega) e^{\psi_{12}(\Omega)} \\
\tilde{A}_{11}(\Omega) &= A_{11}(\Omega) e^{\psi_{11}(\Omega)} \qquad (2.11)
\end{aligned}
$$

we find that, for an order (n) larger than 1, the dispersion is simply given by:

$$
\frac{d^{(n)}k}{d\Omega^{(n)}} = -\left[\frac{d^{(n)}\psi_{12}}{d\Omega^{(n)}} - \frac{d^{(n)}\psi_{11}}{d\Omega^{(n)}}\right]. \qquad (2.12)
$$

---

[4] $\tilde{A}_{11}^{+}(\tau)$ corresponds to the third term $\tilde{A}_{12}^{+}(\tau)$ in Eq. (2.4) taken for identical beams (subscript 1 = subscript 2), not to be confused with the first term $A_{11}(0)$ in that same equation.

Equation (2.12) is not limited to dielectric samples. Instead, any optical transfer function $\tilde{H}$ which can be described by an equation similar to Eq. (1.164), can be determined from such a procedure. For instance, the preceding discussion remains valid for absorbing materials, in which case the wave vector is complex, and Eq. (2.12) leads to a complete determination of the real and imaginary part of the index of refraction of the sample versus frequency. Another example is the response of an optical mirror, as we will see in the following subsection.

2.3. DISPERSION OF INTERFEROMETRIC STRUCTURES

2.3.1. Mirror Dispersion

In optical experiments, mirrors are used for different purposes and are usually characterized only in terms of their reflectivity at a certain wavelength. The latter gives a measure about the percentage of incident light intensity that is reflected. In dealing with femtosecond light pulses, one has, however, to consider the dispersive properties of the mirror [4, 5]. This can be done by analyzing the optical transfer function which, for a mirror, is given by

$$\tilde{H}(\Omega) = R(\Omega)e^{-i\Psi(\Omega)}. \tag{2.13}$$

It relates the spectral amplitude of the reflected field $\tilde{E}_r(\Omega)$ to the incident field $\tilde{E}_0(\Omega)$

$$\tilde{E}_r(\Omega) = R(\Omega)e^{-i\Psi(\Omega)}\tilde{E}_0(\Omega). \tag{2.14}$$

Here $R(\Omega)^2$ is the reflection coefficient and $\Psi(\Omega)$ is the phase response of the mirror. As mentioned earlier a nonzero $\Psi(\Omega)$ in a certain spectral range is unavoidable if $R(\Omega)$ is frequency dependent. Depending on the functional behavior of $\Psi(\Omega)$ (cf. Section 1.3.1), reflection at a mirror not only introduces a certain intensity loss but may also lead to a change in the pulse shape and to chirp generation or compensation. These effects are usually more critical if the corresponding mirror is to be used in a laser. This is because its action is multiplied by the number of effective cavity round trips of the pulse. Such mirrors are mostly fabricated as dielectric multilayers on a substrate. By changing the number of layers and layer thickness, a desired transfer function, i.e., reflectivity and phase response, in a certain spectral range can be realized. As an example, Figure 2.4 shows the amplitude and phase response of a broadband high-reflection

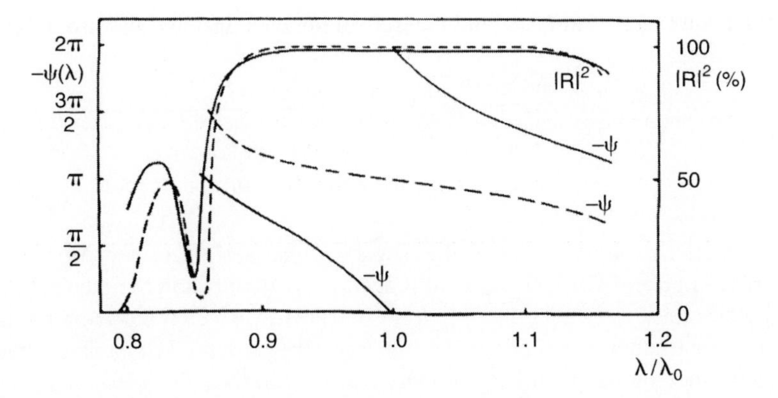

Figure 2.4 Amplitude and phase response for a high reflection multilayer mirror (dashed line) and a weak output coupler (solid line) as a function of the wavelength (Adapted from Dietel *et al.* [5]).

mirror and a weak output coupler. Note that, although both mirrors have similar reflection coefficients around a center wavelength λ_0, the phase response differs greatly. The physical explanation of this difference is that $R(\Omega)$ [or $R(\lambda)$] far from $\omega_0 = 2\pi c/\lambda_0$ (not shown) influences the behavior of $\Psi(\Omega)$ [or $\Psi(\lambda)$] near ω_0.

Before dealing with the influence of other optical components on fs pulses, let us discuss some methods to determine experimentally the mirror characteristics. In this respect the Michelson interferometer is not only a powerful instrument to analyze a sample in transmission, but it can also be used to determine the dispersion and reflection spectrum of a mirror. The interferogram from which the reference spectrum can be obtained is shown on the left of Fig. 2.3. Such a symmetric interference pattern can only be achieved in a well compensated Michelson interferometer (left part of Fig. 2.2) with identical (for symmetry) mirrors in both arms, which are also broadband (to obtain a narrow correlation pattern). For a most accurate measurement, the mirror to be measured should be *inserted* in one arm of the interferometer rather than substituted to one of the reference mirrors. Otherwise, the dispersive properties of that reference mirror cannot be canceled. In Fig. 2.2 (left), a sample mirror is indicated as the dotted line, deflecting the beam (dashed lines) towards a displaced end mirror. As in the example of the transmissive sample, insertion of the reflective sample can in general not be done without losing the relative phase and delay references. The cross-correlation measured after substitution of the sample mirror in one arm of the interferometer (right-hand side of Fig. 2.3) is $\tilde{A}^{+}_{12}(\tau + \tau_f)\exp(i\varphi_f)$, which is the function $\tilde{A}^{+}_{12}(\tau)$ with an unknown phase (φ_f) and delay (τ_f) error. The ratio of the Fourier transforms of both measurements is in analogy with Eq. (2.10):

$$\frac{\tilde{A}_{12}^{+}(\Omega)\, e^{-i(\Omega\tau_f-\varphi_f)}}{\tilde{A}_{11}^{+}(\Omega)\, e^{-i(\Omega\tau_e-\varphi_e)}} = \frac{\tilde{E}_2(\Omega)}{\tilde{E}_1(\Omega)} e^{-i[\Omega(\tau_f-\tau_e)-(\varphi_f-\varphi_e)]}$$

$$= R(\Omega)^2 e^{-i[2\Psi(\Omega)+\Omega(\tau_f-\tau_e)-(\varphi_f-\varphi_e)]}. \tag{2.15}$$

This function is independent of the dispersive and absorptive properties of the reference mirrors. The squared field reflection coefficient and the factor 2 in the phase account for the fact that the beam is reflected twice on the sample mirror. Both the amplitude R and phase Ψ of the transfer function $\tilde{H}(\Omega)$ can be extracted from the measurement, with the limitation that, in general, this measurement will not provide the first two terms of a Taylor expansion of the phase function $\Psi(\Omega)$. Again, this is not a serious limitation, because physically, the undetermined terms are only associated with a phase shift and delay of the fs pulses. Using the notations of Eq. (2.11), the phase shift $-\Psi(\Omega)$ on reflection of the mirror is simply given by:

$$\Psi(\Omega) = -\frac{1}{2}\left[\psi_{12}(\Omega) - \psi_{11}(\Omega) + a + b\Omega\right] \tag{2.16}$$

where a and b are constants that can generally not be determined from such a measurement.

The Michelson interferometer using white light is one of the simplest and most powerful tools to measure the dispersion of transmissive and reflective optics. Knox *et al.* [6] used it to measure directly the group velocity by measuring the delay induced by a sample, at selected wavelengths (the wavelength selection was accomplished by filtering white light). Naganuma *et al.* [7] used essentially the same method to measure group delays, and applied the technique to the measurement of "alpha parameters" (current dependence of the index of refraction in semiconductors) [8]. In fs lasers, the frequency dependence of the complex reflection coefficient of the mirrors contributes to an overall cavity dispersion. Such a dispersion can be exploited for optimal pulse compression, provided there is a mechanism for matched frequency modulation in the cavity. Dispersion will simply contribute to pulse broadening of initially unmodulated pulses, if no intensity or time-dependent index is affecting the pulse phase, as will be discussed later. It is therefore important to diagnose the fs response of dielectric mirrors used in a laser cavity.

A direct method is to measure the change in shape of a fs pulse, after reflection on a dielectric mirror, as proposed and demonstrated by Weiner *et al.* [9]. It is clear in the *frequency domain* that the phases of the various frequency components of the pulse are being scrambled, and therefore the pulse shape should be affected. What is physically happening in the time domain is that the various dielectric layers of the coating accumulate more or less energy at different frequencies, resulting in a delay of some parts of the pulse. Therefore, significant pulse reshaping with broadband coatings occurs only when the coherence length of the pulse length is comparable to the coating thickness. Pulses of less than 30 fs duration were used in Weiner *et al.* [9, 10]. As shown previously, determination of the dispersion in the frequency domain can be made with a simple Michelson interferometer. The latter being a *linear measurement*, yields the same result with incoherent white light illumination or femtosecond pulses of the same bandwidth.

An alternate method, advantageous for its sensitivity, but limited to the determination of the GVD, is to compare glass and coating dispersion inside a fs laser cavity. As will be seen in Chapter 5, an adjustable thickness of glass is generally incorporated in the cavity of mode-locked dye and solid state lasers, to tune the amount of GVD for minimum pulse duration. The dispersion of mirrors can be measured by substituting mirrors with different coatings in one cavity position, and noting the change in the amount of glass required to compensate for the additional dispersion [5, 11]. The method is sensitive, because the effect of the sample mirror is multiplied by the mean number of cycles of the pulse in the laser cavity. It is most useful for selecting mirrors for a particular fs laser cavity.

2.3.2. Fabry–Perot and Gires–Tournois Interferometer

So far we have introduced (Michelson) interferometers only as a tool to split a pulse and to generate a certain delay between the two partial pulses. In general, however, the action of an interferometer is more complex. This is particularly true for multiple-beam devices such as a Fabry–Perot interferometer. Let us consider for instance a symmetric Fabry–Perot, with two identical parallel dielectric reflectors spaced by a distance d. We will use the notations $\tilde{t}_{ij}$ for the field transmission, and $\tilde{r}_{ij}$ for the field reflection, as defined in Figure 2.5.

The complex field transmission function is:

$$\begin{aligned}\tilde{H}(\Omega) &= \tilde{t}_{12}\tilde{t}_{21}e^{-ikd} + \tilde{t}_{12}\tilde{t}_{21}\left(e^{-2ikd'}\cdot \tilde{r}_{21}\tilde{r}_{21}\right)e^{-ikd} \\ &\quad + \tilde{t}_{12}\tilde{t}_{21}e^{-ikd}\left(e^{-2ikd'}\ldots\ \tilde{r}_{21}\tilde{r}_{21}\right)^2 + \cdots \\ &= \tilde{t}_{12}\tilde{t}_{21}e^{-ikd}\frac{1}{1-\tilde{r}_{21}^2e^{-2ikd'}}\end{aligned} \tag{2.17}$$

where $d' = d\cos\theta$.

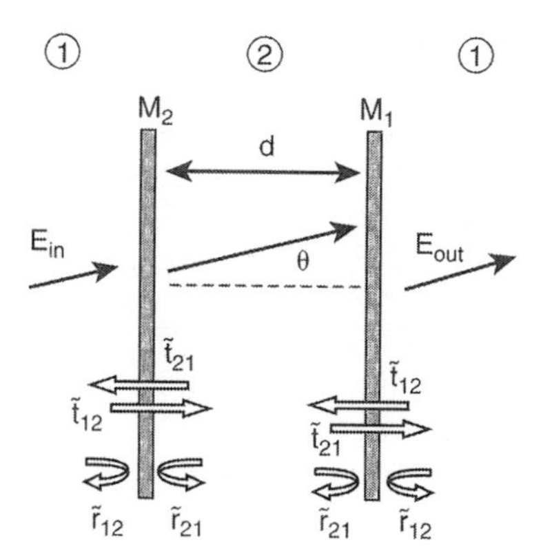

Figure 2.5 Schematic diagram of a Fabry–Perot interferometer. $\tilde{t}_{12}$ is the transmission from outside (1) to inside (2); $\tilde{t}_{21}$ the transmission from inside (2) to outside (1); $\tilde{r}_{12}$ the reflection from outside (1) to inside (2) and $\tilde{r}_{21}$ the reflection from inside (2) to outside (1).

Taking into account the interface properties derived in Appendix B, $\tilde{t}_{12}\tilde{t}_{21} - \tilde{r}_{12}\tilde{r}_{21} = 1$ and $\tilde{r}_{12} = -\tilde{r}^*_{21}$, the field transmission reduces to:

$$\boxed{\tilde{H}(\Omega) = \frac{(1-R)e^{-ikd}}{1-Re^{i\delta}}} \tag{2.18}$$

where,

$$\delta(\Omega) = 2\varphi_r - 2k(\Omega)d\cos\theta \tag{2.19}$$

is the total phase shift of a roundtrip inside the Fabry–Perot, including the phase shift φ_r on reflection on each mirror, θ is the angle of incidence on the mirrors (inside the Fabry–Perot), and $R = |\tilde{r}_{12}|^2$ is the intensity reflection coefficient of each mirror [12].

Similarly, one finds the complex reflection coefficient of the Fabry–Perot:

$$\boxed{\tilde{\mathcal{R}}(\Omega) = \frac{\sqrt{R}\left(e^{i\delta}-1\right)}{1-Re^{i\delta}}.} \tag{2.20}$$

One can easily verify that, if—and only if—kd is real:

$$|\mathcal{R}|^2 + |\mathcal{T}|^2 = 1. \tag{2.21}$$

Equations (2.18) and (2.20) are the transfer functions for the field spectrum. The dependence on the frequency argument Ω occurs through $k = n(\Omega)\Omega/c$ and possibly $\varphi_r(\Omega)$. With $n(\Omega)$ complex, the medium inside the Fabry–Perot is either an absorbing or an amplifying medium, depending on the sign of the imaginary part of the index. We refer to a problem at the end of this chapter for a study of the Fabry–Perot with gain.

The functions $\tilde{H}(\Omega)$ and $\tilde{\mathcal{R}}(\Omega)$ are complex transfer functions, which implies that, for instance, the transmitted field is:

$$\tilde{E}_{out}(\Omega) = \mathcal{T}(\Omega)\tilde{E}_{in}(\Omega) \tag{2.22}$$

where $\tilde{E}_{in}$ is the incident field. Equation (2.22) takes into account all the dynamics of the field and of the Fabry–Perot. In the case of a Fabry–Perot of thickness $d \ll c\tau_p$, close to resonance ($\delta(\Omega) \ll 1$), the transmission function $\tilde{H}(\Omega)$ is a Lorentzian, with a real and imaginary part connected by the Kramers Kronig relation. We refer to a problem at the end of this chapter to show how dispersive properties of a Fabry–Perot can be used to shape a chirped pulse.

In the case of a Fabry–Perot of thickness $d \gg c\tau_p$, the pulse spectrum covers a large number of Fabry–Perot modes. Hence the product (2.22) will represent a frequency comb, of which the Fourier transform is a train of pulses. Intuitively indeed, we expect the transmission and/or reflection of a Fabry–Perot interferometer to consist of a train of pulses of decreasing intensity if the spacing d between the two mirrors is larger than the geometrical pulse length, [Figure 2.6(a)].

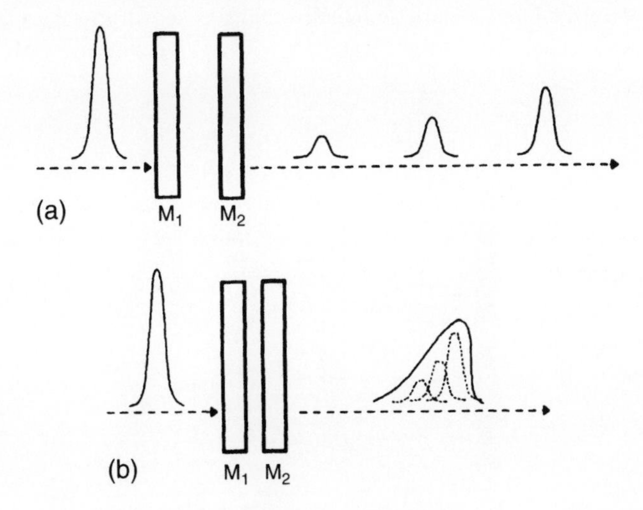

Figure 2.6 Effect of a Fabry–Perot interferometer on a light pulse if the mirror spacing is larger (a) and shorter (b) than the geometrical length of the incident pulse.

The latter condition prevents interference between field components of successive pulses. The free spectral range of the Fabry–Perot interferometer is much smaller than the spectral width of the pulse. On the other hand if d is smaller than the pulse length the output field is determined by interference, as illustrated in Fig. 2.6(b). An example of a corresponding device was the dielectric multilayer mirror discussed before, which can be considered as a sequence of many Fabry–Perot interferometers. Here the free spectral range of one interferometer is much broader than the pulse spectrum, and it is the behavior around a resonance that determines the shape of pulse envelope and phase. The actual pulse characteristics can easily be determined by multiplying the field spectrum of the incident pulse with the corresponding transfer function Eq. (2.18). For a multilayer mirror this function can be obtained from a straightforward multiplication of matrices for the individual layers [13].

Among the various types of interferometers that can be used for pulse shaping, we choose to detail here the Gires–Tournois interferometer [14]. Its striking feature is a high and almost constant amplitude transmission while the spectral phase can be tuned continuously. This device can be used to control the GVD in a fs laser in a similar manner as gratings and prisms. The Gires–Tournois is topologically identical to a ring interferometer, with all mirrors but one being perfect reflectors. The lone transmitting mirror is used as input and output. As with the Gires–Tournois, the output amplitude is unity, whether or not the wave inside the ring is at resonance or not. It is left as an exercise at the end of this chapter to transpose the formulae of the Gires–Tournois to the situation of a ring interferometer.

A sketch of the Gires–Tournois interferometer is shown in Figure 2.7. It is a special type of a Fabry–Perot interferometer with one mirror (mirror M_2) having a

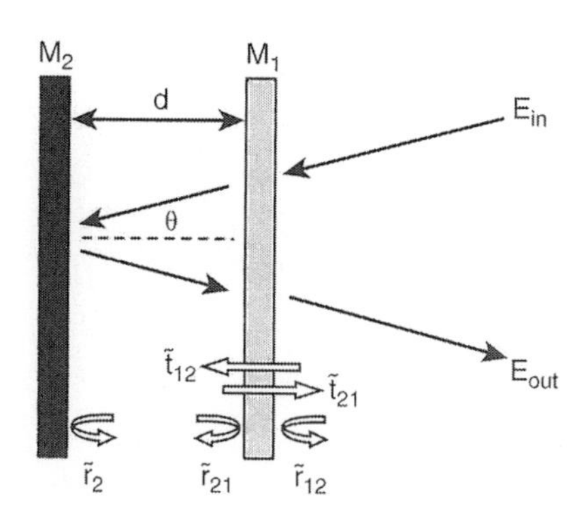

Figure 2.7 Schematic diagram of a Gires–Tournois interferometer.

reflection coefficient of (almost) 1. Consequently the device is used in reflection. In this case the transfer function is given by

$$R(\Omega)e^{-i\Psi(\Omega)} = \frac{-r + e^{i\delta}}{1 - re^{i\delta}}, \tag{2.23}$$

where δ is the phase delay[5] between two successive partial waves that leave the interferometer, and r is the (real) amplitude reflection of M_1 (assumed to be nondispersive). It can easily be shown that the reflectivity of the device is $|R| = 1$, i.e., there is practically no change in the pulse energy. The phase response determined by Eq. (2.23) can be written as

$$\Psi(\Omega) = -\arctan\left[\frac{(r^2 - 1)\sin\delta}{2r - (r^2 + 1)\cos\delta}\right] \tag{2.24}$$

Taking the derivative of both sides of this expression, and dividing by $[\tan^2\Psi + 1]$ yields:

$$\frac{d\Psi}{d\Omega} = \frac{(r^4 - 1) - 2r(r^2 - 1)\cos\delta}{(1 + r^2)^2 + 4r\cos\delta[r\cos\delta - (1 + r^2)]}\frac{d\delta}{d\Omega}. \tag{2.25}$$

It is interesting to find the expression for the group delay at the exact resonances, i.e., the values of Ω that make $\delta = 2N\pi$:

$$\left.\frac{d\Psi}{d\Omega}\right|_{res} = \left(\frac{r + 1}{r - 1}\right)\left.\frac{d\delta}{d\Omega}\right|_{res}. \tag{2.26}$$

The GVD of the device is calculated from the second derivative of the expression Eq. (2.25):

$$\begin{aligned}\frac{d^2\Psi}{d\Omega^2} &= \frac{(r^4 - 1) - 2r(r^2 - 1)\cos\delta}{(1 + r^2)^2 + 4r\cos\delta[r\cos\delta - (1 + r^2)]}\frac{d^2\delta}{d\Omega^2} \\ &+ \frac{2r(r^2 - 1)\sin\delta\left[4r(r^2 + 1)\cos\delta - 4r^2\cos^2\delta - r^2 - 3\right]}{\left\{(1 + r^2)^2 + 4r\cos\delta[r\cos\delta - (1 + r^2)]\right\}^2}\left(\frac{d\delta}{d\Omega}\right)^2.\end{aligned} \tag{2.27}$$

Note that, as pointed out in the problem at the end of this chapter, the same expressions can be derived for the transmission of a ring resonator. At the resonances

[5] In the definition of the phase delay Eq. (2.19) applied to the Gires–Tournois interferometer, θ is the *internal* angle.

of the device, $\delta = 2N\pi$, and the GVD is:

$$\left.\frac{d^2\Psi}{d\Omega^2}\right|_{res} = \left(\frac{r+1}{r-1}\right)\left.\frac{d^2\delta}{d\Omega^2}\right|_{res}. \tag{2.28}$$

As is obvious from Eq. (2.19), the second derivative of δ contains the GVD ($-k''$) of the material inside the interferometer. This material dispersion is enhanced by the factor $(r+1)/(r-1)$ in condition of resonance. This factor can be large in the case of a high finesse resonator ($1 - r \ll 1$).

The GVD given by Eq. (2.27) can be tuned continuously by adjusting δ, which can be either through a change of the mirror separation d or through a change of the external angle of incidence Θ. Gires and Tournois [14] conceived this interferometer to adapt to optical frequencies the pulse compression technique used in radar (sending a frequency modulated pulse through a dispersive delay line). Duguay and Hansen [16] were the first to apply this device for the compression of pulses from a He-Ne laser. Because typical pulse durations were in the order of several hundred ps, the mirror spacing needed to be in the order of few mm. To use the interferometer for the shaping of fs pulses the corresponding mirror spacing has to be on the order of few microns. Heppner and Kuhl [17] overcame this obvious practical difficulty by designing a Gires–Tournois interferometer on the basis of dielectric multilayer systems, as illustrated in Figure 2.8(a). The 100% mirror M_2 is a sequence of dielectric coatings with alternating refractive index deposited on a substrate. A certain spacer of optical thickness d consisting

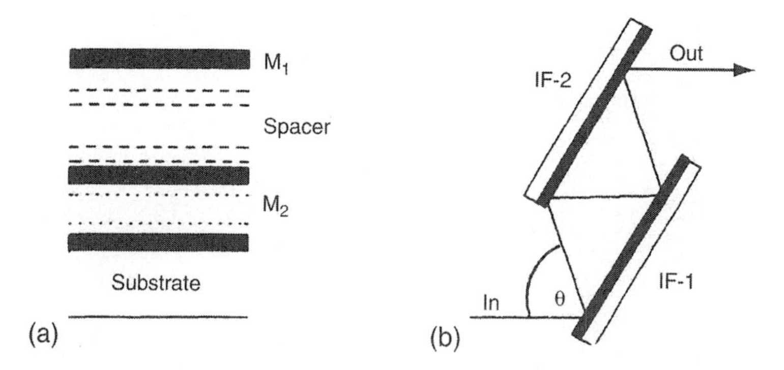

Figure 2.8 Gires–Tournois interferometer for fs light pulses using dielectric multilayers. By rotating two parallel interferometers the overall dispersion can be adjusted through a change of the external angle of incidence Θ and the number of reflections. Note, the beam direction is not changed. The lateral displacement can be compensated by a second pair of interferometers (Adapted from Kuhl and Heppner [15]).

of a series of $\lambda/2$ layers of one and the same material is placed on top of M_2. The partially reflective surface M_1 is realized by one $\lambda/4$ layer of high refractive index. The dispersion of this compact device can be tuned by changing the angle of incidence and/or the number of passes through the interferometer. A possible arrangement which was successfully applied for GVD adjustments in fs lasers is shown in Fig. 2.8(b) [15].

Let us inspect in more detail the actual transfer function of the multilayer Gires–Tournois interferometer taking into account the mirror dispersion. In most general terms, the first reflecting face of the Gires–Tournois is a multilayer dielectric coating, which we will model as an infinitely thin layer with complex reflection coefficient $\tilde{r}_{ij} = r\exp(i\varphi_{r,ij})$ and transmission coefficient $\tilde{t}_{ij} = t\exp(i\varphi_{t,ij})$. The subscripts $i,j = 1,2$ refer to air (1) and spacer dielectric (2). As indicated in Fig. 2.7, $\tilde{t}_{12}$ is the transmission coefficient from air to the spacer, through the multilayer dielectric mirror M_1; $\tilde{r}_{12}$ the reflection coefficient of M_1 to a beam incident from the air, *etc.*... Let us designate by δ the phase shift accumulated by the wave having propagated from the first reflecting layer to the total reflector and back to the first layer:

$$\delta(\Omega) = -2k(\Omega)d\cos\theta + \varphi_{r2} = -\frac{2\Omega n(\Omega)d\cos\theta}{c} + \varphi_{r2}, \tag{2.29}$$

where φ_{r2} is the phase shift on reflection at the totally reflecting layer(s) (mirror M_2), and θ is the internal angle of incidence on the reflecting interfaces. The complex (field) reflection coefficient of the structure is:

$$\begin{aligned} R(\Omega)e^{-i\Psi(\Omega)} &= \frac{\tilde{r}_{12} + (\tilde{t}_{12}\tilde{t}_{21} - \tilde{r}_{12}\tilde{r}_{21})e^{i\delta}}{1 - \tilde{r}_{21}e^{i\delta}} \\ &= \frac{\tilde{r}_{12} + e^{i\delta}}{1 - \tilde{r}_{21}e^{i\delta}}, \end{aligned} \tag{2.30}$$

where the last equality results from the relation between the complex amplitude reflection and transmission derived in Appendix B ($\tilde{t}_{12}\tilde{t}_{21} - \tilde{r}_{12}\tilde{r}_{21} = 1$). The reflectivity of the device is:

$$|R(\Omega)|^2 = \frac{|\tilde{r}_{12}|^2 + 1 + \tilde{r}_{12}e^{-i\delta} + \tilde{r}_{12}^*e^{i\delta}}{|\tilde{r}_{21}|^2 + 1 - \tilde{r}_{21}^*e^{-i\delta} - \tilde{r}_{21}e^{i\delta}}, \tag{2.31}$$

which is only equal to unity if $\tilde{r}_{21} = -\tilde{r}_{12}^*$ and the media are lossless. This relation is consistent with the phase shift upon reflection on a dielectric interface. The expression for the complex reflection of the whole interferometer can be rewritten in terms of the reflection coefficient $\tilde{r}_{21}$:[6]

$$R(\Omega)e^{-i\Psi(\Omega)} = \frac{-\tilde{r}_{21}^{*} + e^{i\delta}}{1 - \tilde{r}_{21}e^{i\delta}}. \tag{2.32}$$

Let us express the reflectivity $\tilde{r}_{21}$ in terms of its amplitude and phase: $\tilde{r}_{21} = r_{21}\exp(i\varphi_{r,21}) = r\exp(i\varphi_r)$. The phase response of the interferometer can now be calculated:

$$\Psi(\Omega) = -\arctan\frac{(r^2-1)\sin\delta(\Omega) - r\sin\varphi_r(\Omega)}{2r\cos\varphi_r(\Omega) - (1+r^2)\cos\delta(\Omega)}, \tag{2.33}$$

which is a generalization of Eq. (2.24) to the more complex multilayer Gires–Tournois structure. Only when $\varphi_r = 0$ and φ_{r2} (in δ) is frequency independent in the range of interest are the dispersions described by Eqs. (2.24) and (2.33) equal. The error may be small in some real situations, as can be seen from the comparison of the approximation Eq. (2.24) with the exact phase $\Psi(\Omega)$ shown in Figure 2.9. The latter functional dependence can be calculated directly through matrix algebra taking into account a certain sequence of dielectric multilayer mirrors [13].

2.3.3. Chirped Mirrors

As mentioned in the previous section, the Gires–Tournois interferometer exhibits a reflectivity close to one over a broad spectrum. This was accomplished by an end mirror of high reflectivity (M_2 in Fig. 2.8). The dispersion on the other hand can be controlled by the spacer and the front mirror. This is expressed in the phase $\delta(\Omega)$ and $\varphi_r(\Omega)$ in Eq. (2.33). The problem is that both mirrors at the same time form a Fabry–Perot structure that has relatively narrow resonances and subsequently a rather complicated dispersion behavior. The most desired alternative would be a process to generate a predefined reflection and phase behavior, $R(\Omega)$ and $\Psi(\Omega)$. Optimization programs applied to dielectric multilayer systems offer such an intriguing and interesting possibility. A dielectric multilayer system consists of a sequence of films characterized by a certain refractive index n_i and thickness d_i. In principle, computer algorithms can be used to find a sequence of (d_i, n_i) combinations representing individual films that come closest to a predefined reflection and phase behavior in a certain spectral range. Of course, there are certain technical constraints that need to be considered: for example the total thickness and number of layers, the manufacturing

---

[6] We recall that $\tilde{r}_{21}$ is the complex field reflection coefficient from inside the Gires–Tournois, on the multilayer dielectric coating, assumed to be infinitely thin.

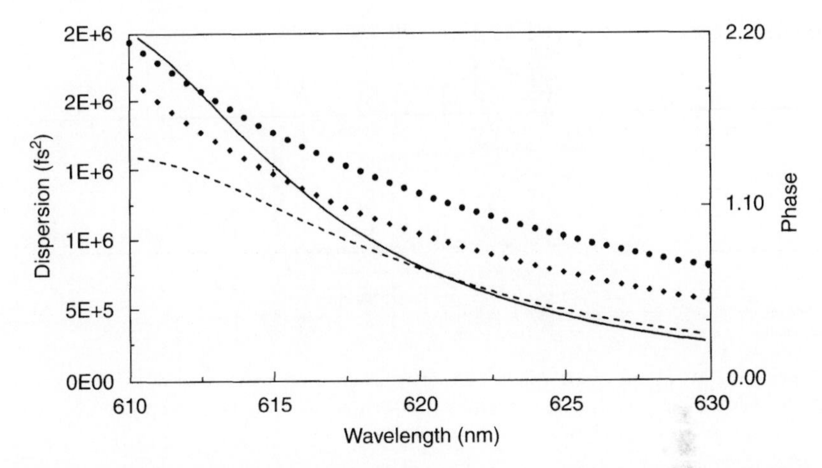

Figure 2.9 Comparison of the exact phase function $\psi(\lambda)$ (diamonds) and its second derivative $d^2\psi/d\Omega^2$ (solid line) of a Gires–Tournois interferometer with the approximation of assuming constant φ_r and φ_{r2} in Eq. (2.33) (circles for ψ, dashed line for the second derivative). The Gires–Tournois interferometer has been designed for a central wavelength $\lambda_0 = 620$ nm. The curves are calculated for an external angle of incidence of 20°. For this particular example, the high reflector is made of 11 layers of TiO_2 (thickness $\lambda_0/4n = 67.4$ nm) alternating with layers of SiO_2 (thickness $\lambda_0/4n = 106.9$ nm) on a glass substrate ($n = 1.5$). The spacer consists of five half wave spacing of SiO_2, for a total thickness of 1068.9 nm. The top reflector ($r = 0.324$) consists of a quarter wave layer of TiO_2 (thickness $\lambda_0/4n = 67.4$ nm). In applying Eq. (2.33), the values of phase shifts upon reflection were $\varphi_r = -0.0192$ (top layer) and $\varphi_{r2} = -0.0952$ (high reflector).

tolerances in n_i and d_i, and the limited choice of available refractive indices n_i (suitable materials). This approach will gain importance in the future as the amplitude and phase responses needed become more and more complicated.

In many cases mirrors are desired that have a constant reflectivity and certain dispersion behavior, for example a constant amount of GVD within a predefined spectral range. This idea was pursued by Szipöks *et al.*, [18] leading to what is now called chirped mirrors. The basic idea is sketched in Figure 2.10. High-reflection mirrors typically consist of stacks of alternating high and low refractive index quarter-wave layers. A chirped mirror is a sequence of those stacks with changing resonance frequency. Wave packets of different center frequency are thus reflected at different depths, making the group delay on reflection a function of frequency.

Unfortunately this is an oversimplified picture that neglects subresonances in particular between the layers and the first air–film interface. This leads to a modulation of the GVD. For this reason computer optimization is necessary to tune the film parameters for a smooth dispersion curve.

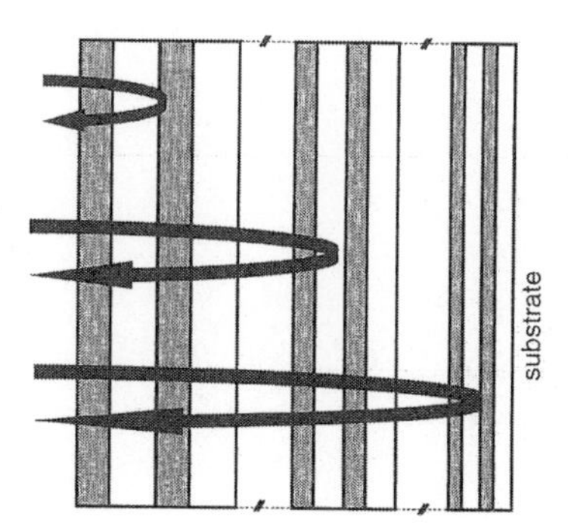

Figure 2.10 Wave packets of different center frequencies are reflected at different depths of a chirped mirror. The mirror consists of stacks of alternating high and low refractive index layers at different resonance frequencies.

Improvements in the initial layer sequence used as a starting point for the final computer optimization have been accomplished, for example, by modulating the ratio of the thickness of the high- and low-index layer of the chirped mirror (double-chirped mirror) [19], by superimposing a quasi-periodic modulation on the linear modulation of the layer thickness [20], and by coating the backside of the substrate [21,22]. As we will see in following chapters such mirrors have made an impressive impact on femtosecond laser source development.

2.4. FOCUSING ELEMENTS

2.4.1. Singlet Lenses

One main function of fs pulses is to concentrate energy in time and space. The ability to achieve extremely high peak power densities partly depends on the ability to keep pulses short in time, and concentrate them in a small volume by focusing. The difference between group and phase velocity in the lens material can reduce the peak intensity in the focal plane by delaying the time of arrival of the pulse front propagating through the lens center relative to the pulse front propagating along peripheral rays. The group velocity dispersion leads to reduction of peak intensity by stretching the pulse in time. As pointed out by Bor [23, 24], when simple focusing singlet lenses are used, the former effect can lead to several picosecond lengthening of the time required to deposit the energy of a fs pulse on focus.

Let us assume a plane pulse and phase front at the input of a spherical lens as sketched in Figure 2.11. According to Fermat's principle, the optical path

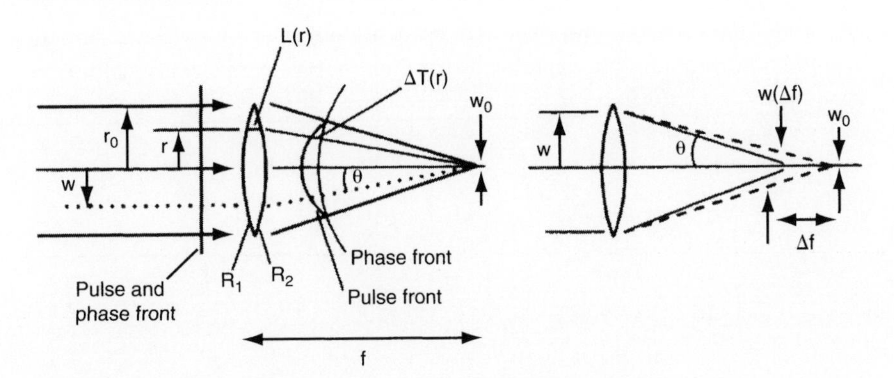

Figure 2.11 Left: delay of the pulse front with respect to the phase front, in the case of a singlet lens. Right: spread of the focal region due to chromatic aberration.

along rays from the input phase front to the focus is independent of the radius coordinate r. The lens transforms the plane phase front into a spherical one which converges in the (paraxial) focus. Assimilating air as vacuum, it is only while propagating through the lens that the pulses experience a group velocity ν_g different from the phase velocity $\nu_p = c/n$. The result is a pulse front that is delayed with respect to the (spherical) phase front, depending on the amount of glass traversed. As we have seen in Chapter 1, the group velocity is:

$$\nu_g = \left(\frac{dk}{d\Omega}\right)^{-1} = \frac{c}{n - \lambda_\ell \frac{dn}{d\lambda}}, \tag{2.34}$$

where λ_ℓ is the wavelength in vacuum. The difference in propagation time between the phase front and pulse front after the lens at radius coordinate r is:

$$\Delta T(r) = \left(\frac{1}{\nu_p} - \frac{1}{\nu_g}\right) L(r), \tag{2.35}$$

where $L(r)$ is the lens thickness. The group delay $\Delta T(r)$ is also the difference of the time of arrival at the focus of pulses traversing the lens at distance r from the axis and peripheral rays touching the lens rim. Pulse parts traveling on the axis ($r = 0$) will arrive delayed in the focal plane of a positive lens compared with pulse parts traversing the lens at $r > 0$. For a spherical thin lens, the thickness L is given by

$$L(r) = \frac{r_0^2 - r^2}{2}\left(\frac{1}{R_1} - \frac{1}{R_2}\right) \tag{2.36}$$

where $R_{1,2}$ are the radii of curvature of the lens, and r_0 is the radius of the lens aperture.[7] Substituting the expressions for the group velocity (2.34) and for the lens thickness (2.36) into Eq. (2.35) yields for the difference in time of arrival between a pulse passing through the lens at the rim and at r:

$$\Delta T(r) = \frac{r_0^2 - r^2}{2c}\left(\frac{1}{R_1} - \frac{1}{R_2}\right)\left(\lambda\frac{dn}{d\lambda}\right)$$
$$= \frac{r_0^2 - r^2}{2c}\lambda\frac{d}{d\lambda}\left(\frac{1}{f}\right) \tag{2.37}$$

where the focal length f has been introduced by $1/f = (n-1)(R_1^{-1} - R_2^{-1})$. Equation (2.38) illustrates the connection between the radius-dependent pulse delay and the chromaticity $d/d\lambda(1/f)$ of the lens. For an input beam of radius r_b the pulse broadening in the focus can be estimated with the difference in arrival time $\Delta T'$ of a pulse on an axial ray and a pulse passing through the lens at r_b:

$$\boxed{\Delta T'(r_b) = \frac{r_b^2}{2c}\lambda\frac{d}{d\lambda}\left(\frac{1}{f}\right).} \tag{2.38}$$

To illustrate the effects of group velocity delay and dispersion, let us assume that we would like to focus a 50 fs pulse at the excimer laser wavelength of 248 nm (KrF) down to a spot size of 0.6 μm, using a fused silica lens (singlet) of focal distance $f = 30$ mm. Let us further assume that the input beam profile is Gaussian. Because the half divergence angle in the focused beam is $\theta = \lambda/(\pi w_0)$, the radius w of the Gaussian beam [radial dependence of the electric field: $\tilde{\mathcal{E}}(r) = \tilde{\mathcal{E}}(0)\exp\{-r^2/w^2\}$] incident on the lens should be approximately $\theta f = (\lambda/\pi w_0)f \approx 4$ mm. To estimate the pulse delay we evaluate $\Delta T'$ at $r_b = w$:

$$\Delta T'(r_b = w) = \frac{w^2}{2c}\lambda\frac{d}{d\lambda}\left(\frac{1}{f}\right)$$
$$= -\frac{w^2}{2cf(n-1)}\left(\lambda\frac{dn}{d\lambda}\right)$$
$$= -\frac{\theta^2 f}{2c(n-1)}\left(\lambda\frac{dn}{d\lambda}\right). \tag{2.39}$$

[7] Regarding sign considerations we will use positive (negative) $R_{1,2}$ for refracting surfaces which are concave (convex) toward the incident side.

For the particular example chosen, $n \approx 1.51$, $\lambda dn/d\lambda \approx -0.17$, and the difference in time of arrival (at the focus) of the rays at $r = 0$ and $r_b = w$ is ≈ 300 fs, which can be used as a rough measure of the pulse broadening.

The effect of the chromaticity of the lens on the spatial distribution of the light intensity near the focal plane is a spread of the optical energy near the focus, because different spectral components of the pulse are focused at different points on axis. For a bandwidth-limited Gaussian pulse of duration $\tau_p = \sqrt{2 \ln 2}\, \tau_{G0}$ with spectral width $\Delta\lambda = 0.441\lambda^2/c\tau_p$, the focus spreads by the amount:

$$\Delta f = -f^2 \frac{d(1/f)}{d\lambda}\Delta\lambda = -\frac{f\lambda^2}{c(n-1)}\frac{0.441}{\tau_p}\frac{dn}{d\lambda}. \tag{2.40}$$

Applying Eq. (2.40) to our example of a 30 mm fused silica lens to focus a 50 fs pulse, we find a spread of $\Delta f = 60$ µm, which is large compared to the Rayleigh range of a diffraction-limited focused monochromatic beam $\rho_0 = w_0/\theta \approx 5$ µm. We can therefore write the following approximation for the broadening of the beam: $w(\Delta f)/w_0 = \sqrt{1 + (\Delta f/2\rho_0)^2} \approx (\Delta f/2\rho_0)$. Substituting the value for Δf from Eq. (2.40):

$$\frac{w(\Delta f)}{w_0} = -\frac{0.44\pi}{\tau_p}\frac{\theta^2 f}{2c(n-1)}\left(\lambda\frac{dn}{d\lambda}\right) \approx -0.44\pi\frac{\Delta T'}{\tau_p}. \tag{2.41}$$

We note that the spatial broadening of the beam because of the spectral extension of the pulse, as given by Eq. (2.41), is (within a numerical factor) the same expression as the group velocity delay [Eq. (2.39)] relative to the pulse duration. In fact, neither expression is correct, in the sense that they do not give a complete description of the spatial and temporal evolution of the pulse near the focus. An exact calculation of the focalization of a fs pulse by a singlet is presented in the subsection that follows.

In addition to the group delay effect, there is a direct temporal broadening of the pulse *in the lens itself* because of GVD in the lens material, as discussed in Section 1.5. Let us take again as an example the fused silica singlet of 30 mm focal length and of 16 mm diameter used to focus a 248 nm laser beam to a 0.6 µm spot size. The broadening will be largest for the beam on axis, for which the propagation distance through glass is $L(r = 0) = d_0 = r_0^2/\{2f(n-1)\} = 2.1$ mm. Using for the second-order dispersion at 250 nm $\lambda d^2n/d\lambda^2 \approx 2.1$ µm^{-1} [23], we find from Eq. (1.125) that a 50 fs (FWHM) unchirped Gaussian pulse on axis will broaden to about 60 fs. If the pulse has an initial upchirp such that the parameter a defined in Eq. (1.33) is $a = -5$, it will broaden on axis to 160 fs. At a wavelength of 800 nm, where the dispersion is much smaller than in the UV (Table 2.1) a bandwidth-limited 50 fs pulse would only broaden to 50.4 fs.

The example illustrates the differences between peak intensity reduction at the focal point of a lens resulting from the difference between group and phase velocity and effects of GVD in the lens material. The latter is strongly chirp dependent, while the former is not. The spread of pulse front arrival times in the focal plane is independent of the pulse duration and is directly related to the spot size that will be achieved (the effect is larger for optical arrangements with a large F-number). The *relative* broadening of the focus, $\propto \Delta T'/\tau_p$, is however larger for shorter pulses. The GVD effect is pulse width dependent, and, in typical materials, becomes significant only for pulse durations well below 100 fs in the VIS and NIR spectral range.

2.4.2. Space–Time Distribution of the Pulse Intensity at the Focus of a Lens

The geometrical optical discussion of the focusing of ultrashort light pulses presented previously gives a satisfactory order of magnitude estimate for the temporal broadening effects in the focal plane of a lens. We showed this type of broadening to be associated with chromatic aberration. Frequently the experimental situation requires an optimization not only with respect to the temporal characteristics of the focused pulse, but also with respect to the achievable spot size. To this aim we need to analyze the space–time distribution of the pulse intensity in the focal region of a lens in more detail. The general procedure is to solve either the wave equation (1.67), or better the corresponding diffraction integral,[8] which in Fresnel approximation was given in Eq. (1.187). However, we cannot simply separate space and time dependence of the field with a product ansatz (1.176) because we expect the chromaticity of the lens to induce an interplay of both. Instead we will solve the diffraction integral for each "monochromatic" Fourier component of the input field $\tilde{E}_0(\Omega)$ which will result in the field distribution in a plane (x, y, z) behind the lens, $\tilde{E}(x, y, z, \Omega)$. The time-dependent field $\tilde{E}(x, y, z, t)$ then is obtained through the inverse Fourier transform of $\tilde{E}(x, y, z, \Omega)$ so that we have for the intensity distribution:

$$I(x, y, z, t) \propto |\mathcal{F}^{-1}\{\tilde{E}(x, y, z, \Omega)\}|^2. \tag{2.42}$$

The geometry of this diffraction problem is sketched in Figure 2.12. Assuming plane waves of amplitude $E_0(\Omega) = E_0(x', y', z' = 0, \Omega)$ at the lens input,

[8] For large F-numbers the Fresnel approximation may no longer be valid, and the exact diffraction integral including the vector properties of the field should be applied.

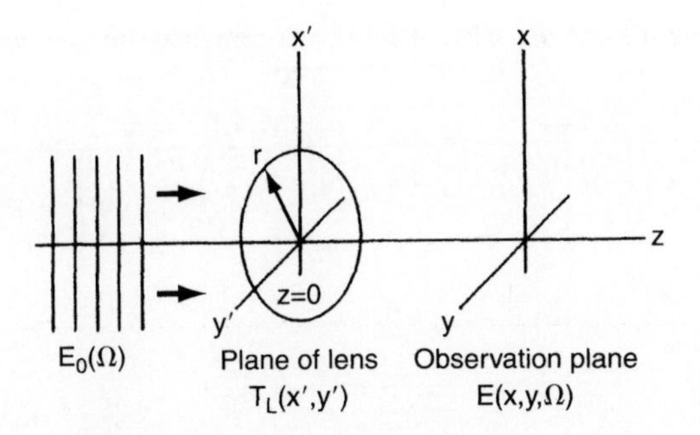

Figure 2.12 Diffraction geometry for focusing.

the diffraction integral to be solved reads, apart from normalization constants:

$$E(x, y, z, \Omega) \propto \frac{\Omega}{c} \iint E_0(\Omega) T_L(x', y') T_A(x', y') e^{-i\frac{k}{2z}[(x'-x)^2+(y'-y)^2]} dx' dy' \tag{2.43}$$

where T_L and T_A are the transmission function of the lens and the aperture stop, respectively. The latter can be understood as the lens rim in the absence of other beam limiting elements. The lens transmission function describes a radially dependent phase delay which in case of a thin, spherical lens can be written:

$$T_L(x', y') = \exp\left\{-i\frac{\Omega}{c}\left[nL(r') + d_0 - L(r')\right]\right\} \tag{2.44}$$

with $r'^2 = x'^2 + y'^2$ and

$$L(r') = d_0 - \frac{r'^2}{2}\left(\frac{1}{R_1} - \frac{1}{R_2}\right) = d_0 - \frac{r'^2}{2(n-1)f}, \tag{2.45}$$

where d_0 is the thickness in the lens center. Note that because of the dispersion of the refractive index n, the focal length f becomes frequency dependent. For a spherical opening of radius r'_0 the aperture function T_A is simply:

$$T_A(r') = \begin{cases} 1 & \text{for} \quad x'^2 + y'^2 = r'^2 \le r_0'^2 \\ 0 & \text{otherwise} \end{cases} \tag{2.46}$$

If we insert Eq. (2.45) into Eq. (2.44) we can rewrite the lens transmission function as:

$$T_L(x', y') = \exp\left\{-i\left[k_g(\Omega)d_0 - \left(k_g(\Omega) - \frac{\Omega}{c}\right)\frac{r'^2}{2}\left(\frac{1}{R_1} - \frac{1}{R_2}\right)\right]\right\}, \quad (2.47)$$

where

$$k_g(\Omega) = \frac{\Omega}{c}n(\Omega) \quad (2.48)$$

is the wave vector in the glass material. Substituting this transmission function in the diffraction integral Eq. (2.43) we find for the field distribution in the focal plane:

$$E(\Omega) \propto \frac{\Omega}{c}e^{-ik_g(\Omega)d_0}\iint T_A E_0(\Omega)\exp\left[i\left(k_g(\Omega) - \frac{\Omega}{c}\right)\frac{r'^2}{2}\left(\frac{1}{R_1} - \frac{1}{R_2}\right)\right]$$
$$\times\, e^{-i\frac{k}{2z}[(x'-x)^2+(y'-y)^2]}dx'dy' \quad (2.49)$$

The exponent of the second exponential function is radially dependent and is responsible for the focusing, while the first one describes propagation through a dispersive material of length d_0. For a closer inspection let us assume that the glass material is only weakly dispersive so that we may expand $k_g(\Omega)$ and $[k_g(\Omega) - \Omega/c]$ up to second order. In both exponential functions this will result in a sum of terms proportional to $(\Omega - \omega_\ell)^m$ $(m = 0, 1, 2)$. According to our discussion in the section about linear elements, optical transfer functions which have the structure $\exp[-ib_1(\Omega - \omega_\ell)]$ give rise to a certain pulse delay. Because b_1 is a function of r' this delay becomes radius dependent, a result which has already been expected from our previous ray–optical discussion. The next term of the expansion $(m = 2)$ is responsible for pulse broadening in the lens material.

A numerical evaluation of Eq. (2.49) and subsequent inverse Fourier transform [Eq. (2.42)] allows one to study the complex space–time distribution of the pulse intensity behind a lens. An example is shown in Figure 2.13. In the aberration-free case we recognize a spatial distribution corresponding to the Airy disc and no temporal distortion. The situation becomes more complex if chromaticity plays a part. We see spatial as well as temporal changes in the intensity distribution. At earlier times the spatial distribution is narrower. This can easily be understood if we remember that pulses from the lens rim (or aperture edge) arrive first in the focal plane and are responsible for the field distribution. At later times pulses from inner parts of the lens arrive. The produced spot becomes larger since the effective aperture size is smaller. If we use achromatic doublets (cf. Section 2.4.3)

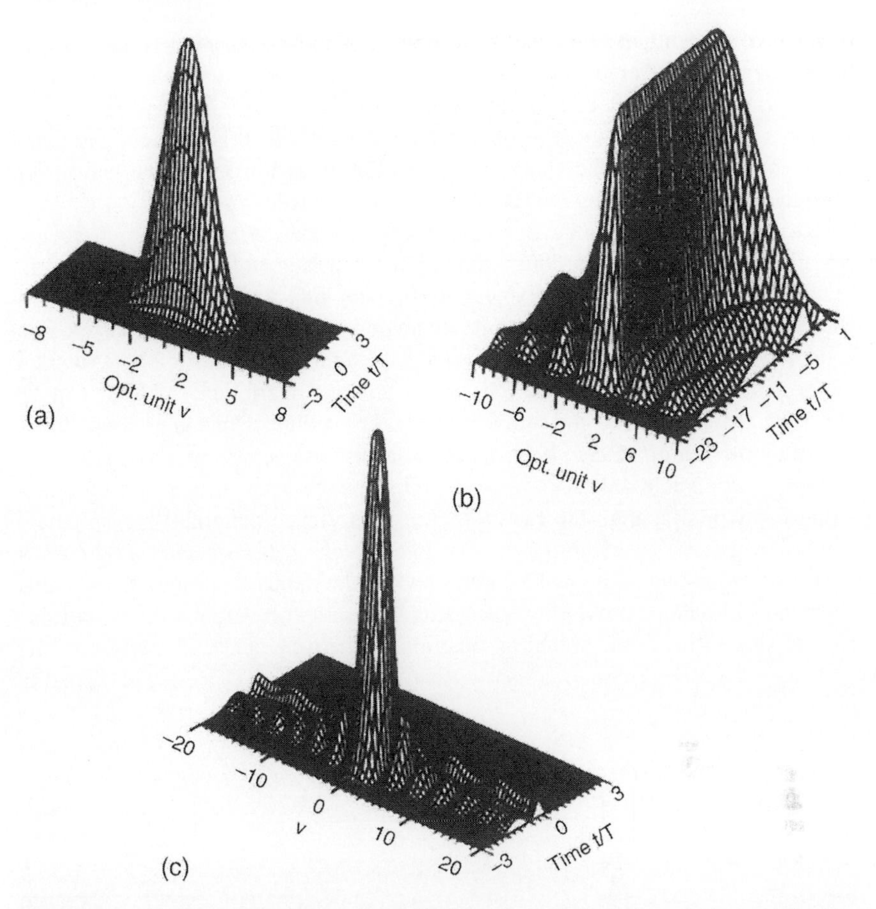

Figure 2.13 Space–time distribution of the pulse intensity in the focal plane of a lens: (a) Focusing without chromatic or spherical aberration, (b) Focusing with chromatic aberration $\tau/T = 20$. The input pulse was chosen to vary as $e^{-(t/T)^2}$. ν is the optical coordinate defined as $\nu = r'_0 k_\ell \sqrt{x^2 + y^2}/f_\ell$ and $\tau = T'(r_0) = \left| \frac{r_0'^2 \lambda}{2fc(n-1)} n'(\lambda) \right|$ is a measure for the dispersion (from Kempe *et al.* [25]). (c) Focusing with chromatic and spherical aberration. The intensity distribution in the plane of the marginal focus is shown (Adapted from Kempe and Rudolph [26]).

the exponential proportional to $(\Omega - \omega_\ell)$ in the corresponding diffraction integral does not appear. The only broadening then is because of GVD in the glass material.

Interesting effects occur if spherical aberration is additionally taken into account which is essential to correctly model strong focusing with singlet lenses [26]. As known from classical optics, spherical aberration results in different focal planes for beams passing through the lens at different r. Because ultrashort pulses passing through different lens annuli experience the same delay, almost no temporal broadening occurs for the light which is in focus, as illustrated in Fig. 2.13(c). The space–time distribution in the focal area can differ substantially from that obtained with a purely chromatic lens.

To measure the interplay of chromatic and spherical aberration in focusing ultrashort light pulses, one can use an experimental setup as shown in Figure 2.14(a). The beam is expanded and sent into a Michelson interferometer. One arm contains the lens to be characterized, which can be translated so as to focus light passing through certain lens annuli onto mirror M_1. Provided the second arm has the proper length, an annular interference pattern can be observed at the output of the interferometer. The radius of this annulus is determined by the setting of Δf. If no spherical aberration is present, an interference pattern is observable only for $\Delta f \approx 0$, and a change of the time delay by translating M_2 would change the radius of the interference pattern. With spherical aberration present, at a certain Δf, an interference pattern occurs only over a delay corresponding to the pulse duration while the radius of the annulus remains constant. This can be proved by measuring the cross-correlation, i.e., by measuring the second harmonic signal as function of the time delay. The width of the

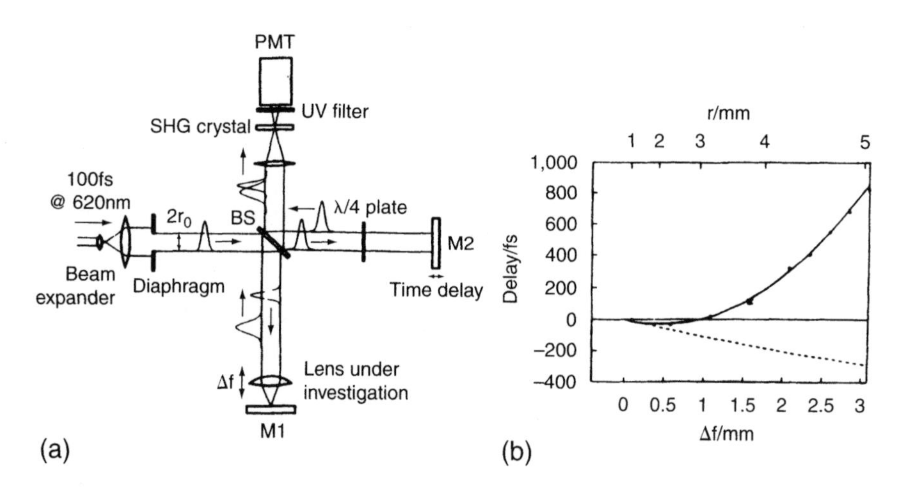

Figure 2.14 (a) Correlator for measuring the effect of chromatic and spherical aberration on the focusing of fs pulses. (b) Measured pulse delay (data points) as a function of the lens position (Δf—deviation from the paraxial focus) and the corresponding radial coordinate r of light in focus. The solid line is obtained with ray–pulse tracing; the dashed curve shows the effect of chromatic aberration only. Lens parameters: $f_0 = 12.7$ mm, BK7 glass (Adapted from Kempe and Rudolph [27]).

cross-correlation does not differ from the width of the autocorrelation which is measured without the lens in the interferometer arm. Figure 2.14(b) shows the position of the peak of the cross-correlation as function of Δf and the corresponding r, respectively. For comparison, the delay associated with chromatic aberration alone is also shown (dashed curve).

2.4.3. Achromatic Doublets

The chromaticity of a lens was found to be the cause for a radial dependence of the time of arrival of the pulse at the focal plane, as was shown by Eq. (2.38). Therefore one should expect achromatic optics to be free of this undesired pulse lengthening. To verify that this is indeed the case, let us consider the doublet shown in Figure 2.15.

The thicknesses of glass traversed by the rays in the media of index n_1 and n_2 are L_1 and L_2 and are given by:

$$L_1 = d_1 - \frac{r^2}{2}\left(\frac{1}{R_1} - \frac{1}{R_2}\right) \tag{2.50}$$

and

$$L_2 = d_2 - \frac{r^2}{2}\left(\frac{1}{R_2} - \frac{1}{R_3}\right) \tag{2.51}$$

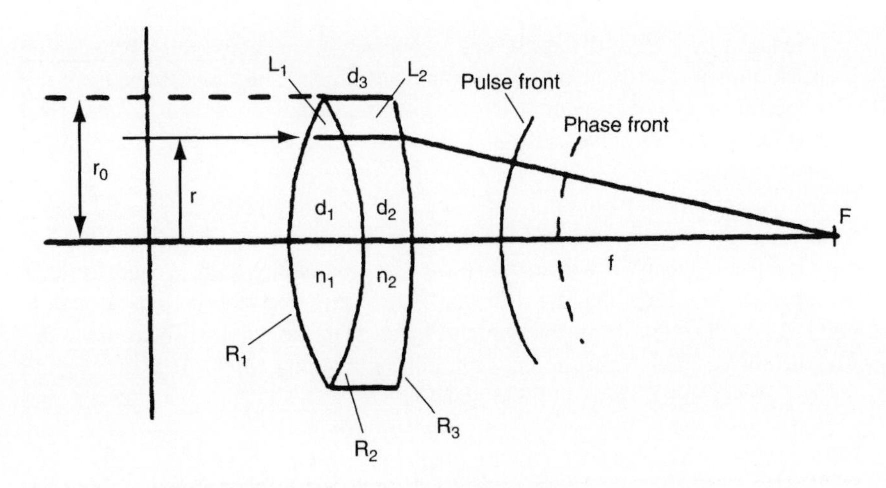

Figure 2.15 Ray tracing in an achromat (Adapted from Bor [23]).

where $d_{1,2}$ is the center thickness of lens 1, 2. The inverse of the focal length of the doublet lens is:

$$\frac{1}{f} = (n_1 - 1)\left(\frac{1}{R_1} - \frac{1}{R_2}\right) + (n_2 - 1)\left(\frac{1}{R_2} - \frac{1}{R_3}\right). \quad (2.52)$$

The condition of achromaticity $\frac{d}{d\lambda}(1/f) = 0$ gives an additional relation between the radii of curvature R_i and the indices n_i. The expression for the transit time in glass in which we have inserted the chromaticity of the doublet is [23]:

$$\boxed{T(r) = \frac{d_1}{c}\left\{n_1 - \lambda\frac{dn_1}{d\lambda}\right\} + \frac{d_2}{c}\left\{n_2 - \lambda\frac{dn_2}{d\lambda}\right\} + \frac{\lambda r^2}{2c}\frac{d}{d\lambda}\left(\frac{1}{f}\right).} \quad (2.53)$$

Equation (2.53) indicates that, for an achromatic doublet for which the third term on the right-hand side vanishes, the transit time has no more radial dependence. The phase front and wave front are thus parallel, as sketched in Fig. 2.15. In this case, the only mechanism broadening the pulse at the focus is GVD. The latter can be larger than with singlet lenses since achromatic doublets usually contain more glass.

2.4.4. Focusing Mirrors

Another way to avoid the chromatic aberration and thus pulse broadening is to use mirrors for focusing. With spherical mirrors and on-axis focusing the first aberration to be considered is the spherical one. The analysis of spherical aberration of mirrors serves also as a basis to the study of spherical aberration applied to lenses.

Let us consider the situation of Figure 2.16, where a plane pulse- and wave-front impinge upon a spherical mirror of radius of curvature R. The reflected rays are the tangents of a caustic—the curve commonly seen as light reflects off a coffee cup. Rays that are a distance r off-axis intersect the optical axis at point T which differs from the paraxial focus F in the paraxial focal plane Σ'. The difference in arrival time between pulses traveling along off-axis rays and on-axis pulses in the paraxial focal plane is:

$$\Delta T = \frac{1}{c}\left[\overline{VQ} - \left(\overline{PS} + \frac{R}{2}\right)\right]. \quad (2.54)$$

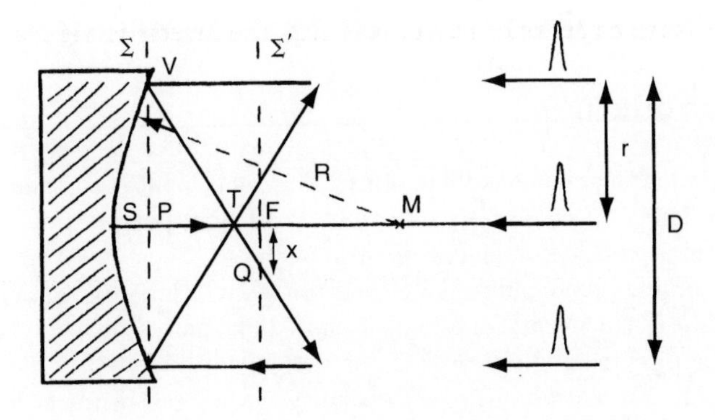

Figure 2.16 Focusing of light pulses by a spherical mirror.

Through simple geometrical considerations one can find an expression for ΔT in the form of an expansion in powers of (r/R). The first nonzero term of that expansion is:

$$\Delta T = \frac{3}{4}\frac{R}{c}\left(\frac{r}{R}\right)^4. \tag{2.55}$$

Likewise, one obtains for the geometrical deviation from the paraxial focus in Σ':

$$x = \frac{R}{2}\left(\frac{r}{R}\right)^3. \tag{2.56}$$

For a beam diameter $D = 3$ mm and a focal length $f = 25$ mm the arrival time difference amounts to only 0.1 fs and the deviation from the paraxial focus $x \approx 1$ µm. The numbers increase rapidly with beam size; $\Delta T \approx 13$ fs, $x \approx 25$ µm for $D = 10$ mm, for example.

In experimental situations where even a small aberration should be avoided, parabolic mirrors can advantageously be used to focus collimated input beams. An example requiring such optics is upconversion experiments where fluorescence with fs rise time from a large solid angle has to be focused tightly, without modifying its temporal behavior. Elliptical mirrors should be used to focus light emerging from a point source. However, because parabolic mirrors are more readily available, a combination of parabolic mirrors may be used in lieu of an ellipsoid.

2.5. ELEMENTS WITH ANGULAR DISPERSION

2.5.1. Introduction

Besides focusing elements there are various other optical components which modify the temporal characteristics of ultrashort light pulses through a change of their spatial propagation characteristics.

Even a simple prism can provide food for thought in fs experiments. Let us consider an expanded parallel beam of short light pulses incident on a prism, and diffracted by the angle $\beta = \beta(\Omega)$, as sketched in Figure 2.17. As discussed in Chapter 1, a Gaussian beam with beam waist w_0 self-diffracts by an angle of approximately $\theta = \lambda/\pi w_0$. In the case of a short pulse (or white light), this diffraction has to be combined with a spectral diffraction, because the light is no longer monochromatic, and different spectral components will be deflected by the prism with a different angle $\beta = \beta(\Omega)$. If the pulse is sufficiently short, both effects are of the same order of magnitude, resulting in a complex space–time problem that can no longer be separated. Throughout this section, whether considering group delays or GVD, we will consider sufficiently broad beams and sufficiently short propagation distances L_p behind the prism. This will allow us to neglect the change in beam diameter because of propagation and spectral diffraction after the prism. In most cases we will also approximate the beam with a flat profile. At the end of this chapter the interplay of propagation and spectral diffraction effects will be discussed for Gaussian beams.

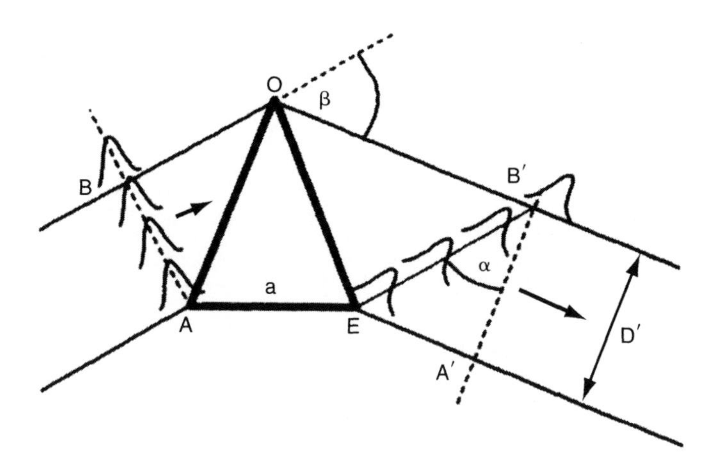

Figure 2.17 Pulse front tilt introduced by a prism. The position of the (plane) wavefronts is indicated by the dashed lines $\overline{AB}$ and $\overline{A'B'}$.

As discussed by Bor [24], the prism introduces a tilt of the pulse front with respect to the phase front. As in lenses, the physical origin of this tilt is the difference between group and phase velocity. According to Fermat's principle the prism transforms a phase front $\overline{AB}$ into a phase front $\overline{A'B'}$. The transit times for the phase and pulse fronts along the marginal ray $\overline{BOB'}$ are equal ($v_p \sim v_g$ in air). In contrast the pulse is delayed with respect to the phase in any part of the ray that travels through a certain amount of glass. This leads to an increasing delay across the beam characterized by a certain tilt angle α. The maximum arrival time difference in a plane perpendicular to the propagation direction is $(D'/c)\tan\alpha$.

Before discussing pulse front tilt more thoroughly, let us briefly mention another possible prism arrangement where the above condition for L_p is not necessary. Let us consider for example the symmetrical arrangement of four prisms sketched in Figure 2.18. During their path through the prism sequence, different spectral components travel through different optical distances. At the output of the fourth prism all these components are again equally distributed in one beam. The net effect of the four prisms is to introduce a certain amount of GVD leading to broadening of an unchirped input pulse. We will see later in this chapter that this particular GVD can be interpreted as a result of angular dispersion and can have a sign opposite to that of the GVD introduced by the glass material constituting the prisms.

2.5.2. Tilting of Pulse Fronts

In an isotropic material the direction of energy flow—usually identified as ray direction—is always orthogonal to the surfaces of constant phase (*wave fronts*) of the corresponding propagating wave. In the case of a beam consisting of ultrashort light pulses, one has to consider in addition planes of constant intensity (*pulse fronts*). For most applications it is desirable that these pulse fronts be parallel to

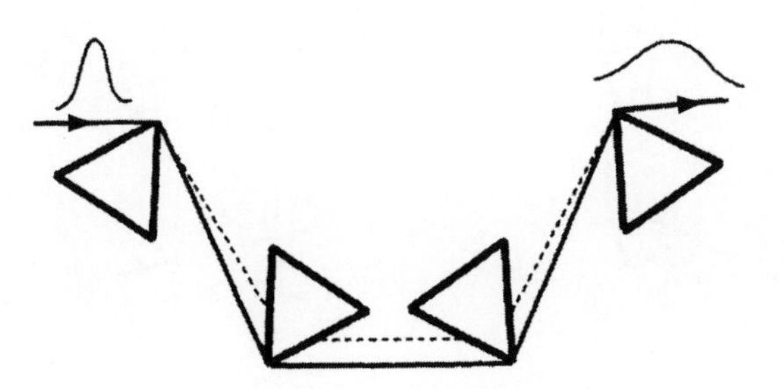

Figure 2.18 Pulse broadening in a four prism sequence.

the phase fronts and thus orthogonal to the propagation direction. In the section on focusing elements we have already seen how lenses cause a radially dependent difference between pulse and phase fronts. This leads to a temporal broadening of the intensity distribution in the focal plane. There are a number of other optical components that introduce a tilt of the pulse front with respect to the phase front and to the normal of the propagation direction, respectively. One example was the prism discussed in the introduction of this section. As a general rule, the pulse front tilting should be avoided whenever an optimum focalization of the pulse energy is sought. There are situations where the pulse front tilt is desirable to transfer a temporal delay to a transverse coordinate. Applications exploiting this property of the pulse front tilt are pulse diagnostics (Chapter 9) and traveling wave amplification (Chapter 7).

The general approach for tilting pulse fronts is to introduce an optical element in the beam path, which retards the pulse fronts as a function of a coordinate transverse to the beam direction. This is schematically shown in Figure 2.19 for an element that changes only the propagation direction of a (plane) wave. Let us assume that a wavefront $\overline{AB}$ is transformed into a wavefront $\overline{A'B'}$. From Fermat's principle it follows that the optical pathlength P_{OL} between corresponding points at the wavefronts $\overline{AB}$ and $\overline{A'B'}$ must be equal:

$$P_{OL}(BB') = P_{OL}(PP') = P_{OL}(AA'). \tag{2.57}$$

Because the optical pathlength corresponds to a phase change $\Delta\Phi = 2\pi P_{OL}/\lambda$, the propagation time of the wavefronts can be written as

$$T_{phase} = \frac{\Delta\Phi}{\omega_\ell} \tag{2.58}$$

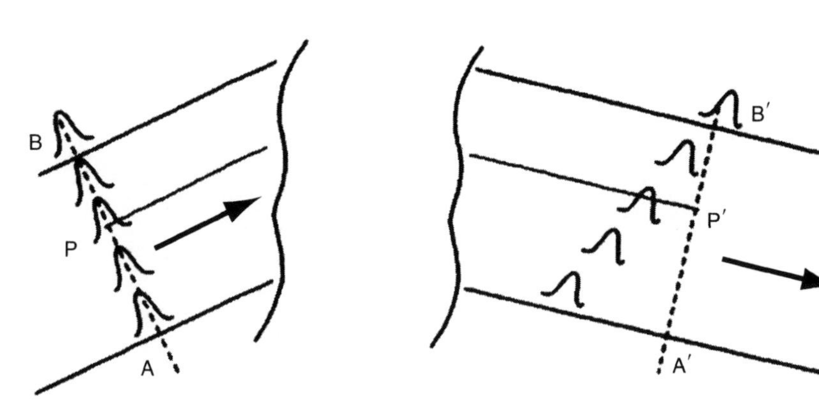

Figure 2.19 Delay of the pulse front with respect to the phase front.

where we referred to the center frequency of the pulse. This phase change is given by

$$\Delta\Phi = \int_P^{P'} k(s)ds = \frac{\omega_\ell}{c}\int_P^{P'} n(s)ds = \omega_\ell \int_P^{P'} \frac{ds}{v_p(s)} \tag{2.59}$$

where s is the coordinate along the beam direction. In terms of the phase velocity the propagation time is

$$T_{phase} = \int_P^{P'} \frac{ds}{v_p(s)}. \tag{2.60}$$

The propagation time of the pulse fronts however, T_{pulse}, is determined by the group velocity

$$T_{pulse} = \int_P^{P'} \frac{ds}{v_g(s)} = \int_P^{P'} \left.\frac{dk}{d\Omega}\right|_{\omega_\ell} ds. \tag{2.61}$$

From Eqs. (2.60) and (2.61) the difference in propagation time between phase front and pulse front becomes

$$\Delta T(P, P') = T_{phase} - T_{pulse} = \int_P^{P'} \left(\frac{1}{v_p} - \frac{1}{v_g}\right) ds = \int_P^{P'} \left[\frac{k_\ell}{\omega_\ell} - \left.\frac{dk}{d\Omega}\right|_{\omega_\ell}\right] ds, \tag{2.62}$$

which can be regarded as a generalization of Eq. (2.35).

A simple optical arrangement to produce pulse front tilting is an interface separating two different optical materials—for instance air (vacuum) and glass (Figure 2.20). At the interface F the initial beam direction is changed by an angle $\beta = \gamma - \gamma'$ where γ and γ' obey Snell's law $\sin\gamma = n(\omega_\ell)\sin\gamma'$. The point A of an incident wavefront $\overline{AB}$ is refracted at time $t = t_0$. It takes the time interval T_{phase} to recreate the wavefront $\overline{A'B'}$ in medium 2, which propagates without distortion with a phase velocity $v_p = c/n(\omega_\ell)$. The time interval T_{phase} is given by

$$T_{phase} = \frac{n\overline{AA'}}{c} = \frac{\overline{BB'}}{c} = \frac{\overline{PF} + n\overline{FP'}}{c} = \frac{D\ \tan\gamma}{c}. \tag{2.63}$$

The beam path from B to B' is through a nondispersive material and thus pulse front and wavefront coincide at B'. In contrast the phase front and pulse front

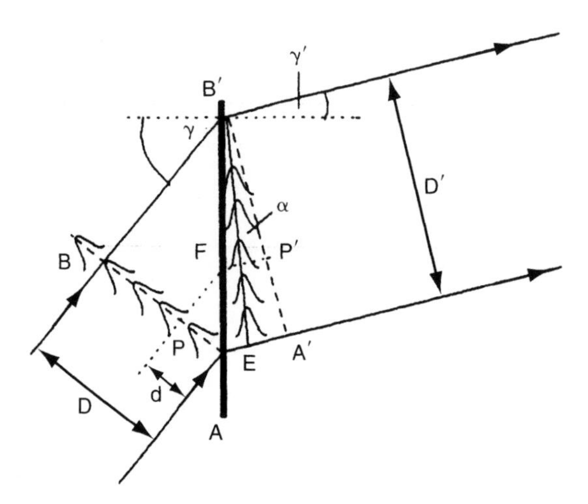

Figure 2.20 Pulse front tilt through refraction at an interface.

propagate different distances during the time interval T_{phase} in medium 2 and thus become separated. Because in (most) optical materials the group velocity is smaller than the phase velocity, the pulse front is delayed with respect to the phase front. In our case this delay increases linearly over the beam cross section. The characteristic tilt angle α between pulse and phase fronts is given by

$$\tan\alpha = \frac{\overline{EA'}}{D'}. \tag{2.64}$$

From simple geometrical considerations we find for the two distances

$$\overline{EA'} = \left(\frac{c}{n} - \nu_g\right) T_{phase} = \left(\frac{c}{n} - \nu_g\right) \frac{D}{c} \tan\gamma \tag{2.65}$$

and

$$D' = D\frac{\cos\gamma'}{\cos\gamma} = D\frac{\sqrt{n^2 - \sin^2\gamma}}{n\cos\gamma}. \tag{2.66}$$

Inserting Eqs. (2.65) and (2.66) in Eq. (2.64) and using the expression for the group velocity, we obtain for α

$$\tan\alpha = \frac{\omega_\ell n'(\omega_\ell)}{\omega_\ell n'(\omega_\ell) + n(\omega_\ell)} \frac{\sin\gamma}{\sqrt{n^2 - \sin^2\gamma}}. \tag{2.67}$$

Following this procedure we can also analyze the pulse front at the output of a prism, cf. Fig. 2.17. The distance $\overline{EA'}$ is the additional pathlength over which the phase has travelled as compared to the pulse path. Thus, we have

$$\overline{EA'} = v_p \left[\frac{a}{v_g} - \frac{a}{v_p} \right] = a\omega_\ell n'(\omega_\ell) \tag{2.68}$$

which results in a tilt angle

$$\tan\alpha = \frac{a}{b}\omega_\ell n'(\omega_\ell) = -\frac{a}{b}\lambda_\ell \left.\frac{dn}{d\lambda}\right|_{\lambda_\ell} \tag{2.69}$$

where $b = D'$ is the beam width.

As pointed out by Bor [24], there is a general relation between pulse front tilt and the angular dispersion $d\beta/d\lambda$ of a dispersive device which reads

$$\boxed{\tan\alpha = \lambda \left| \frac{d\beta}{d\lambda} \right|.} \tag{2.70}$$

The latter equation can be proven easily for a prism, by using the equation for the beam deviation, $d\beta/d\lambda = (a/b)(dn/d\lambda)$, in Eq. (2.69). Similarly to prisms, gratings produce a pulse front tilt, as can be verified easily from the sketch of Figure 2.21. To determine the tilt angle we just need to specify the angular dispersion in Eq. (2.70) using the grating equation.

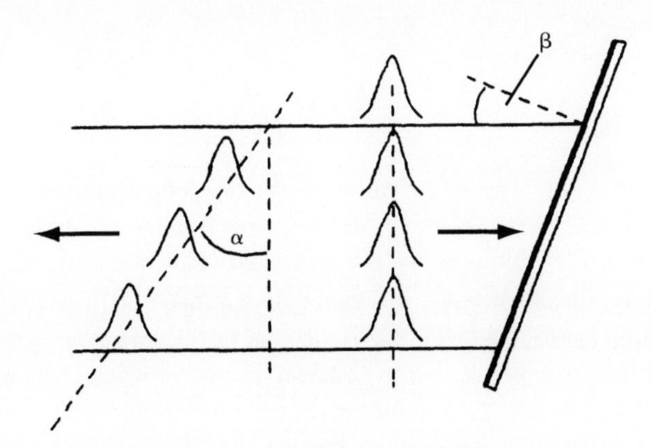

Figure 2.21 Pulse front tilt produced by diffraction at a grating in Littrow configuration.

2.5.3. GVD through Angular Dispersion—General

Angular dispersion has been advantageously used for a long time to resolve spectra or for spectral filtering, utilizing the spatial distribution of the frequency components behind the dispersive element (e.g., prism, grating). In connection with fs optics, angular dispersion has the interesting property of introducing GVD. At first glance this seems to be an undesired effect. However, optical devices based on angular dispersion, which allow for a continuous tuning of the GVD can be designed. This idea was first implemented in Treacy [28] for the compression of chirped pulses with diffraction gratings. The concept was later generalized to prisms and prism sequences [29]. Simple expressions for two and four prism sequences are given in [30, 31]. From a general point of view, the diffraction problem can be treated by solving the corresponding Fresnel integrals [28,32,33]. We will sketch this procedure at the end of this chapter. Another successful approach is to analyze the sequence of optical elements by ray–optical techniques and calculate the optical beam path P as a function of Ω. From our earlier discussion we expect the response of any linear element to be of the form:

$$R(\Omega)e^{-i\Psi(\Omega)} \tag{2.71}$$

where the phase delay Ψ is related to the optical pathlength P_{OL} through

$$\Psi(\Omega) = \frac{\Omega}{c} P_{OL}(\Omega). \tag{2.72}$$

$R(\Omega)$ is assumed to be constant over the spectral range of interest and thus will be neglected.

We know that nonzero terms $[(d^n/d\Omega^n)\Psi \neq 0]$ of order $n \geq 2$ are responsible for changes in the complex pulse envelope. In particular

$$\frac{d^2}{d\Omega^2}\Psi(\Omega) = \frac{1}{c}\left(2\frac{dP_{OL}}{d\Omega} + \Omega\frac{d^2P_{OL}}{d\Omega^2}\right) = \frac{\lambda^3}{2\pi c^2}\frac{d^2P_{OL}}{d\lambda^2} \tag{2.73}$$

is related to the GVD parameter. We recall that, with the sign convention chosen in Eq. (2.71), the phase factor Ψ has the same sign as the phase factor $k_\ell L$. Consistent with the definition given in Eq. (1.117) a positive GVD corresponds to $\frac{d^2\Psi}{d\Omega^2} > 0$. In this chapter, we will generally express $\frac{d^2\Psi}{d\Omega^2}$ in fs^2.

The relation between angular dispersion and GVD can be derived through the following intuitive approach. Let us consider a light ray which is incident onto an optical element at point Q, as in Figure 2.22. At this point we do not specify the element, but just assume that it causes angular dispersion. Thus, different spectral components originate at Q under different angles, within a cone represented by the patterned area in the figure. Two rays corresponding to the center frequency ω_ℓ of

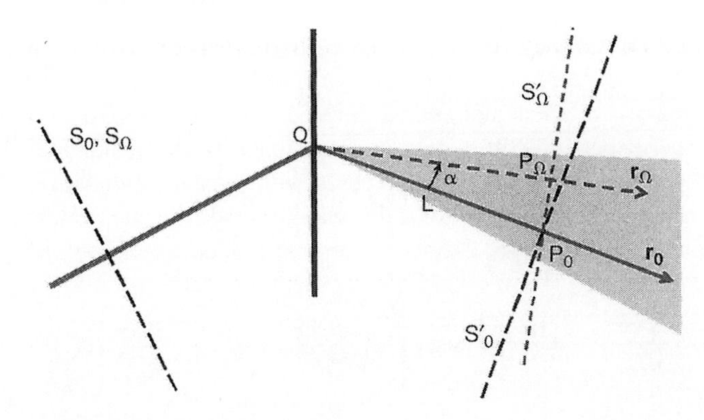

Figure 2.22 Angular dispersion causes GVD. The solid line in the middle of the figure represents the angular dispersive element, providing a frequency-dependent deflection of the beam at the point of incidence Q. The different frequency components of the pulse spread out in the patterned area.

the spectrum, $\vec{r}_0$, and to an arbitrary frequency Ω, $\vec{r}_\Omega$, are shown in Fig. 2.22. The respective wavefronts S are labelled with subscript "0" (for the central frequency ω_ℓ) and "Ω" (for the arbitrary frequency Ω). The planes S_Ω, S_0 and S'_Ω, S'_0 are perpendicular to the ray direction and represent (plane) wave fronts of the incident light and diffracted light, respectively. Let P_0 be our point of reference and be located on $\vec{r}_0$ where $\overline{QP_0} = L$. A wavefront S'_Ω of $\vec{r}_\Omega$ at P_Ω is assumed to intersect $\vec{r}_0$ at P_0. The optical pathlength $\overline{QP_\Omega}$ is thus

$$\overline{QP_\Omega} = P_{OL}(\Omega) = P_{OL}(\omega_\ell)\cos\alpha = L\cos\alpha \tag{2.74}$$

which gives for the phase delay

$$\Psi(\Omega) = \frac{\Omega}{c}P_{OL}(\Omega) = \frac{\Omega}{c}L\cos\alpha \tag{2.75}$$

The dispersion constant responsible for GVD is obtained by twofold derivation with respect to Ω:

$$\left.\frac{d^2\Psi}{d\Omega^2}\right|_{\omega_\ell} = -\frac{L}{c}\left\{\sin\alpha\left[2\frac{d\alpha}{d\Omega} + \Omega\frac{d^2\alpha}{d\Omega^2}\right] + \Omega\cos\alpha\left(\frac{d\alpha}{d\Omega}\right)^2\right\}\Bigg|_{\omega_\ell}$$

$$\approx -\frac{L\omega_\ell}{c}\left(\left.\frac{d\alpha}{d\Omega}\right|_{\omega_\ell}\right)^2 \tag{2.76}$$

where $\sin\alpha = 0$ and $\cos\alpha = 1$ if we take the derivatives at the center frequency of the pulse, $\Omega = \omega_\ell$. The quantity $(d\alpha/d\Omega)|_{\omega_\ell}$, responsible for angular dispersion, is a characteristic of the actual optical device to be considered. It is interesting to note that the dispersion parameter is always negative independently of the sign of $d\alpha/d\Omega$ and that the dispersion increases with increasing distance L from the diffraction point. Therefore angular dispersion always results in negative GVD. Differentiation of Eq. (2.76) results in the next higher dispersion order

$$\begin{aligned}\left.\frac{d^3\Psi}{d\Omega^3}\right|_{\omega_\ell} &= -\frac{L}{c}\left\{\cos\alpha\left[3\left(\frac{d\alpha}{d\Omega}\right)^2 + 3\Omega\frac{d\alpha}{d\Omega}\frac{d^2\alpha}{d\Omega^2}\right]\right.\\ &\quad \left.\left.+\sin\alpha\left[3\frac{d^2\alpha}{d\Omega^2} + \Omega\frac{d^3\alpha}{d\Omega^3} - \Omega\left(\frac{d\alpha}{d\Omega}\right)^3\right]\right\}\right|_{\omega_\ell}\\ &\approx -\frac{3L}{c}\left.\left[\left(\frac{d\alpha}{d\Omega}\right)^2 + \Omega\frac{d\alpha}{d\Omega}\frac{d^2\alpha}{d\Omega^2}\right]\right|_{\omega_\ell},\end{aligned} \tag{2.77}$$

where the last expression is a result of $\alpha(\omega_\ell) = 0$.

The most widely used optical devices for angular dispersion are prisms and gratings. To determine the dispersion introduced by them we need to specify not only the quantity $\alpha(\Omega)$ in the expressions derived previously, but also the optical surfaces between which the path is being calculated. Indeed, we have assumed in the previous calculation that the beam started as a plane wave (plane reference surface normal to the initial beam) and terminates in a plane normal to the ray at a reference optical frequency ω_ℓ. The choice of that terminal plane is as arbitrary as that of the reference frequency ω_ℓ (cf. Section 1.1.1). After some propagation distance, the various spectral component of the pulse will have separated, and a finite size detector will only record a portion of the pulse spectrum.

Therefore, the "dispersion" of an element has only meaning in the context of a particular application, which will associate reference surfaces to that element. This is the case when an element is associated with a cavity, as will be considered in the next section. In the following sections, we will consider combinations of elements of which the angular dispersion is compensated. In that case, a natural reference surface is the normal to the beam.

2.5.4. GVD of a Cavity Containing a Single Prism

Dispersion control is an important aspect in the development of fs sources. The most elementary laser cavity as sketched in Figure 2.23 has an element with angular dispersion. The dispersive element could be the Brewster angle

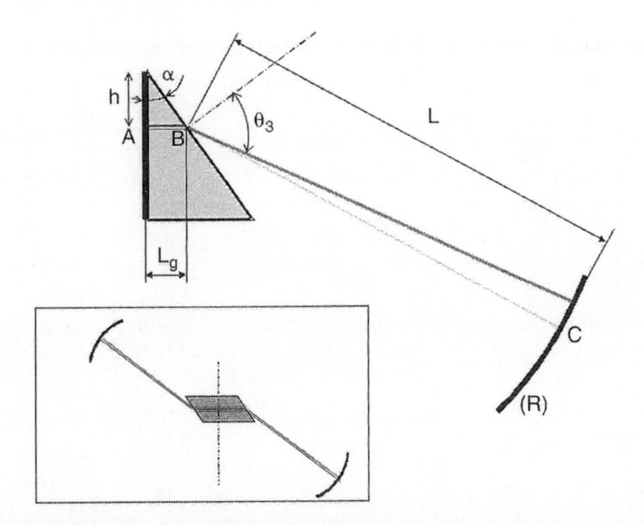

Figure 2.23 Example of a cavity with a single right angle prism. The side of the right angle is an end mirror of the cavity. The cavity is terminated by a curved mirror of radius of curvature R, at a distance L from the Brewster angle exit face of the prism. Stability of the cavity requires that $L + \overline{AB}/n < R$. Translation of the prism allows for an adjustment of the pathlength in glass L_g. The inset shows that this calculation applies to a symmetric cavity with a Brewster angle laser rod and two spherical mirrors.

laser rod itself. The cavity will be typically terminated by a curved mirror. The two reference surfaces to consider are the two end mirrors of the cavity. We have seen that negative GVD is typically associated with angular dispersion, and positive GVD with the propagation through a glass prism or laser rod.[9] One might therefore expect to be able to tune the GVD in the arrangement of Fig. 2.23 from a negative to a positive value. An exact calculation of the frequency dependence presented shows that this is not the case, and that the GVD of this cavity is always positive.

A combination of elements with a tunable positive dispersion can also be desirable in a fs laser cavity. We will consider the case of the linear cavity sketched in Fig. 2.23, whose GVD can be determined analytically.

The cavity is terminated on one end by the plane face of the prism, on the other end by a spherical mirror of curvature R. The prism–mirror distance measured at the central frequency ω_ℓ is L. The beam originates from a distance h from the

[9]It is generally the case—but not always—that optical elements in the visible have positive GVD.

apex of the prism (angle α), such that the pathlength in glass can be written as $L_g = h\tan\alpha$. For the sake of notation simplification, we define:

$$a = \frac{h\tan\alpha}{c}$$

$$b = \frac{L}{2}\left(1 - \frac{L}{R}\right). \tag{2.78}$$

The total phase shift for one half cavity roundtrip is $\Psi(\Omega) = \Psi_{AB}(\Omega) + \Psi_{BC}(\Omega)$. The phase shift through the glass here is simply $-k(\Omega)L_g = -\Psi_{AB}(\Omega)$, with $\Psi_{AB}(\Omega)$ given by:

$$\Psi_{AB}(\Omega) = \Psi_0 + \left.\frac{d\Psi}{d\Omega}\right|_{\omega_\ell} \Delta\Omega + \frac{1}{2}\left.\frac{d^2\Psi}{d\Omega^2}\right|_{\omega_\ell} (\Delta\Omega)^2 + \cdots$$
$$\approx \Psi_0 + a\left[\Omega\frac{dn}{d\Omega} + n(\Omega)\right]_{\omega_\ell} \Delta\Omega + \frac{1}{2}a\left[2\frac{dn}{d\Omega} + \Omega\frac{d^2n}{d\Omega^2}\right]_{\omega_\ell} (\Delta\Omega)^2, \tag{2.79}$$

where $\Delta\Omega = \Omega - \omega_\ell$. For the path in air, we have a phase shift $-k\overline{BC} = -\Psi_{BC}(\Omega)$, with

$$\Psi_{BC}(\Omega) = \frac{\Omega}{c}\left[L + \frac{L}{2}\left(1 - \frac{L}{R}\right)\Delta\theta^2\right] = \frac{\Omega}{c}\left[L + b\Delta\theta^2\right], \tag{2.80}$$

where $\Delta\theta$ is the departure of dispersion angle from the diffraction angle at ω_ℓ. Within the small angle approximation, we have for $\Delta\theta$:

$$\Delta\theta \approx \Delta\Omega\frac{\sin\alpha}{\cos\theta_3}\frac{dn(\Omega)}{d\Omega} = \Delta\Omega\frac{dn(\Omega)}{d\Omega}. \tag{2.81}$$

The last equality ($\sin\alpha = \cos\theta_3$) applies to the case where θ_3 equals the Brewster angle. The GVD dispersion of this cavity is thus:

$$\left.\frac{d^2\Psi}{d\Omega^2}\right|_{\omega_\ell} = \left.\frac{d^2\Psi_{AB}}{d\Omega^2}\right|_{\omega_\ell} + \left.\frac{d^2\Psi_{BC}}{d\Omega^2}\right|_{\omega_\ell} = a\left(2\frac{dn}{d\Omega} + \Omega\frac{d^2}{d\Omega^2}\right)_{\omega_\ell}$$
$$+ \frac{2b\omega_\ell}{c}\left(\left.\frac{dn}{d\Omega}\right|_{\omega_\ell}\right)^2, \tag{2.82}$$

or, using the wavelength dependence of the index of refraction, and taking into account that, for the Brewster prism, $\tan\alpha = 1/n(\omega_\ell)$:

$$\left.\frac{d^2\Psi}{d\Omega^2}\right|_{\omega_\ell} = \frac{h}{nc}\left(\frac{\lambda}{2\pi c}\right)\left.\left(\lambda^2\frac{d^2n}{d\lambda^2}\right)\right|_{\lambda_\ell} + b\frac{\lambda^3}{\pi c^2}\left.\left(\frac{dn}{d\lambda}\right)^2\right|_{\lambda_\ell}. \tag{2.83}$$

The stability of the cavity requires that $R > L$ and that the coefficients a and b be positive. In the visible range, most glasses have a positive GVD ($k'' > 0$ or $d^2n/d\lambda^2 > 0$). Therefore, in a cavity with a single prism as sketched in Fig. 2.23, the GVD is adjustable through the parameter h, but always positive.

The calculation applies to a simple solid state laser cavity as sketched in the inset of Fig. 2.23, with a Brewster angle laser rod. The contributions to the dispersion from each side of the dash–dotted line are additive. Even in this simple example, we see that the total dispersion is not only because of the propagation through the glass, but there is also another contribution because of angular dispersion. It is interesting to compare Eq. (2.76), which gives a general formula associated with angular dispersion, with Eq. (2.83). Both expressions involve the square of the angular dispersion, but with opposite sign.

Femtosecond pulses have been obtained through adjustable GVD compensation with a single prism in a dye ring laser cavity [34]. As in the case of Fig. 2.23, the spectral narrowing that would normally take place because of the angular dispersion of the prism was neutralized by having the apex of the prism at a waist of the resonator. In that particular case, the adjustable positive dispersion of the prism provided pulse compression because of the negative chirp introduced by saturable absorption below resonance, as detailed in Chapter 5.

2.5.5. Group Velocity Control with Pairs of Prisms

2.5.5.1. Pairs of Elements

In most applications, a second element will be associated to the first one, such that the angular dispersion introduced by the first element is compensated, and all frequency components of the beam are parallel again, as sketched in Figure 2.24. The elements will generally be prisms or gratings.

As before, we start from a first reference surface A normal to the beam. It seems then meaningful to chose the second reference surface B at the exit of the system that is normal to the beam. There is no longer an ambiguity in the choice of a reference surface, as in the previous section with a single dispersive element. At any particular frequency, Fermat's principle states that the optical paths are equal from a point of the wavefront A to the corresponding point on the

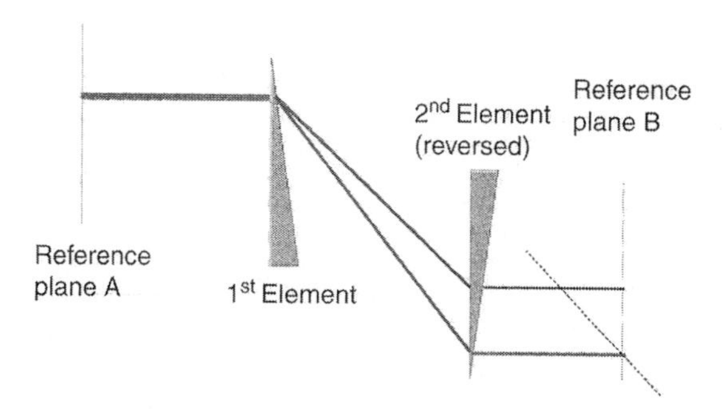

Figure 2.24 Pair of elements with angular dispersion arranged for zero net angular dispersion. The elements are most often prisms or gratings.

wavefront *B*. This is not to say that these distances are not frequency dependent. The spectral components of the beam are still separated in the transverse direction. For that reason, a pair of prisms or gratings provides a way to "manipulate" the pulse spectrum by spatially filtering (amplitude or phase filter) the various Fourier components.

2.5.5.2. Calculation for Matched Isosceles Prisms

One of the most commonly encountered cases of Fig. 2.24, is that where the two angular dispersive elements are isosceles prisms. Prisms have the advantage of smaller insertion losses, which is particularly important with the low gain solid state lasers used for fs applications. To compensate the angular dispersion, the two prisms are put in opposition, in such a way that, to any face of one prism corresponds a parallel face of the other prism (Figure 2.25).

In this section, we consider only the GVD introduced by the prism sequence. The associated pulse front tilt and the effect of beam divergence will be discussed in Section 2.4 using wave optics. There are numerous contributions to the group velocity dispersion that makes this problem rather complex:

(a) GVD because of propagation in glass for a distance L
(b) GVD introduced by the changes in optical path L in each prism, because of angular dispersion
(c) GVD because of the angular dispersion after one prism, propagation of the beam over a distance ℓ, and as a result propagation through different thicknesses of glass at the next prism.

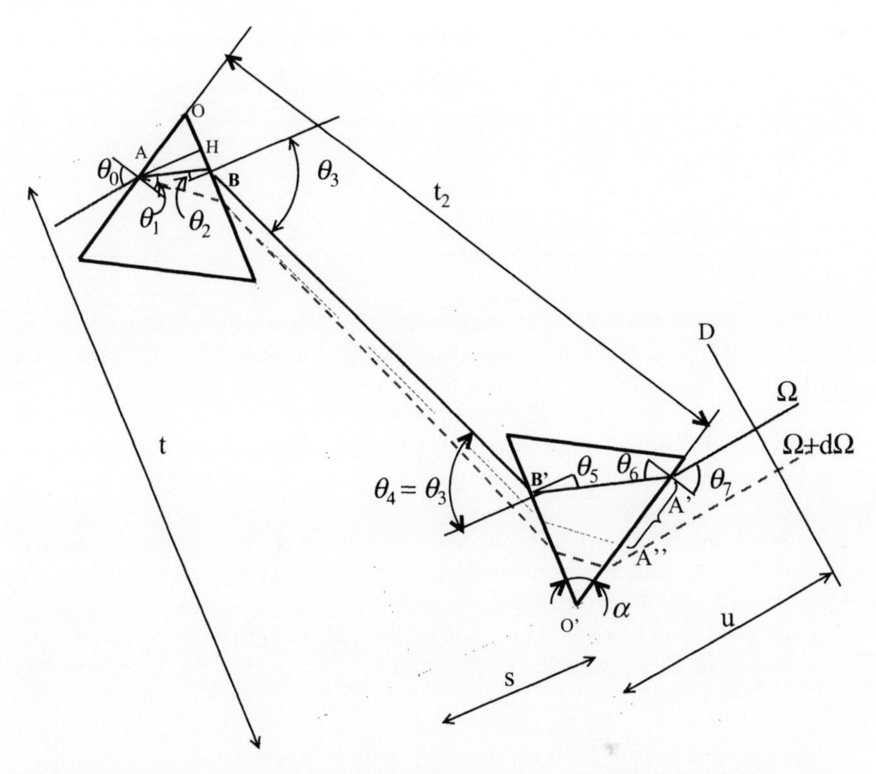

Figure 2.25 Typical two prisms sequence as used in fs laser cavities. The relative position of the prisms is defined by the distance t and the spacing s between the parallel faces OB and $O'B'$. The initial beam enters the prism at a distance $\overline{OA} = a$ from the apex. The distance t_2 between the parallel faces OA and $O'A'$ is $t_2 = t \sin\alpha + s \cos\alpha$. The solid line $ABB'A'D$ traces the beam path at an arbitrary frequency Ω. The beam at the frequency upshifted by $d\Omega$ is represented by the dashed line. The dotted line indicates what the optical path would be in the second prism, if the distance $\overline{BB'}$ were reduced to zero (this situation is detailed in Fig. 2.26). D is a point on the phase front a distance u from the apex O' of the second prism. In most cases we will associate the beam path for a ray at Ω with the path of a ray at the center frequency ω_ℓ.

The optical path $\overline{ABB'A'D}$ at a frequency Ω is represented by the solid line in Fig. 2.25, while the path for a ray upshifted by $d\Omega$ is represented by the dashed line. The successive angles of incidence–refraction are θ_0 and θ_1 at point A, θ_2 and θ_3 at point B, θ_4 and θ_5 at point B', and finally θ_6 and θ_7 at point A'. The two prisms are identical, with equal apex angle α and with pairs of faces oriented parallel as shown in Fig. 2.25. At any wavelength or frequency Ω:

- $\theta_3 = \theta_4$
- $\theta_2 = \theta_5$
- $\theta_1 = \theta_6$
- $\theta_0 = \theta_7$
- $\theta_1 + \theta_2 = \alpha$
- $d\theta_1/d\Omega = -d\theta_2/d\Omega$.

If the prisms are used at minimum deviation at the central wavelength, $\theta_0 = \theta_3 = \theta_4 = \theta_7$. If, in addition to being used at minimum deviation, the prisms are cut for Brewster incidence, the apex angles of both prisms are $\alpha = \pi - 2\theta_0 = \pi - 2\arctan(n)$.

The challenge is to find the frequency dependence of the optical path $\overline{ABB'A'D}$. The initial (geometrical) conditions are defined by

- the distance $a = \overline{OA}$ from the point of impact of the beam to the apex O of the first prisms. For convenience, we will use in the calculations the distance $\overline{OH} = h = \overline{OA}\cos\alpha = a\cos\alpha$,
- the separation s between the parallel faces of the prisms,
- the distance t between the apex O and O', measured along the exit face of the first prism, cf. Fig. 2.25.

The changes in path length because of dispersion can be understood from a glance at the figure, comparing the optical paths at Ω (solid line) and $\Omega + d\Omega$ (dashed line). The contributions that increase the path length are:

1. positive dispersion because of propagation through the prism material of positive dispersion ($\overline{AB}$ and $\overline{B'A'}$)
2. positive dispersion because of the increased path length $\overline{BB'}$ in air (increment $\overline{SB'''}$ in Fig. 2.27)
3. positive dispersion because of the increased path length $\overline{A'D}$ in air (projection of $\overline{A'A'''}$ along the beam propagation direction).

The contributions that decrease the path length (negative dispersion) of the frequency upshifted beam can best be understood with Figures 2.26 and 2.27. Figure 2.26 shows the configurations of the beams if the two prisms were brought together, i.e., $\overline{BB'} = 0$. Figure 2.27 shows an expanded view of the second prism. The negative dispersion contributions emanate from:

1. The shortened path length in glass because of the angular dispersion ($\overline{AA''}$ versus $\overline{AA'}$ in Fig. 2.26),

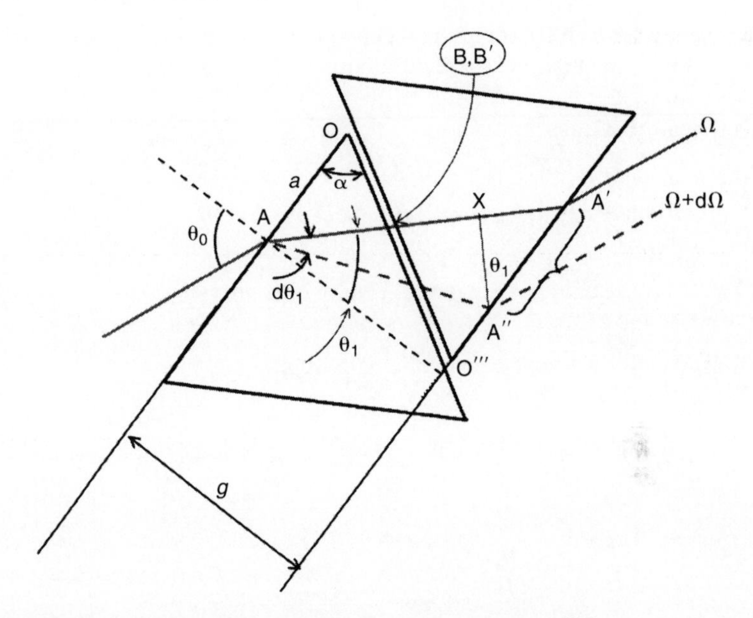

Figure 2.26 Beam passage through the two prisms, when the distance $\overline{BB'}$ (in Fig. 2.25) has been reduced to zero. The distance between the apexes O and O' has been reduced to $\overline{OO'''} = t - \overline{BB'}\sin\theta_3$ (referring to Fig. 2.25). The distance between parallel faces is then $g = \overline{OO'''}\sin\alpha = (t - \overline{BB'}\sin\theta_3)\sin\alpha$.

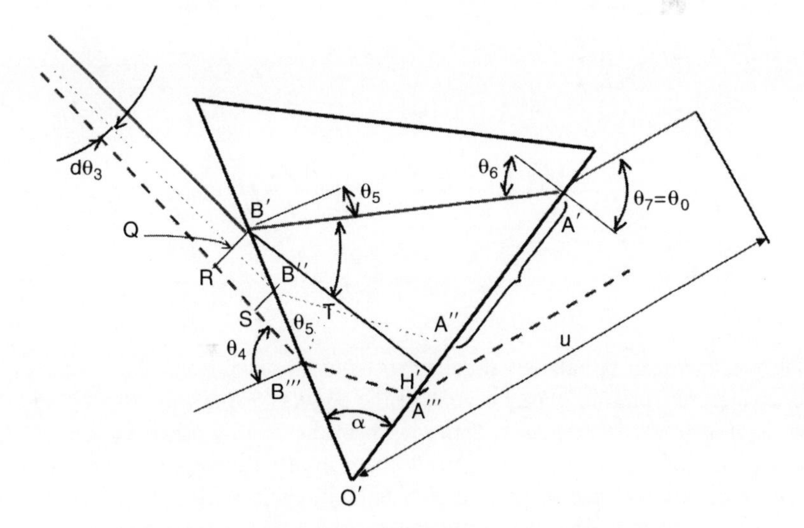

Figure 2.27 Details of the beam passage through the second prism.

2. the shorter path length in the second prism because of the deflection of the beam by the first one (path difference $\overline{B''T}$ in Fig. 2.27).

Path Through Glass

The total path in glass is $L_g = \overline{AB} + \overline{B'A'}$ where $\overline{AB} = a \sin\alpha / \cos\theta_2$ and $\overline{B'A'} = \overline{O'B'} \sin\alpha / \cos(\alpha - \theta_2) = \overline{O'B'} \sin\alpha / \cos\theta_1$, with:

$$\overline{O'B'} = t - s\tan\theta_3 - a(\cos\alpha + \sin\alpha\tan\theta_2). \tag{2.84}$$

We thus have for the total transmitted path in glass:

$$\begin{aligned} L_g = \overline{AB} + \overline{B'A'} &= \frac{a\sin\alpha}{\cos\theta_2} + [t - s\tan\theta_3 - a(\cos\alpha + \sin\alpha\tan\theta_2)]\frac{\sin\alpha}{\cos\theta_1} \\ &= (t - s\tan\theta_3)\frac{\sin\alpha}{\cos\theta_1}. \end{aligned} \tag{2.85}$$

As expected, the total path through glass is independent of the starting position defined by a. If the two prisms are brought together as in Fig. 2.26, they act as a slab of glass with parallel faces, of thickness $g = L_g \cos\theta_1$. There are three contributions to the optical path change from Fig. 2.26: one because of the change in index, a second because of the change in angle, and a third because the path length L_g is frequency dependent. Taking the derivative of $\Omega n L_g / c$ with L_g defined by Eq. (2.85):

$$\begin{aligned} \frac{d(kL_g)}{d\Omega} &= \frac{d}{d\Omega}\left(\frac{n\Omega L_g}{c}\right) \\ &= \frac{L_g}{c}\left(n + \Omega\frac{dn}{d\Omega}\right) + \left(\frac{n\Omega L_g}{c}\tan\theta_1\right)\frac{d\theta_1}{d\Omega} \\ &\quad - \left(\frac{n\Omega s}{c\cos^2\theta_3}\right)\frac{\sin\alpha}{\cos\theta_1}\frac{d\theta_3}{d\Omega} \end{aligned} \tag{2.86}$$

The first term can be attributed solely to material dispersion. The next term is the change in length in glass because of the angular dispersion $d\theta_1/d\Omega$, and the last expresses the reduction in path length in the second prism because of the propagation of the angularly dispersed beam in air. The expression above only partly accounts for the energy tilt associated with the angular dispersion $d\theta_3/d\Omega$. Another contribution arises from the path through air to a reference plane.

Path through Air between and after the Prisms

We have to account for the contributions of the pathlengths $\overline{BB'}$ and $\overline{A'D}$ to the group delay:

$$\frac{d}{d\Omega}\left(\frac{\Omega}{c}\overline{BB'}\right) = \frac{\Omega}{c}\frac{d\overline{BB'}}{d\Omega} + \frac{\overline{BB'}}{c}. \tag{2.87}$$

For the path $\overline{BB'} = s/\cos\theta_3$, there is only a change in length equal to $\overline{SB'''}$, which can be obtained by either differentiating $s/\cos\theta_3$, or simply from geometrical considerations using Fig. 2.25 ($\overline{SB'''} = \overline{SB''}\tan\theta_3 = \overline{BB'}\tan\theta_3 d\theta_3$):

$$\frac{d\overline{BB'}}{d\Omega} = \frac{s}{\cos\theta_3}\tan\theta_3\frac{d\theta_3}{d\Omega}. \tag{2.88}$$

The path in air after the second prism can be expressed as:

$$\overline{A'D} = u - \overline{O'A'}\sin\theta_0. \tag{2.89}$$

Because u is not a function of e, the contribution to the group delay is:

$$\frac{1}{c}\frac{d\left(\Omega\overline{A'D}\right)}{d\Omega} = -\frac{\sin\theta_0}{c}\frac{d}{d\Omega}(\Omega\overline{O'A'}). \tag{2.90}$$

For $\overline{O'A'}$ we find:

$$\begin{aligned}
\overline{O'A'} &= \overline{O'H'} + \overline{H'A'} = \overline{O'B'}\left[\cos\alpha + \sin\alpha\tan(\alpha - \theta_2)\right] \\
&= \left[t - s\tan\theta_3 - a(\cos\alpha + \sin\alpha\tan\theta_2)\right]\left[\cos\alpha + \sin\alpha\tan(\alpha - \theta_2)\right] \\
&= \left[t - s\tan\theta_3 - a\frac{\cos\theta_1}{\cos\theta_2}\right]\left[\cos\alpha + \sin\alpha\tan(\alpha - \theta_2)\right] \\
&= \left[t - s\tan\theta_3\right]\left[\cos\alpha + \sin\alpha\tan\theta_1\right] - a,
\end{aligned} \tag{2.91}$$

where we have used $\cos\alpha + \sin\alpha\tan\theta_2 = \cos(\alpha - \theta_2)\cos\theta_2$. The contribution of $\overline{A'D}$ to the group delay is:

$$\begin{aligned}
-\frac{\sin\theta_0}{c}\frac{d(\Omega\overline{O'A'})}{d\Omega} &= \frac{\overline{O'A'}\sin\theta_0}{c} - \frac{\Omega s\sin\theta_0}{c\cos^2\theta_3}\left[\cos\alpha + \sin\alpha\tan\theta_1\right]\frac{d\theta_3}{d\Omega} \\
&\quad + \frac{\Omega\sin\theta_0}{c}\left[t - s\tan\theta_3\right]\frac{\sin\alpha}{\cos^2\theta_1}\frac{d\theta_1}{d\Omega}
\end{aligned}$$

$$= -\frac{\overline{A'D}}{c} - \frac{n\Omega s}{c\cos^2\theta_3}\left[\cos\alpha\sin\theta_1 + \sin\alpha\frac{\sin^2\theta_1}{\cos\theta_1}\right]\frac{d\theta_3}{d\Omega}$$

$$+ \frac{n\Omega}{c}\left[t - s\tan\theta_3\right]\frac{\sin\alpha\sin\theta_1}{\cos^2\theta_1}\frac{d\theta_1}{d\Omega}. \tag{2.92}$$

In the last equation we used the fact that u is an arbitrary constant, for example zero, so that $\overline{A'D} = -\overline{O'A'}\sin\theta_0$.

Total Path in Glass and Air

After adding all contributions to the total phase

$$\Psi = \frac{\Omega}{c}\left(nL_g + \overline{BB'} + \overline{A'D}\right),$$

we obtain for the group delay using Eqs. (2.86), (2.87), (2.88), (2.90), and (2.92):

$$\frac{d\Psi}{d\Omega} = \frac{d}{d\Omega}\left(\frac{\Omega n L_g}{c}\right) + \frac{d}{d\Omega}\left(\frac{\Omega\overline{BB'}}{c}\right) + \frac{d}{d\Omega}\left(\frac{\Omega\overline{A'D}}{c}\right)$$

$$= \frac{nL_g}{c} + \frac{(\overline{BB'} + \overline{A'D})}{c} + \frac{L_g\Omega}{c}\frac{dn}{d\Omega} + \frac{n\Omega L_g}{c}\tan\theta_1\frac{d\theta_1}{d\Omega}$$

$$+ \left\{-\frac{n\Omega s}{c\cos^2\theta_3}\frac{\sin\alpha}{\cos\theta_1} + \frac{\Omega s}{c\cos\theta_3}\tan\theta_3 + \frac{n\Omega s}{c\cos^2\theta_3}\left[\cos\alpha\sin\theta_1\right.\right.$$

$$\left.\left. + \frac{\sin\alpha\sin^2\theta_1}{\cos\theta_1}\right]\right\}\frac{d\theta_3}{d\Omega} - \frac{n\Omega}{c}\left[t - s\tan\theta_3\right]\frac{\sin\alpha\sin\theta_1}{\cos^2\theta_1}\frac{d\theta_1}{d\Omega}$$

$$= \frac{OPL(ABB'A'D)}{c} + \frac{L_g\Omega}{c}\frac{dn}{d\Omega}$$

$$+ \left(\frac{\Omega s}{c\cos^2\theta_3}\right)\left(-n\sin\alpha\cos\theta_1 + \cos\theta_3\tan\theta_3 + n\cos\alpha\sin\theta_1\right)\frac{d\theta_3}{d\Omega}, \tag{2.93}$$

where we have defined the optical path length $OPL(ABB'A'D) = nL_g + (\overline{BB'} + \overline{A'D})$. The factor preceding $d\theta_3/d\Omega$ cancels, because:

$$\cos\alpha\sin\theta_1 - \sin\alpha\cos\theta_1 + \frac{\sin\theta_3}{n}$$

$$= \cos(\theta_1 + \theta_2)\sin\theta_1 - \sin(\theta_1 + \theta_2)\cos\theta_1 + \sin\theta_2$$

$$= 0.$$

The complete expression for the group delay through the pair of prism reduces to:

$$\boxed{\frac{d\Psi}{d\Omega} = \frac{OPL(ABB'A'D)}{c} + \frac{L_g\Omega}{c}\frac{dn}{d\Omega}.} \tag{2.94}$$

The first terms in the last equation represents the travel delay at the phase velocity:

$$\frac{OPL(ABB'A'D)}{c} = \frac{L_g n}{c} + \frac{s}{c\cos\theta_3} + \frac{\overline{A'D}}{c}. \tag{2.95}$$

The second part of Eq. (2.94) is the carrier to envelope delay caused by the pair of prisms[10]:

$$\tau_{CE}(\Omega) = \frac{\Omega}{c}\frac{d}{d\Omega}OPL(ABB'A'D) = \frac{L_g\Omega}{c}\frac{dn}{d\Omega}. \tag{2.96}$$

The second derivative of the phase, obtained by taking the derivative of Eq. (2.94), is:

$$\begin{aligned}\left.\frac{d^2\Psi}{d\Omega^2}\right|_{\omega_\ell} &= L_g\left[2\left.\frac{dn}{d\Omega}\right|_{\omega_\ell} + \omega_\ell\left.\frac{d^2n}{d\Omega^2}\right|_{\omega_\ell}\right] \\ &\quad - \frac{\omega_\ell}{c\cos\theta_1}\left.\frac{dn}{d\Omega}\right|_{\omega_\ell}\frac{s\sin\alpha}{\cos^2\theta_3}\left.\frac{d\theta_3}{d\Omega}\right|_{\omega_\ell} \\ &\quad + \frac{\omega_\ell}{c}\left.\frac{dn}{d\Omega}\right|_{\omega_\ell}\left(L_g\tan\theta_1\left.\frac{d\theta_1}{d\Omega}\right|_{\omega_\ell}\right).\end{aligned} \tag{2.97}$$

The derivatives with respect to Ω are related. By differentiating Snell's law for the first interface:

$$d\theta_1 = -\frac{\tan\theta_1}{n}dn = -d\theta_2. \tag{2.98}$$

---

[10]We are assuming that the prisms are in vacuum, i.e., the contribution to the dispersion from air is neglected.

For the second interface, taking the previous relation into account, we find:

$$\cos\theta_3 d\theta_3 = n\cos\theta_2 d\theta_2 + \sin\theta_2 dn = -n\cos\theta_2\left(-\frac{\tan\theta_1}{n} + \sin\theta_2\right)dn$$
$$= (\cos\theta_2\tan\theta_1 + \sin\theta_2)\,dn = \frac{\sin\alpha}{\cos\theta_1}dn, \tag{2.99}$$

or:

$$d\theta_3 = \frac{\sin\alpha}{\cos\theta_1\cos\theta_3}dn. \tag{2.100}$$

Therefore, the second-order dispersion Eq. (2.97) reduces to an easily interpretable form:

$$\left.\frac{d^2\Psi}{d\Omega^2}\right|_{\omega_\ell} = \frac{L_g}{c}\left[2\left.\frac{dn}{d\Omega}\right|_{\omega_\ell} + \omega_\ell\left.\frac{d^2n}{d\Omega^2}\right|_{\omega_\ell}\right] - \frac{\omega_\ell}{c}\left(\frac{s}{\cos\theta_3}\right)\left(\left.\frac{d\theta_3}{d\Omega}\right|_{\omega_\ell}\right)^2 - \frac{n\omega_\ell}{c}L_g\left(\left.\frac{d\theta_1}{d\Omega}\right|_{\omega_\ell}\right)^2 \tag{2.101}$$

This equation applies to any pair of identical isosceles prisms in the parallel face configuration represented in Fig. 2.25, for an arbitrary angle of incidence. The GVD is simply the sum of three contributions:

1. The (positive) GVD because of the propagation of the pulse through a thickness of glass L_g.
2. The negative GVD contribution because of the angular dispersion $d\theta_3/d\Omega$ applied to Eq. (2.76) over a distance $\overline{BB'} = s/\cos\theta_3$.
3. The negative GVD contribution because of the angular dispersion $d\theta_1/d\Omega$ (deflection of the beam at the first interface) applied to Eq. (2.76) over a distance L_g in the glass of index n.

In most practical situations it is desirable to write Eq. (2.101) in terms of the input angle of incidence θ_0 and the prism apex angle α. The necessary equations can be derived from Snell's law and Eq. (2.76):

$$\frac{d}{d\Omega}\theta_1 = \left[n^2 - \sin^2(\theta_0)\right]^{-\frac{1}{2}}\left[n\cos\theta_0\frac{d\theta_0}{d\Omega} - \sin\theta_0\frac{dn}{d\Omega}\right]$$

$$\frac{d}{d\Omega}\theta_3 = \left[1 - n^2\sin^2(\alpha+\theta_1)\right]^{-\frac{1}{2}}\left[n\cos(\alpha+\theta_1)\frac{d\theta_1}{d\Omega} + \sin(\alpha+\theta_1)\frac{dn}{d\Omega}\right], \tag{2.102}$$

where $\theta_1 = \arcsin(n^{-1}\sin\theta_0)$ and $d\theta_0/d\Omega = 0$.

For the particular case of Brewster angle prisms and minimum deviation (symmetric beam path through the prism for $\Omega = \omega_\ell$), we can make the substitutions $d\theta_1/dn = -1/n^2$, and $d\theta_3/dn = 2$. Using $\theta_0 = \theta_3 = \theta_4 = \theta_7$, the various angles are related by:

$$\begin{aligned}\tan\theta_0 &= n\\ \sin\theta_0 = \cos\theta_1 &= \frac{n}{\sqrt{1+n^2}}\\ \cos\theta_0 = \sin\theta_1 &= \frac{1}{\sqrt{1+n^2}}\\ \sin\alpha &= \frac{2n}{n^2+1}.\end{aligned} \tag{2.103}$$

The total second-order dispersion in this case becomes:

$$\boxed{\left.\frac{d^2\Psi}{d\Omega^2}\right|_{\omega_\ell} = \frac{L_g}{c}\left[2\left.\frac{dn}{d\Omega}\right|_{\omega_\ell} + \omega_\ell\left.\frac{d^2n}{d\Omega^2}\right|_{\omega_\ell}\right] - \frac{\omega_\ell}{c}\left(4L + \frac{L_g}{n^3}\right)\left(\left.\frac{dn}{d\Omega}\right|_{\omega_\ell}\right)^2}, \tag{2.104}$$

where we have introduced the distance between the two prisms measured along the central wavelength $L = s/\cos\theta_3$. In terms of wavelength:

$$\boxed{\left.\frac{d^2\Psi}{d\Omega^2}\right|_{\omega_\ell} = \frac{\lambda_\ell^3}{2\pi c^2}\left[L_g\left.\frac{d^2n}{d\lambda^2}\right|_{\lambda_\ell} - \left(4L + \frac{L_g}{n^3}\right)\left(\left.\frac{dn}{d\lambda}\right|_{\lambda_\ell}\right)^2\right]}. \tag{2.105}$$

In many practical devices, $L \gg L_g$ and the second term of Eq. (2.105) reduces to $L(dn/d\lambda)^2$.

It is left as a problem at the end of this chapter to calculate the exact third-order dispersion for a pair of prisms. If the angular dispersion in the glass can be neglected ($L \gg L_g$), the third-order dispersion for a Brewster angle prism is:

$$\Psi'''_{\text{tot}}(\omega_\ell) \approx$$

$$\frac{\lambda_\ell^4}{(2\pi c)^2 c}\left[12L\left(n'^2\left[1 - \lambda_\ell n'(n^{-3} - 2n)\right] + \lambda_\ell n' n''\right) - L_g(3n'' + \lambda_\ell n''')\right]. \tag{2.106}$$

To simplify the notation, we have introduced n', n'' and n''' for the derivatives of n with respect to λ taken at λ_ℓ.

The presence of a negative contribution to the GVD because of angular dispersion offers the possibility of tuning the GVD by changing $L_g = g/\sin\theta_0$ (g is the thickness of the glass slab formed by bringing the two prisms together, as shown in Fig. 2.26). A convenient method is to simply translate one of the prisms perpendicularly to its base, which alters the glass path while keeping the beam deflection constant. It will generally be desirable to avoid a transverse displacement of spectral components at the output of the dispersive device. Two popular prism arrangements which do not separate the spectral components of the pulse are sketched in Figure 2.28. The beam is either sent through two prisms, and retro-reflected by a plane mirror, or sent directly through a sequence of four prisms. In these cases the dispersion as described by Eq. (2.101) doubles. The values of Ψ'', Ψ''', etc. that are best suited to a particular experimental situation can be predetermined through a selection of the optimum prism separation $s/\cos\theta_3$, the glass pathlength L_g, and the material (cf. Table 2.1). Such optimization methods are particularly important for the generation of sub-20 fs pulses in lasers [35, 36] that use prisms for GVD control.

In this section we have derived analytical expressions for dispersion terms of increasing order, in the case of identical isoceles prism pairs, in exactly antiparallel configuration. It is also possible by methods of pulse tracing through

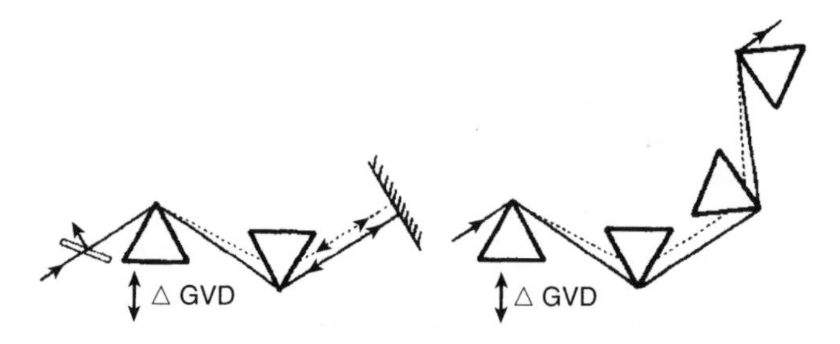

Figure 2.28 Setups for adjustable GVD without transverse displacement of spectral components.

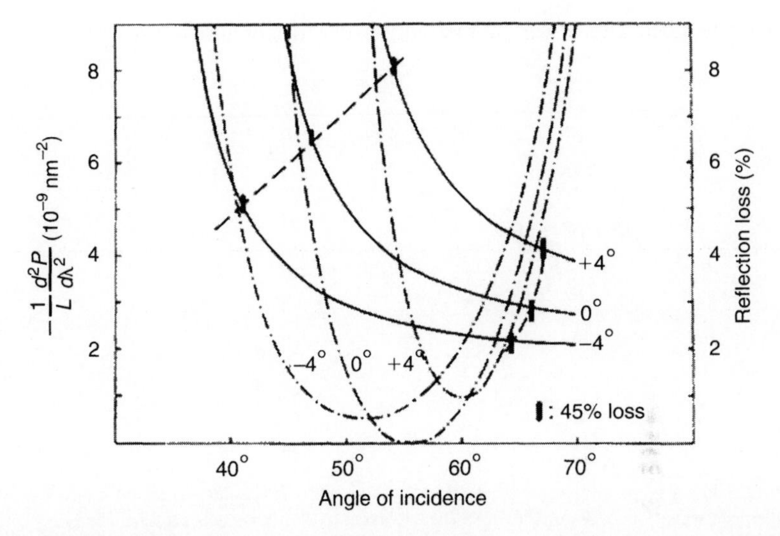

Figure 2.29 Dispersion (solid lines) and reflection losses (dash–dotted lines) of a two-prism sequence (SQ1—fused silica) as a function of the angle of incidence on the first prism surface. Symmetric beam path through the prism at the central wavelength is assumed. Curves for three different apex angles (−4°, 0°, 4°) relative to $\alpha = 68.9°$ (apex angle for a Brewster prism at 620 nm) are shown. The tic marks on the dashed lines indicate the angle of incidence and the dispersion where the reflection loss is 4.5%. (Adapted from Petrov *et al.* [31]).

the prisms to determine the phase factor at any frequency and angle of incidence [30, 31, 37–39]. The more complex studies revealed that the GVD and the transmission factor R [as defined in Eq. (2.71)] depend on the angle of incidence and apex angle of the prism. In addition, any deviation from the Brewster condition increases the reflection losses. An example is shown in Fig. 2.29.

2.5.6. GVD Introduced by Gratings

Gratings can produce larger angular dispersion than prisms. The resulting negative GVD was first utilized by Treacy [28] to compress pulses of a Nd:glass laser. In complete analogy with prisms, the simplest practical device consists of two identical elements arranged as in Figure 2.30 for zero net angular dispersion. The dispersion introduced by a pair of parallel gratings can be determined by tracing the frequency dependent ray path. The optical path length $\overline{ACP}$ between A and an output wavefront $\overline{PP_0}$ is frequency dependent and can be determined

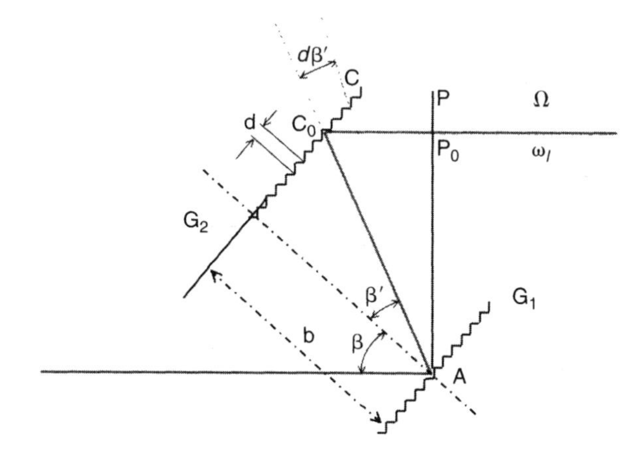

Figure 2.30 Two parallel gratings produce GVD without net angular dispersion. For convenience a reference wavefront is assumed so that the extension of $\overline{PP_0}$ intersects G_1 at A.

with help of Fig. (2.30) to be:

$$\overline{ACP} = \frac{b}{\cos(\beta')}\left[1 + \cos(\beta' + \beta)\right] \tag{2.107}$$

where β is the angle of incidence, β' is the diffraction angle for the frequency component Ω and b is the normal separation between G_1 and G_2. If we restrict our consideration to first-order diffraction, the angle of incidence and the diffraction angle are related through the grating equation

$$\sin\beta' - \sin\beta = \frac{2\pi c}{\Omega d} \tag{2.108}$$

where d is the grating constant. The situation with gratings is however different than with prisms, in the sense that the optical path of two parallel rays out of grating G_1 impinging on adjacent grooves of grating G_2 will see an optical path difference $\overline{CP} - \overline{C_0P_0}$ of $m\lambda$, m being the diffraction order. Thus, as the angle β' changes with wavelength, the phase factor $\Omega\overline{ACP}/c$ increments by $2m\pi$ each time the ray $\overline{AC}$ passes a period of the ruling of G_2 [28]. Because only the relative phase shift across $\overline{PP_0}$ matters, we may simply count the rulings from the (virtual) intersection of the normal in A with G_2. Thus, for first-order diffraction ($m = 1$), we find for $\Psi(\Omega)$:

$$\Psi(\Omega) = \frac{\Omega}{c}\overline{ACP}(\Omega) - 2\pi\frac{b}{d}\tan(\beta'). \tag{2.109}$$

The group delay is given by:

$$\begin{aligned}\frac{d\Psi}{d\Omega} &= \left(\frac{b}{c}\right)\frac{1+\cos(\beta+\beta')}{\cos\beta'} + \frac{\Omega b}{c\cos^2\beta'}\left\{\sin\beta'\left[1+\cos(\beta+\beta')\right]\right. \\ &\quad \left. -\cos\beta'\sin(\beta+\beta')\right\}\frac{d\beta'}{d\Omega} - \frac{2\pi}{d}\frac{b}{\cos^2\beta'}\frac{d\beta'}{d\Omega} \\ &= \left(\frac{b}{c}\right)\frac{1+\cos(\beta+\beta')}{\cos\beta'}. \end{aligned} \tag{2.110}$$

In deriving the last equation, we have made use of the grating equation $\sin\beta' - \sin\beta = 2\pi c/(\Omega d)$. Equation (2.110) shows a remarkable property of gratings, namely that the group delay is simply equal to the phase delay. The carrier to envelope delay is zero. The second-order derivative, obtained by differentiation of Eq. (2.110), is:

$$\begin{aligned}\frac{d^2\Psi}{d\Omega^2} &= \frac{b}{c}\frac{1}{\cos^2\beta'}\left\{\sin\beta'\left[1+\cos(\beta+\beta')\right] - \cos\beta'[\sin(\beta+\beta')]\right\}\frac{d\beta'}{d\Omega} \\ &= \frac{-4\pi^2 bc}{\Omega^3 d^2\cos^3\beta'}, \end{aligned} \tag{2.111}$$

where we have again made use of the grating equation. Evaluating this expression at the central frequency ω_ℓ, and using wavelengths instead of frequencies:

$$\boxed{\left.\frac{d^2\Psi}{d\Omega^2}\right|_{\omega_\ell} = -\frac{\lambda_\ell}{2\pi c^2}\left(\frac{\lambda_\ell}{d}\right)^2\frac{b}{\cos^3\beta'(\lambda_\ell)}.} \tag{2.112}$$

In terms of the distance $L = b/\cos\beta'$ between the gratings along the ray at $\Omega = \omega_\ell$:

$$\left.\frac{d^2\Psi}{d\Omega^2}\right|_{\omega_\ell} = -\frac{\lambda_\ell}{2\pi c^2}\left(\frac{\lambda_\ell}{d}\right)^2\frac{L}{\cos^2\beta'(\omega_\ell)}. \tag{2.113}$$

where $\cos^2\beta'(\omega_\ell) = 1 - [2\pi c/(\omega_\ell d) + \sin\beta]^2$. The third derivative can be written as

$$\left.\frac{d^3\Psi}{d\Omega^3}\right|_{\omega_\ell} = -\frac{3\lambda_\ell}{2\pi c\cos^2\beta'(\omega_\ell)}\left[\cos^2\beta'(\omega_\ell) + \frac{\lambda_\ell}{d}\left(\frac{\lambda_\ell}{d} + \sin\beta\right)\right]\left.\frac{d^2\Psi}{d\Omega^2}\right|_{\omega_\ell}. \tag{2.114}$$

To decide when the third term in the expansion [as defined in Eq. (1.167)] of the phase response of the grating needs to be considered we evaluate the ratio

$$R_G = \left|\frac{b_3(\Omega-\omega_\ell)^3}{b_2(\Omega-\omega_\ell)^2}\right| = \left|\frac{\Psi'''(\omega_\ell)}{3\Psi''(\omega_\ell)}\right| |\Omega-\omega_\ell| \approx \frac{\Delta\omega_p}{\omega_\ell}\left[1 + \frac{\lambda_\ell/d(\lambda_\ell/d + \sin\beta)}{1-(\lambda_\ell/d - \sin\beta)^2}\right] \tag{2.115}$$

where the spectral width of the pulse $\Delta\omega_p$ was used as an average value for $|\Omega - \omega_\ell|$.

Obviously it is possible to minimize (or tune) the ratio of second- and third-order dispersion by changing the grating constant and the angle of incidence. For instance, with $\Delta\omega_p/\omega_\ell = 0.05$, $\lambda_\ell/d = 0.5$ and $\beta = 0^\circ$ we obtain $R_G \approx 0.07$.

Let us next compare Eq. (2.112) with Eq. (2.76), which related GVD to angular dispersion in a general form. From Eq. (2.108) we obtain for the angular dispersion of a grating

$$\left.\frac{d\beta'}{d\Omega}\right|_{\omega_\ell} = -\frac{2\pi c}{\omega_\ell^2 d\cos\beta'} \tag{2.116}$$

If we insert Eq. (2.116) in the general expression linking GVD to angular dispersion, Eq. (2.76), and remember that $L = b/\cos\beta'$, we also obtain Eq. (2.112).

2.5.7. Grating Pairs for Pulse Compressors

For all practical purpose, a pulse propagating from grating G_1 to G_2 can be considered as having traversed a linear medium of length L characterized by a negative dispersion. We can write Eq. (2.112) in the form of:

$$\left.\frac{d^2\Psi}{d\Omega^2}\right|_{\omega_\ell} = k_\ell'' L = -\left\{\frac{\lambda_\ell}{2\pi c^2}\left(\frac{\lambda_\ell}{d}\right)^2 \frac{1}{\cos^2\beta'(\omega_\ell)}\right\} L. \tag{2.117}$$

Referring to Table 1.2, a bandwidth-limited Gaussian pulse of duration τ_{G0}, propagating through a dispersive medium characterized by the parameter k_ℓ'', broadens to a Gaussian pulse of duration τ_G

$$\tau_G = \tau_{G0}\sqrt{1 + \left(\frac{L}{L_d}\right)^2}, \tag{2.118}$$

with a linear chirp of slope:

$$\ddot{\varphi} = \frac{2L/L_d}{1 + (L/L_d)^2} \frac{1}{\tau_{G0}^2} \tag{2.119}$$

where the parameter L_d relates both to the parameters of the grating and to the minimum (bandwidth-limited) pulse duration:

$$L_d = \frac{\tau_{G0}^2}{2|k_\ell''|} = \frac{\pi c^2 d^2 r}{\lambda_\ell^3}\tau_{G0}^2. \tag{2.120}$$

Conversely, a pulse with a positive chirp of magnitude given by Eq. (2.119) and duration corresponding to Eq. (2.118) will be compressed by the pair of gratings to a duration τ_{G0}. A pulse compressor following a pulse stretcher is used in numerous amplifications systems and will be dealt with in Chapter 7. The "compressor" is a pair of gratings with optical path L, designed for a compression ratio $\tau_G/\tau_{G0} = L/L_d$.[11] The ideal compressor of length L will restore the initial (before the stretcher) unchirped pulse of duration τ_0. To a departure x from the ideal compressor length L, corresponds a departure from the ideal unchirped pulse of duration τ_0:

$$\tau_G = \tau_{G0}\sqrt{1 + \frac{x^2}{L_d^2}}. \tag{2.121}$$

This pulse is also given a chirp coefficient (cf. Table 1.2) $\bar{a} = x/L_d$.

In most compressors, the transverse displacement of the spectral components at the output of the second grating can be compensated by using two pairs of gratings in sequence or by sending the beam once more through the first grating pair. As with prisms, the overall dispersion then doubles. Tunability is achieved by changing the grating separation b. Unlike with prisms, however, the GVD is always negative. The order of magnitude of the dispersion parameters of some typical devices is compiled in Table 2.2.

[11] In all practical cases with a pair of gratings, $(L/L_d)^2 \gg 1$.

Table 2.2

Values of second-order dispersion for typical devices.

Device	λ_ℓ (nm)	ω_ℓ (fs^{-1})	Ψ'' (fs^2)
Fused silica ($L_g = 1$ cm)	620	3.04	535
	800	2.36	356
Brewster prism pair, fused silica	620	3.04	−760
$L = 50$ cm	800	2.36	−523
Grating pair $b = 20$ cm; $\beta = 0°$	620	3.04	-9.3×10^4
$d = 1.2$ μm	800	2.36	-3×10^5

The choice between gratings and prism for controllable dispersion is not always a simple one. Prisms pairs have lower losses than gratings (the total transmission through a grating pair usually does not exceed 80%), and are therefore the preferred intracavity dispersive element. Gratings are often used in amplifier chain where extremely high compression and stretching ratios are desired, which implies a small L_d. It should be noted however that L_d is not only determined by the properties of the prism or grating, but is also proportional to τ_{G0}^2 as shown by Eq. (2.120). Therefore, prisms stretcher-compressors are also used in medium power amplifiers for sub-20 fs pulses. The disadvantage of prisms is that the beam has to be transmitted through glass, which, for high power pulses, is a nonlinear medium.

2.5.8. Combination of Focusing and Angular Dispersive Elements

A disadvantage of prism and grating sequences is that for achieving large GVD the length L between two diffraction elements becomes rather large, cf. Eq. (2.76). As proposed by Martinez [40] the GVD of such devices can be considerably increased (or decreased) by using them in connection with focusing elements such as telescopes. Let us consider the optical arrangement of Figure 2.31, where a telescope is placed between two gratings, so that its object focal point is at the grating G_1. According to Martinez, the effect of a telescope can then be understood as follows. Neglecting lens aberrations, the optical beam path between focal points F and F' is independent of α, α'. Therefore, the length which has to be considered for the dispersion reduces from L to

$$z' = L - \overline{FF'} = L - 2(f + f'). \qquad (2.122)$$

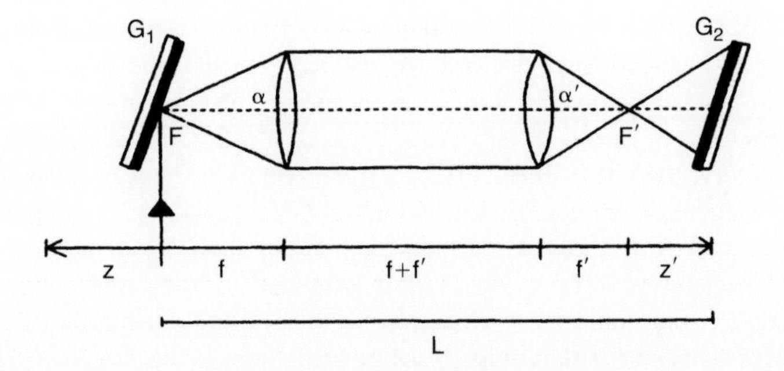

Figure 2.31 Combination of a grating pair and a telescope. In the particular arrangement shown, the object focal point F is at the grating. In general grating 1 can be a distance z away from F. Note: z and z' can be positive as well as negative depending on the relative positions of the gratings and focal points.

In addition, the telescope introduces an angular magnification of $M = f/f'$, which means that the angular dispersion of G_1 is magnified by M to $M(d\alpha/d\Omega)$. For the second grating to produce a parallel output beam its dispersion must be M times larger than that of G_1. With Eq. (2.76) the overall dispersion is now given by

$$\left.\frac{d^2\Psi}{d\Omega^2}\right|_{\omega_\ell} = -\frac{\omega_\ell}{c}\left(\left.\frac{d\alpha}{d\Omega}\right|_{\omega_\ell}\right)^2 z'M^2 = -\frac{\omega_\ell}{c}\left(\left.\frac{d\alpha}{d\Omega}\right|_{\omega_\ell}\right)^2 [L - 2(f + f')]\left(\frac{f}{f'}\right)^2. \tag{2.123}$$

Because it is not practicable to use a second grating of higher dispersion, one can fold the arrangement by means of a roof prism as is done in standard grating compressors. Also, it is not necessary that the object focal point of the telescope coincide with the diffraction point Q at the grating. A possible separation z between Q and F has then to be added to $z'M^2$ to obtain the overall dispersion from Eq. (2.123):

$$\left.\frac{d^2\Psi}{d\Omega^2}\right|_{\omega_\ell} = -\frac{\omega_\ell}{c}\left(\left.\frac{d\alpha}{d\Omega}\right|_{\omega_\ell}\right)^2 \left(z'M^2 + z\right). \tag{2.124}$$

Equation (2.124) suggests another interesting application of telescopic systems. For $z'M^2 + z < 0$ the dispersion changes sign. The largest amount of positive

$d^2\Psi/d\Omega^2$ is achieved for $z = -f$ and $z' = -f'$. Because the sign of the angular dispersion is changed by the telescope we have to tilt the second grating to recollimate the beam. For the folded geometry we can use a mirror instead of a roof prism.

In summary, the use of telescopes in connection with grating or prism pairs allows us to increase or decrease the amount of GVD as well as to change the sign of the GVD. As will be discussed later, interesting applications of such devices include the recompression of pulses after long optical fibers and extreme pulse broadening (>1000) before amplification [40,41]. A more detailed discussion of this type of dispersers, including the effects of finite beam size, can be found in Martinez [33].

2.6. WAVE-OPTICAL DESCRIPTION OF ANGULAR DISPERSIVE ELEMENTS

Because our previous discussion of pulse propagation through prisms, gratings, and other elements was based on ray–optical considerations, it failed to give details about the influence of a finite beam size. These effects can be included by a wave-optical description which is also expected to provide new insights into the spectral, temporal, and spatial field distribution behind the optical elements. We will follow the procedure developed by Martinez [33], and use the characteristics of Gaussian beam propagation, i.e., remain in the frame of paraxial optics.

First, let us analyze the effect of a single element with angular dispersion as sketched in Figure 2.32. The electric field at the disperser can be described by a complex amplitude $\tilde{U}(x, y, z, t)$ varying slowly with respect to the spatial and

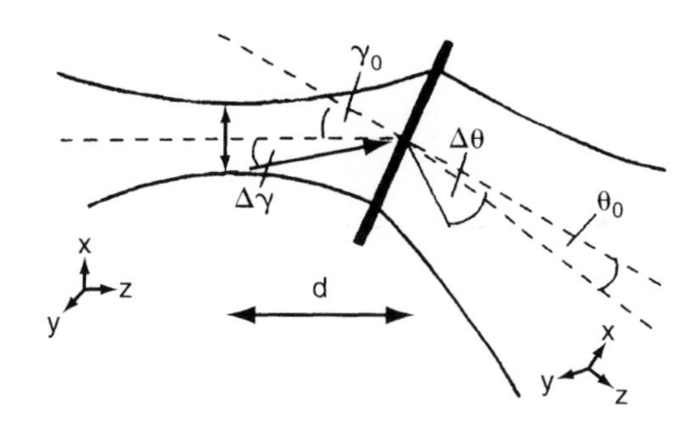

Figure 2.32 Interaction of a Gaussian beam with a disperser.

temporal coordinate:

$$E(x, y, z, t) = \frac{1}{2}\tilde{U}(x, y, z, t)e^{i(\omega_\ell t - k_\ell z)} + c.c. \tag{2.125}$$

Using Eq. (1.185) the amplitude at the disperser can be written as

$$\tilde{U}(x, y, t) = \tilde{\mathcal{E}}_0(t)\exp\left[-\frac{ik_\ell}{2\tilde{q}(d)}(x^2 + y^2)\right] = \tilde{U}(x, t)\exp\left[\frac{-ik_\ell y^2}{2\tilde{q}(d)}\right] \tag{2.126}$$

where $\tilde{q}$ is the complex beam parameter, d is the distance between beam waist and disperser, and $\tilde{\mathcal{E}}_0$ is the amplitude at the disperser. Our convention shall be that x and y refer to coordinates transverse to the respective propagation direction z. Further, we assume the disperser to act only on the field distribution in the x direction, so that the field variation with respect to y is the same as for free space propagation of a Gaussian beam. Hence, propagation along a distance z changes the last term in Eq. (2.126) simply through a change of the complex beam parameter $\tilde{q}$. According to Eq. (1.186) this change is given by

$$\tilde{q}(d + z) = \tilde{q}(d) + z. \tag{2.127}$$

To discuss the variation of $\tilde{U}(x, t)$ it is convenient to transfer to frequencies $\bar{\Omega}$ and spatial frequencies ρ applying the corresponding Fourier transforms

$$\tilde{U}(x, \bar{\Omega}) = \int_{-\infty}^{\infty} \tilde{U}(x, t)e^{-i\bar{\Omega}t}dt \tag{2.128}$$

and

$$\tilde{U}(\rho, \bar{\Omega}) = \int_{-\infty}^{\infty} \tilde{U}(x, \bar{\Omega})e^{-i\rho x}dx. \tag{2.129}$$

A certain spatial frequency spectrum of the incident beam means that it contains components having different angles of incidence. Note that $\bar{\Omega}$ is the variable describing the spectrum of the envelope (centered at $\bar{\Omega} = 0$), while $\Omega = \bar{\Omega} + \omega_\ell$ is the actual frequency of the field. In terms of Fig. 2.32 this is equivalent to a certain angular distribution $\Delta\gamma$. The spatial frequency ρ is related to $\Delta\gamma$ through

$$\Delta\gamma = \frac{\rho}{k_\ell}. \tag{2.130}$$

For a plane wave, $\tilde{U}(\rho, \bar{\Omega})$ exhibits only one nonzero spatial frequency component which is at $\rho = 0$. The disperser not only changes the propagation direction

$(\gamma_0 \to \theta_0)$ but also introduces a new angular distribution $\Delta\theta$ of beam components which is a function of the angle of incidence γ and the frequency $\bar{\Omega}$

$$\begin{aligned}\Delta\theta &= \Delta\theta(\gamma, \Omega)\\ &= \left.\frac{\partial\theta}{\partial\gamma}\right|_{\gamma_0} \Delta\gamma + \left.\frac{\partial\theta}{\partial\Omega}\right|_{\omega_\ell} \bar{\Omega}\\ &= \alpha\Delta\gamma + \beta\bar{\Omega}.\end{aligned} \tag{2.131}$$

The quantities α and β are characteristics of the disperser and can easily be determined, for example, from the prism and grating equations.[12] By means of Eq. (2.130) the change of the angular distribution $\Delta\gamma \to \Delta\theta$ can also be interpreted as a transformation of spatial frequencies ρ into spatial frequencies $\rho' = \Delta\theta k_\ell$ where

$$\rho' = \alpha k_\ell \Delta\gamma + k_\ell \beta\bar{\Omega} = \alpha\rho + k_\ell\beta\bar{\Omega}. \tag{2.132}$$

Just behind the disperser we have an amplitude spectrum $\tilde{U}_T(\rho', \bar{\Omega})$ given by

$$\tilde{U}_T(\rho', \bar{\Omega}) = C_1 \tilde{U}\left(\frac{1}{\alpha}\rho' - \frac{k_\ell}{\alpha}\beta\bar{\Omega}, \bar{\Omega}\right) \tag{2.133}$$

where C_1 and further constants C_i to be introduced are factors necessary for energy conservation that shall not be specified explicitly. In spatial coordinates the field distribution reads

$$\begin{aligned}\tilde{U}_T(x, \bar{\Omega}) &= \int_{-\infty}^{\infty} \tilde{U}_T(\rho', \bar{\Omega}) e^{i\rho' x} d\rho'\\ &= C_1 \int_{-\infty}^{\infty} \tilde{U}\left(\frac{1}{\alpha}\rho' - \frac{k_\ell}{\alpha}\beta\bar{\Omega}, \bar{\Omega}\right) e^{i\rho' x} d\rho'\\ &= C_1 \int_{-\infty}^{\infty} \tilde{U}(\rho, \bar{\Omega}) e^{i\alpha\rho x} e^{ik_\ell\beta\bar{\Omega}x} d(\alpha\rho)\\ &= C_2 e^{ik_\ell\beta\bar{\Omega}x} \tilde{U}(\alpha x, \bar{\Omega}).\end{aligned} \tag{2.134}$$

---

[12] For a Brewster prism adjusted for minimum deviation we find $\alpha = 1$ and $\beta = -(\lambda^2/\pi c)(dn/d\lambda)$. The corresponding relations for a grating used in diffraction order m are $\alpha = \cos\gamma_0/\cos\theta_0$ and $\beta = -m\lambda^2/(2\pi c d \cos\theta_0)$.

The disperser introduces a phase factor $\exp(ik_\ell \beta \bar{\Omega} x)$ and a magnification factor α. For the overall field distribution we obtain with Eqs. (2.126), (2.127), and (2.134)

$$\tilde{U}_T(x, y, \bar{\Omega}) = C_3 \tilde{\mathcal{E}}_0(\bar{\Omega}) e^{ik_\ell \beta \bar{\Omega} x} \exp\left[-\frac{ik_\ell}{2\tilde{q}(d)}\left(\alpha^2 x^2 + y^2\right)\right]. \tag{2.135}$$

The field a certain distance L away from the disperser is connected to the field distribution Eq. (2.135) through a Fresnel transformation which describes the free space propagation. Thus, it can be written as

$$\tilde{U}_T(x, y, L, \bar{\Omega}) = C_4 \exp\left[\frac{-ik_\ell y^2}{2\tilde{q}(d + L)}\right] \int \tilde{\mathcal{E}}_0(\bar{\Omega}) e^{ik_\ell \beta \bar{\Omega} x'} \times \exp\left[-i\frac{k_\ell}{2\tilde{q}(d)}\alpha^2 x'^2\right] \exp\left[-\frac{i\pi}{L\lambda}(x - x')^2\right] dx'. \tag{2.136}$$

Solving this integral yields an analytical expression for the spectral amplitude

$$\tilde{U}_T(x, y, L, \bar{\Omega}) = C_5 \tilde{\mathcal{E}}_0(\bar{\Omega}) \exp\left[-ik_\ell \frac{x^2}{2L}\right] \times \exp\left[-ik_\ell \frac{y^2}{2\tilde{q}(d + L)}\right] \exp\left\{\frac{ik_\ell L}{2} \frac{\tilde{q}(d)}{\tilde{q}(d + \alpha^2 L)}\left[\frac{x^2}{L^2} + 2\frac{\beta x}{L}\bar{\Omega} + \beta^2 \bar{\Omega}^2\right]\right\}. \tag{2.137}$$

The phase term proportional to $\bar{\Omega}^2$ is responsible for GVD according to our discussion in the section on linear elements. As expected from our ray–optical treatment, this term increases with increasing distance L and originates from angular dispersion β. The term linear in $\bar{\Omega}$ varies with the transverse coordinate x. It describes a frequency variation across the beam and accounts for different propagation directions of different spectral components. We know that exponentials proportional to $b_1 \bar{\Omega}$ result in a pulse delay, as discussed previously following Eq. (1.169). Because $b_1 \propto x$ the pulse delay changes across the beam—a feature which we have called tilt of pulse fronts. This proves the general connection between angular dispersion and pulse front tilting introduced earlier in a more intuitive way.

For a collimated input beam and $\alpha = 1$ we can estimate

$$\frac{\tilde{q}(d)}{\tilde{q}(d + \alpha^2 L)} \approx 1 \tag{2.138}$$

and the temporal delay becomes $b_1 = k_\ell \beta x$. Looking at the beam at a certain instant the corresponding spatial delay is $k_\ell \beta x c$. Thus, we find for the tilt angle α

$$|\tan\alpha| = \left|\frac{d}{dx}(k_\ell \beta x c)\right| = \omega_\ell \left|\left.\frac{d\theta}{d\bar{\Omega}}\right|_{\bar{\Omega}=0}\right| = \lambda_\ell \left|\left.\frac{d\theta}{d\lambda}\right|_{\lambda_\ell}\right| \tag{2.139}$$

which confirms our previous results, cf. Eq. (2.70). With the same approximation we obtain for the GVD term:

$$\frac{d^2\Psi}{d\bar{\Omega}^2} = 2b_2 = -k_\ell L \beta^2 = -\frac{L\omega_\ell}{c}\left(\left.\frac{d\theta}{d\Omega}\right|_{\omega_\ell}\right)^2 \tag{2.140}$$

in agreement with Eq. (2.76).

For compensating the remaining angular dispersion we can use a properly aligned second disperser which has the parameters $\alpha' = 1/\alpha$ and $\beta' = \beta/\alpha$. According to our general relation for the action of a disperser (2.134) the new field distribution after this second disperser is given by

$$\begin{aligned}\tilde{U}_F &= C_2 e^{ik_\ell \frac{\beta}{\alpha}\bar{\Omega}x}\tilde{U}_T(\frac{x}{\alpha}, y, L, \bar{\Omega}) \\ &= C_6 \tilde{\mathcal{E}}_0(\bar{\Omega}) e^{\frac{i}{2}k_\ell \bar{\Omega}^2\beta^2 L} \exp\left\{\frac{-ik_\ell}{2}\left[\frac{(x+\alpha\beta\bar{\Omega}L)^2}{\tilde{q}(d+\alpha^2 L)} + \frac{y^2}{\tilde{q}(d+L)}\right]\right\},\end{aligned} \tag{2.141}$$

which again exhibits the characteristics of a Gaussian beam. Hence, to account for an additional propagation over a distance L', we just have to add L' in the arguments of $\tilde{q}$. As discussed by Martinez [33] $\alpha \neq 1$ gives rise to astigmatism (the position of the beam waist is different for the x and y directions) and only for a well-collimated input beam does the GVD not depend on the travel distance L' after the second disperser. For $\alpha = 1$ and $q(d + L + L') \approx q(0)$ (collimated input beam) the field distribution becomes

$$\boxed{\tilde{U}_F(x, y, L, \bar{\Omega}) = C_6 \tilde{\mathcal{E}}_0(\bar{\Omega}) e^{\frac{i}{2}k_\ell \beta^2 L \bar{\Omega}^2} \exp\left\{-\frac{(x+\beta\bar{\Omega}L)^2 + y^2}{w_0^2}\right\}.} \tag{2.142}$$

The first phase function is the expected GVD term. The $\bar{\Omega}$ dependence of the second exponential indicates the action of a frequency filter. At constant position x, its influence increases with increasing $(\beta\bar{\Omega}L/w_0)^2$, i.e., with the ratio of the lateral displacement of a frequency component Ω and the original beam waist. The physics behind is that after the second disperser, not all frequency components can interfere over the entire beam cross-section, leading to an effective bandwidth reduction and thus to pulse broadening. If the experimental situation requires even this to be compensated, the beam can be sent through an identical second pair of dispersers (e.g., prisms). Within the approximations introduced previously we just have to replace L by $2L$ in Eqs. (2.141) and (2.142). For a well-collimated beam ($\beta\bar{\Omega}L/w_0 \ll 1$) this results in

$$\tilde{U}_{F2}(x, y, L, \bar{\Omega}) = C_7\tilde{\mathcal{E}}_0(\bar{\Omega})e^{ik_\ell\beta^2L\bar{\Omega}^2}e^{-(x^2+y^2)/w_0^2}. \tag{2.143}$$

In this (ideal) case the only modification introduced by the dispersive element is the phase factor leaving the beam characteristics unchanged.

It is quite instructive to perform the preceding calculation with a temporally chirped input pulse as in Eq. (1.33) having a Gaussian spatial as well as temporal profile [33]:

$$\tilde{\mathcal{E}}_0(t) = \mathcal{E}_0e^{-(1+ia)(t/\tau_G)^2}e^{-(x^2+y^2)/w_0^2}. \tag{2.144}$$

The (temporal) Fourier transform yields

$$\tilde{\mathcal{E}}_0(\bar{\Omega}) = C_8\exp\left(i\frac{\bar{\Omega}^2\tilde{\tau}a}{4}\right)\exp\left(-\frac{\bar{\Omega}^2\tilde{\tau}^2}{4}\right)\exp\left(-\frac{x^2+y^2}{w_0^2}\right) \tag{2.145}$$

with $\tilde{\tau}^2 = \tau_G^2/(1+a^2)$, where according to our discussion following Eq. (1.39) $\tau_G/\tilde{\tau}$ is the maximum possible shortening factor after chirp compensation. This pulse is to travel through an ideal two-prism sequence described by Eq. (2.142) where βL has to be chosen so as to compensate exactly the quadratic phase term of the input pulse Eq. (2.145). Under this condition the insertion of Eq. (2.145) into Eq. (2.142) yields

$$\tilde{U}_F(x, y, \bar{\Omega}, L) = C_9e^{-\bar{\Omega}^2\tilde{\tau}^2/4}\exp\left[-\frac{(x+\beta\bar{\Omega}L)^2+y^2}{w_0^2}\right]. \tag{2.146}$$

The time-dependent amplitude obtained from Eq. (2.146) after inverse Fourier transform is

$$\tilde{\mathcal{E}}(t) = C_{10}\exp\left(\frac{-t^2}{\tilde{\tau}^2(1+u^2)}\right)\exp\left[\left(\frac{-x^2}{(1+u^2)w_0^2}-\frac{y^2}{w_0^2}\right)\right]\exp\left[\frac{-iu^2xt}{(1+u^2)\beta L}\right], \tag{2.147}$$

where $u = 2\beta L/(\tilde{\tau}w_0)$. The last exponential function accounts for a frequency sweep across the beam which prevents the different frequency components from interfering completely. As a result, the actual shortening factor is $\sqrt{1+u^2}$ times smaller than the theoretical one, as can be seen from the first exponential function. The influence of such a filter can be decreased by using a large beam size. A measure of this frequency filter, i.e., the magnitude of the quantity $(1+u^2)$, can be derived from the second exponent. Obviously the quantity $(1+u^2)$ is responsible for a certain ellipticity of the output beam which can be measured.

2.7. OPTICAL MATRICES FOR DISPERSIVE SYSTEMS

In Chapter 1 we pointed out the similarities between Gaussian beam propagation and pulse propagation. Even though this fact has been known for many years [28, 42], it was only recently that optical matrices have been introduced to describe pulse propagation through dispersive systems [43–47] in analogy to optical ray matrices. The advantage of such an approach is that the propagation through a sequence of optical elements can be described using matrix algebra. Dijaili [46] defined a 2×2 matrix for dispersive elements which relates the complex pulse parameters (cf. Table 1.2) of input and output pulse, $\tilde{p}$ and $\tilde{p}'$, to each other. Döpel [43] and Martinez [45] used 3×3 matrices to describe the interplay between spatial (diffraction) and temporal (dispersion) mechanisms in a variety of optical elements, such as prisms, gratings and lenses, and in combinations of them. The advantage of this method is the possibility to analyze complicated optical systems such as femtosecond laser cavities with respect to their dispersion—a task of increasing importance, as attempts are being made to propagate ultrashort pulses near the bandwidth limit through complex optical systems. The analysis is difficult since the matrix elements contain information pertaining to both the optical system and of pulse.

One of the most comprehensive approaches to describe ray and pulse characteristics in optical elements by means of matrices is that of Kostenbauder [47]. He defined 4×4 matrices which connect the input and output ray and pulse coordinates to each other. As in ray optics, all information about the optical system is carried in the matrix while the spatial and temporal characteristics of the pulse are represented in a ray–pulse vector $(x, \Theta, \Delta t, \Delta \nu)$. Its components are defined by position x, slope Θ, time t, and frequency ν. These coordinates have to be understood as *difference quantities* with respect to the coordinates of a *reference pulse*. The spatial coordinates are similar to those known from the ordinary $\begin{pmatrix} A & B \\ C & D \end{pmatrix}$ ray matrices. However, the origin of the coordinate system is defined now by the path of a diffraction limited reference beam at the average pulse frequency. This reference pulse has a well-defined arrival time at any reference plane; the coordinate Δt, for example, is the difference in arrival time of the pulse under investigation. In terms of such coordinates and using a 4×4 matrix, the action of an optical element can be written as

$$\begin{pmatrix} x \\ \Theta \\ \Delta t \\ \Delta \nu \end{pmatrix}_{out} = \begin{pmatrix} A & B & 0 & E \\ C & D & 0 & F \\ G & H & 1 & I \\ 0 & 0 & 0 & 1 \end{pmatrix} \begin{pmatrix} x \\ \Theta \\ \Delta t \\ \Delta \nu \end{pmatrix}_{in} \tag{2.148}$$

where A, B, C, D are the components of the ray matrix and the additional elements are

$$E = \frac{\partial x_{out}}{\partial \Delta \nu_{in}}, \quad F = \frac{\partial \Theta_{out}}{\partial \Delta \nu_{in}}, \quad G = \frac{\partial \Delta t_{out}}{\partial x_{in}}, \quad H = \frac{\partial \Delta t_{out}}{\partial \Theta_{in}}, \quad I = \frac{\partial \Delta t_{out}}{\partial \Delta \nu_{in}}. \tag{2.149}$$

The physical meaning of these matrix elements is illustrated by a few examples of elementary elements in Figure 2.33. The occurrence of the zero elements can easily be explained using simple physical arguments, namely (a) the center frequency must not change in a linear (time invariant) element and (b) the ray properties must not depend on t_{in}. It can be shown that only six elements are independent of each other and therefore three additional relations between the nine nonzero matrix elements exists [47]. They can be written as

$$\begin{aligned} AD - BC &= 1 \\ BF - ED &= \lambda_\ell H \\ AF - EC &= \lambda_\ell G. \end{aligned} \tag{2.150}$$

Using the known ray matrices [48] and Eq. (2.149), the ray–pulse matrices for a variety of optical systems can be calculated. Some example are shown in Table 2.3.

A system matrix can be constructed as the ordered product of matrices corresponding to the elementary operations (as in the example of the prism constructed from the product of two interfaces and a propagation in glass). An important feature of a system of dispersive elements is the frequency dependent optical beam path P, and the corresponding phase delay Ψ. This information is sufficient for geometries that do not introduce a change in the beam parameters. Examples which have been discussed in this respect are four-prism and four-grating sequences illuminated by a well-collimated beam.

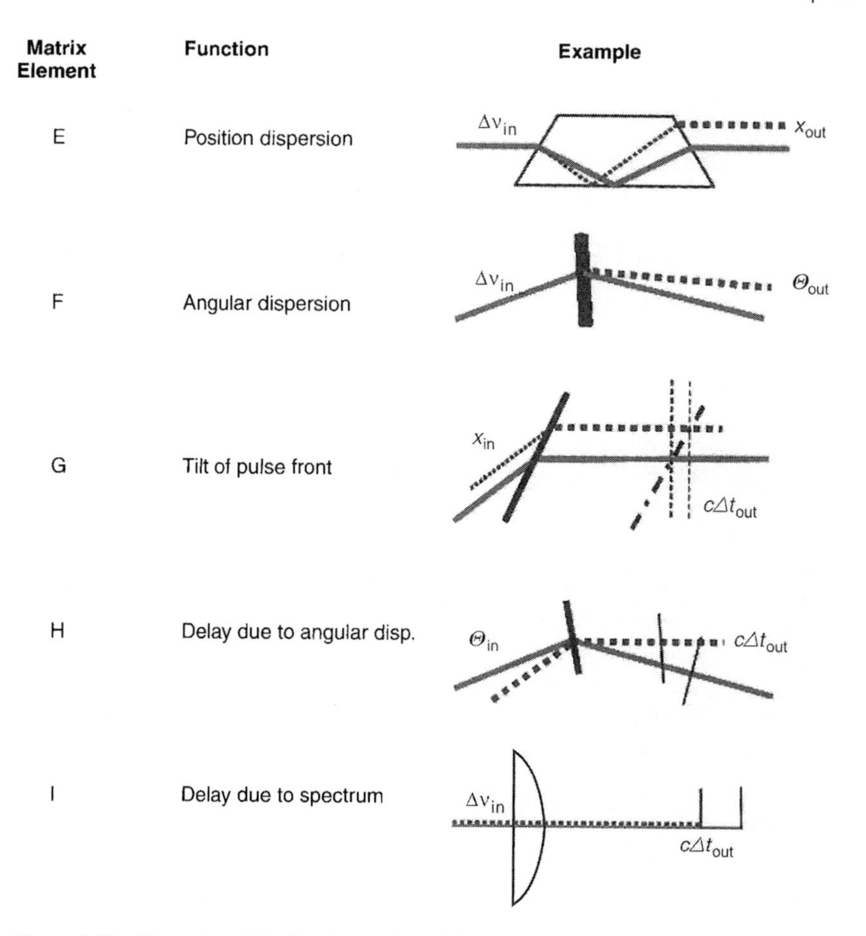

Figure 2.33 Illustration of the function performed by the matric elements E, F, G, H, and I. The path of the reference beam at the central wavelength is represented by the solid line, while the dotted line indicates the displaced path caused by Θ_{in}, x_{in}, or $\Delta\nu_{in}$. A dispersive prism introduces a transverse wavelength dependent displacement of the beam, x_{out}. To a change in optical frequency $\Delta\nu_{in}$ from the central frequency ν_ℓ corresponds an angular deviation Θ_{out} at a dispersive interface. At the same dispersive interface, to a transverse displacement x_{in} left of the interface corresponds an energy front tilt $\Delta t_{out} = Gx_{in}$ right of the interface. There is also a contribution to the energy front tilt associated with the angular dispersion, which is $\Delta t_{out} = H\Theta_{in}$. Finally, on axis of a lens which has chromatic aberration, the displaced wavelength suffers a delay $\Delta t_{out} = I\Delta\nu_{in}$.

As shown in Kostenbauder, [47], Ψ can be expressed in terms of the coordinates of the system matrix as

$$\Psi = \frac{\pi\Delta\nu^2}{B}(EH - BI) - \frac{\pi}{B\lambda_\ell}Q(\Delta\nu) \qquad (2.151)$$

Table 2.3

Examples of ray–pulse matrices.

Lens or mirror ($\mathcal{M}_L$)	**Brewster prism** ($\mathcal{M}_{BP}$)
$\begin{pmatrix} 1 & 0 & 0 & 0 \\ -1/f & 1 & 0 & 0 \\ 0 & 0 & 1 & 0 \\ 0 & 0 & 0 & 1 \end{pmatrix}$	$\begin{pmatrix} 1 & L_g/n^3 & 0 & -\frac{SL_g}{n^3} \\ 0 & 1 & 0 & -2S \\ -\frac{2S}{\lambda_\ell} & -\frac{L_g S}{n^3\lambda_\ell} & 1 & \frac{L_g S^2}{n^3\lambda_\ell} + 2\pi L_g k''_\ell \\ 0 & 0 & 0 & 1 \end{pmatrix}$
f, focal length	$S = 2\pi \left.\frac{\partial n}{\partial \Omega}\right\rvert_{\omega_\ell}$, L_g, mean glass path
Dispersive slab ($\mathcal{M}_{DS}$)	**Grating** ($\mathcal{M}_G$)
$\begin{pmatrix} 1 & L/n & 0 & 0 \\ 0 & 1 & 0 & 0 \\ 0 & 0 & 1 & 2\pi L k''_\ell \\ 0 & 0 & 0 & 1 \end{pmatrix}$	$\begin{pmatrix} -\frac{\cos\beta'}{\cos\beta} & 0 & 0 & 0 \\ 0 & -\frac{\cos\beta}{\cos\beta'} & 0 & \frac{c(\sin\beta' - \sin\beta)}{\lambda_\ell \sin\beta'} \\ \frac{\sin\beta - \sin\beta'}{c\sin\beta} & 0 & 1 & 0 \\ 0 & 0 & 0 & 1 \end{pmatrix}$
$k''_\ell = \left.\frac{d^2k}{d\Omega^2}\right\rvert_{\omega_\ell}$, L_g, thickness of slab	β, angle of incidence; β', diffraction angle

where

$$Q(\Delta\nu) = \begin{pmatrix} x_{in} & x_{out} \end{pmatrix} \begin{pmatrix} A & -1 \\ -1 & D \end{pmatrix} \begin{pmatrix} x_{in} \\ x_{out} \end{pmatrix} + 2 \begin{pmatrix} E & \lambda_0 H \end{pmatrix} \begin{pmatrix} x_{in} \\ x_{out} \end{pmatrix} \tag{2.152}$$

and x_{in}, x_{out} are the position coordinates of the input and output vectors, respectively. The argument $\Delta\nu$ of Q and Ψ is the cyclic frequency coordinate relative to the pulse central frequency $\Delta\nu = (\Omega - \omega_\ell)/2\pi$. The calculations according to Eq. (2.151) have to be repeated for a set of frequencies to obtain $\Psi(\nu)$. From $\Psi(\nu)$ we can then determine chirp and temporal behavior of the output pulses using the relation [Eq. (1.165)] for linear elements without losses. For pulses incident on-axis ($x_{in} = 0$), Eq. (2.152) yields for the phase response

$$\Psi_M(\Delta\nu) = \frac{1}{4\pi B}\left[\left(EH - BI - \frac{1}{\lambda_\ell}DE^2\right)\Delta\nu^2 - 4\pi EH\Delta\nu\right], \tag{2.153}$$

where the index M is to express the derivation of the phase response from the ray–pulse matrix. Information about the temporal broadening can also be gained directly from the matrix element I because $\Delta t_{out} = \Delta t_{in} + \Delta\nu I$. Wave packets centered at different frequencies need different times to travel from the input to the exit plane which gives an approximate broadening of $I\Delta\nu$ for a bandwidth-limited input pulse with a spectral width $\Delta\omega_\ell = 2\pi\Delta\nu_\ell$. For on-axis propagation ($x_{in} = x_{out} = 0$) we find $Q(\Delta\nu) = 0$ and the dispersion is given by the first term in Eq. (2.151). For a dispersive slab, for example, we find from Table 2.3:

$$\Psi_M = \frac{1}{2} L_g k_\ell''(\Omega - \omega_\ell)^2 \tag{2.154}$$

which agrees with Eq. (1.172) and the accompanying discussion.

As another example let us discuss the action of a Brewster prism at minimum deviation and analyze the ray–pulse at a distance L_a behind it. The system matrix is the product of $(\mathcal{M}_{BP})$ and $(\mathcal{M}_{DS})$ for free space, which is given by

$$\begin{pmatrix} 1 & B+L_a & 0 & E+FL_a \\ 0 & 1 & 0 & F \\ G & H & 1 & I \\ 0 & 0 & 0 & 1 \end{pmatrix}. \tag{2.155}$$

For the sake of simplicity the elements of the prism matrix have been noted $A, B, \ldots, H$. For the new position and time coordinate we obtain

$$x_{out} = x_{in} + (B + L_a)\Theta_{in} + (E + FL_a)\Delta\nu \tag{2.156}$$

and

$$\Delta t_{out} = Gx_{in} + H\Theta_{in} + \Delta t_{in} + I\Delta\nu. \tag{2.157}$$

Let us next verify the tilt of the pulse fronts derived earlier. The pulse front tilt can be understood as an arrival time difference Δt_{out} which depends on the transverse beam coordinate x_{out}. The corresponding tilt angle α' is then

$$\tan\alpha' = \frac{\partial(c\Delta t_{out})}{\partial x_{out}} = c\frac{\partial \Delta t_{out}}{\partial x_{in}}\frac{\partial x_{in}}{\partial x_{out}} = cG. \tag{2.158}$$

After we insert G for the Brewster prism, the tilt angle becomes (cf. Table 2.3):

$$\tan\alpha' = -2\omega_\ell \left.\frac{\partial n}{\partial \Omega}\right|_{\omega_\ell} = 2\lambda_\ell \left.\frac{\partial n}{\partial \lambda}\right|_{\lambda_\ell}. \tag{2.159}$$

This result is equivalent to Eq. (2.69) if we use $a/b = 2$, which is valid for Brewster prisms. The different signs result from the direction of the x-axis chosen here.

As a final example we want to apply the matrix formalism to discuss the field distribution behind a two-prism sequence used for pulse compression, such as the one sketched in Fig. 2.25. We assume that one prism is traversed at the apex while the second is responsible for a mean glass path L_g. The corresponding system matrix is obtained by multiplying matrix (2.155) from the left with the transposed[13] matrix of a Brewster prism. The result is

$$\begin{pmatrix} 1 & \frac{L_g}{n^3} + L_a & 0 & -S\left[\frac{L_g}{n^3} + 2L_a\right] \\ 0 & 1 & 0 & 0 \\ 0 & \frac{S}{\lambda_\ell}\left[\frac{L_g}{n^3} + 2L_a\right] & 1 & -\frac{S^2}{\lambda_\ell}\left[\frac{L_g}{n^3} + 4L_g\right] - 2\pi k_\ell'' L_g \\ 0 & 0 & 0 & 1 \end{pmatrix}. \tag{2.160}$$

To get a simplified expression, we make the assumption that $L_g \ll L_a$, which allows us to neglect terms linear in L_g in favor of those linear in L_a, whenever they appear in a summation. For the second derivative of the phase response (2.153) we find

$$\Psi''(\omega_\ell) = L_g k_\ell'' - \frac{8\pi}{\lambda_\ell} L_a \left(\left. \frac{dn}{d\Omega} \right|_{\omega_\ell} \right)^2. \tag{2.161}$$

which is consistent with the exact solution Eq. (2.101), within the approximation of $L_g \ll L_a$, implying negligible angular dispersion inside the prisms.

It is well-known that ray matrices can be used to describe Gaussian beam propagation, e.g., [48]. The beam parameter of the output beam is connected to the input parameter by

$$\tilde{q}_{out} = \frac{A\tilde{q}_{in} + B}{C\tilde{q}_{in} + D}. \tag{2.162}$$

Kostenbauder [47] showed that, in a similar manner, the ray–pulse matrices contain all information which is necessary to trace a generalized Gaussian beam through the optical system. Using a 2×2 complex "beam" matrix $(\tilde{Q}_{in})$, the amplitude of a generalized Gaussian beam is of the form

$$\exp\left[-\frac{i\pi}{\lambda_\ell} \begin{pmatrix} x_{in} & x_{out} \end{pmatrix} \left(\tilde{Q}_{in}\right)^{-1} \begin{pmatrix} x_{in} \\ t_{in} \end{pmatrix} \right] \tag{2.163}$$

[13] Note that the second prism has an orientation opposite to the first one.

which explicitly varies as

$$\exp\left[-\frac{i\pi}{\lambda_\ell}\left(\tilde{Q}^r_{xx}x_{in}^2 + 2\tilde{Q}^r_{xt}x_{in}t_{in} - \tilde{Q}^r_{tt}t_{in}^2\right)\right]$$

$$\times \exp\left[\frac{\pi}{\lambda_\ell}\left(\tilde{Q}^i_{xx}x_{in}^2 + 2\tilde{Q}^i_{xt}x_{in}t_{in} - \tilde{Q}^i_{tt}t_{in}^2\right)\right], \tag{2.164}$$

where $\tilde{Q}^r_{ij}$, $\tilde{Q}^i_{ij}$ are the real and imaginary coordinates of the matrix $(\tilde{Q}_{in})^{-1}$ and $\tilde{Q}_{xt} = -\tilde{Q}_{tx}$. The first factor in Eq. (2.164) expresses the phase behavior and accounts for the wave front curvature and chirp. The second term describes the spatial and temporal beam (pulse) profile. Note that unless $\tilde{Q}^{r,i}_{xt} = 0$ the diagonal elements of $(\tilde{Q}_{in})$ do not give directly such quantities as pulse duration, beam width, chirp parameter, and wave front curvature. One can show that the field at the output of an optical system is again a generalized Gaussian beam where in analogy to (2.162) the generalized beam parameter $(\tilde{Q}_{out})$ can be written as

$$(\tilde{Q}_{out}) = \frac{\begin{pmatrix} A & 0 \\ G & 1 \end{pmatrix}(\tilde{Q}_{in}) + \begin{pmatrix} B & E/\lambda_\ell \\ H & I/\lambda_\ell \end{pmatrix}}{\begin{pmatrix} C & 0 \\ 0 & 0 \end{pmatrix}(\tilde{Q}_{in}) + \begin{pmatrix} D & F/\lambda_\ell \\ 0 & 1 \end{pmatrix}}. \tag{2.165}$$

The evaluation of such matrix equations is quite complex since it generally gives rather large expressions. However, the use of advanced algebraic formula manipulation computer codes makes this approach practicable.

2.8. NUMERICAL APPROACHES

The analytical and quasi-analytical methods to trace pulses give much physical insight but fail if the optical systems become too demanding and/or many dispersion orders have to be considered.

There are commercial wave and ray tracing programs available that allow one to calculate not only the geometrical path through the system but also the associated phase. Thus complete information on the complex field distribution (amplitude and phase) in any desired plane is retrievable.

2.9. PROBLEMS

1. Dispersion affects the bandwidth of wave plates. Calculate the maximum pulse duration for which a 10th order quarter wave plate can be made of crystalline quartz, at 266 nm, using the parameters given with Eq. (2.2).

We require that the quarter-wave condition still be met with 5% accuracy at $\pm$ $(1/\tau_p)$ of the central frequency. What is the thickness of the wave plate?

2. We consider here a Fabry–Perot cavity containing a gain medium. To simplify, we assume the gain to be linear and uniform in the frequency range around a Fabry–Perot resonance of interest. Consider this system to be irradiated by a tunable probe laser of frequency ν_p.

(a) Find an expression for the transmission and reflection of this Fabry–Perot with gain as a function of the frequency of the probe laser.
(b) Find the gain for which the expression for the transmission tends to infinity. What does it mean?
(c) Describe how the gain modifies the transmission function of the Fabry–Perot (linewidth, peak transmission, peak reflection). Sketch the transmission versus frequency for low and high gain.
(d) With the probe optical frequency tuned to the frequency for which the empty (no gain) Fabry–Perot has a transmission of 50%, find its transmission factor for the value of the gain corresponding to lasing threshold.

3. Calculate the transmission of pulse propagating through a Fabry–Perot interferometer. The electric field of the pulse is given by $E(t) = \mathcal{E}(t)e^{i\omega_\ell t}$, where $\mathcal{E}(t) = \exp(-|t|/\tau)$ and $\tau = 10$ ns. The Fabry–Perot cavity is 1 mm long, filled with a material of index $n_0 = 1.5$, and both mirrors have a reflectance of 99.9%. The wavelength is 1 μm. What is the transmission linewidth (FWHM) of this Fabry–Perot? Find analytically the shape (and duration) of the pulse transmitted by this Fabry–Perot, assuming exact resonance.

4. Consider the same Fabry–Perot as in the previous problem, on which a Gaussian pulse (plane wave) is incident. The frequency of the Gaussian pulse is 0.1 ns^{-1} below resonance. Calculate (numerically) the shape of the pulse transmitted by this Fabry–Perot, for various values of the pulse chirp a. The pulse envelope is:

$$\tilde{\mathcal{E}}(t) = e^{-(1+ia)\left(\frac{t}{\tau_G}\right)^2}.$$

Is there a value of a for which the pulse transmitted has a minimum duration?

5. Consider the Gires–Tournois interferometer. (a) As explained in the text, the reflectivity is R = constant = 1, while the phase shows a strong variation with frequency. Does this violate the Kramers–Kronig relation? Explain your answer. (b) Derive the transfer function [Eq. (2.30)].

6. Derive an expression for the space–time intensity distribution of a pulse in the focal plane of a chromatic lens of focal length $f(\lambda)$. To obtain an analytical formula make the following assumptions. The input pulse is bandwidth-limited and exhibits a Gaussian temporal and transverse spatial profile. The lens has an infinitely large aperture, and the GVD can be neglected. [*Hint:* You can apply Gaussian beam analysis for each spectral component to obtain the corresponding field in the focal plane. Summation over spectral contributions (Fourier back-transform) gives then the space–time field distribution.]
7. Calculate the third-order dispersion for a pair of isosceles prisms, not necessarily used at the minimum deviation angle, using the procedure that led to Eq. (2.101). Compare with Eq. (2.106).
8. Calculate the optimum pair of prisms to be inserted into the cavity of a femtosecond pulse laser at 620 nm. The criterium is that the prism pair should provide a 20% GVD tunability around—800 fs^2, and the next higher-order dispersion should be as small as possible. With the help of Table 2.1 choose a suitable prism material, calculate the apex angle of the prisms for the Brewster condition at symmetric beam path, and determine the prism separation. If needed, assume a beam diameter of 2 mm to estimate a minimum possible glass path through the prisms.
9. Derive the ray–pulse matrix (2.160) for a pair of Brewster prisms. Verify the second-order dispersion given in relation [Eq. (2.101)], without the assumption of $L_g \ll L_a$.
10. Derive the delay and aberration parameter of a spherical mirror as given in Eqs. (2.54) and (2.55). Explain physically what happens if a parallel input beam impinges on the mirror with a certain angle α.
11. A parallel beam with plane pulse fronts impinges on a circular aperture with radius R centered on the optic axis. The pulse is unchirped and Gaussian. Estimate the frequency shift that the diffracted pulse experiences if measured with a detector placed on the optic axis. Give a physical explanation of this shift. Make a numerical estimate for a 100 fs and a 10 fs pulse. Can this effect be used to obtain ultrashort pulses in new spectral regions by placing diffracting apertures in series? [*Hint:* You can start with Eq. (2.49) and take out the lens terms. For mathematical ease you can let $R \rightarrow \infty$.] Note that a frequency shift (of the same origin) occurs when the on-axis pulse spectrum of a Gaussian beam is monitored along its propagation path.
12. Consider the three-mirror ring resonator sketched in Figure 2.34. Two of the mirrors are flat and 100% reflecting, while one mirror of field reflectivity $r = 0.9999\%$ and 60 cm curvature, serves as input and output of this resonator. We are operating at a wavelength of 800 nm. The perimeter of the ring is 60 cm. A beam with a train of pulses, of average incident

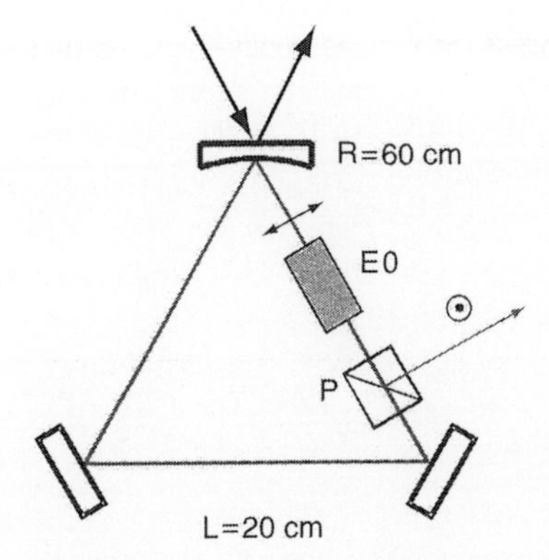

Figure 2.34 Ring resonator. Consider the electro-optic switch (EO, Pockel's cell) and the polarizing beam splitter only for part (d). The polarization of the beam circulating in the cavity gets rotated from the plane of the ring into the orthogonal direction when an electrical pulse is applied to the Pockel's cell, and extracted from the cavity by a polarizing beam splitter. The rise time of the electrical pulse is short compared to the cavity roundtrip time.

power of $P_0 = 1$ mW is sent, properly aligned, into the input path of this resonator.

(a) Calculate the size and location of the beam waist w_0 of the fundamental mode of this resonator, and the size of the beam (w) at the output mirror. Explain why the output power P_2 does not depend on the wavelength.

(b) Derive an expression for the field inside the resonator E_i as a function of the input field E_0.

(c) Consider this passive cavity being irradiated from the outside by a train of femtosecond pulses, for its use as a photon storage ring. Show that two conditions need to be fulfilled for this cavity to be exactly resonant, which may not always be simultaneously met.

(d) Let us assume next that the train of pulses, with a wavelength near 800 nm, corresponds to exactly a "resonance" of this resonator, both in frequency and repetition rate. A fast electro-optic switch is included in the ring, such that it directs the electromagnetic wave out of the resonator for a roundtrip time of the cavity, every N roundtrip times (cavity dumping). The switch opens in a short time compared to the roundtrip time. Explain how this device can be used to create short output pulses with a larger single-pulse energy than the incident pulses. What energy could be obtained in the case of (i) $N = 100$ and (ii) $N = 5000$.

BIBLIOGRAPHY

[1] V. G. Dmitriev, G. G. Gurzadyan, and D. N. Nikogosyan. *Handbook of Nonlinear Material*. Springer-Verlag, Berlin, Germany, 1999.

[2] F. J. Duarte. Dye lasers. In F. J. Duarte, Ed. *Tunable Lasers Handbook*. Academic Press, Boston, MA, 1994.

[3] W. H. Press, B. P. Flannery, S. E. Teukolsky, and W. P. Vetterling. *Numerical Recipes*. Cambridge University Press, New York, 1986.

[4] S. DeSilvestri, P. Laporta, and O. Svelto. The role of cavity dispersion in cw mode-locked lasers. *IEEE Journal of Quantum Electronics*, QE-20:533–539, 1984.

[5] W. Dietel, E. Dopel, K. Hehl, W. Rudolph, and E. Schmidt. Multilayer dielectric mirrors generated chirp in femtosecond dye ring lasers. *Optics Communications*, 50:179–182, 1984.

[6] W. H. Knox, N. M. Pearson, K. D. Li, and C. A. Hirlimann. Interferometric measurements of femtosecond group delay in optical components. *Optics Letters*, 13:574–576, 1988.

[7] K. Naganuma, K. Mogi, and H. Yamada. Group delay measurement using the fourier transform of an interferometric cross-correlation generated by white light. *Optics Letters*, 7:393–395, 1990.

[8] K. Naganuma and H. Yasaka. Group delay and alpha-parameter measurement of 1.3 micron semiconductor traveling wave optical amplifier using the interferometric method. *IEEE Journal of Quantum Electronics*, 27:1280–1287, 1991.

[9] A. M. Weiner, J. G. Fujimoto, and E. P. Ippen. Femtosecond time-resolved reflectometry measurements of multiple-layer dielectric mirrors. *Optics Letters*, 10:71–73, 1985.

[10] A. M. Weiner, J. G. Fujimoto, and E. P. Ippen. Compression and shaping of femtosecond pulses. In D. H. Auston and K. B. Eisenthal, Eds. *Ultrafast Phenomena IV*, Springer-Verlag, NY, 1984, pp. 11–15.

[11] J.-C. Diels, W. Dietel, E. Dopel, J. J. Fontaine, I. C. McMichael, W. Rudolph, F. Simoni, R. Torti, H. Vanherzeele, and B. Wilhelmi. Colliding pulse femtosecond lasers and applications to the measurement of optical parameters. In D. H. Auston and K. B. Eisenthal, Eds. *Ultrafast Phenomena IV*, Springer-Verlag, NY, 1984, pp. 30–34.

[12] M. Born and E. Wolf. *Principles of Optics—Electromagnetic theory of propagation, interference and diffraction of light*. Pergamon Press, Oxford, New York, 1980.

[13] M. V. K. and Thomas E. Furtak. *Optics*. John Wiley & Sons, New York, 1986.

[14] J. Desbois, F. Gires, and P. Tournois. A new approach to picosecond laser pulse analysis and shaping. *IEEE Journal of Quantum Electronics*, QE-9:213–218, 1973.

[15] J. Kuhl and J. Heppner. Compression of fs optical pulses with dielectric multi-layer interferometers. *IEEE Journal of Quantum Electronics*, QE-22:182–185, 1986.

[16] M. A. Duguay and J. W. Hansen. Compression of pulses from a mode-locked He-Ne laser. *Applied Physics Letters*, 14:14–15, 1969.

[17] J. Heppner and J. Kuhl. Intracavity chirp compensation in a colliding pulse mode-locked laser using thin-film interferometers. *Applied Physics Letters*, 47:453–455, 1985.

[18] R. Szipöks, K. Ferenz, C. Spielman, and F. Krausz. Chirped multilayer coatings for broadband dispersion control in femtosecond lasers. *Optics Letters*, 19:201–203, 1994.

[19] F. X. Kärtner, N. Matuschek, T. Schibli, U. Keller, H. A. Haus, C. Heine, R. Morf, V. Scheuer, M. Tilsch, and T. Tschudi. Design and fabrication of double chirped mirrors. *Optics Letters*, 22:831–833, 1997.

[20] G. Tempea, F. Krausz, C. Spielman, and K. Ferenz. Dispersion control over 150 THz with chirped dielectric mirrrors. *IEEE Selected Topics in Quantum Electronics*, 4:193–196, 1998.

[21] G. Tempea, V. Yakovlev, B. Bacovic, F. Krausz, and K. Ferenz. Tilted-front interface chirped mirrors. *Journal of the Optical Society of America B*, 18:1747–1750, 2001.

[22] N. Matuschek, L. Gallmann, D. H. Shutte, G. Steinmeyer, and U. Keller. Back-side coated chirped mirrors with smooth ultra-broadband dispersion characteristics. *Applied Physics B*, 71:509–522, 2000.

[23] Z. Bor. Distortion of femtosecond laser pulses in lenses and lens systems. *Journal of Modern Optics*, 35:1907–1918, 1988.

[24] Z. Bor. Distorsion of femtosecond laser pulses in lenses. *Optics Letters*, 14:119–121, 1989.

[25] M. Kempe, W. Rudolph, U. Stamm, and B. Wilhelmi. Spatial and temporal transformation of femtosecond laser pulses by lenses and lens systems. *Journal of the Optical Society of America B*, 9:1158–1165, 1992.

[26] M. Kempe and W. Rudolph. The impact of chromatic and spherical aberration on the focusing of ultrashort light pulses by lenses. *Optics Letters*, 18:137–139, 1993.

[27] M. Kempe and W. Rudolph. Femtosecond pulses in the focal region of lenses. *Physics Review* A, 48:4721–4729, 1993.

[28] E. B. Treacy. Optical pulse compression with diffraction gratings. *IEEE Journal of Quantum Electronics*, QE-5:454–460, 1969.

[29] F. J. Duarte and J. A. Piper. Dispersion theory of multiple prisms beam expanders for pulsed dye lasers. *Optics Communications*, 43:303–307, 1982.

[30] R. L. Fork, O. E. Martinez, and J. P. Gordon. Negative dispersion using pairs of prism. *Optics Letters*, 9:150–152, 1984.

[31] V. Petrov, F. Noack, W. Rudolph, and C. Rempel. Intracavity dispersion compensation and extracavity pulse compression using pairs of prisms. *Exp. Technik der Physik*, 36:167–173, 1988.

[32] C. Froehly, B. Colombeau, and M. Vampouille. Shaping and analysis of picosecond light pulses. *Progress of Modern Optics*, Vol. XX:115–125, 1981.

[33] O. E. Martinez. Grating and prism compressor in the case of finite beam size. *Journal of the Optical Society of America B*, 3:929–934, 1986.

[34] W. Dietel, J. J. Fontaine, and J. -C. Diels. Intracavity pulse compression with glass: A new method of generating pulses shorter than 60 femtoseconds. *Optics Letters*, 8:4–6, 1983.

[35] P. F. Curley, C. Spielmann, T. Brabec, F. Krausz, E. Wintner, and A. J. Schmidt. Operation of a fs Ti:sapphire solitary laser in the vicinity of zero group-delay dispersion. *Optics Letters*, 18:54–57, 1993.

[36] M. T. Asaki, C. P. Huang, D. Garvey, J. Zhou, H. Kapteyn, and M. M. Murnane. Generation of 11 fs pulses from a self-mode-locked Ti:sapphire laser. *Optics Letters*, 18:977–979, 1993.

[37] O. E. Martinez, J. P. Gordon, and R. L. Fork. Negative group-velocity dispersion using diffraction. *Journal of the Optical Society of America A*, 1:1003–1006, 1984.

[38] F. J. Duarte. Generalized multiple-prism dispersion theory for pulse compression in ultrafast dye lasers. *Optics and Quantum Electronics*, 19:223–229, 1987.

[39] C. P. J. Barty, C. L. Gordon III, and B. E. Lemoff. Multiterawatt 30-fs Ti:sapphire laser system. *Optics Letters*, 19:1442–1444, 1994.

[40] O. E. Martinez. 3000 times grating compressor with positive group velocity dispersion: Application to fiber compensation in the 1.3–1.6 μm region. *IEEE Journal of Quantum Electronics*, 23:59–64, 1987.

[41] M. Pessot, P. Maine, and G. Mourou. 1000 times expansion compression of optical pulses for chirped pulse amplification. *Optics Communications*, 62:419, 1987.

[42] S. A. Akhmanov, A. P. Sukhorov, and A. S. Chirkin. Stationary phenomena and space-time analogy in nonlinear optics. *Translation of Pis'ma V Zhurnal Eksperimentalnoi i Teoreticheskoi Fiziki*, 28:748–757, 1969.

[43] E. Doepel. Matrix formalism for the calculation of group delay and group-velocity dispersion in linear optical elements. *Journal of Modern Optics*, 37:237–242, 1990.

[44] O. E. Martinez. Matrix formalism for pulse compressors. *IEEE Journal of Quantum Electronics*, QE-24:2530–2536, 1988.

[45] O. E. Martinez. Matrix formalism for dispersive cavities. *IEEE Journal of Quantum Electronics*, QE-25:296–300, 1989.

[46] S. P. Dijaili, A. Dienes, and J. S. Smith. ABCD matrices for dispersive pulse propagation. *IEEE Journal of Quantum Electronics*, QE-26:1158–1164, 1990.

[47] A. G. Kostenbauder. Ray-pulse matrices: A rational treatment for dispersive optical systems. *IEEE Journal of Quantum Electronics*, QE-26:1148–1157, 1990.

[48] A. E. Siegman. *Lasers*. University Science Books, Mill Valley, CA, 1986.

3

Light–Matter Interaction

The generation and application of fs light pulses revolve around light–matter interaction. In the preceding chapter we considered situations in which the propagating field does not change the material response. The result was a linear dependence of the output field on the input field, a feature attributed to linear optical elements. The medium could be described by a transfer function $\tilde{H}(\Omega)$, which for a material, is completely determined by a complex dielectric constant $\tilde{\epsilon}(\Omega)$. The real part of $\tilde{\epsilon}(\Omega)$ is responsible for dispersion, determining phase and group velocity, for example. The imaginary part describes (frequency-dependent) loss or gain. In many cases these linear optical media can be considered as host materials for sources of nonlinear polarization. It is the latter which will be discussed in this section.

Let us consider a pulse propagating through a (linear) optical material, for instance, glass. In addition to the processes described above, we will have to consider nonlinear optical interaction if the electric field strength is high. This can result from high pulse power and/or tight focusing. The decomposition of the polarization according to Eq. (1.68), $P = P^L + P^{NL}$, accounts then for different optical properties of one and the same material. Another possible situation is that the host material affects the pulse through the linear optical response only, but contains additional substances interacting nonlinearly with the pulse. This, for example, can occur if the light pulse is at resonance with a dopand material, which considerably increases the interaction strength. Examples of such composite materials are glass doped with ions, dye molecules dissolved in a (transparent) solvent, and resonant gas molecules surrounded by a (nonresonant) buffer gas.

In these cases the contributions, P^L and P^{NL}, belong to different materials.[1] Finally, there can be situations where the nonresonant (host) material as well as a resonant (second) material produce a nonlinear polarization. Here P^{NL} may be decomposed into a nonresonant $P^{N,NL}$ and resonant part $P^{R,NL}$:

$$P^{NL} = P^{N,NL} + P^{R,NL}. \tag{3.1}$$

3.1. DENSITY MATRIX EQUATIONS

The semiclassical treatment has been most widely applied to discuss the interaction of short light pulses with matter. The electromagnetic field is used as a classical quantity, whereas the matter is described in the frame of quantum mechanics. This both allows a simple interpretation of the results in terms of measurable quantities and accounts for the quantum properties of matter. We will restrict ourselves to atomic particles (atoms, molecules, etc.) with a transition at resonance with the incident light pulse and will idealize the resonant medium as an ensemble of two-level systems. By "resonant" we mean that there is sufficient overlap of the pulse spectrum with the transition profile, for the pulse to experience absorption or gain. If not specified differently, we will refer to a volume element at a certain space coordinate z. This allows us to consider only the time dependence of the sample response.

In addition we have to take into account the interaction of the resonant particles with each other and/or with surrounding nonresonant particles. The physical nature of such interactions can be manifold; for instance, collisions or changing local fields because of particle motion may contribute. Generalizing, one says that the resonant system is coupled to a dissipative system which in turn is regarded as a reservoir with a large number of degrees of freedom. The result is a time dependence of certain physical properties of the resonant particle, which implies a time dependence of the strength of interaction with an incident electric field.

We should keep in mind that, typically, the light pulse interacts with a large number of particles ($\approx 10^5$–10^{23} cm^{-3}). We can therefore exclude the possibility of calculating the dynamic behavior of each individual ensemble member and summing up for the macroscopic sample response. But do we really need to consider individual particles? The answer to this question depends on the time scale on which we look at the medium and, fortunately, in many cases is no. The physical reason is the random character of the forces acting on the resonant atoms, as illustrated in Figure 3.1, which is quite understandable if we think of the Brownian molecular motion. Mathematically, this force can be represented

[1]Of course, resonant matter also may interact purely linearly with the light pulse if the field strength is sufficiently small (linear loss and gain).

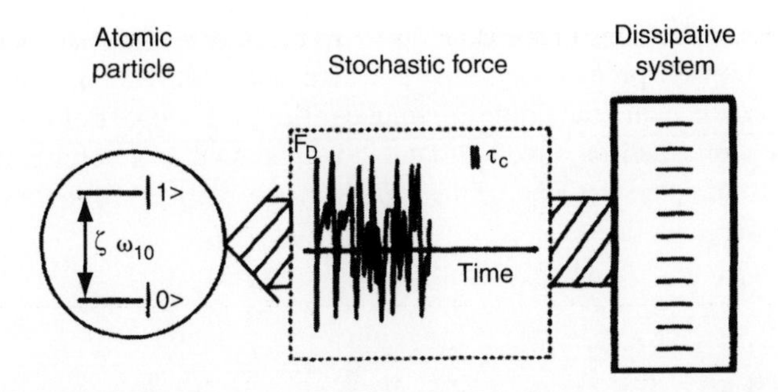

Figure 3.1 Resonant atomic particle under the influence of a dissipative system exerting a stochastic force with correlation time τ_c.

by a stochastic function $F_D(t)$ characterized by a certain correlation time τ_c. The latter means that the correlation of first order

$$A_F(\tau) = \int_{-\infty}^{+\infty} F_D(t')F_D(t' - \tau)\,dt' \tag{3.2}$$

has appreciable values only for $|\tau| < \tau_c$. In other words, there is no net force on the particle under investigation on a time scale $t > \tau_c$; a situation we will refer to as stationary state. If we assume homogeneity of the medium, all resonant particles will experience similar distortions by the dissipative system. To decide whether the particles have identical properties, we have to perform a measurement, for example, to determine the absorptivity or reflectivity. Because each measurement requires a certain time T_M, the result is a sample property, $\bar{X}$, averaged over T_M:

$$\bar{X}(t) = \frac{1}{T_M}\int_{t-T_M/2}^{t+T_M/2} X(t')\,dt'. \tag{3.3}$$

This averaging makes evident the difficulties of defining identical properties. For measuring times (e.g., given by the interaction time with a light pulse) $T_M > \tau_c$, we expect to see identical behavior for all ensemble members. For $T_M < \tau_c$ the stochastic influence of the dissipative system does not average to zero and the atomic particles show "individuality." The magnitude of τ_c depends on the actual medium. In condensed matter, at room temperature, it is mostly shorter than 10^{-14} s.

In general, it is impossible to calculate the behavior of the resonant particle under the influence of the dissipative system exactly, no matter what time scale is of interest. The reasons are the complexity of the dissipative system and its large number of degrees of freedom. In terms of quantum mechanics we have to deal with a weakly prepared system [1] which can be described favorably by its density operator $\hat{\rho}$ and the corresponding density matrix equations, respectively. If the action of the dissipative system can be regarded as a perturbation for the two-level system, the elements of the density matrix obey the following equations of motion, see e.g., [2]:

$$\frac{d}{dt}\rho_{10}(t) = -i\omega_{10}\rho_{10}(t) - \frac{1}{T_2}\rho_{10}(t) \tag{3.4}$$

$$\frac{d}{dt}\rho_{11}(t) = -\frac{1}{T_1}[\rho_{11}(t) - \rho_{11}^{(e)}] \tag{3.5}$$

$$\rho_{01}(t) = \rho_{10}^{*}(t) \tag{3.6}$$

$$\rho_{00}(t) = 1 - \rho_{11}(t) \tag{3.7}$$

where T_1, T_2 are the energy (longitudinal) and phase (transverse) relaxation time, respectively. This system of equations describes the evolution of the two-level system toward a stationary state (e.g., thermal equilibrium) denoted by the superscript (e). The elements of the density matrix, ρ_{kl}, can be calculated by means of the eigenstates of the (unperturbed) two-level system $|0\rangle$ and $|1\rangle$ and read

$$\rho_{kl} = \langle k|\hat{\rho}|l\rangle \qquad k, l = 0, 1. \tag{3.8}$$

With the density operator $\hat{\rho}$ given any measurable quantity X can be determined as an expectation value of the corresponding observable $\hat{X}$

$$X = Tr(\hat{\rho}\hat{X}). \tag{3.9}$$

Assuming a homogeneous distribution of two-level systems of number density $\bar{N}$, the occupation number density of level k is

$$N_k = \bar{N}Tr(\hat{\rho}\hat{N}_k) = \bar{N}\rho_{kk} \qquad k = 0, 1 \tag{3.10}$$

and the polarization can be written as

$$P = \bar{N}Tr(\hat{\rho}\hat{p}) = \bar{N}(p_{10}\rho_{01} + p_{01}\rho_{10}). \tag{3.11}$$

Here $\hat{N}_{0,1}$ is the occupation number operator and $\hat{p}$ is the atomic dipole operator with the matrix elements p_{10} and p_{01}. The diagonal elements of the density matrix

ρ_{00} and ρ_{11} determine the probability of finding the system in state $|0\rangle$ and $|1\rangle$, respectively.

Let us assume that at $t = 0$ the system can be described by the initial conditions $N_{0,1}(0)$ and $P(0)$. For $t > 0$ the evolution of the polarization and the occupation numbers according to Eqs. (3.4)–(3.7) is then given by

$$\boxed{P(t) \propto P(0)e^{-2t/T_2}\cos(\omega' t)} \tag{3.12}$$

and

$$\boxed{N_{0,1}(t) = N_{0,1}^{(e)} + \left[N_{0,1}(0) - N_{0,1}^{(e)}\right]e^{-t/T_1}.} \tag{3.13}$$

As can be seen the polarization behaves like a damped harmonic oscillator where T_2^{-1} plays the role of a damping constant and $\omega' = \sqrt{\omega_{10}^2 + T_2^{-2}} \approx \omega_{10}$ is the oscillation frequency.

The relaxation of the polarization is not necessarily associated with a relaxation of the energy of the two-level system. Instead, T_2 can be regarded as the phase memory time of the two-level system, i.e., as the time interval in which the two-level system remembers the phase of the oscillation at $t = 0$ (excitation event). It is T_1 which describes the recovery of the occupation numbers toward their equilibrium values $N_{0,1}^{(e)}$ and thus determines an energy relaxation. The convenience of the density matrix Eqs. (3.4)–(3.7) is obvious; the dissipative system enters through two characteristic relaxation constants only. However, to derive Eqs. (3.4)–(3.7), it is necessary to perform an averaging over time intervals τ_c which limits the range of validity. Therefore, with the view on ultrashort pulse interaction, we have to be aware that the temporal resolution with which we can trace the dynamics of the two-level system with Eqs. (3.4)–(3.7) is always worse than τ_c.

What determines these relaxation rates and what are typical values? For the reasons mentioned above, an exact calculation of T_1, T_2 is rather difficult if not impossible for many systems of practical relevance. These parameters have a complex dependence on particle density, type of interaction, particle velocity etc. As will be discussed later, measurements with ultrashort light pulses can yield the desired information directly. A limiting case occurs for isolated two-level systems which are not influenced by surrounding particles. The only dissipative system acting on the atomic particle is the vacuum field, causing natural line broadening which implies $T_2 = 2T_1$. Generally speaking, whenever the interaction with the dissipative system is such that each event leads to an occupation change, $T_2 = 2T_1$ holds. Here the dephasing time is determined by the energy relaxation time. However, there are many other interaction processes which do not change

Table 3.1

Typical energy and phase relaxation times and interaction cross sections of some materials.

Medium	T_1 (s)	T_2 (s)	$\sigma_{01}^{(0)}$ (cm^2)
Solids doped with resonant atomic systems	10^{-3}–10^{-6}	10^{-11}–10^{-14}	10^{-19}–10^{-21}
Dye molecules solved in an organic solvent	10^{-8}–10^{-12}	10^{-13}–10^{-14}	10^{-16}
Semiconductors	10^{-4}–10^{-12}	10^{-12}–10^{-14}	–

the occupation numbers, but which may affect the phase of the oscillations. These additional relaxation channels for the phase are responsible for the fact that most often (particularly in condensed matter) $T_2 < (\ll) T_1$. Table 3.1 offers some examples. It should also be noted that the transition profile resulting from the steady state solution of Eqs. (3.4)–(3.7) implies a Lorentzian line shape with a FWHM given by $2T_2^{-1}$ (which is consistent with the Fourier transform of the exponential decay law), Eq. (3.12).

So far we have considered an ensemble of identical atoms, which enabled us to calculate the macroscopic quantities by multiplying the contribution of one particle by the particle number (density), as shown in Eqs. (3.10) and (3.11). The result is a transition profile which has exactly the same shape and width as those of a single atom. Such a medium is referred to as *homogeneously broadened.* However, in many real situations, different particles from an ensemble of identical atoms have (slightly) different resonance frequencies ω'_{10}. Among the various causes for such a distribution of frequencies, the most common are Doppler shift in gases, and different local surroundings in solids (e.g., defects, impurities). The total response of the medium is now given by the sum of the responses of all subensembles with polarization $P'(\omega'_{10})$ and occupation numbers $N'_{0,1}(\omega'_{10})$. A sub-ensemble characterized by a certain transition frequency ω'_{10} is to contain particles with resonance frequencies which fall within a homogeneous linewidth. Thus, for the total polarization we obtain

$$P(t) = \sum_{\omega'_{10}} P'(t, \omega'_{10}), \tag{3.14}$$

and likewise the total number density of particles in state $|0\rangle$ or $|1\rangle$ is

$$N_{0,1}(t) = \sum_{\omega'_{10}} N'_{0,1}(t, \omega'_{01}). \tag{3.15}$$

Frequently the distribution of atomic species with a certain transition frequency can be described by a distribution function $g_{inh}(\omega'_{10} - \omega_{ih})$ centered at ω_{ih} and having a FWHM of $\Delta\omega_{inh}$. This line shape is referred to as *inhomogeneous* (as opposed to the homogeneous broadening cited previously). The polarization and occupation number densities which result from particles having resonance frequencies in the interval $(\omega'_{10}, \omega'_{10} + d\omega'_{10})$ can be written as

$$P'(t, \omega'_{10}) = \bar{N} g_{inh}(\omega'_{10} - \omega_{ih})[p_{01}\rho_{10}(t, \omega'_{10}) + p_{10}\rho_{01}(t, \omega'_{10})]d\omega'_{10} \tag{3.16}$$

and

$$N'_{0,1}(t, \omega'_{10}) = \bar{N} g_{inh}(\omega'_{10} - \omega_{ih})\rho_{00,11}(t, \omega'_{10})d\omega'_{10} \tag{3.17}$$

where g_{inh} must satisfy the normalization condition

$$\int_0^\infty g_{inh}(\omega'_{10} - \omega_{ih})d\omega'_{10} = 1. \tag{3.18}$$

The density matrix elements can be determined from the set of density matrix Eqs. (3.4)–(3.7) for each frequency ω'_{10}. In terms of g_{inh} the total polarization and occupation numbers are

$$\begin{aligned} P(t) &= \int_0^\infty P'(t, \omega'_{10})d\omega'_{10} \\ &= \int_0^\infty \bar{N} g_{inh}(\omega'_{10} - \omega_{ih})[p_{01}\rho_{10}(t, \omega'_{10}) + p_{10}\rho_{01}(t, \omega'_{10})]d\omega'_{10} \end{aligned} \tag{3.19}$$

and

$$\begin{aligned} N_{0,1}(t) &= \int_0^\infty N'(t, \omega'_{10})d\omega'_{10} \\ &= \int_0^\infty \bar{N} g_{inh}(\omega'_{10} - \omega_{ih})\rho_{00,11}(t, \omega'_{10})d\omega'_{10}. \end{aligned} \tag{3.20}$$

The use of Eqs. (3.20) and (3.19) in connection with Eqs. (3.4)–(3.7), and (3.9) for the determination of $P(t)$ and $N_{0,1}(t)$ requires that the composition of the subensembles remains unchanged during the time range of interest. In other words, each atom possesses a certain (constant) transition frequency ω'_{10}. Surely this concept of "static" inhomogeneous broadening is justified where the inhomogeneity is caused by a time-independent crystal disorder etc.; it becomes questionable when the resonance frequency ω'_{10} is a function of time. The latter, for example, is true if ω'_{10} is influenced by particle motion as is the case with

Doppler broadening associated with velocity changing collisions. A similar situation arises in liquid dye solutions, where the absorbing molecules move through changing local fields, which (through Stark shifts) also causes random drifts of the resonance frequency. Such statistical changes of phase because of drifts in resonance frequency are generally represented by a "cross relaxation time" T_3. The equation of motion for the occupation numbers of an ensemble of particles with resonance frequency $\bar{\omega}'_{10}$ then becomes:

$$\frac{d}{dt}N'_{0,1}(t,\bar{\omega}'_{10}) = -\frac{1}{T_1}[N'_{0,1}(t,\bar{\omega}'_{10}) - N^{(e)}_{0,1}(\bar{\omega}'_{10})] - \frac{1}{T_3}[N'_{0,1}(t,\bar{\omega}'_{10}) - N_{0,1}] \tag{3.21}$$

where

$$N_{0,1} = g_{inh}(\bar{\omega}'_{10} - \omega_{ih})d\omega'_{10} \int_0^{\infty} \bar{N} g_{inh}(\omega'_{10} - \omega_{ih})\rho_{00,11}(t,\omega'_{10})d\omega'_{10}. \tag{3.22}$$

Note that the integral in the last equation represents the total number of particles in state $|0\rangle$ and $|1\rangle$, respectively, at a certain instant t. Therefore the second term in the right-hand side of Eq. (3.21) is responsible for the relaxation of a disturbed particle distribution toward an equilibrium distribution given by the inhomogeneous line shape.

For illustration, let us assume that at $t = t_0$ an initial Gaussian distribution [Figure 3.2(a)] of particles in state $|0\rangle$ having a certain resonance frequency is distorted into (b) by excitation ($|0\rangle$ to $|1\rangle$) of a subensemble with resonance frequency $\bar{\omega}'_{10}$. If $\Delta\omega_{inh}$ is much larger than the homogeneous linewidth $\Delta\omega_h = 2/T_2$ a hole of width $\Delta\omega_h$ will be burnt in the distribution $N'_0(\omega'_{10})$ around the excitation frequency $\bar{\omega}'_{10}$ [Fig. 3.2(b)]. On a time scale $\Delta t > T_3$, these excited particles change their resonance frequency under the influence of the linewidth determining processes, and the Gaussian distribution is re-established (c). For $T_3 \ll T_1$

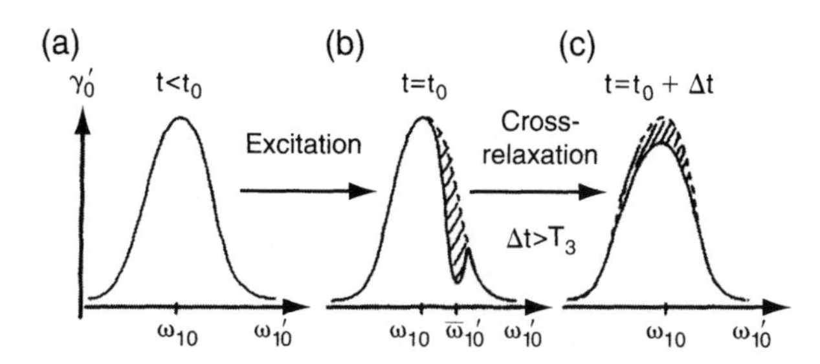

Figure 3.2 Cross-relaxation of an inhomogeneously broadened medium (notation: $\omega_{ih} = \omega_{10}$ and $\gamma'_0 = N'_0$).

the hatched areas in Fig. 3.2, representing the number of excited particles, are equal. The initial distribution is then reached after relaxation of the two-level systems into the ground state. The question arises as to what happens if the measuring time T_M is much larger than the cross-relaxation time T_3? We expect then to measure a distribution similar to curve (c) in Fig. 3.2. The sample behaves as if it were homogeneously broadened. This shows that the classification of homogeneous and inhomogeneous line broadening depends on the time scale of interest.

Let us next turn to the interaction of an electric field (light pulse) with the two-level system. For the model described by Eqs. (3.4)–(3.7) to be valid, the pulse duration τ_p should be (much) larger than the correlation time of the dissipative system τ_c. Consistently with this requirement, we can simply "add" the field interaction terms to Eqs. (3.4)–(3.7). In the case of dipole interaction, the Hamiltonian is given by $\hat{p}E$, and the density matrix equations read [2]

$$\frac{d}{dt}\rho_{10}(t) + i\omega_{10}\rho_{10}(t) + \frac{1}{T_2}\rho_{10}(t) = i\frac{p}{\hbar}[\rho_{00}(t) - \rho_{11}(t)]E(t) \tag{3.23}$$

$$\rho_{01}(t) = \rho_{10}^*(t) \tag{3.24}$$

$$\frac{d}{dt}\rho_{11}(t) + \frac{1}{T_1}[\rho_{11}(t) - \rho_{11}^{(e)}] = i\frac{p}{\hbar}[\rho_{10}(t) - \rho_{01}(t)]E(t) \tag{3.25}$$

$$\frac{d}{dt}\rho_{00}(t) + \frac{1}{T_1}[\rho_{00}(t) - \rho_{00}^{(e)}(t)] = i\frac{p}{\hbar}[\rho_{01}(t) - \rho_{10}(t)]E(t). \tag{3.26}$$

For simplicity, we have introduced $p_{01} = p_{10} = p$. The total population is conserved:

$$\rho_{00}(t) + \rho_{11}(t) = 1. \tag{3.27}$$

For calculating the change of the pulse (electric field) as it propagates through the medium we have to evaluate the wave equation (1.67). Assuming that the resonant atomic particles are embedded in a (transparent) host material with dielectric constant ϵ and applying the SVEA [Eq. (1.92)], the pulse envelope obeys Eq. (1.88) complemented by the resonant polarization. As we have treated the electric field in Eq. (1.83), the polarization of the two-level system is also decomposed into a slowly varying envelope $\tilde{\mathcal{P}}$ and a rapidly oscillating contribution

$$P(t,z) = \frac{1}{2}\tilde{\mathcal{P}}(t,z)e^{i(\omega_\ell t - k_\ell z)} + c.c. \tag{3.28}$$

With this ansatz the propagation equation becomes

$$\frac{\partial}{\partial z}\tilde{\mathcal{E}} - \frac{i}{2}k_\ell''\frac{\partial^2}{\partial t^2}\tilde{\mathcal{E}} + \mathcal{D} = i\frac{\mu_0}{2k_\ell}\left(\frac{\partial^2}{\partial t^2}\tilde{\mathcal{P}} + 2i\omega_\ell\frac{\partial}{\partial t}\tilde{\mathcal{P}} - \omega_\ell^2\tilde{\mathcal{P}}\right), \tag{3.29}$$

where we have transferred to local coordinates. In the frame of the SVEA, the right-hand side of this equation can be approximated by the last term. We will see in Section 3.3.3 that propagation of short pulses requires to include also the next before last term. Because in this section we shall concentrate on the effects of nonlinear light–matter interaction, we assume the host material to be weakly dispersive and neglect GVD and higher-order dispersion ($k_\ell'' = \mathcal{D} = 0$). In later chapters we will study the interplay of dispersion with various nonlinearities. Hence, we obtain the following propagation equation

$$\boxed{\frac{\partial}{\partial z}\tilde{\mathcal{E}}(t,z) = -i\frac{\mu_0\omega_\ell^2}{2k_\ell}\tilde{\mathcal{P}}(t,z) = -i\frac{\mu_0\omega_\ell c}{2n}\tilde{\mathcal{P}}(t,z).} \tag{3.30}$$

The polarization needed to solve this equation must be derived from the density matrix equations according to Eq. (3.11) or Eq. (3.19), depending on the kind of line broadening.

Propagation problems which are governed by Eq. (3.30) are associated with pulse shaping and play a decisive role in the pulse generation in lasers and in pulse amplification and shaping. We therefore give this equation a close scrutiny.

First, however, let us discuss briefly some of the important features of the resonant light–matter interaction, under the assumption that the pulse duration is much longer than all relaxation constants, which allows us to use a stationary solution for Eq. (3.23). In addition we will consider propagation only over a small distance Δz. A formal solution of Eq. (3.30) yields for the change of the complex pulse envelope

$$\Delta\tilde{\mathcal{E}}(t,z) = -i\frac{\mu_0\omega_\ell^2}{2k_\ell}\tilde{\mathcal{P}}(t,z)\Delta z. \tag{3.31}$$

As will be discussed in the next subsection in detail, using Eq. (3.11) and the stationary solution of Eqs. (3.23) and (3.26), we obtain for the polarization

$$\tilde{\mathcal{P}}(t,z) = i\frac{p^2T_2/\hbar}{iT_2(\omega_\ell - \omega_{10}) + 1}\;[\rho_{11}(t) - \rho_{00}(t)]\bar{N}\tilde{\mathcal{E}}(t,z). \tag{3.32}$$

This relation inserted into Eq. (3.31) gives

$$\Delta\tilde{\mathcal{E}}(t,z) = \frac{\mu_0\omega_\ell^2 p^2}{2\hbar k_\ell}\frac{T_2}{1+iT_2(\omega_\ell-\omega_{10})}[\rho_{11}(t)-\rho_{00}(t)]\bar{N}\tilde{\mathcal{E}}(t,z)\Delta z, \quad (3.33)$$

which has the formal solution

$$\tilde{\mathcal{E}}(t,z+\Delta z) = \tilde{\mathcal{E}}(t,z)e^{A(t)+iB(t)}, \quad (3.34)$$

where

$$A = \frac{\mu_0\omega_\ell^2 p^2 T_2/(2\hbar k_\ell)}{T_2^2(\omega_{10}-\omega_\ell)^2+1}[\rho_{11}(t)-\rho_{00}(t)]\bar{N}\Delta z \quad (3.35)$$

and

$$B = A(\omega_{10}-\omega_\ell)T_2. \quad (3.36)$$

The real part A of the exponential factor in Eq. (3.34) represents absorption or amplification. The imaginary part B has the structure of a phase modulation term. According to our discussion in Chapter 1 [Eq. (1.175)] we can interpret B as resulting from a time-dependent optical path length (difference) $(2\pi/\lambda)\Delta z[n(t)-1]$, where the time dependence of B and n, respectively, is determined by the inversion density $\Delta\rho = \rho_{11}(t) - \rho_{00}(t)$. Equating both relations yields an expression for the refractive index

$$n(t) = 1 + \frac{Ac}{\omega_\ell\Delta z}(\omega_\ell-\omega_{10})T_2 \quad (3.37)$$

in the vicinity of the resonance frequency ω_{10}. In addition to the explicit frequency dependence (ω_ℓ) in Eq. (3.37), the index depends on the laser frequency (ω_ℓ) through A in Eq. (3.35). These dependencies are shown in Figure 3.3 for an absorbing sample for two different (negative) values of $\Delta\rho$. A smaller $|\Delta\rho|$ results in a smaller absorption coefficient regardless of ω_ℓ while for n we find a decrease for $\omega_\ell < \omega_{10}$ and an increase for $\omega_\ell > \omega_{10}$. At exact resonance ($\omega_\ell = \omega_{10}$) the refractive index does not change at all if $\Delta\rho$ is varied.

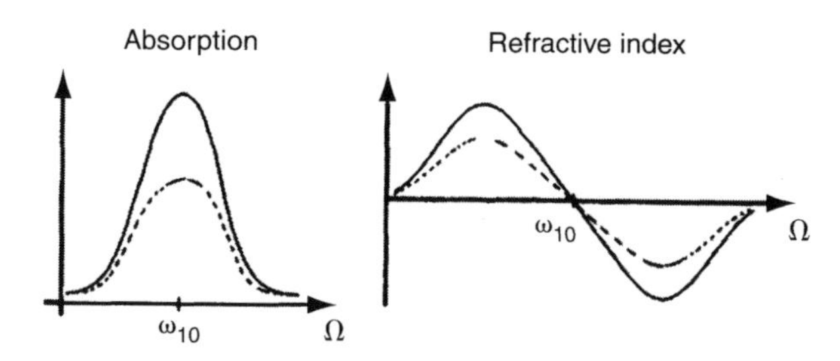

Figure 3.3 Absorption coefficient and refractive index in the vicinity of an optical resonance for two values of the population inversion $\Delta\rho = \rho_{11} - \rho_{00}$.

3.2. PULSE SHAPING WITH RESONANT PARTICLES

3.2.1. General

Let us consider a pulse traveling through a medium of certain length consisting of particles with number density $\bar{N}$ at resonance with the light. How is the pulse being modified by the medium? In answering this question we assume that only the resonant particles influence the pulse, i.e., we neglect GVD of the host material and nonresonant nonlinear contributions. Depending on how the medium is prepared we expect light to be absorbed or amplified. Absorption can occur when $\Delta\rho = (\rho_{11}-\rho_{00}) < 0$, i.e., if the majority of resonant particles interacting with the pulse are in the ground state. Amplification is expected in the opposite case where $\Delta\rho = (\rho_{11}-\rho_{00}) > 0$. To reach this situation the particles have to be excited by an appropriate pump mechanism.[2]

At first glance absorption (amplification) seems only to decrease (increase) the pulse energy by a certain factor while leaving the other pulse characteristics unchanged. We will see this to be true only under specific conditions. In the general case the pulse at the output exhibits a different envelope (shape) as well as a different time-dependent phase, as compared to its input characteristics. The leading part of the pulse changes the properties of the medium, which will then act in a different manner on the trailing part. In terms of amplification and absorption one can say that these quantities become time dependent because of the change of the inversion density $\Delta\rho(t)$. We expect the pulse to be more heavily absorbed (amplified) at its leading part which, of course, modifies the envelope shape. According to our discussion previously, a time-dependent occupation number means that different pulse parts "see" different optical path lengths, resulting in

[2]Other energy levels have to be taken into account to describe the pumping.

a time-dependent phase change (chirp). To discuss the actual pulse distortion we need to determine the temporal behavior of the occupation numbers by means of Eqs. (3.23)–(3.26). In general, numerical methods are required to analyze this problem. An analytical description can only be made for some limiting cases. The physics of the pulse–matter interaction depends strongly on the ratio of the pulse duration and the characteristic response time of the medium, as well as on the pulse intensity and energy. We will start with the approximation that the phase relaxation time is much shorter than the pulse duration, generally referred to as rate equation approximation. Femtosecond pulses do challenge this often used approximation. Because the duration of fs pulses can be comparable or shorter than the phase memory of the medium, the polarization oscillations excited by the leading part of the pulse can interfere coherently with subsequent pulse parts. We will discuss first this situation approximately as a perturbation of the *rate equation approximation* (REA). Phenomena related to the dominant influence of coherent light–matter interaction will be dealt with in the following chapter.

To illustrate the action of the phase memory, let us inspect the equation of motion for the polarization. Within the SVEA, and assuming $|\omega_\ell - \omega_{10}| \ll \omega_{10}$, ω_ℓ as well as $\omega_{10}T_2, \omega_\ell T_2 \gg 1$, Eqs. (3.11), (3.23), (3.24), and (3.28) yield a first-order differential equation for the slowly varying envelope component $\tilde{\mathcal{P}}$:

$$\frac{d}{dt}\tilde{\mathcal{P}} + \left[\frac{1}{T_2} + i(\omega_\ell - \omega_{10})\right]\tilde{\mathcal{P}} = i\frac{\bar{N}p^2}{\hbar}(\rho_{11} - \rho_{00})\tilde{\mathcal{E}}. \tag{3.38}$$

The solution of Eq. (3.38) can be written formally in the form

$$\tilde{\mathcal{P}}(t) = i\frac{\bar{N}p^2}{\hbar}\int_{-\infty}^{t} \tilde{\mathcal{E}}(t')\Delta\rho(t')e^{[i(\omega_\ell-\omega_{10})+1/T_2](t'-t)}dt' \tag{3.39}$$

where $\Delta\rho = \rho_{11} - \rho_{00}$ is the population inversion. Obviously the polarization at time t depends on values of the electric field (modulus and phase) and population numbers for all $t' \leq t$, a dependence weighted by the function $e^{-(t-t')/T_2}$. The latter implies that the memory time is T_2. For later reference we will derive two other representations of the solution of Eq. (3.38). Continuous application of partial integration of Eq. (3.39) leads to

$$\tilde{\mathcal{P}}(t) = i\frac{\bar{N}p^2T_2}{\hbar}\tilde{L}(\omega_\ell - \omega_{10})\Bigg\{\tilde{\mathcal{E}}(t)\Delta\rho(t) + \sum_{n=1}^{\infty}\tilde{L}^n(\omega_\ell - \omega_{10})\left(T_2\frac{d}{dt}\right)^n [\tilde{\mathcal{E}}(t)\Delta\rho(t)]\Bigg\}, \tag{3.40}$$

where

$$\tilde{L}(\omega_\ell - \omega_{10}) = \frac{1}{i(\omega_\ell - \omega_{10})T_2 + 1} \tag{3.41}$$

is the complex line shape factor. Fourier transforming Eq. (3.40) yields

$$\tilde{\mathcal{P}}(\Omega) = i\frac{\bar{N}p^2T_2}{\hbar}\tilde{L}(\omega_\ell - \omega_{10})\Big[\mathcal{F}\{\tilde{\mathcal{E}}(t)\Delta\rho(t)\}$$
$$+ \sum_{n=1}^{\infty}\tilde{L}^n(\omega_\ell - \omega_{10})(i\Omega T_2)^n\mathcal{F}\{\tilde{\mathcal{E}}(t)\Delta\rho(t)\}\Big]. \tag{3.42}$$

Note that in Eq. (3.40) and in Eq. (3.42) the first summand corresponds to the zero-order term.

3.2.2. Pulses Much Longer Than the Phase Relaxation Time ($\tau_p \gg T_2$)

If the pulse duration is much longer than the phase relaxation time of the medium there is no coherent superposition of polarization and electric field oscillations. The memory of the medium is only through the change of the occupation numbers. Assuming that the fastest component of the dynamics of the occupation inversion is determined by the pulse under discussion, $\tau_p \gg T_2$ implies that $|T_2(d/dt)| \ll 1$ in Eq. (3.40). In this case we may neglect all terms with $n \geq 1$, a procedure called rate equation approximation, and the polarization reduces to

$$\tilde{\mathcal{P}}(t) = i\frac{p^2T_2}{\hbar}\tilde{L}(\omega_\ell - \omega_{10})\tilde{\mathcal{E}}(t)\Delta N(t) \tag{3.43}$$

where we have introduced the population inversion density

$$\Delta N = \bar{N}(\rho_{11} - \rho_{00}) = \bar{N}\Delta\rho = N_1 - N_0. \tag{3.44}$$

Note that $\tilde{\mathcal{P}}$ given by Eq. (3.43) corresponds to the stationary solution of Eqs. (3.38) and (3.23), respectively. In terms of Eq. (3.42) the rate equation approximation requires $|(i\Omega T_2)^n\mathcal{F}\{\tilde{\mathcal{E}}(t)\Delta\rho(t)\}| \ll 1$, which means that the spectral width of the pulse is much smaller than $1/T_2$, hence much smaller than the spectral width of the transition. Strictly speaking, this approximation requires a monochromatic wave interacting with an extended transition profile. It does, however, yield satisfactory results in numerous practical cases, even when the above condition is only marginally satisfied.

To study the behavior of modulus and phase of a pulse in propagating through the resonant medium we have to solve the propagation equation (3.30) with the polarization given by Eq. (3.43). Thus, we have

$$\frac{\partial}{\partial z}\tilde{\mathcal{E}} = \frac{p^2 T_2 \mu_0 \omega_\ell^2}{2k_\ell \hbar}\tilde{L}(\omega_\ell - \omega_{10})\Delta N \tilde{\mathcal{E}}. \tag{3.45}$$

The equation of motion for the population inversion density at location z, ΔN, can be obtained from Eqs. (3.23)–(3.26) and reads

$$\boxed{\frac{\partial}{\partial t}\Delta N + \frac{1}{T_1}(\Delta N - \Delta N^{(e)}) = -\frac{p^2 T_2}{\hbar^2}|\tilde{L}(\omega_\ell - \omega_{10})|^2 \Delta N |\tilde{\mathcal{E}}|^2,} \tag{3.46}$$

where $\Delta N^{(e)}$ is the equilibrium value of ΔN before the pulse arrives. This equation is often referred to as the rate equation for the population inversion. Using the amplitude and phase representation $\tilde{\mathcal{E}} = \mathcal{E}\exp(i\varphi)$ for the complex pulse envelope, Eq. (3.45) yields for the modulus

$$\boxed{\frac{\partial}{\partial z}\mathcal{E} = \frac{1}{2}\sigma_{01}^{(0)}|\tilde{L}(\omega_\ell - \omega_{10})|^2 \Delta N \mathcal{E}} \tag{3.47}$$

and for the phase

$$\boxed{\frac{\partial}{\partial z}\varphi = -\frac{1}{2}\sigma_{01}^{(0)}|\tilde{L}(\omega_\ell - \omega_{10})|^2(\omega_\ell - \omega_{10})T_2 \Delta N.} \tag{3.48}$$

The quantity

$$\sigma_{01}^{(0)} = \frac{p^2 T_2 \omega_{10}}{\epsilon_0 c n \hbar} \tag{3.49}$$

is the interaction cross-section at the center of the transition ($\omega_\ell = \omega_{10}$). The interaction cross-section at frequency ω_ℓ is given by

$$\sigma_{01} = \sigma_{01}(\omega_\ell - \omega_{10}) = \sigma_{01}^{(0)}|\tilde{L}(\omega_\ell - \omega_{10})|^2 = \frac{\sigma_{01}^{(0)}}{1 + T_2^2(\omega_\ell - \omega_{10})^2} \tag{3.50}$$

and can be considered as a measure of the interaction strength or of the probability that an absorption (emission) process takes place. Typical values for the interaction cross sections were listed in Table 3.1. The temporal change of the population inversion, cf. Eq. (3.46), does not depend on the phase of the pulse but on $|\tilde{\mathcal{E}}|^2$. This suggests the convenience of rewriting this equation in terms of the photon flux density F defined in Eq. (1.23). We find

$$\frac{\partial}{\partial t}\Delta N + \frac{1}{T_1}(\Delta N - \Delta N^{(e)}) = -2\sigma_{01}\Delta N F \tag{3.51}$$

and by means of Eq. (3.47) the photon flux density is found to obey

$$\frac{\partial}{\partial z}F = \sigma_{01}\Delta N F. \tag{3.52}$$

From Eq. (3.48) we see that a phase change can only occur if

$$\Delta N(\omega_\ell - \omega_{10})T_2 \neq 0,$$

which states that the laser frequency does not coincide with the center frequency of the transition. A phase modulation $(d\varphi/dt)$ requires in addition that $(d/dt)\Delta N \neq 0$, i.e., the population difference must change during the interaction with the pulse.

For a quantitative discussion we may proceed as follows. We solve Eqs. (3.51) and (3.52) to obtain the pulse envelope change and $\Delta N(t,z)$, which then allows us to determine the phase modulation with Eq. (3.48). In general this requires numerical means. Here we will deal with the two limiting cases where the pulse duration is much shorter and much longer than the energy relaxation time T_1. Combining Eqs. (3.51) and (3.52) to eliminate ΔN leads to:

$$\frac{\partial^2}{\partial t \partial z}\ln F + 2\sigma_{01}\frac{\partial}{\partial z}F + \frac{1}{T_1}\left(\frac{\partial}{\partial z}\ln F - \sigma_{01}\Delta N^{(e)}\right) = 0 \tag{3.53}$$

with the boundary condition $F(z = 0, t) = F_0(t)$ and the initial condition $(\partial/\partial z)\ln F(t \to -\infty, z) = \sigma_{01}\Delta N^{(e)}(z)$.

For $\tau_p \ll T_1$ the third term in Eq. (3.53) can be neglected and the remaining differential equation can be integrated with respect to z, which yields

$$\frac{\partial}{\partial t}\ln\frac{F(z,t)}{F_0(t)} + 2\sigma_{01}[F(z,t) - F_0(t)] = 0. \tag{3.54}$$

This equation has a solution of the form [3]:

$$\boxed{F(z,t) = F_0(t)\frac{e^{2\sigma_{01}\bar{W}_0(t)}}{e^{-a} - 1 + e^{2\sigma_{01}\bar{W}_0(t)}}} \tag{3.55}$$

where $\bar{W}_0(t) = \int_{-\infty}^{t} F_0(t')dt' = 1/(\hbar\omega_\ell)\int_{-\infty}^{t} I_0(t')dt'$ (I_0 intensity of the incident pulse), cf. Eqs. (1.21), (1.22), and

$$a = \sigma_{01}\Delta N^{(e)}z \tag{3.56}$$

is the absorption ($\Delta N^{(e)} < 0$) or amplification ($\Delta N^{(e)} > 0$) coefficient corresponding to a sample of length z. $\bar{W}_0(t)$ is a measure of the incident pulse energy (area) density until time t in units of (photons)/cm^2. The total incident energy density is $\hbar\omega_\ell\bar{W}_0(t = \infty) = \hbar\omega_\ell\bar{W}_{0,\infty} = W_0$. The transmitted energy density $W(z,t) = \hbar\omega_\ell\bar{W}(z,t)$ is obtained by integrating Eq. (3.55) with respect to time and can be written as

$$\boxed{W(z,t) = \hbar\omega_\ell\int_{-\infty}^{t} F(z,t')dt' = W_s\ln\left[1 - e^a\left(1 - e^{W_0(t)/W_s}\right)\right],} \tag{3.57}$$

where $W_s = \hbar\omega_\ell/(2\sigma_{01})$ is the saturation energy density of the medium. With Eq. (3.51), in the limit $\tau_p \ll T_1$, we can express the population inversion as

$$\boxed{\Delta N(z,t) = \Delta N^{(e)}e^{-2\sigma_{01}\bar{W}(z,t)} = \frac{\Delta N^{(e)}}{1 - e^a[1 - e^{W_0(t)/W_s}]}.} \tag{3.58}$$

It is obvious from Eqs. (3.55) and (3.58) that a modification of the pulse shape is related to a change in the population inversion. The latter is controlled by the magnitude of $2\sigma_{01}\bar{W}_0$. A characteristic quantity is the ratio s of pulse energy density to saturation energy density of the transition:

$$s = 2\sigma_{01}\bar{W}_{0,\infty} = \frac{W_0}{W_s}. \tag{3.59}$$

Changes of the occupation numbers and the pulse shape become significant with increasing saturation parameter s. For large s, the population inversion approaches zero during the pulse (saturation), cf. Eq. (3.58), and the pulse distortion at the trailing part becomes small. The result is that a saturable absorber attenuates the leading edge more than the trailing edge, leading to pulse steepening which is associated with a pulse shortening. In an amplifier the leading part of the pulse experiences higher gain than the trailing part. This can result in pulse shortening as well as broadening, depending on the steepness of the incident pulse. Our qualitative discussion can easily be proven quantitatively by evaluating Eq. (3.55). Figure 3.4 shows some examples. Possible applications of this kind of pulse shaping in lasers and in pulse amplifiers will be discussed later.

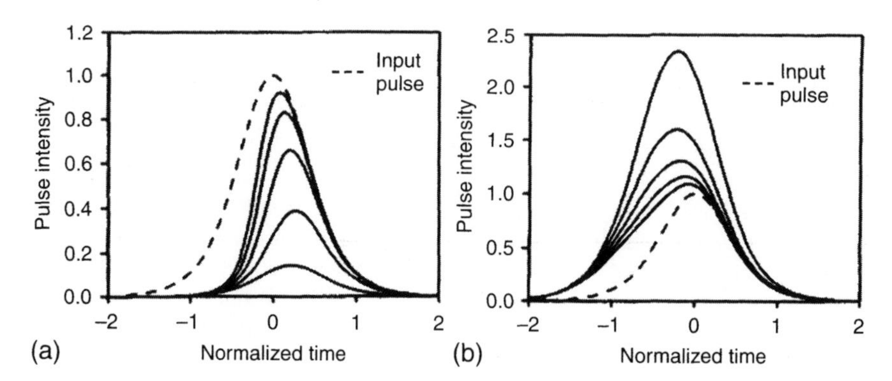

Figure 3.4 Pulse shaping through saturable absorption and amplification. (a) absorber: $a = -10$, s varies from 2 to 8 (increment 1) in the order of increasing intensities. (b) amplifier: $a = 2$, s varies from 0.5 to 8 (increment 0.5) in the order of decreasing intensities, For the input pulse we assumed $F_0(t) = \cosh^{-2}(1.76t/\tau_p)$.

It should be mentioned that the steepening, which can be achieved by a saturable absorber is limited, although this limitation is not included in the model used here. The reason is that for Eq. (3.55) to remain valid (as the pulse shortens) the rise time of the shaped pulse must still be (much) longer than T_2, which, in this manner, sets a lower limit.

Let us turn next to the second case where $\tau_p \gg T_1$. It should be noted that dealing with fs light pulses this situation is less probable than the previous one. We can now neglect the term with the temporal derivation with respect to the term containing $1/T_1$ in Eq. (3.53). The resulting differential equation is:

$$2\sigma_{01}T_1\frac{\partial}{\partial z}F + \frac{\partial}{\partial z}\ln F = \sigma_{01}\Delta N^{(e)} = \alpha. \tag{3.60}$$

It can be rewritten as

$$\frac{d}{dz}F = \frac{\alpha F}{1 + F/F_s} \tag{3.61}$$

or in terms of intensities as

$$\boxed{\frac{d}{dz}I = \frac{\alpha I}{1 + I/I_s},} \tag{3.62}$$

where we have defined a saturation flux density $F_s = (2\sigma_{01}T_1)^{-1}$ and a saturation intensity $I_s = \hbar\omega_\ell/(2\sigma_{01}T_1)^{-1}$. Equations (3.61) and (3.62) represent Beer's law for a saturable medium. Their integration yields

$$\ln \frac{F(z,t)}{F_0(t)} + \frac{F(z,t) - F_0(t)}{F_s} = a \tag{3.63}$$

and

$$\boxed{\ln \frac{I(z,t)}{I_0(t)} + \frac{I(z,t) - I_0(t)}{I_s} = a,} \tag{3.64}$$

where $a = \alpha z$, positive for amplifying media. Equation (3.63) contains the unknown $F(z,t)$ and $I(z,t)$ implicitly. To get an insight into the pulse distortion we will assume $|a| \ll 1$, from which we expect $F = F_0 + \Delta F$ with $|\Delta F/F_0| \ll 1$. Inserting this into Eq. (3.63) and expanding the logarithmic function gives the following relation for the flux at the output of the medium of length z:

$$F(z,t) = F_0(t) \left[1 + \frac{a}{1 + F_0(t)/F_s} \right]. \tag{3.65}$$

The absorption (or amplification) term becomes time dependent, where its value is now controlled by the instantaneous photon flux density rather than by the energy density. A characteristic quantity of the medium is now F_s or I_s. The result of this kind of saturation is that the pulse peak where the intensity takes on a maximum is less absorbed (amplified) than the wings, as illustrated in Figure 3.5.

3.2.3. Phase Modulation by Quasi-Resonant Interactions

According to our previous discussion, the change of the occupation numbers (saturation) results in a change of the refractive index and we expect a phase modulation to occur. Using Eqs. (3.48) and (3.52) the time-dependent frequency change

$$\delta\omega(t) = \frac{\partial \varphi}{\partial t}$$

can be written as

$$\delta\omega(t) = -\frac{(\omega_\ell - \omega_{10})T_2}{2} \sigma_{01}^{(0)} |\tilde{L}|^2 \int_0^z \frac{\partial}{\partial t} \Delta N dz = -\frac{(\omega_\ell - \omega_{10})T_2}{2} \frac{\partial}{\partial t} \ln \frac{F(z,t)}{F_0(t)}. \tag{3.66}$$

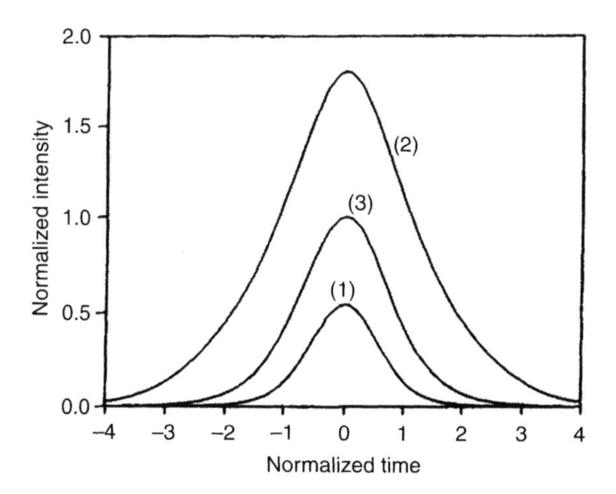

Figure 3.5 Pulse shaping in a saturable absorber (1) and depletable amplifier (2) for $\tau_p \gg T_1$, $a = \mp 3$ and $F_0/F_s = 1.5$ according to Eq. (3.63). For the input pulse (3) we assumed $F_0(t) = \cosh^{-2}(1.76t/\tau_p)$.

The sign of $\delta\omega$ depends on the sign of $(\omega_\ell - \omega_{10})T_2$, i.e., on whether the interaction takes place above or below resonance, and on the sign of $\ln(F/F_0)$. The latter is positive (negative) for $F > (<) F_0$ which is true for an amplifier (absorber).

For $\tau_p \ll T_1$ the pulse energy controls the dynamics of ΔN, and by means of Eq. (3.55), $\delta\omega(t)$ becomes

$$\delta\omega(t) = -(\omega_\ell - \omega_{10})T_2\sigma_{01}\frac{e^{-a}-1}{e^{-a}-1+e^{2\sigma_{01}\bar{W}_0(t)}}F_0(t), \tag{3.67}$$

or equivalently, in terms of the intensity and saturation energy density W_s:

$$\delta\omega(t) = -\frac{(\omega_\ell - \omega_{10})T_2}{2}\left(\frac{e^{-a}-1}{e^{-a}-1+e^{W(t)/W_s}}\right)\frac{I(t)}{W_s}. \tag{3.68}$$

The optically thin medium approximation ($|a| \ll 1$) of Eq. (3.67) is:

$$\delta\omega(t) \simeq (\omega_\ell - \omega_{10})T_2 a e^{-2\sigma_{01}\bar{W}_0(t)}\sigma_{01}F_0(t). \tag{3.69}$$

The other limiting case ($\tau_p \gg T_1$) is associated with a $\delta\omega(t)$ given by

$$\delta\omega(t) = (\omega_\ell - \omega_{10})T_2\sigma_{01}T_1\frac{\partial}{\partial t}\left[F(t,z) - F_0(t)\right], \tag{3.70}$$

which for small $|a|$ results in

$$\delta\omega(t) = \frac{(\omega_\ell - \omega_{10})T_2 a}{2} \frac{1}{(1 + F_0(t)/F_s)^2} \frac{\partial}{\partial t}\left(\frac{F_0(t)}{F_s}\right), \tag{3.71}$$

as can easily be verified by inserting Eq. (3.65) into Eq. (3.70). Figure 3.6 shows the time-dependent frequency change described by Eq. (3.67) for an amplifier. Equation (3.71) indicates a frequency change toward the leading part of the pulse with increasing saturation. A similar behavior occurs if the pulse passes through an absorber. As a result the frequency change across the FWHM of the pulses becomes maximum at a certain level of saturation.

To get an idea about the order of magnitude of the frequency change let us estimate $\delta\bar{\omega} = \delta\omega(t = -\tau_p/2) - \delta\omega(t = \tau_p/2)$ for a sech^2 pulse using Eq. (3.69) for an absorber. For $\bar{W}_{0,\infty}/\bar{W}_s = W_0/W_s = 1$ we find

$$\tau_p \delta\bar{\omega} \simeq (\omega_\ell - \omega_{10})T_2 a \tag{3.72}$$

which for $a = -0.1$ and $(\omega_\ell - \omega_{10})T_2 = 1$ yields $\tau_p\delta\bar{\omega} \approx -0.1$. If we compare this with the pulse duration–bandwidth product, cf. Table 1.1, $\tau_p\Delta\omega = 2.8$, we hardly expect that the induced frequency change is of importance. This is usually true for a single passage; however, it can play a significant role in lasers where the pulse passes through the media many times before it is coupled out [4].

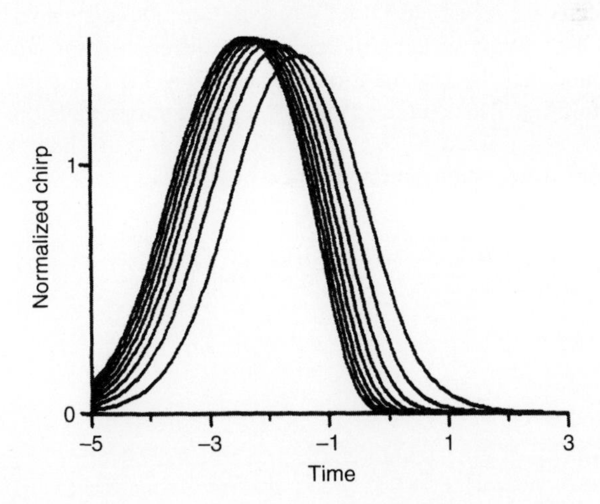

Figure 3.6 Normalized chirp $\tau_p\delta\omega$ versus time after passage through an amplifier according to Eq. (3.67) for a sech^2 input pulse, $e^{|a|} = 100$, $s = 0.5, \ldots, 4$ ($\Delta s = 0.5$, from right to left), $(\omega_\ell - \omega_{10})T_2 = -1$.

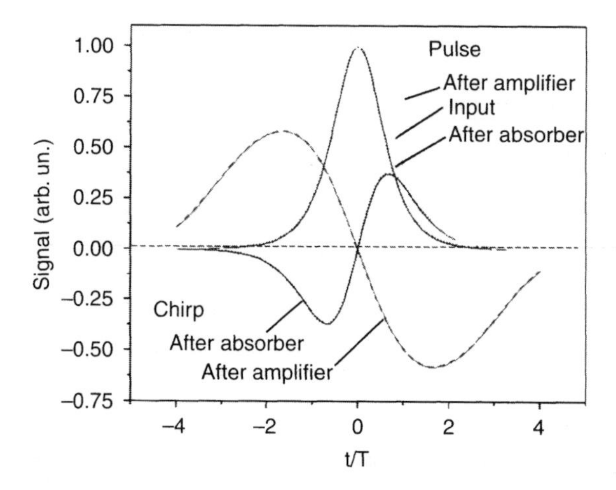

Figure 3.7 Normalized chirp and pulse shapes at the output of a saturable absorber and a depletable amplifier for $\tau_p/T_1 \gg 1$ [Eqs. (3.63) and (3.71)] and a $\text{sech}^2(t/T)$ input pulse. $e^{|a|} = 100$, $s = 1$, $(\omega_\ell - \omega_{10})T_2 = 1$.

Chirp introduced by a fast ($T_1 \ll \tau_p$) absorber and amplifier is shown in Figure 3.7 for comparison. It should be noted that our discussion can easily be expanded to include effects of multilevel systems [5]. This, for example, is necessary to model dye molecules more accurately.

For completeness let us now briefly discuss an inhomogeneously broadened sample, i.e., a distribution of particles having different resonance frequencies ω'_{10}. We assume that the preconditions for applying the rate equation approximations are fulfilled. The source terms in the propagation equations for the field components $\mathcal{E}, \varphi$, cf. Eqs. (3.47), and (3.48), have to be integrated over the inhomogeneous distribution function g_{inh}. This yields

$$\begin{aligned}\frac{\partial}{\partial z}\mathcal{E} &= \frac{1}{2}\mathcal{E}\int_0^\infty \sigma_{01}^{(0)}|\tilde{L}(\omega_\ell - \omega_0)|^2 \Delta N'(t, \omega'_{10})d\omega'_{10} \\ &= \frac{1}{2}\mathcal{E}\int_0^\infty \sigma_{01}(\omega_\ell - \omega'_{10})g_{inh}(\omega'_{10})\bar{N}[\rho_{11}(\omega'_{10}) - \rho_{00}(\omega'_{10})]d\omega'_{10} \qquad (3.73)\end{aligned}$$

and

$$\begin{aligned}\frac{\partial}{\partial z}\varphi = &-\frac{1}{2}\int_0^\infty (\omega_\ell - \omega'_{10})T'_2\sigma_{01}(\omega_\ell - \omega'_{10})g_{inh}(\omega'_{10})\bar{N} \\ &\times [\rho_{11}(t, \omega'_{10}) - \rho_{00}(t, \omega'_{10})]d\omega'_{10}. \qquad (3.74)\end{aligned}$$

The matrix elements ρ_{kk} can be obtained from the density matrix equations. A detailed discussion of chirp generation in such samples can be found in [6].

3.2.4. Pulse Durations Comparable with or Longer Than the Phase Relaxation Time ($\tau_p \geq T_2$)

If the incident intensity varies on a time scale which is comparable with the phase relaxation time of the medium, we cannot apply the rate equation approximation as was done before but have to solve the complete set of density matrix equations (3.4)–(3.7) and the wave equation (3.30). The physics behind this case is that the pulse "feels" the spectral properties of the resonant medium, that is, different spectral components of the pulse experience different modification during the interaction. This becomes obvious and simple if the interaction is weak, and we may neglect saturation and a change in the occupation numbers, respectively. For $\Delta\rho =$ constant, we find for the polarization after Fourier transforming Eq. (3.38)

$$\tilde{\mathcal{P}}(\Omega) = i\frac{\bar{N}p^2 T_2 \Delta\rho}{\hbar} \frac{T_2}{1 + iT_2(\Omega + \omega_\ell - \omega_{10})} \tilde{\mathcal{E}}(\Omega). \tag{3.75}$$

Substituting the polarization in the Fourier transformed propagation equation (3.30), the transmitted amplitude spectrum can be found after integration with respect to z

$$\tilde{\mathcal{E}}(\Omega, z) = \tilde{\mathcal{E}}(\Omega, 0) e^{\frac{1}{2}a(\Omega)} e^{-\frac{i}{2}a(\Omega)(\Omega + \omega_\ell - \omega_{10})T_2} \tag{3.76}$$

where $a(\Omega)$ is the frequency dependent coefficient of the small signal gain (absorption) which is given by

$$a(\Omega) = \sigma_{01}^{(0)} \bar{N} \Delta\rho z \frac{1}{1 + T_2^2(\Omega + \omega_\ell - \omega_{10})^2}. \tag{3.77}$$

The corresponding relation for the power spectrum reads

$$S(\Omega, z) = S(\Omega, 0) e^{a(\Omega)}. \tag{3.78}$$

As can be seen from Eq. (3.78), different spectral components experience different absorption–gain. Thus the sample acts as filter. For off-resonant interaction ($\omega_\ell \neq \omega_{10}$), the filter also introduces a spectral phase represented by the last term in Eq. (3.76). Therefore, in addition to a change in pulse shape the output pulse can be chirped.

Generally, numerical integration of the density matrix and wave equations is required to deal with pulse–matter interaction in the presence of varying population number changes. For the limiting case of small saturation [$s < (<) 1$], a small absorption–gain coefficient [$|a| < (<) 1$], and pulse durations being still longer than T_2, we may utilize a perturbation approach [7]. This gives for the pulse amplitude at the output of such an absorber–amplifier

$$\tilde{\mathcal{E}}(t,z) = \left\{1 + \frac{1}{2}a^{(0)}\tilde{L}\left[1 - 2\sigma_{01}\bar{W}_0(t) + \frac{1}{2}\left(2\sigma_{01}\bar{W}_0(t)\right)^2 + T_2\tilde{L}(2\sigma_{01})F_0(t)\right]\right.$$
$$\left. - \frac{1}{2}a^{(0)}\tilde{L}^2\left[1 - 2\sigma_{01}\bar{W}_0(t)\right]T_2\frac{d}{dt} + \frac{1}{2}a^{(0)}\tilde{L}^3T_2^2\frac{d^2}{dt^2}\right\}\tilde{\mathcal{E}}_0(t) \quad (3.79)$$

where $a^{(0)} = \sigma_{01}^{(0)}\Delta N^{(e)}z$ is the absorption–gain coefficient at the resonance frequency of the transition. For $T_2 \rightarrow 0$, we obtain a relation which corresponds to Eq. (3.55) if we expand it up to terms linear in a and quadratic in $(2\sigma_{01}\bar{W}(t))$. The additional terms in Eq. (3.79) come into play if $T_2(d/dt)\tilde{\mathcal{E}}_0(t)$ is not vanishingly small, that is if the pulse duration is of the same order of magnitude as T_2. Then the medium not only remembers the number of absorbed–amplified photons but also the phase of the electric field over a time period T_2.

3.3. NONLINEAR, NONRESONANT OPTICAL PROCESSES

3.3.1. General

Nonresonant optical processes are particularly useful in femtosecond phenomena because they can lead to conversion of optical frequencies with minimum losses. Nonlinear nonresonant phenomena are currently exploited to make use of the most efficient laser sources, which are only available at few wavelengths, to produce shorter pulses at different wavelengths (nonlinear frequency conversion and compression) and amplify them (parametric amplification). In contrast to the previous section where the interaction was dominated by a resonance, we will be dealing with situations where the light frequency is far away from optical resonances. Nonlinear crystals lend themselves nearly ideally to frequency conversion with ultrashort pulses because their nonlinearity is electronic and typically nonresonant from the near UV through the visible to the near IR spectral region. Therefore, the processes involved respond (nearly) instantaneously on the time scale of even the shortest optical pulse. There appears to be no limit in the palette of frequencies that can be generated through nonlinear optics, from dc (optical rectification) to infrared (difference frequency generation and optical parametric generation and amplification), to visible and UV (sum frequency generation). The shorter the pulse, the higher the peak intensity for a given pulse energy (and thus the more efficient the nonlinear process).

For cw light of low intensity, a medium with a nonresonant nonlinearity appears completely transparent and merely introduces a phase shift. For pulses, as discussed in Chapter 1, dispersion has to be taken into account, which can lead to pulse broadening and shortening depending on the input chirp, and to phase modulation effects. The light–matter interaction is linear, i.e., there is a linear relationship between input and output field, which results in a constant spectral intensity. A typical example is the pulse propagation through a piece of glass. The situation becomes much more complex if the pulse intensity is large, which can be achieved by focusing or/and using amplified pulses. The high electric field associated with the propagating pulse is no longer negligibly small as compared to typical local fields inside the material such as inner atomic (inner molecular) fields and crystal fields. The result is that the material properties are changed by the incident field and thus depend on the pulse. The induced polarization which is needed as source term in the wave equation is formally described by the relationship

$$
\begin{aligned}
P &= \epsilon_0 \chi(E) E = \epsilon_0 \chi^{(1)} E + \epsilon_0 \chi^{(2)} E^2 + \epsilon_0 \chi^{(3)} E^3 + \cdots + \epsilon_0 \chi^{(n)} E^n + \ldots \\
&= P^{(1)} + P^{(2)} + \cdots + P^{(n)} + \ldots.
\end{aligned} \tag{3.80}
$$

The quantities $\chi^{(n)}$ are known as the nonlinear optical susceptibilities of nth order where $\chi^{(1)}$ is the linear susceptibility introduced in Eq. (1.70). The ratio of two successive terms is roughly given by

$$
\left| \frac{P^{(n+1)}}{P^{(n)}} \right| = \left| \frac{\chi^{(n+1)} E}{\chi^{(n)}} \right| \approx \left| \frac{E}{E_{mat}} \right| \tag{3.81}
$$

where E_{mat} is a typical value for the inherent electrical field in the material. For simplicity we have taken both E and P as scalar quantities. Generally, $\chi^{(n)}$ is a tensor of order $(n+1)$ which relates an n-fold product of vector components E_j to a certain component of the polarization of nth order,[3] $P^{(n)}$; see, for example, [2, 8, 9].

[3]Note that this product can couple up to n different input fields depending on the conditions of illumination.

3.3.2. Noninstantaneous Response

For Eq. (3.80) to be valid in the time domain, we must assume that the sample responds instantaneously to the electric field; in other words, it does not exhibit a memory. The polarization at an instant $t = t_0$ must depend solely on field values at $t = t_0$. As discussed in the previous section for resonant interaction, a noninstantaneous response and memory effects, respectively, are a result of phase and energy relaxation processes. They become noticeable if they proceed on a time scale of the pulse duration or longer. Fortunately, in nonresonant light–matter interaction many processes are well described by an instantaneous response even when excited by pulses with durations of the order of 10^{-14} s. This is generally true for nonlinear effects of electronic origin. Often however, the motion of the much heavier atomic nuclei and molecules contribute to the material response. In such a case, memory effects are likely to occur on a fs time scale, and the nth-order polarization depends on the history of the field:

$$P^{(n)}(t) = \epsilon_0 \int\int \cdots \int \chi^{(n)}(t_1, t_2, ..., t_n) E(t - t_1) E(t - t_1 - t_2) \ldots \times E(t - t_1 - \cdots - t_n) dt_1 dt_2 \cdots dt_n, \tag{3.82}$$

which illustrates the influence of the electric field components at earlier times.

Let us discuss the meaning of a memory of the nonlinear polarization for the case of $n = 2$. The nonlinear polarization of second order is responsible for second harmonic generation or frequency mixing or parametric amplification:

$$P^{NL}(t) = P^{(2)}(t) = \epsilon_0 \int\int \chi^{(2)}(t_1, t_2) E_1(t - t_1) E_2(t - t_1 - t_2) dt_1 dt_2, \tag{3.83}$$

where E_1 and E_2 are optical fields, which can be identical and $\chi^{(2)}$ is the susceptibility of second order. Note that, even though the expression (3.83) is a time convolution, its Fourier transform is not a simple product, but also a convolution in the frequency domain. This convolution takes a simple form in the case of an instantaneous nonlinearity:

$$\chi^{(2)}(t_1, t_2) = \chi_0^{(2)} \delta(t - t_1) \delta(t - t_2). \tag{3.84}$$

In the time domain, the corresponding nonlinear polarization is:

$$P^{NL}(t) = \epsilon_0 \chi_0^{(2)} E_1(t) E_2(t), \tag{3.85}$$

By taking directly the Fourier transform of this expression, we find that the nonlinear polarization in the frequency domain is a convolution:

$$P^{NL}(\Omega) = \int P^{NL}(t)e^{-i\Omega t}dt = \epsilon_0 \chi_0^{(2)} \int E_1(\Omega - \Omega')E_2(\Omega')d\Omega'. \tag{3.86}$$

For monochromatic waves and long pulses, where the fields can be approximated by δ-functions in the frequency domain, Eq. (3.86) reduces to a product.

Equation (3.86) fails as soon as the nonlinear response can no longer be considered to be instantaneous. We will now show how one can find the general expression for a nonlinear polarization of second order, cf. Eq. (3.83), in the frequency domain. Fourier-transforming Eq. (3.83) yields:

$$P^{NL}(\Omega) = \epsilon_0 \int\int \chi^{(2)}(t_1, t_2)\left[\int E_1(t - t_1)E_2(t - t_1 - t_2)e^{-i\Omega t}dt\right] dt_1 dt_2 \tag{3.87}$$

where we have changed the order of integration. The expression in brackets, $C(t_1, t_2, \Omega)$, is the Fourier transform of a product, which can be written as the convolution of the FT's of the factors, $e^{-i\Omega t_1}E_1(\Omega)$ and $e^{-i\Omega(t_1+t_2)}E_2(\Omega)$:

$$C(t_1, t_2, \Omega) = \int e^{-i\Omega'(t_1+t_2)}E_2(\Omega')e^{-i(\Omega-\Omega')t_1}E_1(\Omega - \Omega')d\Omega'. \tag{3.88}$$

After inserting Eq. (3.88) into Eq. (3.87) the polarization in the frequency domain reads

$$P^{NL}(\Omega) = \epsilon_0 \int\int \chi^{(2)}(t_1, t_2)\,[C(t_1, t_2, \Omega)]\,dt_1 dt_2. \tag{3.89}$$

Inserting Eq. (3.88) into Eq. (3.89) and changing the order of integration, we find

$$P^{NL}(\Omega) = \epsilon_0 \int E_2(\Omega')E_1(\Omega - \Omega')\chi^{(2)}(\Omega, \Omega')d\Omega', \tag{3.90}$$

where

$$\chi^{(2)}(\Omega, \Omega') = \int\int \chi^{(2)}(t_1, t_2)e^{-i\Omega' t_2}e^{-i\Omega t_1}dt_1 dt_2. \tag{3.91}$$

The result of Eq.(3.90) is easily generalized to higher-order susceptibilities. If the susceptibility is not frequency dependent we reproduce the result of Eq. (3.86). This is again a manifestation of the fact that an instantaneous response (no memory) is characterized by nondispersive material properties.

3.3.3. Pulse Propagation

To study pulse propagation in a nonlinear optical medium we can proceed as in the previous section. To the linear wave equation for the electric field, which contains the $\chi^{(1)}$ contribution, we add the nonresonant nonlinear polarization. As a result we obtain Eq. (3.29) again, but with the nonlinear polarization as source term:

$$\left(\frac{\partial}{\partial z}\tilde{\mathcal{E}} - \frac{i}{2}k_\ell''\frac{\partial^2}{\partial t^2}\tilde{\mathcal{E}} + \mathcal{D}\right)e^{i(\omega_\ell t - k_\ell z)} + c.c. = i\frac{\mu_0}{k_\ell}\frac{\partial^2}{\partial t^2}P^{NL}. \tag{3.92}$$

The polarization appearing on the right-hand side can be instantaneous, or be the solution of a differential equation as in the case of most interactions with resonant atomic or molecular systems, see previous section and Chapter 4. If we represent the polarization as a product of a slowly varying envelope $\tilde{\mathcal{P}}$ and a term oscillating with an optical frequency ω_p, $e^{i\omega_p t}$, the right-hand side of Eq. (3.92) can be written as

$$\frac{\partial^2}{\partial t^2}\left(\tilde{\mathcal{P}}e^{i\omega_p t} + c.c.\right) = \left(\frac{\partial^2}{\partial t^2}\tilde{\mathcal{P}} + 2i\omega_p\frac{\partial}{\partial t}\tilde{\mathcal{P}} - \omega_p^2\tilde{\mathcal{P}}\right)e^{i\omega_p t} + c.c. \tag{3.93}$$

To compare the magnitude of the individual terms we approximate $(\partial/\partial t)\tilde{\mathcal{P}}$ with $\tilde{\mathcal{P}}/\tau_p$ which yields for the ratio of two successive members of the sum in the brackets $\omega_p\tau_p$. Therefore, if the pulse duration is (much) longer than an optical period, that is $\omega_p\tau_p = 2\pi\tau_p/T_p \gg 1$, we may neglect the first two terms in favor of $\omega_p^2\tilde{\mathcal{P}}$. This will simplify the further evaluation of Eq. (3.92) significantly.

As pointed out previously the SVEA becomes questionable if the pulses contain only few optical cycles. Brabec and Krausz [10] derived a propagation equation under less stringent conditions. If

$$|(\partial/\partial z)\tilde{\mathcal{E}}| \ll k_\ell|\tilde{\mathcal{E}}| \tag{3.94}$$

and

$$|(\partial/\partial t)\tilde{\mathcal{E}}| \ll \omega_\ell|\tilde{\mathcal{E}}| \tag{3.95}$$

or

$$\left|1 - \frac{\nu_p}{\nu_g}\right| \ll 1 \tag{3.96}$$

are satisfied pulse propagation in the presence of a nonlinear polarization of slowly varying amplitude $\mathcal{P}^{(NL)}$ and dispersion can be described by

$$\left[\frac{\partial}{\partial z} + \frac{i}{2k_\ell}\left(1 - \frac{i}{\omega_\ell}\frac{\partial}{\partial t}\right)^{-1} \nabla_\perp^2 - \frac{\alpha_0}{2} + i\hat{D}\right]\tilde{\mathcal{E}} = -i\frac{\omega_\ell c \mu_0}{2n_0}\left(1 - \frac{i}{\omega_0}\frac{\partial}{\partial t}\right)\tilde{\mathcal{P}}^{(NL)}, \tag{3.97}$$

where

$$\hat{D} = \frac{\alpha_1}{2}\frac{\partial}{\partial t} + \sum_{m=2}^{\infty}\frac{1}{m!}\left(k_m + i\frac{\alpha_m}{2}\right)\left(-i\frac{\partial}{\partial t}\right)^m$$

with

$$\alpha_m = \frac{\partial^m}{\partial \Omega^m}\left[\operatorname{Im} k(\Omega)\right]_{\omega_\ell} \quad \text{and} \quad k_m = \frac{\partial^m}{\partial \Omega^m}\left[\operatorname{Re} k(\Omega)\right]_{\omega_\ell},$$

see Appendix C. The coordinates z, t refer to a frame moving with the group velocity of the pulse. The additional time derivatives of the nonlinear polarization and the diffraction term ($\nabla_\perp^2$) become important for extremely short pulses. This propagation equation was termed "slowly evolving envelope equation." In many materials phase and group velocity are not much different and condition (3.96) is satisfied. Conditions (3.94) and (3.96) can be combined to the "slowly evolving wave approximation" [10]

$$\left|\frac{\partial}{\partial z}E\right| \ll k_\ell |E|, \tag{3.98}$$

which states that the amplitude and phase of the electric field must not change significantly over a propagation distance of the order of a wavelength.

In general, when a nonlinear polarization is involved, there will not be just one propagation equation of the form of Eq. (3.93), but as many as the number of waves that participate in the nonlinear optical process. For instance, a third-order polarization excited by a field at frequency ω_ℓ will create a polarization at $3\omega_\ell = \omega_\ell + \omega_\ell + \omega_\ell$, and a polarization at $\omega_\ell = \omega_\ell - \omega_\ell + \omega_\ell$. The first process is generation of a third harmonic field, and the second is either two-photon absorption or a nonlinear index of refraction, depending on the phase of the nonlinear susceptibility. The generated field at $3\omega_\ell$ will propagate, and interfere with the field at $3\omega_\ell$ produced at a different location by the fundamental. The third harmonic field may also lead to the generation of other frequencies, through the third-order process. For instance, there will be regeneration of the fundamental frequency through the third-order process $\omega_\ell = 3\omega_\ell - \omega_\ell - \omega_\ell$, and the latter field will also interfere with the propagated fundamental. The third harmonic may also create a ninth harmonic through the nonlinear susceptibility. At a minimum, there will be at least two differential equations of the form Eq. (3.93), with a third-order susceptibility, corresponding to the fundamental and third harmonic fields. More equations have to be added if more frequencies are generated.

It is beyond the scope of the book to give a detailed description of the various possible nonlinear effects and excitation schemes. The reader is referred to the standard texts on nonlinear optics, for example Schubert and Wilhelmi [2], Boyd [8], Bloembergen [9], and Shen [11]. Here we shall restrict ourselves to a nonlinearity of second order that is responsible for second harmonic generation, optical parametric amplification (OPA), and to a nonlinearity of third order describing (self-) phase modulation [(S)PM].

The tensor character of the nonlinear susceptibility describes the symmetry properties of the material. For all substances with inversion symmetry, $\chi^{(2n)} = 0$ $(n = 1, 2 \ldots)$ holds, and therefore no second harmonic processes can be observed in isotropic materials and centrosymmetric crystals for example. In contrast, third-order effects are always symmetry allowed. However, even in isotropic materials, the tensor character of the nonlinear susceptibility should not be ignored. The electric field of the light itself can break the symmetry, leading to interesting polarization rotation effects.

In the following sections we will discuss various examples of nonlinear optical processes with short light pulses. The propagation of the corresponding wave packets at carrier frequency ω_i is described by a group velocity ν_i for which

$$\boxed{\frac{1}{\nu_i} = \frac{n(\omega_i)}{c} + \frac{\omega_i}{c}\left.\frac{dn}{d\Omega}\right|_{\omega_i}} \tag{3.99}$$

holds. Sometimes it will also be necessary to specify the polarization direction, $\hat{e}_j$, of the waves participating in the nonlinear process.

Unless stated otherwise we will assume that the nonlinear susceptibility is much faster than the time scale of interest (pulse duration). This will allow us to simplify the derivations by applying the concept of an instantaneous material response. Also, to simplify the discussion on effects typical for the conversion of short light pulses, we will usually neglect any change in intensity because of focusing effects; an approximation, which generally holds for nonlinear materials shorter than the Rayleigh range. An exception is when self-focusing occurs, a nonlinear effect discussed in Section 3.8.

3.4. SECOND HARMONIC GENERATION (SHG)

Second harmonic generation has gained particular importance in ultrashort pulse physics as a means for frequency conversion and nonlinear optical correlation. Owing to the characteristics of ultrashort pulses, a number of new features unknown in the conversion of cw light have to be considered [12–16]. We will examine first the relatively simple case of type I SHG, in which the fundamental wave propagates as an ordinary (o) or extraordinary (e) wave, producing an extraordinary or ordinary second harmonic (SH) wave, respectively. We will briefly discuss at the end of this section the more complex case of type II SHG, in which the nonlinear polarization, responsible for the generation of a second harmonic propagating as an e wave, is proportional to the product of the e and o components of the fundamental. We will see that group velocity mismatch between the fundamental and the SH leads generally to a reduced conversion efficiency and pulse broadening. Under certain circumstances, however, it is possible to have simultaneously high conversion efficiency and efficient compression of the second harmonic in presence of group velocity mismatch. Second harmonic is only a particular case of sum frequency generation. Therefore, in some of the subsections to follow, we will treat in parallel SHG and the more general case of sum frequency generation.

3.4.1. Type I Second Harmonic Generation

Let us assume a light pulse incident on a second harmonic generating crystal. The electric field propagating inside the material consists of the original (fundamental) field (subscript $i = 1$) and the second harmonic field (subscript $i = 2$). The total field obeys a wave equation similar to Eq. (3.92) with a nonlinear polarization of second order as source term:

$$\left[\left(\frac{\partial}{\partial z}+\frac{1}{\nu_1}\frac{\partial}{\partial t}-\frac{ik_1''}{2}\frac{\partial^2}{\partial t^2}\right)\tilde{\mathcal{E}}_1+\mathcal{D}_1\right]e^{i(\omega_1 t-k_1 z)}$$
$$+\frac{k_2}{k_1}\left[\left(\frac{\partial}{\partial z}+\frac{1}{\nu_2}\frac{\partial}{\partial t}-\frac{ik_2''}{2}\frac{\partial^2}{\partial t^2}\right)\tilde{\mathcal{E}}_2+\mathcal{D}_2\right]e^{i(\omega_2 t-k_2 z)}+c.c.=i\frac{\mu_0}{k_1}\frac{\partial^2}{\partial t^2}P^{(2)} \tag{3.100}$$

where the second-order polarization can be written as

$$P^{(2)}=\epsilon_0\chi^{(2)}\frac{1}{4}\left[\tilde{\mathcal{E}}_1 e^{i(\omega_1 t-k_1 z)}+\tilde{\mathcal{E}}_2 e^{i(\omega_2 t-k_2 z)}+c.c.\right]^2. \tag{3.101}$$

Because the group velocities ν_1 and ν_2 are not necessarily equal there is no coordinate frame in which both the fundamental and SH pulses are at rest. Therefore z and t are the (normal) coordinates in the laboratory frame. With the simplifications introduced above for the polarization, we obtain two coupled differential equations for the amplitude of the fundamental wave

$$\boxed{\left(\frac{\partial}{\partial z}+\frac{1}{\nu_1}\frac{\partial}{\partial t}-\frac{ik_1''}{2}\frac{\partial^2}{\partial t^2}\right)\tilde{\mathcal{E}}_1+\mathcal{D}_1=-i\chi^{(2)}\frac{\omega_1^2}{4c^2k_1}\tilde{\mathcal{E}}_1^*\tilde{\mathcal{E}}_2e^{i\Delta kz}} \tag{3.102}$$

and for the SH wave

$$\boxed{\left(\frac{\partial}{\partial z}+\frac{1}{\nu_2}\frac{\partial}{\partial t}-\frac{ik_2''}{2}\frac{\partial^2}{\partial t^2}\right)\tilde{\mathcal{E}}_2+\mathcal{D}_2=-i\chi^{(2)}\frac{\omega_2^2}{4c^2k_2}\tilde{\mathcal{E}}_1^2e^{-i\Delta kz},} \tag{3.103}$$

where $\Delta k = 2k_1(\omega_1) - k_2(\omega_2)$ is the wave vector mismatch calculated with the wave vector values at the carrier frequency of the fundamental and second harmonic. Because k_1, k_2 are functions of the orientation of the wave vector with respect to the crystallographic axis, it is often possible to find crystals, beam geometry and beam polarizations, for which $\Delta k = 0$ (phase matching) is achieved [2,8,9]. Note that in the case of ultrashort pulses the wave vectors vary over the bandwidth of the pulse. This variation caused by the linear polarization has already been taken into account by the time derivatives on the left-hand sides of Eqs. (3.102) and (3.103), cf. Eq. (1.89).

Type I—Small Conversion Efficiencies

Small conversion efficiencies occur at low input intensities and/or small length of the nonlinear medium and nonlinear susceptibility. Under these circumstances we may assume that the fundamental pulse does not suffer losses. If we assume in addition that $k_1'' = k_2'' = \mathcal{D}_1 = \mathcal{D}_2 \approx 0$ we find for the fundamental pulse, using Eq. (3.102), $\tilde{\mathcal{E}}_1(t,z) = \tilde{\mathcal{E}}_1(t - z/\nu_1)$. The fundamental pulse travels distortionless in a frame moving with the group velocity ν_1. This expression can be inserted into the generating equation for the SH, Eq. (3.103). Integration with respect to the propagation coordinate yields for the SH at $z = L$:

$$\tilde{\mathcal{E}}_2\left(t-\frac{L}{\nu_2},L\right)=-i\frac{\chi^{(2)}\omega_2^2}{4c^2k_2}\int_0^L\tilde{\mathcal{E}}_1^2\left[t-\frac{z}{\nu_2}+\left(\frac{1}{\nu_2}-\frac{1}{\nu_1}\right)z\right]e^{-i\Delta kz}dz. \tag{3.104}$$

Using the correlation theorem, Eq. (3.104) can be transformed into the frequency domain:

$$\tilde{\mathcal{E}}_2(\Omega,L)=-i\frac{\chi^{(2)}\omega_2^2}{4c^2k_2}\int\tilde{\mathcal{E}}_1(\Omega')\tilde{\mathcal{E}}_1(\Omega-\Omega')d\Omega'\int_0^L e^{i\left[(\nu_2^{-1}-\nu_1^{-1})\Omega-\Delta k\right]z}dz. \tag{3.105}$$

After integration with respect to the propagation coordinate we obtain for the SH field

$$\tilde{\mathcal{E}}_2(\Omega, L) = -i\frac{\chi^{(2)}\omega_2^2 L}{4c^2 k_2}\text{sinc}\left\{\left[\left(\frac{1}{\nu_2} - \frac{1}{\nu_1}\right)\Omega - \Delta k\right]\frac{L}{2}\right\}$$

$$\times \int \tilde{\mathcal{E}}_1(\Omega')\tilde{\mathcal{E}}_1(\Omega - \Omega')d\Omega' \quad (3.106)$$

and for the spectral intensity of the SH (apart from the conversion factor from field squared to intensity):

$$|\tilde{\mathcal{E}}_2(\Omega, L)|^2 = \left(\frac{\chi^{(2)}\omega_2^2 L}{4c^2 k_2}\right)^2 \text{sinc}^2\left\{\left[\left(\frac{1}{\nu_2} - \frac{1}{\nu_1}\right)\Omega - \Delta k\right]\frac{L}{2}\right\}$$

$$\times \left|\int \tilde{\mathcal{E}}_1(\Omega')\tilde{\mathcal{E}}_1(\Omega - \Omega')d\Omega'\right|^2. \quad (3.107)$$

Maximum conversion is achieved for zero group velocity mismatch ($\nu_1 = \nu_2$) and zero phase mismatch ($\Delta k = 0$).

The term $(\nu_2^{-1} - \nu_l^{-1})z$ in the argument of $\tilde{\mathcal{E}}_1$ in Eq. (3.104) describes the walk-off between the second harmonic pulse and the pulse at the fundamental wavelength owing to the different group velocities. The result is a broadening of the second harmonic pulse, as can be seen from Figure 3.8. Only for crystal lengths

$$L \ll L_D^{SHG} = \frac{\tau_{p1}}{|\nu_2^{-1} - \nu_1^{-1}|} \quad (3.108)$$

can the influence of the group velocity mismatch on the shape of the SH pulse be neglected. In this case the SHG intensity varies with the square of the product of crystal length and intensity of the fundamental, cf. Eq. (3.104). Because of this quadratic dependence, the second harmonic pulse is shorter than the fundamental pulse (by a factor $\sqrt{2}$ for Gaussian pulses). For $L \gg L_D^{SHG}$ the pulse duration is determined by the walk-off and approaches a value of $L \times |\nu_2^{-1} - \nu_1^{-1}|$, the peak power remains constant, and the energy increases linearly with L. Of course, one needs to avoid this regime if short second harmonic pulses are required. The group velocity mismatch between the fundamental and SH pulse is listed in Table 3.2 for some typical crystals used for SHG. Similar conclusions can be

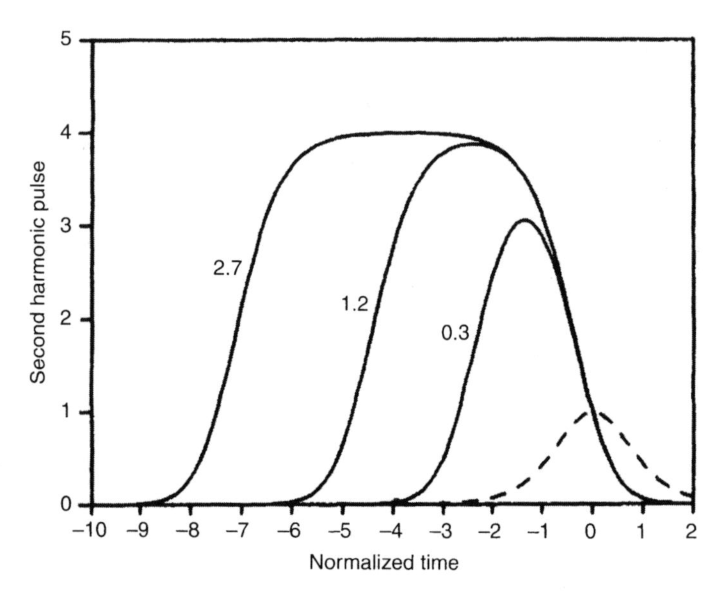

Figure 3.8 SH pulse at different, normalized crystal lengths, L/L_D^{SHG} according to Eq. (3.104). (– – – input sech - pulse; the intensity is not to scale.)

Table 3.2

Phase matching angle θ and group velocity mismatch $(\nu_2^{-1} - \nu_1^{-1})$ for type I phase matching (oo-e) in some negative uni-axial crystals. The data were obtained from Sellmeier equations, see Zernicke [17], Choy and Byer [18], and Kato [19].

Crystal	λ (nm)	θ (°)	$(\nu_2^{-1} - \nu_1^{-1})$ (fs/mm)
KDP	550	71	266
	620	58	187
	800	45	77
	1000	41	9
$LiIO_3$	620	61	920
	800	42	513
	1000	32	312
BBO	500	52	680
	620	40	365
	800	30	187
	1000	24	100
	1500	20	5

drawn from the frequency domain solution for the SH pulse. The group velocity mismatch causes the SHG process to act as a frequency filter, cf. Eq. (3.107). The bandwidth becomes narrower with increasing crystal length. In addition, the sinc^2 term in Eq. (3.107) introduces a modulation of the spectrum of the second harmonic. The period of that modulation can serve to estimate the group velocity mismatch $(\nu_2^{-1} - \nu_1^{-1})$ of the particular crystal used.

It is interesting to note what happens when the phase matching condition is not satisfied ($\Delta k \neq 0$). The introduction of $\exp(-i\Delta kz)$ in the integrand of Eq. (3.104) produces a second harmonic output that varies periodically with the propagation distance. The periodicity length is given by

$$L_P^{SHG} = \frac{2\pi}{\Delta k} \tag{3.109}$$

if group velocity mismatch can be neglected. In such cases it is recommended to work with crystal lengths $L < L_P^{SHG}$.

Type I—Large Conversion Efficiencies

The simple approach of the previous section does no longer apply to conversion efficiencies larger than a few tens of percent. We have to consider the depletion of the fundamental pulse as the SH pulse grows according to the complete system of differential equations (3.102), (3.103). In the phase and group velocity matching regime, the SH energy approaches its maximum value asymptotically. Because of their lower intensities, the pulse wings reach this "saturation" regime later and the SH pulse duration τ_{p2} broadens until it reaches a value that is approximately given by the duration of the fundamental pulse τ_{p1}. Therefore, even a moderate energy conversion requires very high conversion efficiencies for the peak intensities. Figure 3.9 shows schematically the conversion efficiencies in various regimes for zero group velocity mismatch (long pulses).

With the inclusion of group velocity and phase mismatch, the processes involved in SHG become complex. Numerical studies of Eqs. (3.102) and (3.103) in Karamzin and Sukhorukov [20], Eckardt and Reintjes [21], and Kothari and Carlotti [22] reveal pulse splitting and a periodical behavior of the conversion efficiency with propagation length under certain circumstances. The complexity results partly from the fact that the phase of the fundamental wave becomes dependent on the conversion process. For cw light, the phase of the fundamental wave can easily be obtained from Eqs. (3.102) and (3.103) and reads [21]

$$\varphi_1(z) = \frac{1}{2} \arccos\left[\frac{c^2 k_1 \tilde{\mathcal{E}}_2(z)}{\chi^{(2)} \omega_1^2 \tilde{\mathcal{E}}_1^2(z)} \Delta k\right] - \frac{\pi - \Delta kz}{4}. \tag{3.110}$$

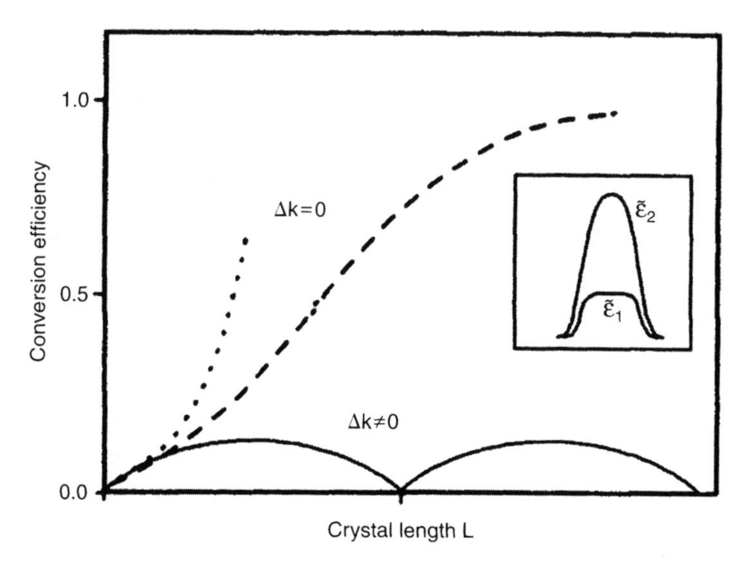

Figure 3.9 Conversion efficiencies neglecting (· · ·) and taking into account (— — —) depletion of the fundamental wave. The inset illustrates the shaping of the SH and fundamental pulse in the crystal.

This phase is responsible for a new phase mismatch $\Delta k_{\text{eff}} z = \varphi_2(z) - 2\varphi_1(z)$ which, as opposed to the Δk introduced earlier, is a function of the field amplitudes. The result is that the conversion efficiency drops more rapidly for spectral components for which $\Delta k \neq 0$. Thus, the SH process acts like an intensity dependent spectral filter for short pulses, reducing the conversion efficiency and leading to distortions of the temporal profile. As shown experimentally by Kuehlke and Herpers, [23] an optimum input intensity can exist for maximum energy conversion of fs pulses. Usually these conversion efficiencies do not exceed a few tens of percent.

There is another interesting consequence of the nonlinear phase. In most cases, for both the SH and the fundamental fields, we have to consider a certain dependence of the field amplitudes on the transverse spatial coordinate (beam profile). According to Eq. (3.110) this leads to a phase $\phi_1(x, y, z)$, which can result in focusing or defocusing of the fundamental wave [24]. This lensing effect is similar to self-focusing based on the Kerr effect that will be discussed in Section 3.8. There, the self-lensing will be introduced as the result of a nonlinear polarization of third order, $P^{(3)}$, as opposed to the former case of Eq. (3.110), which is derived from a second-order nonlinear polarization.

Type I—Compensation of the Group Velocity Mismatch

A nonzero group velocity mismatch limits the frequency doubling efficiency of femtosecond light pulses to a few tens of percents. It is interesting to note that the group velocity mismatch is equivalent to the fact that the phase matching condition does not hold over the entire pulse spectrum. We want to leave the actual workout of this fact to one of the problems at the end of this chapter. Generally, it is not possible to match the group velocities by choosing suitable materials while keeping the phase matching condition for the center frequencies, $\Delta k = 0$, as indicated in Table 3.2. However, because phase matching is achieved most often by angular tuning, simultaneous phase matching of an extended spectrum is feasible by realizing different angle of incidence for different spectral components. Corresponding practical arrangements for fs light pulses were suggested in Szabo and Bor [25] and Martinez [26], and implemented for sum frequency generation to 193 nm [27]. In Figure 3.10, two gratings, G_1 and G_2, are used to disperse and recollimate the beam, respectively. Two achromatic lenses (or telescopes [26]) image A onto the crystal and onto B to ensure zero group velocity dispersion. The combination of L_1 and G_1 enables different angles of incidence for different spectral components. The desired magnification of lens L_1 is determined by the angular dispersion of the first grating $a_1 = d\beta/d\Omega$ and the derivative of the phase matching angle $a_2 = d\theta/d\Omega$

$$M_1 = \frac{d\theta/d\Omega}{d\beta/d\Omega}. \tag{3.111}$$

The magnification of the second lens has to be chosen likewise, taking into account the doubled frequency at the output of the crystal.

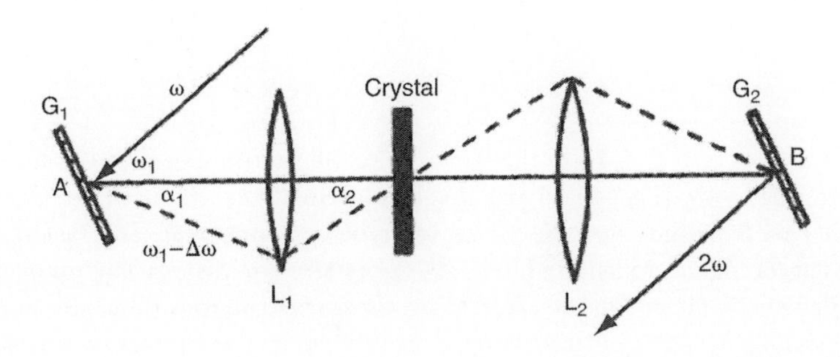

Figure 3.10 Frequency doubler for ultrashort (broadband) light pulses. (Adapted from Szabo and Bor [25].)

3.4.2. Second Harmonic Type II: Equations for Arbitrary Phase Mismatch and Conversion Efficiencies

Treatment in the Time Domain

As pointed out before there is no analytical solution to the general problem of SH generation. Numerical procedures have to be used to describe the propagation of the fundamental and the SH pulses under the combined action of (linear) dispersion and nonlinear effects. The possible effects are particularly complex and interesting in type II SH generation.

Type II SH generation involves the interaction of three waves, the SH, and an ordinary (o) and extraordinary (e) fundamental. Group velocity mismatch of these three waves does not always lead to pulse broadening. In the case of SHG type II, it is possible to achieve significant pulse compression at either the fundamental or the up-converted frequency.

To describe type II frequency conversion we extend the system of equations (3.102) and (3.103). We choose a retarded time frame of reference traveling with the second harmonic signal at its group velocity ν_2. The fundamental pulse has a component $\tilde{\mathcal{E}}_o(t)\exp[i(\omega_1 t - k_o z)]$ propagating as an ordinary wave (subscript o) at the group velocity ν_o, and a component $\tilde{\mathcal{E}}_e(t)\exp[i(\omega_1 t - k_e z)]$ propagating as an extraordinary wave (subscript e) at the group velocity ν_e. The system of equations describing the evolution of the fundamental pulses $\tilde{\mathcal{E}}_o$ and $\tilde{\mathcal{E}}_e$, and the generation of the SH wave $\tilde{\mathcal{E}}_2$ is:

$$\left[\frac{\partial}{\partial z} + \left(\frac{1}{\nu_o} - \frac{1}{\nu_2}\right)\frac{\partial}{\partial t}\right]\tilde{\mathcal{E}}_o + \mathcal{D}_o = -i\chi^{(2)}\frac{\omega_1^2}{4c^2 k_o}\tilde{\mathcal{E}}_e^*\tilde{\mathcal{E}}_2 e^{i\Delta kz} \tag{3.112}$$

$$\left[\frac{\partial}{\partial z} + \left(\frac{1}{\nu_e} - \frac{1}{\nu_2}\right)\frac{\partial}{\partial t}\right]\tilde{\mathcal{E}}_e + \mathcal{D}_e = -i\chi^{(2)}\frac{\omega_1^2}{4c^2 k_e}\tilde{\mathcal{E}}_o^*\tilde{\mathcal{E}}_2 e^{i\Delta kz} \tag{3.113}$$

$$\frac{\partial}{\partial z}\tilde{\mathcal{E}}_2 + \mathcal{D}_2 = -i\chi^{(2)}\frac{\omega_1^2}{c^2 k_2}\tilde{\mathcal{E}}_e\tilde{\mathcal{E}}_0 e^{-i\Delta kz}, \tag{3.114}$$

where $\Delta k = k_o + k_e - k_2$ is the wave vector mismatch calculated at the pulse carrier frequency. The phase matching condition $\Delta k_0 = k_o(\omega_1) + k_e(\omega_1) - k_2(2\omega_1) = 0$ implies that the *phase velocities* are matched. The fact that the waves at ω_1 and ω_2 remain in phase does not necessarily imply that *pulses* reach simultaneously the end of the crystal. The three wave packets propagate at *group velocities* ν_o, ν_e, and ν_2 that, in general, are different. The expression Eq. (3.106) found for type-I SH generation without pump depletion can be regarded a special solution of Eqs. (3.112–3.114).

SHG for Short Pulses—Treatment in the Frequency Domain

When dealing with the conversion of short pulses, it is not sufficient to include dispersion only up to first order, that is $\mathcal{D} = 0$. For $\mathcal{D} \neq 0$, however, the system of equations (3.112)–(3.114) contains higher-order time derivatives whose treatment is numerically difficult. The problem can be stated more clearly in the frequency domain, using the complete functional dependence of the k vectors (or the indices of refraction), rather than power series. We start from the wave equation for the electric field

$$\left(\frac{\partial^2}{\partial z^2} - \frac{1}{c^2}\frac{\partial^2}{\partial t^2}\right)\tilde{\mathbf{E}}(t,z) = \mu_0 \frac{\partial^2}{\partial t^2}\left[\tilde{\mathbf{P}}^L(t,z) + \tilde{\mathbf{P}}^{NL}(t,z)\right] \tag{3.115}$$

where the electric field is the sum of the three participating waves

$$\tilde{\mathbf{E}}(t,z) = \hat{\mathbf{e}}_{\mathbf{o}}\tilde{E}_o(t,z) + \hat{\mathbf{e}}_{\mathbf{e}}\tilde{E}_e(t,z) + \hat{\mathbf{e}}_{\mathbf{e}}\tilde{E}_2(t,z) \tag{3.116}$$

and the nonlinear polarization

$$\tilde{\mathbf{P}}^{NL}(t,z) = \epsilon_0\chi^{(2)}\left[\tilde{E}_o(t,z)\tilde{E}_e(t,z)\hat{\mathbf{e}}_{\mathbf{e}} + \tilde{E}_2(t,z)\tilde{E}_o^*(t,z)\hat{\mathbf{e}}_{\mathbf{e}} + \tilde{E}_2(t,z)\tilde{E}_e^*(t,z)\hat{\mathbf{e}}_{\mathbf{o}}\right]. \tag{3.117}$$

Without loss of generality, we have assumed that the SH field $\tilde{\mathbf{E}}_2$ propagates as an extraordinary wave with polarization vector $\hat{\mathbf{e}}_{\mathbf{e}}$. The nonlinear polarization terms are responsible for the evolution of the SH, the fundamental e-wave and the fundamental o-wave, respectively.

Following the same procedure as in Section 1.2, we take the Fourier transform of Eq. (3.115):

$$\left[\frac{\partial^2}{\partial z^2} + \mu_0\Omega^2\epsilon(\Omega)\right]\tilde{\mathbf{E}}(\Omega,z) = -\mu_0\Omega^2\tilde{\mathbf{P}}^{NL}(\Omega,z) \tag{3.118}$$

where we used the expressions (1.70) for the linear polarization and Eq. (1.73) for the dielectric constant. The nonlinear polarization in the frequency domain is a sum of three convolution integrals; the first member of the sum, for example, is $\epsilon_0\chi^{(2)}\hat{e}_e \int E_o(\Omega',z)E_e(\Omega-\Omega',z)d\Omega'$.

For the electric field components we make the ansatz

$$\tilde{E}_q(\Omega,z) = \frac{1}{2}\tilde{a}_q(\Omega,z)e^{-ik_q(\Omega)z}. \tag{3.119}$$

where the subscript q stands for o, e, or 2. The amplitudes $\tilde{a}_q(\Omega,z)$ peak at the central frequencies of the corresponding pulse. The ansatz is a solution of the linear wave equation [$\tilde{\mathbf{P}}^{NL}(\Omega, z) = 0$ in Eq. (3.118)]. Hence $k_q(\Omega) = \Omega n_q(\Omega)/c = \Omega\sqrt{\mu_0\epsilon(\Omega)}$. Inserting the ansatz into Eq. (3.118) and separating out the three field components according to polarization and frequency yields

$$\frac{\partial}{\partial z}\tilde{a}_2(\Omega, z) = \frac{-i\Omega^2\chi^{(2)}}{4c^2 k_2(\Omega)}\int \tilde{a}_o(\Omega', z)\tilde{a}_e(\Omega - \Omega', z)e^{i[-k_o(\Omega')-k_e(\Omega-\Omega')+k_2(\Omega')]z}d\Omega' + \frac{i}{2k_2(\Omega)}\frac{\partial^2}{\partial z^2}\tilde{a}_2(\Omega, z) \tag{3.120}$$

$$\frac{\partial}{\partial z}\tilde{a}_o(\Omega, z) = \frac{-i\Omega^2\chi^{(2)}}{4c^2 k_o(\Omega)}\int \tilde{a}_2(\Omega', z)\tilde{a}_e^*(\Omega - \Omega', z)e^{i[k_o(\Omega')+k_e(\Omega-\Omega')-k_2(\Omega')]z}d\Omega' + \frac{i}{2k_o(\Omega)}\frac{\partial^2}{\partial z^2}\tilde{a}_o(\Omega, z) \tag{3.121}$$

$$\frac{\partial}{\partial z}\tilde{a}_e(\Omega, z) = \frac{-i\Omega^2\chi^{(2)}}{4c^2 k_e(\Omega)}\int \tilde{a}_2(\Omega', z)\tilde{a}_o^*(\Omega - \Omega', z)e^{i[k_e(\Omega')+k_o(\Omega-\Omega')-k_2(\Omega')]z}d\Omega' + \frac{i}{2k_e(\Omega)}\frac{\partial^2}{\partial z^2}\tilde{a}_e(\Omega, z). \tag{3.122}$$

The sum of the three equations (3.120) is equivalent to the second-order wave equation (3.118) or (3.115). There is however an approximation involved in splitting the single wave equation in a system of three equations, namely that the spectral range of the pulses E_o and E_e centered at ω_ℓ and $2\omega_\ell$ do not overlap. For numerical calculations, it is more convenient to consider field amplitudes shifted to zero frequency:

$$\begin{aligned}\tilde{\mathcal{E}}_2(\Delta\Omega = \Omega - 2\omega_\ell) &= \tilde{a}_2(\Omega, z)\\ \tilde{\mathcal{E}}_o(\Delta\Omega = \Omega - \omega_\ell) &= \tilde{a}_o(\Omega, z)\\ \tilde{\mathcal{E}}_e(\Delta\Omega = \Omega - \omega_\ell) &= \tilde{a}_e(\Omega, z),\end{aligned} \tag{3.123}$$

where we have dropped for simplicity of notation the variable z in the argument of the field amplitudes on the left of Eq. (3.123). These fields are the Fourier transforms of the envelopes defined in Chapter 1, Eq. (1.10). The envelope was defined in Eq. (1.10) with a similar ansatz as Eq. (3.119), hence not involving any SVEA. In situations where the SVEA applies, these shifted amplitudes [Eq. (3.123)] become identical to the spectral amplitudes defined in Eq. (1.10).

The set of equations (3.120) can easily be written for the spectral field envelopes defined in Eq. (3.123).

The set of coupled equations (3.120) with (3.123) is convenient for a numerical treatment. No other assumption or approximation has been made, except that the spectra of the fundamental and second harmonic do not overlap, to be able to split a single Maxwell's second-order propagation equation into three coupled differential equations. The dispersion of the material is contained in the frequency dependence of the wave vectors $k_q^2(\Omega) = \Omega^2 n^2(\Omega)/c^2$. The second derivative of the envelope with respect to z can generally be neglected, unless the spectral envelope of the field changes on length scales of the wavelength.

It should be noted that no moving frame of reference has been adopted in this section. Hence, the fields are propagating at their respective group velocities. A more convenient representation of the solution uses a frame of reference propagating at one of the group velocities, for instance that of the second harmonic (see also Section 3.4.1). The temporal (complex) envelopes in this frame of reference moving at the velocity ν_2 are obtained from the solutions $\tilde{\mathcal{E}}_i(\Delta\Omega, z)$ through the transformation:

$$\tilde{\mathcal{E}}_2(t, z) = \int_{-\infty}^{\infty} \tilde{\mathcal{E}}_2(\Delta\Omega, z) e^{-i[k_2(\Delta\Omega) - \frac{\Delta\Omega}{\nu_2}]z} e^{-i\Delta\Omega t} d\Delta\Omega \tag{3.124}$$

$$\tilde{\mathcal{E}}_o(t, z) = \int_{-\infty}^{\infty} \tilde{\mathcal{E}}_o(\Delta\Omega, z) e^{-i[k_o(\Delta\Omega) - \frac{\Delta\Omega}{\nu_2}]z} e^{-i\Delta\Omega t} d\Delta\Omega \tag{3.125}$$

$$\tilde{\mathcal{E}}_e(t, z) = \int_{-\infty}^{\infty} \tilde{\mathcal{E}}_e(\Delta\Omega, z) e^{-i[k_e(\Delta\Omega) - \frac{\Delta\Omega}{\nu_2}]z} e^{-i\Delta\Omega t} d\Delta\Omega. \tag{3.126}$$

3.4.3. Pulse Shaping in Second Harmonic Generation (Type II)

In this section we will describe the situation where group velocity mismatch can be utilized to shape (shorten) ultrashort light pulses as a result of nonlinear frequency conversion.

Akhmanov *et al.* [28] analyzed the situation where an SH pulse and a fundamental pulse are simultaneously incident on a nonlinear crystal with $\nu_2 > \nu_1$. If a short SH pulse is launched in the trailing edge of a long fundamental pulse the SH will extract energy from various parts of the fundamental while moving through the fundamental pulse because of the group velocity mismatch. High peak powers of the second harmonic and considerable pulse shortening were predicted.

Similar effects can be expected in type II phase matching. Here another degree of freedom, the group velocity mismatch $(\nu_o - \nu_e)$ of the ordinary and extraordinary fundamental, can be adjusted. Consider a phase matched situation $(\Delta k = 0)$ where again a short SH pulse, of sufficient intensity to deplete the fundamental, is seeded at the trailing edge of a longer fundamental pulse at the input of the nonlinear crystal. This situation leading to pulse compression, [16, 29] was implemented with subpicosecond pulses in Wang and Davies [30] and Heinz *et al.* [31]. Subsequently, pulse compression through second harmonic generation in very long (5–6 cm) KDP (KH_2PO_4) and KD*P (KD_2PO_4) crystals was predicted [29] and demonstrated [32]. We will describe in a subsequent section a similar compression mechanism in synchronously pumped optical parametric oscillators [33–36].

Let us now discuss the situation where the group velocity of the second harmonic is intermediate between the two fundamental waves, $\nu_o < \nu_2 < \nu_e$ and assume that the e-wave pulse enters the crystal delayed with respect to the o-wave pulse. A second harmonic will be generated at the temporal overlap between the two pulses, as sketched in Figure 3.11(a). In a frame of reference moving at the group velocity of the second harmonic, the two fundamental pulses will travel toward each other. After some propagation distance, the overlap of the fundamentals increases, and so does the second harmonic intensity (Fig. 3.11(b)). However, if the fundamental pulses are sufficiently intense as sketched in Figs. 3.11(c) and (d), the intensity of the second harmonic may be large enough to deplete the fundamentals. As a result, the spatial overlap of the two (depleted) fundamentals remains small, as they move into each other (Fig. 3.11(d)).

An interesting situation arises when the walk-off lengths for the two fundamental pulses, $L_e = \tau_p/(\nu_e^{-1} - \nu_2^{-1})$ and $L_o = \tau_p/(\nu_o^{-1} - \nu_2^{-1})$, are equal and opposite in sign, and only three or four times longer than the crystal length. This case was analyzed in detail by Wang and Dragila [16] and Stabinis *et al.* [29]. The crystal angle θ; the walk-off lengths for a 12 ps pulse at 1.06 μm, and their ratio m are listed in Table 3.3 for SHG type II in KDP and DKDP. To generate compressed SH pulses, the faster e wave is sent delayed with respect to the o wave into the crystal. An SH seed originates from the short overlap region between the two e and o fundamental pulses. As this SH propagates through the crystal it is amplified, while the overlap between all three pulses increases. Because of its faster group velocity, the second harmonic always sees an undepleted o wave at its leading edge. Compression of the second harmonic results from the differential amplification of the leading edge with respect to the trailing edge. Implementation of this compression requires an accurate control of the pulse intensity, hence a well defined temporal profile, and a square or super Gaussian beam profile. Compression factors in excess of 30 have been observed [32, 37].

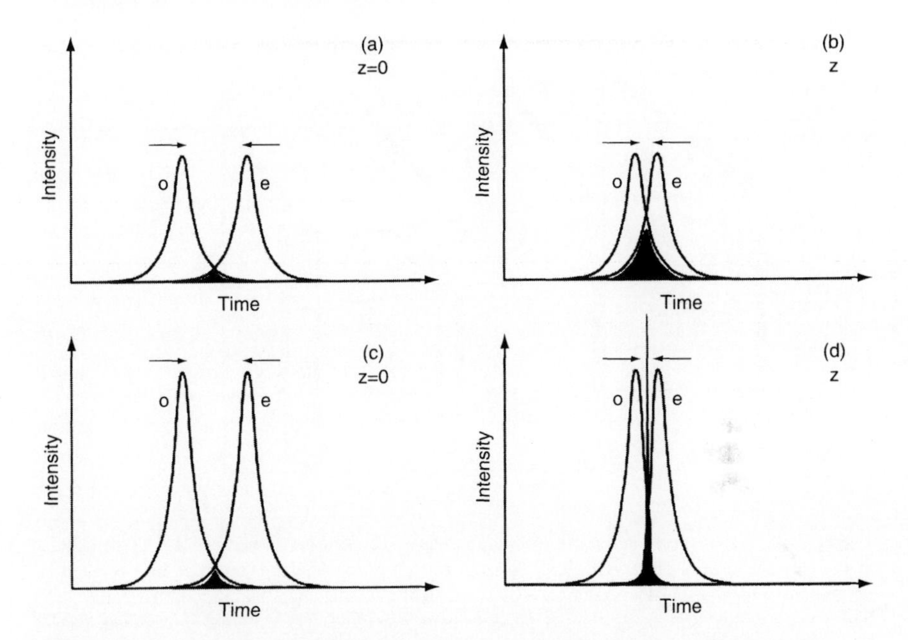

Figure 3.11 Pulse shaping in type II SHG. The three interacting pulses—fundamental ordinary, fundamental extraordinary, and second harmonic—are represented in a temporal frame of reference moving with the group velocity of the second harmonic.

Table 3.3

Relevant constants for SHG type II in KDP and DKDP at 1.06 μm. *m* is the ratio of the walk-off lengths.

Crystal	θ	L_o (cm)	L_e (cm)	$\|m\|$
KDP	59.2^o	20.9	−15.8	1.32
DKDP	53.5^o	28.1	−20.0	1.40

3.4.4. Group Velocity Control in SHG through Pulse Front Tilt

The condition of SH group velocity intermediate between the ordinary and extraordinary fundamentals cannot in general be met at any wavelength, for any crystal. It is however possible to adjust the group velocity through a tilt of the energy front with respect to the phase front. Figure 3.12 illustrates the basic principle in the case of a degenerate type I process. As shown in the lower part of

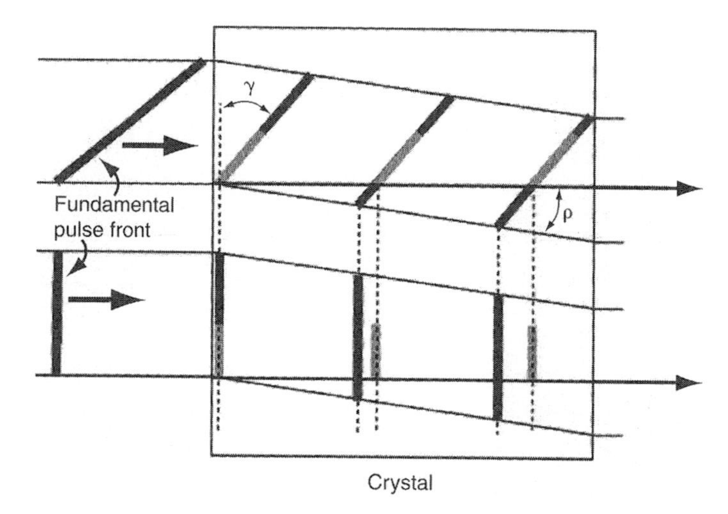

Figure 3.12 SHG using tilted pulse fronts. The wave vector of the (o-wave) SH is perpendicular to the entrance face of the crystal. Top—the pulse front of the fundamental (dark bar) is tilted by an amount compensating the group velocity mismatch of fundamental and second harmonic (grey bar) pulse exactly. Bottom—the input pulse front is not tilted. The different group velocities of fundamental and SH pulse lead to walk-off.

the figure, the temporal overlap of the interacting pulses decreases because of the lower velocity of the SH pulse relative to the fundamental pulse. Furthermore, the spatial overlap decreases because of the walk-off (ρ in the figure) of the fundamental wave from the SH wave. These two negative effects can however be used in conjunction with pulse front tilt to match the relative velocities of the two pulses, as illustrated in the upper part of Fig. 3.12. Loosely speaking: seen in the frame of reference of the SH wave, the lateral walk-off of the fundamental beam decrease the component of the pulse velocity along the direction of propagation of the SH just to match its (group) velocity.

Unfortunately, for femtosecond pulses, it is not practical to generate a large pulse front tilt. For instance, a pulse front tilt of the order of 40° would be required for SHG type II and compression in BBO of an 800 nm pulse of a Ti:Sapphire laser in collinear interaction [38]. A dispersive element like a prism with a ratio of beam diameter to base length of 20 (in the case of SF10 glass) would be needed to achieve the required energy front tilt of 40°. Dispersion in the glass would lead to large pulse broadening and phase modulation.

A better approach for group velocity matching in SHG as well as in parametric three-wave interactions of fs pulses is to use a noncollinear geometry [39, 40]. Table 3.4 shows how the group velocities of the participating waves can be changed from the collinear to the noncollinear case (here for an internal angle of 2°) leading to conditions for compression. The sketch of the interaction

Table 3.4

Group velocities for the *o*, *e* and SH *e* waves. Values are calculated for type II SHG of 800 nm radiation for collinear and noncollinear (internal angle is 2°) interaction in BBO.

Geometry	Fundamental (o) ν_o (10^8 m/s)	Second harmonic (e) ν_2 (10^8 m/s)	Fundamental (e) ν_e (10^8 m/s)
Collinear	1.780	1.755	1.843
Noncollinear	1.798	1.905	1.934

geometry in Figure 3.13 shows that it is possible to obtain the respective group velocities through manipulation of the angles of incidence in the crystal and the energy tilt produced by a prism. Figure 3.13 pertains to the case of noncollinear interacting plane waves inside a 250 μm long BBO crystal, cut for type II SHG of laser pulses at 800 nm.

The situation sketched in Fig. 3.13 was simulated with the system of equations Eq. (3.120), (3.121), and (3.122), with the substitution of Eqs. (3.123) for the

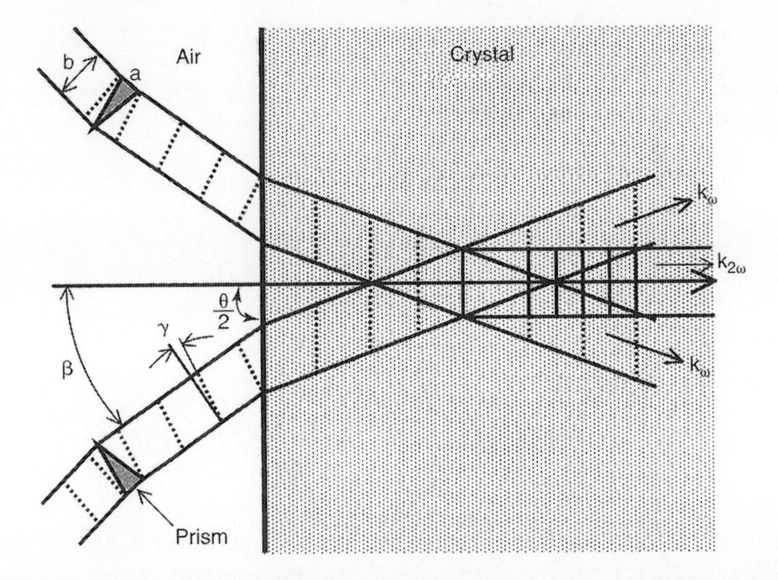

Figure 3.13 Geometry for precompensation the pulse front tilt resulting from propagation through an interface (the pulse fronts are depicted as dotted lines) using prisms. For the chosen type II SHG of 800 nm pulses: $\theta = 2°$, which is the internal angle of the fundamental beams for noncollinear interaction. To match the energy fronts of both fundamentals, an external pulse front tilt of $\gamma = 0.6°$ for $\beta = 1.6°$ is required. Also sketched in the figure are thin SF10 prisms (a/b = 0.2) to realize the desired γ.

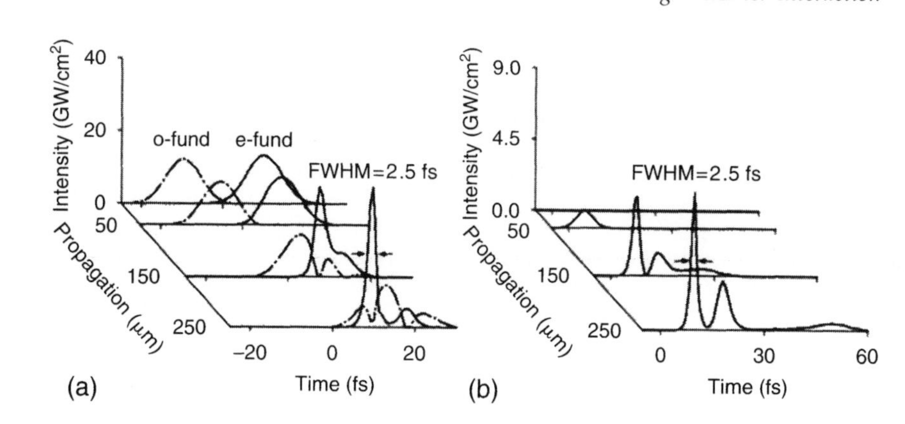

Figure 3.14 Calculated fundamental (a) and SH waves (b) as they propagate and interact inside a 250 μm long BBO crystal (note the different scales for (a) and (b)). Group velocity mismatch leads to a compressed SH pulse with an FWHM of 2.5 fs (b), for 10 fs fundamental input pulses. Furthermore, considerable pulse reshaping can be seen for the extraordinary fundamental in (a), leading to a shortened FWHM from 10 fs to 2.5 fs with a shoulder. Pulse energies were 400 μJ for each fundamental pulse, at an FWHM of 10 fs with a predelay of 20 fs of the *e*-fundamental with respect to the *o*-fundamental. (Adapted from Biegert and Diels [38].)

case of a 10 fs fundamental pulse with a 20 fs predelay of the fundamental *e* with respect of the fundamental *o*. Successive intensity profiles are plotted in Figure 3.14. The 10 fs fundamental pulse gives rise to a 2.5 fs second harmonic. Even in this extreme case of pulse compression, the maximum value for $\partial^2\tilde{\mathcal{E}}/\partial z^2$ is 200 times smaller than $2k\,\partial\tilde{\mathcal{E}}/\partial z$, and the second-order partial derivatives can therefore be neglected in Eqs. (3.120), (3.121), and (3.122). Figure 3.14(b) shows a peak intensity of 9 GW/cm^2 for the 2.5 fs second-harmonic for initial peak intensities for the fundamental pulses of roughly 13 GW/cm^2. This indicates 70% conversion in intensity.

3.5. OPTICAL PARAMETRIC INTERACTION

3.5.1. Coupled Field Equations

Similar considerations as in the previous section can be made for a number of other nonlinear processes of second order used for generating pulses at new frequencies. Figure 3.15 shows schematically three possible situations. In parametric up conversion two pulses of frequencies ω_1 and ω_2, respectively, are sent through a nonlinear medium (crystal) and produce a pulse of frequency $\omega_3 = \omega_1 + \omega_2$.

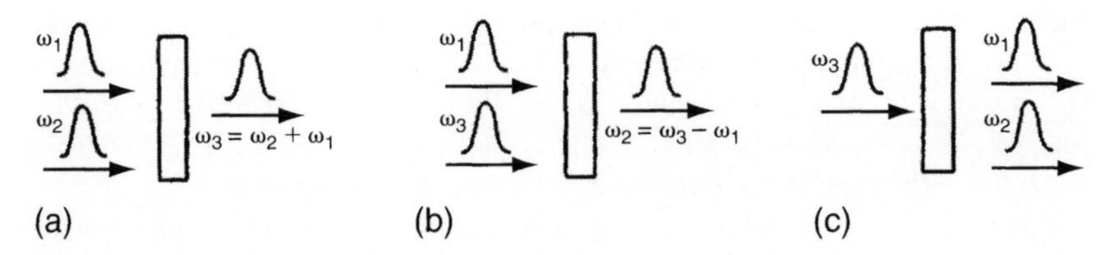

Figure 3.15 Nonlinear optical processes of second order for generating pulses of new frequencies. (a) Parametric up conversion, (b) parametric down conversion, (c) parametric oscillation.

In parametric down conversion a pulse with the difference frequency is generated. In parametric oscillation, a single pulse of frequency ω_3 generates two pulses of frequency ω_1 and ω_2 such that $\omega_1 + \omega_2 = \omega_3$. Which process occurs depends on the realization of the phase matching condition. In principle, to obtain a fs output pulse through up or down conversion, it is sufficient to have only one fs input pulse. The second input can be a longer pulse or even cw light. Mokhtari *et al.*, [41] for example, mixed 60 fs pulses at 620 nm from a dye laser with 85 ps pulses from an Nd:YAG laser (1064 nm) to obtain up converted fs pulses at 390 nm. Parametric frequency mixing of two fs input pulses were reported, for example, in Elsaesser and Nuss [42] and Jedju and Rothberg [43]. If the input pulse (pump pulse) is sufficiently strong, two pulses of frequencies ω_1 and ω_2, for which the phase matching condition is satisfied, can arise. In this case, noise photons which are always present in a broad spectral range can serve as seed light. This process is known as optical parametric oscillation; the generated pulses are called idler and signal pulses—the usual convention being that the signal is the generated radiation with the shorter wavelength.

With similar assumptions that allowed us to derive the equations for SHG, we obtain three coupled differential equations for the interaction of the three optical fields as shown in Fig. 3.15:

$$\left(\frac{\partial}{\partial z} + \frac{1}{\nu_1}\frac{\partial}{\partial t}\right)\tilde{\mathcal{E}}_1 - \frac{i}{2}k_1''\frac{\partial^2}{\partial t^2}\tilde{\mathcal{E}}_1 + \mathcal{D}_1 = -i\chi^{(2)}\frac{\omega_1^2}{4c^2k_1}\tilde{\mathcal{E}}_2^*\tilde{\mathcal{E}}_3 e^{i\Delta kz} \tag{3.127}$$

$$\left(\frac{\partial}{\partial z} + \frac{1}{\nu_2}\frac{\partial}{\partial t}\right)\tilde{\mathcal{E}}_2 - \frac{i}{2}k_2''\frac{\partial^2}{\partial t^2}\tilde{\mathcal{E}}_2 + \mathcal{D}_2 = -i\chi^{(2)}\frac{\omega_2^2}{4c^2k_2}\tilde{\mathcal{E}}_1^*\tilde{\mathcal{E}}_3 e^{i\Delta kz} \tag{3.128}$$

$$\left(\frac{\partial}{\partial z} + \frac{1}{\nu_3}\frac{\partial}{\partial t}\right)\tilde{\mathcal{E}}_3 - \frac{i}{2}k_3''\frac{\partial^2}{\partial t^2}\tilde{\mathcal{E}}_3 + \mathcal{D}_3 = -i\chi^{(2)}\frac{\omega_3^2}{4c^2k_3}\tilde{\mathcal{E}}_1\tilde{\mathcal{E}}_2 e^{-i\Delta kz} \tag{3.129}$$

where $\Delta k = k_1(\omega_1) + k_2(\omega_2) - k_3(\omega_3)$, and the higher-order dispersion terms defined in Eqs. (1.88), and (1.89) have been included. The conversion efficiencies are maximum if the phase matching condition, $\Delta k = 0$, is satisfied.

The description of the various processes in Fig. 3.15 by Eqs. (3.127)–(3.129) differs only in the initial conditions, that is the field amplitudes at the crystal input. This system of equations is analogous to the ones encountered for SHG. For relatively weak pulses, conversion efficiencies are low, and the group velocity dispersion contributes to a broadening of the generated radiation.

As in the case of SHG, there is a particularly interesting regime which combines the complexities of short pulses, high intensity, and long interaction lengths. The pulses have to be sufficiently short that simultaneous phase matching cannot be achieved over the pulse bandwidth. The crystal length and pulse intensities are sufficiently high for regeneration of the pump to occur. These conditions are also referred to as "giant pulse regime," or sometimes "nonlinear parametric generation."

3.5.2. Synchronous Pumping

Higher efficiencies can be obtained by placing the nonlinear crystal in an optical resonator for the signal pulse. The crystal is then pumped by a sequence of pulses whose temporal separation exactly matches the resonator round trip time. In this manner, the signal pulse passes through the amplifying crystal many times before it is coupled out. Laenen *et al.* [44] pumped the crystal with a train of 800 fs pulses from a frequency doubled Nd:glass laser and produced 65–260 fs signal pulses which were tunable over a range from 700 to 1800 nm.

Intracavity pumping of an optical parametric oscillator (OPO) is also possible to take advantage of the high intracavity pulse power for the pumping. This technique was demonstrated by Edelstein *et al.*, [45] by placing the OPO crystal in the resonator of a fs dye laser and building a second resonator around the crystal for the signal pulse. At repetition frequencies of about 80 MHz, the mean output power of the parametric oscillator was on the order of several milliwatts. A high gain, short lifetime laser such as a dye or semiconductor laser is desirable, because the mode-locking of such a laser is less sensitive to feedback from the faces of the crystal. Intracavity pumping of an optical parametric oscillator has however also been demonstrated in a Ti:sapphire laser [46] which has a long gain lifetime.

3.5.3. Chirp Amplification

So far we have been concerned with amplitude modulation effects in nonlinear mixing. Phase modulation introduces an element of complexity that one generally tries to avoid, in particular in the conditions of giant pulse compression discussed previously. It has been recognized however that second-order interactions can be used to generate or amplify a phase modulation. We will consider here as an example chirp amplification that can take place in parametric processes.

The frequencies of the three interacting pulses obey the relation

$$\omega_3(t) = \omega_1(t) + \omega_2(t). \tag{3.130}$$

If we substitute for the time dependent frequencies $\omega_i(t) = \omega_i + \dot{\varphi}_i(t)$ $(i = 1, 2, 3)$, we obtain for the phases

$$\dot{\varphi}_3(t) = \dot{\varphi}_2(t) + \dot{\varphi}_1(t). \tag{3.131}$$

In addition, for efficient parametric oscillation, the phase matching condition must be satisfied, which now implies

$$k_3[\omega_3(t)] = k_2[\omega_2(t)] + k_1[\omega_1(t)]. \tag{3.132}$$

For $|\dot{\varphi}_i| \ll \omega_i$ and linearly chirped pump pulses a Taylor expansion of Eq. (3.132) yields

$$\left.\frac{dk_3}{d\Omega}\right|_{\omega_3} \dot{\varphi}_3(t) = \left.\frac{dk_2}{d\Omega}\right|_{\omega_2} \dot{\varphi}_2(t) + \left.\frac{dk_1}{d\Omega}\right|_{\omega_1} \dot{\varphi}_1(t). \tag{3.133}$$

The chirps at the three frequencies are thus related by the *group* velocities ν_i. From Eqs. (3.131) and (3.133), a relation between the chirp of idler and signal pulses and pump pulse can be found:

$$\dot{\varphi}_1 = p\dot{\varphi}_3 \tag{3.134}$$

$$\dot{\varphi}_2 = (1 - p)\dot{\varphi}_3, \tag{3.135}$$

where p is the chirp enhancement coefficient:

$$p = \left(\frac{\nu_3^{-1} - \nu_2^{-1}}{\nu_1^{-1} - \nu_2^{-1}} \right), \tag{3.136}$$

and ν_i are the group velocities at the respective frequencies ω_i.

Equation (3.136) indicates that chirp amplification is most pronounced in a condition of degeneracy where $\omega_1 \approx \omega_2$. The mechanism of chirp amplification can also be understood by means of the tuning curves. For instance, Fig. 3.16 shows the tuning curves for phase matching (type I) in KDP. A small change in pump wavelength λ_3 results in a strong change of signal wavelength. As a result, a slightly chirped pump pulse generates signal and idler with enhanced (and opposite) chirp [47]. The signal pulses with enhanced chirp can be compressed in a grating pair compressor. Using the natural chirp of frequency doubled

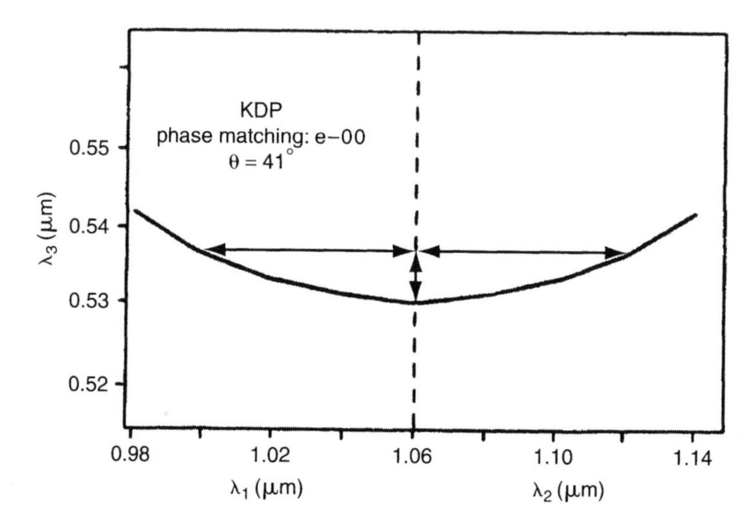

Figure 3.16 Tuning curves of a KDP optical parametric oscillator, pumped at a wavelength of $\lambda_3 = 0.53$ μm. (Adapted from Jankauskas *et al.* [47].)

pulses from an Nd:glass laser, pulses of 50 fs at 920 nm were obtained by this technique [47].

3.6. THIRD-ORDER SUSCEPTIBILITY

3.6.1. Fundamentals

The third-order contribution to the nonlinear polarization in an isotropic medium is [4]:

$$\begin{aligned}\tilde{P}_x^{(3)} &= \epsilon_0 \chi^{(3)} \sum_{j=x,y} \left[\tilde{E}_x \tilde{E}_j \tilde{E}_j^* + \tilde{E}_j \tilde{E}_x \tilde{E}_j^* + \tilde{E}_j \tilde{E}_j \tilde{E}_x^* \right] \\ &= \epsilon_0 \chi^{(3)} \left\{ 2 \left[|\tilde{E}_x|^2 + |\tilde{E}_y|^2 \right] \tilde{E}_x + \tilde{E}_x |\tilde{E}_x|^2 + \tilde{E}_y \tilde{E}_y \tilde{E}_x^* \right\} \\ &= 3\epsilon_0 \chi^{(3)} \left\{ \left[|\tilde{E}_x|^2 + \frac{2}{3} |\tilde{E}_y|^2 \right] \tilde{E}_x + \frac{1}{3} (\tilde{E}_x^* \tilde{E}_y) \tilde{E}_y \right\},\end{aligned} \qquad (3.137)$$

[4] We consider here only the terms in the polarization oscillating at the same optical frequency as the driving field.

where we consider the $\hat{x}$ component of the polarization, under the influence of a light pulse propagating along z, with field components along $\hat{x}$ and $\hat{y}$. If there is only a component of the field along x:

$$\tilde{P}_x^{(3)} = 3\epsilon_0 \chi^{(3)} |\tilde{E}_x|^2 \tilde{E}_x. \tag{3.138}$$

In the transparent region of many materials, $\chi^{(3)}$ can be approximated by a real quantity, and the nonlinear polarization results in an index of refraction that depends nonlinearly on the propagating field. Usually the lowest order of this dependence is expressed by one of the following equivalent relations

$$\boxed{\begin{aligned} n &= n_0 + n_2 |\tilde{\mathcal{E}}(t)|^2 \\ &= n_0 + 2n_2 \langle E^2(t) \rangle \\ &= n_0 + \bar{n}_2 I(t), \qquad (3.139) \end{aligned}}$$

where $\bar{n}_2 = 2n_2/(\epsilon_0 c n_0)$. The quantity n_2 is called nonlinear index coefficient and describes the strength of the coupling between the electric field and the refractive index. The most often quoted quantities are n_2 in esu units, and $\bar{n}_2$ in cm^2/W. The conversion factor between the two quantities is:

$$\bar{n}_2\ (\text{cm}^2/\text{W}) = \frac{2}{(300)^2 n_0} \sqrt{\frac{\mu_0}{\epsilon_0}} n_2\ (\text{esu}) \approx \frac{8.378}{n_0} \cdot 10^{-3} n_2\ (\text{esu}) \tag{3.140}$$

Many different physical processes can account for an intensity-dependent change in index of refraction because of a third-order nonlinearity. Table 3.5 gives some examples. As a rule of thumb, the larger the nonlinearity, the longer the corresponding response time. For fs pulse excitation, it is only a (nonresonant) nonlinearity of electronic origin that can be considered to be without inertia. The corresponding nonlinear refractive index can be described by relations (3.139), and is a result of an optical nonlinearity of third order. If only one pulse is incident

Table 3.5

Examples of nonlinear refractive index parameters.

Origin	Example	$\bar{n}_2$ (cm^2/W)	Response time (s)
Electronic			
Nonresonant	Glass	10^{-16}–10^{-15}	10^{-15}–10^{-14}
	Air	10^{-18}–10^{-19}	10^{-14}–10^{-13}
Resonant	Semiconductor doped glass	10^{-10}	10^{-11}
Molecular motion	CS_2	10^{-12}	10^{-12}

on the sample the complete (including terms not oscillating at ω_ℓ) polarization reads

$$\tilde{P}^{(3)} = \epsilon_0 \chi^{(3)} E^3 = \epsilon_0 \chi^{(3)} \left(\frac{3}{8} |\tilde{\mathcal{E}}|^2 \tilde{\mathcal{E}} e^{i\omega_\ell t} + \frac{1}{8} \tilde{\mathcal{E}}^3 e^{3i\omega_\ell t} \right) + c.c. \tag{3.141}$$

assuming an instantaneous response. The terms with $3\omega_\ell$ in the argument of the exponential function describe third harmonic generation. In the cases where this latter process is sufficiently weak not to impact the propagation of the wave at ω_ℓ, one can show from Eq. (3.141) that

$$n_2 = \frac{3\chi^{(3)}}{8n_0}, \tag{3.142}$$

(see Problem 3 at the end of this chapter). The intensity dependence of n implies a refractive index varying in time and space. The temporal variation, as discussed in Chapter 1 [Eq. (1.175)], leads to a pulse chirp. The (transverse) spatial refractive index dependence leads to lensing effects. These processes are called self-phase modulation (SPM) and self-focusing (SF), respectively. In most cases self-focusing is undesirable and is avoided by minimizing the sample length and/or working with uniform beam profiles whenever possible.

To describe SPM we can substitute Eq. (3.141) into the wave equation (3.92). Let us first recall the approximations needed to derive the simple expression for the second derivative of the polarization used in the previous subsection. These approximations lead to an estimate of the first correction terms. If the nonlinearity is not perfectly inertialess (i.e., does not respond instantaneously to the electric field), we have to compute the polarization using the integral expression (3.82). For response times smaller than the pulse duration, the Fourier transform of Eq. (3.82) can be expanded into a Taylor series about ω_ℓ. Termination of the series after the second term and back transformation into the time domain yields for the polarization, in terms of the field envelope:

$$\tilde{P}^{(3)}(t) = \frac{3}{8}\epsilon_0 \left[\chi^{(3)} |\tilde{\mathcal{E}}|^2 \tilde{\mathcal{E}} + i \left. \frac{\partial \chi^{(3)}}{\partial \Omega} \right|_{\omega_\ell} \frac{\partial}{\partial t} \left(|\tilde{\mathcal{E}}|^2 \tilde{\mathcal{E}} \right) \right] e^{i(\omega_\ell t - k_\ell z)} + c.c. \tag{3.143}$$

The expression (3.143) restates once more that a nonzero response time τ_r leads to a frequency dependence of the susceptibility. The critical parameter is the spectral variation of this susceptibility over the frequency range covered by the pulse spectrum. To study nonlinear propagation problems, the SVEA is generally applied to the polarization, as expressed in Eq. (3.93). Inserting the expansion (3.143) into the second derivative of the polarization (3.93) and keeping terms up to the first temporal derivative lead to:

$$i\frac{\mu_0}{k_\ell}\frac{d^2}{dt^2}P^{(3)} = \left[-i\frac{n_2 k_\ell}{n_0}|\tilde{\mathcal{E}}|^2\tilde{\mathcal{E}} - \beta\frac{\partial}{\partial t}\left(|\tilde{\mathcal{E}}|^2\tilde{\mathcal{E}}\right) + \ldots\right]e^{i(\omega_\ell t - k_\ell z)} + c.c. \quad (3.144)$$

where

$$\beta = \frac{n_2}{c}\left(2 - \frac{\omega_\ell}{\chi^{(3)}}\left.\frac{\partial\chi^{(3)}}{\partial\Omega}\right|_{\omega_\ell}\right). \quad (3.145)$$

It should be remembered that the temporal derivative in Eq. (3.144) becomes important if the light period is not negligibly short compared to the pulse duration. Equations (3.144) and (3.145) indicate that the first-order correction to the SVEA and the finite response time of the nonlinear susceptibility (the two summands forming β) have the same action on pulse propagation. Corrections to the SVEA may also be important in the spatial (transverse) propagation of the beam as indicated in Eq. (3.97).

Pulse propagation through transparent media which is affected by SPM and dispersion has played a crucial role in fiber optics. For a review we refer to the monograph by Agrawal [48]. In this chapter we will discuss the physics behind SPM and neglect GVD. In Chapter 8 we will describe effects associated with the interplay of SPM and GVD.

3.6.2. Short Samples with Instantaneous Response

For interaction lengths much shorter than the dispersion length L_D, and for an instantaneous nonlinearity, the wave equation (3.92) with the source term (3.144) simplifies to

$$\frac{\partial}{\partial z}\tilde{\mathcal{E}}(z,t) = -i\frac{3\omega_\ell^2\chi^{(3)}}{8c^2 k_\ell}|\tilde{\mathcal{E}}^2|\tilde{\mathcal{E}} = -i\frac{n_2 k_\ell}{n_0}|\tilde{\mathcal{E}}|^2\tilde{\mathcal{E}} \quad (3.146)$$

after applying the SVEA to the polarization term. For real $\chi^{(3)}$, substituting $\tilde{\mathcal{E}} = \mathcal{E}\exp(i\varphi)$ into Eq. (3.146) and separating the real and imaginary parts result in an equation for the pulse envelope

$$\frac{\partial}{\partial z}\mathcal{E} = 0 \quad (3.147)$$

and for the pulse phase

$$\frac{\partial \varphi}{\partial z} = -\frac{n_2 k_\ell}{n_0}|\mathcal{E}|^2. \tag{3.148}$$

Obviously the pulse amplitude $\mathcal{E}$ is constant in the coordinate system traveling with the group velocity, that is, the pulse envelope remains unchanged, $\mathcal{E}(t, z) = \mathcal{E}(t, 0) = \mathcal{E}_0(t)$. Taking this into account, we can integrate Eq. (3.148) to obtain for the phase

$$\boxed{\varphi(t, z) = \varphi_0(t) - \frac{k_\ell n_2}{n_0} z \mathcal{E}_0^2(t)} \tag{3.149}$$

which results in a phase modulation given by

$$\frac{\partial \varphi}{\partial t} = \frac{d\varphi_0}{dt} - \frac{n_2 k_\ell}{n_0} z \frac{d}{dt} \mathcal{E}_0^2(t). \tag{3.150}$$

This result can be interpreted as follows. The refractive index change follows the pulse intensity instantaneously. Thus, different parts of the pulse "feel" different refractive indices, leading to a phase change across the pulse. Unlike the phase modulation associated with GVD, this SPM produces new frequency components and broadens the pulse spectrum. To characterize the SPM it is convenient to introduce a nonlinear interaction length

$$L_{NL} = \frac{n_0}{n_2 k_\ell \mathcal{E}_{0m}^2} \tag{3.151}$$

where $\mathcal{E}_{0m}$ is the peak amplitude of the pulse. The quantity z/L_{NL} represents the maximum phase shift which occurs at the pulse peak, as can be seen from Eq. (3.150). Figure 3.17 shows some examples of the chirp and spectrum of self-phase modulated pulses. Because n_2 is mostly positive far from resonances (see Problem 4 at the end of this chapter), upchirp occurs in the pulse center. We also see that SPM can introduce a considerable spectral broadening. This process as well as some nonlinear processes of higher order can be used to generate a white light continuum, as discussed later in this chapter.

For an order of magnitude estimate let us determine the frequency change $\delta\omega$ over the FWHM of a Gaussian pulse. Substituting $\mathcal{E}_{0m} e^{-2\ln 2(t/\tau_p)^2}$ for $\mathcal{E}_0(t)$ in Eq. (3.150) yields

$$\delta\omega\tau_p = \frac{8 \ln 2}{\sqrt{2}} \frac{z}{L_{NL}}. \tag{3.152}$$

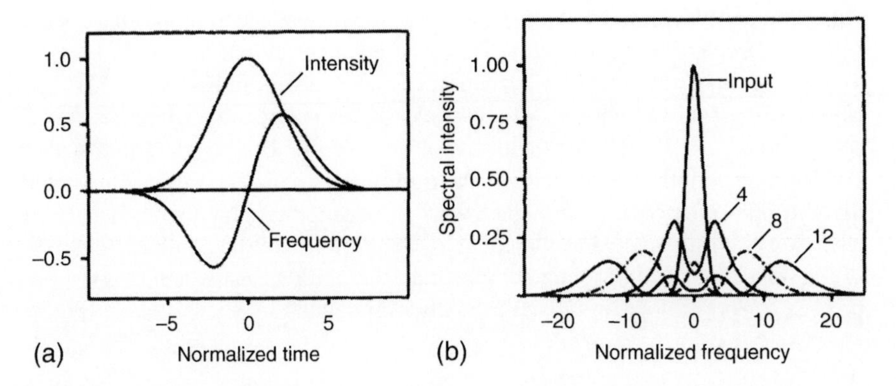

Figure 3.17 Frequency modulation and spectrum of self-phase modulated Gaussian pulses for different propagation lengths z/L_{NL}.

For $z = L_{NL}$ the normalized frequency sweep is $\delta\omega\tau_p \approx 4$. Note that the original pulse duration–bandwidth product of the unchirped Gaussian input pulse was $\Delta\omega\tau_p \approx 3$.

A pulse can also be phase modulated in the field of a second pulse if both pulses interact in the medium. The phase of pulse 1 is then determined by

$$\frac{\partial \varphi_1}{\partial z} = -\frac{n_2 k_\ell}{n_0}\left(|\tilde{\mathcal{E}}_1|^2 + 2|\tilde{\mathcal{E}}_2|^2\right)\tilde{\mathcal{E}}_1. \tag{3.153}$$

This offers the possibility of phase modulating weak pulses by means of a strong pulse. Both pulses can differ in their wavelength, duration, polarization state, etc. This process is known as cross-phase modulation and has found several interesting applications [49]. For example, it is possible to transfer information from one pulse train to another by induced spectral changes [50].

3.6.3. Short Samples and Noninstantaneous Response

For short pulses and/or a noninstantaneous sample response, the source term given by Eq. (3.144) with $\beta \neq 0$ must be incorporated in the wave equation. The pulse propagation is now governed by

$$\frac{\partial}{\partial z}\tilde{\mathcal{E}} + i\frac{n_2 k_\ell}{n_0}|\tilde{\mathcal{E}}|^2\tilde{\mathcal{E}} + \beta\frac{\partial}{\partial t}\left(|\tilde{\mathcal{E}}|^2\tilde{\mathcal{E}}\right) = 0. \tag{3.154}$$

Comparison with Eqs. (1.87) and (1.88) suggests that the term with a time derivative can be interpreted as an intensity dependent group velocity. For $\beta > (<)\ 0$ the pulse center is expected to travel slower (faster) than the trailing edge. This causes a steepening of the trailing (leading) edge of the pulse, known as self-steepening. It is similar to the formation of shock waves in acoustics. It is not necessarily associated with a slow response of the polarization, because $\beta = 2$ for $\partial\chi^{(3)}/\partial\omega\big|_{\omega_\ell} = 0$, from the definition of Eq. (3.145). This shock term can also be identified with the second term of the right-hand side in Eq. (3.97). One notes that this first-order correction to the SVEA is a loss term, rather than a dispersive term. It can be interpreted as the energy required to drive the nonlinear index.

To solve Eq. (3.154) we again substitute the complex pulse amplitude by a product of an envelope and a phase function to obtain

$$\frac{\partial}{\partial z}\mathcal{E} + 3\beta\mathcal{E}^2\frac{\partial}{\partial t}\mathcal{E} = 0 \tag{3.155}$$

and

$$\frac{\partial}{\partial z}\varphi + \beta\mathcal{E}^2\frac{\partial}{\partial t}\varphi = -\frac{n_2 k_\ell}{n_0}\mathcal{E}^2. \tag{3.156}$$

The equation for the envelope can be solved independently of the phase equation. Its solution can formally be written as [51]

$$\mathcal{E}(z,t) = \mathcal{E}\left(z = 0, t - 3\beta z\mathcal{E}^2(t,z)\right). \tag{3.157}$$

For a Gaussian input pulse, $\mathcal{E}_{0m}e^{-(t/\tau_G)^2}$, we find

$$\mathcal{E}(z,t) = \mathcal{E}_{0m}\exp\left\{-\left[\frac{t - 3\beta z\mathcal{E}^2(z,t)}{\tau_G}\right]^2\right\}, \tag{3.158}$$

which contains the envelope implicitly. Figure 3.18 shows the pulse envelope for different values of $3\beta z/\tau_G$. To observe the shock pulse with $|d\mathcal{E}/dt| \to \infty$, the shock term in the propagation should dominate other terms such as dispersion (neglected here for the sake of simplicity). This, for example, is the case for filamentation (self-focusing) in air and is responsible for the associated white light or "conical emission" (Sections 3.7 and 13.2.2).

The envelope function can be inserted in Eq. (3.156) to obtain an implicit (analytical) solution for the phase, [51] which, however, is rather complex. A numerical evaluation of the pulse spectrum $|\mathcal{F}\{\tilde{\mathcal{E}}(t,z)\}|^2$ reveals an asymmetric behavior, which is expected, because we are now dealing with SPM of asymmetric pulses. Such spectra were reported by Knox *et al.*, [52] for example, as result of SPM of 40 fs pulses in glass. Rothenberg and Grischkowsky [53] directly measured the steepening of pulses in propagating through optical glass fibers.

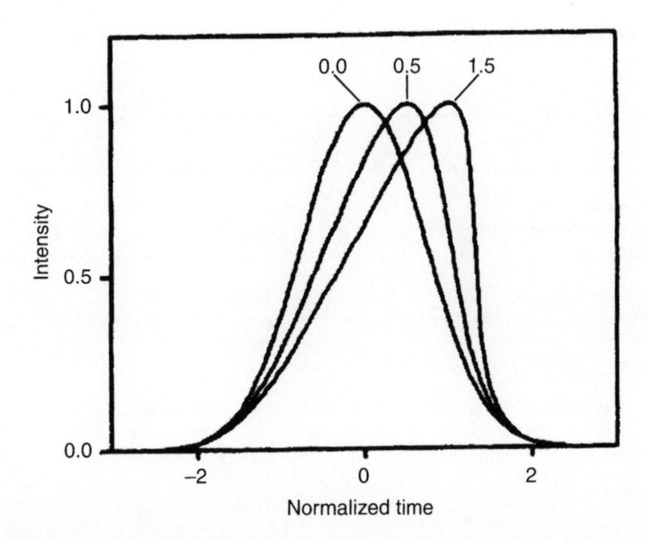

Figure 3.18 Pulse envelope according to Eq. (3.158) for different values of $3\beta z/\tau_G$.

It should also be noted that SPM in connection with saturation, which was discussed previously, is a particular example of a time-dependent sample response. In the case of saturation, there is a memory effect associated with the change of occupation numbers, with a characteristic time determined by the corresponding (energy) relaxation time.

3.6.4. Counter-Propagating Pulses and Third-Order Susceptibility

In a laser cavity, we may encounter the situation where counter-propagating pulses meet in a nonlinear medium. The standing wave formed at the intersection of the colliding pulses creates a phase (real third-order susceptibility) or an amplitude grating (purely imaginary third-order susceptibility, resulting in two photon absorption). The grating formed by the nonlinear interaction will diffract each beam into the direction of the other. As illustrated in Figure 3.19, the transmitted forward propagating field (affected by the nonlinear susceptibility) is combined with the part of the counter-propagating field that is diffracted to the right. Experimentally, these two contributions being undistinguishable, we observe only the total forward propagating pulse at the right of the sample as a "transmitted" beam. When the incident counter-propagating intensities are different, it appears as if the medium had a different transfer function for the forward and backward propagating beams, as will be explained. In what follows we will assume that the medium is much thinner than the pulse length, $d \ll \tau_p c$.

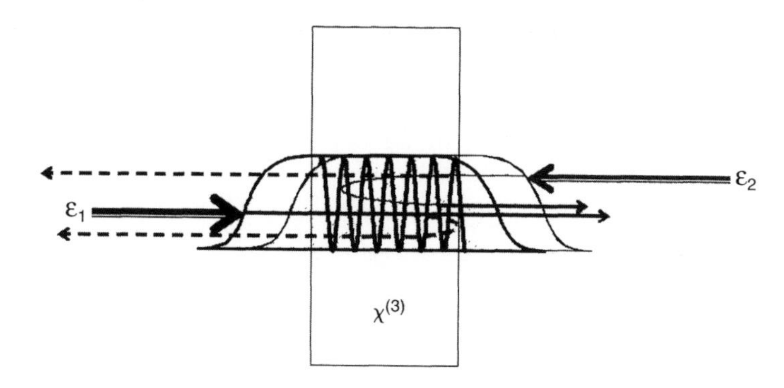

Figure 3.19 Geometry of the interaction of counter-propagating pulses in a medium with a third-order nonlinear susceptibility. The standing wave field produces a grating that diffracts (reflects) the waves. What is observed exiting the sample is the sum of the transmitted and reflected fields.

Let us assume two counter-propagating beams $\tilde{E}_1 = (1/2)\tilde{\mathcal{E}}_1 \exp[i(\omega_\ell t - kz)]$ and $\tilde{E}_2 = (1/2)\tilde{\mathcal{E}}_2 \exp[i(\omega_\ell t + kz)]$. From Eq. (3.138), we find for the third-order polarization responsible for phase modulation and two photon absorption:

$$\begin{aligned} P^{(3)} &= 3\epsilon_0\chi^{(3)}(\tilde{E}_1 + \tilde{E}_2)(\tilde{E}_1 + \tilde{E}_2)^*(\tilde{E}_1 + \tilde{E}_2) + c.c. \\ &= 3\epsilon_0\chi^{(3)}\left\{|\tilde{E}_1|^2 + |\tilde{E}_2|^2 + \tilde{E}_1\tilde{E}_2^* + \tilde{E}_1^*\tilde{E}_2\right\}\left\{\tilde{E}_1 + \tilde{E}_2\right\} + c.c. \\ &= \frac{3}{8}\epsilon_0\chi^{(3)}\left\{\left(|\tilde{\mathcal{E}}_1|^2 + 2|\tilde{\mathcal{E}}_2|^2 + \tilde{\mathcal{E}}_1\tilde{\mathcal{E}}_2^* e^{-2ik_\ell z}\right)\tilde{\mathcal{E}}_1 e^{i(\omega_\ell t - k_\ell z)}\right. \\ &\quad \left. + \left(2|\tilde{\mathcal{E}}_1|^2 + |\tilde{\mathcal{E}}_2|^2 + \tilde{\mathcal{E}}_1^*\tilde{\mathcal{E}}_2 e^{2ik_\ell z}\right)\tilde{\mathcal{E}}_2 e^{i(\omega_\ell t + k_\ell z)}\right\} + c.c. \end{aligned} \tag{3.159}$$

We recognize in Eq. (3.159) a forward and a backward propagating field, and two terms involving higher-order spatial Fourier components. Even in the case of a homogeneous medium, the terms in $\exp(\pm 2ikz)$ lead to a coupling through a higher-order (fifth-order) process, which we will neglect. We will see in the exercise at the end of the chapter that these terms become important if the nonlinear susceptibility has a periodic structure on the scale of the wavelength. The forward propagating nonlinear polarization is:

$$\tilde{P}^{(3)}_{\text{forward}} = \frac{1}{2}\tilde{\mathcal{P}}^{(3)}_1 e^{i(\omega_\ell t - k_\ell z)} = \frac{3}{8}\epsilon_0\chi^{(3)}\left(|\tilde{\mathcal{E}}_1|^2 + 2|\tilde{\mathcal{E}}_2|^2\right)\tilde{\mathcal{E}}_1 e^{i(\omega_\ell t - k_\ell z)}. \tag{3.160}$$

In the retarded frame of reference, and in the SVEA, the field amplitude varies as, cf. Eq. (3.30):

$$\frac{\partial \tilde{\mathcal{E}}_1}{\partial z} = -i\frac{\mu_0 \omega_\ell c}{2n_0}\tilde{\mathcal{P}}_1^{(3)}. \tag{3.161}$$

Up to this point, no assumption has been made for the complex third-order susceptibility, which we will assume to be of the form:

$$\chi^{(3)} = \chi_r^{(3)} - i\chi_i^{(3)}. \tag{3.162}$$

The real part of the third-order susceptibility leads to the nonlinear index n_2 as was expressed in Eq. (3.142):

$$n_2 = \frac{3\chi_i^{(3)}}{8n_0}$$

For the case of real $\chi^{(3)}$, inserting Eq. (3.160) into Eq. (3.161) results in the following propagation equation for $\tilde{\mathcal{E}}_1$

$$\begin{aligned}\frac{\partial \tilde{\mathcal{E}}_1}{\partial z} &= -i\frac{\omega_\ell}{c}\frac{3}{8}\frac{\chi_r^{(3)}}{n_0}\left(|\tilde{\mathcal{E}}_1|^2 + 2|\tilde{\mathcal{E}}_2|^2\right)\tilde{\mathcal{E}}_1 \\ &= -i\frac{\omega_\ell}{c}n_2\left(|\tilde{\mathcal{E}}_1|^2 + 2|\tilde{\mathcal{E}}_2|^2\right)\tilde{\mathcal{E}}_1 = -ik_{NL}\tilde{\mathcal{E}}_1,\end{aligned} \tag{3.163}$$

where k_{NL} represents a nonlinear propagation constant. This equation describes self- and cross-phase modulation. The factor of 2 in front of $|\tilde{\mathcal{E}}_2|^2$ leads to the asymmetry in the induced phase mentioned in the introduction if the two counter-propagating beams have different amplitudes. There is no energy exchange between the two beams in case of a real susceptibility $\chi^{(3)}$.

Case of Two Photon Absorption

The imaginary part of the third-order susceptibility is responsible for two-photon absorption. For purely imaginary $\chi^{(3)}$ the propagation equation for the pulse amplitude becomes

$$\frac{\partial \tilde{\mathcal{E}}_1}{\partial z} = -\frac{\omega_\ell}{c}\frac{3}{8}\frac{\chi_i^{(3)}}{n_0}\left(|\tilde{\mathcal{E}}_1|^2 + 2|\tilde{\mathcal{E}}_2|^2\right)\tilde{\mathcal{E}}_1 = -\frac{\beta_2}{2}\left(|\tilde{\mathcal{E}}_1|^2 + 2|\tilde{\mathcal{E}}_2|^2\right)\tilde{\mathcal{E}}_1, \tag{3.164}$$

which leads to the definition of the two photon absorption coefficient β_2:

$$\beta_2 = \frac{3}{4}\frac{\omega_\ell}{cn_0}\chi_i^{(3)}. \tag{3.165}$$

In terms of intensities, the expression for the attenuation of beam 1 in presence of beam 2 is:

$$\frac{dI_1}{dz} = -\beta_2 (I_1 + 2I_2) I_1. \tag{3.166}$$

Again, it is only when the two counter-propagating beams have equal intensity that the relative attenuation of both beams is equal.

3.7. CONTINUUM GENERATION

One of the most impressive (and simplest) experiments with ultrashort light pulses is the generation of a white light continuum. At the same time continuum generation with laser pulses is one of the most complex and difficult to analyze processes as it combines spatial and temporal effects and their interplay. This is one reason while the spectral supercontinuum has remained an area of active theoretical and experimental research for a long time. For reviews on this subject see Alfano, [54] and the special issue Zheltikov, [55] for a summary of research.

Provided the pulse is powerful enough, focusing into a transparent material results in a substantial spectral broadening. The output pulse appears on a sheet of paper as a white light flash, even if the exciting pulse is in the near IR or near UV spectral range. This is often accompanied by colors distributed in rings. Continuum generation was first discovered with ps pulses by Alfano and Shapiro [56] and has since been applied to numerous experiments. One of the most attractive fields of application is time-resolved spectroscopy, where the continuum pulse is used as an ultrafast spectral probe.

Spectral super broadening was observed in many different (preferably transparent) materials including liquids, solids, and gases. Essential processes contributing to the continuum generation are common to all. Figure 3.20 shows as an example a white light continuum generated in a solid with near IR fs pulses [57] and in gas with UV fs pulses [58]. As can be seen, the continuum does not have a "flat" uniform spectrum. A broad palette of fs laser sources is still desirable to create a continuum with a maximum energy concentration in any particular wavelength range. Continuum generation is a rather complex

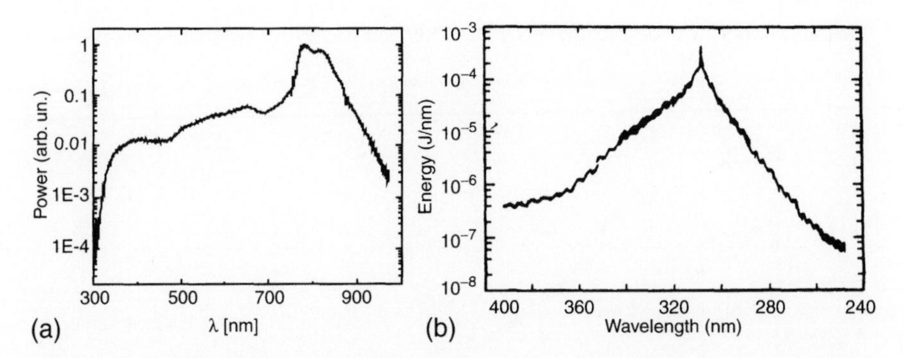

Figure 3.20 Femtosecond spectral super broadening. (a) In a 0.5-mm CaF_2 sample with 20 fs pulses at 800 nm. (Adapted from Zeller *et al.* [57].) (b) In a 60 cm long Ar cell (40 atm), the pump pulses (4 mJ, 308 nm, 160 fs) were focused with a 50 cm lens. (Adapted from Glownia *et al.* [58].)

issue which involves changes in the temporal and spatial beam characteristics. With fs pulses, the dominant process and the starting mechanism leading to spectral super broadening is SPM because of an intensity-dependent refractive index. However, a number of other nonlinear effects play a role as well. The interplay of self-focusing (see Section 5.6.6) and various nonlinear processes make the exact treatment of the continuum generation with short pulses extremely complex. Indeed, an inspection of Fig. 3.20 shows that the spectral features cannot be explained by the action of SPM alone. Other nonlinear effects that are likely to contribute are parametric four wave mixing and Raman scattering. The strong anti Stokes component visible in Fig. 3.20 is likely because of multiphoton excitation of the dielectric material followed by avalanche ionization [59]. The resulting electron plasma in the conduction band produces a fast rise of a negative refractive index component that can explain the dominant broadening toward the shorter wavelengths. As will be explained in the next section SPM is associated with self-focusing leading to extremely high intensities where the beam collapses. It is at this point where the continuum generating nonlinear processes are most effective.

The continuum pulse at the sample output is chirped. This is because of the time dependence of the nonlinear optical processes, which produces various spectral components at different parts within the pulse. Another origin of chirp is the propagation of the generated continuum from the point of beam collapse through the medium and all optical elements (including the path through air) leading to the detector. The chirp of the continuum was measured by fs frequency domain interferometry [60] and transient grating diffraction [57] for example. The dispersive processes account for most of the chirp of the continuum for pulses <50 fs. This is illustrated in Figure 3.21.

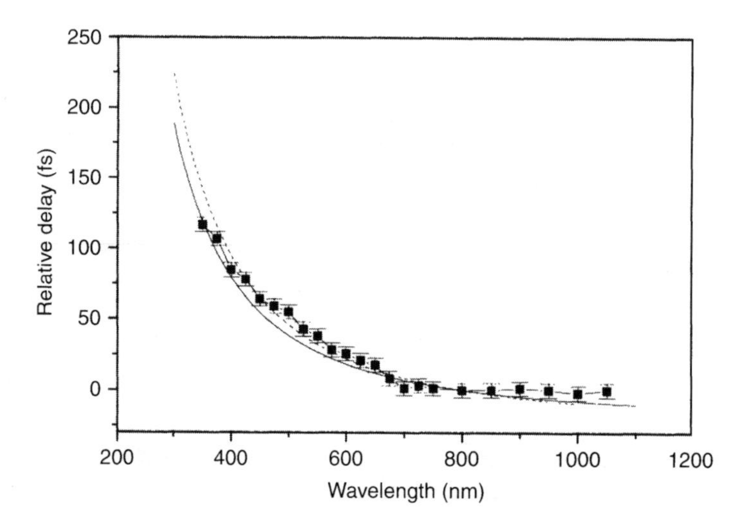

Figure 3.21 Spatio-temporal distribution of a fs white light continuum after propagation through 1 m air (640 Torr). Solid line: dispersion expected from the 1 m path through the atmosphere; dotted line: atmospheric path and 0.11 mm of CaF_2 (the distance from the beam collapse to the exit face of the crystal). (Adapted from Zeller *et al.* [57].)

Sending the continuum pulse through a proper element with GVD can result in pulse compression (Chapter 8). The "ideal" fs continuum pulse is thus a nearly bandwidth-limited fs pulse which is considerably shorter than the original pump pulse. Such continuum pulses enabled optical spectroscopy with time resolution better than 10 fs [61, 62].

Traditionally continua were generated in bulk materials with amplified fs pulses. Dispersion and the associated pulse broadening together with the finite propagation length until beam collapse limited the effective material length. This changed with the introduction of microstructured fibers [63] and tapered fibers [64]. These fibers either shift the wavelength of zero dispersion to regions were fs oscillator pulses were readily available and/or reduce the dispersion while increasing the nonlinearity. Because of the possible large propagation lengths at constant beam diameter (guided modes) the overall nonlinear interaction length can be greatly increased, which allowed the generation of continua with nJ and sub-nJ pulses directly from oscillators, and even using cw light (e.g., [65]). Using pulses directly from Ti:sapphire oscillators, continua covering almost two decades from $\approx$380 nm to 1600 nm were obtained.

An interesting application of the continuum is in metrology. Under certain circumstances, that will be discussed in Chapter 5, the continuum extends the regular mode structure of a mode-locked laser, making it possible to perform frequency mixing experiments over more than one octave in frequency.

3.8. SELF-FOCUSING

The nonlinearities of a medium affect both the temporal and spatial dependence of the electric field of the light. In the previous sections we have avoided this difficulty by assuming a uniform beam profile or neglecting nonlinear space–time coupling effects. However, any nonlinear interaction strong enough to affect the pulse temporal profile will also affect its transverse profile. One example is SHG, slightly off-exact phase matching condition (large nonlinear phase shifts), or at large conversion efficiencies. As sketched in the inset of Fig. 3.9, an initially Gaussian temporal profile will be depleted predominantly in the center, resulting in a flattened shape. The same interaction will also transform an initially Gaussian beam into a profile with a flat top.

In this section we will consider as an important example the problem of self-focusing. The intensity-dependent index of refraction causes an initially collimated beam to become focused in a medium with $n_2 > 0$. It is the same intensity dependence of the refractive index that causes SPM of a fs pulse, as we have seen in the previous section.

3.8.1. Critical Power

The action of a nonuniform intensity distribution across the beam profile on a nonlinear refractive index results in a transverse variation of the index of refraction, leading either to focusing or defocusing. Let us assume a cw beam with a Gaussian profile $I = I_0 \exp(-2r^2/w^2)$, and a positive $\bar{n}_2$, as is typical for a nonresonant electronic nonlinearity. The refractive index decreases monotonically from the beam center with increasing radial coordinate. One can define a "self-trapping" power $P_{cr,1}$ as the power for which the wavefront curvature (on axis) because of diffraction is exactly compensated by the change in the wavefront curvature because of the self-lensing over a small propagation distance Δz. We assume that the waist of the Gaussian beam is at the input boundary of the nonlinear medium ($z = 0$). Within the paraxial approximation, diffraction results in a spherical curvature of the wavefront at a small distance Δz from the beam waist:

$$\varphi_{\text{diff}}(\Delta z) = -\frac{k_\ell \Delta z}{2\rho_0^2} r^2, \tag{3.167}$$

where ρ_0 is the Rayleigh range. This result is obtained by approximating $1/R \approx \Delta z/\rho_0^2$ in Eq. (1.181). The action of the nonlinear refractive index results in a radial dependence of the phase after a propagation distance Δz:

$$\varphi_{sf}(r, \Delta z) = -\bar{n}_2 \frac{2\pi}{\lambda} \Delta z I_0 e^{-(2r^2/w_0^2)} \approx -\bar{n}_2 \frac{2\pi}{\lambda_\ell} \Delta z I_0 \left(1 - 2\frac{r^2}{w_0^2}\right), \quad (3.168)$$

where the last equation is an approximation of the wavefront near the beam center. This equation follows from Eq. (3.149) after replacing $n_2 \mathcal{E}_0^2$ by $\bar{n}_2 I(r)$. The input beam has the critical power $P_{cr,1} = \int 2\pi r I(r) dr$ when the radial parts of Eqs. (3.167) and (3.168) compensate each other:

$$P_{cr,1} = I_0 \frac{\pi w_0^2}{2} = \frac{\lambda^2}{8\pi n_0 \bar{n}_2} \quad (3.169)$$

where we have made use of $\rho_0 = \pi w_0^2 n_0/\lambda$. One says that the beam is self-trapped because neither diffraction nor focusing seems to occur. This value of critical power is also derived by Marburger [66] by noting that the propagation equation is equivalent to that describing a particle moving in a one-dimensional potential. The condition for which the potential is "attractive" (leads to focusing solutions) is $P \geq P_{cr,1}$. One can define another critical power $P_{cr,2}$ as the power for which the phase factor on-axis of the Gaussian beam, $\arctan(z/\rho_0)$, exactly compensates the nonlinear phase shift $\bar{n}_2 I_0$:

$$P_{cr,2} = \frac{\lambda_\ell^2}{4\pi n_0 \bar{n}_2}. \quad (3.170)$$

This second value of the critical power is also obtained by assuming that the beam profile remains Gaussian and deriving an expression for "scale factor" $f(z) = w(z)/w_0$ as a function of distance z [67, 68]. The function $f(z)$ reaches zero after a finite distance if the power exceeds the value $P_{cr,2}$.

Another common approach to defining a self trapping power is to approximate the radial beam profile by a flat top of diameter d [8]. The refractive index inside the tube of diameter d is $n = n_0 + \bar{n}_2 I_0$. The critical angle for total internal reflection, α, is determined by $\sin\alpha = n_0/(n_0 + \bar{n}_2 I_0)$. The beam is trapped inside the tube if the diffraction angle $\theta_d \approx 1.22\lambda/(2n_0 d)$ is equal to $\theta_{cr} = \pi/2 - \alpha$. From this condition and using the fact that $\bar{n}_2 I_0 \ll n_0$ one can derive the critical power $P_{cr,3} = I_0 \pi d^2/4$:

$$P_{cr,3} = \frac{(1.22)^2 \pi \lambda_\ell^2}{32 n_0 \bar{n}_2} \quad (3.171)$$

Table 3.6

Three approaches to defining a critical power for self-focusing. They differ by the coefficient *a* in Eq. (3.172).

Phase on-axis	Wavefront curv.	Waveguide
$\frac{2\pi}{\lambda_\ell}\bar{n}_2 I_0 z = \arctan\frac{z}{\rho_0}$	$\frac{2\pi}{\lambda_\ell}\bar{n}_2 I_0 = -\frac{k_\ell r^2}{2R}$	$\theta_{cr} = \theta_d$
$P_{cr} = \frac{1}{4\pi}\frac{\lambda_\ell^2}{n_0\bar{n}_2}$	$P_{cr} = \frac{1}{8\pi}\frac{\lambda_\ell^2}{n_0\bar{n}_2}$	$P_{cr} = \frac{(1.22)^2\pi}{32}\frac{\lambda_\ell^2}{n_0\bar{n}_2}$

These three approaches, summarized in Table 3.6, define a critical power of the form:

$$P_{cr} = a\frac{\lambda_\ell^2}{n_0\bar{n}_2} \tag{3.172}$$

The point of agreement between these different definition is the existence of a critical *power* rather than a critical *intensity*. This result is not surprising, because, for a given power, both the diffraction to be compensated and the lensing effect (nonlinear index) are inversely proportional to the beam diameter. It is not possible without numerical calculation to predict exactly the evolution of the beam at powers close to any of these critical powers. The Gaussian characteristic of the beam will be altered. Only numerical calculation can determine the fate of the beam over long propagation distances. Calculations made in steady state conditions have demonstrated the existence of a critical power $P_{cr} \approx 3.77P_{cr,1} \approx 1.03P_{cr,2}$, corresponding to the value of $a \approx 0.142$ in Eq. (3.172) [66].

A laser beam whose power exceeds the critical power reaches a focus after a finite propagation distance—the self-focusing length z_{SF}. Even for cw beams the exact treatment of beam propagation in an n_2 medium can only be done numerically. According to such a simulation

$$z_{SF} \approx \frac{0.183\rho_0}{\sqrt{(\sqrt{P/P_{cr}} - 0.852)^2 - 0.0219}} \tag{3.173}$$

see, for example, Marburger [66]. Here ρ_0 is the Rayleigh range of the original beam assumed to have a beam waist at the sample input.

With some restrictive assumptions one can derive an approximate self-focusing length analytically, which shows a similar structure as Eq. (3.173). We will sketch this approach at the end of Section 3.9.1.

3.8.2. The Nonlinear Schrödinger Equation

Let us consider the propagation of a laser beam along the direction z, in a medium characterized by a linear index of refraction n, a third-order nonlinear polarization and a linear loss–gain coefficient α [cf. Eq. (3.141)]. The nonlinear polarization

$$\tilde{P}_{NL}^{(3)} = \frac{3}{8}\epsilon_0 \chi^{(3)} |\tilde{\mathcal{E}}|^2 \tilde{\mathcal{E}} e^{i(\omega_\ell t - kz)}. \tag{3.174}$$

is to be substituted into the wave equation (1.67), with the field given by Eq (1.83). If we assume a steady state condition (no time dependence) the equation describing the spatial dependence of the electric field is:

$$\left[\frac{\partial}{\partial z} + \frac{i}{2k_\ell}\left(\frac{\partial^2}{\partial x^2} + \frac{\partial^2}{\partial y^2}\right) + i\frac{3\omega_\ell^2}{8c^2 k_\ell}\chi^{(3)}|\mathcal{E}|^2 - \frac{\alpha}{2}\right]\tilde{\mathcal{E}} = 0 \tag{3.175}$$

The prefactor in front of the nonlinear term can also be written as $3\omega_\ell^2 \chi^{(3)}/(8c^2 k_\ell) = n_2 k_\ell / n_0$. One recognizes in Eq. (3.175) a three-dimensional generalization of the nonlinear Schrödinger equation [69]:

$$i\frac{\partial \psi}{\partial z} - a\frac{\partial^2 \psi}{\partial x^2} - b|\psi|^2\psi + c\psi = 0. \tag{3.176}$$

The last term of Eq. (3.175) is a linear gain or absorption associated with the imaginary part of the linear index of refraction. In one dimension, these equations were shown by Zakharov and Shabat [70] to have steady state solutions labelled "solitons." These solitons correspond to a balance between the self-focusing and the diffraction. The physical reality is however more complex than can be included in the nonlinear Schrödinger Eq. (3.176). Indeed, once the nonlinearity exceeds the threshold to overcome diffraction, the beam collapses to a point (see next section). To obtain a dynamically stable solution in the transverse dimension, it is necessary to include a higher-order nonlinearity in the polarization to prevent this collapse, as will be demonstrated.

If we consider a short pulse propagating as a plane wave through an infinite medium with the nonlinear polarization of Eq. (3.174), the temporal evolution of the field is given by a similar nonlinear Schrödinger equation:

$$\left[\frac{\partial}{\partial z} - \frac{ik''}{2}\frac{\partial^2}{\partial t^2} + i\frac{n_2 k_\ell}{n_0}|\mathcal{E}|^2 - \frac{\alpha}{2}\right]\tilde{\mathcal{E}} = 0, \tag{3.177}$$

where the independent variables are now z and t. The soliton solution to Eq. (3.177) results from a balance between dispersion and SPM caused by the nonlinear polarization. The condition for the existence of a temporal soliton is that (anomalous) dispersion $\frac{ik''}{2}\frac{\partial^2}{\partial t^2}$ balances self-temporal lensing (positive SPM) $\frac{in_2k_\ell}{n_0}|\mathcal{E}|^2\tilde{\mathcal{E}}$. Normal dispersion and a negative SPM can also lead to soliton solutions [71].

3.9. BEAM TRAPPING AND FILAMENTS

Once the beam power is sufficient for self-focusing to overcome diffraction, the beam collapses to a point. In general, after the beam has collapsed, it diffracts. However, numerous experiments have shown self-guiding of high peak power infrared femtosecond pulses through the atmosphere [72–77]. Similar observations were made in the UV [78, 79]. After reaching the focus, the light appeared to trap itself in self-induced waveguides or "filaments" of the order of 100 μm diameter. Before addressing problems specific to ultrashort pulses we will discuss a steady state model to illustrate the possibility of beam self-trapping.

3.9.1. Beam Trapping

We start with the time-free wave equation

$$\left[\frac{\partial}{\partial z}+\frac{i}{2k_\ell}\left(\frac{\partial^2}{\partial x^2}+\frac{\partial^2}{\partial y^2}\right)\right]\tilde{\mathcal{E}}(x,y,z)=-ik_{NL}\tilde{\mathcal{E}}(x,y,z). \tag{3.178}$$

For the nonlinear propagation constant on the right-hand side we assume a nonlinear refractive index because of a Kerr nonlinearity and a contribution of next order.

$$k_{NL}=\frac{\omega_\ell}{c}\left(n_2|\tilde{\mathcal{E}}|^2+n_3|\tilde{\mathcal{E}}|^3\right) \tag{3.179}$$

A physical system that can give rise to a negative n_3 will be introduced later. A general solution to Eq. (3.178) is only possible by numerical means. To illustrate the possibility of beam trapping we will analyze the term on the right-hand side of the wave equation near in the vicinity of the beam center. Assuming a Gaussian beam profile, $\tilde{\mathcal{E}}=\mathcal{E}_0(w_0/w)\exp(-r^2/w^2))$, over a propagation distance Δz, the medium introduces a phase factor

$$\phi_{NL}(r)=k_{NL}\Delta z=\frac{\omega_\ell}{c}\left[n_2\frac{\mathcal{E}_0^2w_0^2}{w^2}e^{-2r^2/w^2}+n_3\frac{\mathcal{E}_0^3w_0^3}{w^3}e^{-3r^2/w^2}\right]\Delta z. \tag{3.180}$$

The curvature of the r dependent phase on-axis determines the focusing characteristics of a slice of thickness Δz. For $n_3 = 0$ this is the phase factor that was discussed in Section 3.8.1 and was found responsible for self-focusing. The curvature of the phase term in the vicinity of the beam center is

$$\frac{d^2}{dr^2}\phi_{NL}(r)\bigg|_{r\approx 0} = -Q\left[1 + \tilde{\mathcal{E}}_0 w_0 \left(\frac{n_3}{wn_2}\right)\right] \tag{3.181}$$

where $Q = 4\omega_\ell n_2 \tilde{\mathcal{E}}_0^2 w_0^2 \Delta z/(cw^4)$. For $n_2 > 0$ as is the case in most materials and $n_3 < 0$ we have the situation that the term in brackets can change sign depending on the ratio $n_3/(n_2 w)$ for given input beam $(\tilde{\mathcal{E}}_0 w_0)$. For $n_2 w > (<)\, n_3$ the material will act like a positive (negative) lens. A positive lens tends to decrease w on propagation until at some point the sign of the phase term reverses leading to negative lensing. This in turn increases w until the process is reversed again. This suggests the possibility of a periodically changing beam diameter (trapped beam) even if diffraction is included. The effect of the latter is that the phase curvature should have a certain positive value depending on w before the beam actually contracts. Similar beam trapping can be expected from contributions that are of order $m > 2$ if the sign of n_m is negative. The nonlinear refractive indices have their origin in nonlinear susceptibilities of order $m+1$. The Kerr effect, $\chi^{(3)}$ producing a nonlinear index n_2, is one example, which we discussed in detail previously.

Let us now briefly describe a physical system that can produce a negative n_3. Let us assume that the beam propagates through a gas that can be ionized by a three photon absorption. The free electrons (density N_e) can recombine with the positive ions and a steady state will be reached.

$$\frac{d}{dt}N_e = \sigma^{(3)}|\tilde{\mathcal{E}}|^6 N_0 - \beta_{ep}N_e^2 = 0 \tag{3.182}$$

Here $\sigma^{(3)}$ is a three photon absorption cross section, N_0 is the number density of atoms, and β_{ep} is the two-body recombination constant. From Eq. (3.182) we can obtain the steady state density of free electrons as a function of the laser field

$$N_{eq} = \sqrt{\frac{\sigma^{(3)}N_0}{\beta_{ep}}}|\tilde{\mathcal{E}}|^3. \tag{3.183}$$

The (small) change of the refractive index $\Delta\tilde{n}$ associated with the laser generated free electrons can be estimated with the Drude model:

$$\tilde{n}^2 = (1 + \Delta\tilde{n})^2 \approx 1 + 2\Delta\tilde{n} = 1 + \frac{\omega_p^2}{\omega_\ell^2}\left(1 - i\frac{\gamma}{\omega_\ell}\right). \tag{3.184}$$

where $\omega_p^2 = N_{eq}^2 e^2/(m\epsilon_0)$ is the plasma frequency and γ is the dephasing rate determined by collisions. With the equilibrium electron density from Eq. (3.183) the change of the refractive index becomes:

$$\Delta\tilde{n} = \sqrt{\frac{\sigma^{(3)}N_0}{\beta_{ep}}} \frac{e^2}{2\omega_\ell^2 m\epsilon_0} \left(1 - i\frac{\gamma}{\omega_\ell}\right) |\tilde{\mathcal{E}}|^3. \tag{3.185}$$

The index change is complex; the imaginary part accounts for free electron absorption, the real part determines the nonlinear index n_3 used in Eq. (3.179).

Numerical solutions of Eq. (3.178) with the nonlinear k-vector (3.179) and n_3 because of three photon ionization [80] show several features that are similar to those observed with fs filamented pulses. The power loss with distance plotted in Figure 3.22 is remarkably low, after an initial drop. The explanation for the low losses can be found in the plot of electron density and beam size $w(z)$ of Fig. 3.22. As the beam size decreases toward its minimum value w_{min}, the electron density reaches a peak, before falling back to an insignificant value as the beam expands. The beam waist w appears to "ricochet" at every period on the minimum value. In most cases, the loss mechanism is effective only for a small fraction of the period of oscillation of the beam. The loss decreases with distance because that fraction of period spent at the shortest dimension w decreases with distance. This phenomenon is the steady state analogue of "dynamic replenishment" observed for fs filaments in numerical simulations by Mlejnek *et al.* [81].

Equation (3.178) with an n_2 nonlinearity allows one to derive an approximate expression for the self-focusing length z_{SF}, see for example Shen [11]. To this

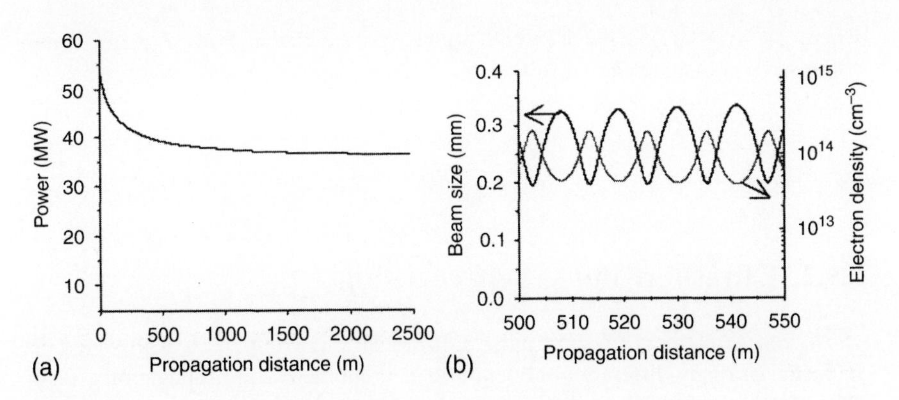

Figure 3.22 Numerical simulation of the nonlinear propagation of a 60 MW cw Gaussian UV (250 nm) beam in air. Left: plot of beam power versus distance. Right: beam size and electron density versus propagation distance [80].

end we insert the ansatz of a Gaussian beam profile

$$\tilde{\mathcal{E}} = \mathcal{E}_0[w_0/w(z)]\exp(-r^2/w(z)^2)$$

and assume that the Gaussian beam shape is maintained throughout the propagation (focusing). The beam waist w_0 is at the sample input ($z = 0$). Within the frame of paraxial optics we approximate the nonlinear term by

$$n_2|\tilde{\mathcal{E}}|^2 \approx n_2\mathcal{E}_0^2 \frac{w_0^2}{w^2(z)}\left[1 - 2\frac{r^2}{w^2(z)}\right]. \tag{3.186}$$

Sorting the powers of r and setting the prefactors to zero results in a second-order differential equation for the beam waist:

$$\frac{d^2}{dz^2}w(z) = -\frac{4}{k_\ell^2}\left(\frac{P}{P_{cr}} - 1\right)\frac{1}{w^3(z)}, \tag{3.187}$$

where P is the (constant) beam power and P_{cr} is the critical power defined as $P_{cr,1}$ in Eq. (3.169). The solution to this differential equation is

$$w^2(z) = w_0^2 - \frac{w_0^2}{\rho_0^2}\left(\frac{P}{P_{cr}} - 1\right)z^2 \tag{3.188}$$

where $\rho_0 = kw_0^2/2$ is the Rayleigh range. Provided that $P > P_{cr}$, the beam collapses at a distance equal to the self-focusing length $z = z_{SF}$, where

$$z_{SF} = \frac{\rho_0}{\sqrt{P/P_{cr} - 1}}. \tag{3.189}$$

3.9.2. Ultrashort Pulse Self-Focusing

An exact treatment of short pulse self-focusing is complex as it involves the inclusion of many different nonlinear effects combined with propagation effects. The general approach is to start with Eq. (3.97) and specify the nonlinear polarization of the material in which the pulse propagates. For example, by specifying the nonlinear polarization as the Kerr effect and the Raman effect, one arrives at a generalized nonlinear Schrödinger equation often used to describe pulse propagation in fibers and in bulk materials leading to filaments:

$$\begin{aligned}\frac{\partial \tilde{\mathcal{E}}}{\partial z} &= \frac{\alpha_0}{2}\tilde{\mathcal{E}} + i\hat{D} - \frac{i}{2k_\ell}\left(1 - \frac{i}{\omega_\ell}\frac{\partial}{\partial_z}\right)^{-1} \nabla_\perp^2 \tilde{\mathcal{E}} \\ &- i\gamma(1-f_R)\left(1 - \frac{i}{\omega_\ell}\frac{\partial}{\partial_t}\right)|\tilde{\mathcal{E}}|^2\tilde{\mathcal{E}} \\ &+ i\gamma f_R\left[1 - \frac{i}{\omega_\ell}\frac{\partial}{\partial t}\right]\left\{\tilde{\mathcal{E}}\int_0^\infty h_R(t')|\tilde{\mathcal{E}}(t-t')|^2 dt'\right\}\end{aligned} \tag{3.190}$$

where f_R indicates the fraction of the nonlinearity that is a delayed Raman contribution, as opposed to the "instantaneous" electronic contribution. $\gamma = \pi\bar{n}_2 n_0/(377\lambda)$ is the effective nonlinearity. The Raman response function is given by $h_R(t)$. When applied to fibers [48], the transverse Laplacian term is no longer relevant.

Numerical solutions of Eq. (3.190) exist for different propagation problems (material parameters), see for example references [82–88]. The results vary because of the complexity of the equation and the fact that some of the material parameters in Eq. (3.190) are not well known. Comparison of the numerical results with experiments is hampered by the difficulties encountered when measuring the parameters of the propagating pulse and the filaments. We will discuss some of the properties of self-focusing associated with short pulses and leading to filaments in more detail in Chapter 13.

3.10. PROBLEMS

1. Verify the temporal behavior of the polarization and occupation number as given in Eqs. (3.12) and (3.13).
2. Estimate the absolute value of the refractive index change at a frequency off resonance by $\Delta\omega_F/2$ which is caused by saturation of a homogeneously broadened, absorbing transition with Lorentzian profile. The change in the absorption at resonance at the pulse center was measured to be 50%. The pulse duration τ_p is much larger than the energy relaxation time T_1. The small signal absorption coefficient is α_0.
3. Show that the nonlinear refractive index coefficient is related to the third-order susceptibility through $n_2 = 3\chi^{(3)}/(8n_0)$ [cf. Eq. (3.142)].
4. Starting from the density matrix equations of a two-level system, find an approximate expression for n_2 which is valid for ω_ℓ far from a single

resonance. In particular, comment on the statement following Eq. (3.151) that n_2 is mostly positive.

5. Chirp enhancement through parametric interaction provides an interesting possibility to compress pulses. To illustrate this, let us consider the following simplified model. An initially unchirped Gaussian pulse is sent through a group velocity dispersive element, e.g., a block of glass of length L and GVD parameter k'', leading to linearly chirped output pulses (chirp parameter a). This pulse now serves as a pump in a parametric process producing a (GAUSSIAN) output pulse of the same duration but with an increased chirp parameter (of opposite sign), $a' = -Ra$. A second piece of glass of suitable length can be used to compress this pulse. Calculate the total compression factor in terms of L and R. Note that this configuration would allow us to control the achievable compression factor simply by changing the GVD (e.g., L) of the first linear element that controls the initial chirp.
6. Consider the situation of Section 3.6.4 relating to the interaction of two counter-propagating pulses in a nonlinear medium characterized by a third-order nonlinear susceptibility $\chi^{(3)}$. The purpose of this problem is to compare the counter-propagating interaction in the case of the homogeneous medium with the case of a set of Multiple Quantum Wells (MQWs) separated by a wavelength, as sketched in Figure 3.23. For the homogeneous medium, the nonlinear susceptibility is uniform and equal to $\chi^{(3)}$. The stratified medium is assumed to be made of N (infinitely thin) quantum wells separated by half a wavelength, each quantum well having

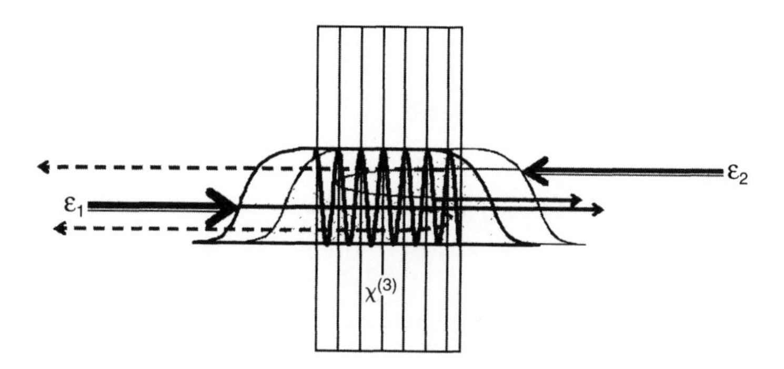

Figure 3.23 Geometry of counter-propagating wave interaction in a medium with a third-order nonlinear susceptibility concentrated in quantum wells. We assume the spacing between quantum wells to have a negligible nonlinear susceptibility. The quantum wells are located at the antinodes of the standing wave field, resulting in a maximum interaction between the nonlinear medium and the light field.

a susceptibility $\chi^{(3)}\delta(kz - j\pi)/N$. The medium susceptibility $\chi^{(NL)}$ can thus be represented by:

$$\chi^{(NL)}(z) = \sum_{j=1}^{N} \frac{\chi^{(3)}}{N}\delta(kz - j\pi) \tag{3.191}$$

Insert this susceptibility in Eq. (3.159), and average (integrate) over the thickness of the medium, to find an average third-order polarization. Show that the coupling term in this polarization, which in an homogeneous medium would average to zero, has now a contribution of the same order as the other terms.

It is interesting to contrast the result from the MQW with that of the homogeneous medium. First, the nonlinear polarization is larger in the case of the MQW sample: if the fields are equal, $\mathcal{P}_1^{(3)} = 4\epsilon_0\chi^{(3)}|\mathcal{E}_1^2|\mathcal{E}_1$ to be compared with $\mathcal{P}_1^{(3)} = 3\epsilon_0\chi^{(3)}|\mathcal{E}_1^2|\mathcal{E}_1$ in the homogeneous case. Second, the "nonreciprocity" because of cross-phase modulation, which appears in the homogeneous case, is not present in the MQW geometry. Show that, for the change in index, instead of $\Delta n_{\text{nl}} = n_2(\tilde{\mathcal{E}}_1^2 + 2\tilde{\mathcal{E}}_2^2)$ for the forward beam, in the homogeneous case—a consequence of Eq. (3.163)—we have

$$\Delta n_{\text{nl}} = n_2\left(\tilde{\mathcal{E}}_1^2 + \tilde{\mathcal{E}}_2^2 + 2\mathcal{E}_1\mathcal{E}_2\right) = \frac{3\chi^{(3)}}{8n_0}\left(\tilde{\mathcal{E}}_1^2 + \tilde{\mathcal{E}}_2^2 + 2\mathcal{E}_1\mathcal{E}_2\right), \tag{3.192}$$

an expression that is the same for both directions of propagation. This is basically a result from the fact that the emission of layers of dipoles (spaced by a wavelength) in the forward and backward directions is equal [89].

BIBLIOGRAPHY

[1] O. Hittmair. *Quantum Mechanics*. Thiemig, Munich, Germany, 1972.

[2] M. Schubert and B. Wilhelmi. *Nonlinear Optics and Quantum Electronics*. John Wiley & Sons, New York, 1978.

[3] L. M. Frantz and J. S. Nodvik. Theory of pulse propagation in a laser amplifier. *Journal of Applied Physics*, 34:2346–2349, 1963.

[4] D. Kuehlke, W. Rudolph, and B. Wilhelmi. Calculation of the colliding pulse mode locking in cw dye ring lasers. *IEEE Journal of Quantum Electronics*, QE-19:526–533, 1983.

[5] V. Petrov, W. Rudolph, and B. Wilhelmi. Chirping of femtosecond light pulses passing through a four-level absorber. *Optics Communications*, 64:398–402, 1987.

[6] V. Petrov, W. Rudolph, and B. Wilhelmi. Chirping of ultrashort light pulses by coherent interaction with samples with an inhomogeneously broadened line. *Soviet Journal of Quantum Electronics*, 19:1095–1099, 1988.

[7] W. Rudolph. Calculation of pulse shaping in saturable media with consideration of phase memory. *Optics and Quantum Electronics*, 16:541–550, 1984.

[8] W. Boyd. *Nonlinear Optics*. Academic Press, New York, 1991.

[9] N. Bloembergen. *Nonlinear Optics*. World Scientific Publishing Company, Singapore, 1996.

[10] T. Brabec and F. Krausz. Nonlinear optical pulse propagation in the single-cycle regime. *Physical Review Letters*, 78:3282–3284, 1997.

[11] Y. R. Shen. *The Principles of Nonlinear Optics*. John Wiley & Sons, New York, 1984.

[12] R. C. Miller. Second harmonic generation with a broadband optical maser. *Physics Letters*, A26:177–178, 1968.

[13] J. Comly and E. Garmire. Second harmonic generation from short light pulses. *Applied Physics Letters*, 12:7–9, 1968.

[14] S. A. Akhmanov, A. S. Chirkin, K. V. Drabovich, A. I. Kovrigin, R. V. Khokhlov, and A. P. Sukhorukov. Nonstationary nonlinear optical effects and ultrashort light pulse formation. *IEEE Journal of Quantum Electronics*, 4:598–605, 1968.

[15] W. H. Glenn. Second harmonic generation by ps optical pulses. *IEEE Journal of Quantum Electronics*, QE-5:281–290, 1969.

[16] Y. Wang and R. Dragila. Efficient conversion of picosecond laser pulses into second-harmonic frequency using group-velocity-dispersion. *Physical Review A*, 41:5645–5649, 1990.

[17] F. Zernicke. Refractive indices of ADP and KDP between 200 nm and 1500 nm. *Journal of the Optical Society of America*, 54:1215–1220, 1964.

[18] M. Choy and R. L. Byer. Accurate second-order susceptibility measurements of visible and infrared optical pulses. *Physical Review*, B14:1693–1706, 1976.

[19] K. Kato. Second harmonic generation to 2058 å in β-bariumborate. *IEEE Journal of Quantum Electronics*, QE-22:1013–1014, 1986.

[20] Y. N. Karamzin and A. P. Sukhorukov. Limitation on the efficiency of frequency doublers of picosecond light pulses. *Soviet Journal of Quantum Electronics*, 5:496–500, 1975.

[21] R. C. Eckardt and J. Reintjes. Phase matching limitations of high efficiency second harmonic generation. *IEEE Journal of Quantum Electronics*, QE-20:1178–1187, 1984.

[22] N. C. Kothari and X. Carlotti. Transient second harmonic generation: Influence of effective group velocity dispersion. *Journal of the Optical Society of America*, 5:756–764, 1988.

[23] D. Kuehlke and U. Herpers. Limitation of the second harmonic conversion of intense femtosecond pulses. *Optics Communications*, 69:75–80, 1988.

[24] L. A. Ostrovski. Self-action of light in crystals. *Translation of Pis'ma V Zhurnal Eksperimentalnoi i Teoreticheskoi Fiziki*, 5:272–274, 1967.

[25] G. Szabo and Z. Bor. Broadband frequency doubler for femtosecond light pulses. *Applied Physics B*, B50:51–54, 1990.

[26] O. E. Martinez. Achromatic phase matching for second harmonic generation of femtosecond light pulses. *IEEE Journal of Quantum Electronics*, QE-25:2464–2468, 1989.

[27] T. Hofmann, K. Mossavi, F. K. Tittel, and G. Szabo. Spectrally compensated sum-frequency mixing scheme for generation of broadband radiation at 193 nm. *Optics Letters*, 17:1691–1693, 1992.

[28] S. A. Akhmanov, A. P. Sukhorukov, and A. S. Chirkin. Nonstationary phenomena and space-time analogy in nonlinear optics. *Soviet Physics JETP*, 28:748–757, 1969.

[29] A. Stabinis, G. Valiulis, and E. A. Ibragimov. Effective sum frequency pulse compression in nonlinear crystals. *Optics Communication*, 86:301–306, 1991.

[30] Y. Wang and B. L. Davies. Frequency-doubling pulse compressor for picosecond high power neodymium laser pulses. *Optics Letters*, 17:1459–1461, 1992.

[31] P. Heinz, A. Laubereau, A. Dubietis, and A. Piskarskas. Fiberless two-step parametric compression of sub-picosecond laser pulses. *Lithuanian Physical Review*, 33:314–317, 1993.

[32] A. Umbrasas, J. C. Diels, G. Valiulis, J. Jacob, and A. Piskarskas. Generation of femtosecond pulses through second harmonic compression of the output of a Nd:YAG laser. *Optics Letters*, 20:2228–2230, 1995.

[33] A. Umbrasas, J. C. Diels, J. Jacob, and A. Piskarskas. Parametric oscillation and compression in KTP crystals. *Optics Letters*, 19:1753–1755, 1994.

[34] J. D. V. Khaydarov, J. H. Andrews, and K. D. Singer. Pulse compression in a synchronously pumped optical parametric oscillator from group-velocity mismatch. *Optics Letters*, 19:831–833, 1994.

[35] J. D. V. Khaydarov, J. H. Andrews, and K. D. Singer. 20-fold pulse compression in a synchronously pumped optical parametric oscillator. *Applied Physics Letters*, 65:1614–1616, 1994.

[36] L. Lefort, S. D. Butterworth, Y. P. Svirko, K. Puech, D. C. Hanna, and D. H. Jundt. Generation of fs pulses from order of magnitude pulse compression in a cw synchronously pumped optical parametric oscillator. In *CLEO '98*, OSA, San Francisco, CA, 1998.

[37] J. Biegert, V. Kubecek, and J.-C. Diels. Pulse compression: Type II second harmonic pulse compression. In J. G. Ed., Webster, *Encyclopedia of Electrical and Electronics Engineering (#17)*, volume 17, Wiley's, New York, NY, IEEE, 1998, pp. 446–454.

[38] J. Biegert and J. C. Diels. Compression of pulses of a few optic cycles through harmonic conversion. *Journal of Optical Society B*, 18:1218–1226, 2001.

[39] T. R. Zhang, H. R. Choo, and M. C. Downer. Phase and group velocity matching for second harmonic generation of femtosecond pulses. *Applied Optics*, 29:3927–3933, 1990.

[40] P. DiTrapani, A. Andreoni, C. Solcia, P. Foggi, R. Danielius, A Dubietis, and A. Piskarskas. Matching of group velocities in three-wave parametric interaction with femtosecond pulses and application to traveling-wave generators. *Journal of the Optical Society of America B*, 12:2237–2244, 1995.

[41] A. Mokhtari, A. Chebira, and J. Chesnoy. Subpicosecond fluorescence dynamics of dye molecules. *Journal of the Optical Society of America B*, B-7:1551–1557, 1990.

[42] T. Elsaesser and M. C. Nuss. Femtosecond pulses in the mid-infrared generated by down conversion of a traveling wave dye laser. *Optics Letters*, 16:411–413, 1991.

[43] T. M. Jedju and L. Rothberg. Tunable femtosecond radiation in the mid-infrared for time resolved absorption in semiconductors. *Applied Optics*, 27:615–618, 1988.

[44] R. Laenen, H. Graener, and A. Laubereau. Broadly tunable femtosecond pulses generated by optical parametric oscillation. *Optics Letters*, 15:971–973, 1990.

[45] D. C. Edelstein, E. S. Wachman, and C. L. Tang. Broadly tunable high repetition rate femtosecond optical parametric oscillator. *Applied Physics Letters*, 54:1728–1730, 1989.

[46] X. Meng, R. Quintero, and J. C. Diels. Intracavity pumped optical parametric bidirectional ring laser as a differential interferometer. *Optics Communications*, 233:167–172, 2004.

[47] A. Jankauskas, A. Piskarskas, D. Podenas, A. Stabinis, and A. Umbrasas. New nonlinear optical methods of powerful femtosecond pulse generation and dynamic interferometry. In Z. Rudzikas, A. Piskarskas, and R. Baltramiejunas, Eds. *Proceedings of the V international symposium on Ultrafast Phenomena in Spectroscopy*, Vilnius, A World Scientific, 1987, pp. 22–32.

[48] G. P. Agrawal. *Nonlinear Fiber Optics*. Academic Press, Boston, MA, 1995.

[49] R. R. Alfano, Q. X. Li, T. Jimbo, J. T. Manassah, and P. P. Ho. Induced spectral broadening of a weak ps pulse in glass produced by an intense ps pulse. *Optics Letters*, 14:626–628, 1986.

[50] W. Rudolph, J. Krueger, P. Heist, and W. Wilhelmi. Ultrafast wavelength shift of light induced by light. In *ICO-15*, p.1319. SPIE, Bellingham, WA, 1990.

[51] D. Anderson and M. Lisak. Nonlinear asymmetric self phase modulation and self steepening of pulses in long waveguides. *Physical Review A*, 27:1393–1398, 1983.

[52] W. H. Knox, R. L. Fork, M. C. Downer, R. H. Stolen, and C. V. Shank. Optical pulse compression to 8 fs at a 5 kHz repetition rate. *Applied Physics Letters*, 46:1120–1121, 1985.

[53] J. E. Rothenberg and D. Grischkowsky. Observation of the formation of an optical intensity shock and wave breaking in the nonlinear propagation of pulses in optical fibers. *Phys. Rev. Lett.*, 62:531–533, 1989.

[54] R. R. Alfano, Ed., *The Supercontinuum Laser Source*. Springer, New York, 1989.

[55] A. Zheltikov, Ed., Special issue on Supercontinuum Generation. *Applied Physics B*, 77:143–376, 2003.

[56] R. R. Alfano and S. L. Shapiro. Observation of self-phase modulation and small-scale filaments in crystals and glasses. *Physics Review Letters*, 24:592–594, 1970.

[57] J. Zeller, J. Jasapara, W. Rudolph, and M. Sheik-Bahae. Spectro-temporal characterization of a fs white light continuum by transient-grating diffraction. *Optics Communication*, 185:133–137, 2000.

[58] J. H. Glownia, J. Misewich, and P. P. Sorokin. Ultrafast ultraviolet pump probe apparatus. *Journal of the Optical Society of America B*, B3:1573–1579, 1986.

[59] A. Brodeur and S. L. Chin. Ultrafast white-light continuum generation and self-focusing in transparent condensed media. *Journal of the Optical Society of America B*, 16:637–650, 2000.

[60] E. Tokunaga, Q. Terasaki, and T. Kobayashi. Induced phase modulation of chirped continuum pulses studied with a femtosecond frequency-domain interferometer. *Optics Letters*, 18:370–372, 1992.

[61] W. H. Knox, M. C. Downer, R. L. Fork, and C. V. Shank. Amplifier femtosecond optical pulses and continuum generation at 5 kHz repetition rate. *Optics Letters*, 9:552–554, 1984.

[62] R. L. Fork, C. H. Cruz, P. C. Becker, and C. V. Shank. Compression of optical pulses to six femtoseconds by using cubic phase compensation. *Optics Letters*, 12:483–485, 1987.

[63] J. K. Ranka, R. S. Windeler, and A. J. Stentz. Visible continuum generation in air-silica microstructure optical fibers with anomalous dispersion at 800 nm. *Optics Letters*, 25:25–27, 2000.

[64] T. A. Birks, W. J. Waldsorth, and P. S. Russel. Supercontinuum generation in tapered fibers. *Optics Letters*, 25:1415–1417, 2000.

[65] J. W. Nicholson, A. K. Abeeluck, C. Headley, M. F. Yan, and C. G. Jorgensen. Pulsed and continuous-wave supercontinuum generation in highly nonlinear, dispersion-shifted fibers. *Applied Physics B*, 77:211–218, 2003.

[66] J. H. Marburger. In J. H. Sanders and S. Stendholm, Eds., *Progress in Quantum Electronics*, volume 4, 1977, p. 35.

[67] S. A. Akhmanov, A. P. Sukhorukov, and R. V. Khokhlov. Self focusing and self trapping of intense light beams in a nonlinear medium. *Translation of Zhurnal Eksperimentalnoi i Teoreticheskoi Fizik*, 23:1025–1033, 1966.

[68] S. A. Akhmanov, A. P. Sukhorukov, and R. V. Khokhlov. Development of an optical waveguide in the propagation of light in a nonlinear medium. *Translation of Zhurnal Eksperimentalnoi i Teoreticheskoi Fizik*, 24:198–201, 1967.

[69] E. Infeld and G. Rowlands. *Nonlinear waves, solitons and chaos*. Cambridge University Press, New York, 1990.

[70] V. E. Zakharov and A. B. Shabat. Exact theory of two-dimensional self-focusing and one-dimensional self-modulation of waves in nonlinear media. *Translation of Zhurnal Eksperimentalnoi i Teoreticheskoi Fizik*, 34:62–69, 1972.

[71] J.-C. Diels, W. Dietel, J. J. Fontaine, W. Rudolph, and B. Wilhelmi. Analysis of a mode-locked ring laser: Chirped-solitary-pulse solutions. *Journal of the Optical Society of America B*, 2:680–686, 1985.

[72] A. Braun, G. Korn, X. Liu, D. Du, J. Squier, and G. Mourou. Self-channeling of high-peak-power fs laser pulses in air. *Optics Letters*, 20:73–75, 1994.

[73] E. T. J. Nibbering, P. F. Curley, G. Grillon, B. S. Prade, M. A. Franco, F. Salin, and A. Mysyrowicz. Conical emission from self-guided femtosecond pulses in air. *Optics Letters*, 21:62–64, 1996.

[74] A. Braun, G. Korn, X. Liu, D. Du, J. Squier, and G. Mourou. Self-channeling of high-peak-power femtosecond laser pulses in air. *Optics Letters*, 20:73–75, 1995.

[75] L. Woeste, S. Wedeking, J. Wille, P. Rairouis, B. Stein, S. Nikolov, C. Werner, S. Niedermeier, F. Ronneberger, H. Schillinger, and R. Sauerbrey. Femtosecond atmospheric lamp. *Laser und Optoelektronic*, 29:51–53, 1997.

[76] B. La Fontaine, F. Vidal, Z. Jiang, C. Y. Chien, D. Comtois, A. Desparois, T. W. Johnston, J.-C. Kieffer, and H. Pepin. Filamentation of ultrashort pulse laser beams resulting from their propagation over long distances in air. *Physics of Plasmas*, 6:1615–1621, 1999.

[77] P. Rairoux, H. Schillinger, S. Niedermeier, M. Rodriguez, F. Ronneberger, R. Sauerbrey, B. Stein, D. Waite, C. Wedeking, H. Wille, L. Woeste, and C. Ziener. Remote sensing of the atmosphere using ultrashort laser pulses. *Applied Physics B*, 71:573–580, 2000.

[78] X. M. Zhao, P. Rambo, and J.-C. Diels. Filamentation of femtosecond UV pulses in air. In *QELS, 1995*, volume 16, OSA, Baltimore, MD, 1995, p. 178.

[79] J. Schwarz, P. K. Rambo, J.-C. Diels, M. Kolesik, E. Wright, and J. V. Moloney. UV filamentation in air. *Optics Communications*, 180:383–390, 2000.

[80] J. Schwarz and J.-C. Diels. Analytical solution for UV filaments. *Physical Review A*, 65:013806-1–013806-10, 2001.

[81] M. Mlejnek, E. M. Wright, and J. V. Moloney. Dynamic spatial replenishment of femtosecond pulse propagating in air. *Optics Letters*, 23:382–384, 1998.

[82] G. Fibich and G. C. Papanicolaou. Self-focusing in the perturbed and unperturbed nonlinear Schrödinger equation in critical dimension. *SIAM Journal of Applied Mathematics*, 60:183–240, 1998.

[83] G. Fibich and A. L. Gaeta. Critical power for self-focusing in bulk media and in a hollow waveguides. *Optics Letters*, 29:1772–1774, 2004.

[84] A. L. Gaeta. Catastrophic collapse of ultrashort pulses. *Physical Review Letters*, 84:3582–3585, 2000.

[85] N. Aközbek, C. M. Bowden, A. Talebpour, and S. L. Chin. Femtosecond pulse propagation in air: Variational analysis. *Physical Review B*, 61:4540–4549, 2000.

[86] G. Fibich, S. Eisenmann, B. Ilan, and A. Zigler. Control of multiple filamentation in air. *Optics Letters*, 25:335–337, 2000.

[87] A. Couairon and L. Bergé. Light filaments in air for ultraviolet and infrared wavelengths. *Physical Review Letters*, 88:13503-1–13503-4, 2002.

[88] P. Bennett and A. B. Aceves. Parallel numerical integration of Maxwell's full-vector equations in nonlinear focusing media. *Physica D*, 184:352–375, 2003.

[89] Paul Pulaski and J.-C. Diels. Demonstration of a nonreciprocal multiple-quantum-well structure leading to unidirectional operation of a ring laser. In *CLEO'98*, Optical Society of America, San Francisco, CA, 1998.

4

Coherent Phenomena

This chapter reviews some aspects of coherent interactions between light and matter. By coherent interaction, it is meant that the exciting pulse is shorter than the phase memory time of the excited medium. Femtosecond pulses have made quite an impact in this field (some selected examples of application are presented in Chapters 10 and 11), because the phase-relaxation time of absorbing transitions in condensed matter is generally in the fs range.

Experiments in this field have somewhat contradictory requirements: a source of high coherence, but ultrashort duration. The experimentalist has to walk a tightrope to meet the coherence requirements for pulses sometimes shorter than 10 fs.

To be classified in the field of coherent interactions, the result of a particular experiment should depend on the coherence of the source. This brings us to the question of what constitutes an *incoherent* source? To make a radiation source incoherent, should random phase fluctuations be applied in the time domain or in the frequency domain? Is the temporal coherence always linked to the spatial coherence? These questions are discussed in the first few sections of this chapter.

Coherent interactions with near resonant two-level systems constitute the nucleus of this chapter (Section 3). Section 4 extends the theory of coherent interactions to multilevel molecular or atomic systems.

4.1. FROM COHERENT TO INCOHERENT INTERACTIONS

The highest level of coherence is easy to define: monochromatic radiation with a δ-function spectrum, emitted from a point source, has an infinite coherence time and coherence length. Most interesting phenomena of coherent interaction, however, occur with broad bandwidth radiation, involving temporal variations of either the field amplitude or its phase, which are faster than the dephasing time of the system of atoms or molecules being excited. Therefore, a short pulse will be defined as coherent if its bandwidth–duration product satisfies the minimum uncertainty relation for that particular pulse shape, as defined in Table 1.1. The lack of coherence can be deterministic. Such is the case for chirped pulses, as discussed in Chapter 1. It can also be of aleatory origin. In the latter case, incoherence can be introduced through either

- statistical fluctuations in the time domain; or
- statistical fluctuations in the frequency domain.

In the first approach, given a fixed amount of time during which the radiation is applied, one can introduce incoherence through statistical (temporal) fluctuations of the field *amplitude or phase*. The bandwidth of the radiation is increased as a result of the decrease in coherence. In the second approach, the statistical fluctuations are introduced in the amplitude and phase of the field in the frequency domain. The bandwidth of the radiation is fixed. The distinction between incoherence in the time and frequency domains can best be understood through the experimental implementation of these two concepts. Coherence is a parameter in the interaction of radiation with matter. The influence of coherence on a particular process will be different if incoherence is defined as statistical fluctuations in time or frequency. To illustrate this concept, let us look at the influence of coherence on multiphoton photoionization of atoms. To study the influence of coherence on two- and three-photon resonant ionization of Cs and other alkali, Lecompte *et al.* [1] have adjusted the bandwidth of the cavity of a Q-switched Nd:glass laser used to excite a resonant transition. By changing the number of modes let to oscillate simultaneously, the temporal structure of the pulse is also modified. Because the relative phase and amplitude of the oscillating modes are random (the laser is Q-switched, not mode-locked), one has introduced statistical fluctuations in the Q-switched pulse as illustrated in Figure 4.1(a). For the particular interaction being investigated, it was shown in De Bethune [2] that, for all practical purposes, the radiation can be considered to be incoherent when 20 or more modes are allowed to oscillate simultaneously. Clearly, the bandwidth of the radiation is modified.

Multiphoton interactions are typically intensity-dependent processes. It is not surprising therefore that an increase in multiphoton interaction is observed, because the fluctuations result in larger peak intensities, and hence an increase in nonlinear transition rates. Such an approach is appropriate for the study

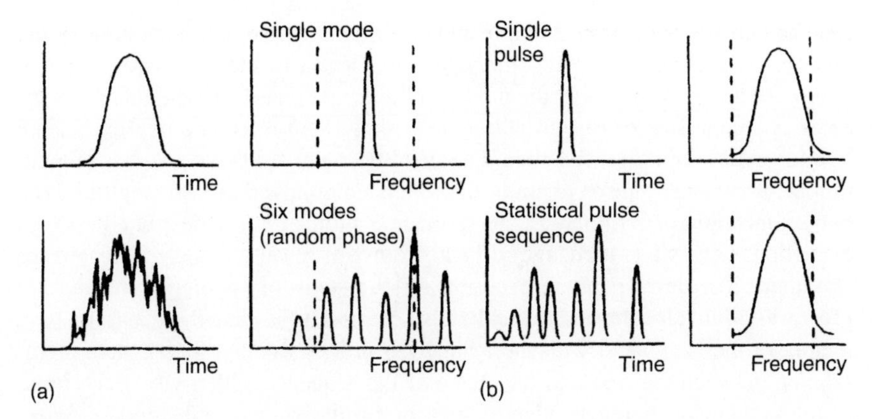

Figure 4.1 Illustration of two different ways to act on the coherence of an optical signal. On the left sides of (a) and (b), the time domain representation is shown. On the right sides, the frequency representation is depicted. Two material resonances (single photon or multiphoton) are represented by dashed lines. (a) Modifying coherence by changing the number of oscillating modes of a Q-switched laser. On top, a single oscillating mode, shown off resonance. As the number of oscillating modes, of random amplitude and phase, is increased, resonances can occur (bottom). The corresponding situation in the time domain (left) shows increasing amplitude and phase fluctuations within the pulse duration. The radiation is never completely random, even in the time domain, because the regularity of mode spacing is seen as a repeating pattern of fluctuations. (b) Modifying coherence in the frequency domain by creating a random pulse sequence in the time domain. On the left side, the time domain shows an increasing number of ultrashort pulses of random phase. On the right side, the Fourier amplitude remains within the single pulse bandwidth. The two Fourier transforms have different amplitude and/or phase fluctuations (not shown) under the same envelope. In case (b), with increasing incoherence, the detuning from resonance with the material transitions (dashed lines) is not modified.

of nonresonant interactions. However, if the multiphoton process has sharp resonances, such as indicated by the dashed lines in Fig. 4.1(a), then the problem of finding the influence of coherence on the resonance curves is ill-defined because the resonance and coherence conditions are not independent. Indeed, in the presence of a sharp resonance, it is not possible to modify independently the parameter "detuning" and the parameter coherence [Fig. 4.1(a)]. If the bandwidth of the laser is increased, to allow more modes to oscillate simultaneously, the resonance condition may be modified (if one of the modes coincides with one resonance).

For the study of the influence of source incoherence on near resonant light–matter interaction, it would be desirable to have the radiation confined in a fixed bandwidth. The experimental procedure consists of exciting the sample with a random sequence of pulses [Fig. 4.1(b)]. Each single pulse has the same envelope in the frequency domain, but a different phase corresponding to the random time of arrival in the time domain. Thus, in the frequency domain, within the spectral envelope, the field amplitude and phase become random variables. A comparison of Figs. 4.1(a) and (b) shows that these two approaches are Fourier transforms of each other. The condition of validity of the use of multimode Q-switched pulses to study the influence of coherence on (multiphoton) photo ionization is that there be no resonance within the broadest bandwidth of excitation. Otherwise, a change in bandwidth will mean a change in resonance condition. The corresponding criterion for the study of coherence in resonant processes with ultrashort statistical pulse sequences is that there has to be a memory time associated with the resonance(s), long enough to establish a correlation between the first and last pulse of the sequence. Otherwise, each pulse of the sequence acts independently, and the result of a two pulse measurement is simply the sum of the results for each individual pulse. This "memory time" is the phase-relaxation time of the resonance process, such as was introduced for single photon processes in the previous chapter. As an example, we refer to Diels and Stone [3] for a study of the influence of coherence on multiphoton ionization.

The sequence of pulses represented in Fig. 4.1(b) cannot be defined as random out of context. The set of numbers representing the relative phases and delays between pulses are chosen out of a particular statistical distribution, of which the parameters define the coherence of the source. A large number of experiments must be averaged to establish the response of a process to the light statistics chosen. Sets of relative phases and delays can be predetermined to match any selected statistical distribution. There are two possible techniques for generating such random pulse sequences. The first one involves simply splitting and delaying pulses with polarizers and mirrors. The second one consists of filtering the Fourier transform of a single pulse.

Should white light be seen as an ensemble of monochromatic sources statistically distributed in amplitude and phase in the frequency domain or as a random sequence of ultrashort pulses? The question is not purely academic. One or the other aspect dominates, depending on the particular experimental situation. It is the second definition that generally applies to the conditions of coherent interactions discussed in this chapter.

Light is defined as coherent in time if there is a well-defined phase relation between the radiation at times t and $t + \delta t$. The typical instrument to measure this type of coherence is the Michelson interferometer. Young's double slit experiment will determine the degree of spatial coherence of a source [4]. A definition of temporal coherence for electromagnetic radiation is the normalized first-order correlation $\langle \tilde{E}(t)\tilde{E}^*(t+\tau)\rangle / \sqrt{\langle I(t)I(t+\tau)\rangle}$. A bandwidth-limited ultrashort pulse has thus a coherence of unity. A similar definition can be applied to spatial coherence. A single source of diameter $d \ll \lambda$ has perfect spatial coherence.

If the source is an atom, spherical waves are emitted with identical fluctuations at all points at equal distance from the source. Macroscopically, we are used to looking at a volume average of randomly distributed dipoles. Some new sources have the emitted dipoles arranged in a regular fashion. For instance, multiple quantum well lasers are made of "sheets" of emitting atoms which are thinner than the wavelength of the light. This particular arrangement of emitters has important implications for the macroscopic properties of the source [5, 6].

4.2. COHERENT INTERACTIONS WITH TWO-LEVEL SYSTEMS

4.2.1. Maxwell–Bloch Equations

Whether we are dealing with molecular or atomic transitions, the situation can arise where the ultrashort duration of the optical pulse becomes comparable with—or even less than—the phase-relaxation time of the excitation. In the frequency domain, the pulse spectrum is broader than the homogeneous linewidth defined in the first section of Chapter 3. If the pulse is so short that its spectrum becomes much larger than the inhomogeneous linewidth, the medium response becomes similar to that of a single atom. It may seem like a simplified situation when the excitation occurs in a time shorter than all interatomic interaction. It is in fact quite to the contrary; in dealing with longer pulses, the faster phase-relaxation time of the induced excitation simplifies the light–matter response. One is accustomed to dealing with a steady state rather than the "transient" response of light–matter interaction.

We will start from the semiclassical equations for the interaction of near resonant radiation with an ensemble of two-level systems inhomogeneously broadened around a frequency ω_{ih}. The extension to multilevel systems will be discussed in the next section. We refer to the book by Allen and Eberly [7] for more detailed developments.

We summarize briefly the results of Chapter 3 for a two-level system, of ground state $|0\rangle$ and upper state $|1\rangle$, excited by the field $E(t)$. The density matrix equation for this two-level system is:

$$\dot{\rho} = \frac{1}{i\hbar}[H_0 - pE, \ \rho], \tag{4.1}$$

where H_0 is the unperturbed Hamiltonian, and p the dipole moment that is parallel to the polarization direction of the field. Introducing the complex field through $E = \tilde{E}^{+} + \tilde{E}^{-}$ in Eq. (4.1) leads to the following differential equations for the diagonal and off-diagonal matrix elements:

$$\dot{\rho}_{11} - \dot{\rho}_{00} = \frac{2p}{\hbar}\left[i\rho_{01}\tilde{E}^{-} - i\rho_{10}\tilde{E}^{+}\right] \tag{4.2}$$

$$\dot{\rho}_{01} = i\omega_0\rho_{01} + \frac{ip\tilde{E}^{+}}{\hbar}\left[\rho_{11} - \rho_{00}\right], \tag{4.3}$$

where ω_0 is the resonance frequency of the two-level system. It is generally convenient to define a complex "pseudo-polarization" amplitude $\tilde{Q}$ by

$$i\rho_{01}p\bar{N} = \frac{1}{2}\tilde{Q}\exp(i\omega_\ell t) \tag{4.4}$$

where $\bar{N} = \bar{N}_0 g_{inh}(\omega_0 - \omega_{ih})$ and $\bar{N}_0$ is the total number density of the two-level systems. The real part of $\tilde{Q}$ will describe the attenuation (or amplification for an initially inverted system) of the electric field. Note that $\tilde{Q} = i\tilde{\mathcal{P}}$ where $\tilde{\mathcal{P}}$ is the slowly varying polarization envelope defined in Eq. (3.28). Further we introduce a normalized population inversion:

$$w = p\bar{N}(\rho_{11} - \rho_{00}). \tag{4.5}$$

The complete system of interaction and propagation equations can now be written as:

$$\dot{\tilde{Q}} = i(\omega_0 - \omega_\ell)\tilde{Q} - \kappa\tilde{\mathcal{E}}w - \frac{\tilde{Q}}{T_2} \tag{4.6}$$

$$\dot{w} = \frac{\kappa}{2}[\tilde{Q}^*\tilde{\mathcal{E}} + \tilde{Q}\tilde{\mathcal{E}}^*] - \frac{w - w_0}{T_1} \tag{4.7}$$

$$\frac{\partial\tilde{\mathcal{E}}}{\partial z} = -\frac{\mu_0\omega_\ell c}{2n}\int_0^\infty \tilde{Q}(\omega_0')g_{inh}(\omega_0' - \omega_{ih})d\omega_0'. \tag{4.8}$$

The quantity $\kappa\mathcal{E}$ with $\kappa = p/\hbar$ is the Rabi frequency. T_1 and T_2 are respectively the energy and phase-relaxation times. Most of the energy conserving relaxations are generally lumped in the phase-relaxation time T_2. Equation (4.8) has been obtained from Eq. (3.30) by integrating over the polarization of subensembles with resonance frequency ω_0'. The set of Eqs. (4.6)–(4.8) is generally designated as Maxwell–Bloch equations.

Another common set of notations to describe the light–matter interaction uses only real quantities, such as the in-phase (v) and out-of-phase (u) components of the pseudo-polarization $\tilde{Q}$, and, for the electric field $\tilde{\mathcal{E}}$, its (real) amplitude $\mathcal{E}$ and its phase φ. Defining

$$\tilde{Q} = (iu + \nu)e^{i\varphi} \tag{4.9}$$

and substituting in the above system of equations leads to the usual form of Bloch equations[1] for the subensemble of two-level systems having a resonance frequency ω_0.

$$\dot{u} = (\omega_0 - \omega_\ell - \dot{\varphi})\nu - \frac{u}{T_2} \tag{4.10}$$

$$\dot{v} = -(\omega_0 - \omega_\ell - \dot{\varphi})u - \kappa\mathcal{E}w - \frac{\nu}{T_2} \tag{4.11}$$

$$\dot{w} = \kappa\mathcal{E}\nu - \frac{w - w_0}{T_1} \tag{4.12}$$

where the initial value for w at $t = -\infty$ is

$$w_0 = p\bar{N}(\rho_{11}^{(e)} - \rho_{00}^{(e)}). \tag{4.13}$$

The propagation equation, [Eq. (4.8)] in terms of $\tilde{\mathcal{E}}$ and φ, becomes

$$\frac{\partial \mathcal{E}}{\partial z} = -\frac{\mu_0 \omega_\ell c}{2n} \int_0^\infty \nu(\omega_0')g_{inh}(\omega_0' - \omega_{ih})d\omega_0' \tag{4.14}$$

$$\frac{\partial \varphi}{\partial z} = -\frac{\mu_0 \omega_\ell c}{2n} \int_0^\infty \frac{u(\omega_0')}{\mathcal{E}} g_{inh}(\omega_0' - \omega_{ih})d\omega_0'. \tag{4.15}$$

The vector representation of Feynman *et al.* [9] for the interaction equations is particularly useful in the description of coherent phenomena. The representation is a cinematic representation of the set of equations (4.10), (4.11), and (4.12). For simplicity, we consider first an undamped isolated two-level system ($T_1 = T_2 = T_3 = \infty$), and construct a fictitious vector $\vec{\mathcal{P}}$ of components (u, v, w), and a pseudo-electric field vector $\vec{\mathcal{E}}$ of components $(\kappa\mathcal{E}, 0, -\Delta\omega)$. The detuning is defined as $\Delta\omega = \omega_0 - \omega_\ell - \dot{\varphi}$. The system of Eqs. (4.10)–(4.12) are then the cinematic equations describing the rotation of a pseudo-polarization vector $\vec{\mathcal{P}}$ rotating around the pseudo-electric vector $\vec{\mathcal{E}}$ with an angular velocity

[1]These equations are the electric dipole analog of equations derived by F. Bloch [8] to describe spin precession in magnetic resonance.

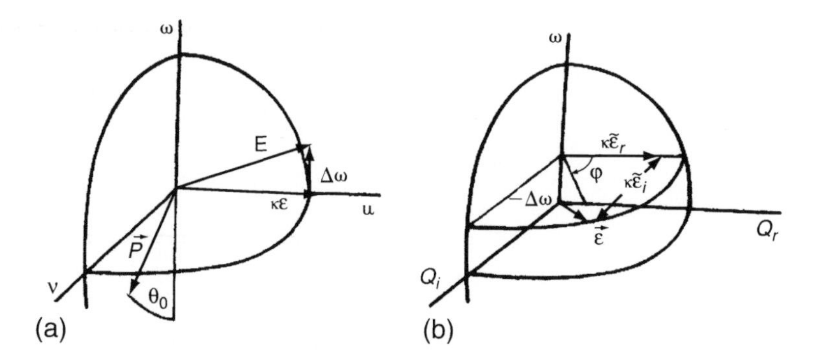

Figure 4.2 Vector model for Bloch's equations. (a) The motion of the pseudo-polarization vector $\vec{\mathcal{P}}$ (initially pointing downward along the w axis) is a rotation around the pseudo-electric field vector $\vec{\mathcal{E}}$ with an angular velocity proportional to the amplitude of that vector. (b) In the complex amplitude representation, the phase of the electric field determines the particular vertical plane containing the pseudo-electric field vector $\vec{\tilde{\mathcal{E}}}$.

given by the amplitude of the vector $\vec{\mathcal{E}}$ [Figure 4.2(a)]. The vectorial form of Eqs. (4.10)–(4.12) is thus:

$$\boxed{\partial\vec{\mathcal{P}}/\partial t = \vec{\mathcal{E}} \times \vec{\mathcal{P}}} \tag{4.16}$$

Depending on whether the two-level system is initially in the ground state or inverted, the pseudo-polarization vector is initially pointing down or up. Because we have assumed no relaxation, the length of the pseudo-polarization vector is a constant of the motion, and the tip of the vector moves on a sphere. The conservation of length of the pseudo-polarization vector can be verified directly from the set of Bloch's equations. Indeed, the sum of each equation (4.10), (4.11) and (4.12) multiplied by u, ν, and w, respectively, yields after integration:

$$u^2 + \nu^2 + w^2 = w_0^2 \tag{4.17}$$

which is satisfied for each subensemble of two-level systems. As shown in Fig. 4.2(a), a resonant excitation ($\Delta\omega = 0$) will tip the pseudo-polarization vector by an angle $\theta_0 = \int_{-\infty}^{\infty} \kappa\mathcal{E}dt$ in the (v, w) plane. For a sufficiently intense pulsed excitation, it is possible to achieve complete population inversion when $\theta_0 = \pi$. The effect of phase relaxation (homogeneous broadening) is to shrink the pseudo-polarization vector as it moves around. To take into account inhomogeneous broadening, we have to consider an ensemble of pseudo-polarization vectors, each corresponding to a different detuning $\Delta\omega$.

A similar representation can be made for the system of Eqs. (4.6)–(4.7). The pseudo-polarization vector is then the vector $\vec{Q}(Q_i, Q_r, w)$ rotating around a pseudo-electric field vector $\vec{\mathcal{E}}(\kappa\tilde{\mathcal{E}}_r, \kappa\tilde{\mathcal{E}}_i, -\Delta\omega)$ [Fig. 4.2(b)]. Physically, the first two components of the pseudo-polarization vector $\vec{Q}$ represent the dipolar resonant field that opposes the applied external field (and is thus responsible for absorption).

4.2.2. Rate Equations

If the light field envelope is slowly varying with respect to T_2, Bloch's equations reduce to the standard rate equations. For pulses longer than the dephasing time T_2, the two first Bloch equations, (4.10) and (4.11) are stationary on the time scale of the pulse. Solving these equations for u, ν, and substituting ν into the third equation (4.12) for the population difference, leads to the rate equation:

$$\dot{w} = -\frac{\mathcal{E}^2(\kappa^2 T_1 T_2)}{1 + \Delta\omega^2 T_2^2}\frac{w}{T_1} - \frac{w - w_0}{T_1}. \tag{4.18}$$

We note that this equation is identical to the rate equation (3.46) introduced in Chapter 3 in terms of ΔN ($\Delta N = w/p$). Equation (4.18) defines a saturation field at resonance $\tilde{\mathcal{E}}_{s0} = 1/(\kappa\sqrt{T_1 T_2})$. Off-resonance, a larger field, $\tilde{\mathcal{E}}_s = \tilde{\mathcal{E}}_{s0}\sqrt{1 + \Delta\omega^2 T_2^2}$ is required to saturate the same transition.

In Chapter 3 we discussed various cases of pulse propagation through resonant media resulting from the rate equations. For example, for pulses much shorter than the energy relaxation time $\tau_p \ll T_1$ and purely homogeneously broadened media the rate equation (4.18) can be integrated together with the propagation equation (4.8), which yields for the transmitted intensity

$$I(z,t) = I_0(t)\frac{e^{W(t)/W_s}}{e^{-a} - 1 + e^{W(t)/W_s}}. \tag{4.19}$$

In this last equation $W(t) = \int_{-\infty}^{t} I_0(t)dt$, and $a = \sigma_{01}^{(0)} w_0 z/p$ is the linear gain–absorption coefficient. Equation (4.19) corresponds to Eq. (3.55) which was written for the photon flux F.

Femtosecond pulse propagation through a homogeneously broadened saturable medium in the limit of $T_2 \ll \tau_p \ll T_1$ is completely determined by two parameters: the saturation energy density W_s and the linear absorption (gain) coefficient a. Equation (4.19) is particularly useful in calculating pulse propagation in amplifiers, as shown in Chapter 3, and further detailed in Chapter 7.

4.2.3. Evolution Equations

Energy Conservation

The total energy in the resonant light–matter system should be conserved if the pulses are shorter than the energy relaxation time T_1, because no energy is dissipated into the bath. The pulse energy density was defined in Eq. (1.22):

$$W = \frac{1}{2}\epsilon_0 cn \int_{-\infty}^{\infty} \mathcal{E}^2 dt = \frac{1}{2}\sqrt{\epsilon/\mu_0} \int_{-\infty}^{\infty} \mathcal{E}^2 dt. \tag{4.20}$$

A simple energy conservation law can be derived by integrating Eq. (4.14) over time, after multiplying both sides by $\mathcal{E}$ and using the third Bloch equation (4.12):

$$\begin{aligned} \frac{dW}{dz} &= \sqrt{\frac{\epsilon}{\mu_0}} \int_{-\infty}^{\infty} \mathcal{E}\frac{\partial \mathcal{E}}{\partial z} dt \\ &= -\frac{\mu_0 \omega_\ell\, c}{2}\sqrt{\frac{\epsilon}{\mu_0}} \int_{-\infty}^{\infty}\int_0^{\infty} \nu(\omega_0')\mathcal{E} g_{inh}(\omega_0' - \omega_{ih}) d\omega_0'\, dt \\ &= -\frac{\hbar\omega_\ell}{2p}\int_0^{\infty} \left[w_\infty(\omega_0') - w_0(\omega_0')\right] g_{inh}(\omega_0' - \omega_{ih}) d\omega_0'. \end{aligned} \tag{4.21}$$

The population difference (per unit volume) $(w_\infty - w_0)/p$ integrated over the inhomogeneous transition is a measure of the energy stored in the medium, as a consequence of the energy lost by the pulse, dW/dz.

Area Theorem

There are other conservation and evolution laws that can be derived for certain parameters associated with pulses of arbitrary shape. An essential physical parameter for single photon coherent interactions is the *pulse area* θ_0, defined as the tipping angle of the pseudo-polarization vector (at resonance) as illustrated in Fig. 4.2(a):

$$\theta_0 = \int_{-\infty}^{\infty} \kappa\mathcal{E}(t)dt. \tag{4.22}$$

For a pulse at resonance, the area θ_0 fully describes in which state the medium is left. It can be seen directly from the vector model of Fig. 4.2(a), or by direct integration of Bloch's equations (4.10), (4.11), and (4.12) for exact resonance $[\omega_\ell = \omega_0$ and $\dot{\varphi}(t) = 0]$ and negligible relaxation ($T_1 \approx \infty$ and $T_2 \approx \infty$), that:

$$w_\infty = w(t = \infty) = w_0 \cos\theta_0. \tag{4.23}$$

Similarly, one finds that the polarization (absorptive component) is proportional to $\sin\theta_0$. A "π pulse" is thus a pulse that will completely invert a two-level system, leaving it in a pure state with no macroscopic polarization. A "2π pulse" will leave the system in the ground state, having completed a cycle of population inversion and return to ground state.

Because the area involves the integral of the electric field amplitude rather than the pulse intensity, higher energy densities will be required to achieve the same area with shorter pulses. Therefore, experiments of single photon coherent resonant interactions with fs pulses require intensities at which higher-order nonlinear optical effects may have to be taken into account. Another consequence of the electric field amplitude dependence of the area is, that in an absorbing medium, it is possible to have the area conserved, or even growing with distance, although the energy is decreasing. Such a situation arises when the pulse duration increases with distance.

For inhomogeneously broadened media an "area theorem" can be derived which tells us exactly how the pulse area evolves with propagation distance. With the assumptions that the pulses are at resonance ($\omega_\ell = \omega_{ih}$), and shorter than both the energy relaxation time T_1 and the phase-relaxation time T_2, a time integration of Eq. (4.14), taking into account Bloch's equations (4.10)–(4.12), yields the *area theorem* [10]:

$$\boxed{\frac{d\theta_0}{dz} = \frac{\alpha_0}{2}\sin\theta_0,} \tag{4.24}$$

where

$$\alpha_0 = \frac{\pi\mu_0\omega_\ell\, cp}{\hbar n} w_0(\omega_{ih}) = \frac{\pi\kappa^2\hbar\omega_\ell}{\epsilon_0 cnp} w_0(\omega_{ih}) \tag{4.25}$$

is the linear absorption coefficient (at resonance) for the inhomogeneously broadened transition. $w_0(\omega_{ih})$ is the initial inversion density at the transition center ($\omega_0' = \omega_{ih}$).

The area theorem applies to amplifying ($w_0 > 0$) as well as to absorbing media ($w_0 < 0$). The derivation is straightforward if we assume a square pulse at resonance, and neglect the reshaping of the pulse. However, Eq. (4.24) can be proven to be quite general and applies to any pulse shape. A graphical representation of the solution of Eq. (4.24) is shown in Figure 4.3.

The low intensity limit of Eq. (4.24) ($\sin\theta_0 \approx \theta_0$) is the exponential attenuation law (Beer's law). A remarkable feature of Eq. (4.24) is that it points to *area conserving pulses*. Indeed, for any pulse of area equal to a multiple of π, the area is conserved. This is true, for instance, for a "zero area" pulse, which is not necessarily a zero energy pulse. A signal consisting of two pulses π out of phase has an area equal to zero, but an energy equal to twice the single pulse energy.

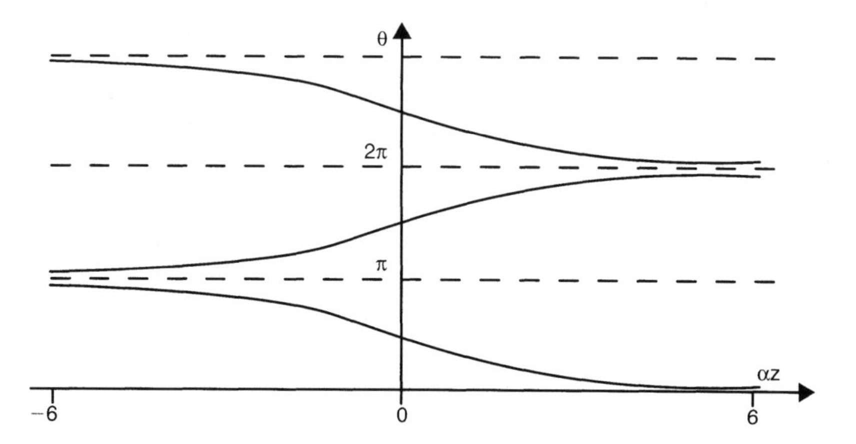

Figure 4.3 Graphic solution of the area theorem equation. The area is plotted as a function of distance (in units of linear absorption length (α_0^{-1})) for an absorbing medium. There is only one point on the set of curves that corresponds to any given area. The origin of distance is the abscissa corresponding to the initial area. From this initial point, a pulse will propagate to the right in an absorbing medium, to the left in an amplifier. Any initial area will evolve asymptotically with distance toward the limits 0, $2\pi, \ldots, 2N\pi$ (N integer) in an absorbing medium. The asymptotic limits are $\pi, 3\pi, \ldots, (2N+1)\pi$ in an amplifier.

Such a pulse sequence propagates without loss of "area" through a resonant medium.

In addition the area theorem tells us which of the steady-state areas are stable solutions. For a pulse with an initial area smaller than π, the right side of Eq. (4.24) is negative, and the area as well as the pulse energy will decay with distance as the pulse is absorbed. On the other hand, if the pulse area is initially between π and 2π, θ *increases* with distance. With such a pulse, a population inversion has been achieved, and the pulse tail is amplified. The stimulated emission at the pulse tail results in pulse stretching, and hence an increase in pulse area with distance. The pulse energy, however, still decreases with distance in the reshaping process. Pulse reshaping proceeds until the area reaches the value of 2π. Once the pulse reshaping is completed, the electric field envelope has acquired a well-defined and stable hyperbolic secant (sech) shape and propagates without further distortion or attenuation through the resonant absorbing medium. This phenomenon is called *self-induced transparency*. Pulses of initial area $2n\pi$ break up into n "2π" pulses. It can easily be verified, by substitution into Eqs. (4.10)–(4.12), that the envelope given by:

$$\mathcal{E}(t) = \frac{2}{\kappa\tau_s}\operatorname{sech}\left(\frac{t}{\tau_s} - \frac{z}{\tau_s v_e}\right), \tag{4.26}$$

is a solution of Bloch's equations. The pulse given by Eq. (4.26) has an area of 2π and a duration (FWHM) of $1.763\tau_s$. This solution is valid on- and off-resonance. In Eq. (4.26), $\nu_e \ll c$ is the envelope velocity of the 2π pulse. For a pulse duration short compared to the inverse (inhomogeneous) linewidth of the absorber, the envelope velocity is given by $\nu_e = 2/(\alpha_0 \tau_s)$. This slow velocity essentially expresses that the first half ($\tau_s/2$) of the pulse is absorbed in a distance α_0^{-1}, to be restored to the second half by stimulated emission.

The subscript in θ_0 indicates that the amplitude of the electric field of a pulse at frequency ω_ℓ is used in the previous definition (4.22). In Eq. (4.22) the pulse envelope θ_0 was introduced by a time integral. Another definition for the area is related to the amplitude of the Fourier transform of the pulse defined in Eq. (1.6):

$$\theta = \kappa|\tilde{\mathcal{E}}(\Omega - \omega_0 = 0)|. \tag{4.27}$$

Because $\kappa\mathcal{E}(t)$ has the dimension of a frequency, the Fourier transform of that quantity is dimensionless. It is left as a problem at the end of this chapter to show that the definitions (4.22) and (4.27) are equivalent for unchirped pulses at resonance. Chirped or nonresonant pulses of the same energy or of the same area θ_0 will have a smaller area θ, because of the broadening of the Fourier spectrum because of the phase modulation. The distinction is important in determining the threshold for nonlinear propagation phenomena such as *self-induced transparency* [11]. Chirped off- or on-resonance pulses will evolve toward a pure unmodulated 2π pulse at the original pulse frequency ω_ℓ, provided the initial area θ is larger than π, as has been demonstrated by numerous computer simulations [11].

The coherent absorber is therefore an ideal filter for phase and amplitude fluctuations for pulses of initial area larger than π. Because of the slow propagation velocity of the 2π pulse, all fast phase and amplitude noise propagates ahead of the pulse as a "precursor" that is ultimately absorbed [11–13]. An example of the evolution of a small Gaussian disturbance (90° out-of-phase) initially superimposed on top of a 2π hyperbolic secant pulse is illustrated in Figure 4.4. The amplitude and phase of the electric field are represented as a function of time and distance. The propagation distance is expressed in units of the linear absorption length. The initial "noise pulse" on top of the 2π pulse is indicated by the dashed circle around the top of the main pulse. The phase of the perturbation is indicated above the amplitude disturbance. The phase modulation is seen to be continuously amplified with propagation distance. This is consistent with the discussion on the average frequency of a pulse off-resonance being shifted further away from resonance with distance. In the spectral domain (not shown), this perturbation appears as little "bumps" on the wings of the pulse spectrum. The amplification of the phase modulation corresponds to the Fourier components of the perturbation being rejected farther away from resonance. Numerical simulations have shown that the phase

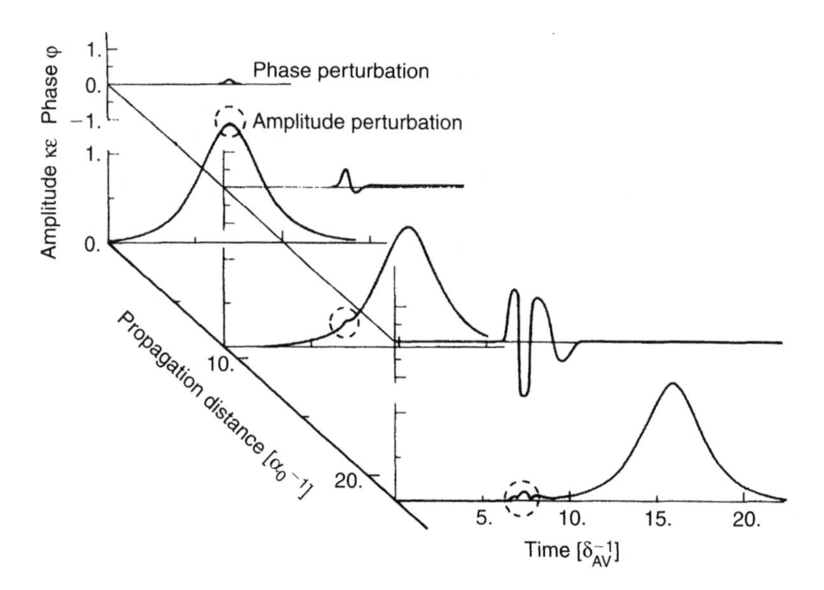

Figure 4.4 Propagation of a 2π pulse at resonance, showing the "filtering" of a small 90° out-of-phase Gaussian disturbance (shown in the dashed circles). The medium is inhomogeneously broadened with the line profile $g_{inh}(\Delta\omega) = (1/\sqrt{\pi})\exp[-(\Delta\omega/\delta_{AV})^2]$. The initial amplitude is $\kappa\mathcal{E}(t) = (2/1.5)\text{sech}[(t-7)/1.5] + (0.1i/0.3\sqrt{\pi})\exp\{-[(t-7)/0.3]^2\}$, where the time t is expressed in units of δ_{AV}^{-1}. Three steps of propagation are shown. The amplitude perturbation is indicated by the dashed circle. The initial modulation is seen to be amplified in amplitude and phase. Because of its faster group velocity as compared to the envelope velocity of the 2π pulse, it separates in time from the latter. The amplification of the phase fluctuation indicates that the perturbation also separates in frequency from the main pulse (Adapted from Diels and Hahn [12]).

perturbations become amplitude and phase perturbations, which propagate at the normal group velocity of the medium, rather than the low velocity ν_e of the 2π pulse. When the 2π pulse has separated in time from the noise signal, it has reshaped into a lower energy sech pulse broadened by 0.7% in the particular example shown.

Pulse Frequency Evolution

Bloch's equations describe the transient response of the complex polarization for a two-level system. As mentioned in the introduction, the pulse frequency is no longer a conserved quantity in the presence of a transient polarization. The medium considered in this section can be represented by a collection of two-level systems, each with a homogeneous broadening of T_2^{-1}, with a frequency distribution represented by a function $g_{inh}(\omega_0' - \omega_{ih})$ (inhomogeneous broadening).

In this section, we will derive expressions from Bloch's equations that will allow us to establish quantitatively the evolution of the average carrier frequency of a pulsed signal as it propagates through the medium. We have seen in Chapter 1 that the frequency modulation $\dot{\varphi}$ has to be averaged to define a mean pulse frequency [Eq. 1.18]. We will introduce the notation $\langle\dot{\varphi}\rangle$ for the average phase derivative (average deviation of the frequency from ω_ℓ):

$$\langle\dot{\varphi}\rangle = \frac{\int_{-\infty}^{\infty} \mathcal{E}^2 \dot{\varphi} dt}{\int_{-\infty}^{\infty} \mathcal{E}^2 dt} = \frac{1}{2W}\left[\sqrt{\frac{\epsilon}{\mu_0}} \int_{-\infty}^{\infty} \mathcal{E}^2 \dot{\varphi} dt\right]. \tag{4.28}$$

Multiplying both sides by W and taking the derivative we obtain:

$$\begin{aligned} \frac{d(W\langle\dot{\varphi}\rangle)}{dz} &= W\frac{d\langle\dot{\varphi}\rangle}{dz} + \langle\dot{\varphi}\rangle\frac{dW}{dz} \\ &= \frac{1}{2}\sqrt{\frac{\epsilon}{\mu_0}}\frac{d}{dz}\int_{-\infty}^{\infty} \mathcal{E}^2 \dot{\varphi} dt. \end{aligned} \tag{4.29}$$

The last term of Eq. (4.29) can be directly calculated using the Maxwell–Bloch equations. In particular, we can insert the time derivative of Eq. (4.15) in this equation:

$$\begin{aligned} &\frac{1}{2}\sqrt{\frac{\epsilon}{\mu_0}}\int_{-\infty}^{\infty}\left(\mathcal{E}^2\frac{\partial\dot{\varphi}}{\partial z} + \dot{\varphi}\frac{\partial\mathcal{E}^2}{\partial z}\right)dt \\ &= -\frac{\omega_\ell}{4}\int_{-\infty}^{\infty} dt \int_0^{\infty} d\omega_0'\, g_{inh}(\omega_0' - \omega_{ih})\left[\dot{u}\mathcal{E} - u\dot{\mathcal{E}} + 2\nu\mathcal{E}\dot{\varphi}\right] \\ &= -\frac{\omega_\ell}{2}\int_0^{\infty} d\omega_0' \int_{-\infty}^{\infty} dt\, g_{inh}(\omega_0' - \omega_{ih})\left[\dot{u}\mathcal{E} + \dot{\varphi}\nu\mathcal{E}\right] \\ &= -\frac{\omega_\ell}{2}\int_0^{\infty} d\omega_0' \int_{-\infty}^{\infty} dt\, g_{inh}(\omega_0' - \omega_{ih})\left[(\omega_0' - \omega_\ell)\nu\mathcal{E} - \frac{u\mathcal{E}}{T_2}\right] \end{aligned} \tag{4.30}$$

where we have made use of the first Bloch equation (4.10). We have already derived an expression for the evolution of the pulse energy density W. Combining Eqs. (4.21), (4.29), and (4.30) yields the following expression for the evolution with propagation distance of the pulse carrier frequency:

$$\boxed{\frac{d\langle\dot{\varphi}\rangle}{dz} = \frac{\omega_\ell}{2\kappa W}\int_0^{\infty} g_{inh}(\omega_0' - \omega_{ih})\left[\omega_0' - \omega_\ell - \langle\dot{\varphi}\rangle\right](w_\infty - w_0)\, d\omega_0' + \frac{2\langle k\rangle}{T_2}.} \tag{4.31}$$

In analogy to the definition of the average frequency in Chapter 1 [cf. Eq. (1.18)], we have introduced the average contribution to the propagation vector because of the resonant dispersion of the two-level system:

$$\langle k \rangle = \frac{\int_{-\infty}^{\infty} \mathcal{E}^2 (\partial\varphi/\partial z) dt}{\int_{-\infty}^{\infty} \mathcal{E}^2\, dt}$$
$$= \frac{\omega_\ell}{4W} \int_0^\infty d\omega_0' \int_{-\infty}^{\infty} dt\, g_{inh}(\omega_0' - \omega_{ih}) u\mathcal{E} \quad (4.32)$$

The polarization amplitude u—and hence the resonant contribution to the wave vector $\langle k \rangle$—will shrink with time in presence of phase relaxation (finite T_2). The corresponding temporal modulation of the polarization is responsible for the second term of the right-hand side of Eq. (4.31). For short pulses, however, ($\tau_p \ll T_2$), this second term can be neglected.

The frequency shift is proportional to the overlap integral of the lineshape $g_{inh}(\omega_0' - \omega_{ih})$ with the (frequency-dependent) change of the inversion $(w_\infty - w_0)$ times the detuning $(\omega_0' - \omega_\ell - \langle\dot{\varphi}\rangle)$. The ratio of absorbed energy (which is proportional to $(w_\infty - w_0)$) to the pulse energy W is maximum in the weak pulse limit ($\theta \ll 1$). Therefore, the frequency pushing as described by Eq. (4.31) is important in the weak pulse limit, and for narrow lines [$T_2 \to \infty$; $g_{inh}(\omega_0' - \omega_{ih}) \approx \delta(0)$].

Bloch's equations can be solved analytically in the weak, short pulse limit, i.e., for pulses that do not induce significant changes in population and have a duration short compared to the phase relaxation time T_2. The interaction equation (4.6) can be written in the integral form:

$$\tilde{Q}(t) = \int_{-\infty}^{t} \kappa \mathcal{E} w e^{-i[(\omega_0' - \omega_\ell)t' - \varphi(t')]} dt'. \quad (4.33)$$

For weak pulses ($w \approx w_0$) and the right hand-side of Eq. (4.33) at $t = \infty$ is proportional to the Fourier transform of $\kappa\tilde{\mathcal{E}}w$. Thus we have:

$$|\tilde{Q}|^2 = u^2 + v^2 = \kappa^2 w_0^2 |\tilde{\mathcal{E}}(\omega_0' - \omega_\ell)|^2$$
$$\approx -2w_0(w_\infty - w_0) \quad (4.34)$$

where $\tilde{\mathcal{E}}(\omega_0' - \omega_\ell)$ is the amplitude of the Fourier transform of the field envelope at the line frequency ω_0'. The last equality results from the conservation of the length of the pseudo-polarization vector ($u^2 + v^2 + w^2 = w_0^2 =$ constant). The approximation is made that the change in population is small, $w_\infty^2 = [w_0 + (w_\infty - w_0)]^2 \approx w_0^2 + 2w_0(w_\infty - w_0)$. Insertion of the approximation equation (4.34) into the energy evolution equation (4.21) leads to the expected result that the absorbed energy is proportional to the overlap of the pulse and line spectra:

$$\frac{dW}{dz} = \frac{\kappa\omega_\ell\, w_0}{4}\int_0^\infty |\tilde{\mathcal{E}}(\omega_0' - \omega_\ell)|^2 g_{inh}(\omega_0' - \omega_{ih})d\omega_0'. \tag{4.35}$$

A similar expression is found by inserting the approximation equation (4.34) into the frequency evolution equation (4.31):

$$\frac{d\langle\dot{\varphi}\rangle}{dz} = \frac{\kappa\omega_\ell w_0}{2W}\int_0^\infty (\omega_0' - \omega_\ell - \langle\dot{\varphi}\rangle)|\tilde{\mathcal{E}}(\omega_0' - \omega_\ell)|^2 g_{inh}(\omega_0' - \omega_{ih})d\omega_0' \tag{4.36}$$

Comparing expressions (4.35) and (4.36) leads to the conclusion that the frequency shift with distance is largest for sharp lines (compared to the pulse bandwidth), and pulses off-resonance by approximately their bandwidth. The maximum possible shift is about half a pulse bandwidth per absorption length.

The initial frequency pushing with distance given by Eq. (4.36) has an interesting simple dependence on the initial pulse area, in the case of a sech^2-shaped pulse. It can be verified that the functional dependence is exactly:

$$\frac{d\langle\dot{\varphi}\rangle}{dz} = \frac{\kappa\omega_\ell\, w_0}{4\epsilon_0 cn}\frac{1-\cos\theta_0}{\theta_0^2}\Gamma_F(\omega_\ell), \tag{4.37}$$

where $\Gamma_F(\omega_\ell)$ is the spectral overlap function:

$$\Gamma_F(\omega_\ell) = \frac{\int_0^\infty \kappa^2\mathcal{E}^2(\omega_0' - \omega_\ell)g_{inh}(\omega_0' - \omega_{ih})d\omega_0'}{\int_{-\infty}^\infty \kappa^2\mathcal{E}(t)dt}. \tag{4.38}$$

The above expression pertains to sech^2-shaped pulses. However, Figure 4.5 illustrates that Eq. (4.37) still provides a reasonable approximation even for asymmetric pulse shapes.

The frequency shift with distance is a manifestation of the fact that the response time of matter is finite. We alluded to this in Chapter 1, when we wrote the polarization as a convolution integral, expressing that the instantaneous polarization is not an immediate function of the electric field of the light, but a function of the history of the field [Eq. (1.71)]. Such convolution expressions can be obtained directly by Fourier transformation of Bloch's equations [for instance Eq. (4.6)].

To appreciate physically why the carrier frequency should not be a conserved quantity, let us consider a square pulse sent through a near resonant absorbing medium at the average carrier frequency ω_ℓ where $\omega_\ell < \omega_{ih}$. High frequency components within the pulse spectrum are closer to the center of the inhomogeneous transition and, thus, are absorbed more than low frequency components.

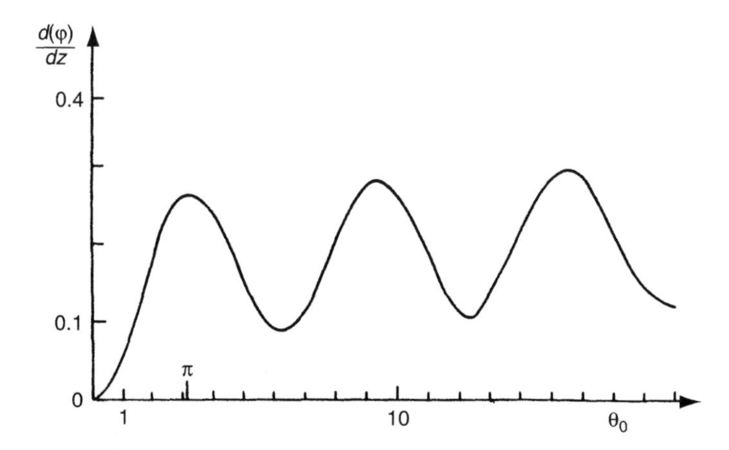

Figure 4.5 Frequency pushing with distance for an asymmetric pulse applied one linewidth off-resonance, as a function of initial pulse area. The medium is inhomogeneously broadened with the line profile $g_{inh}(\Delta\omega) = (1/\sqrt{\pi})\exp[-(\Delta\omega/\delta_{AV})^2]$. The pulse shape is given by $\kappa\mathcal{E}(t) = 1/[\exp(t/2) + \exp(t/0.8)]$. The frequency (time) is normalized to the (inverse of the) inhomogeneous broadening linewidth.

The result is a continuous shift of the pulse spectrum to lower frequencies. Essentially the same phenomenon has been discussed in Chapter 1 (Section 1.2.2). There, the cause was a frequency dependent imaginary part of the dielectric constant. It is left as a problem to find the connection between both descriptions. Another way to describe the frequency pushing is in the time domain. Initially—i.e., prior to the arrival of the light pulse—there were no induced dipoles, because there was no field present to induce them. The initial value of the susceptibility χ (or, using the notations of this chapter, the pseudo-polarization Q or u, v) is thus 0. As the leading edge of the pulse enters the medium, it excites the electronic dipoles, resulting in a change of χ and in a change of the refractive index, respectively. As we have seen before this results in a pulse chirp which can manifest itself in a shift of the average pulse frequency. Such an effect can be simply visualized with help of Figure 4.6, which is a three-dimensional representation of the electric field versus time and distance.

Because the frequency pushing with distance is a general manifestation of the response time of matter, it will occur whenever there is a transient in the polarization because of a change in applied field. Even continuous radiation will be frequency shifted if there are random fluctuations of the phase of the field. Analytical expressions can be derived for the shift of average frequency with distance for a phase fluctuating dc field [11].

Given a symmetric line shape, is the frequency pushing symmetric above and below resonance? There is a blue shift equal to the red shift, only within

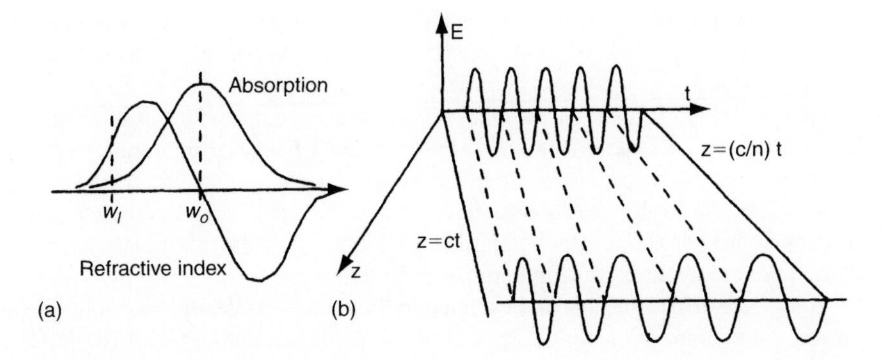

Figure 4.6 Sketch illustrating the evolution of the average frequency of a short pulse as it propagates through matter. Figure (a) shows the relative frequencies of the radiation (ω_ℓ) and material resonance (ω_0). These are however steady-state values. Prior to applying the field (and for the first optical cycle(s) of the radiation), the electric dipoles have not yet been induced, and the abscissa is the correct representation of both the absorption and dispersion. If a square pulse is applied, the value of the absorption and dispersion evolve with time from the initial condition (0) to the value on the corresponding curve at ω_ℓ. A representation of the evolution of the wave packet in time and space is shown in (b). The leading edge of the square pulse is not affected, because the dipoles have not yet been induced. The time varying index leads to a shift of the average pulse frequency to lower frequencies.

the framework of the SVEA. Numerical simulations indicate that the material response is larger below resonance than above, leading to a larger red shift than blue shift. This may not come as a surprise, because the response of a system with a resonance at ω_0 will "follow" a low frequency ($\omega_\ell \ll \omega_0$) excitation, but has zero response at frequencies far beyond its resonance ($\omega_\ell \gg \omega_0$).

4.2.4. Steady-State Pulses

A steady-state pulse is a pulse for which the envelope $\mathcal{E}(t - z/\nu_e)$ is conserved along propagation. The quantity ν_e is the velocity of the pulse envelope. Usually it depends on properties of the medium and the pulse and differs from the group and phase velocity. If the pulse envelope remains constant, various pulse parameters such as the pulse duration, energy, and area should also be conserved.

In the case of inhomogeneously broadened media, the area theorem applies. From Eq. (4.24), a necessary condition for the existence of a steady-state pulse is that $d\theta/dz = (\alpha_0/2)\sin\theta_0 = 0$. Obviously there are several values of the area which do not change with distance; they are $\theta_0 = 0, \pi, 2\pi, 3\pi, \ldots$. A quick glance at the graphical representation (Fig. 4.3) of Eq. (4.24) indicates that in an absorbing medium ($\alpha < 0$) the areas $\theta_0 = 0, 2\pi, 4\pi \ldots$ are stable solutions, although for amplifying media, the stable areas correspond to uneven numbers of π. At resonance, steady-state pulses correspond to the π pulse. As mentioned previously, in the case of an absorber, only the area 2π corresponds to a stable stationary pulse. In the amplifier, however, even with constant area, the pulse energy will tend to grow to infinity unless balanced by a loss mechanism. We will use a scattering coefficient σ_s to describe such a loss mechanism in searching for steady-state pulses in amplifiers. It is the same coefficient representing linear losses as the κ_1 introduced in Eq. (1.142)—the notation σ_s is chosen here to avoid confusion with the Rabi coefficient κ.

Transverse effects have been neglected in the search for steady-state solutions and in the evolution equations derived in the previous sections. In free space, the plane wave approximation will only hold within the confocal parameter of a beam of finite size. Even within this limit, transverse effects in coherent interaction may contribute to self-focusing and defocusing of the beam [14]. As a rule of thumb, the plane wave approximation for self-induced transparency may be considered to hold for approximately five linear absorption lengths. There are however two important types of confinements for which the single dimensional approach remains valid over long distances:

- optical fibers and
- optical cavities.

The search for steady-state solutions in conditions of coherent interactions has more than purely academic interest. The main motivations for these studies relate to

- stability,
- minimizing energy losses, and
- maximizing energy extraction.

We have seen in the previous section that the 2π pulse in absorbing media acts as a filter, "cleaning" the signal from amplitude and phase fluctuations. The steady-state pulse in an absorber has minimum energy loss, because it returns the absorbing two-level system to the ground state. In an amplifier or in a laser, steady-state pulses can be found that bring the two-level system from inversion to the ground state—and hence extract the maximum energy possible from the gain medium.

Steady-State Pulses in Amplifiers

For finite T_2, stable steady-state pulses do not exist in absorbing media. The loss of coherence leads to irreversible energy losses, causing to the collapse of an initial 2π pulse. In the case of an amplifier, however, steady-state pulses can be found even in the presence of homogeneous broadening, because there is a gain mechanism to compensate for the losses. Because there is no mechanism to "push back" in time, the leading edge of the pulse as in the case of "2π pulse propagation," the envelope velocity will generally be close to the phase velocity in an amplifier. In searching for steady-state solutions off-resonance, it is important to take into consideration the correct *dispersive* response, and the corresponding *phase velocity* v_p. It is this same near-resonant phase velocity—as opposed to the phase velocity in the host medium c/n—that is used in the definition of the slowly varying components of a propagating physical quantity $\tilde{X}$:

$$X = \frac{1}{2}\tilde{X}\exp[i\omega_\ell(t - \frac{z}{v_p})]. \tag{4.39}$$

Following the procedure of Petrov and Rudolph, [15] we start directly from the second-order Maxwell equation for a linearly polarized plane wave:

$$\frac{\partial^2 E}{\partial z^2} - \frac{1}{c^2}\frac{\partial^2 E}{\partial t^2} = \mu_0\left[\frac{\partial^2 P}{\partial t^2} + \sigma_s\frac{\partial E}{\partial t}\right]. \tag{4.40}$$

We are looking for solutions that satisfy the steady-state condition for the slowly varying envelopes, describing "form-stable" pulse propagation. In a frame of reference moving with the (yet unknown) envelope velocity v_e, any function $\tilde{X}$ associated with the pulse (i.e., pulse envelope, polarization, medium inversion) should remain unchanged:

$$\frac{\partial}{\partial z}\tilde{X}(t - z/v_e, z) = 0. \tag{4.41}$$

It can be shown that, within the SVEA, $v_e \approx v_p$. Substituting the slowly varying field and pseudo-polarization amplitudes (4.6) and (4.7) defined in the section on Maxwell–Bloch equations into Eq. (4.40), we find:

$$\beta\tilde{\mathcal{E}} + ig\tilde{Q} - i\gamma\tilde{\mathcal{E}} = 0 \tag{4.42}$$

where we have defined:

$$\beta = \omega_\ell^2\left(\frac{1}{v_p^2} - \frac{1}{c^2}\right) \tag{4.43}$$

$$g = \mu_0\omega_\ell^2 \tag{4.44}$$

$$\gamma = \mu_0\omega_\ell\sigma_s. \tag{4.45}$$

Equation (4.42) expresses that, at steady-state, the "slippage" of the pulse envelope with respect to the wave (term β) results from a balance between gain (second term) and scattering losses (third term). According to Eq. (4.42), the steady-state pseudo-polarization $\tilde{Q} = Q\exp(i\vartheta)$ is proportional to the field $\mathcal{E}$:

$$\tilde{\mathcal{E}} = \tilde{A}\tilde{Q} \tag{4.46}$$

where

$$\tilde{A} = \frac{g\gamma - i\,g\beta}{\beta^2 + \gamma^2}. \tag{4.47}$$

Substituting in Eq. (4.6) for the time evolution of the pseudo-polarization yields:

$$\dot{Q} + i\dot{\vartheta}Q = -i(\omega_\ell - \omega_0)Q - \kappa w\tilde{A}Q - \frac{Q}{T_2}. \tag{4.48}$$

The real and imaginary parts of Eq. (4.48) and Bloch's equation (4.7) for the population difference w form a complete set:

$$\dot{Q} = -\frac{g\gamma\kappa}{\beta^2 + \gamma^2}wQ - \frac{Q}{T_2} \tag{4.49}$$

$$\dot{\vartheta} = -(\omega_\ell - \omega_0) + \frac{\kappa g\beta}{\beta^2 + \gamma^2}w \tag{4.50}$$

$$\dot{w} = \frac{\kappa g\gamma}{\beta^2 + \gamma^2}Q^2. \tag{4.51}$$

Taking the time derivative of the first of these equations [Eq. (4.49)], and substituting $\dot{w}$ from the last Eq. (4.51), we find a (relatively) simple equation which is known to have a sech solution for functions Q that tend to zero at both ends of the time scale [16]:

$$Q\ddot{Q} - \dot{Q}^2 + \left[\frac{\kappa g\gamma}{\beta^2 + \gamma^2}\right]^2 Q^4 = 0. \tag{4.52}$$

Substituting the sech solution $\tilde{\mathcal{E}}(t) = \mathcal{E}_0\text{sech}(t/\tau_s)\exp(i\varphi)$ into Eq. (4.46), and hence a form $\tilde{Q}(t) = Q_0\text{sech}(t/\tau_s)\exp(i\vartheta)$ in Eq. (4.52), yields a relation between β and the field amplitude $\mathcal{E}_0$. If we choose the carrier frequency ω_ℓ to be the average pulse frequency, we impose that the average phase derivative over the pulse duration is zero ($\int_{-\infty}^{\infty}\dot{\varphi}|\mathcal{E}_0|^2 = 0$). Equation (4.49) provides an additional relation to determine the steady-pulse parameters [pulse duration τ_s (time normalization factor of sech pulse shape, as defined in Table 1.1), phase φ, amplitude $\mathcal{E}_0$,

Table 4.1

Main parameters associated with steady-state pulses in homogeneously broadened amplifiers. The pulse duration τ_s is defined in Table 1.1.

		Steady-State pulse
Envelope	$\mathcal{E}$	$\mathcal{E}_0 \text{sech}(t/\tau_s) e^{i\varphi(t)}$
Field amplitude	$\mathcal{E}_0$	$[1/(\kappa\tau_s)]\sqrt{1 + (\omega_\ell - \omega_0)^2 T_2^2}$
Pulse duration	τ_s	$T_2 \left\{[\alpha_0/(\sigma_s \kappa^2 \mathcal{E}_0^2 \tau_s^2)] - 1\right\}$
Average carrier frequency	ω_ℓ	ω_ℓ
Chirp	$\dot{\varphi}(t)$	$\tau_s^{-1}[T_2(\omega_\ell - \omega_0)]^2 \tanh(t/\tau_s)$
Pulse area	θ_0	$\pi\sqrt{1 + (\omega_\ell - \omega_0)^2 T_2^2}$
Pulse energy density	W	$\mathcal{E}_0^2 \tau_s \sqrt{\frac{\epsilon}{\mu_0}}$

and frequency ω_ℓ]. The relations yielding the steady-state pulse parameters are summarized in Table 4.1.

The larger the amount off-resonance, the larger the field amplitudes. A larger "power broadening" is required to extract the same energy from the gain medium off-resonance than on-resonance. This class of solutions include the "π pulse" solution of Arecchi and Bonifacio [17, 18] and the "$\pi\sqrt{2}$" solution derived in Diels and Hahn [11, 13]. These types of coherent steady-state pulses are optimizing the energy extraction from the amplifier.

4.3. MULTIPHOTON COHERENT INTERACTION

4.3.1. Introduction

The high intensities associated with fs pulses lead to a nonlinear response of the real and imaginary parts of the polarization. The imaginary part of the nonlinear polarization is associated for instance with multiphoton transitions and will exhibit a n-photon resonance when two levels of an atomic or molecular system can be connected by n optical quanta. As for single photon transition, there will be a linewidth and dephasing time associated with the higher-order resonance. When the fs pulses are shorter than the multiphoton phase-relaxation time, we are dealing with *coherent transient resonant multiphoton interaction*. This situation is as complex as its name, because we are cumulating the problems associated with nonlinear optics, coherent phenomena, transient absorption, transient dispersion, and propagation. In fact it is so complex that few have addressed this problem either experimentally or theoretically. One might therefore wonder whether there is any benefit in even considering such situations.

Before we lay down the theoretical framework of multiphoton coherent resonant interaction, we show the potential benefits of fs coherent excitation with a few simple examples (an encouragement to the reader to proceed to the more tedious theoretical subsection).

There is a basic property of coherent transient interaction that is common to single photon and multiphoton transitions. There is no longer a saturating intensity that tends to equalize the population of the levels connected by the radiation. Radiation coherently interacting with two-level systems can transfer all the population from one level into the other. If the two levels are initially inverted (amplifier), the total energy stored in the two-level system is transferred to the radiation (π pulse amplification). In the case of a system initially in the ground state (absorber), lossless transmission can be observed (2π pulse propagation). A typical application is harmonic generation in the presence of multiphoton resonances. The resonant enhancement leads to a better conversion efficiency, which saturates partly because of multiphoton absorption. Multiphoton coherent propagation effects can be exploited to achieve larger penetration depths—and hence better conversion.

Transitions always proceed via intermediate levels. In some cases, the effect of numerous intermediate levels far off-resonance can be combined into one or more "virtual" intermediate state(s). The multilevel system of equations then reduces to coupled equations involving only the two extreme upper and lower resonant levels. This approximation is typical of atomic systems where only a few levels are directly connected within the pulse bandwidth at low quantum numbers. In these systems, if two levels are connected by an n-photon transition, there is generally no intermediate resonance (of order $m < n$) within the pulse bandwidth.

The situation is different in molecules. There are generally so many levels around that one can describe the n-photon process as a "ladder" of single photon transitions. In the presence of a large number of levels—all with a different detuning with respect to the radiation—maintaining some degree of coherence requires a particular excitation tailored to the transfer function of the system. Pulse shaping techniques have reached such a level of sophistication that it should be possible to provide a fs excitation with a shape designed to produce a pure inversion or frequency selective excitation of a particular level. The example of the multiphoton excitation of CH_3F by a *group of pulses* that will be considered in Section 4.4.2 shows that even in the case of such a complex anharmonic vibrational–rotational ladder, highly *selective* excitation can be achieved with *broadband* pulses. This complex problem has a solution nearly as simple as that of the unfortunate "Romeo" ape of Figure 4.7(a) faced with an anharmonic ladder. In that analogy also, getting a group of friends to make a multiape coherent transition could succeed in getting the system in the upper state.

We will start in Section 4.4.2 with the more general (and complex) systems comprising a large number of levels (such as molecules). In the following

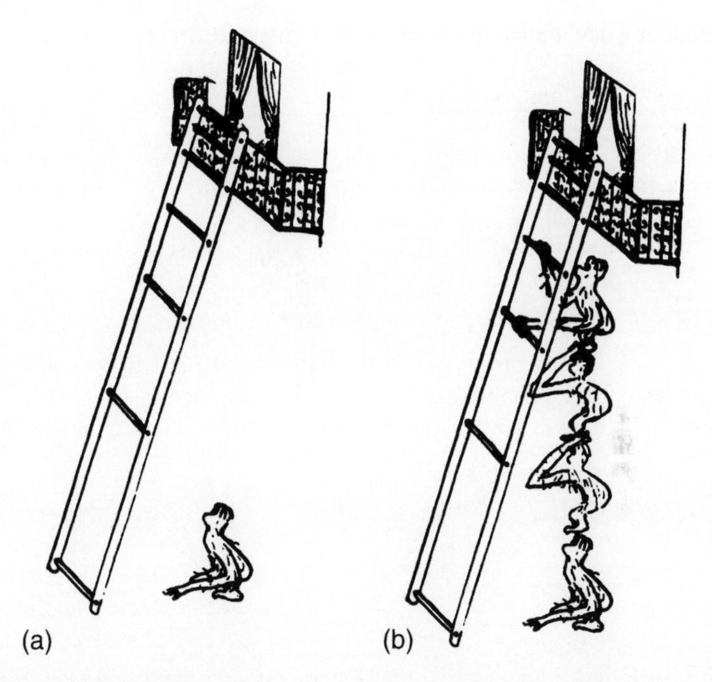

Figure 4.7 The problem of reaching a high level with a harmonic ladder (a), and the multiple-pulse multiphoton coherent solution (b).

section, Section 4.4.3. we will proceed from this more general situation to the particular approximation that leads to the multiphoton analog of Bloch's optical equations.

4.3.2. Multiphoton Multilevel Transitions

General Formalism

The complex atomic–molecular system is represented by the unperturbed Hamiltonian H_0 to which corresponds a set of eigenstates ψ_k of energy $\hbar\omega_k$. In the presence of an electric field E, the state of the atomic–molecular system is described by the wavefunction ψ, a solution of the time–dependent Schrödinger equation:

$$H\psi = i\hbar\frac{\partial\psi}{\partial t}, \tag{4.53}$$

with the total Hamiltonian given by:

$$H = H_0 + H' = H_0 - p \cdot E(t) \tag{4.54}$$

where p is the dipole moment. In the standard technique for solving time-dependent problems, the wave function ψ is written as a linear combination of the basis functions ψ_k:

$$\psi(t) = \sum_k a_k(t)\psi_k. \tag{4.55}$$

This expression for ψ is inserted in the time-dependent Schrödinger equation (4.53). Taking into account the normalization conditions for the basis functions ψ_k, one finds the coefficients a_k have to satisfy the following set of differential equations:

$$\begin{aligned}\frac{da_k}{dt} &= -\frac{i}{\hbar}\hbar\omega_k a_k + \sum_j \frac{i}{\hbar} p_{k,j}\mathcal{E}(t)\cos[\omega_\ell t + \varphi(t)]a_j \\ &= -i\omega_k a_k + \sum_j \frac{i}{2\hbar} p_{k,j}[\tilde{\mathcal{E}}e^{i\omega_\ell t} + \tilde{\mathcal{E}}^* e^{-i\omega_\ell t}]a_j\end{aligned} \tag{4.56}$$

where p_{kj} are the components of the dipole coupling matrix[2] for the transition $k \to j$, and a_k are the amplitudes of the eigenstates.

Apart from the basic assumption that the time scale is short enough for all phase and amplitude relaxation to be negligible, Eq. (4.56) is of a quite general nature, and it is ideally suited to numerical integration. It can be used to solve the most complex problem of light–matter interaction, because no assumptions have been made as to the transitions, resonances, detuning, or degeneracies. However, the complete numerical treatment leaves little room for physical insight. We will therefore consider some simplified cases for which general trends emerge from the solution. Even though numerical analysis is still required, it seems possible to gain some intuition as to the response of the system.

As a first simplifying assumption, we will consider only the levels that can be connected directly by a single photon at ω_ℓ. The separations of the successive energy levels k from the ground state 0 are $\hbar\omega_{0k}$. We designate with an index k a state at an energy close to $\hbar k\omega_\ell$ (the ground state being thus labeled with the index 0). The important levels to consider will be those sketched in Figure 4.8(a) for which the detuning:

$$\Delta_k = \omega_{0k} - k\omega_\ell \tag{4.57}$$

[2] We will use both the notation p_{ik} and $p_{i,k}$ depending on the complexity of the subscripts.

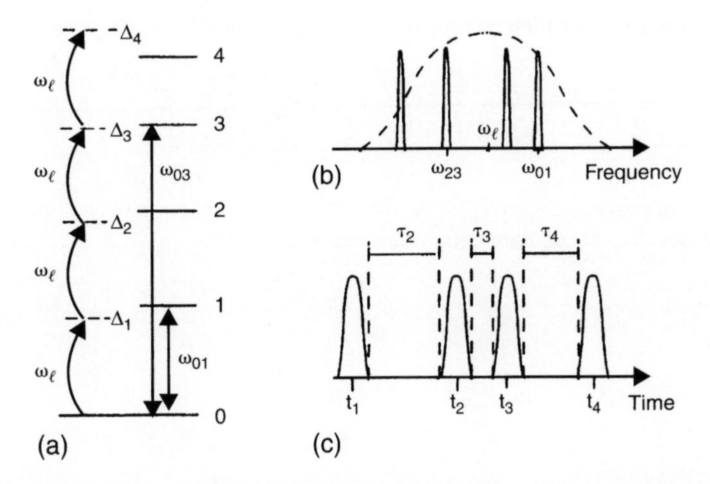

Figure 4.8 (a) Energy level diagram, (b) spectral representation, and (c) sequence of exciting pulses.

is at most comparable to the sum of the radiation bandwidth and the transition rate. Such a near-resonance condition is illustrated by the sketch of Fig. 4.8(b) where the dashed line indicates the pulse spectrum overlapping the successive single photon transitions. Consistent with this approximation, we replace the set of coefficients a_k, which have temporal variations at optical frequencies, by the "slowly varying" set of coefficients c_k, using the transformation:

$$a_k = e^{-ik\omega_\ell t} c_k. \tag{4.58}$$

This is the same type of transformation to a "rotating set of coordinates" as we applied to the set of density matrix equations (4.2) and (4.3) in the section on two-level systems.

Selecting from the sum in Eq. (4.56) the particular pairs of levels that matches best the resonance condition for an incident field at frequency ω_ℓ, leads to the simpler form:

$$\frac{dc_k}{dt} = -i\Delta_k c_k + \frac{i}{2\hbar} p_{k-1,k} \tilde{\mathcal{E}}^*(t) c_{k-1} + \frac{i}{2\hbar} p_{k,k+1} \tilde{\mathcal{E}}(t) c_{k+1}. \tag{4.59}$$

The use of amplitudes c_k rather than elements of the density matrix is more convenient in dealing with multilevel systems. It is easier, however, to get a physical picture from a density matrix representation. The density matrix elements can be calculated from $\rho_{ii} = a_i a_i^*$ and $\rho_{ij} = \varrho_{ij} e^{i\omega_\ell t} = a_i a_j^* e^{i\omega_\ell t}$ for $i \neq j$. To verify that the two descriptions are equivalent, let us look at the equations for a two-level system which can be written in the form:

$$\frac{d}{dt}\begin{pmatrix} c_0 \\ c_1 \end{pmatrix} = \begin{pmatrix} 0 & i\frac{p_{01}\tilde{\mathcal{E}}}{2\hbar} \\ i\frac{p_{01}\tilde{\mathcal{E}}^*}{2\hbar} & -i\Delta_1 \end{pmatrix}\begin{pmatrix} c_0 \\ c_1 \end{pmatrix}. \tag{4.60}$$

Multiplying by c_i^* $(i = 0, 1)$, we find the following set of equations for the amplitudes $c_i^* c_j$ of the elements of the density matrix:

$$\dot{\rho}_{11} - \dot{\rho}_{00} = \frac{d}{dt}(c_1 c_1^* - c_0 c_0^*) = \frac{ip_{01}}{\hbar}[\tilde{\mathcal{E}}^*(c_0 c_1^*) - \tilde{\mathcal{E}}(c_0 c_1^*)^*] \tag{4.61}$$

$$\dot{\varrho}_{01} = \frac{d}{dt}(c_0 c_1^*) = i\Delta_1 c_0 c_1^* + \frac{ip_{01}}{2\hbar}\tilde{\mathcal{E}}(\rho_{11} - \rho_{00}). \tag{4.62}$$

The substitutions:

$$\kappa\tilde{\mathcal{E}} = \frac{p_{01}}{\hbar}\tilde{\mathcal{E}} \tag{4.63}$$

$$\frac{1}{2}\tilde{Q} = ic_0 c_1^* p_{01}\bar{N} \tag{4.64}$$

$$\rho_{11} - \rho_{00} = \frac{w}{p_{01}\bar{N}}, \tag{4.65}$$

lead identically to the equations of motion for the pseudo-polarization vector [Eqs. (4.6) and (4.12)] derived earlier in this chapter. Consequently, some form of Bloch vector model can be used whenever the multilevel system can be reduced to a two-level system. It will be shown in the next subsection that such a simplification can be made even for multiphoton processes and n levels under certain conditions of detuning of the intermediate levels.

To simplify the notation, we will use complex Rabi frequencies to represent the electric field of the radiation:

$$\tilde{E}_{k+1} = \frac{i}{2\hbar}p_{k+1,k}\tilde{\mathcal{E}} = \frac{i}{2\hbar}p_{k+1,k}\tilde{\mathcal{E}}e^{\varphi_k}. \tag{4.66}$$

This particular notation is only used in the following sections of this chapter. The complex quantity $\tilde{E}_{k+1}$ represents the complex electric field envelope, expressed in units of frequency (the Rabi frequency for the transition $k \to k+1$). This definition involves an arbitrary choice of phase. In general, the initial cycles (near $t = 0$) of the applied electromagnetic field are taken to have zero phase $[\varphi_k(t \approx 0) = 0]$. The definition (4.66) corresponds to transitions from level k to level $k + 1$ in Eq. (4.59). The same definition applied to transitions from the lower level [$k-1 \rightarrow k$ in Eq. (4.59)] yields $\frac{i}{2\hbar} p_{k-1,k} \tilde{\mathcal{E}}^*(t) c_{k-1} = -\tilde{E}^*_{k-1}$. Let us consider the harmonic oscillator approximation, for which the dipole moments are proportional to the square root of the ratio of the indices of the successive levels [19]: $p_{k+1,k} \propto \sqrt{(k+1)/k}$. It results from this proportionality that the Rabi frequencies [defined by Eq. (4.66)] are proportional to the Rabi frequency of the first transition $0 \rightarrow 1$: $\tilde{E}_k = \sqrt{k}\,\tilde{E}_1$, where k is the level index. Let us assume in addition a constant anharmonicity: Each successive transition frequency is reduced by the same frequency χ:

$$\begin{aligned} \omega_{0,j} - \omega_{0,j-1} &= \omega_{0,j-1} - \omega_{0,j-2} - \chi = \ldots \\ &= \omega_{0,1} - (j-1)\chi. \end{aligned} \tag{4.67}$$

As an illustrative example, let us consider a four-level system excited by a single pulse of average frequency ω_ℓ. With the above mentioned assumptions and notations, the system of equations (4.56) reduces to:

$$\frac{d}{dt}\begin{pmatrix} c_0 \\ c_1 \\ c_2 \\ c_3 \end{pmatrix} = \begin{pmatrix} 0 & \tilde{E}_1 & 0 & 0 \\ -\tilde{E}_1^* & -i\Delta_1 & \tilde{E}_2 & 0 \\ 0 & -\tilde{E}_2^* & -i\Delta_2 & \tilde{E}_3 \\ 0 & 0 & -\tilde{E}_3^* & -i\Delta_3 \end{pmatrix} \begin{pmatrix} c_0 \\ c_1 \\ c_2 \\ c_3 \end{pmatrix}. \tag{4.68}$$

Equation (4.59) and the particular four-level example Eq. (4.68) were established assuming the presence of levels nearly equally spaced (cf. Fig. 4.8), and a single frequency source of ultrashort pulses. We leave it as a problem at the end of this chapter to generalize this treatment to a polychromatic source of several pulses, matching stepwise transitions of a discrete level system, such as that of a simple atom. It can be shown that Eqs. (4.53) and (4.68) still apply, with an appropriate redefinition of the complex Rabi frequencies $\tilde{E}_i$ and detunings Δ_i.

Climbing the Ladder

One of the main interests in coherent interactions with ultrashort pulses is efficient creation of a population inversion. We will use the example of an anharmonic multilevel system to illustrate the high efficiency of optical pumping that can be achieved in conditions of coherent interactions.

The set of interaction equations (4.68) is particularly convenient for the study of the response of the multilevel system to a series of identical pulses (random sequence or regular train). Let us consider, for instance, the simple level structure of Fig. 4.8. We will assume that only the first four transitions participate in the excitation process. The system is excited by the sequence of four ultrashort pulses of Fig. 4.8(c), with amplitude:

$$\tilde{\mathcal{E}}(t-t_1)+\tilde{\mathcal{E}}(t-t_2)e^{i\varphi_1}+\tilde{\mathcal{E}}(t-t_3)e^{i\varphi_2}+\tilde{\mathcal{E}}(t-t_4)e^{i\varphi_3} \tag{4.69}$$

where t_i denotes the pulse delay. The interaction with a single pulse can be represented by a matrix that transforms any set of initial $c_i(t_0)$ into a set of coefficients $c_i(t_1^+)$ (where t_1^+ designates the time just following the pulse). The phase of each pulse corresponds to a rotation of that matrix. Between pulses, all the $\tilde{E}_i$ in Eq. (4.68) are zero, and the interaction matrix is diagonal. The solution of the system of equations (4.68) becomes in this case particularly simple. The eigenvalues are the successive detunings given by Eq. (4.57). For the four-level system, the coefficients c_i after the first pulse will evolve according to:

$$c_0(t)=c_0(t_1^+)$$
$$c_i(t)=c_i(t_1^+)e^{-\Delta_i(t-t_1^+)} \quad \text{for } i=1,2,3. \tag{4.70}$$

The c_i after a delay τ_2 [in Fig. 4.8(c)] are the initial conditions for the second pulse and so forth.

We have seen in the previous section how a π pulse excitation can completely invert a two-level system, while incoherent excitation or continuous irradiation (or irradiation with pulses longer than the phase-relaxation time T_2) can at most equalize the population of the upper and lower states. Similarly, "coherent pumping" can also be used to complete inversion of multilevel systems. The situation is, however, more complex because of the plurality of Rabi frequencies and detunings involved. There is no simple analytical solution to this problem, which is to find a waveform $\tilde{\mathcal{E}}(t)$ which transforms the initial population state (1, 0, ..., 0) into a final state (0, 0, ..., 1). A practical method leading to the multilevel inversion is to excite the system with sequences of pulses, applied in a time short compared with the phase-relaxation time. The general procedure is to optimize the relative phases and delays between pulses to maximize the transfers of population toward higher energy levels.

Before illustrating this concept by a few simple examples, let us examine first whether this problem should, in general, have a solution. Each pulse has set in motion a few near-resonant harmonic oscillators in the molecular system, of amplitude given by the matrix elements $c_i c_j^*$. After passage of the first pulse, the matrix element $c_i c_j^*$ is left with a phase $\varphi_{ij}^{(1)}$, which, between pulses, will increase linearly with time as $(\Delta_j-\Delta_i)(t-t_1^+)$ [Eq. (4.59)]. Two parameters can be adjusted for the next (second) pulse:

- the interpulse spacing τ_2, to ensure that several matrix elements $c_i^* c_j$ are "caught" in phase by the next (second) pulse, and

- the phase of the next pulse, φ_2, equal or opposite to that of the various matrix elements that were in phase after the delay τ_2.

For the sake of illustration, let us suppose that we want the second pulse to start in phase with $c_0c_1^*, c_1c_2^*$, and $c_2c_3^*$. The common phase has to satisfy simultaneously the equations:

$$\begin{aligned}\varphi_{01}^{(1)} + \Delta_1\tau_2 + m_1 \cdot 2\pi &= \varphi_2 \\ \varphi_{12}^{(1)} + (\Delta_2 - \Delta_1)\tau_2 + m_2 \cdot 2\pi &= \varphi_2 \\ \varphi_{23}^{(1)} + (\Delta_3 - \Delta_2)\tau_2 + m_3 \cdot 2\pi &= \varphi_2.\end{aligned} \tag{4.71}$$

The three equations (4.71) can be solved to determine τ_2, φ_2, m_1, m_2, and m_3. Because the m_i are integers, the system of equations (4.71) will, in general, have only approximate rather than exact solutions. Among the various approximate solutions within certain tolerance limits of φ_2, the one corresponding to the smallest delay τ_2 will be selected.

The procedure of selecting the proper delays and phases of the pulse train can be explained by means of a simple picture in the frequency domain. A sequence of equal pulses ($t_j - t_i$ is the delay between successive pulses) exhibits a modulated spectrum where the envelope is given by the single pulse spectrum. If the pulse delays and phases are chosen properly the modulated spectrum matches the transition profile of the energy ladder. Of course, finding the optimum pulse sequence is more complex than a simple spectral overlap of the optical excitation and the transition frequencies, because of the nonlinearity of the interaction.

The various phases and elements ($c_ic_i^* = \rho_{ii}$, and $c_ic_j^* = \tilde{\varrho}_{ij}\exp[i\varphi_{ij}]$) of the density matrix can be represented, for convenience, at any stage during the excitation by the pulse sequence in the form of a matrix:

$$(\mathcal{M}) = \begin{pmatrix} \rho_{00} & 2|\tilde{\varrho}_{01}| & 2|\tilde{\varrho}_{02}| & 2|\tilde{\varrho}_{03}| \\ \dfrac{\varphi_{10}}{2\pi} & \rho_{11} & 2|\tilde{\varrho}_{12}| & 2|\tilde{\varrho}_{13}| \\ \dfrac{\varphi_{20}}{2\pi} & \dfrac{\varphi_{21}}{2\pi} & \rho_{22} & 2|\tilde{\varrho}_{23}| \\ \dfrac{\varphi_{30}}{2\pi} & \dfrac{\varphi_{31}}{2\pi} & \dfrac{\varphi_{32}}{2\pi} & \rho_{33} \end{pmatrix}.$$

Let us now apply the procedure outlined above to a specific multilevel system. The parameters chosen for this particular example are taken from the ν_3 C–F stretch mode of vibration of the molecule CH_3F. For the first few levels of the vibrational ladder the transition frequencies and detunings are:

$$\frac{\omega_{01}}{2\pi} = 197.66 \text{ ps}^{-1} \qquad \Delta_1 = 4.45 \text{ ps}^{-1}$$
$$\frac{\omega_{02}}{2\pi} = 392.34 \text{ ps}^{-1} \qquad \Delta_2 = 5.92 \text{ ps}^{-1}$$
$$\frac{\omega_{03}}{2\pi} = 584.42 \text{ ps}^{-1} \qquad \Delta_3 = 4.79 \text{ ps}^{-1}.$$

We assume that level 4 and higher energy levels are far off-resonance so that they can be neglected. Our intention is to show that indeed a suitable pulse sequence can lead to an almost complete inversion of the system (which is initially in the ground state). For the excitation of this model four-level system, we use four Gaussian pulses of 1 ps FWHM at 1025 cm^{-1} (or an angular frequency of $\omega_\ell = 193.20 \text{ ps}^{-1}$) [cf. Fig. 4.8(c)]. The pulse peak amplitudes (in terms of the complex Rabi frequency) are $|\tilde{E}_i(0)| = 2 \text{ ps}^{-1}$. This pulse amplitude and duration, for a dipole moment of $p_{01} = 0.21$ Debye,[3] corresponds to a pulse energy density of 5 J/cm^2. The pulse phases and delays are:

$$\varphi_1 = \varphi_2 = 0 \qquad \varphi_3 = \varphi_4 = \pi/3$$
$$\tau_2 = \tau_3 = 1.8 \text{ ps} \qquad \tau_3 = \tau_4 = 0 \text{ ps}. \tag{4.72}$$

The successive density matrices, $(\mathcal{M}_1)$ at a time τ_2 after application of the first pulse and just before arrival of the second pulse, $(\mathcal{M}_2)$ at time τ_3 (following the second pulse), and $(\mathcal{M}_4)$ at time τ_4 (following the fourth pulse) are:

$$(\mathcal{M}_1) = \begin{pmatrix} 0.71 & 0.57 & 0.44 & 0.56 \\ 0.21 & 0.11 & 0.18 & 0.23 \\ 0.24 & 0.03 & 0.07 & 0.17 \\ -0.18 & 0.11 & 0.08 & 0.11 \end{pmatrix}$$

$$(\mathcal{M}_2) = \begin{pmatrix} \mathbf{0.46} & 0.12 & 0.49 & \mathbf{0.87} \\ 0.08 & 0.01 & 0.06 & 0.11 \\ 0.04 & -0.04 & 0.13 & 0.45 \\ 0.25 & 0.17 & 0.21 & \mathbf{0.40} \end{pmatrix}$$

$$(\mathcal{M}_4) = \begin{pmatrix} \mathbf{0.01} & 0.05 & 0.04 & 0.22 \\ -0.02 & \mathbf{0.05} & 0.07 & 0.42 \\ -0.10 & -0.08 & \mathbf{0.03} & 0.31 \\ -0.05 & -0.03 & 0.05 & \mathbf{0.91} \end{pmatrix}.$$

[3] 1 Debye = 10^{-18} esu = (1/3) 10^{-29} C· m

We note that after the first pulse, the diagonal terms of the density matrix indicate a distribution of 29% of the population among the three excited states. The population density in the middle of the pulse sequence (after excitation by the second pulse, see matrix ($\mathcal{M}_2$)) is particularly interesting, for having the population nearly equally distributed among the two extreme (ground and upper) states, and a maximum value for the off-diagonal matrix element $|\tilde{\varrho}_{03}|$. The subsequent pair of exciting pulses (pulse 3 and 4) is identical to the first one, except that it is phase–shifted by $\pi/3$. That second pulse sequence results in a 91% inversion of the four-level system.

This type of excitation is the multiphoton analogue of a nonresonant "zero area pulse excitation" [20] for single photon transitions, which consists of a sequence of two pulses π out-of-phase. We recall that, in the case of single photon transitions, referring to the area theorem Eq. (4.24), a signal consisting of successive equal positive and negative (phase $= \pi$) half will have a zero area θ_0, and conserve its zero area during propagation. The Fourier transform of such a waveform is zero at its resonance frequency but has sidebands that can produce significant excitation of absorbing lines off-resonance.

In the preceding example of multistep excitation, each of the two successive pair of pulses acts in an analogous way on the four-level system. We have seen that a vector model can be constructed for two-level systems. It will be shown in the next section under which conditions such a model can be extended to a multiphoton resonance. In general, however, the submatrix $\begin{pmatrix} \rho_{00} & 2|\tilde{\varrho}_{0n}| \\ \ldots & \rho_{nn} \end{pmatrix}$ can lead to such a description at times τ_2, τ_3 and τ_4 [Fig. 4.8 (c)], but not *during* the application of the pulses, when all states are mixed by the electromagnetic field. The general strategy that emerged from a study of anharmonic systems [21] is to design a pulse (or a sequence of pulses) $\tilde{\mathcal{E}}_s(t)$ leading to a matrix $\begin{pmatrix} \rho_{00} & 2|\tilde{\varrho}_{0n}| \\ \ldots & \rho_{nn} \end{pmatrix} = \begin{pmatrix} 0.5 & 1 \\ \ldots & 0.5 \end{pmatrix}$. Repeating the same excitation, with a dephasing of π/n, i.e., applying $\tilde{\mathcal{E}}(t)\exp(i\pi/n)$, generally leads to a good approximation of a pure inversion $\begin{pmatrix} 0 & 0 \\ \ldots & 1 \end{pmatrix}$, as in the case of the four-level system considered previously.

The theory presented applies to any physical system that can be modeled adequately by a set of isolated levels as sketched in Fig. 4.8. It can be generalized to a step ladder excitation in atomic systems, where radiation pulses containing different frequencies $\omega_{\ell,i}$ matching the ladder of energy levels are applied.

A Molecular System

The basic approximation made so far is that we have discrete, isolated levels $(1, 2, 3, \ldots)$ separated by about the photon energy of the exciting pulse(s).

This, however does not apply to most molecular systems. In considering transitions of a real molecule, the model should be corrected to include all the vibrational–rotational transitions that fall within the excitation spectrum. That is, any level previously labelled by 1, 2, 3, ..., corresponds now to a certain vibrational band which contains many rotational levels. We will show in this section that this additional element of complexity can be handled numerically, and that the general conclusion—namely that a total inversion can be achieved with properly shaped input signals—remains valid. Throughout this section we will neglect transitions between different electronic levels.

Depending on the molecular structure the actual vibration–rotational spectra can be complex, and its detailed derivation is beyond the scope of this book. The reader is referred to the monograph by Herzberg [22] for example. In the case of molecular transitions, the wave function has to be expanded in a series of energy eigenfunctions corresponding to the quantum numbers υ (vibrational quantum number) and J (rotational quantum number). A dipole allowed transition can only occur between two vibrational bands that satisfy the selection rules $\Delta J = \pm 1$ and $\Delta\upsilon = \pm 1$. Therefore the initial population of rotational states determines essentially the number of rotational levels that needs to be considered for each vibrational state during the excitation.

The definition of the detuning has to be extended to take into account the rotational level structure. Instead of Eq. (4.57), we define the detuning of the level labeled with the quantum number (υ, J) as:

$$\Delta_{\upsilon,J} = \Delta_{\upsilon,0} + B\,J(J-1) \tag{4.73}$$

where

$$\Delta_{\upsilon,0} = \omega_{\upsilon,0} - \upsilon\omega_\ell \tag{4.74}$$

is the detuning of the molecule in the rotational ground state, and $\Delta_{0,0} = 0$. B is the rotational constant. To simplify the notations, we will use for the matrix elements of the dipole moment:

$$p_{\upsilon,+} = \langle \upsilon, J-1|p|J, \upsilon+1\rangle \tag{4.75}$$

$$p_{\upsilon,-} = \langle \upsilon, J+1|p|J, \upsilon+1\rangle. \tag{4.76}$$

The J dependence of the dipole moments $p_{\upsilon,+}$ and $p_{\upsilon,-}$ is given by [23, 24]:

$$p_{\upsilon,+} = p_\upsilon\sqrt{\frac{J+1}{2J+3}} \quad \text{and} \quad p_{\upsilon,-} = \sqrt{\frac{J}{2J-1}}. \tag{4.77}$$

For this simplified model of a vibrational–rotational level structure, the system of differential equations for the coefficients c_{kJ} takes a form similar to Eq. (4.59). The first two equations are:

$$\begin{aligned}
\frac{d}{dt}c_{0J} &= -i\Delta_{0J}c_{0J} + \tilde{E}_1[p_{0,+}c_{1,J-1} + p_{0,-}c_{1,J+1}] \\
\frac{d}{dt}c_{1J} &= -\tilde{E}_1^*[p_{0,-}c_{0,J+1} + p_{0,+}c_{0,J-1}] - i\Delta_{1J}c_{1J} \\
&\quad - \tilde{E}_2[p_{1,+}c_{2,J-1} + p_{1,-}c_{2,J+1}].
\end{aligned} \tag{4.78}$$

As for the previous Eq. (4.56), the frame of reference is rotating at the angular velocity ω_ℓ.

The system of equations (4.78) can be solved numerically starting with an initial (Boltzmann) distribution for the population of the rotational levels in the vibrational state ($\upsilon = 0$):

$$c_{0J}c_{0J}^* = A(J+1)e^{[-BJ(J+1)/k_BT]} \tag{4.79}$$

where k_B is Boltzmann's constant, T is the absolute temperature, and A a normalization factor (total population = 1). The numerical calculations were applied to the situation sketched in Figure 4.9 and will be detailed. A particular result is shown in Figure 4.10.

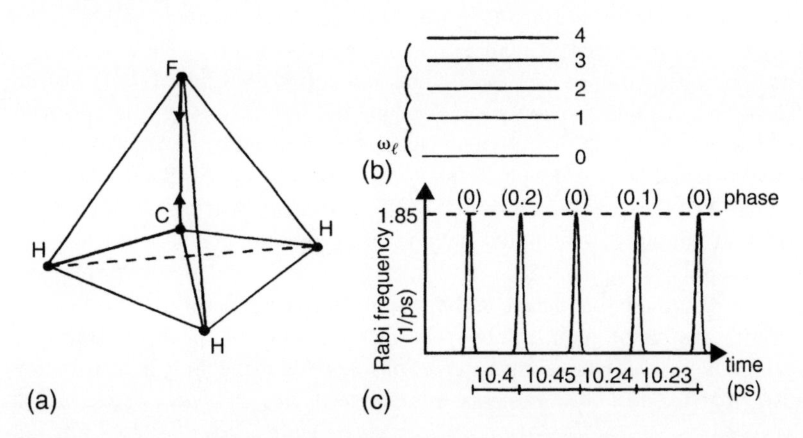

Figure 4.9 Multiphoton excitation of the CH_3F molecule. Sketch of the molecule identifying the C–F stretch mode (a), and corresponding energy level diagram (b). Optimized five pulse sequence leading to selective excitation of the fifth ($\upsilon = 4$) vibrational band (c). The absolute phase of each pulse (in radians) and the relative delays between pulses are indicated.

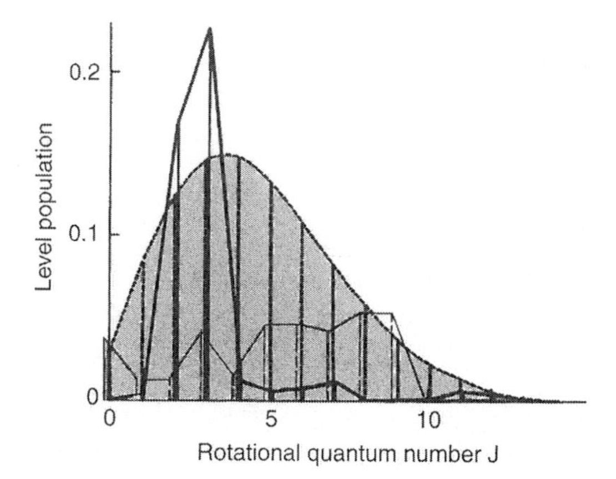

Figure 4.10 Selective pumping of the CF stretch mode of a CH_3F molecule by a sequence of five Gaussian pulses, each 1 ps FWHM and with a peak Rabi angular frequency of 1.85 ps^{-1}. With the first pulse taken as reference for phase and time, the successive pulses are at t = 10.4, 20.85, 31.09, and 41.32 ps, with a phase of 0.2, 0, 0.1, 0 radian, respectively. The ground state population versus rotational quantum number is represented before (dotted lines, shaded area) and after (thin dashed line) the five pulse excitation, as a function of rotational quantum number J. The heavy lines indicate the population distribution among the rotational lines after the four pulse excitation. The radiation wavelength is 9.7713 μm. (Adapted from [21].)

Let us choose again as an example the excitation of the C–F stretch mode of the CH_3F molecule (Fig. 4.9), but now taking into account rotational levels. The transition frequencies are $\omega_{1,0}/(2\pi c) = 1048.6$ cm^{-1}, $\omega_{2,0}/(2\pi c) = 2081.4$ cm^{-1}, $\omega_{3,0}/(2\pi c) = 3100.4$ cm^{-1}, and $\omega_{4,0}/(2\pi c) = 4104.4$ cm^{-1}. The excitation consists of a sequence of five identical Gaussian pulses for which $1/\lambda_\ell = \omega_\ell/2\pi c = 1023.4$ cm^{-1}, slightly below exact four photon resonance with the upper vibrational level (Fig. 4.9). To compute the interaction with the train of pulses, a system of circa 50×4 coupled equations (4.78) has to be solved numerically.[4] As a result of the excitation by a first pulse, a complex distribution of populations is created among the rotational lines in each vibrational state. When a second identical pulse is applied, the modification of the population distribution is a function of the relative delays and phases of the two pulses.

The question of main interest—in fact the main purpose of dealing with the complexities of coherent interactions—is: can one find a distribution of pulses that excites *only a single rotation level in the upper vibrational state*

[4] The exact number of equations needed depends on the number of rotational states initially populated, incremented by the number of additional states being accessed by stepwise $J \rightarrow J+1$ excitations.

(number 4)? The transfer function of this problem is too complex to solve the general inverse problem—i.e., find the pulse shape, intensity, phase, and spacing that would provide optimal selective excitation of the vibrational state $\upsilon = 4$. Numerical algorithms, however, exist to solve such problems in particular cases. With some intuition and luck, a trial and error method can, after a few iterations, lead to quite satisfactory solutions. The general guidelines are as follows:

- Each pulse should be short and intense enough to meet the bandwidth requirements of the anharmonicity. As indicated in Fig. 4.8(b), the spectrum of a single pulse (dashed line) should cover the frequencies of the successive transitions to be excited. We choose 1 ps pulses, of intensity corresponding to a peak Rabi frequency of $|\tilde{E}(0)_1| = 1.85\ 10^{12}\ \mathrm{s}^{-1}$ (notation defined by Eq. (4.66).
- The interpulse delay should be close to $\pi/2B$ (10.45 ps for CH_3F), which is one half of the period of the difference frequency between adjacent rotational lines.

Because there are two upward transition possible for any particular level with quantum numbers υ, J ($\upsilon, J \rightarrow \upsilon + 1, J + 1$ and $\upsilon, J \rightarrow \upsilon + 1, J - 1$), we can expect the initial Boltzman distribution of population in the ground vibrational state to be transposed as a broader distribution in the excited states (rotational line heating). The particular choice of interpulse delay ($\pi/2B$) should minimize this effect. Indeed, let us consider an isolated pair of vibrational levels resonant with the laser radiation, in a frame of reference rotating at the angular velocity corresponding to the (forbidden) $J \leftrightarrow J$ transition. A first pulse creates off-diagonal elements for the transitions $J \rightarrow J + 1$ and $J \rightarrow J - 1$. The precession after a delay $\pi/2B$ brings these two matrix elements exactly in opposite phase; hence, they cancel each other. The transitions to the next vibrational level will populate the sublevel J from a pure state at $J - 1$ and $J + 1$.

Systematic calculations of the excitation to the fifth level, for pulse sequences of various relative phases and delays, lead to the optimal five-pulse sequence sketched in Fig. 4.9(c). The population distribution created by the pulse sequence in the upper level ($\upsilon = 4$) is shown in Fig. 4.10 (heavy line), as a function of rotational quantum number. For comparison the Boltzmann distribution taken as initial condition for the ground state $\upsilon = 0$ (at a temperature of $T = 250^{\circ}$K) is indicated by the light gray area under the dashed curve. The thin broken line shows the population distribution in the ground state after the pulse sequence. It is remarkable to note to which extent the choice of interpulse delay of $\pi/2B$ resulted in a cooling of the rotational temperature *in the excited state*. Indeed, the pulse sequence has essentially resulted in transferring most of the populations in the $J = 2$ and $J = 3$ rotational state of the upper vibrational level. A population inversion has been achieved for the vibrational levels (higher total population in $v = 4$ than in $v = 0$).

4.3.3. Simplifying a N-Level System to a Two-Level Transition

Taking into account the detailed level structure of a complex molecule, following the procedure of the previous section is an arduous task. The atomic or molecular system can be assimilated to a two-level system (the upper and lower levels) if the intermediate levels are sufficiently off-resonance. To quantify the conditions under which multiphoton transitions—as opposed to the cascade of single photon transitions discussed in the previous sections—will be observed, two problems have to be solved. First, one has to determine the detuning $\Delta_n = \omega_{n0} - n\omega_\ell$ that provides the maximum amount of mixing (i.e., the largest amount of population exchange for a given optical field) between the extreme levels. Second, we have to determine the minimum detuning of the intermediate levels for which the perturbation of the intermediate levels is negligible. For simplicity of the analysis, we will approximate the pulse by a square wave.

Two main assumptions are that (a) the detuning Δ_k of the intermediate level k is larger than the Rabi frequencies $\tilde{E}_k$ defined in the previous section, and that (b) there is no resonance between any pair of intermediate levels (i.e., $\omega_{ij} \neq \omega_\ell$).

The approach in the following paragraph requires familiarity with properties of matrices, such as can be found in Franklin [25]. It is not essential to the understanding of the remainder of this book. The same final result has also been obtained by a totally different approach [26].

We start from the time-dependent Schrödinger equation (4.53) for the n-level system. The wave function $\psi(t)$ is expanded the usual way [cf. Eq. (4.55)] in terms of the eigenfunctions ψ_k, corresponding to the eigenvalues $\hbar\omega_k$ of the unperturbed Hamiltonian H_0 [defined in Eq. (4.54)]. We proceed also to the same rotating wave approximation to represent the interaction by the set of differential equation put in matrix form in Eq. (4.68) (the latter equation is readily generalized to a n-level system).

Because we have assumed the field to be constant during the interaction (square pulse approximation), the coefficients $c_k(t)$ of this expansion can be found exactly through a calculation of the eigenvalues and eigenvectors of the matrix of this set of differential equations. The coefficients $c_k(t)$, solution of Eq. (4.68), are given by:

$$c_k(t) = \sum_{0}^{n-1} x_{kj} e^{\lambda_j t}, \tag{4.80}$$

where x_{kj} are the eigenvectors corresponding to the eigenvalues λ_j of the interaction matrix. Because of the particular form of the interaction matrix $\mathcal{H}$ [the $n \times n$ square matrix in Eq. (4.68)], some recurrence relations exist to systematically search for the eigenvalues and eigenvectors. Indeed, the matrix $\mathcal{H}$ belongs to the class of tri-diagonal matrices, so-called because the only nonzero elements are of the type a_{ii}, $a_{i,i\pm1}$ and $a_{i\pm1,i}$. Properties of these matrices are given in Franklin, [25] pages 251–253. Let Ψ_{n-1} be the determinant of the characteristic equation of the interaction matrix $\mathcal{H}$, and $\Psi_{k-1}(\lambda)$ be the determinant of the submatrix containing the first k rows and columns:

$$\Phi_{k-1}(\lambda) = (H_{k-1,k-1} - \lambda)\Phi_{k-2}(\lambda) - H_{k-1,k-2}H_{k-2,k-1}\Phi_{k-3}(\lambda). \tag{4.81}$$

The characteristic equation $\Phi_{k-1}(\lambda) = 0$ has in general k solution which are the eigenvalues $\lambda_0, \lambda_1, \ldots \lambda_{k-1}$. There are in general k eigenvalues. To each solution λ_j corresponds a set of solutions or eigenvector of the eigenvalue equation:

$$\begin{pmatrix} H_{00}+\lambda & H_{01} & 0 & \ldots & \ldots \\ H_{10} & H_{11}+\lambda & H_{12} & 0 & \ldots \\ 0 & H_{21} & H_{22}+\lambda & H_{23} & \ldots \\ \ldots & 0 & \ldots & \ldots & \ldots \end{pmatrix} \begin{pmatrix} x_0 \\ \vdots \\ x_{k-1} \end{pmatrix} = 0. \tag{4.82}$$

Recurrence relations can be found in Franklin [25] for the eigenvectors of matrix $\mathcal{H}$:

$$\begin{aligned} x_{k-1} &= \Phi_{k-2}(\lambda); \\ x_\nu &= (-1)^{k-\nu} H_{\nu,\nu+1} \ldots H_{k-1,k}\Phi_{\nu-1}(\lambda), \end{aligned} \tag{4.83}$$

where ν takes any value $< (k-1)$, and λ can take any of the λ_j values. The sum of the roots is equal to the sum of the diagonal elements $\sum_0^{k-1} H_{ii}$ and is the coefficient of the term of the characteristic equation in λ^{k-1} where k is the order of the submatrix. The product of the roots is equal to the value of the determinant and is equal and opposite to the constant term of the characteristic equation.

These recurrence relations are sufficient to determine, for a given n-level system, the n roots λ_i of the characteristic equation, and the n^2 numbers $x_k(\lambda_i)$ that constitute the n eigenvectors of the characteristic equation, needed to determine the coefficient c_k according to Eq. (4.80). As stated previously, it is desirable to be able to reduce the n-level system to an equivalent two-level system, between which the radiation induces multiphoton transitions. We outline the procedure of this reduction for a three-level system, leaving the generalization to a n-level as a problem.

In the particular case of the three-level system, the characteristic equation is:

$$\Phi_2(\lambda) = \begin{vmatrix} \lambda & \tilde{E}_1 & 0 \\ -\tilde{E}_1^* & -i\Delta_1 + \lambda & \tilde{E}_2 \\ 0 & -\tilde{E}_2^* & -i\Delta_2 + \lambda \end{vmatrix} = 0$$

$$= \lambda^3 - i(\Delta_1 + \Delta_2)\lambda^2 - (\Delta_1\Delta_2 - |\tilde{E}_1|^2 - |\tilde{E}_2|^2)\lambda - i\Delta_2|\tilde{E}_1|^2. \quad (4.84)$$

According to the recurrence expressions (4.83), the solutions $x_k(\lambda_i)$ $(k, i = 0, 1, 2)$ should be proportional to:

$$x_0 = \tilde{E}_1\tilde{E}_2 \quad (4.85)$$

$$x_1 = -\lambda\tilde{E}_2 \quad (4.86)$$

$$x_2 = |\tilde{E}_1|^2 - \lambda(i\Delta_1 - \lambda), \quad (4.87)$$

where λ takes any of the three values, solution, of the characteristic equation (4.84). Consistent with the original assumption that $\Delta_1 \gg \Delta_2$, $|\tilde{E}_1|$, $|\tilde{E}_2|$, one approximate solution to the characteristic equation (4.84) is $\lambda = i\Delta_1$. With a large detuning of the intermediate level, we will have thus one eigenvalue $\lambda_1 \approx i\Delta_1 \gg \lambda_0, \lambda_2$. Consistent with the absence of population transfer to and from level 1, the coefficient c_1 is proportional to $\exp(i\Delta_1 t)$. One of the eigenvalues (λ_1) is much larger than the two others, because according to Eq. (4.86), the values $x_1(\lambda_i)$ are proportional to λ.

Because our goal is to reduce the three-level system to an equivalent two-level system, the characteristic equation (reduced to second order) should have the same form as that of a two-level system, which can be seen by reducing the determinant in Eq. (4.84) to be:

$$\Phi_1(\lambda) = \lambda^2 - i\Delta\lambda + |\tilde{E}_1|^2, \quad (4.88)$$

where $|\tilde{E}_1|$ is the Rabi frequency and Δ the detuning of the two-level system. For the similarity between the equation for the three-level system and the two-level model [Eq. (4.88)] to hold, the detuning of the upper state should be redefined as:

$$\Delta = \Delta_2 - \frac{|\tilde{E}_1|^2}{\Delta_1} - \frac{|\tilde{E}_2|^2}{\Delta_1}. \quad (4.89)$$

The correction to the original detuning Δ_2, proportional to the square of the optical electric field, is a "Stark shift."

We note that in the case of the two-level system, the frequency of the Rabi cycling at exact resonance is determined by the difference between the two eigenvalues. "Exact resonance" in the case of our tri-level system implies [from Eq. (4.89)] that $\Delta_2 = [|\tilde{E}_1|^2 + |\tilde{E}_2|^2]/\Delta_1$. Substituting that value in the characteristic equation, and after some algebraic manipulations, one finds for the Rabi cycling frequency of the equivalent two-level system:

$$E_{\text{Rabi}} = \lambda_2 - \lambda_0 = \frac{|\tilde{E}_1||\tilde{E}_2|}{\Delta_1}. \tag{4.90}$$

We conclude thus that the three-level system, within the approximation of intermediate level far off-resonance, is equivalent to a two-level system. The resonance condition is modified by the Stark shift given by Eq. (4.89). A two photon Rabi frequency can be defined, proportional to the product of the transition frequencies $\tilde{E}_1$ and $\tilde{E}_2$, hence proportional to the square of the electric field amplitude.

The preceding considerations extend to an arbitrary number of levels. Quite generally, the multilevel system can, under certain conditions of large detuning of the intermediate levels, be reduced to a two-level system. The generalized Rabi frequency and the Stark shift can be incorporated into the density matrix equations of a two-level system, or into the vector model.

For the n-level system, a crucial approximation is that the detuning of the intermediate levels be large compared with the Rabi frequencies $|\tilde{E}_k|$. This is a more complex condition than might appear at first glance; there should be no accidental resonance of order smaller than n for any pair of levels in the system being considered.

We refer the interested reader to a derivation of the generalized multilevel Rabi frequencies. One finds that the multilevel Rabi frequency E_{Rabi}, as in the case of the three-level system, is the small difference of order $|\tilde{E}_k^{n-1}|/\Delta_j^{n-2}$ between two roots of the characteristic equation, and is

$$E_{\text{Rabi}} = 2\frac{\tilde{E}_1\tilde{E}_2 \ldots \tilde{E}_{n-1}}{\Delta_1\Delta_2 \ldots \Delta_{n-2}} = \kappa_{n-1}\tilde{\mathcal{E}}^{n-1}. \tag{4.91}$$

$\kappa_{n-1}\tilde{\mathcal{E}}^{n-1}$ is the generalized Rabi frequency, with the scale factor for the multilevel Rabi cycling given by:

$$\kappa_{n-1} = \frac{p_{01} \cdots p_{n-2,n-1}}{(\Delta_1 \ldots \Delta_{n-2})(2\hbar)^2}. \tag{4.92}$$

The Stark shift is the detuning $\delta\omega_s$ of the upper level Δ_{n-1} consistent with these solutions.

$$\delta\omega_s = \Delta_{n-1} = \frac{|\tilde{E}_1|^2}{\Delta_1} - \frac{|\tilde{E}_{n-1}|^2}{\Delta_{n-2}} \tag{4.93}$$

$$= \left[\left(\frac{p_{01}}{2\hbar\Delta_1}\right)^2 - \left(\frac{p_{n-1,n-2}}{2\hbar\Delta_{n-2}}\right)^2\right]|\mathcal{E}|^2. \tag{4.94}$$

Because the multilevel system is equivalent to a two-level system, a simple system of two density matrix equations should apply (or a modified Bloch vector model). Using the notation $\tilde{\varrho}_{ij} = c_i c_j^*$ for the off-diagonal matrix element in the rotating frame, we can write for the two-level system approximation:

$$\dot{\tilde{\varrho}}_{0,n-1} = i(\Delta_{0,n-1} - \delta\omega_s) + i\kappa_{n-1}\tilde{\mathcal{E}}^{n-1}(\rho_{n-1,n-1} - \rho_{00}),$$

$$\dot{\rho}_{n-1,n-1} - \dot{\rho}_{00} = 2\kappa_{n-1}\left[\tilde{\mathcal{E}}^{(n-1)*}\tilde{\varrho}_{0,n-1} + \tilde{\mathcal{E}}^{(n-1)}\tilde{\varrho}^*_{0,n-1}\right]. \tag{4.95}$$

The reduction of multiphoton interaction to a coherent model involving only the two extreme levels may seem like a coarse approximation. It has the value of making Bloch's vector model applicable to this complex problem. In particular, it has been shown to be valuable in designing pulse sequences leading to a complete population inversion in multilevel systems [21]. The validity of the two-level approximation has been tested numerically by computing the response of three- to five-level systems excited by sequences of pulses of different phase [28].

In the following subsection, we consider as an example of higher-order two-level system four photon resonance in mercury.

4.3.4. Four Photon Resonant Coherent Interaction

In an atomic system, the level density is much smaller than in molecules. Therefore, in the case of a multiphoton resonance, one will generally not find intermediate resonances between a pair of levels and the light field. The two-level approximation introduced above is generally appropriate. We will discuss in this section harmonic generation by multiphoton resonant coherent propagation in atomic vapors.

We have considered in Chapter 3 nonresonant nonlinear polarization, leading to self-lensing, parametric generation, and harmonic generation. The nonlinear susceptibility responsible for these effects can be enhanced by the proximity of a resonance. The presence of a resonance, on the other hand, complicates the interaction and introduces losses.

Coherent interaction processes take increasing importance the higher the order (n) of the nonlinear process[5] is. As we have seen in the previous sections, the transition rate between the two levels (a generalized Rabi frequency) is proportional to the nth power of the field envelope $\mathcal{E}$. Therefore, the parameter characteristic of the interaction is the pulse "area," proportional to the integral of the multiphoton Rabi frequency defined in Eq. (4.91), $\theta_n = \int \kappa_n \mathcal{E}^n(t)dt$. For a two photon transition, the area is proportional to the pulse energy. In the case of higher-order processes, for the same pulse energy, shorter pulses have a larger area, hence favor the higher-order nonlinear process. With the progress in pulse shaping techniques (see Chapter 8), femtosecond pulses could be used to establish multiphoton coherences, manipulate population transfers, and generate with high efficiency higher-order harmonics. This is a rather unexplored area of nonlinear optics, but extremely promising to generate efficiently short wavelength radiation, or store a large amount of energy in an atom or molecule.

Let us consider for instance the case of resonant n^{th} harmonic generation, where the fundamental radiation is in a kth ($k < n$) photon resonance with a particular atomic transition. At moderate powers, the conversion efficiency is orders of magnitude larger than in the nonresonance case, because of the intermediate resonance. However, as the power is increased, in an attempt to achieve higher conversion efficiencies, the k–photon absorption can deplete the fundamental radiation, before significant conversion is achieved. The energy of the fundamental radiation is transferred to the atomic system rather than to the harmonic radiation. A solution to this problem is to minimize the energy lost to the atomic system by the fundamental by making use of pulse shapes or sequences that propagate through the (multiphoton) resonant medium with minimum absorption losses.

This technique has been used successfully to increase the third harmonic conversion in the case of a two photon resonant transition in lithium vapor [28–30]. A first pulse excites (via a two photon transition) the $4s$ level of lithium. For a second pulse, following the first one with a delay short compared to the phase-relaxation time of the two photon transition, and sent through the same medium $\pi/2$ out-of-phase, two photon stimulated emission occurs. The second pulse recovers the energy lost by the first one to the medium. If conditions of phase matching (for third harmonic generation) are not met, the peak intensity of the second pulse increases with propagation distance (two photon stimulated emission) at the expense of the first pulse which is depleted. If conditions for phase matched third harmonic generation are met, however, energy is continuously transferred from the second pulse to the third harmonic field. As a result,

[5]In this subsection n is the *order* of the process, or the number of photons required to make the transition between the two levels, and should not be confused with the n representing the number of levels in the previous subsection.

instead of growing at the expense of the first pulse, the second pulse propagates with near constant peak intensity over a long distance. Efficient third harmonic generation occurs during that second pulse.

Among the multiphoton processes of various orders n, two photon resonant processes are somewhat unique, because, as long as the pulses are shorter than the inverse linewidth of the resonances, the result of the interaction is independent of the pulse duration and depends only on the pulse energy. This is mainly because, as we have seen before, the rate of population transfer (Rabi frequency) and level motion (Stark shift) are all proportional to the pulse intensity. For instance, if a 10 ps pulse of 1 mJ/cm^2 is a π pulse producing complete inversion, a 10 fs pulse of 1 mJ/cm^2, having a 10^3 times higher intensity, will produce exactly the same result.

The situation is quite different with higher-order processes, as pointed out at the beginning of this section. We will therefore chose the example of a four photon resonance to illustrate the potential benefits, as well as some of the difficulties associated with higher-order resonances. Specifically, we will consider four photon coherent resonant excitation of mercury, as sketched in Figure 4.11. If its intensity is sufficiently high, a fs pulse of amplitude $\tilde{\mathcal{E}}(t)$ at ω_ℓ (fundamental) will be attenuated through four photon absorption. The excited level (6^1D_2 in Fig. 4.11) can be seen as a "springboard" for third (frequency $3\omega_\ell$, field $\tilde{\mathcal{E}}_3$) and fifth (frequency $5\omega_\ell$, field $\tilde{\mathcal{E}}_5$) harmonic generation through Stokes and anti-Stokes Raman processes. The motivation for studying such a complex system

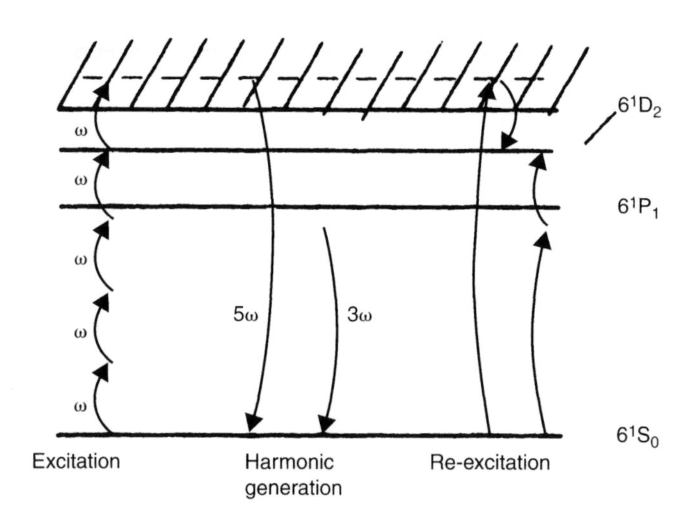

Figure 4.11 Energy levels of mercury and some of the possible excitation and harmonics generation mechanisms. The third and fifth harmonic can be generated by a Raman process involving the 6^1D_2 state for example. The main loss mechanisms are four photon absorption and five photon ionization.

is to generate efficiently third and/or fifth harmonic, using the four photon resonance to enhance the nonlinear susceptibility, and coherent propagation effects to minimize energy transfer (absorption) to the atomic system. We use this problem as an example to illustrate the various interesting phenomena that complicate significantly the study of higher-order processes.

Because a resonant level has been excited by four photon absorption (6^1D_2 in Fig. 4.11), generation of the third harmonic frequency is possible through a Raman process. Energy losses occur because of direct (four photon) absorption, and subsequent absorption of another photon leading to (five photon) ionization.

As discussed in the previous section, the system of interaction equations can be reduced to a system of three equations resembling Bloch's equations [cf. Eq. (4.95)]. One can define a fourth-order Rabi frequency proportional to the fourth power of the complex electric field amplitude $\tilde{\mathcal{E}}(t)$. The off-diagonal matrix element representing the four photon coherent resonant excitation is a quantity oscillating at the frequency $4\omega_\ell$, with a complex amplitude of the pseudo-polarization $\tilde{Q}_4$. As shown in Appendix D, third and fifth harmonics $\tilde{\mathcal{E}}_3$ and $\tilde{\mathcal{E}}_5$ are generated by the combinations $\tilde{Q}_4\tilde{\mathcal{E}}^*$ and $\tilde{Q}_4\tilde{\mathcal{E}}$, respectively which appear as source terms in the wave equation for the electric field amplitudes. These coupling terms describe stimulated Raman processes. There is also a nonresonant contribution to the harmonic fields. In particular, the nonresonant third harmonic generation source term is proportional to the third power of the field [of the form $\chi^{(3)}\tilde{\mathcal{E}}^3(t)$] and may not be negligible compared to the Raman term $\tilde{Q}_4\tilde{\mathcal{E}}^*$. Combinations of these fields ($\tilde{\mathcal{E}}_1^4$, $\tilde{\mathcal{E}}_1\tilde{\mathcal{E}}_3$, $\tilde{\mathcal{E}}_1^*\tilde{\mathcal{E}}_5$ and $\tilde{\mathcal{E}}_1^2\tilde{\mathcal{E}}_5\tilde{\mathcal{E}}_3^*$) are the source terms in the interaction equations (D.1 and (D.2) given in Appendix D.[6] The upper level of the transition is pumped by four photons at the fundamental frequency, but also by the sum of a third harmonic photon and a fundamental photon. If these two excitation processes have opposite phase, there is no more interaction between the resonant radiation and the medium (interaction quenching).

The complexity of the problem is because of the number of combinations of fields which give rise to an interaction of order four or lower. Second-order Stark shifts can sweep intermediate levels through resonance, creating "transient resonances."

The example of four photon resonant transient coherent interaction in Hg vapor points to three puzzling effects typical of higher-order systems:

1. interference between resonant and nonresonant harmonic generation;
2. transient Stark shifts; and
3. interaction quenching.

---

[6]In Eqs. D.1 and D.2, the field amplitudes $\tilde{\mathcal{E}}_i$ are replaced by corresponding Rabi frequencies $\tilde{V}_i$ defined in Eq. D.2.

All these effects are particularly dramatic when an intermediate level approaches resonance with an harmonic of the field. This is in particular the case with the $6S$–$6D$ four photon resonance of Hg, where the $6P$ state approaches a three photon resonant condition, as shown in Fig. 4.11. This near-resonance accounts for the three above effects. First, the nonresonant susceptibility $\chi^{(3)}$ given in Eq. (D.8) is enhanced, resulting in a nonresonant contribution to the third harmonic, which can interfere with the resonant term in Eq. (D.9).

For the particular levels of Hg shown in Fig. 4.11, the four photon area is:

$$\theta_4 = 0.0175 \frac{W^2}{\tau_p} \tag{4.96}$$

where W is the energy density in J/cm^2 of a pulse of duration τ_p (in ps). This example shows indeed that higher-order coherent interactions require fs pulses. An area of $\theta_4 = 0.7$ can be achieved with a 100 fs pulse of 2 J/cm^2 [200 μJ in a cross section of (0.1 mm)^2].

Because of the proximity of the $6P$ level (indicated in Fig. 4.11), the main contribution to the Stark shift of the resonant transition is a shift of the upper $6D$ level:

$$\delta\omega_2 \approx \frac{1}{\sqrt{r_{04}}} \left[\frac{|p_{04}|^2}{(\omega_{34} - \omega_\ell)} \right] |\tilde{V}_1|^2, \tag{4.97}$$

where the index 0, 3, and 4 have been given to the ground, the $6P$ and the $6D$ levels, respectively. Equation (4.97) is the Stark shift expression Eq. (D.5) in which the dominant term of the susceptibility [Eq. (D.6)] has been inserted. The symbol $|\tilde{V}_1(t)|$ represents the electric field in units of $s^{-1/4}$, defined through the four photon Rabi frequency $\tilde{V}_1^4(t) = r_{04}\tilde{\mathcal{E}}_1^4(t)$. The various coefficients are defined in Appendix D.

As the field intensity of the fs pulse increases, the level $6D$ (level 4) is shifted in accordance with Eq. (4.97). The resulting change in the denominator of Eq. (4.97) implies a redefinition of the detuning $\delta\omega$. Thus, with increasing field amplitude, there is a nonlinear self-induced detuning effect, reducing the resonant coherent interaction with ultrashort pulses. This effect has been predicted within the framework of the approximation used in Refs. [30, 32]. It should be noted that in this particular example the field induced shifts are too large and too fast to be consistent with the adiabatic approximation used in the elaboration of the four photon resonant interaction equations (see Appendix D).

The last mentioned effect of interaction quenching is probably the most challenging—yet unsolved—dilemma of resonant multiphoton coherent interaction. When intermediate resonances make the generation of an harmonic field particularly efficient, the source terms for the four photon coherent excitation through the harmonic and through the fundamental are out-of-phase—hence cancel each other. In the density matrix equations for this interaction (as outlined in Appendix D), the "coherent excitation" is represented by an off-diagonal matrix element $\tilde{\rho}_{04}$ or its complex amplitude defined as $\tilde{Q}_4 = 2i\tilde{\rho}_{04}e^{-4i\omega_\ell t}$. The sequence of events that unfolds by solving the system of interaction equations (D.1) can be told through phenomenological arguments. As a strong pulse propagates through the medium, coherent excitation (of amplitude $\tilde{Q}_4$) is created through a source term proportional to the fourth power of the fundamental field $\tilde{\mathcal{E}}_1^4$, resulting also in some population of the level 4 (6^1D_2 in Fig. 4.11). The mixing of the four photon excitation and the fundamental leads to the generation of a third harmonic field ($\omega_{3\ell} = 3\omega_\ell$; $\tilde{\mathcal{E}}_3 \propto \tilde{Q}_4\tilde{\mathcal{E}}_1^*$) and a fifth harmonic field ($\omega_{3\ell} = 5\omega_\ell$; $\tilde{\mathcal{E}}_5 \propto \tilde{Q}_4\tilde{\mathcal{E}}_1$). Both the third and fifth harmonic field are themselves source terms for the coherent four photon excitation through second-order processes: difference frequency generation $\omega_{5\ell} - \omega_\ell$ (source term for $\tilde{Q}_4 \propto \tilde{\mathcal{E}}_5\tilde{\mathcal{E}}_1^*$) and sum frequency generation $\omega_{3\ell} + \omega_\ell$ (source term for $\tilde{Q}_4 \propto \tilde{\mathcal{E}}_3\tilde{\mathcal{E}}_1$). The phase relation between the three source terms for the coherent excitation is such that they cancel each other. For all practical purposes, the radiation does not interact with the resonant system anymore. This effect is illustrated in the simulation of Figure 4.12, where the generated third harmonic and the five-photon ionization are plotted as a function of propagation distance. Because of the proximity of the 6^1P_1 level to a three photon resonance condition, the third harmonic generation dominates the fifth harmonic in this particular example. The incident wavelength is at 566.7 nm, or 6 nm above the weak field resonance (560.7 nm) to compensate

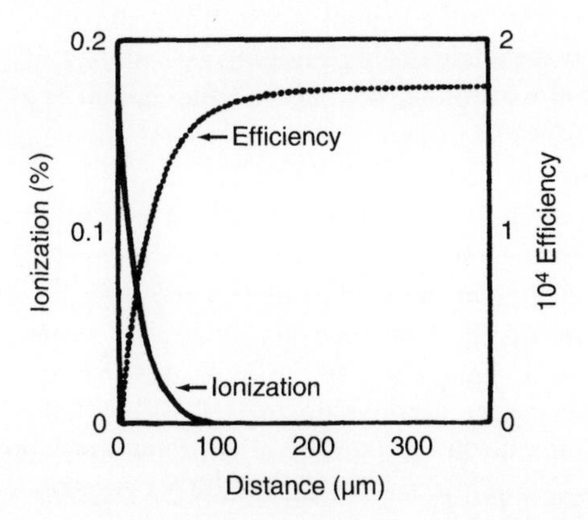

Figure 4.12 Five photon ionization (in %; solid line) and third harmonic peak field conversion factor η (dashed line) versus propagation distance (in μm) in mercury vapor. The vapor pressure is 10 torr. The incident pulse shape is the Gaussian $\tilde{\mathcal{E}}_1(t) = \tilde{\mathcal{E}}_{10}\exp[-(t/\tau_G)^2]$, with $\tau_G = 3$ ps. The pulse energy density is 20 J/cm^2. (Adapted from [29].)

for the Stark shift. As the pulse propagates through the mercury vapor, the third harmonic generation is seen to saturate after a short distance, while the ionization drops to zero. The third harmonic being generated with a phase opposite to that of the fundamental, the interaction vanishes, because the second order $(\omega_{3\ell} + \omega_\ell)$ and fourth order $(4\omega_\ell)$ mechanisms of four photon resonant excitation [Eq. (D.1)] cancel each other. This problem could be solved by adjusting the relative phase of the fundamental and third harmonic and by using phased pulse sequences to control the relative phase of the coherent excitation (term of amplitude $\tilde{Q}_4$) and that of the radiation.

Multiphoton resonances reduce the energy requirements on higher-order harmonic generation by enhancing the nonlinear susceptibilities. Energy losses associated with the inevitable multiphoton absorption associated with the resonance can be minimized by exploiting the property of reversibility of light–matter energy transfers in coherent multiphoton resonant interactions with fs pulses. The above example, however, illustrates the complexity of multiphoton coherent effects. Mastering the theory and being able to generate the fs pulses shapes dictated by numerical simulations is the price to pay for high nonlinear conversion with low energy fs pulses.

These techniques are of increasing importance for the generation of femtosecond pulses in the UV, where one has to work with gases rather than crystals.

4.3.5. Miscellaneous Applications

Because fs pulses can be shorter than dephasing times even in condensed matter, the reversibility of coherent interactions can be exploited in various applications. All experiments using photons as a source of momentum can be made more efficient with ultrashort pulses. When an atom of mass M absorbs a photon, the ratio of the recoil energy $M\nu^2/2$ to the photon energy is the minute quantity $\hbar\omega_\ell/(2Mc^2) \leq 10^{-19}$.

To illustrate the latter point, let us consider the situation sketched in Figure 4.13, where the intracavity beam path of a mode-locked laser crosses an atomic beam at a right angle. Let us assume that the laser is tuned to an absorbing transition of the atom, and the intracavity mode-locked pulse intensity is such as to be a π pulse for the atom. At the first passage of the pulse, m atoms will have been given a momentum $M\nu$, while the pulse has lost m photons. If the number of photons in the pulse is $n \gg m$, the pulse reflected back by the mirror will still be a π pulse, which will return the m atoms to the ground state, thereby restituting its original energy to the mode-locked pulse and imparting an addition momentum $M\nu$ to the m atoms *in the same direction*. The net result is that the number of photons is conserved, but the atoms have been given a kinetic

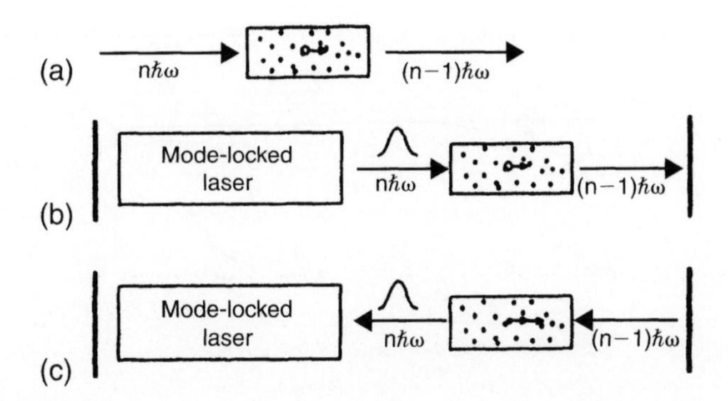

Figure 4.13 Momentum transfers between a collection of atoms and an optical pulse. In (a), a pulse of n photons loses one photon to the atomic system, providing a small recoil to the absorbing atom. (b) and (c) show two successive steps of intracavity interaction of a π pulse with an atomic beam. In a first passage, the atoms are inverted, resulting in a recoil to the right. The sketch assumes only one atom interacting with the beam. That atom absorbs one photon from the pulse and is imparted a kinetic energy $Mv^2/2 = (\hbar\omega_\ell/2Mc^2)\hbar\omega_\ell$ (b). The reflected pulse recovers its energy by stimulated emission at the second passage, while giving an additional recoil to the right (c).

energy of $2Mv^2$. It is left as a problem at the end of this chapter to determine how the energy is conserved in this problem. A simple mechanical analogy for this problem is sketched in Figure 4.14.

Many optical interactions that involve optical pumping can be performed more efficiently with ultrashort pulses. However, it may come as a surprise that, with these ultrashort broad bandwidth pulses, *frequency selective* excitation is also possible. The frequency selectivity can be obtained from a combination of one or more pulses. For instance, it can easily be seen that a zero area pulse has no Fourier component at its average frequency. Such a pulse can be designed to produce complete inversion off-resonance and no excitation at resonance.

Similarly, a Gaussian pulse can be designed to be a π pulse at resonance and a 2π pulse off-resonance. The case of a square pulse is the most obvious. For a square π pulse at resonance (ω_0) with a line to be excited $\theta_0 = \kappa\mathcal{E}\tau_p = \pi$. If we want to leave a line at frequency $\omega_1 = \omega_0 - \Delta\omega$ unexcited, the condition for the pulse duration and amplitude is $\tau_p\sqrt{\kappa^2\mathcal{E}^2 + \Delta\omega^2} = 2\pi$ or $\kappa\mathcal{E} = \sqrt{3}\,\Delta\omega$ [as can be seen from the vector model sketched in Fig. 4.2 (a)]. These properties can be exploited for selective optical excitation, for instance, isotope separation [33, 34]. It can be shown that, in the case of Doppler broadened transitions, the "π–2π" (π pulse on resonance; 2π pulse off-resonance) selective excitation scheme is more selective and efficient with Gaussian shaped than with square pulses [20]. If the pulse duration and amplitude are appropriately chosen,

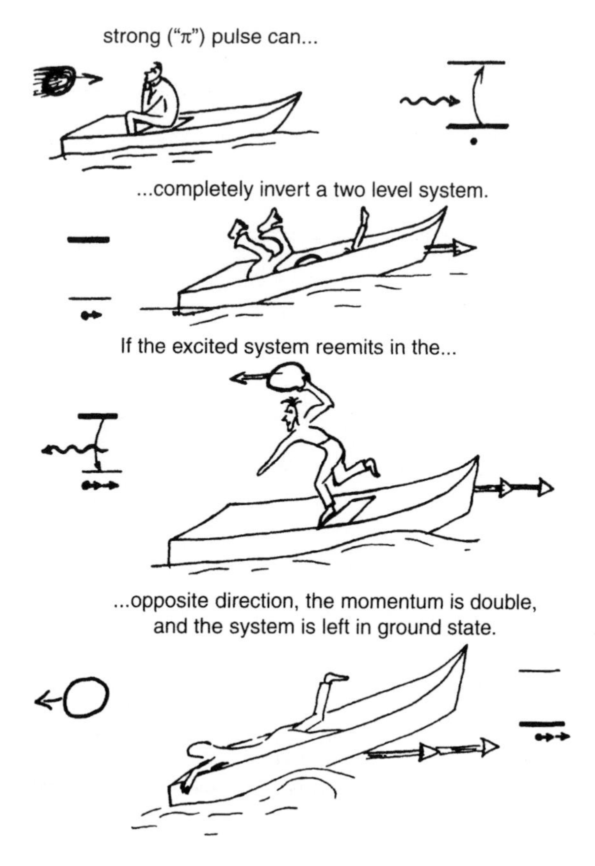

Figure 4.14 Momentum transfers between a short pulse and a two-level system; the "mechanical" analogy.

the selected line can be completely inverted in a single shot, while the transition to be left undisturbed is completely returned to ground state. The selectivity is high, even though the pulse is short enough to interact with both lines. There is a simple mechanical analogy to this "π–2π" excitation. The analog of the two absorbers at different frequencies are two pendulae of different lengths. The element to be selected is the longer pendulum (Fig. 4.15), while the "unwanted" transition is the shorter toddler. The longer pendulum has a lower frequency. The potential energy is the analogue of the population difference (taking as reference the axis of the pendulum). For the same kick (light pulse) applied to both pendulae, it is possible to create a complete inversion for the selected transition, while the unwanted element has gone through a whole cycle and returned to the ground state.

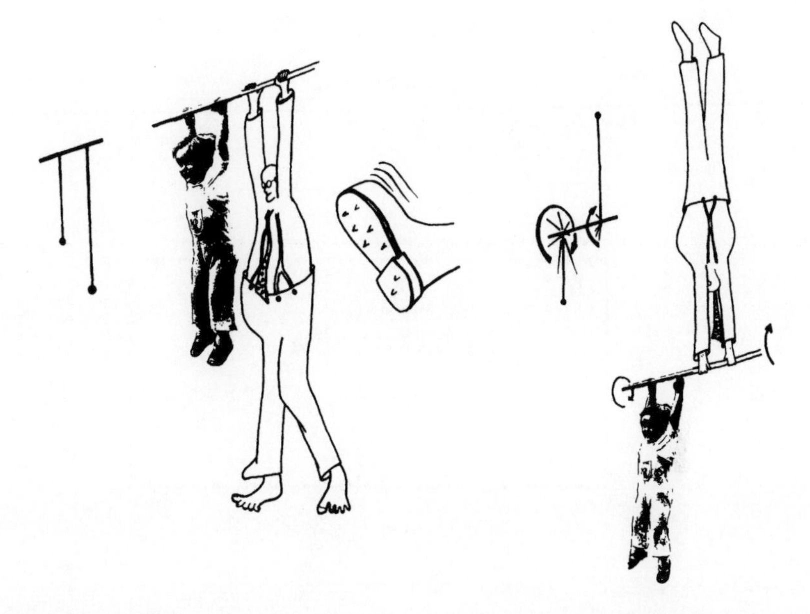

Figure 4.15 Selective optical excitation with a π pulse on resonance and a 2π pulse off-resonance: the pendulum analogy.

Another application of coherent interaction involves the use of an atomic resonance as a local oscillator, to extract phase information of a chirped pulse [35] with frequency $\omega_\ell + \dot{\varphi}(t)$. The chirped pulse to be analyzed is sent through a cell containing sodium vapor with a sharp resonance at ω_0. The transmitted pulse interferes with the resonant reradiation from the atomic line, resulting in a field component modulated at a frequency $\omega_\ell + \dot{\varphi}(t) - \omega_0$. This modulation carries the information about the chirp and is measured by cross-correlation with a shorter fs pulse. The result of such a measurement is illustrated in Figure 4.16. The chirp was induced on a 5.4 ps pulse from a synchronously pumped dye laser by passage through a polarization preserving fiber (3 m long). The pulse bandwidth was increased by SPM from 1 Å to 50 Å. A fraction of the original self-modulated pulse is sent through a two-stage compression (as described in Chapter 6) to create the 22 fs pulses needed to cross-correlate the signal transmitted by the sodium cell [34]. The upper trace (a) in Fig. 4.16 is the cross-correlation of the input to the sodium cell with the 22 fs pulse. The lower figure (b) shows the modulation induced in the sodium cell (optical thickness $a = 95$). Because the phase of the pulse changes by exactly π from one extremum to the next, a plot of the phase versus time can easily be made. Such a measurement is particularly useful to test the linearity of frequency chirping techniques.

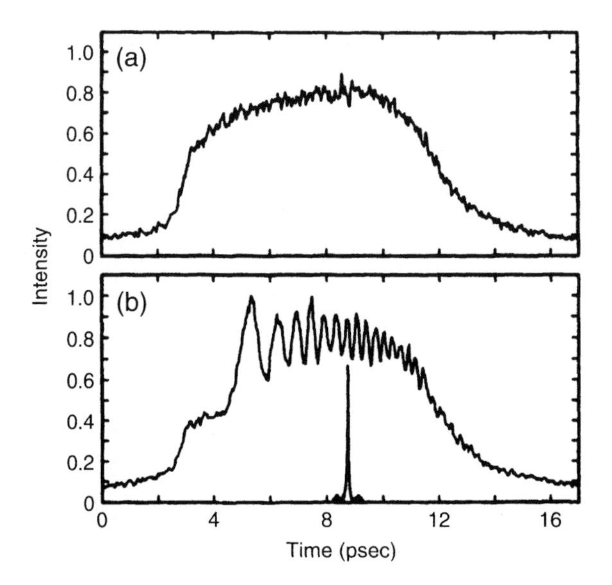

Figure 4.16 Transmission of a frequency swept pulse through a sodium cell. (a) Cross-correlation of the input pulse with a 22 fs probe pulse. (b) Cross-correlation of the output pulse with a 22 fs probe pulse. The modulation indicates that the frequency of the pulse transmitted through the fiber varies with time (chirp). The 22 fs probing pulse's autocorrelation (FWHM 34 fs) is also shown. (Adapted from [34].)

4.4. PROBLEMS

1. Compare the population transfer when a two-level system is excited by (a) a step function dc field (zero carrier frequency) and (b) a step function electric field resonant with the transition frequency.
2. Find the steady-state solution of Bloch's equations valid for monochromatic, cw incident radiation, and a homogeneously broadened two-level system. Calculate the results of an absorption measurement for high incident field intensity (i.e., calculate the absorption versus wavelength and light intensity). The calculated width of the absorption profile will turn out to be a function of the incident intensity. However, if you perform a pump probe experiment—i.e., you saturate the line with a field at frequency ω_1, and measure the absorption profile by tuning the frequency ω_ℓ of a weak probe beam—you find that the linewidth is $1/T_2$ independently of the intensity of the pump at ω_1. Explain.
3. Demonstrate the area theorem for a square pulse at resonance, in an inhomogeneously broadened medium, where g_{ih} can be approximated by a square function of width $\Delta\omega_{ih}$.

4. Show that a pulse with zero area has no Fourier component at its average frequency.

5. Show that the definitions (4.22) and (4.27) are equivalent for unchirped pulses at resonance.

6. Discuss the connection between the frequency pushing (mentioned in Section 1.2.1) and $\langle \dot{\varphi} \rangle$ obtained in Chapter 4 (Section 4.3.3) for the weak pulse limit. *Hint:* Define a dielectric constant in terms of u, ν, and w.

7. A short (weak) Gaussian pulse is sent through a resonant absorber with $T_2 \gg \tau_G$ (no inhomogeneous broadening). Being much narrower than the pulse spectrum, the absorbing line should act as a frequency filter. It is therefore a broadened pulse (in time) that should emerge from the absorber. Yet, according to the area theorem, the pulse area should decrease. Resolve this apparent contradiction in both frequency and time domains.

8. Calculate the initial frequency shift with distance $d\langle \dot{\varphi} \rangle / dz$ for a Gaussian pulse, off-resonance by $1/\tau_G$ with an absorbing (homogeneously broadened) transition with $T_2 = 100\tau_G$. Express your answer in terms of the linear attenuation.

9. Find an expression for $d\langle \dot{\varphi}^2 \rangle / dz$ for a pulse propagating through an ensemble of two-level systems. $\langle \dot{\varphi}^2 \rangle$ is defined as $\int \dot{\varphi}^2 \mathcal{E}^2 dt / \int \mathcal{E}^2 dt$. *Hint:* Use a similar procedure as for the derivation of the expression for the frequency shift with distance, writing first an expression for the space derivative of $W\langle \dot{\varphi}^2 \rangle$, and using Maxwell-Bloch's equations to evaluate each term of the right-hand side.

10. Referring to Fig. 4.8(a), let us consider a stepwise excitation with a polychromatic pulse given by the sum $\tilde{\mathcal{E}}_1(t) e^{i\omega_{\ell,1} t} + \tilde{\mathcal{E}}_2(t) e^{i\omega_{\ell,2} t} + \tilde{\mathcal{E}}_3(t) e^{i\omega_{\ell,3} t} + \ldots$. The frequencies $\omega_{\ell,1}, \omega_{\ell,2}, \omega_{\ell,3} \ldots$, are nearly resonant with the successive transitions $(0 \rightarrow 1)$, $(1 \rightarrow 2)$, $(2 \rightarrow 3), \ldots$. Derive Eq. (4.59) for this situation. *Hint:* Instead of Eq. (4.57), the detunings are now $\Delta_1 = \omega_{01} - \omega_{\ell,1}$, $\Delta_2 = \omega_{02} - (\omega_{\ell,1} + \omega_{\ell,2})$; $\Delta_3 = \omega_{03} - (\omega_{\ell,1} + \omega_{\ell,2} + \omega_{\ell,3}), \ldots$. The complex Rabi frequencies are to be defined as $\tilde{E}_{k+1} = \frac{i}{2\hbar} p_{k+1,k} \tilde{\mathcal{E}}_{k+1}$.

BIBLIOGRAPHY

[1] C. Lecompte, G. Mainfray, C. Manus, and F. Sanchez. Laser temporal coherence effects on multiphoton ionization processes. *Physical Review A*, A-11:1009, 1975.

[2] J.-L. De Bethune. Quantum correlation functions for fields with stationary mode. *Nuovo Cimento*, B 12:101–117, 1972.

[3] J.-C. Diels and J. Stone. Multiphoton ionization under sequential excitation by coherent pulses. *Physical Review A*, A31:2397–2402, 1984.

[4] E. Hecht and A. Zajac. *Optics*. Addison-Wesley, Menlo Park, CA, 1987.

[5] M. Lai and J.-C. Diels. Interference between spontaneous emission in different directions. *American Journal of Physics*, 58:928–930, 1990.

[6] M. Lai and J.-C. Diels. Ring-laser configuration with spontaneous noise reduced by destructive interference of the laser outputs. *Physical Review A*, A42:536–542, 1990.

[7] L. Allen and J. H. Eberly. *Optical Resonances and Two-level Atoms*. John Wiley & Sons, 1975.

[8] F. Bloch. Magnetic resonances. *Physical Review*, 70:460, 1946.

[9] R. P. Feynman, F. L. Vernon, and R. W. Hellwarth. Geometrical representation of the Schrödinger equation for solving maser problems. *Journal of Applied Physics*, 28:49–52, 1957.

[10] S. McCall and E. L. Hahn. Self-induced transparency. *Physical Review*, 183:457, 1969.

[11] J.-C. Diels and E. L. Hahn. Carrier-frequency distance dependence of a pulse propagating in a two-level system. *Physical Review A*, A-8:1084–1110, 1973.

[12] J.-C. Diels and E. L. Hahn. Phase modulation propagation effects in ruby. *Physical Review A*, A-10:2501–2510, 1974.

[13] J.-C. Diels and E. L. Hahn. Pulse propagation stability in absorbing and amplifying media. *IEEE Journal of Quantum Electronics*, QE-12:411–416, 1976.

[14] F. P. Mattar and M. C. Newstein. Transverse effects associated with the propagation of coherent optical pulses in resonant media. *IEEE Journal of Quantum Electronics*, QE-13:507, 1977.

[15] V. Petrov and W. Rudolph. The chirped steady state pulse in an amplifier with loss. *Physics Letters A*, 145:192–194, 1990.

[16] E. Kamke. *Differentialgleichungen*. Geest and Portig, Leipzig, Germany, 1969.

[17] F. T. Arecchi and R. Bonifacio. Theory of optical maser amplifiers. *IEEE Journal of Quantum Electronics*, QE-1:169–178, 1965.

[18] J. A. Armstrong and E. Courtens. π pulses in homogeneously broadened amplifiers. *IEEE Journal of Quantum Electronics*, QE-4:411–416, 1968.

[19] C. Cohen-Tannoudji, B. Diu, and F. Laloe. *Quantum Mechanics*. John Wiley & Sons, NY, 1977.

[20] J-C. Diels. Efficient selective optical excitation for isotope separation using short laser pulses. *Physical Review A*, 13:1520–1526, 1976.

[21] J.-C. Diels and S. Besnainou. Multiphoton coherent excitation of molecules. *Journal of the Chemical Physics*, 85:6347–6355, 1986.

[22] G. Herzberg. *Molecular Spectra and Molecular Structure: I. Spectra of Diatomic Molecules*. Van Nostrand, NY, 1950.

[23] C. H. Townes and A. L. Schawlow. *Microwave Spectroscopy*. McGraw-Hill, NY, 1955.

[24] W. Gordy and R. L. Cook. *Microwave Molecular Spectra*. John Wiley & Sons, NY, 1970.

[25] J. N. Franklin. *Matrix Theory*. Prentice Hall, Englewood Cliffs, N. J., 1968.

[26] J. Biegert. *Polychromatic multiphoton coherent excitation of sodium*. Der Andere Verlag, Osnabrueck, Germany, 2001.

[27] J.-C. Diels, J. Stone, S. Besnainou, M. Goodman, and E. Thiele. Probing the phase coherence time of multiphoton excited molecules. *Optics Communications*, 37:11–14, 1981.

[28] J.-C. Diels and A. T. Georges. Coherent two-photon resonant third and fifth harmonic VUV generation in metal vapors. *Physical Review A*, 19:1589–1591, 1979.

[29] J.-C. Diels, N. Nandini, and A. Mukherjee. Multiphoton coherences. *Physica Acta*, T23:206–210, 1988.

[30] A. Mukherjee, N. Mukherjee, J.-C. Diels, and G. Arzumanyan. Coherent multiphoton resonant interaction and harmonic generation. In G. R. Fleming and A. E. Siegman, editors, *Picosecond Phenomena V*, pages 166–168, Springer-Verlag, Berlin, 1986.

[31] N. Mukherjee, A. Mukherjee, and J.-C. Diels. Four-photon coherent resonant propagation and transient wave mixing: Application to the mercury atom. *Physical Review A*, A38: 1990–2004, 1988.

[32] J.-C. Diels. Two-photon coherent propagation, transmission of 90° phase shifted pulses, and application to isotope separation. *Optics and Quantum Electronics*, 8:513, 1976.

[33] J.-C. Diels. Application of coherent interactions to isotope separation. In Plenum Publishing Co., editor, *Proceedings of the 4th Rochester conference on coherence and quantum optics*, page 707, 1978.

[34] J. E. Rothenberg and D. Grischkowsky. Subpicosecond transient excitation of atomic vapor and the measurement of optical phase. *Journal of Optical Society B*, 4:174–179, 1987.

[35] J. E. Rothenberg. Self-induced heterodyne: The interaction of a frequency-swept pulse with a resonant system. *IEEE Journal of Quantum Electronics*, QE-22:174–179, 1986.

5

Ultrashort Sources I: Fundamentals

5.1. INTRODUCTION

The standard source of ultrashort pulses is a mode-locked laser. Fundamental properties of the radiation emitted by such a source, both in time and frequency domains, are presented in this first section. Section 5.2 exposes the main theoretical models to predict the shape of the pulses generated in such a laser. General considerations about the evolution of the pulse energy are given in Section 5.3. Section 5.4 is dedicated to the analysis of the main components of the laser, outlining the mechanism of pulse shaping of each element (or groups of elements). Of course, the laser resonator itself has its role in the mode-locked operation. The remainder of this chapter, Section 5.5 is therefore dealing with the properties of the laser cavity.

5.1.1. Superposition of Cavity Modes

Central to the generation of ultrashort pulses is the laser cavity with its longitudinal and transverse modes. A review of the mode spectrum of a laser cavity is contained in Section 5.5.1. Mode-locked operation requires a well-defined mode structure. As will be shown, mode-locking refers to establishing a phase relationship between longitudinal modes. A transverse mode structure will generally contribute to amplitude noise (at frequencies corresponding to the differences between mode frequencies). Most fs lasers operate in a single TEM_{00} transverse mode. A typical laser cavity can support a large number of longitudinal modes. In the absence of transverse mode structure, we can consider that the laser can operate on any of the longitudinal modes of index m, whose frequency ν_m satisfies the condition

$$\nu_m = \frac{mc}{2\sum_i n_i(\nu_m)L_i} \equiv \frac{mc}{2n(\nu_m)L} \tag{5.1}$$

where m is a positive integer and $n_i(\nu_m)L_i$ is the optical pathlength at the frequency ν_m of the cavity element i of length L_i. The total pathlength $OL = \sum_i n_i(\nu_m)L_i$ is the sum of the optical pathlengths of all cavity elements. We will formally write $OL = n(\nu)L$, where L is the geometrical cavity length and n is an effective average refractive index. We will first consider the ideal textbook case where the mode spacing $\Delta = \nu_{m+1} - \nu_m = c/(2nL)$ is constant, which implies that n is nondispersive for frequencies within the laser gain bandwidth.

The electric field of a laser that oscillates on M adjacent longitudinal modes of frequency $\omega_m = 2\pi\nu_m = \omega_\ell + 2\pi m\Delta$ with equal field amplitude $\mathcal{E}_0$ can be written as

$$\tilde{E}^+(t) = \frac{1}{2}\tilde{\mathcal{E}}(t)e^{i\omega_\ell t} = \frac{1}{2}\mathcal{E}_0 e^{i\omega_\ell t} \sum_{m=(1-M)/2}^{(M-1)/2} e^{i(2m\pi\Delta t + \phi_m)}, \tag{5.2}$$

where we now count m from $(1 - M)/2$ to $(M - 1)/2$. Here ϕ_m is the phase of mode m, which is random for a free-running laser. The mode spectrum is centered about a cavity mode of frequency $\omega_\ell = 2\pi p\Delta$, where p is a large positive integer. The laser field, except for a phase factor, is a repeating pattern with a periodicity of $1/\Delta$, because, for any integer q,

$$\tilde{E}^+\left(t + \frac{1}{\Delta}q\right) = \tilde{E}^+(t)e^{i\omega_\ell q/\Delta} \tag{5.3}$$

as can be verified using Eq. (5.2). This periodicity is the cavity round-trip time $\tau_{RT} = 1/\Delta$. Figure 5.1 compares the laser output for random and constant phase ϕ_m.

For any particular distribution of phases, the time-dependent laser power can be written as:

$$P(t) \propto \mathcal{E}_0^2[M + f(t)]. \tag{5.4}$$

Here $|f(t)| < M$ carries the information on the time dependence of the periodic laser output. This follows from the fact that the length of the sum vector of M unit vectors of random phase is equal to $\sqrt{M}$ (random walk). Note that each member of the sum in Eq. (5.2) represents such unit vector. For random phases

ϕ_m, the average laser intensity $\langle I\rangle = M\mathcal{E}_0^2/(2\sqrt{\mu_0/\epsilon})$ is the sum of the intensity of the individual modes.

Forcing all the modes to have an equal phase ϕ_0—a procedure called mode-locking—implies in the time domain that all the waves of different frequency will add constructively at one point, resulting in an intense and short burst of light [Figure 5.1(c)]. For the case of M oscillating modes of equal amplitude the sum in Eq. (5.2) can be calculated analytically and the total electric field is

$$\tilde{E}^+(t) = \frac{1}{2}\tilde{\mathcal{E}}(t)e^{i\omega_\ell t} = \frac{1}{2}\mathcal{E}_0 e^{i\phi_0}e^{i\omega_\ell t}\frac{\sin(M\pi\Delta t)}{\sin(\pi\Delta t)}. \tag{5.5}$$

For large M this corresponds to a train of single pulses spaced by $\tau_{RT} = 1/\Delta$. The duration of one of such burst, τ_p, can be estimated from Eq. (5.5):

$$\boxed{\tau_p \approx \frac{1}{M\Delta}}. \tag{5.6}$$

If we identify $M\Delta$ with the spectral width $\Delta\nu$ of the laser output we recognize the relationship $\tau_p \approx 1/\Delta\nu$ from Chapter 1. The ratio τ_{RT}/τ_p is thus a measure of the number of longitudinal modes oscillating in phase. For example, to produce a train of 10-fs pulses with a period of 10 ns about 10^6 modes are required.

At the pulse peak the contribution of the M modes add constructively to produce a field amplitude $\mathcal{E}(t = t_{peak}) = M\mathcal{E}_0$. Unlike the case of the random superposition of modes, the peak intensity is now equal to the product of the intensity of a single mode and the square of the number of modes: $I(t) = \mathcal{E}^2/(2\sqrt{\mu_0/\epsilon}) = M^2\mathcal{E}_0^2/(2\sqrt{\mu_0/\epsilon})$. Note that both the free-running and

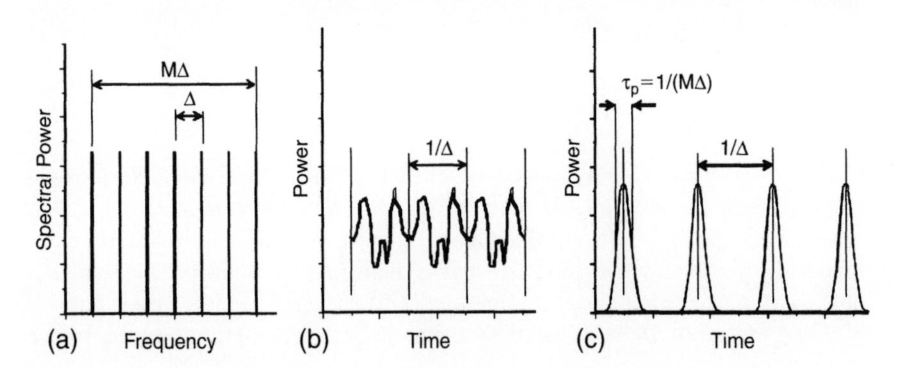

Figure 5.1 Spectral amplitude of a set of equally spaced cavity modes (a), and possible field amplitudes in the time domain, (b) and (c), that belong to this spectrum. In (b), the modes have a random phase distribution. In (c), all modes are "locked" to the same phase.

mode-locked laser described here have identical mode combs given by Eq. (5.2) (sketched in Fig. 5.1). Within that description, the average power of the free-running and mode-locked train are both equal to M times the power of each mode.[1] The peak power of the pulsed output exceeds that of the cw free-running laser by a factor M.

5.1.2. Cavity Modes and Modes of a Mode-Locked Laser

The above description assumes equally spaced modes. This is no longer true if the refractive index is a function of frequency, $n = n(\nu)$, which is the case for a typical laser cavity. The mode spacing at frequency ν can be estimated by

$$\Delta(\nu) \approx \frac{c}{2Ln(\nu)}\left[1 + \frac{\nu}{n(\nu)}\frac{dn}{d\nu}\right]^{-1}. \tag{5.7}$$

To derive this result we approximated $n(\nu_{m+1}) \approx n(\nu_m) + \frac{dn}{d\nu}\Delta$ in Eq. (5.1). The actual frequency dependence of the mode spacing depends on the cavity dispersion that was lumped into $n(\nu)$. The Fourier transform of a nonuniform frequency comb is a nonperiodic, nonuniform pulse train. To illustrate this point we consider a cavity dispersion that leads to a mode spacing that varies linearly with the mode index. The frequency of cavity mode m can then be written as

$$\omega_m = \omega_\ell + 2\pi(1 + m\gamma_d)m\Delta, \tag{5.8}$$

where γ_d is the dispersion coefficient. The superposition of M cavity modes of equal amplitude and phase produces an electric field

$$\tilde{E}^+(t) = \frac{1}{2}\tilde{\mathcal{E}}(t)e^{i\omega_\ell t} = \frac{1}{2}\tilde{\mathcal{E}}_0 e^{i\omega_\ell t}\sum_{m=(1-M)/2}^{(M-1)/2} e^{i2\pi(1+m\gamma_d)m\Delta t}. \tag{5.9}$$

Unlike in the case of equally spaced modes shown in Eqs. (5.2) and (5.3), the field amplitude, the intensity and the power $P(t)$ are not periodic. At $t = 0$ all cavity modes are in phase producing the maximum possible field amplitude. After one round-trip the modes still interfere mostly constructively and produce another pulse with a somewhat smaller amplitude. This scenario persists over a number of round-trips until, roughly speaking, the mode at the end of the spectrum [mode

[1] This is a coarse approximation that does not consider the interaction of the light with the gain medium. In an actual laser, the emission of each mode is not the same in cw or mode-locked operation.

index $(M-1)/2]$ becomes π out-of-phase with the central mode. From Eq. (5.9) one expects this to happen after about $q_M \approx 2(M^2\gamma_d)^{-1}$ round-trips. Note that we can interpret the term quadratic in m in Eq. (5.9) as a phase term for mode m that changes with time. Depending on the actual value of Δ and γ_d further round trips lead to random superposition of modes with different phases. As a result the individual pulses become broader, have unequal spacing and fluctuating amplitudes.

Figure 5.2 shows, as an example, the maximum of the field amplitude during one round trip as a function of the round-trip number for $M = 101$ and $\gamma_d = 10^{-5}$.

So far we have referred to modes of a hypothetical cavity and their superposition. In a mode-locked laser, even though $n = n(\nu)$, the pulse spacing (in time) and the mode spacing (in frequency) are constants.[2] The theory of the mode-locked laser shows this transformation of the unequal mode spacing of the passive cavity into a perfect comb to be the result of an interplay of dispersion and nonlinear optical processes. Before elaborating on this surprising result, let us describe two sets of experiments that further demonstrate the difference between cavity modes and the Fourier transform of a mode-locked laser. The experiments were performed on a standard Ti:sapphire linear laser as depicted in the left of Figure 5.3. In a first experiment the laser was operated in cw mode. The bandwidth and center frequency of that laser can be controlled by

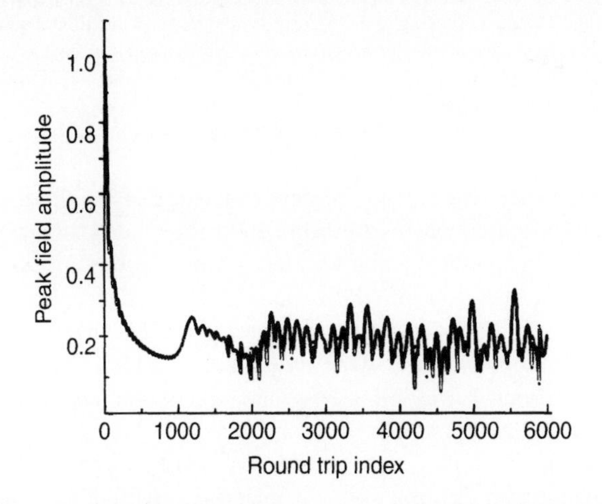

Figure 5.2 Pulse train produced by a set of $M = 101$ cavity modes whose frequencies are not equally spaced, $\omega_m = \omega_\ell + 2\pi(1 + m\gamma_d)m\Delta$, where $\gamma_d = 10^{-5}$.

[2]The Fourier transform of a regular pulse train is a comb of frequency spikes. Here the term mode spacing refers to the spacing between the teeth of that comb, and not between the actual longitudinal modes of the laser cavity.

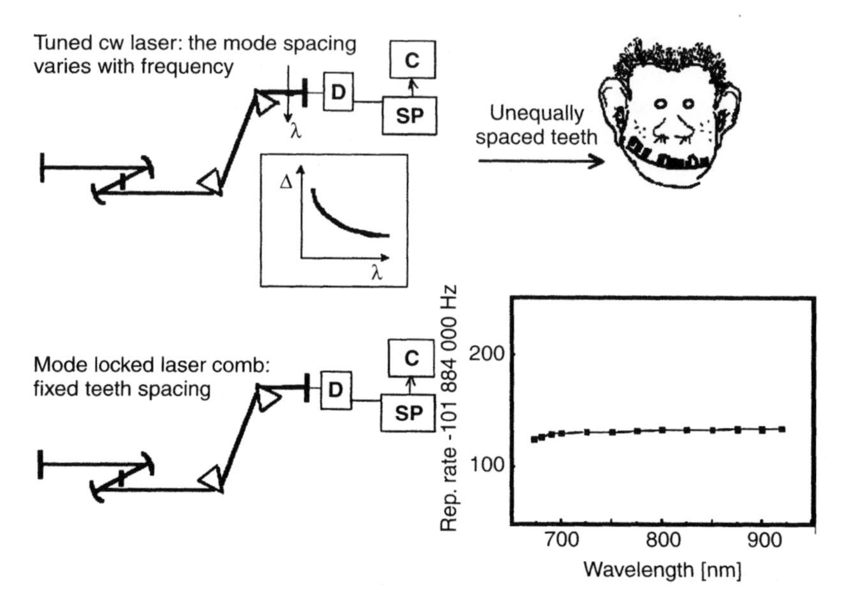

Figure 5.3 Top left: a standard Ti:sapphire laser operated in cw mode. As the wavelength (optical frequency) is being tuned, the beat note between adjacent modes changes because of dispersion of the cavity. Bottom left: the same laser is mode-locked, and portions of the output spectrum are selected with a spectrometer. The mode spacing of the fs comb is recorded as a function of wavelength (bottom right) either with a frequency counter or spectrum analyzer. The small change of the repetition rate arises from a thermal expansion of the cavity during the measurement. (D, detector; C, counter or spectrum analyzer; SP, spectrometer.)

translating intracavity slits [1,2]. The beat frequency of longitudinal modes that oscillate in a narrow frequency spectrum equals the mode spacing frequency Δ. It was found that the mode spacing frequency is not a constant across the tuning range of the laser, but follows the expected frequency dependence given by Eq. (5.7). The inverse mode spacing frequency would correspond to the pulse repetition frequency if the laser were mode-locked and the spectrum limited to a narrow range about ν. Mode-locking that laser is similar to the orthodontist intervention on the mode comb.

One can perform a similar measurement on the same laser, mode-locked after opening the bandwidth limiting slit, and selecting a particular wavelength range of the spectrum with a spectrometer, as sketched on the lower left of Fig. 5.3. The laser emits a train of pulses of 9 fs duration, spanning a 200 nm broad spectrum. A 0.2 nm bandwidth of the output spectrum of the laser is selected with a spectrometer and sent onto a fast photodiode. The signal was recorded with a frequency counter and a spectrum analyzer [3] as shown on the lower right of Fig. 5.3. The sensitivity of the spectrum analyzer (spectral resolution 1 Hz) allows the measurement to extend far into the wings of the mode-locked spectrum, greater than 50 dB down from the center peak. The repetition rate of the different wavepackets does not vary as a function of frequency or wavelength over a total span of 250 nm. It is the nonlinear phase shift because of the mode-locking mechanism (in this case the Kerr modulation) that compensates for the GVD. Such a result is expected, because the different wavepackets formed with any group of modes should all travel at the same group delay, or they will not produce a pulse that "stays together" after several round-trips. It is also consistent with the Fourier transform of an infinite train of equally spaced pulses, which produces a comb of equally spaced spectral components.

A recent experiment with a stabilized laser has confirmed that "teeth" of the frequency comb, which is the Fourier transform of the pulse train, are equally spaced throughout the pulse bandwidth to 3.0 parts in 10^{17} [4].

5.1.3. The "Perfect" Mode-Locked Laser

The perfect mode-locked laser produces a continuous train of *identical* pulses at a constant repetition rate. Such a perfect mode-locked laser has to be stabilized to minimize, for example, length fluctuations because of thermal expansion and vibrations.

A mode-locked fs laser requires a broadband gain medium, which will typically sustain over 100,000 longitudinal modes. The *train* of pulses results from the leakage (outcoupling) of a single pulse traveling back and forth in a cavity of constant length. The round-trip time of the cavity is thus a constant, implying a perfectly regular comb of pulses in the time domain. The frequency spectrum of such a pulse train is a perfect frequency comb, with equally spaced teeth, at variance with the unequal comb of longitudinal modes of a nonmode-locked cw laser.

The historical and standard textbook definition of mode-locking presented in the previous section originates from the description of the laser in the frequency domain, where the emission is considered to be made up of the sum of the radiation of each of these (longitudinal) modes. This description can still be applied to the ideal mode-locked laser considered in this section, if a fictitious perfect comb with equal tooth spacing is substituted to the real longitudinal modes of the cavity. This frequency description of mode-locking is equivalent to having, in the time domain, a continuous single frequency carrier, sampled at equal time intervals τ_{RT} by an envelope function, as shown in the top part of Figure 5.4.

Our ideal mode-locked laser emits a train of equally spaced pulses with a period τ_{RT}, which corresponds to a comb of modes in the spectral domain whose spacing is constant, $\Delta = 1/\tau_{RT}$. Consequently the mode frequency can be

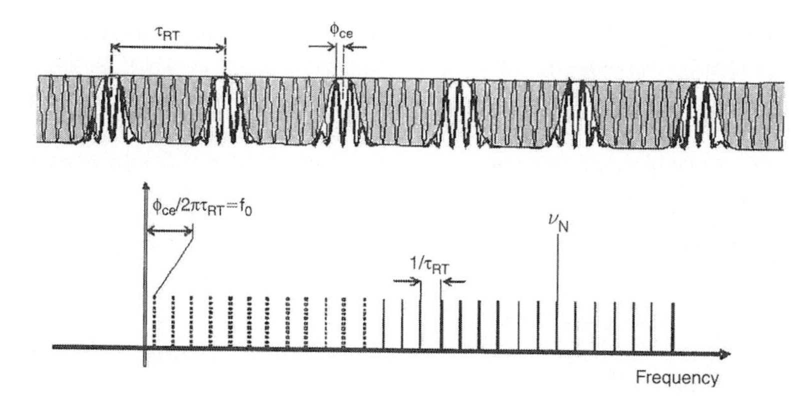

Figure 5.4 Top: a pure carrier at a frequency ν_N is modulated periodically by envelopes, at regular time intervals τ_{RT}. Bottom: the corresponding frequency picture. A comb of δ functions in frequency, is extended to near zero frequency. The frequency f_0 of the first mode is the carrier to envelope offset.

expressed as

$$\nu_m = f_0 + m\Delta = f_0 + \frac{m}{\tau_{RT}}, \tag{5.10}$$

where m is the mode index that now starts at $m = 0$. Note that $f_0 < \Delta$ is nonzero in general. This is different from the cold cavity referred to in the introduction of this chapter, where the mode frequencies are solely determined by the optical pathlength of the cavity $Ln(\nu)$. In cases where the index can be approximated by a constant over the gain bandwidth, the group velocity is equal to the phase velocity, and the mode frequencies are integer multiples of Δ.

While the pulse envelope peaks again exactly after one round-trip time τ_{RT} the phase of a mode with index m changes by

$$2\pi\nu_m\tau_{RT} = 2\pi f_0\tau_{RT} + 2\pi m\Delta\tau_{RT} = 2\pi f_0\tau_{RT} + 2\pi m. \tag{5.11}$$

Apart from multiples of 2π each mode acquires an additional phase with respect to the pulse envelope

$$\phi_{CE} = 2\pi f_0\tau_{RT}. \tag{5.12}$$

This is illustrated in Fig. 5.4. Because the phase shift ϕ_{CE} is independent of the mode index it leads to a slippage of the phase of the carrier frequency with respect to the pulse envelope. The frequency f_0 responsible for this slippage is called carrier to envelope offset (CEO). One can also interpret the relative shift of envelope and carrier as the result of the difference of phase and group velocity. An average group velocity can be defined as $\bar{v}_g = 2L/\tau_{RT}$. The time a phase front of a mode of index N needs to complete one round-trip ($2L$) is N/ν_N, which suggest to define an average phase velocity $\bar{v}_p = 2L\nu_n/N$. The delay between the pulse envelope and an arbitrary point on the phase front can now be written as

$$\tau_{CE} = 2L\left(\frac{1}{\bar{v}_g} - \frac{1}{\bar{v}_p}\right) = (\tau_{RT} - N/\nu_N), \tag{5.13}$$

which yields for the phase

$$\phi_{CE} = 2\pi\nu_N\tau_{CE} = 2\pi(\tau_{RT}\nu_N - N). \tag{5.14}$$

It is only when $f_0 = 0$ that the repetition rate is an integer number of optical cycles of an oscillating mode, cf. Eqs. (5.14) and (5.12).

The ability to measure (or control) f_0 implies that one is able to establish a link between the optical frequencies of the mode comb (ν_m) and the radio frequency ($1/\tau_{RT}$). Let us assume for instance that one optical mode at ν_N of the laser is linked to an optical frequency standard and that $f_0 = 0$ so that there are N optical cycles $1/\nu_N$ within the pulse period τ_{RT}. Under these conditions, the repetition rate can be considered to be a radio frequency standard with a relative linewidth, $\Delta\nu/\nu$, N times narrower than that of the optical reference.

The existence of a perfectly regular frequency comb has revolutionized the field of metrology. Such a comb can be used as a ruler to measure the spacing between any pairs of optical frequencies ν_1 and ν_2. The technique is similar to a standard measurement of length with a ruler. One measures the beat note $\Delta\nu_1$ between the source at ν_1 and the closest tooth—assigned the index m_1—of the frequency comb, as well as the beat note $\Delta\nu_2$ between the source at ν_2 and the neighboring tooth m_2 of the frequency comb. The frequency difference between the two sources is $\nu_2 - \nu_1 = \Delta\nu_2 - \Delta\nu_1 + (m_2 - m_1)/\tau_{RT}$.

We will discuss the frequency rulers and the mode-locked laser as time standard in Chapter 13. Details on stabilization techniques as well as frequency standards can be found in Ye and Cundiff [5].

5.1.4. The "Common" Mode-Locked Laser

The expression mode-locking suggests equidistant longitudinal modes of the laser cavity emitting in phase. As mentioned in the previous section, this frequency description of mode-locking is equivalent to having, in the time domain, a continuous single frequency carrier, sampled at equal time intervals by an envelope function. Unless sophisticated stabilization techniques as described in Section 13.4 are used, an ordinary mode-locked laser does not at all fit the above description. We shall use the term common mode-locked laser when the cavity length is not stabilized across the spectrum. In such a common situation, each cavity mirror is subject to vibrational motions. A typical mechanical resonance is around 100 Hz, with a motion amplitude ΔL of up to 1 μm. Because of that motion, the position of the longitudinal modes of the cavity is not fixed in time. As the cavity length L drifts, so does the mode frequency ν_m and the repetition rate $1/\tau_{RT}$. From Eq. (5.10), we can express the change in mode frequency $\Delta\nu_m$ because of a change in cavity length ΔL:

$$\Delta\nu_m = \left(\frac{df_0}{dL} + m\frac{d\Delta}{dL}\right)\Delta L = \left(\frac{df_0}{dL} - \frac{m}{\tau_{RT}^2}\frac{d\tau_{RT}}{dL}\right)\Delta L. \tag{5.15}$$

It depends on the specifics of the mode-locked laser how the CEO f_0 and the roundtrip time (group velocity) vary individually with L.

Pulse Train Coherence

Because of this change of the carrier frequency, the repetition rate and the carrier to envelope offset, one can no longer talk of an output pulse train made of identical pulses. The difference between the properties of the radiation from an ultrastable "frequency comb" as opposed to the common mode-locked laser can be established in a coherence measurement. Coherence can be measured with a Mach–Zehnder interferometer, as sketched in Figure 5.5. In the case of a single pulse, the interference contrast approaches zero for optical delays Δx of the interferometer exceeding the coherence length of the pulse. The interferogram will resemble that shown in Fig. 2.3. In the case of a pulse train from a perfect mode-locked laser, as the delay of the interferometer is being scanned, an identical fringe pattern reappears at delays equal to an integer multiple q of the pulse spacing τ_{RT}. In a common mode-locked laser, the visibility of these reoccurring fringes will decay with increasing q. To explain this loss in fringe contrast let us assume that at each delay Δx we measure a signal from N pulse pairs. The signal at the detector

$$S(q, \Delta x) = \eta^2 \sum_{i=1}^{N} \left\langle \mathcal{E}^2(t)\left[\cos(\omega_\ell t + \phi_i) + \cos(\omega_\ell t + \phi_{i+q} + k\Delta x)\right]^2 \right\rangle. \tag{5.16}$$

Here ϕ_i is the relative phase of the carrier with respect to the peak of the pulse envelope and $\langle\rangle$ denotes time integration over the pulse envelope and carrier

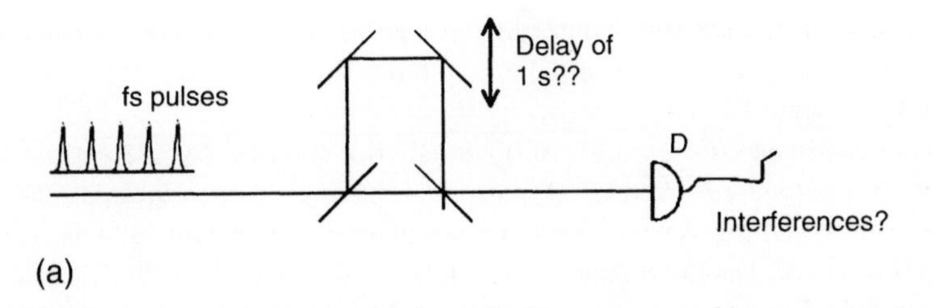

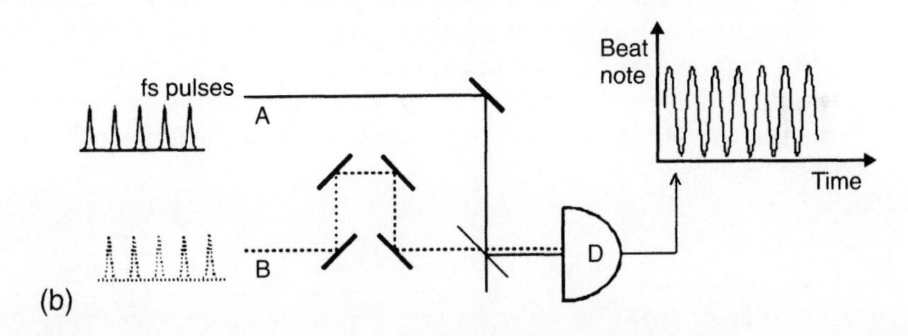

Figure 5.5 Pulse train coherence measured by a Mach–Zehnder interferometer. (a) A fringe pattern is observed around delays that are multiples of the pulse period, $m\tau_{RT}$. The fringe contrast deteriorates with increasing m for trains from common mode-locked lasers. (b) The coherence of an unknown source A can be measured by optical beating with an ideal reference source B, provided both sources have the same repetition rate. An optical delay is required to ensure that the pulse of each train interfere at the detector. The bandwidth of the beat note carries the information on the coherence properties.

period. After performing the time integration we obtain

$$S(q, \Delta x) = W_0 \sum_{i=1}^{N} \left[1 + \cos(\phi_i - \phi_{i+q} - k\Delta x)\right], \tag{5.17}$$

where W_0 is the energy of one pulse pair. Because N is typically a large number $\sum \cos(\phi_i - \phi_{i+q} - k\Delta x) \approx 0$ if the phase difference $\delta\phi_i = \phi_i - \phi_{i+q}$ is random, which results in zero fringe contrast.

The change in cavity length ΔL will result in a mode shift $\Delta \nu_m$ given by Eq. (5.15), resulting in a total phase shift $q\Delta\nu_m\tau_{RT}$. Let us use as an estimate for $\Delta\nu_m$ the value of $\nu_m \Delta L/L$. The fringes will disappear for the value of ΔL that makes this phase shift of the order of unity. Because $\Delta\nu_m$ is an optical frequency, it takes only a few round-trips in a typical laser to reach that value. It has been possible, however, to stabilize lasers to have a pulse train coherent over delays 14 orders of magnitude larger than the single pulse duration or $q \approx 10^8$, as will be shown in Chapter 13.

The method discussed previously is obviously not practical for measuring the coherence of a pulse train over many interpulse spacings as the required optical delay line can exceed hundreds of km. Another method to measure the coherence of a common source A is to compare it with a perfectly coherent source B (lower part of Fig. 5.5). Let us assume that both mode-locked lasers are locked[3] to the same repetition rate $1/\tau_{RT}$. From Eq. (5.15), it results that the mode frequency fluctuations are all equal to carrier-to-envelope fluctuations $d\nu_m/dL = df_0/dL$ if τ_{RT} is constant. Let us assume that the CEO of source (B), $f_{B,0}$, is kept constant by a control unit while the carrier frequency of source (A), $f_{A,0}$, is let to fluctuate. If the two pulse trains are made to interfere on a detector, a beat note will be observed. Using Eq. (5.10) to define the frequencies of the mode combs we find for the beat frequency

$$f_b(t) = |f_{B,0} - f_{A,0}(t)|. \tag{5.18}$$

The observation of $f_b(t)$ over a certain time period allows one to measure the bandwidth of this beat note. The inverse of the bandwidth is the coherence time of the source (B). A convolution of the bandwidths of each source is involved if the CEOs of both sources fluctuate.

As an example, Fig. 5.6 shows the beat note and its spectrum produced by two pulse trains of equal repetition rate. In this particular case of an unstabilized laser source, mechanical vibrations constantly change the cavity length, resulting in excursions of the cavity mode frequency of the order of one MHz. The beat note bandwidth however can be extremely narrow ($\approx$1 Hz) if, as is the case in Fig. 5.6, identical cavity length fluctuations exist in the two resonators from which the interfering pulse trains originate. This large degree of mutual coherence indicated by the narrow beat note bandwidth proves that the cavity does have an influence on the pulse train.

There is a simple method to generate two pulse trains that have the same repetition rate and are subject to the same cavity fluctuations. The method consists in constructing a single ring or linear resonator that emits two pulse trains of the same repetition rate but different mode frequencies, as will be described in Chapter 13. The two pulse trains from such a source produced the beat note shown in Fig. 5.6. A bandwidth of less than 1 Hz is observed because of the fact

[3]This locking does not necessarily imply stabilization: a synchronously pumped optical parametric oscillator and its pump laser have by design the same repetition rate. So does the two outputs of a bidirectional mode-locked ring laser.

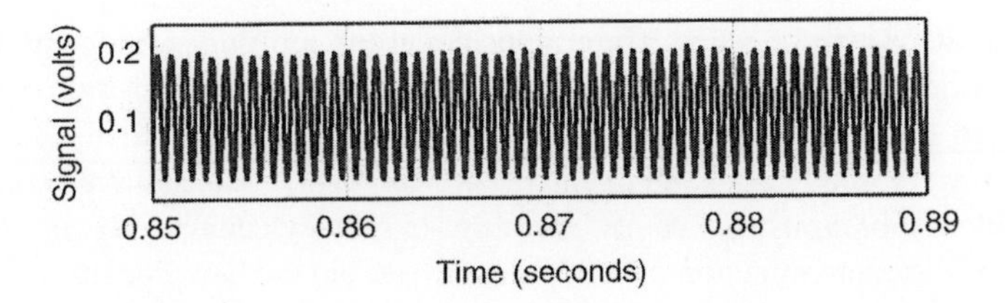

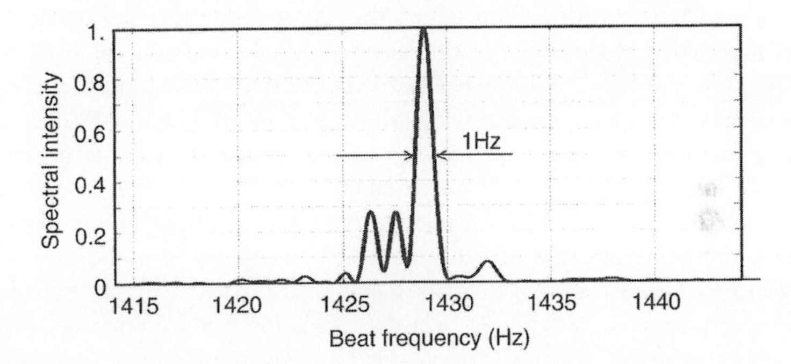

Figure 5.6 Top: portion of a 1.5 second long beat note recorded between two femtosecond pulse trains of the same repetition rate. The two pulse trains are issued from a common cavity. Lower figure: the Fourier transform of the full 1.5 s recording. The bandwidth of the central peak of the beat note spectrum is only 1 Hz wide. The side peaks are because of the fact that the beat note drifts over longer periods of time, because of a small rocking motion of the optical table (gyroscopic response) and/or air currents (Fresnel drag).

that the cavity length fluctuations change the mode comb of the two interfering pulses in a similar way.

To illustrate this let us assume one of the mirrors moves with a constant velocity ν relative to the cavity axis. In the time domain picture, both counter-propagating pulses reflected off this mirror experience a Doppler shift $\Delta\nu_{dop} = 2(\nu/c)\nu_N$. This leaves the beat note (difference frequency) unchanged. It also is instructive to analyze the mode comb of a cavity, which for simplicity we assume to be empty, under the condition of a moving mirror. The cavity length changes according to $L_0 + vt$. From Eq. (5.1), the mode comb frequencies can be expected to change according to

$$\nu_m(t) = \frac{mc}{2(L_0 + \nu t)}. \tag{5.19}$$

For small velocities, the frequency change during one round trip τ_{RT}

$$\Delta\nu_m = \nu_m(t) - \nu_m(t + \tau_{RT}) \approx \frac{2\nu}{c}\nu_m(t) = \Delta\nu_{dop} \tag{5.20}$$

is equal to the Doppler shift. These considerations are only valid within the approximations that the change in cavity length at each round-trip is much smaller than the wavelength, and that the relative change in velocity during a cavity lifetime is small compared to unity. Thus for all practical situations relating to mechanical vibrations of a cavity, the pulse frequency changes because of the Doppler shift at each reflection on a moving mirror, but the Doppler shifted mode frequency ν_N remains resonant with the cavity. A consequence of the equality between the Doppler shift and the cavity resonance shift expressed by Eq. (5.19) is that the cavity modes follow the same temporal evolution for both senses of circulation in a ring cavity.[4] This fact explains why pulse trains generated in opposite sense of circulation in ring cavities can be mutually coherent. Figure 5.6 is thus also a demonstration that the pulse train emitted by an unstabilized laser possesses properties pertinent to the resonator. This fact in itself is remarkable, considering that radiation should completely fill a cavity to define the cavity modes, and that a femtosecond pulse occupies only one part in a million of the cavity length. The beat note of Fig. 5.6 is an indication that the femtosecond pulse has started from noise distributed over the whole laser, noise that contained the mode structure of the cavity and maintained it through the compression process, shaping and evolution toward the fs pulse. The 1 Hz bandwidth of the beat note in Fig. 5.6 signifies that the mode structure of the cavity is remembered over at least 10^8 round-trips.

Time Domain Versus Frequency Domain Description of a Mode-Locked Laser

There are two basic approaches to describe the operational principle of a perfect mode-locked laser: the frequency and the time domain approach. The frequency domain picture that we have stressed so far considers the oscillation of a number of equally spaced (by Δ) longitudinal modes of amplitude $\mathcal{E}_m$ and phase ϕ_0, whose frequencies ν_m are given by Eq. (5.10). Some mechanism is then introduced to lock the relative phases of the modes to each other so that their coherent superposition produces a periodic pulse train in the time domain

$$\tilde{\mathcal{E}}(t)e^{i\omega_\ell t} = \sum_{m=1}^{M} \mathcal{E}_m e^{i\phi_0} e^{i2\pi\nu_m t}. \tag{5.21}$$

This locking can be accomplished through active, passive, and a combination of those techniques. Most femtosecond lasers utilize some kind of passive mode-locking, where intensity-dependent loss and/or dispersion mechanisms

[4]In that particular case the repetition rate is locked to the same value, and both terms in Eq. (5.15) are equal for both intracavity pulses.

favor pulsed over continuous radiation. The problem with the frequency domain picture is the difficulty to treat the various processes in the laser including the coupling between the large number of modes to predict the pulse parameters $M, \mathcal{E}_m, \nu_m$, and ϕ_0. Recall that the mode frequencies ν_m are not given by the dispersion of the cold cavity, cf. Eq. (5.7), but establish themselves in the process of mode-locking. Therefore the simple picture presented in the beginning of this chapter can only serve as a qualitative description of the mode-locking process.

We have seen in the previous section that the frequency domain picture, in which all longitudinal modes within the gain bandwidth oscillate in phase, is an oversimplification. The ratio of the laser cavity length to the pulse duration would be a measure of the number of modes oscillating in phase. Typically, for a meter long laser producing a train of 100 fs pulses, there would be over 100,000 longitudinal modes contributing to the pulse bandwidth. Because of the dispersion of the intracavity elements, the longitudinal modes are not equidistant over that range. Moreover, because a real laser resonator is not infinitely rigid, one cannot even talk of a fixed set of modes. Therefore, the most common approach to model a mode-locked laser is to analyze, in the time domain, the shaping mechanisms of one (sometimes more) pulse(s) traveling back and forth in a linear cavity, or circulating in a ring cavity. This is the description that will generally be followed in this book, and in particular in this chapter, Sections 5.2 and 5.3. In this picture the function of the cavity is not to establish a comb of modes but rather to force the circulating field to interact periodically with the cavity elements. Most analyses follow one of two main routes—(a) the evolution of the pulse from noise (spontaneous emission) and (b) the characterization of a steady state where the circulating pulse reproduces itself after an integer number (ideally one) of round trips. In either case, the result is the complex pulse envelope $\tilde{\mathcal{E}}(t)$ rather than the frequency domain parameters M, ν_m, and $\mathcal{E}_m$.

In the case of passive mode-locking, some intensity dependent loss or dispersion mechanism is used to favor operation of pulsed over continuous radiation. Another type of mode-locking mechanism is active: A coupling is introduced between cavity modes, "locking" them in phase. Between these two classes are "hybrid" and "doubly mode-locked" lasers in which both mechanisms of mode-locking are used. In parallel to this categorization in "active" and "passive" lasers, one can also classify the lasers as being modulated inside (the most common approach) or outside (usually in a coupled cavity) the resonator.

5.1.5. Basic Elements and Operation of a fs Laser

There are a few basic elements essential to a fs laser:

- a broadband ($\Delta\nu_g \gg 1$ THz) gain medium,
- a laser cavity,

- an output coupler,
- a dispersive element,
- a phase modulator, and
- a gain–loss process controlled by the pulse intensity or energy.

The items listed above refer more to a function than to physical elements. For instance, the gain rod in a Ti:sapphire laser can cumulate the functions of gain (source of energy), phase modulator (through the Kerr effect), loss modulation (through self-lensing), and gain modulation. To reach femtosecond pulse durations, there is most often a dispersive mechanism of pulse compression present, with phase modulation to broaden the pulse bandwidth, and dispersion (positive or negative, dependent on the sign of the phase modulation) to eliminate the chirp and compress the pulse. It is the dispersion of the whole cavity that has to be factored in the calculation of pulse compression. This interplay of nonlinear and dispersion processes is responsible for the perfect (equally spaced) mode comb of the mode-locked laser even if the dispersion of the cold cavity calls for unequally spaced cavity modes.

The radiation builds up from noise as in any oscillator. In continuously pumped lasers, the noise is because of spontaneous emission from the active medium. The evolution from noise to a regular train of pulses has been the object of numerous theories and computer simulations since the first mode-locked laser was operated (see, for instance, [6]). The pump power has to exceed a given threshold P_{th} for this transition from noise to pulsed operation to occur. This threshold is sometimes higher than the power required to sustain mode-locking: A mode-locked laser will not always restart if its operation has been interrupted. Such a hysteresis is sometimes observed with dye lasers and is common with Ti:sapphire lasers. Generally, it is a loss or gain modulation that is at the origin of the pulse formation.

Emergence of a pulse from noise is only the first stage of a complex pulse evolution. Subsequently, the pulse—which may contain sub-fs noise spikes and be as long as the cavity round-trip time—will be submitted to several compression mechanisms that will bring it successively to the ps and fs range. Progress in ultrashort pulse generation has resulted from the understanding of compression mechanisms that can act at the shortest time scale. Most of this chapter will be devoted to the analysis of the most common compression schemes. First, a loss (saturable absorption) and gain (synchronous pumping, gain saturation) mechanism will steepen the leading and trailing edges of the pulse, reducing its duration down to a few ps. Dispersive mechanisms—such as SPM and compression—take over from the ps to the fs range.

There are mechanisms of pulse broadening that prevent pulse compression from proceeding indefinitely in the cavity. The most obvious and simple broadening arises from the bandwidth limitation of the cavity (bandwidth of the difference spectrum of gains and losses). The bandwidth limit of the amplifier medium has been reached in some (glass lasers, Nd:YAG lasers), but not all, lasers. Other pulse width limitations arise from higher-order dispersion of optical components and nonlinear effects (four wave mixing coupling in dye jets, two photon absorption, Kerr effect, etc...).

The pulse evolution in a cw pumped laser leads generally to a steady state, in which the pulse reproduces itself after an integer number of cavity round trips (ideally one). The pulse parameters are such that gain and loss, compression and broadening mechanisms, as well as shaping effects, balance each other.

5.2. CIRCULATING PULSE MODEL

5.2.1. General Round-Trip Model

As mentioned in the previous section, the simplest model for a practical mode-locked laser is that of a pulse circulating in the cavity. The pulse travels successively through the different resonator elements, each contributing to the pulse shaping in a particular manner. The block diagram of Figure 5.7 is the basis for the most commonly used theoretical description of such lasers. Which elements need to be considered and in which order will depend on the type of laser to be modeled. Each block of the diagram of Fig. 5.7 can represent a real physical element or a function rather than a physical element. For instance, the "saturable loss" in Fig. 5.7 can represent either a saturable absorber, or the contribution of all elements that give rise to an intensity or energy dependent transmission.

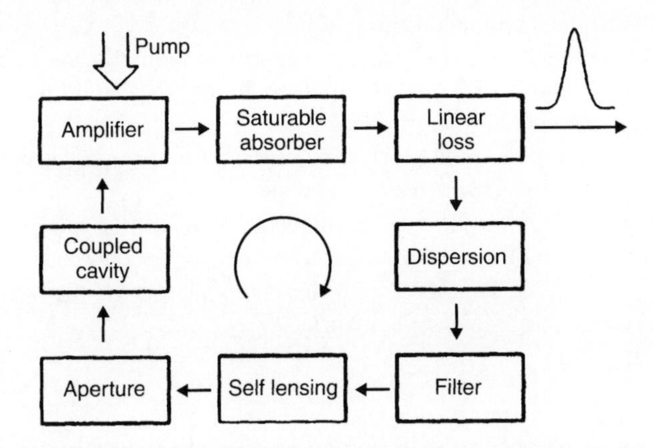

Figure 5.7 Schematic representation of the circulating pulse model describing a fs laser.

If we describe symbolically the action of each resonator element by an operator function T_i, the field after the $(n+1)$-th round-trip can be written in terms of the field before that round-trip as

$$\tilde{\mathcal{E}}^{(n+1)}(t) = (T_N T_{N-1} \dots T_2 T_1)\,\tilde{\mathcal{E}}^{(n)}(t), \tag{5.22}$$

where we have numbered the resonator elements from 1 to N. If the parameters of the laser elements are suitably chosen, the fields $\tilde{\mathcal{E}}^{(i)}$ will evolve toward a steady-state pulse, which reproduces itself (apart from a constant phase factor ϕ_0) after subsequent round trips, i.e., $\tilde{\mathcal{E}}^{(n+1)} = \tilde{\mathcal{E}}^{(n)}$ for n being large enough. The resulting steady-state condition

$$\tilde{\mathcal{E}}(t)e^{i\phi_0} = (T_N T_{N-1} \dots T_2 T_1)\,\tilde{\mathcal{E}}(t), \tag{5.23}$$

has been the basis for numerous analytical models.[5] These models usually assume certain beam and pulse shapes with parameters that are determined from Eq. (5.23). In most cases the operators have to be suitably approximated to allow for analytical treatment. We will discuss this procedure in detail in the next sections. While the analytical or semianalytical solutions give much insight into the physical mechanisms involved in fs lasers the complexity of the processes often calls for numerical modeling.

The round trip model illustrated in Fig. 5.7 is well-suited for a numerical treatment. Starting from noise (spontaneous emission) the field is traced through each cavity round-trip. The main advantages of this approach are

- Ease of incorporating various processes and optical elements with complicated transfer functions, leading to the modeling of virtually any laser.
- There is no need to make restrictive approximations for the transfer functions. This allows one, for example, to follow the evolution of both the temporal and spatial field profile.
- The evolution of the mode-locked pulse from noise can be predicted as can the response of the laser to external disturbances.
- One is not limited to the time or frequency domain. By using Fast Fourier Transforms (FFT), one can chose to model any phenomena in the most appropriate frame (for instance, phase modulation in the time domain, dispersion in the frequency domain).

---

[5] A more general definition expresses that the pulse reproduces itself every m round trips, i.e., $\tilde{\mathcal{E}}^{(n+m)}e^{i\phi_0} = \tilde{\mathcal{E}}^{(n)}$.

Even though the transient evolution toward steady state may take thousands of round-trips, the modeling can generally be implemented with a personal computer. The main disadvantage of a computer model that includes a plethora of processes is that it is difficult to get a clear physical picture of the laser operation.

5.2.2. Continuous Model

If the change in electric field introduced by each element of the cavity, at each round-trip, is small, the pulse evolution can be modeled by a differential equation. An additional simplification is to assume that the pulse evolves along the mode of a stable cavity, and thus the spatial pulse evolution is decoupled from the temporal evolution. Because the change per element and per round-trip is assumed to be infinitesimal, the order of the elements in the cavity does not matter, and the laser is equivalent to an infinitely long medium, in which the resonator elements are uniformly distributed (Figure 5.8). This "continuous model" in which the resonator elements are replaced by a uniform medium is similar to the propagation of a pulse in a fiber.

The continuous model is aimed at searching for a stationary pulse, which is a shape-preserving signal propagating through this model medium. Such a pulse is called a soliton of first order or fundamental soliton. Pulses that reproduce after a certain periodicity length are labeled solitons of higher order. In an actual laser, the "higher-order solitons" will reproduce after a given number of resonator round trips. We have seen some examples of solitons in Chapters 3 and 4. One of the simplest cavity model leading to solitons is that of a laser with linear gain balancing linear losses and a combination of SPM and dispersion. This particular soliton model and the related equations are discussed in more detail in the following section.

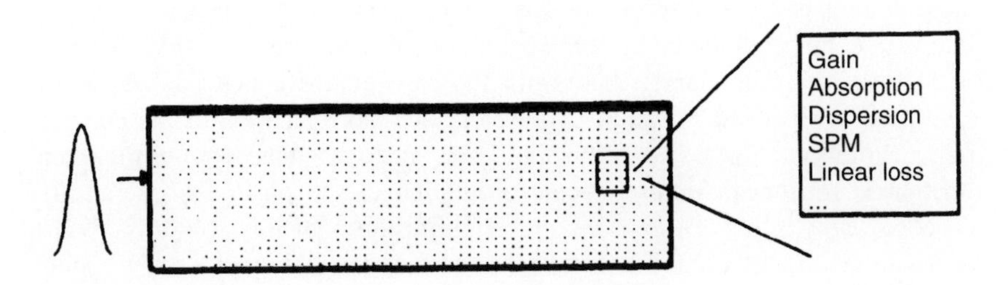

Figure 5.8 Representation of a fs laser as an infinitely long medium with the distributed properties of the cavity.

Solitons in Femtosecond Lasers

The existence of solitons is related to a particular structure of the equations governing propagation through the composite medium. No exact soliton solutions have been found in any model incorporating most of the resonator elements. However, several subsystems have been found to lead to soliton solutions. For instance, considering *only* the amplifier and an absorber as cavity elements, soliton solutions can be found [8] which are related to the π and 2π pulse propagation, as detailed in Chapter 4. Another subsystem that has been used considers the laser to consist only of a GVD and Kerr medium. It is this latter model, where absorption and gain are assumed to balance each other exactly, that will be discussed here in more detail.

The elegance of the theory is at the expense of simplifying assumptions that are not quite compatible with a pulse *formation* mechanism. Because the only compression mechanism assumed in the model presented later is purely dispersive, there is no intensity-dependent mechanism that could preferably amplify the noise fluctuations of the laser to start the pulse operation. Even though the Kerr effect is taken into account, one assumes that the self-focusing associated with it has no influence.

The ring laser model reduces to a product of two operations, as sketched in Figure 5.9: phase modulation in the time domain (upper part of the figure) and dispersion in the frequency domain. Both operations are combined into a single equation (see problem at the end of this chapter) shown in a square box in the middle of the figure.

This resulting expression is the nonlinear Schrödinger equation, which has been analyzed in detail by Zakharov and Shabat [7], using the inverse scattering method [9]. The problem is reduced to a search for eigenvalues of coupled differential equations. The soliton is the eigenfunction associated with that eigenvalue. The order of the soliton is the number of poles associated with that solution. A soliton of order 1 is a sech-shaped pulse. It exhibits a stable pulse shape, propagating without distortion. Solitons of order n are periodic solutions with n characteristic frequencies. Periodic evolution of the pulse train has been observed in some dye lasers [10, 11]. Salin *et al.* [12] interpreted the periodicity in pulse evolution of a fs laser as a manifestation of a soliton of order larger than 1. In the case of a dye laser, however, the nonlinear Schrödinger equation is only a crude approximation of the complex pulse evolution. For lasers with large gain, such that the continuity approximation is no longer valid, periodic oscillation of the pulse energy can also be observed [13, 14]. Despite the oversimplifications of this soliton model, it appears to describe certain features of the stationary operation of a fs Ti:sapphire laser [15, 16]. For instance, the unchirped pulse that returns identical to itself after each round-trip is associated with the soliton of order 1. With some minor

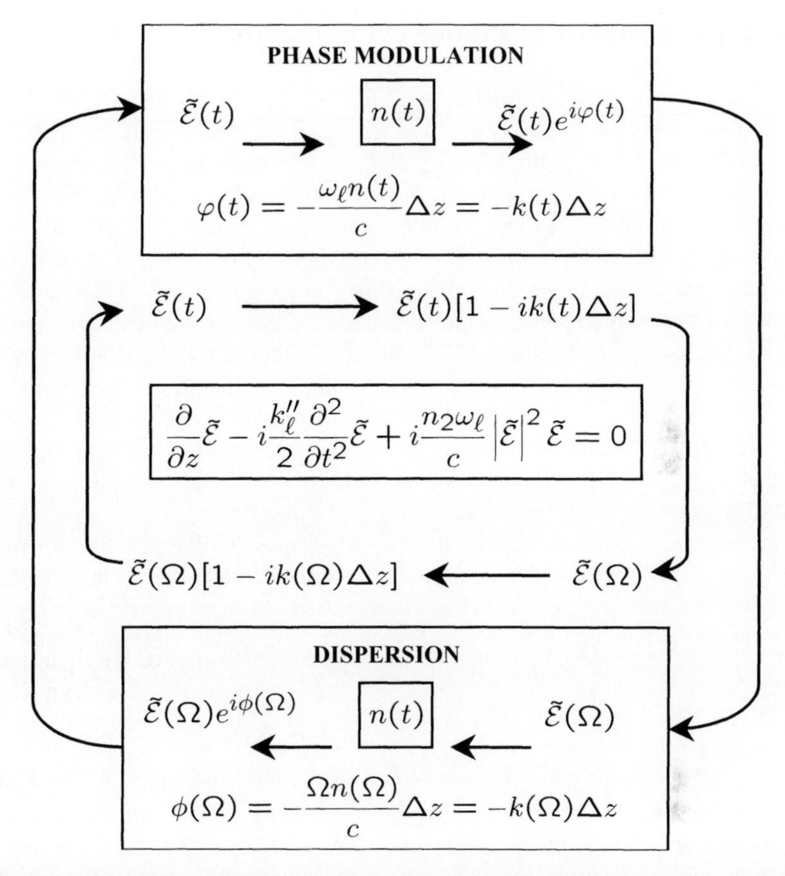

Figure 5.9 Model used to describe a "soliton" laser. Only infinitesimal SPM (top part) and dispersion (bottom part of the figure) are applied on the pulse circulating in the cavity. Phase modulation leads to a time-dependent wave vector $k(t) = \omega_\ell n(t)/c = \omega_\ell n_2|\tilde{\mathcal{E}}|^2/c$. The combined operation of phase modulation and dispersion results in the equation written in the center of the picture for which soliton solutions are known (nonlinear Schrödinger equation in Zakharov and Shabat [7]). A steady state in which gain and loss compensate is assumed.

changes in alignment, a periodicity is observed in the pulse train. If this periodicity contains m frequencies, it is often possible to represent the pulse by a soliton of order $n = m - 1$.

Before we give a detailed description of the evolution of various pulse parameters and the role of the most common cavity elements, in the next two sections, the basic features of a numerical model and an analytical approach will be explained.

5.2.3. Elements of a Numerical Treatment

Because of its central importance in the description of the pulse evolution let us first elaborate on the successive treatment of processes in the time and frequency domains. If a single cavity element represents several processes it is often convenient to divide it into thin slices. The term thin means that the change in the complex pulse envelope caused by one slice is small. In this case the order of processes considered in one slice is unimportant. The choice about which processes are treated in which domain (time, frequency, or spatial frequency) is made based on numerical or analytical convenience and feasibility. Gain and the Kerr effect are typically dealt with in the time domain ((x, y, z, t) space) while free-space propagation and dispersion are usually treated in the frequency domain ((k_x, k_y, z, t) or (x, y, z, Ω) space). For example, if gain and dispersion occur in one element, for each slice one has to solve a differential equation in the time domain to deal with the gain and subsequently treats the effect of the dispersion and diffraction in the frequency domain.

Figure 5.10 illustrates the procedure for the sequence of an element (or slice) with gain, free-space propagation, phase modulation, and dispersion. The illustration starts with an electric field $\tilde{E}_1(x, y, z, t) = \tilde{\mathcal{E}}_1(x, y, z, t)e^{i\omega_\ell t}$ entering the gain medium. This could be noise if we want to simulate the pulse evolution. This first step where all processes but the gain are neglected can formally be written using a transfer operator T_g.

$$\tilde{\mathcal{E}}_2(x, y, t) = T_g(t)\tilde{\mathcal{E}}_1(x, y, t). \tag{5.24}$$

In practice one solves the differential equations derived in Chapters 3 and 4 for a medium with population inversion. The next step involves propagation over a distance L_P. This diffraction problem is best described in the frequency domain. FFT algorithms are applied to obtain $\tilde{E}_3(\Omega, x, y)$. In Fresnel approximation the

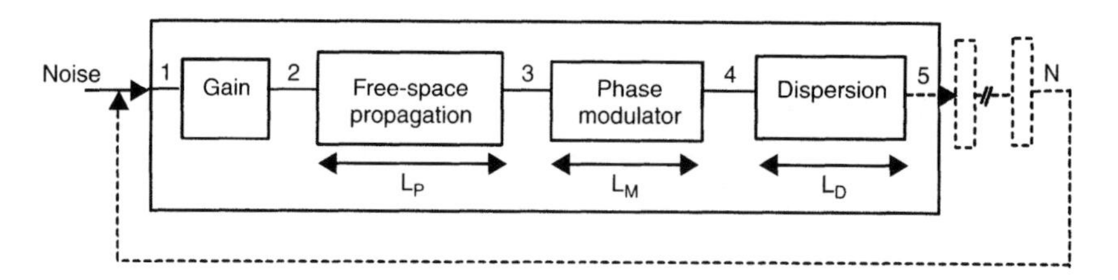

Figure 5.10 Illustration of some of the main elements and processes in a circulation model that describes pulse evolution in a laser. The numbers refer to the subscript of the electric field before and after a certain element or process.

propagation step through free space is described by Eq. (1.203):

$$\tilde{E}_3(x,y,\Omega) = \frac{i\Omega}{2\pi c L_P} e^{-i\Omega L_P/c} \int\int dx' dy' \tilde{E}_2(x',y',\Omega) \times \exp\left\{-\frac{i\Omega}{2L_P c}\left[(x'-x)^2 + (y'-y)^2\right]^{1/2}\right\}. \tag{5.25}$$

Inverse FFT then produces the output in the time domain $\tilde{E}_3(x,y,t)$. Except for pulses of a few optical cycles or shorter the approximation $\Omega \approx \omega_\ell$ can be made in the terms preceding the integral and in the exponent of the integrand. As explained in Chapter 1 this is equivalent to separating the space and time effects on propagation.

The next element introduces a phase modulation. Let us assume that through some effect the (nondispersive) refractive index of the material is modulated in time and/or space, $n = n(x,y,t)$. Its effect on the pulse is advantageously described in the time picture

$$\tilde{\mathcal{E}}_4(x,y,t) = \tilde{\mathcal{E}}_3(x,y,t) \exp\left[-i\frac{\omega_\ell}{c} n(x,y,t) L_M\right]. \tag{5.26}$$

Note that the pulse envelope $|\mathcal{E}(t)|$ does not change while the pulse spectrum and spatial frequency spectrum are modified because of the action of such a phase modulator.

As detailed in Chapter 1, cf. Eq. (1.166), a dispersive element is characterized by its (linear) transfer function, which for a dispersive path of length L_D is simply the propagator $\exp[-ik(\Omega)L_D]$ with $k = \Omega n(\Omega)/c$. Thus

$$\tilde{E}_5(x,y,\Omega) = \tilde{E}_4(x,y,\Omega) \exp\left[-i\frac{\Omega}{c} n(\Omega) L_D\right]. \tag{5.27}$$

The necessary input field is obtained after FFT of the output of the phase modulator.

This procedure is continued until all resonator elements are taken into account. The final output pulse $\tilde{\mathcal{E}}_N(x,y,t)$ is then coupled back into the first element (gain in our case) to start the next round-trip.

As pointed out previously, the treatment of a single cavity element may require a procedure as just described. This, for example, is true for the gain crystal in a Kerr lens mode-locked laser. This element is responsible for gain, dispersion, self-lensing and diffraction (beam propagation). The procedure is exemplified in Figure. 5.11. The crystal is divided into slices of thickness Δz and the various effects are dealt with one at a time in each slice. At the beginning of each slice the pulse properties are defined by the complex amplitude

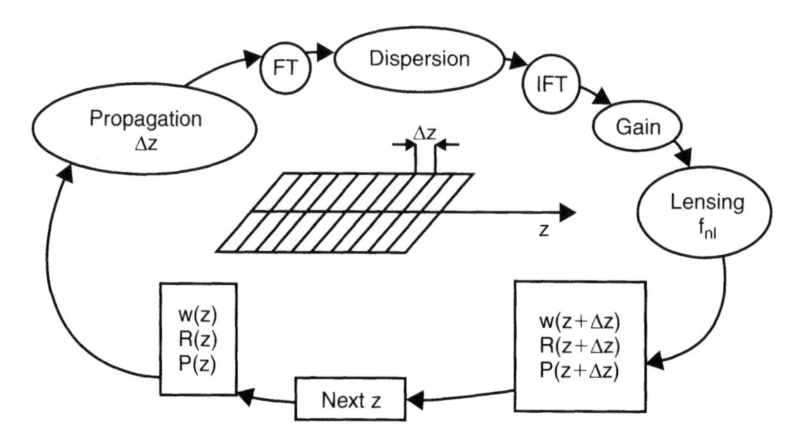

Figure 5.11 Successive calculations to be made to propagate a pulse through each slice Δz of a gain crystal. FT and IFT stand for Fourier transform and inverse Fourier transform and indicate that the treatment of dispersion can conveniently be done in the frequency domain.

$\tilde{\mathcal{E}}(z, r, t) = \mathcal{E}(z, r, t) \exp(\varphi(z, r, t))$. At the end of each slice we obtain $\tilde{\mathcal{E}}(z + \Delta z, r, t)$ which acts as the input for the next slice, $z + \Delta z \to z$.

To study the switch on dynamics of a fs laser one starts from noise. The noise bandwidth is roughly given by the width of the fluorescence spectrum of the amplifier while its magnitude corresponds to the light emitted spontaneously into the solid angle defined by the cavity modes. In most cases the particular noise features vanish after few round-trips, and the final results are independent of the field originally injected. Figure 5.12 shows as an example the development of the pulse envelope, instantaneous frequency, and energy as a function of round-trips completed after the switch on of the laser mode-locked with a slow saturable absorber. Obviously the pulse parameters become stationary after several hundred round-trips, which amounts to several microseconds.

5.2.4. Elements of an Analytical Treatment

A number of approximate analytical procedures has been developed by New [18,19] and Haus [20] to describe the steady-state regime. The problem often reduces to finding a complex pulse envelope $\tilde{\mathcal{E}}(t)$ that satisfies the steady-state condition [20]:

$$\tilde{\mathcal{E}}(t + h) = \prod_{i=1}^{N} T_i \tilde{\mathcal{E}}(t). \tag{5.28}$$

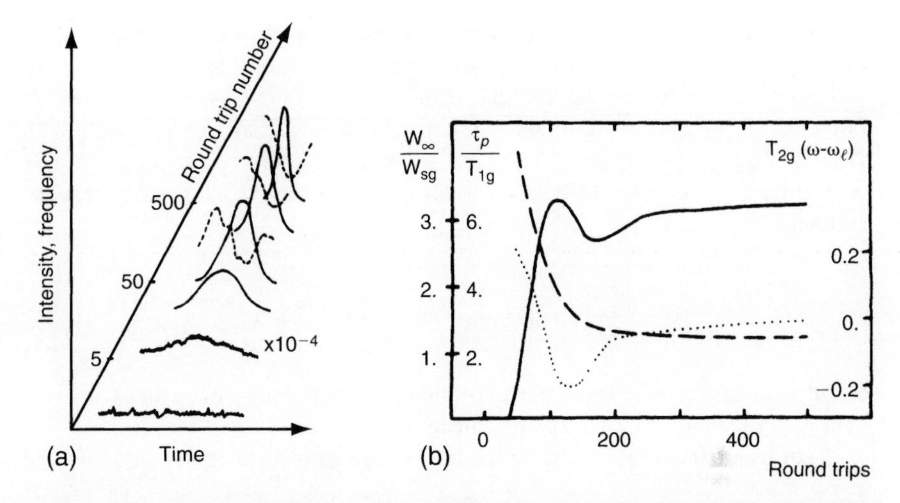

Figure 5.12 (a) Evolution of pulse envelope (solid line) and instantaneous frequency (dashed line) after switch on of the laser, and (b) corresponding steady-state pulse energy (solid line), pulse duration (dashed line), and frequency (dotted line). The active media were described by the density matrix equations introduced in Chapter 3. All pulse parameters including the average frequency develop as a result of the interplay of resonator elements. No extra frequency selective element was necessary to limit the pulse duration (normalized to the spectral width of the gain transition $2/T_{2g}$). (Adapted from Petrov [17].)

Equation (5.28) states that the pulse envelope reproduces itself after each round-trip, except for a temporal translation h including a constant pase shift. The main challenge is to find appropriate operator functions for the different resonator elements that are amenable to an analytical evaluation of Eq. (5.28). A convenient approximation is to assume that the modification introduced by each resonator element is small, which allows one to terminate the expansion of the corresponding operator functions after a few orders. Another consequence of that approximation is that the order of the resonator elements is no longer relevant. We will briefly describe this approach here with a small number of possible resonator elements and processes. Some of the most frequently used operators representative of resonator elements are derived below.

The transformation of the pulse envelope by a saturable loss–gain can be expressed as

$$\boxed{\tilde{\mathcal{E}}_{out}(t) = \left\{1 + \frac{1}{2}a_a^{(0)}\tilde{L}\left[1 - \frac{W(t)}{W_{sa}} + \frac{1}{2}\left(\frac{W(t)}{W_{sa}}\right)^2\right]\right\}\tilde{\mathcal{E}}_{in}(t).} \tag{5.29}$$

Equation (5.29) is the rate equation approximation ($T_2 \to 0$) of Eq. (3.79). The expansion parameters are the small signal absorption coefficient ($a_a < 0$) and the ratio of pulse energy density to the saturation density of the transition. Equation (5.29) applies to a gain medium with the substitutions $a_a \to a_g$ and $W_{sa} \to W_{sg}$.

The transfer function of a GVD element can be derived from Eq. (1.172) and reads

$$\boxed{\tilde{\mathcal{E}}_{out}(t) = \left\{1 + ib_2 \frac{d^2}{dt^2}\right\} \tilde{\mathcal{E}}_{in}(t).} \tag{5.30}$$

For this expansion to be valid, the dispersion parameter b_2 has to be much smaller than τ_p^2. In the case of a transparent medium of thickness d, $b_2 = k''d/2$.

A corresponding expression for the action of a Kerr medium of length d is

$$\boxed{\tilde{\mathcal{E}}_{out}(t) = \left\{1 - i\frac{k_\ell n_2 d}{n_0}|\tilde{\mathcal{E}}_{in}(t)|^2\right\} \tilde{\mathcal{E}}_{in}(t)} \tag{5.31}$$

which can easily be derived from Eq. (3.146).

A linear loss element, which for example represents the outcoupling mirror, can be modeled according to

$$\boxed{\tilde{\mathcal{E}}_{out}(t) = \left\{1 - \frac{1}{2}\gamma\right\} \tilde{\mathcal{E}}_{in}(t)} \tag{5.32}$$

where γ is the intensity loss coefficient (transmission coefficient), for which $\gamma \ll 1$ is assumed.

Each resonator contains frequency selective elements which can be used to tune the frequency. Such elements are for example prisms, Lyot filters, and mirrors with a certain spectral response. Together with the finite gain profile, they ultimately restrict the bandwidth of the pulse in the laser. Let us assume a Lorentzian shape for the filter response in the frequency domain $\tilde{H} = [1 + i(\Omega - \omega_\ell)/\Delta\omega_F]^{-1}$ where the FWHM $\Delta\omega_F$ is much broader than the pulse spectrum. After expansion up to second order and retransformation to the time domain:

$$\boxed{\tilde{\mathcal{E}}_{out}(t) = \left\{1 - \frac{2}{\Delta\omega_F}\frac{d}{dt} + \frac{4}{\Delta\omega_F^2}\frac{d^2}{dt^2}\right\} \tilde{\mathcal{E}}_{in}(t).} \tag{5.33}$$

If all passive elements are chosen to have an extremely broad frequency response, the finite transition profiles of the active media act as effective filters. This can be taken into account by using the operator defined in Eq. (3.79) for the media instead of Eq. (5.29) derived from the rate equations.

The operator describing the pulse change at each round-trip is obtained by multiplying the transfer functions of all elements of the cavity, neglecting products of small quantities. To evaluate the steady state (5.28), $\tilde{\mathcal{E}}(t+h)$ can be conveniently written as $(1 + h\frac{d}{dt} + \ldots)\tilde{\mathcal{E}}(t)$, leading to an integro-differential steady-state equation for the complex pulse envelope. The type of laser to be modeled determines the actual elements (operators) that need to be included. A parametric approach is generally taken to solve the steady-state equation. An analytical expression is chosen for the pulse amplitude and phase, depending on a number of parameters. This ansatz is substituted in the steady-state equation, leading to a set of algebraic equations for the unknown pulse parameters. Several types of mode-locked fs lasers have been modeled by this approach [21–24]. Changes in the beam profile because of the self-lensing effect have been incorporated [25, 26]. The transverse dimension is included through a modification of the pulse matrices introduced in Chapter 2 to include the action of the various active resonator elements [26].

5.3. EVOLUTION OF THE PULSE ENERGY

Before proceeding with a discussion of the various processes of pulse formation and compression, we will consider only the evolution of pulse energy in the presence of saturable gain and nonlinear losses. Based on the continuous model we will derive rate equations that describe the evolution of the pulse energy on time scales of the cavity round-trip time and longer. The rate equations will be written in terms of derivatives with respect to time. To relate this to the spatial derivatives used in the continuous model we apply

$$\frac{d}{dz} \approx \frac{1}{\nu_g}\frac{d}{dt} = \frac{\tau_{RT}}{2L}\frac{d}{dt}, \tag{5.34}$$

where L is the cavity length. Through most of this section we will neglect the transverse variation of the beam intensity (flat top beam) and diffraction effects. If we assume a beam cross section area A we may refer either to the total pulse energy $\mathcal{W}$ or the energy density $W = \mathcal{W}/A$.

We will concentrate first on parameters that may lead to a *continuous* mode-locked pulse train, as opposed to *Q-switched mode-locking*. Most broadband solid-state laser media being used for short pulse operation have a long gain lifetime. As a result, there is a tendency for the intracavity pulse to grow until the gain has been depleted. The laser operation thereafter ceases until the gain is recovered, which takes a time of the order of the gain material lifetime (typically microseconds). The output of such a laser consists in bursts of Q-switched mode-locked pulse trains.

This first subsection is dedicated to straightforward linear cavities. The case of ring cavities and some linear cavities with two pulses per cavity round-trip is more complex because it involves mutual coupling between counter propagating pulses in an absorber or nonlinear loss element.

5.3.1. Rate Equations for the Evolution of the Pulse Energy

Nonlinear Element

The hypothetical laser to be considered here consists of a gain and a loss medium whose parameters vary with the intensity and the energy of the evolving pulse, depending on the time constant of the nonlinearity. Examples are saturable gain and loss as described in detail in Chapter 3. A nonlinear element will be said to provide *negative feedback* if it enhances the net cavity losses with increasing energy or intensity. The reverse (cavity losses decreasing with intensity or energy) occurs for a nonlinear element that provides *positive feedback*. "Saturable absorption" is an example of positive feedback: the loss decreases with increasing intensity. Positive feedback is needed for the establishment of a pulse train. It is generally desirable to have a positive feedback dominating the nonlinearities of the cavity at higher intensities. Examples of negative feedback are two photon absorption and intracavity SHG. It will be shown in Section 5.4 that Kerr lensing contains both types of feedback. Another important example of positive and negative feedback is found with semiconductor absorbers, as discussed in Section 6.5.

We will consider in this section a combination of positive and negative passive feedback nonlinearities. The nonlinear losses can be expressed through their dependence on the pulse energy density W. We assume that at a certain energy, a negative feedback takes over, i.e., the loss start increasing with energy. The simplest form of nonlinear loss that will show a transition from positive to negative feedback is:

$$L(\mathcal{W}) = L_L + a(\mathcal{W} - \mathcal{W}_0)^2, \tag{5.35}$$

where $\mathcal{W}_0$ defines the energy at which the nonlinear losses switch over from saturable losses (positive feedback) to induced losses (negative feedback). As we will see when discussing specific examples of cavities, Eq. (5.35) is a second-order fit for the actual energy dependence of the losses, hence L_L is not simply a sum of the linear losses but may also contain a contribution from the nonlinear elements.

The saturable gain is another factor that determines the dynamics of the pulse evolution in the cavity. We will show in an example of saturable absorption and intracavity two photon absorption (cf. Section 5.3.2) how the parameters L_L, a, and $\mathcal{W}_0$ are related to those material parameters. In the case of mode-locking dominated by self-lensing, we will show in Section 5.4.3 the connection between the phenomenological parameters L_L, a, and $\mathcal{W}_0$ and properties such as the magnitude of the nonlinearity, the transverse dimension of the beam, the length, and position of the nonlinear element.

Rate Equations

In the present derivation of the evolution of the pulse energy we will use a rate equation approximation for the gain medium. Referring for instance to Eq. (4.18) for a two-level system, we can write for the population difference ΔN:

$$\frac{d\Delta N}{dt} = -\frac{I(t)\Delta N}{I_s T_1} - \frac{\Delta N - \Delta N_0}{T_1} - \frac{\Delta N + \Delta N_0}{2} R \tag{5.36}$$

where $I(t)$ is the laser intensity, R is a constant pumping rate,[6] and $W_s = I_s T_1$ is the saturation energy density. ΔN_0 is the equilibrium population difference in the absence of the pump and laser field. For most gain media, the energy relaxation time T_1 is longer than the pulse duration. The preceding equation is equivalent to the rate equation often used to model a gain medium:

$$\frac{d\Delta N}{dt} = -\frac{I\Delta N}{I_s' T_p} - \frac{\Delta N - \Delta N_0}{T_p} + R' \tag{5.37}$$

which has a constant pump rate $R' = -R\Delta N_0$, and where the energy relaxation time T_1 has been replaced by a shorter characteristic constant T_p given by:

$$\frac{1}{T_p} = \frac{1}{T_1} + \frac{R}{2}. \tag{5.38}$$

A modified saturation intensity was introduced as $I_s' = (T_1/T_p)I_s$. Without laser field ($I = 0$) the population difference, according to Eq. (5.37), approaches an

[6] The pumping term is proportional to the population of the ground state which is $N_1 = (\Delta N + \Delta N_0)/2$. R is an effective (assumed to be constant) pump rate that also contains the properties of a third energy level involved in the pumping process.

equilibrium value

$$\Delta N_e = T_p R' + \Delta N_0, \tag{5.39}$$

which can be positive for a sufficiently large pumping $R' > -\Delta N_0/T_p$, or $R > 2/T_1$.[7] In terms of ΔN_e Eq. (5.37) can be written as

$$\frac{d\Delta N}{dt} = -\frac{I\Delta N}{W_s} - \frac{\Delta N - \Delta N_e}{T_p}. \tag{5.40}$$

Because we are neglecting effects of pulse shape and will be considering gain media with relaxation times much longer than the cavity round-trip time, the effect of a short pulse on depleting the gain is equivalent to that of a constant intensity filling the cavity for a round-trip time τ_{RT}. Defining a gain factor $G = \sigma \Delta N \ell$, where σ is the cross section for stimulated emission, and ℓ is length of the gain medium traversed per round trip, an equivalent form for Eq. (5.40) is:

$$\frac{dG}{dt} = -\frac{G - G_e}{T_p} - \frac{G\mathcal{W}}{\mathcal{W}_{es}T_p}. \tag{5.41}$$

In Eq. (5.41), $\mathcal{W}_{es} = W_s \times A_g \times \tau_p/T_p$ is an effective saturation energy in the gain medium, where A_g is the cross section of the beam at that location. The physical meaning of the energy $\mathcal{W}_{es}$ is obvious from the steady state ($dG/dt = 0$) solution of Eq. (5.41):

$$G(\mathcal{W}) = \frac{G_e}{1 + \frac{\mathcal{W}}{\mathcal{W}_{es}}}. \tag{5.42}$$

Next we need a rate equation for the pulse energy. We assume that the laser consists only of the gain medium and the nonlinear loss element that was introduced in Eq. (5.35). The combined effect of gain and loss for the pulse energy per round-trip is $d\mathcal{W}/(dt/\tau_{RT}) = (G - L)\mathcal{W}$ or:

$$\frac{d\mathcal{W}}{dt} = \frac{G - L_L - a(\mathcal{W} - \mathcal{W}_0)^2}{\tau_{RT}}\mathcal{W}. \tag{5.43}$$

The system of Eqs. (5.41) and (5.43) describes the evolution of the energy of a single pulse.

[7] In the condition leading to population inversion, the recovery rate $1/T_p$ as defined by Eq. (5.38) is dominated by the pumping rate R.

$$\frac{d\mathcal{W}}{dt} = \frac{G - L_L - a(\mathcal{W} - \mathcal{W}_0)^2}{\tau_{RT}}\mathcal{W} \quad (5.44)$$

$$\frac{dG}{dt} = -\frac{G - G_e}{T_p} - \frac{G\mathcal{W}}{\mathcal{W}_{es}T_p}. \quad (5.45)$$

It is useful to investigate first under which condition there is a steady-state solution for the evolution equations. Steady state, $d\mathcal{W}/dt = 0$, is reached when $G(\mathcal{W}) = L_L + a(\mathcal{W} - \mathcal{W}_0)^2$, cf. Eq. (5.44). With Eq. (5.42) for the gain coefficient, this condition becomes:

$$\frac{G_e}{1 + \frac{\mathcal{W}}{\mathcal{W}_{es}}} = L_L + a(\mathcal{W} - \mathcal{W}_0)^2 \quad (5.46)$$

The existence of a real solution for the pulse energy $\mathcal{W}$ indicates that a steady-state regime of the laser is possible.[8] This is exemplified in the two examples discussed below for a short and a long lifetime gain medium.

Immediately after the gain is turned on, the energy is given by the spontaneous emission into the lasing mode, $\mathcal{W}_{sp} \ll \mathcal{W}_0$. At this early time the evolution of energy and gain can be calculated using $\mathcal{W} = 0$ and $G = G_e$ in the right-hand side of Eq. (5.44), and $\mathcal{W} = \mathcal{W}_{sp}$ and $G = G_e$ in Eq. (5.45):

$$\mathcal{W}(t) = \mathcal{W}_{sp} \exp\left(\frac{G_e - L_L - a\mathcal{W}_0^2}{\tau_{RT}} t\right) \quad (5.47)$$

$$G(t) = G_e \exp\left(-\frac{\mathcal{W}_{sp}}{\mathcal{W}_{es}T_p} t\right). \quad (5.48)$$

The laser is self-starting (the energy increases) if the gain exceeds the loss ($G_e > L_L + a\mathcal{W}_0^2$). The gain decreases from its initial value G_e because of saturation.

Case of a Laser with a Short Lifetime Gain Medium

The gain medium of a dye or a semiconductor laser is characterized by an energy relaxation time in the nanosecond range (typically a few nanoseconds), and a large gain cross section leading to a saturation energy density of the order of a few mJ/cm^2. Therefore, all the time constants in the system of Eqs. (5.44)

[8]Note that Eqs. (5.44) and (5.45) do not distinguish between mode-locked and cw laser. As such they can only be used to discuss the evolution of the pulse energy or cw power ($\mathcal{W}/\tau_{RT}$).

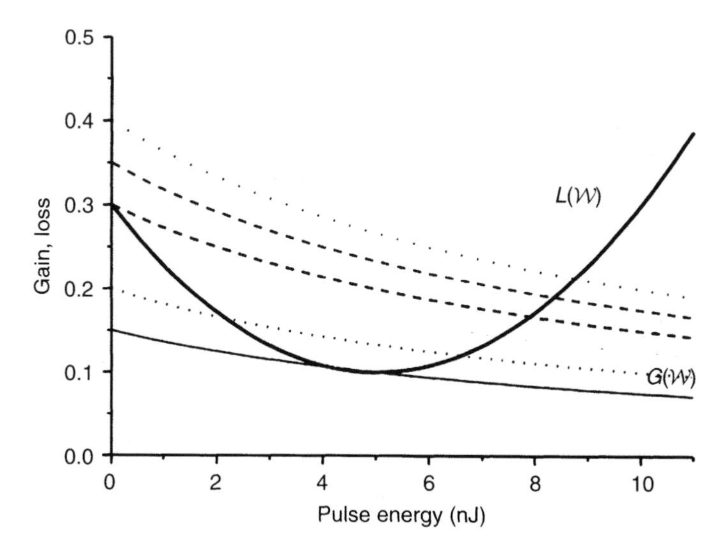

Figure 5.13 Saturated gain and loss of a laser with a short gain life time. The solid line describes the loss $L(\mathcal{W})$. The other lines represent the gain $G(\mathcal{W})$ for different small signal gain values $G_e = G(\mathcal{W} = 0)$. The laser parameters are defined in the text.

and (5.45) are of the same order of magnitude. The saturable absorber saturates at an energy smaller than the gain medium. The case of a strong positive–negative feedback with $a = 0.008\ \text{nJ}^{-2}$, with a turnover energy $\mathcal{W}_0 = 5$ nJ and a larger gain saturation energy of $\mathcal{W}_{es} = 10$ nJ (this proportion would be typical in a dye laser) is illustrated in Figure 5.13. The heavier solid line represents the losses $L(\mathcal{W})$ (linear loss $L_L = 0.1$) of Eq. (5.35). The other succession of curves represents the saturated gain $G(\mathcal{W})$, cf. Eq. (5.42), for various levels of pumping, i.e., the unsaturated gain G_e varies from 0.15 to 0.4, with increments of 0.5. Steady-state solutions $\mathcal{W}$ exist where the loss and the gain curves intersect, which represents a solution of Eq. (5.46). Obviously, the laser is self-starting for values of unsaturated gain larger than 0.3. Steady-state solutions can be expected only for $G_e > 0.15$.

The steady-state pulse energy should correspond to the highest energy intersection of the saturated gain curves with the loss curve. The low-energy intersection is an unstable equilibrium; a small positive excursion of the pulse energy from this value will drive the system toward the high-energy intersection point. For values of initial gain less than 0.3, an initial energy larger than the first intersection of the saturated curve with the loss curve is required. For instance, if the unsaturated gain is 0.25, an initial pulse energy larger than 1 nJ is required to have evolution toward the steady-state energy of 7.35 nJ.

Case of a Laser with a Long Lifetime Gain Medium

In most solid-state crystalline lasers, such as Ti:sapphire, Li:CAF, Nd:YAG, or Nd:vanadate, the energy relaxation time of the upper lasing level is orders of magnitude larger than the cavity round-trip time. The loss modulation is small $[a(\mathcal{W}_0-\mathcal{W})^2 \ll 1]$, with a turnover point for the nonlinear loss curve much higher than the gain saturation $\mathcal{W}_0 \gg \mathcal{W}_{es}$. In most solid-state laser crystals used for fs pulse generation, the gain lifetime being in the microsecond range, the two Eqs. (5.41) and (5.43) operate on totally different time constants. Once the gain G_e has been switched on, the pulse energy reaches a value that corresponds to the steady state of Eq. (5.44). Because of the long lifetime T_p of the upper state, the gain [Eq. (5.45)] evolves on a much longer time scale of thousands of round-trips. Thus one can assume that the pulse energy derived from the steady-state solution of Eq. (5.44) follows the slowly evolving gain adiabatically. Substituting the steady-state solution $\mathcal{W}$ of Eq. (5.44) into Eq. (5.45) yields:

$$\frac{dG}{dt} = -\frac{G - G_e}{T_p} - \frac{G\mathcal{W}_0}{T_p\mathcal{W}_{es}} \mp \sqrt{\frac{G - L_L}{a}}\frac{G}{T_p\mathcal{W}_{es}}. \tag{5.49}$$

A plot of the function dG/dt versus gain G is shown in Figure 5.14. The steady-state condition corresponds to the intersection of these curves with the abscissa, $dG/dt = 0$. As initial condition, the pulse energy $\mathcal{W}$ is small, and the gain has its

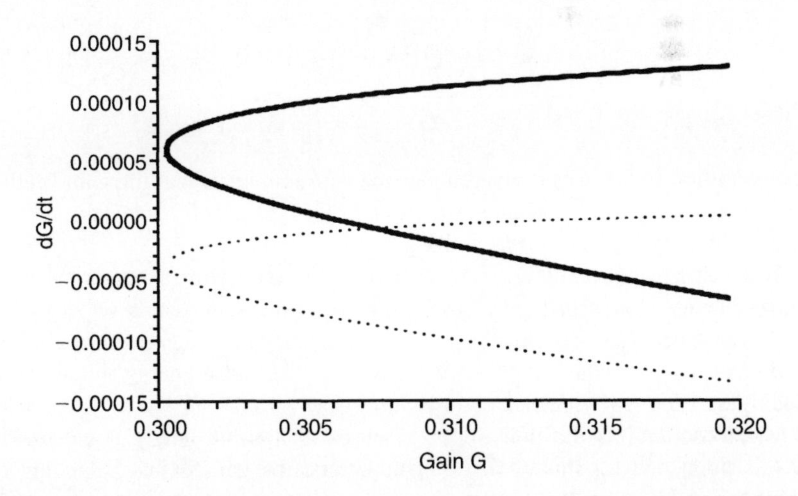

Figure 5.14 Plot of the function dG/dt from Eq. (5.49), taken for two sets of parameters leading either to cw mode-locking (solid line) or to Q-switched mode-locking (dotted line). The parameters are $\mathcal{W}_0 = 25$ nJ, $\mathcal{W}_s = 13$ nJ, Ge = 1 (solid line); and $L_L = 0.3$, $T_p = 2000$ ns and $G_e = 0.8$ (dotted line).

maximum value equal to the linear gain G_e, cf. Eqs. (5.47) and (5.48). The laser starts necessarily on the lower branch of any of these curves, where dG/dt is negative because the gain decreases as the laser power builds up. Referring to the solid line, the gain is expected to decrease until $dG/dt = 0$, which happens at the lower branch of this curve. The point $dG/dt = 0$ describes a stable steady state, because an increase in pulse energy leading to a further decrease in gain leads to a positive dG/dt, hence an increase in gain and return toward the $dG/dt = 0$ point. This indicates the possibility of the existence of a stable pulse train.

The situation is different with the parameter set leading to the dotted curve. Here, dG/dt is still negative at the minimum gain value $G = L_L = 0.3$. At this point there are no real solutions to Eq. (5.49). Reducing the gain below L_L as required by the negative dG/dt drives the laser below threshold. Hence the laser will turn itself off, until the gain can recover to its small signal value, and the laser can start again. This describes the scenario of Q-switching.

CW-Mode-Locking Versus Q-Switching

The condition for a cw regime (or a stable pulse train) is thus that at the minimum value $G = L_L$, the gain derivative as shown in Eq. (5.49) be positive, $\left.\frac{dG}{dt}\right|_{G=L_L} > 0$. This can be expressed as:

$$\boxed{\epsilon_g = G_e - L_L\left(1 + \frac{\mathcal{W}_0}{\mathcal{W}_{es}}\right) > 0.} \tag{5.50}$$

The condition (5.50) was derived under the approximation that the gain lifetime is infinite. It remains a good approximation for the typical solid-state laser cavities with a round-trip time of 10 ns and a gain lifetime of 2 μs.

To illustrate this point Eqs. (5.44) and (5.45) were solved numerically and the results are plotted in Figure 5.15. Figure (a) corresponds to a set of parameters leading to $\epsilon_g \approx 0$ [Eq. (5.50)], and shows that steady-state continuous mode-locking is reached after a few transients. The pulse energy initially rises quickly as Eq. (5.44) reaches equilibrium at the initial gain value. After a few transient oscillations the pulse energy settles to a value nearly equal to $\mathcal{W}_0$. At this point and for this choice of parameters, the gain is just balancing the linear losses. Figure 5.15(b) corresponds to a slightly smaller gain (gain reduced from $G_e = 0.87$ to $G_e = 0.8$). Substituting the values in Eq. (5.50), we find a negative value for $\epsilon = -0.07$, which indicates Q-switching. Indeed, a periodic burst of mode-locked pulse trains is seen.

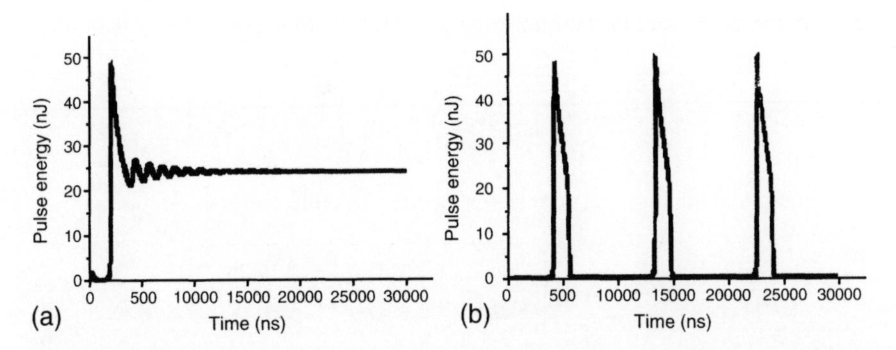

Figure 5.15 Pulse energy versus round-trip index. The lifetime of the upper laser level is 2000 ns. The round-trip time is 10 ns. The linear loss parameter is $L_L = 0.3$. The turnover energy is $\mathcal{W}_0 =$ 25 nJ. The saturation energy is $\mathcal{W}_s = 13$ nJ. The amplitude coefficient for the nonlinear interaction is $a = 0.0006$ nJ^{-2}. The figure on the left (a) is for a linear gain of $G_e = 0.87$. The figure on the right (b) is for a linear gain of $G_e = 0.8$.

5.3.2. Connection of the Model to Microscopic Parameters

In the previous sections we found a condition for continuous mode-locked operation, as opposed to Q-switched operation. The model was based on a simple model for the nonlinear intracavity losses of a hypothetical element [cf. Eq. (5.35)]. Although the exact functional behavior of the losses at the transition from positive to negative feedback depends on the actual cavity elements and processes involved, the general trend described by Eq. (5.35) is quite general. To illustrate this point we identify the terms of that expression with physical quantities for a specific example.

We consider here the case where the nonlinear losses are caused by a saturable absorber of thickness d_1 and a two photon absorber of thickness d_2. Each element is traversed once per round-trip, and we assume that the change in energy per round-trip is small ($\Delta\mathcal{W}/\mathcal{W} \ll 1$). For a saturable absorber whose relaxation time is much longer than the pulse duration we derived a relation between the input and output pulse energy in Chapter 3, cf. Eq. (3.57):

$$\mathcal{W}_{out} = \mathcal{W}_{sa} \ln\left[1 - e^{-\alpha_0 d_1}\left(1 - e^{\mathcal{W}_{in}/\mathcal{W}_{sa}}\right)\right], \tag{5.51}$$

where $\mathcal{W}_{sa}$ is the saturation energy for the absorber. After expanding this expression into a Taylor series up to first order in the small signal absorption coefficient, $\alpha_0 d_1$, and up to third order in $(\mathcal{W}_{in}/\mathcal{W}_{sa})$ we obtain

$$\mathcal{W}_{out} = \mathcal{W}_{in} - \mathcal{W}_{sa}\alpha_0 d_1 \left[\frac{\mathcal{W}_{in}}{\mathcal{W}_{sa}} - \frac{1}{2}\left(\frac{\mathcal{W}_{in}}{\mathcal{W}_{sa}}\right)^2 + \frac{1}{6}\left(\frac{\mathcal{W}_{in}}{\mathcal{W}_{sa}}\right)^3 \right]. \tag{5.52}$$

The energy attenuation per round-trip for the saturable absorber becomes:

$$\frac{d\mathcal{W}}{d(t/\tau_{RT})} = -\mathcal{W}_{sa}\alpha_0 d_1 \left[\frac{\mathcal{W}}{\mathcal{W}_{sa}} - \frac{1}{2}\left(\frac{\mathcal{W}}{\mathcal{W}_{sa}}\right)^2 + \frac{1}{6}\left(\frac{\mathcal{W}}{\mathcal{W}_{sa}}\right)^3 \right], \tag{5.53}$$

where we have replaced $\mathcal{W}_{in}$ by $\mathcal{W}$. Clearly, the transmission increases with increasing pulse energy (saturation). The opposite can be expected from a two photon absorber. Here, a beam of intensity I is attenuated according to

$$\frac{dI}{dz} = -\beta_2 I^2, \tag{5.54}$$

where β_2 is the two photon absorption cross section. Spatial integration relates the input to the output intensity

$$\frac{1}{I_{in}} - \frac{1}{I_{out}} = -\beta_2 d_2. \tag{5.55}$$

Assuming small changes per pass ($\Delta I = |I_{out} - I_{in}| \ll I_{in}, I_{out} \approx I$) we can approximate

$$\frac{dI}{d(t/\tau_{RT})} = -\beta_2 d_2 I^2. \tag{5.56}$$

Within our approximation of a fixed pulse shape (duration τ_p) and flat top beam of area A we can obtain from Eq. (5.56) the rate of change for the pulse energy $\mathcal{W} = I\tau_p A$

$$\frac{d\mathcal{W}}{d(t/\tau_{RT})} = -\beta_w d_2 \mathcal{W}^2, \tag{5.57}$$

where $\beta_w = \beta_2/(A\tau_p)$ is the effective two photon absorption cross section for the pulse energy.[9]

[9] We have made the approximation in this section that reshaping and focusing effects are negligible.

If we add linear losses L_0 to the effect of the one photon absorber [Eq. (5.53)] and the two photon absorber [Eq. (5.57)], we find for the total beam attenuation per round trip:

$$\frac{d\mathcal{W}}{d(t/\tau_{RT})} = -\left\{L_0 + \alpha_0 d_1 - \frac{\alpha_0 d_1 \mathcal{W}}{2\mathcal{W}_{sa}} + \frac{\alpha_0 d_1 \mathcal{W}^2}{6\mathcal{W}_{sa}^2} + \beta_w d_2 \mathcal{W}\right\} \mathcal{W}. \tag{5.58}$$

This equation can be compared to the formal expression for the nonlinear loss introduced in Eq. (5.35):

$$L(\mathcal{W}) = L_L + a(\mathcal{W} - \mathcal{W}_0)^2. \tag{5.59}$$

We find the correspondences

$$L_L = L_0 + \alpha_0 d_1 - a\mathcal{W}_0^2, \tag{5.60}$$

where the turnover energy is given by:

$$\mathcal{W}_0 = \frac{3\mathcal{W}_{sa}}{2}\left(1 - \frac{2\beta_w d_2 \mathcal{W}_{sa}}{\alpha_0 d_1}\right). \tag{5.61}$$

The amplitude of the nonlinear losses is characterized by the parameter a:

$$a = \frac{\alpha_0 d_1}{6\mathcal{W}_{sa}^2}. \tag{5.62}$$

Colliding Pulses in the Loss Element

In some cases several pulses can circulate in the cavity. This is the case for a bidirectional ring laser and a linear laser where several pulses exist during one round-trip. If the cavity consists of a saturable gain and a saturable loss medium the pulses of a ring laser will collide in the loss medium if the attenuation decreases with energy. This favors optimum pulse overlap in the absorber because each pulse feels an absorption that is saturated by twice the energy. The same situation can occur in linear resonators where two (or more) pulses oscillate. If the absorber is in the cavity center equal pulse spacing results. Asymmetric pulse spacings have been observed that result in colliding pulses in nonlinear elements placed off center in linear cavities (see for example Lai *et al.* [27]). The evolution of such regimes is complex; at this point we want to sketch the modifications necessary for the nonlinear loss element only.

For a thin saturable absorber the transmitted pulse energy, according to Eq. (5.52), is:

$$\mathcal{W}_{out} = \mathcal{W}_{in} \left\{ 1 - \alpha_0 d_1 \left[1 - \frac{\beta}{2} \frac{\mathcal{W}_{in}}{\mathcal{W}_{sa}} + \frac{\theta}{6} \left(\frac{\mathcal{W}_{in}}{\mathcal{W}_{sa}} \right)^2 \right] \right\}. \tag{5.63}$$

We have introduced the coefficients β and θ to describe the colliding pulse effect on the saturation. If the absorber is geometrically thin compared to the pulse length and the pulses interact incoherently, for example because of crossed polarizations, $\mathcal{W}_{in}$ should be replaced by $2\mathcal{W}_{in}$ and $\theta = \beta^2 = 4$.

In the case of coherent overlap the two counter-propagating pulses produce an intensity grating in the absorber.

$$I(z) = I_0 \left[1 + \cos(2kz) \right]. \tag{5.64}$$

While at the nodes of the intensity modulation there is no saturation, the saturation at the maxima corresponds to an energy density of $4\mathcal{W}_{in}$. An analysis of the grating and its effect on the propagating field gives $\beta = 3$ and $\theta = 5$ in Eq. (5.63) [28].

5.4. PULSE SHAPING IN INTRACAVITY ELEMENTS

In any description of a laser that follows the round-trip model, either numerically or analytically, each element is taken in sequence, represented either by a matrix or a scalar function. Essential intracavity elements are analyzed in this section. In the sections that follow and in Appendix E, we will derive expressions for the most essential combinations of intracavity elements. The term element refers here more to a function than a physical element, because each component of a laser will have generally a plurality of properties which are most easily treated separately. For instance, the Ti:sapphire crystal in a laser may serve simultaneously as a gain medium, dispersive element, nonlinear nonresonant element, astigmatism compensator. In each subsection characterizing an element, we will give expressions for its function at various levels of approximation, either in the time domain or in the frequency domain, as appropriate.

The various elements are organized in resonant, nonresonant passive elements and active elements. Under resonant elements we include saturable absorbers and gain, because they are generally associated with a near-resonant transition.

The organization of this section is as follows:

1. Saturable absorbers and gain
2. Nonlinear nonresonant elements

   (a) SPM and cross-phase modulation
   (b) Polarization coupling and rotation
   (c) Two photon absorption
3. Self-lensing
4. Summary of compression mechanisms
5. Dispersion

5.4.1. Saturation

Most fs mode-locked lasers involve some intensity dependent loss mechanism. Saturation of an absorber is the first one that comes to mind and has been used for mode-locking. The typical passive mode-locking element favors pulsed over cw operation by reducing the cavity losses for high intensities. Through the sheer mechanism of saturation, an absorber–gain element can also produce phase modulation and coupling between counter-propagating beams.

Because there should be at least one pulse per cavity round-trip time τ_{RT}, the recovery time τ_r of the device should not exceed that time: $\tau_r \leq \tau_{RT}$. Within that constraint, there is still room for a distinction between "slow" and "fast" intensity dependent elements. A slow element—such as the saturable absorber of a fs laser—will recover in a time long compared with the pulse duration. A fast element—such as a Kerr lens—will have its time constant(s) even shorter than the fs pulse.

The most commonly used saturable absorbers are dyes and semiconductors—specifically MQW's. When used in a free flowing dye jet, the saturable absorber dye has the advantage of a continuously adjustable optical density (through its concentration) and a high damage threshold. The latter is because of the tiny interaction volume (a tight focal spot) is replenished in microseconds. To ensure a good stability and optical quality of the jet, a viscous solvent—typically ethylene glycol—is used. Saturable absorber dyes have a saturation energy density of the order of 1 mJ/cm^2, and an energy relaxation time ranging from 1 ps to several ns. The obvious disadvantage of a dye is the inconvenience of having to deal with carcinogenic solutions and noisy, bulky circulation systems. There is a widespread effort to replace dyes by MQW saturable absorbers.

MQW's provide the substitute saturable absorber with the smaller saturation energy required to mode-lock semiconductor lasers [29]. Measurements performed at room temperature with cw radiation and a 5 μm spot size indicate a saturation intensity of less than 1 kW/cm^2 for MQW, against 10 MW/cm^2 for pure GaAs [30]. It has been possible to achieve even more control on the parameters of the saturable absorber (in particular its saturation intensity) by inserting a MQW in a Fabry–Perot used in antiresonance [31, 32]. Such a device can be added to an end mirror of a mode-locked laser. Because of the antiresonance condition, the material inside the Fabry–Perot is subjected to a smaller field than the one present in the laser cavity, hence a better damage threshold and higher saturation intensity for the device than for the MQW used directly. Such a device has therefore been successfully applied to most cw mode-locked solid state lasers.

The general case of saturable gain and absorption was treated in Chapters 3 and 4. Here we want to summarize the main results for a fast and a slow element in the rate equation approximation.

Fast Absorber or Amplifier

The propagation equation for the intensity through a fast saturable material

$$\frac{d}{dz}I = \frac{\alpha I}{1 + I/I_s}, \tag{5.65}$$

cf. Eq. (3.62), can be integrated and one obtains after a distance Δz an implicit solution for the output intensity

$$\ln \frac{I(\Delta z, t)}{I_0(t)} + \frac{I(\Delta z, t) - I_0(t)}{I_s} = \alpha \Delta z = a. \tag{5.66}$$

Here $a = \alpha \Delta z$ is the small signal absorption ($a < 0$) or gain ($a > 0$) coefficient. For optically thin elements or a slice Δz of an arbitrary element, $|a| \ll 1$, the change in intensity becomes

$$\Delta I(t) \simeq \frac{a I_0(t)}{1 + I_0(t)/I_s}. \tag{5.67}$$

The phase modulation associated with saturation and interaction away from the line center, according to Eq. (3.71), is

$$\varphi(\Delta z, t) \simeq -\frac{1}{2}(\omega_\ell - \omega_{10}) T_2 a \ln \frac{I(\Delta z, t)}{I_0(t)}. \tag{5.68}$$

These fast elements can follow the pulse instantaneously. The saturation and consequently the pulse shaping is therefore controlled by the intensity. As explained in detail in Chapter 3 this leads to pulse shortening in an absorber and broadening in an amplifier.

Slow Absorber or Amplifier

In slow elements the saturation is controlled by the pulse energy. The medium at a given time t accumulates the changes in the occupation numbers induced by all parts of the pulse arriving prior to t. From Eq. (3.55) we obtain for the intensity after such an element

$$I(z,t) = I_0(t)\frac{e^{W_0(t)/W_s}}{e^{-a} - 1 + e^{W_0(t)/W_s}}. \tag{5.69}$$

For weak absorption or gain ($|a| \ll 1$) this expression can be simplified and the change of pulse intensity

$$\Delta I(t) \simeq aI_0(t)e^{-W(t)/W_s}. \tag{5.70}$$

In this limit the phase modulation is given by Eq. (3.67) and reads:

$$\varphi(t) \simeq -\frac{1}{2}(\omega_\ell - \omega_{10})T_2 a e^{-W_0(t)/W_s}. \tag{5.71}$$

The pulse shaping in these elements is a result of (unsaturated) attenuation or gain in the leading part of the pulse while the trailing part is less affected (saturated transition).

5.4.2. Nonlinear Nonresonant Elements

(a) Self-Phase Modulation

Some elements impress a nonlinear phase on the propagating pulse. As detailed in Chapter 3, this phase is the result of a nonlinear process of third order and characterized by a nonlinear polarizability $\chi^{(3)}$. In the limit of a fast nonlinearity the response is instantaneous and is usually described by an intensity-dependent refractive index. Acting only on the phase, such an element leaves the pulse envelope, $\mathcal{E}_0(t)$, unchanged. From Eq. (3.149)

$$\varphi(t,z) = \varphi_0(t) - \frac{k_\ell n_2}{n_0} z\mathcal{E}_0^2(t) = \varphi_0(t) - \frac{k_\ell \bar{n}_2}{n_0} zI_0^2(t). \tag{5.72}$$

If the actual profile of the incident beam is taken into account the index change becomes a function of the transverse coordinate, which leads to self-lensing effects. The general mechanism is described in Chapter 3; the effect of such an element in a fs laser is discussed in the next section.

(b) Polarization Coupling and Rotation

Nonlinear effects can also act on the polarization state of the laser pulse. This effect is used in some lasers (for instance in fiber lasers [33]) to produce mode-locking. Let us consider a pulse with arbitrary polarization, with complex amplitudes $\tilde{\mathcal{E}}_x(t)$ and $\tilde{\mathcal{E}}_y(t)$ along the principal axis characterized by the unit vectors $\hat{x}$ and $\hat{y}$:

$$\mathbf{E} = \frac{1}{2}\left(\hat{x}\tilde{\mathcal{E}}_{0x}(t) + \hat{y}\tilde{\mathcal{E}}_{0y}(t)\right) e^{i(\omega_\ell t - k_\ell z)} + c.c. \tag{5.73}$$

The propagation of such a field through the nonlinear material leads to a coupling of the two polarization components. One can calculate, see [33], the nonlinear index change probed by polarizations along $\hat{x}$ and $\hat{y}$:

$$\Delta n_{\mathrm{nl},x} = n_2\left[|\tilde{\mathcal{E}}_{0x}|^2 + \frac{2}{3}|\tilde{\mathcal{E}}_{0y}|^2\right]$$

$$\Delta n_{\mathrm{nl},y} = n_2\left[|\tilde{\mathcal{E}}_{0y}|^2 + \frac{2}{3}|\tilde{\mathcal{E}}_{0x}|^2\right]. \tag{5.74}$$

In an element of thickness d_m, this induced birefringence leads to a phase change between the x and y components of the field vector

$$\Delta\Phi(t) = \frac{2\pi}{\lambda_\ell}\left(\Delta n_{\mathrm{nl},x} - \Delta n_{\mathrm{nl},y}\right) = \frac{2\pi n_2 d_m}{3\lambda_\ell}\left[|\tilde{\mathcal{E}}_{0x}(t)|^2 - |\tilde{\mathcal{E}}_{0y}(t)|^2\right]. \tag{5.75}$$

The phase shift is time dependent and, in combination with another element, can represent an intensity-dependent loss element.

To illustrate this further let us consider a sequence of such a birefringent element and a linear polarizer. We assume that the incident pulse, $\mathcal{E}_0\cos(\omega t)$, is linearly polarized with components

$$\mathcal{E}_{0x}(t) = \mathcal{E}_0(t)\cos\alpha$$

$$\mathcal{E}_{0y}(t) = \mathcal{E}_0(t)\sin\alpha. \tag{5.76}$$

The pass direction of the polarizer is at $\alpha + 90°$ resulting in zero transmission through the sequence for low-intensity light ($\Delta\Phi \approx 0$). Neglecting a common phase the field components at the output of the nonlinear element are

$$\mathcal{E}'_x(t) = [\mathcal{E}_0(t)\cos\alpha]\cos(\omega_\ell t)$$

$$\mathcal{E}'_y(t) = [\mathcal{E}_0(t)\sin\alpha]\cos[\omega_\ell t + \Delta\Phi(t)]. \tag{5.77}$$

Next the pulse passes through the linear polarizer. The total transmitted field is the sum of the components from $\mathcal{E}'_x(t)$ and $\mathcal{E}'_y(t)$ along the polarizer's path direction

$$\mathcal{E}_{out}(t) = \mathcal{E}_0(t) \cos\alpha \sin\alpha \{\cos(\omega_\ell t) + \cos[\omega_\ell t + \Delta\Phi(t)]\}. \tag{5.78}$$

The total output intensity $I_{out}(t) = \langle \mathcal{E}^2(t) \rangle$ is

$$I_{out}(t) = I_{in}(t)\frac{1}{2}[1 - \cos\Delta\Phi(t)]\sin^2(2\alpha). \tag{5.79}$$

Let us now assume a Gaussian input pulse $I_{in} = I_0 \exp\left[2(t/\tau_G)^2\right]$ and parameters of the nonlinear element so that for the pulse center the phase difference

$$\Delta\Phi(t=0) = \frac{2\pi n_2 d_m}{3\lambda_\ell}\mathcal{E}_0^2(t=0)\left(\sin^2\alpha - \cos^2\alpha\right) = \pi. \tag{5.80}$$

For this situation we obtain a transmitted pulse

$$I_{out}(t) = \frac{1}{2} I_{in}(t)\left\{1 - \cos\left[\pi e^{-2(t/\tau_G)^2}\right]\right\}. \tag{5.81}$$

The transmission is maximum (= 1) where the nonlinear element acts like a half-wave plate that rotates the polarization by 90°, lining it up with the pass direction of the polarizer. For the parameters chosen here this happens at the pulse center (t = 0). The phase shift $\Delta\Phi$ is smaller away from the pulse center producing elliptically polarized output and an overall transmission that approaches zero in the pulse wings. Thus this sequence of elements can give rise to an intensity dependent transmission similar to a fast absorber.

(c) Two Photon Absorption

In the case of an imaginary susceptibility of third order $[\chi^{(3)}]$ there is a resonant transition at twice the photon energy of the incident wave. As explained in Chapter 3 this may lead to two photon absorption, which is governed by the propagation equation for the pulse intensity

$$\frac{d}{dz} I(t,z) = -\beta_2 I^2(t,z). \tag{5.82}$$

Integrating this equation over a propagation distance d yields for the output pulse $I_{out}(t)$ in terms of the input $I_{in}(t)$

$$I_{out}(t) - I_{in}(t) = -\beta_2 d I_{in}(t) I_{out}(t). \quad (5.83)$$

If the pulse modification introduced by this element is small the change in pulse intensity can be approximated:

$$\Delta I(t) = -\beta_2 d I^2(t). \quad (5.84)$$

For counter-propagating pulses of intensities I_1 and I_2 in an optically and geometrically thin ($d \ll \tau_p c$) absorber the induced change is

$$\Delta I_1(t,d) = -\beta_2 d \left[I_1^2(t,0) + 2I_2^2(t,0) \right]. \quad (5.85)$$

This follows directly from integrating Eq. (3.166) using the approximations for thin absorbers.

5.4.3. Self-Lensing

An intensity-dependent index of refraction results in spatial phase modulation, because of the transverse variation of the intensity, as well as in temporal phase modulation through the time-dependent intensity of the pulses. We will consider here laser beams with an intensity profile that peaks on-axis. The radial intensity distribution causes a variation in index resulting in a wavefront curvature. Therefore, the nonlinear element can be adequately represented by a lens with an intensity-dependent focal distance. Self-lensing can be caused either by the Kerr effect (nonresonant nonlinearity), or by an off-resonance saturation (resonant nonlinearity). The calculations presented in this section will take as an example the Kerr nonlinearity. In Chapter 3, Eq. (3.168), we derived an expression for the radial dependence of the phase in the vicinity of the beam center, assuming a Gaussian beam profile

$$\varphi(r,t) = \frac{2\pi d}{\lambda_\ell} \Delta n_{\mathrm{nl}}(r,t) = -\bar{n}_2 \frac{2\pi}{\lambda_\ell} d I_0(t) e^{-(2r^2/w_0^2)} \approx -\bar{n}_2 \frac{2\pi}{\lambda_\ell} d I_0 \left(1 - 2\frac{r^2}{w_0^2}\right). \quad (5.86)$$

Here $I_0(t)$ is the intensity on axis ($r = 0$) and w_0 is the beam waist located at the input of a thin sample of thickness d. This expression should be compared to the phase factor that is introduced by a thin lens of focal length f, for example [34],

$$T(r) = \exp\left(ik_\ell \frac{r^2}{2f}\right). \tag{5.87}$$

Obviously the nonlinear element acts like a lens of focal length

$$f = \frac{w_0^2}{4\bar{n}_2 d I_0}. \tag{5.88}$$

Note that $f = f(t)$ is controlled by the time dependence of the pulse envelope $I_0(t)$. A similar expression applies to any nonlinear change in index that results in a parabolic radial phase dependence $\varphi(r) = Br^2$. The generalized expression for the focal length is

$$f = \frac{k_\ell}{2B}. \tag{5.89}$$

Another example is off-resonance interaction with a saturable absorber or amplifier.

Let us now consider the transmission of a pulse with a Gaussian beam and temporal profile,

$$I(r,t) = \hat{I} \exp\left[2(t/\tau_G)^2\right] \exp\left[-2r^2/w_0^2\right],$$

through a sequence of a nonlinear lens element and an aperture of radius R a distance z away. To explain the time dependence of the transmission analytically we will make certain restrictive assumptions. One of these assumptions is that the beam remains Gaussian after the nonlinear element. This requires to consider the element as a thin lens of certain focal length f. Strictly speaking, the latter is only true in the vicinity of the beam center. It will be obvious that similar effects occur in the general case; its treatment, however, requires a numerical approach. The waist of the incident Gaussian beam with Rayleigh range $\rho_0 = \pi w_0^2/\lambda$ is placed at the nonlinear element ($z = 0$). After a lens of focal length f the waist of the Gaussian beam develops as, see for example [34],

$$w^2(z) = w_0'^2 \left[1 + \frac{(z_m - z)^2}{\rho_0'^2}\right] \tag{5.90}$$

where

$$w_0'^2 = w_0^2 \frac{f^2}{f^2 + \rho_0^2} \tag{5.91}$$

is the beam waist after the lens, which occurs at a distance

$$z = z_m(f) = f \frac{\rho_0^2}{f^2 + \rho_0^2}. \tag{5.92}$$

$\rho_0' = \pi w_0'^2/\lambda$ is the Rayleigh range of the beam after the nonlinear element. By way of Eq. (5.88) we can write the focal length of the nonlinear element

$$f(t) = f_0 \exp\left[2\left(\frac{t}{\tau_G}\right)^2\right], \tag{5.93}$$

where $f_0 = w_0^2/(4d\bar{n}_2\hat{I})$. We will consider the behavior of the lens aperture sequence in the vicinity of the pulse center, for which we can approximate

$$f(t) \approx f_0 \left[1 + 2(t/\tau_G)^2\right]. \tag{5.94}$$

The aperture is placed in the plane of the beam waist produced by the pulse center, that is, at $z = z_{m0} = z_m(f_0)$. The power transmitted through the aperture is then

$$\begin{aligned} P_{out} &= \int_0^R rdr \int_0^{2\pi} d\phi \; \frac{w_0^2}{w^2(z_{m0})} \hat{I} \exp\left[-2\left(\frac{r}{w(z_{m0})}\right)^2\right] \\ &= \left[1 - e^{-2R^2/w^2(z_{m0})}\right] P_{\text{in}} \approx \frac{2R^2}{w_0'^2(z_{m0})} P_{in}, \end{aligned} \tag{5.95}$$

where the input power $P_{in} = \hat{I}\pi w_0^2/2$, and $R \ll w(z_{m0})$ was assumed to derive the last equation. Inserting Eqs. (5.90) through (5.92) with $f(t)$ from Eq. (5.94) into Eq. (5.95) yields for the time-dependent transmission through the lens–aperture sequence

$$\frac{P_{out}}{P_{in}} \approx \frac{2R^2\left(f_0^2 + \rho_0^2\right)}{w_0^2 f_0^2} \left(1 - \frac{2\rho_0^2}{f_0^2 + \rho_0^2} a t^2\right), \tag{5.96}$$

where, consistent with Eq. (5.94), we have kept expansion terms up to t^2 only. The transmission is time dependent with the maximum at the pulse center.

This reflects the fact that the shortest focal length is produced at the intensity maximum leading to the smallest beam size at the position of the aperture. Obviously, such a lens aperture sequence is just another example of a fast element whose overall transmission is controlled by the pulse intensity.

Laser pulse induced lensing does take place in nearly all ultrashort pulse mode-locked lasers. In femtosecond pulse lasers, there is always a pulse shaping mechanism in the cavity, involving a balance of dispersion and temporal SPM. The latter effect implies necessarily a spatial modulation of the wavefront, hence self-lensing. As a result of self-lensing, the size of the cavity modes is modified, leading to an increase or reduction of losses because (a) there is a change in transmission through an aperture (hard aperture) or (b) there is a change in spatial overlap between the cavity mode and the pump beam in the gain medium (soft aperture).

5.4.4. Summary of Compression Mechanisms

Figure 5.16 summarizes the compression mechanisms and the associated pulse shaping that were discussed in the previous sections.

5.4.5. Dispersion

The effect of dispersion is most simply treated in the frequency domain. Using the notations of Chapter 2, a dispersive element is characterized by a frequency dependent phase factor $\psi(\Omega)$. In the particular case where the dispersion is because of propagation through a thickness d of a homogeneous medium of index $n(\Omega)$, the dispersive phase factor is simply $\psi(\Omega) = k(\Omega)d$. The most rigorous procedure to model dispersion is to take the temporal Fourier transform of the pulse, $\tilde{\mathcal{E}}_{in}(\Omega)$, and multiply by the dispersion factor, to find the field $\tilde{\mathcal{E}}_{out}(\Omega)$ after the dispersive element:

$$\tilde{\mathcal{E}}_{out}(\Omega) = \tilde{\mathcal{E}}_{in}(\Omega)e^{-i\psi(\Omega)}. \tag{5.97}$$

An inverse Fourier transform will lead to the field $\tilde{\mathcal{E}}_{out}(t)$ in the time domain.

When analytical expressions are sought to model the evolution of a pulse in a cavity and the dispersion per round-trip is small, one can use an approximate analytical solution in the time domain. We approximate $\psi(\Omega) \approx \psi''\big|_{\omega_\ell} (\Omega - \omega_\ell)^2/2$. This is the lowest order of a Taylor expansion that produces a change in pulse shape. Expanding the exponential function $\exp[-i\,\psi''\big|_{\omega_\ell} (\Omega - \omega_\ell)^2/2]$ up to first order and Fourier transform to the time

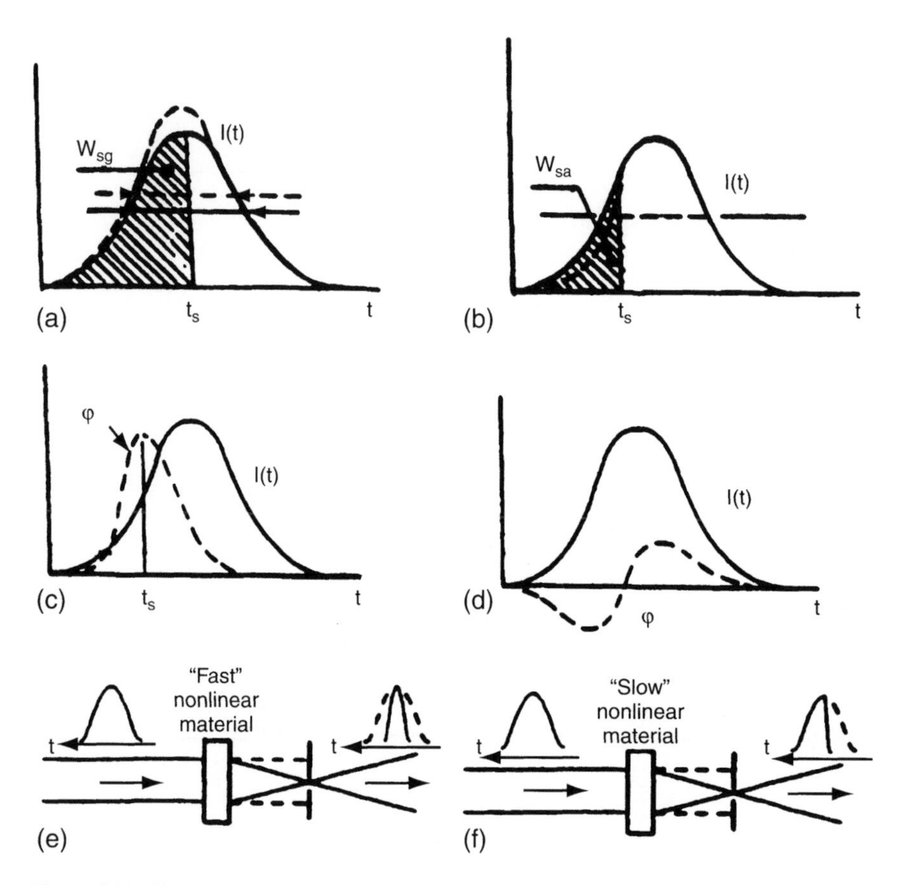

Figure 5.16 Various compression mechanisms. (a) Gain saturation: The original pulse and its pulse width are indicated by the solid line. The leading edge of the pulse is amplified, until the accumulated energy equals the saturation energy density W_{sg} at time t_s, as indicated by the dashed area. In the case of the figure, the pulse tail is not amplified, resulting in a shorter amplified pulse (dashed line). (b) Saturable absorption: The leading edge of the pulse is attenuated, until the saturation energy density W_{sa} is reached at t_s. (c) Frequency modulation (dotted line) because of saturation peaks at the time t_s when the saturation energy density W_{sa} is reached. For $t > t_s$, the pulse experiences a downchirp, if the carrier frequency of the pulse is smaller than the resonance frequency of the absorber. (d) Frequency modulation (dotted line) produced by the Kerr effect. The central part of the pulse (intensity profile indicated by the solid line) experiences an upchirp. (e) Self-focusing by a fast nonlinearity combined with an aperture, leads to a compression by attenuating both leading and trailing edges. (f) In the case of self-focusing by a slow nonlinearity combined with an aperture, only the trailing edge is trimmed.

domain yields

$$\tilde{\mathcal{E}}_{out}(t) = \tilde{\mathcal{E}}_{in}(t) - \frac{i}{2}\,\psi''\Big|_{\omega_\ell}\frac{\partial^2}{\partial t^2}\tilde{\mathcal{E}}_{in}(t), \tag{5.98}$$

which is a special case of Eq. (5.30). Note that this treatment of dispersion is equivalent to solving the differential equation (1.93) for an incremental step Δz. There the dispersion was that of an optical material, $\psi(\Omega) = k_\ell''(\Omega)\Delta z$.

5.5. CAVITIES

Resonators are an essential part of any laser. We will review first the mode spectrum of a laser cavity and the standard ABCD-matrix cavity analysis. Because mirrors are used as focusing elements, astigmatism complicates significantly the calculation and design (optical positioning of the elements) of the resonator. Finally, we will analyze the effect of a Kerr lens in a laser cavity.

5.5.1. Cavity Modes and ABCD Matrix Analysis

Mode-locked operation requires a well-defined mode structure. It is generally understood that the longitudinal modes are locked in phase. A transverse mode structure will generally contribute to amplitude noise (at frequencies corresponding to the differences between mode frequencies). Most fs lasers operate in a single TEM_{00} transverse mode. We will see, however, that some multiple transverse mode lasers have the same longitudinal mode structure as the fundamental TEM_{00}.

This subsection reviews standard ABCD matrix calculations of the stability of laser resonators. Most fs laser cavities have at least one beam waist (for instance, one for the gain medium, and possibly one for a passive mode-locking element).

A beam with an electric field amplitude having a Gaussian radial dependence is uniquely defined by its complex $\tilde{q}$ parameter defined by Eq. (1.186). The phase variation on axis ($r = 0$) of the beam is determined by the phase angle $\Theta(z) = \arctan z/\rho_0$ according to Eq. (1.183).

The parameter of the fundamental Gaussian beam that can reproduce itself in a cavity can be determined by the standard technique using ABCD matrices [35]. Let A, B, C, and D be the elements of the 2×2 system matrix obtained by calculating the product of the matrices of all cavity components, defining a complete round-trip from a point P. The complex $\tilde{q}$ parameter of the Gaussian beam at the point P is given by:

$$\boxed{\frac{1}{\tilde{q}} = -\frac{A-D}{2B} - i\frac{\sqrt{1-(A+D)^2/4}}{|B|}}. \tag{5.99}$$

Using the definition of the $\tilde{q}$ parameter, Eq. (1.186), leads to the beam characteristics at point P:

$$R = -\frac{2B}{A-D} \tag{5.100}$$

$$\frac{\pi w^2}{\lambda_\ell} = \frac{|B|}{\sqrt{1-(A+D)^2/4}}. \tag{5.101}$$

The modes of a cavity are determined by the condition that, after one round-trip from point P to point P, the total phase variation is a multiple of 2π. For the fundamental TEM_{00} mode, and a simple cavity consisting of two concave mirrors at distance d_1 and d_2 from the beam waist (which we will assume here to be inside the resonator), the phase variation for the half round-trip should be a multiple of π:

$$\begin{aligned} l\pi &= k(d_1+d_2) + \Theta(d_1) + \Theta(d_2) \\ &= k(d_1+d_2) + \Delta\Theta \\ &= k(d_1+d_2) + \arctan\left(\frac{d_1}{\rho_0}\right) + \arctan\left(\frac{d_2}{\rho_0}\right), \end{aligned} \tag{5.102}$$

where $k = 2\pi\nu/c$. If $d_{1,2} \ll \rho_0$, the mode spacing frequency $\Delta\nu = \nu_{l+1} - \nu_l = c/[2(d_1+d_2)]$.

The Fourier transform of the output of a continuously mode-locked laser is a comb of frequency spikes having a maximum overlap with the longitudinal mode structure represented by Eq. (5.102). Apertures are often inserted in the cavity of mode-locked lasers to avoid the complication introduced by a transverse mode structure. In some lasers, a saturable absorber acts as an aperture. The small cross section of the inverted region in the gain medium can also act as a mode limiting aperture.

The fundamental TEM_{00} Gaussian beam is not necessarily the only existing mode in a fs laser. Let us consider the influence that the transverse mode structure can have on the operation of a mode-locked laser. If the laser is multimode, the electric field can be expressed as an expansion of Hermite–Gaussian modes (see, for instance, [35]):

$$E(x, y, z, t) = \sum_n \sum_m c_{nm} u_n(x) u_m(y) e^{i(\omega t - kz)} + c.c., \tag{5.103}$$

where

$$u_s(x) = \left(\frac{2}{\pi}\right)^{\frac{1}{4}} \sqrt{\frac{\exp[i(2s+1)\Theta(z)]}{2^s s! w(z)}}$$

$$\times H_s\left(\frac{\sqrt{2}x}{w(z)}\right) \exp\left\{-\left[i\frac{kx^2}{2R(z)} + \frac{x^2}{w^2(z)}\right]\right\}. \tag{5.104}$$

Here H_s are Hermite–Gaussian polynomials of order s. The subscript m and n are the transverse mode indices.

For the simple cavity consisting of the two curved mirrors introduced above, the spacing between transverse modes is:

$$\Delta\nu = \frac{c}{2(d_1 + d_2)}(\Delta n + \Delta m)\left[\arctan\left(\frac{d_1}{\rho_0}\right) + \arctan\left(\frac{d_2}{\rho_0}\right)\right] \tag{5.105}$$

where Δn and Δm are the difference in transverse mode numbers [35]. If the various transverse modes oscillate independently with random phase, the output of the laser will have a noise component corresponding to the beating of the various transverse modes. This noise component will be periodic if the mode spacing is equal in the two transverse directions. In some cases, [36] the transverse modes can be locked in phase, resulting in a periodic spatial scanning of the beam. The noise contribution corresponding to the transverse mode beating will be low frequency if $d_{1,2} < \rho_0$.

Calculation of the exact stability diagram, position, and size of the beam waists, is a tedious but essential task in the design of fs lasers. Because it is not essentially different from the design of any laser, we will refer to the appropriate literature for details, for example [35]. A few general criteria to consider in the design of the laser cavity are:

- minimal losses.
- a flexibility of varying the ratio of beam waists in various resonator elements in large proportions. The saturation level in the gain medium will be a factor in determining the intracavity power. In the case of Kerr lensing, it is important to identify the location where the beam size variation because of self-lensing will be maximum (positioning of an aperture).

In the case of passive mode-locking, the ratio of beam waists in the absorber–amplifier is important, because this is the parameter that determines the relative saturation of the gain and absorber.

- a small spot size in the amplifier may be desirable for efficient pumping and heat removal.
- a round spot is desirable in the passive mode-locking element (for instance, the amplifier rod in the case of a Ti:sapphire laser, or the saturable absorber jet in a dye laser) to have the most uniform possible wavefront across the beam, because it is a region of the cavity which contributes to the phase modulation.

Mirrors or lenses can be used to create beam waists in a laser cavity. Because of the requirement of minimum losses and dispersion, one will generally choose reflective optics over lenses.

5.5.2. Astigmatism and Its Compensation

It is not always an easy task to create a waist of minimal size with off-axis reflective optics. Indeed, let us consider the typical focusing geometry sketched in Figure 5.17. The smaller the focal spot in A, the larger the diameter w of the incident beam on the mirrors, hence the larger the clearance angle θ required to have the focal point fall outside of the incident beam cross section. However, the astigmatism caused by a large angle of incidence θ will make it impossible to obtain the desired small focal spot with a cylindrically symmetric Gaussian beam incident from the left.

Because a tight focusing is required in the nonlinear elements of a mode-locked laser, there is clearly a need for minimizing or reducing the astigmatism. There are

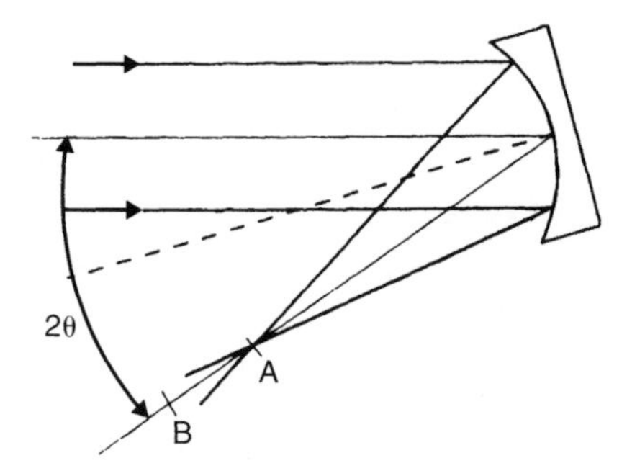

Figure 5.17 Off-axis focusing of a Gaussian beam leading to astigmatism. In the plane of the figure the focal distance of the mirror is $(R/2)\cos\theta$, where R is the radius of curvature of the mirror. The first focalization is therefore a line perpendicular to the plane of the figure originating from A. In the orthogonal plane the focal distance of the mirror is $(R/2)/\cos\theta$. There will therefore be a focal line in the plane of the figure at B.

some exceptions. A large astigmatism may sometimes be desirable in the gain medium. This is the case when it is desirable to take maximum advantage of a self-lensing effect. Another example of such a need is to match the elongated shape of the gain region of semiconductor lasers.

Let us choose as transverse coordinate y for the plane of incidence (the plane of the figure in Fig. 5.17), and x for the orthogonal direction. The locations of the two focal lines corresponding respectively to the plane of the figure and to the orthogonal plane are:

$$f_y = f\cos\theta$$

$$f_x = \frac{f}{\cos\theta}, \tag{5.106}$$

where θ is the angle of incidence and $f = R/2$ is the focal length of the mirror (see for example Kogelnik *et al.* [37]).

Other elements in a cavity, such as Brewster plates, also have astigmatic properties that can limit the performance of the system. The gain medium of a Ti:sapphire laser or of a dye laser is generally a plane parallel element put at Brewster's angle. Kogelnik *et al.* [37] have shown under which condition the astigmatism can be compensated by such elements.

To analyze the astigmatism of such elements let us consider the propagation of a Gaussian beam through a plate of thickness d and refractive index n put at the Brewster angle directly next to the beam waist (size w_0) at $z = 0$, as sketched in Figure 5.18.

On entering the medium, the beam waist takes the values:

$$w_{0x} = w_0$$

$$w_{0y} = w_0\frac{\cos\theta_r}{\cos\theta_B} = w_0\frac{\sin\theta_B}{\cos\theta_B} = nw_0, \tag{5.107}$$

where $\theta_B = \arctan n$ is the Brewster angle, and $\theta_r = 90° - \theta_B$ is the angle of refraction. To a thickness d, there corresponds a propagation distance

$$\chi = \frac{d}{\cos\theta_r} = d\frac{\sqrt{1+n^2}}{n}. \tag{5.108}$$

Applying the propagation law (1.182) for the beam waist across the thickness d yields:

$$w_x = w_0\sqrt{1+\left(\frac{\lambda_\ell\chi}{n\pi w_0^2}\right)^2} = w_0\sqrt{1+\left(\frac{\lambda_\ell}{\pi w_0^2}\frac{d\sqrt{1+n^2}}{n^2}\right)^2} \tag{5.109}$$

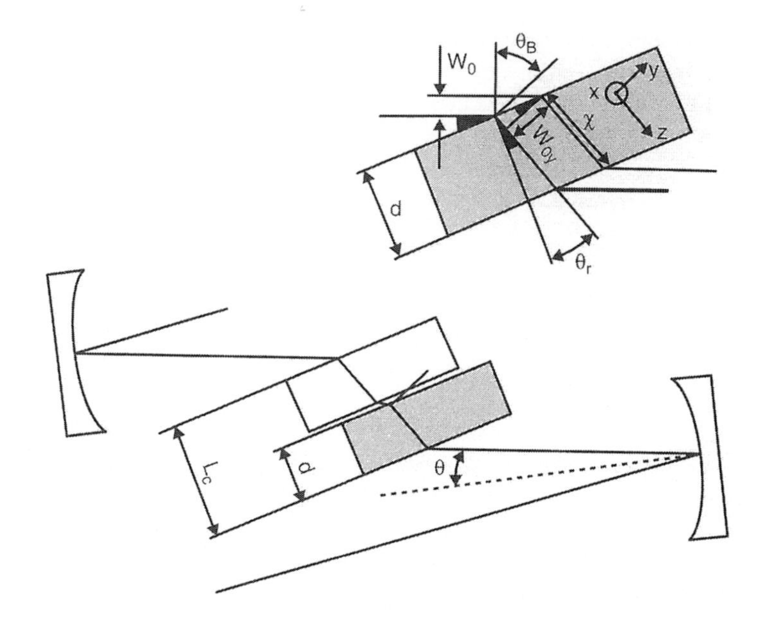

Figure 5.18 Brewster plate geometry for the calculation of astigmatism compensation. On the top right: a Brewster plate of thickness d is placed close to a beam waist (w_0). The propagation through the plate is calculated along z, for the plane of the figure (coordinate y) and orthogonal to the plane of the figure. Bottom part of the figure: a Brewster cut laser crystal of length L_c inserted in an X or Z configuration (the Z configuration is shown) between two curved mirrors can be considered to be made of two halves of thickness $d = L_c/2$.

$$w_y = nw_0 \sqrt{1 + \left(\frac{\lambda_\ell \chi}{n^3 \pi w_0^2}\right)^2} = nw_0 \sqrt{1 + \left(\frac{\lambda_\ell}{\pi w_0^2} \frac{d\sqrt{1+n^2}}{n^4}\right)^2}. \tag{5.110}$$

On exiting the crystal, the beam waists take the values w_x and w_y/n, along x and y, respectively. Therefore, the Brewster plate can be seen from Eqs. (5.109) and (5.110) to be equivalent to propagation in free space of distances equal to

$$d_x = d\frac{\sqrt{1+n^2}}{n^2} \tag{5.111}$$

$$d_y = d\frac{\sqrt{1+n^2}}{n^4} \tag{5.112}$$

in the planes xz and yz, respectively. The beam issued from the waist w_0 after passing through the Brewster plate will be again collimated by a mirror of radius R

at an angle θ if the difference between the two distances d_x and d_y compensates the difference in focal distances f_x and f_y. Using Eqs. (5.106), (5.111), and (5.112), this condition is equivalent to:

$$\frac{2d}{R}\frac{\sqrt{n^4-1}}{n^4} = \frac{\sin^2\theta}{\cos\theta}. \tag{5.113}$$

In the case of a typical Ti:sapphire laser, the crystal of length $L_c = 2d$ is inserted near the focal point between two curved mirrors. Each half crystal of thickness $d = L_c/2$ can compensate the astigmatism of each of the mirrors, provided the angle of incidence is chosen to be a solution of:

$$\frac{L_c}{R}\frac{\sqrt{n^4-1}}{n^4} = \frac{\sin^2\theta}{\cos\theta}. \tag{5.114}$$

For a 9 mm long Ti:sapphire crystal inserted between two mirrors of 10 cm curvature, Eq. (5.114) indicates that astigmatism compensation occurs at an angle of $\theta = 9.5°$.

In the case of dye lasers, we find that for a typical jet thickness of 200 μm and a mirror curvature of 5 cm, the compensated angle is less than 2.5°. Compensation is impossible for tight focusing in a saturable absorber jet of typical thickness of 50 μm. Two options are available in such a situation:

- insert a glass window at Brewster angle between the two curved mirrors, of a thickness sufficient to compensate the astigmatism caused by the mirrors, or
- use the astigmatism of another part of the cavity, to obtain a minimum astigmatism-free spot between the two mirrors.

The latter approach is only possible in situations where there are two or more waists in the cavity. Calculations show that large angles can actually result in astigmatism compensation, and, even with angles of incidence on the focusing mirrors exceeding 10°, large stability ranges have been found [38]. Resonators for dye lasers have been designed to provide a round and minimum size spot in the absorber jet, using the cavity geometry (relative location of the components) to compensate the astigmatism, without a need for inserting additional "compensating elements." The price to pay for an optimal focusing at one waist of the cavity, is a large astigmatism at another part of the cavity. Returning to Fig. 5.17, the beam incident from the left will focus first on a line originating from A, perpendicular to the plane of the figure, next in a line at B in the plane of the figure. However, if the beam incident from the left is collimated in the plane of the figure, but convergent in the orthogonal direction, the focal line B

will recede towards the focus A. The incident beam parameters can be adjusted such as to create a tight round focal spot.

To assess the importance of astigmatism, let us consider a simple ring cavity with two beam waists. We assume that one beam waist is formed by two lenses of 15 mm focal distance, spaced by 30 mm. The other waist is formed by two mirrors of focal distance $f = 25$ mm separated by a distance d. The distances between the two waists are 1 m and 3 m in a 4 m perimeter ring cavity. From the expression for the ABCD matrix for this cavity one finds that the stability range is $f < d < 1.01f$. If astigmatism because of an angle of incidence θ on the two mirrors of $f = 25$ mm is taken into account, there will be a different stability condition corresponding to each of the two focal distances $f_x = f/\cos\theta$ and $f_y = f\cos\theta$. The cavity is stable if the two stability ranges overlap. For an angle $\theta \geq 5.7°$, this cavity is no longer stable. In this situation other degrees of freedom, such as beam propagation out of plane and/or the insertion of additional elements are options.

5.5.3. Cavity with a Kerr Lens

In this section we will treat the effect of a Kerr lens on the beam parameters in a laser cavity perturbatively based on Gaussian beam analysis. A simple, geometrical optics description of a nonlinear lens aperture sequence is presented in Appendix E.

General Approach

It is convenient to use the ABCD matrix approach to evaluate the intensity-dependent losses introduced by Kerr lensing in a laser cavity [39, 40]. The ABCD matrix of the resonator is calculated starting from a reference plane at the position of the Kerr medium. Let $\mathcal{M}_1 = \begin{pmatrix} A_1 & B_1 \\ C_1 & D_1 \end{pmatrix}$ be the ABCD matrix for low intensity (negligible Kerr effect). At high intensity, the nonlinear lensing effect modifies this matrix as follows:

$$\mathcal{M} = \begin{pmatrix} 1 & 0 \\ -\frac{1}{f_{nl}} & 1 \end{pmatrix} \begin{pmatrix} A_1 & B_1 \\ C_1 & D_1 \end{pmatrix} = \begin{pmatrix} A_1 & B_1 \\ C_1 + \delta C & D_1 + \delta D \end{pmatrix}$$

$$= \begin{pmatrix} A & B \\ C & D \end{pmatrix}, \tag{5.115}$$

where $\delta C = -A_1/f_{nl}$; $\delta D = -B_1/f_{nl}$, and f_{nl} is the time dependent focal length of the Kerr medium, cf. Eq. (5.88). A Gaussian beam is uniquely characterized by its complex beam parameter $\tilde{q}$ (cf. Chapter 1 and [39]). In the absence of

Kerr lensing, the eigenmode of the empty cavity is characterized by a complex beam parameter $\tilde{q}_1$, at the location the system matrix calculation was started. The inverse of this beam parameter is:

$$\tilde{s}_1 = \frac{1}{\tilde{q}_1} = \frac{1}{R_1} - i\frac{\lambda_\ell}{\pi w_1^2}$$
$$= \frac{C_1 + D_1\tilde{s}_1}{A_1 + B_1\tilde{s}_1}, \tag{5.116}$$

Solving the eigenvalue equation (5.116) yields the complex beam parameter at the location of the nonlinear crystal (Kerr medium):

$$w_1^2 = \frac{|B_1|\lambda}{\pi} \times \sqrt{\frac{1}{1 - [(A_1 + D_1)/2]^2}} \tag{5.117}$$

$$R_1 = \frac{2B_1}{D_1 - A_1}. \tag{5.118}$$

In the presence of the Kerr lensing, the eigenvalue $\tilde{s}$ is the solution of the eigenmode equation for the complete round-trip ABCD matrix:

$$\tilde{s} = \frac{C + D\tilde{s}}{A + B\tilde{s}}. \tag{5.119}$$

For $\tilde{s}$ we can make the ansatz:

$$\tilde{s} = \tilde{s}_1 + \delta\tilde{s} \tag{5.120}$$

where $\delta\tilde{s}$ is the small change in $\tilde{s}$ produced by the Kerr lens that we will determine next. For this we multiply both sides of Eq. (5.119) by $A + B\tilde{s}$, substitute Eq. (5.120) for $\tilde{s}$, and replace all four matrix elements X by $X_1 + \delta X$. Keeping only terms up to first order we obtain:

$$A_1\tilde{s}_1 + B_1\tilde{s}_1^2 + (2D_1\tilde{s}_1 + A_1)\delta\tilde{s} = C_1 + D_1\tilde{s}_1 + \delta C + \tilde{s}_1\delta D + D_1\delta\tilde{s}, \tag{5.121}$$

where we have made use of the fact that $\delta A = \delta B = 0$, cf. Eq. (5.115). Solving for $\delta\tilde{s}$ and using δC and δD from Eq. (5.115), and using Eq. (5.116) yields:

$$\delta\tilde{s} = \frac{1}{f_{\text{nl}}}\left[\frac{-(A_1 + B_1\tilde{s}_1)}{A_1 + 2B_1\tilde{s}_1 - D_1}\right]. \tag{5.122}$$

The change in $\delta\tilde{s}$ implies that the lensing effect results in a change in beam size at any location in the cavity. Typically, an aperture is used at a particular location of the cavity, where, ideally, the self-lensing results in the largest reduction in beam size. Let $\mathcal{M}_m = \begin{pmatrix} A_m & B_m \\ C_m & D_m \end{pmatrix}$ be the ABCD matrix that connects the reference point of the cavity (location of the Kerr lens) to the position of the aperture. The complex parameter $\tilde{s}_m$ at the aperture is:

$$\tilde{s}_m = \frac{C_m + D_m\tilde{s}}{A_m + B_m\tilde{s}}. \tag{5.123}$$

The relative change in beam size at the aperture $\delta w_m/w_m$ because of the Kerr lens is related to the change in $\tilde{s}_m$:

$$\frac{\delta w_m}{w_m} = -\frac{1}{2}\frac{\mathrm{Im}(\delta\tilde{s}_m)}{\mathrm{Im}(\tilde{s}_m)}. \tag{5.124}$$

The change in beam parameter at the aperture $\delta\tilde{s}_m$ can be inferred from the change in beam parameter $\delta\tilde{s}$ at the point of reference. Let us make the ansatz that $\tilde{s}_m = \tilde{s}_{m0} + \delta\tilde{s}_m$, where $\tilde{s}_{m0}$ is the complex s parameter without the Kerr lens [$\tilde{s}_{m0}$ is given by Eq. (5.123) for $\tilde{s}_1 = \tilde{s}$]. Inserting these expressions for $\tilde{s}_m$ and $\tilde{s} = \tilde{s}_1 + \delta\tilde{s}$ into Eq. (5.123) and keeping only terms up to first order, we find:

$$\delta\tilde{s}_m = \frac{D_m - B_m\delta\tilde{s}_{m0}}{A_m + B_m\tilde{s}_1}\delta\tilde{s}. \tag{5.125}$$

In this equation, $\tilde{s}_{m0}$ can be substituted from Eq. (5.123) where $\tilde{s}$ has been replaced by $\tilde{s}_1$, yielding:

$$\delta\tilde{s}_m = \frac{\delta\tilde{s}}{(A_m + B_m\tilde{s}_1)^2}. \tag{5.126}$$

Finally we substitute $\delta\tilde{s}$ with Eq. (5.122) to obtain:

$$\delta\tilde{s}_m = \frac{1}{f_{\mathrm{nl}}}\left[\frac{-(A_1 + B_1\tilde{s}_1)}{(A_1 + 2B_1 s_1 - D_1)(A_m + B_m\tilde{s}_1)^2}\right]. \tag{5.127}$$

The last equation (5.127) contains all the information necessary to estimate the effect of Kerr induced lensing on a cavity. Let us consider an aperture of radius w_a.

Using Eq. (5.95), we can estimate the ratio of the energy loss through the aperture with Kerr effect ($\Delta\mathcal{W}$) to the loss without Kerr effect ($\Delta\mathcal{W}_0$):

$$\frac{\Delta\mathcal{W}}{\Delta\mathcal{W}_0} = \exp\left\{-2\left(\frac{w_a}{w_m + \delta w_m}\right)^2 + 2\left(\frac{w_a}{w_m}\right)^2\right\} = \exp\left\{4\left(\frac{w_a}{w_m}\right)^2 \frac{\delta w_m}{w_m}\right\} \tag{5.128}$$

where δw_m is determined by Eqs. (5.124) and (5.127).

5.6. PROBLEMS

1. By simple energy conservation arguments, find (in the approximation $T_1 \rightarrow \infty$) an expression for the energy gained (lost) per unit distance for an amplifier (absorber) $\frac{dW}{dz}$ as function of the change in population difference and of the photon energy. Introduce into that expression the linear gain (absorption) coefficient, and combine with the rate equation to derive the evolution equation for the pulse energy density:

$$\frac{dW}{dz} = -\alpha_0 W_s \left[1 - e^{-W/W_s}\right], \tag{5.129}$$

where α_0 is the linear gain (absorption) coefficient, and W_s is the saturation energy density.

2. Derive an evolution equation for the pulse energy in a mode-locked laser ring cavity consisting of (a) the sequence output mirror (reflectivity r)—gain—saturable absorber; (b) the sequence gain—mirror—absorber; and (c) the sequence absorber—gain—mirror. Neglect reshaping of the pulse and integrate Eq. (5.129) to yield the energy $\mathcal{W}_{out} = W_{out}A_a$ at the end of a loss element of thickness d_a, as a function of the input energy $\mathcal{W}_{in} = W_{in}A_a$. A_a is the cross section of the beam in this particular element. Show that the output pulse of energy $\mathcal{W}_{out}$, after transmission of a pulse of input energy $\mathcal{W}_{in}$ through a saturable absorber is:

$$\mathcal{W}_{out} = A_a W_{sa} \ln\left[1 - e^{\alpha_a d_a}\left(1 - e^{\mathcal{W}_{in}/A_a W_{sa}}\right)\right]. \tag{5.130}$$

3. Consider the elementary round-trip model of Figure 5.19. Choose a reference point just after the saturable loss, where the beam cross section is A_a.

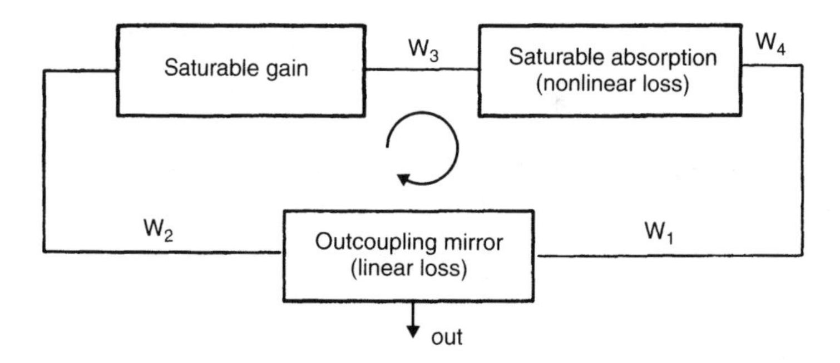

Figure 5.19 Simplified round-trip model for a passively mode-locked laser, showing the evolution of the pulse energy density, W.

Show that the energy density W_4 at the end of series loss-gain-nonlinear-loss is related to the energy density W_1 entering this sequence by:

$$1 + e^{-a_a}\left[e^{W_4/W_{sa}} - 1\right] = \left\{1 + e^{a_g}\left[e^{RW_1A_a/(W_{sg}A_g)} - 1\right]\right\}^{\frac{W_{sg}A_g}{W_{sa}A_a}}, \tag{5.131}$$

where R is the output coupler (intensity) reflectivity, A_a and A_g are the beam cross sections in the absorber and gain elements, respectively; W_{sa} and W_{sg} the saturation energy densities in the absorber and gain media. Note that in a steady state, $\mathcal{W}_1 = \mathcal{W}_4$, and Eq. (5.131) can be used to calculate the pulse energy.

Solve Eq. (5.131) in the approximation $W_4/W_{sa} \ll 1$ and $W_1/W_{sg} \ll 1$. Show that, even for a laser below threshold for cw operation ($R = r^2 < \exp[-(a_a + a_g)]$, two solutions can be found for W_1. The first solution is the minimum intracavity energy required to start mode-locked oscillation. The second solution is the steady-state intracavity energy. Discuss the stability of both solutions. Under which condition can the first solution be small compared to the steady-state intracavity energy?

4. Derive the equation for the soliton laser, following the procedure sketched in Fig. 5.9.
5. Calculate the spectrum of a continuously mode-locked Ti:sapphire laser emitting a continuous train of identical Gaussian pulses, 40 fs FWHM each, at a repetition rate of 80 MHz. The laser cavity is linear, with two prisms separated by 60 cm [$(dn/d\lambda)^2 = 10^{-9}$ nm^{-2}]. Neglecting all other contributions to the dispersion, calculate the longitudinal mode spectrum

of the cavity. Is the latter spectrum identical to the Fourier amplitude (squared) of the train of pulses? If not, explain the difference(s).

6. Three mode-locked lasers generate pulses of 40 fs duration at the wavelengths of 620, 700, and 780 nm, respectively. Find under which condition the output of these lasers could be combined to provide a train of much shorter pulses. What are the practical problems to be solved? Assuming perfect technology, how short would these pulses be?

7. Consider a passively mode-locked dye laser consisting of a saturable absorber, a depletable amplifier, an output coupler and a bandwidth limiting element. Following the approach described in Section 5.2.4, write down the equation for the steady-state pulse envelope assuming both absorber and amplifier are traversed at resonance. The result should be an integrodifferential equation with differentials up to second order (d^2/dt^2) and containing integrals of the type $W(t)$ and $[W(t)]^2$ where $W(t) = \int_{-\infty}^{t} F(t')dt'$.

8. In short cavity and distributed feedback lasers short pulse generation is possible in the single (longitudinal) mode regime. At first sight this seems to be in contradiction to the necessity of a broad laser spectrum. Make a quantitative estimate of the essential laser parameters required for the generation of a 500 fs pulse (*Hint:* consult the paragraph on miniature dye lasers in Section 6.8.1).

9. A Gaussian beam passes through a sequence of a Kerr medium of thickness d and an aperture of diameter D. The beam waist $w_0 = 25$ μm is at the position of the Kerr medium ($d = 1$ mm, $\bar{n}_2 = 3 \cdot 10^{-16}$ cm^2/W). Find the distance from the Kerr medium to the aperture L, and the diameter of the aperture D, such that the overall transmission changes by 1% if the input illumination is switched from cw (no Kerr lens) to a 25 fs, 10 nJ pulse.

10. Recall Eq. (5.28) that determines the field envelope, $\mathcal{E}(t)$ at the output of a saturable absorber or amplifier, valid for small induced changes. Use this equation to derive an equation for the output pulse energy in terms of the input energy, cf. Eq. (5.51).

BIBLIOGRAPHY

[1] W. H. Knox. *In situ* measurement of complete intracavity dispersion in an operating Ti:sapphire femtosecond laser. *Optics Letters*, 17:514–516, 1992.

[2] W. H. Knox and J. P. Gordon. Frequency-domain dispersion measurements in tunable mode-locked lasers. *Journal of the Optical Society of America B*, 10:2071–2079, 1993.

[3] R. J. Jones, J.-C. Diels, J. Jasapara, and W. Rudolph. Stabilization of the frequency, phase, and repetition rate of an ultra-short pulse train to a Fabry–Perot reference cavity. *Optics Communications*, 175:409–418, 2000.

[4] T. Udem, J. Reichert, R. Holzwarth, and T. W. Hänsch. Accurate measurement of large optical frequency differences with a mode-locked laser. *Optics Letters*, 24:881–883, 1999.

[5] J. Ye and S. Cundiff. *Femtosecond Optical Frequency Comb: Principle, Operation and Applications*. Springer, New York, 2005.

[6] S. V. Chekalin, P. G. Kryukov, Y. A. Matveetz, and O. B. Shatherashvili. The processes of formation of ultrashort laser pulses. *Opto-Electronics*, 6:249–261, 1974.

[7] V. E. Zakharov and A. B. Shabat. Exact theory of two-dimensional self-focusing and one-dimensional self-modulation of waves in nonlinear media. *Translation of Zhurnal Eksperimentalnoi i Teoreticheskoi Fiziki*, 34:62–69, 1972.

[8] V. Petrov, W. Rudolph, and B. Wilhelmi. Evolution of chirped light pulses and the steady-state regime in passively mode-locked femtosecond dye lasers. *Review of Physical Applications*, 22:1639–1650, 1987.

[9] C. S. Gardner, G. Green, M. Druskal, and R. Miura. Method for solving the Korteweg-deVries equation. *Physical Review Letters*, 19:1095, 1967.

[10] J.-C. Diels and H. Sallaba. Black magic with red dyes. *Journal of the Optical Society of America*, 70:629, 1980.

[11] J.-C. Diels, J. Menders, and H. Sallaba. Generation of coherent pulses of 60 optical cycles through synchronization of the relaxation oscillations of a mode-locked dye laser. In R. M. Hochstrasser, W. Kaiser, and C. V. Shank, Eds., *Picosecond Phenomena II*, Springer-Verlag, Berlin, Germany 1980, p. 41.

[12] F. Salin, P. Grangier, G. Roger, and A. Brun. Observation of high-order solitons directly produced by a femtosecond ring laser. *Physical Review Letters*, 56:1132–1135, 1986.

[13] T. Tsang. Observation of higher order solitons from a mode-locked Ti:sapphire laser. *Optics Letters*, 18:293–295, 1993.

[14] J.-C. Diels. Femtosecond dye lasers. In F. Duarte and L. Hillman, Eds., *Dye Laser Principles: With Applications*, Academic Press, Boston, MA, 1990, pp. 41–132.

[15] Y. Ishida, N. Sarukura, and H. Nakano. Soliton like pulse shaping in a passively mode locked Ti:sapphire laser. In C. B. Harris, E. P. Ippen, G. A. Mourou, and A. H. Zewail, Eds., *Ultrafast Phenomena VII*, Springer-Verlag, Berlin, Germany, 1990, pp. 75–77.

[16] P. F. Curley, Ch. Spielmann, T. Brabec, F. Krausz, E. Wintner, and A. J. Schmidt. Operation of a fs Ti:sapphire solitary laser in the vicinity of zero group-delay dispersion. *Optics Letters*, 18:54–57, 1993.

[17] V. Petrov, W. Rudolph, and B. Wilhelmi. Computer simulation of passive mode locking of dye lasers with consideration of coherent light-matter interaction. *Optical and Quantum Electronics*, 19:377–384, 1987.

[18] G. H. C. New. Mode-locking of quasi-continuous lasers. *Optics Communications*, 6:188–193, 1972.

[19] G. H. C. New. Pulse evolution in mode-locked quasi-continuous lasers. *IEEE Journal of Quantum Electronics*, QE-10:115–124, 1974.

[20] H. A. Haus. Theory of mode-locking with a slow saturable absorber. *IEEE Journal of Quantum Electronics*, QE-11:736–746, 1975.

[21] J.-C. Diels, J. J. Fontaine, I. C. McMichael, W. Rudolph, and B. Wilhelmi. Experimental and theoretical study of a femtosecond laser. *Soviet Journal of Quantum electronics*, 13:1562–1569, 1983.

[22] W. Rudolph and B. Wilhelmi. Calculation of light pulses with chirp in passively mode-locked lasers taking into account the phase memory of absorber and amplifier. *Applied Physics*, B-35:37–44, 1984.

[23] J.-C. Diels, W. Dietel, J. J. Fontaine, W. Rudolph, and B. Wilhelmi. Analysis of a mode-locked ring laser: chirped-solitary-pulse solutions. *Journal of the Optical Society of America B*, 2:680–686, 1985.

[24] H. A. Haus and Y. Silberberg. Laser mode-locking with addition of nonlinear index. *IEEE Journal of Quantum Electronics*, QE-22:325–331, 1986.

[25] X. M. Zhao and J.-C. Diels. Stability study of fs dye laser with self-lensing effects. *Journal of the Optical Society America B*, 10(7):1159–1165, 1993.

[26] J. L. A. Chilla and O. E. Martinez. Spatial-temporal analysis of the self-mode-locked Ti:sapphire laser. *Journal of the Optical Society B*, 10:638–643, 1993.

[27] M. Lai, J. Nicholson, and W. Rudolph. Multiple pulse operation of a femtosecond Ti:sapphire laser. *Optics Communications*, 142:45–49, October 1997.

[28] D. Kuehlke, W. Rudolph, and B. Wilhelmi. Calculation of the colliding pulse mode locking in cw dye ring lasers. *IEEE Journal of Quantum Electronics*, QE-19:526–533, 1983.

[29] U. Keller, G. W. 'tHooft, W. H. Knox, and J. E. Cuninham. Femtosecond pulses from a continously self-starting passively mode-locked Ti:sapphire laser. *Optics Letters*, 16:1022–1024, 1991.

[30] P. W. Smith, Y. Silberberg, and D. Q. B. Miller. Mode-locking of semiconductor diode lasers using saturable excitonic nonlinearities. *Journal of the Optical Society America*, B-2:1228–1235, 1985.

[31] U. Keller. Ultrafast all-solid-state technology. *Applied Physics B*, 58:347–363, 1994.

[32] L. R. Brovelli, U. Keller, and T. H. Chiu. Design and operation of antiresonant Fabry–Perot saturable semiconductor absorbers for mode-locked solid-state lasers. *Journal of the Optical Society America B*, 12:311–322, 1995.

[33] G. P. Agrawal. *Nonlinear Fiber Optics*. Academic Press, ISBN 0-12-045142-5, Boston, 1995.

[34] Miles V. Klein and Thomas E. Furtak. *Optics*. John Wiley and Sons, ISBN 0-471-87297-0, New York, 1986.

[35] Anthony E. Siegman. *Lasers*. University Science Books, Mill Valley, CA, 1986.

[36] S. L. Shapiro, editor. *Ultrashort Light Pulses*. Springer Verlag, Berlin, Heidelberg, New York, 1977.

[37] H. W. Kogelnik, E. P. Ippen, A. Dienes, and C. V. Shank. Astigmatically compensated cavities for cw dye lasers. *IEEE Journal of Quantum Electronics*, QE-8:373–379, 1972.

[38] N. Jamasbi, J.-C. Diels, and L. Sarger. Study of a linear femtosecond laser in passive and hybrid operation. *Journal of Modern Optics*, 35:1891–1906, 1988.

[39] H. W. Kogelnik and T. Li. Laser beams and resonators. *Applied Optics*, 5:1550–1567, 1966.

[40] A. Agnesi and G. C. Reali. Analysis of unidirectional operation of Kerr lens mode-locked ring oscillators. *Optics Communications*, 110:109–114, 1994.

6

Ultrashort Sources II: Examples

In the previous chapter the elements of passive mode-locking and their function for pulse shaping were described in detail. Analytical and numerical methods of characterizing mode-locked lasers were presented. Passive mode-locking is indeed the most widely applied and successful technique to produce pulses whose bandwidth approaches the limits imposed by the gain medium of dye and solid-state lasers including fiber lasers. Passive mode-locking was the technique of choice to produce sub 50-fs pulses in dye lasers and, today, is routinely applied in solid-state and fiber lasers. Sub 5-fs pulses have been obtained from Ti:sapphire lasers without external pulse compression [1] using this method.

In this chapter we will review additional techniques of mode-locking and discuss examples of mode-locked lasers. The purely active or synchronous mode-locking will be covered first, followed by the hybrid passive–active technique. Other techniques not discussed in the previous chapter are additive mode-locking, methods based on second-order nonlinearities, and passive negative feedback. For their important role as saturable absorbers we will review the relevant properties of semiconductor materials. The later part of this chapter is devoted to specific examples of popular lasers.

6.1. SYNCHRONOUS MODE-LOCKING

A simple method to generate short pulses is to excite the gain medium at a repetition rate synchronized with the cavity mode spacing. This can be done by using a pump that emits pulses at the round-trip rate of the cavity to be pumped. One of the main advantages of synchronous mode-locking is that a much broader range of gain media can be used than in the case of passive mode-locking. This includes semiconductor lasers and, for instance, laser dyes such as styryl 8, 9, and 14, which have too short a lifetime to be practical in cw operation, but are quite efficient when pumped with short pulses.

Ideally, the gain medium in a synchronously pumped laser should have a short lifetime, so that the duration of the inversion is not larger than that of the pump pulse. An extreme example is the case of optical parametric oscillators (OPO) where the gain lives only for the time of the pump pulse.

Synchronous pumping is sometimes used in situations that do not meet this criterion, just as starting mechanism. This is the case in some Ti:sapphire lasers, where the gain medium has a longer lifetime than the cavity round-trip time, and therefore synchronous pumping results in only a small modulation of the gain. The small modulation of the gain coefficient $\alpha_g(t)$ is sufficient to start the pulse formation and compression mechanism by dispersion and SPM [2]. The initial small gain modulation grows because of gain saturation by the modulated intracavity radiation, resulting in a shortening of the function $\alpha_g(t)$, and ultimately ultrashort pulses.

The simple considerations that follow, neglecting the influence of saturation, show the importance of cavity synchronism. If the laser cavity is slightly longer than required for exact synchronism with the pump radiation (train of pulses), stimulated emission and amplified spontaneous emission will constantly accumulate at the leading edge of the pulse, resulting in pulse durations that could be even longer than the pump pulse. Therefore, to avoid this situation, the cavity length should be slightly shorter than that required for exact synchronism with the pump radiation. Let us assume first perfect synchronism. The net gain factor per round-trip is

$$G(t) = e^{[\alpha_g(t)d_g - L]}, \tag{6.1}$$

where L is the natural logarithm of the loss per cavity round-trip. After n round-trips, the initial spontaneous emission of intensity I_{sp} has been amplified sufficiently to saturate the gain α_g, and thus the pulse intensity is approximately $I(t) \approx I_{sp} \times \left\{e^{[\alpha_{g0}(t)d_g - L]}\right\}^n = I_{sp} \times [G_0(t)]^n$. The pulse is thus $\sqrt{n}$ times narrower than the unsaturated gain function $G_0(t)$.

For a cavity shorter than required for exact synchronism, in a frame of reference synchronous with the pulsed gain $\alpha_g(t)$, the intracavity intensity of the jth round-trip is related to the previous one by:

$$I_j(t) = I_{j-1}(t + \delta)e^{[\alpha_g(t)d_g - L]}, \tag{6.2}$$

where δ is the mismatch between cavity round-trip time and the pump pulse spacing. The net gain for the circulating pulse $e^{[\alpha_g(t)d_g - L]}$ exists in the cavity for

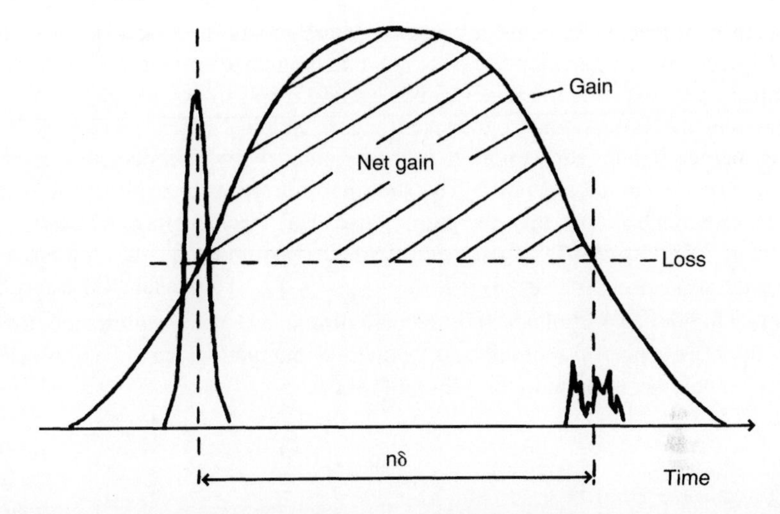

Figure 6.1 Net gain (gain minus loss) temporal profile as it appears at each periodicity of the pump pulse. If the round-trip time of the laser cavity is slightly shorter (by δ) than the pump period, radiation emitted at the right edge of the gain profile will reappear shifted to the left by that amount δ at each successive round-trip. A pulse will experience gain for a maximum of n passages, given by the ratio of the duration of the net gain to the mismatch δ.

a time $n\delta$ only, as can be seen from Figure 6.1. The laser oscillation will start from a small noise burst $I_{sp}(t)$. The intracavity pulse after n round trips can be approximated by:

$$I(t) = I_{sp}(t + n\delta)\left[e^{(\alpha_{av}d_g - L)}\right]^n, \tag{6.3}$$

where $I_{sp}(t + n\delta)$ is the spontaneous emission noise present in the cavity in the time interval $(n-1)\delta \rightarrow n\delta$, and $\alpha_{av} = \frac{1}{n\delta}\int \alpha_g(t)dt$ is a gain coefficient averaged over the n round-trips.

These simple considerations indicate that in the absence of any spectral filtering mechanism and neglecting the distortion of the gain curve $\alpha_g(t)$ by saturation, the pulse should be roughly $\sqrt{n}$ times shorter than the duration of the gain window. The timing mismatch δ is an essential parameter of the operation of a synchronously mode-locked laser. The shape of the autocorrelation (see Chapter 9) is typically a double-sided exponential, which—as pointed out by Van Stryland [3]—is a signature for a possible random distribution of pulse duration in the train. The interferometric autocorrelation also indicates a random (Gaussian) distribution of pulse frequencies [4]. These fluctuations in pulse duration and frequency have also been observed in theoretical simulations by New and Catherall [5] and Stamm [6].

Gain saturation—neglected in the elementary model discussed so far—does play an essential role in pulse shaping and compression for synchronously pumped lasers. We refer to a paper by Nekhaenko *et al.* [7] for a detailed review of the various theories of synchronous pumping. In a typical synchronously pumped laser, the net gain (at each round-trip) is "terminated" by gain depletion at each passage of the circulating pulse. The shortening of the gain period results in a laser pulse much shorter than the pump pulse. This mechanism was analyzed in detail by Frigo *et al.* [8]. It has been verified experimentally that the shortest pulse duration is approximately $\tau_p \approx \sqrt{\tau_{pump} T_{2g}}$ [9]. This result illustrates the fact that the finite spectral width of the gain profile, $\delta\nu_g \approx T_{2g}^{-1}$, ultimately limits the shortest obtained pulse duration. Numerical simulations have been made to relate the number of round-trips required to reach steady state to the single-pass gain [10].

Regenerative Feedback

As we have seen at the beginning of the previous section, the laser cavity should never be longer than the length corresponding to exact synchronism with the pump radiation to generate pulses shorter than the pump pulse. This implies strict stability criteria for the pump laser cavity, its mode-locking electronics, and the laser cavity (invar or quartz rods were generally used for synchronously pumped dye laser cavities). Considerations of thermal expansion of the support material and typical cavity lengths clearly shows the need for thermal stability. Indeed, the thermal expansion coefficient of most rigid materials for the laser support exceeds 10^{-5}/°C. Because the cavity length approaches typically 2 m, even a temperature drift of 0.5°C would bring the laser out of its stability range. However, because it is the *relative* synchronism of the laser cavity with its pump source that is to be maintained, a simpler and efficient technique is to use the noise (longitudinal mode beating) of the laser itself, to drive the modulator of the pump laser [11] if the latter is actively modelocked. This technique, sometimes called "regenerative feedback," has been applied to some commercial synchronously pumped mode-locked lasers, and even to a Ti:sapphire laser [2].

Seeding

Even if somewhat oversimplified, the representation of Fig. 6.1 gives a clue to an important source of noise in the synchronously pumped dye laser. The seed $I_{sp}(t)$ has a complex electric field amplitude $\tilde{\epsilon}(t)$ with random phase. As pointed out in Catherall and New [12] and in Stamm [6] it is this spontaneous emission source that is at the origin of the noise of the laser. Could the noise be reduced by adding to $\tilde{\epsilon}$ a minimum fraction $\eta E(t)$ of the laser output, just large enough so that the phase of $\eta E(t) + \tilde{\epsilon}(t)$ is equal to the phase of the output

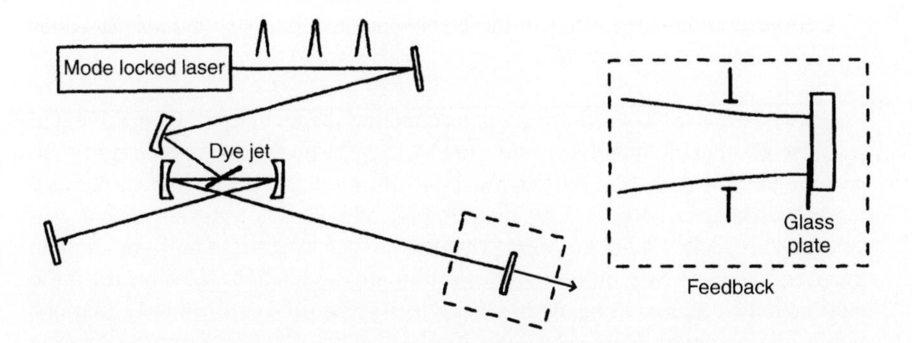

Figure 6.2 Typical synchronously pumped dye laser. The length of the dye laser cavity has to be matched to the repetition rate of the pump pulses. The noise in a synchronously pumped laser can be reduced by reinjection of a portion of the output *ahead* of the main intracavity pulse. A thin glass plate on the output mirror intercepts and reflects part of the beam into the cavity, with the desired advance. The fraction of energy reflected (of the order of 10^{-6}) is determined by the overlap of the aperture and the glass plate. (Adapted from Peter *et al.* [13].)

fields $E(t)$ (which essentially implies $\eta E(t) \gg \tilde{\epsilon}$)? Both calculation and experiment have demonstrated a dramatic noise reduction by seeding the cavity with a small fraction of the pulse *in advance* of the main pulse [13]. The emphasis here is on small; only a fraction of the order of 10^{-7} (not exceeding 10^{-5}) of the output power should be reinjected. A possible implementation would consist of reflecting back a fraction of the output pulse delayed by slightly less than a cavity round-trip. This amounts to a weakly coupled external cavity. A much simpler implementation demonstrated by Peter *et al.* [13] consists in inserting a thin glass plate (microscope cover glass for instance) in front of the output mirror (Figure 6.2). The amount of light reinjected is adjusted by translating the glass plate in front of the beam. The timing of the reinjected signal is determined by the thickness of the plate.

6.2. HYBRID MODE-LOCKING

Synchronous pumping alone can be considered as a good source of ps rather than fs pulses. The disadvantages of this technique, as compared to passive mode-locking, are:

- a longer pulse duration,
- larger amplitude and phase noise,
- the duration of the pulses of the train are often randomly distributed, [3] and

- when attempting to achieve the shortest pulse durations, the pulse frequencies are randomly distributed [4].

One solution to these problems is to combine the techniques of passive and active mode-locking in a hybrid system [14,15]. Depending on the optical thickness of the absorber, the hybrid mode-locked laser is either a synchronously mode-locked laser perturbed by the addition of saturable absorption or a passively mode-locked laser pumped synchronously. The distinction is obvious to the user. The laser with little saturable absorption modulation will have the noise characteristics and cavity length sensitivity typical of synchronously pumped lasers, but a shorter pulse duration. The laser with a deep passive modulation (concentrated saturable absorber for a dye, or a large number of MQW for a semiconductor saturable absorber) shows intensity autocorrelation traces identical to those of the passively mode-locked laser [16]. The sensitivity of the laser to cavity detuning decreases. The reduction in noise can be explained as being related to the additional timing mismatch introduced by the absorber, which partially compensates the pulse advancing influence of the gain and spontaneous emission [17].

6.3. ADDITIVE PULSE MODE-LOCKING

6.3.1. Generalities

There was in the late 1980s a resurrection of interest in developing additive pulse mode-locking (APML), a technique involving coupled cavities. One of the basic ideas—to establish the mode coupling outside the main laser resonator—was suggested in 1965 by Foster *et al.* [18] and applied to mode-locking a He-Ne laser [19]. In that earlier implementation, an acousto-optic modulator is used to modulate the laser output at half the intermode spacing of the laser. The frequency shifted beam is reflected back through the modulator, resulting in a first-order diffracted beam, which is shifted in frequency by the total mode spacing, and reinjected into the laser cavity through the output mirror. The output mirror of the laser forms, with the mirror used to reinject the modulated radiation, a cavity with the same mode spacing as the main laser cavity. If the laser is close to threshold, a small extracavity modulation fed back into the main cavity can be sufficient to lock the longitudinal modes.

Unlike this technique more recent APML implementations are based on passive methods. In the purely dispersive version, pulses from the coupled cavity are given some phase modulation, such that the first half of the pulse fed back into the laser adds in phase with the intracavity pulse, while the second half

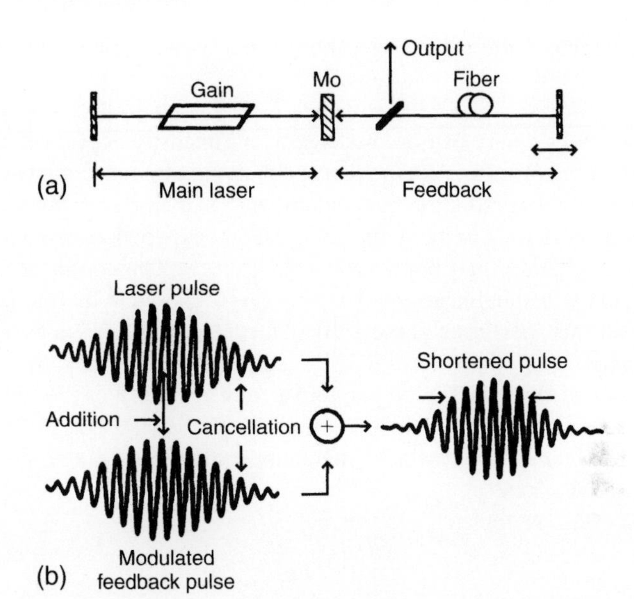

Figure 6.3 A typical additive pulse mode-locked laser (a). At the output mirror M_0, the pulse of the main cavity [(b), top left] adds coherently to the pulse of the auxiliary cavity [(b), bottom left], to result in a shortened pulse [(b), right]. (Courtesy E. Wintner.)

has opposite phase [20]. At each round-trip, the externally injected pulse thus contributes to compress the intracavity pulse, by adding a contribution to the leading edge and subtracting a certain amount from the trailing edge, as sketched in Figure 6.3. This technique has first been applied to shortening pulses generated through other mode-locking mechanisms. A reduction in pulse duration by as much as two orders of magnitudes was demonstrated with color-center lasers [21–24] and with Ti:sapphire lasers [25].

It was subsequently realized that the mechanism of pulse addition through a nonlinear coupled cavity is sufficient to passively mode-lock a laser. This technique has been successfully demonstrated in a Ti:sapphire laser, [26] Nd:YAG [27,28] Nd:YLF [29,30], Nd:glass [31], and KCl color-center lasers [32]. A detailed description of the coherent addition of pulses from the main laser and the extended cavity which takes place in the additive pulse mode-locking has been summarized by Ippen *et al.* [33].

Coherent field addition is only one aspect of the coupled cavity mode-locked laser. The nonlinearity from the coupled cavity can be, for example, an amplitude modulation, as in the "soliton" laser [34], or a resonant nonlinear reflectivity via a quantum well material [35].

6.3.2. Analysis of APML

Analysis of APML [23,33] has shown that the coupling between a laser and an external nonlinear cavity can be modeled as an intensity-dependent reflectivity of the laser end mirror. Let r be the real (field amplitude) reflection coefficient of the output mirror. The radiation transmitted through that mirror into the auxiliary (external) cavity returns to the main cavity having experienced a field amplitude loss γ ($\gamma < 1$) and a total phase shift $-[\phi + \Phi(t)]$. The nonlinear phase shift $\Phi(t)$ induced by the nonlinearity is conventionally chosen to be zero at the pulse peak [33] so that the linear phase shift ϕ includes a bias because of the peak nonlinear phase shift. Therefore, if $r\tilde{\mathcal{E}}(t)$ is the field reflected at the mirror, the field transmitted through the output mirror, the auxiliary cavity (loss γ) and transmitted a second time through the output mirror is $(1 - r^2)\gamma e^{-i\phi}\tilde{\mathcal{E}}(t)e^{-i\Phi(t)}$. If d is the length of the nonlinear medium, and assuming a $\bar{n}_2$ nonlinearity, according to Eq. (3.149):

$$\Phi(t) = \frac{2\pi\bar{n}_2}{\lambda_\ell}[I_{ax}(t) - I_{ax}(0)]\,d \tag{6.4}$$

where $I_{ax}(t)$ is the intensity of the field in the auxiliary cavity. For a qualitative discussion we determine the total reflection by adding the contribution of the reinjected field from the auxiliary cavity to the field reflection r of the output mirror, which leads to a time-dependent complex "reflection coefficient" $\tilde{\Gamma}$:

$$\tilde{\Gamma}(t) = r + \gamma(1 - r^2)e^{-i\phi}[1 - i\Phi(t)]. \tag{6.5}$$

In Eq. (6.5), it has been assumed that Φ is small, allowing us to substitute for the phase factor $e^{-i\Phi}$ its first-order expansion. There is a differential reflectivity for different parts of the pulses. If one sets $\phi = -\pi/2$, then $|\tilde{\Gamma}|$ has a maximum value at the pulse center where $\Phi = 0$, and smaller values at the wings:

$$\tilde{\Gamma}(t) = r + \gamma(1 - r^2)\Phi(t) + i[\gamma(1 - r^2)]. \tag{6.6}$$

The reflection is thus decreasing when Φ becomes negative in the wings of the pulse, which is the "coherent field subtraction" sketched in Fig. 6.3. The compression factor is determined by the ratio of $\gamma(1 - r^2)$ to r, which can be related to the ratio of energy in the auxiliary cavity to that in the main cavity [note that $\gamma(1 - r^2)$ is the maximum amount of energy that can be subtracted from the pulse in the main cavity at each round-trip].

This dynamic reflectivity can be adjusted for pulse shortening at each reflection, until a steady-state balance is achieved between the pulse shortening and pulse spreading because of bandwidth limitation and dispersion.

6.4. MODE-LOCKING BASED ON NONRESONANT NONLINEARITY

Various techniques of mode-locking using second-order nonlinearities have been developed. A first method is a direct extension of Kerr lens mode-locking, which has been analyzed in the previous chapter. A giant third-order susceptibility can be found near phase matching conditions in SHG, not unlike the situation encountered with a third-order susceptibility, which is seen to be enhanced near a two photon resonance [36,37]. In this method, called cascaded second-order nonlinearity mode-locking, the nonlinear crystal is used in mismatched conditions with a mirror that reflects totally both the fundamental and SH waves. The cascade of sum and difference frequency generation induces a transverse focusing of the fundamental beam in a way similar to Kerr self-focusing. This method has been applied to solid-state lasers by Cerullo *et al.* [38] and Danailov *et al.* [39]. The resonance condition (the phase matching bandwidth) implied in this method does not make it applicable to the fs range.

Another technique was introduced by Stankov, [40,41] who demonstrated passive mode-locking in a Q-switched laser by means of a nonlinear mirror consisting of a second harmonic generating crystal and a dichroic mirror. Dispersion between the crystal and the dichroic mirror is adjusted so that the reflected SH is converted back to the fundamental.

A third method, based on polarization rotation occurring with type II second harmonic generation, is the equivalent of Kerr lens mode locking in fiber lasers. It has also been applied to some solid-state lasers. The last two methods will be discussed in more detail in the following subsections.

6.4.1. Nonlinear Mirror

The principle of operation of the nonlinear mirror can be understood with the sketch of Figure 6.4, showing the end cavity elements that provide the function of nonlinear reflection. A frequency doubling crystal in phase matched orientation

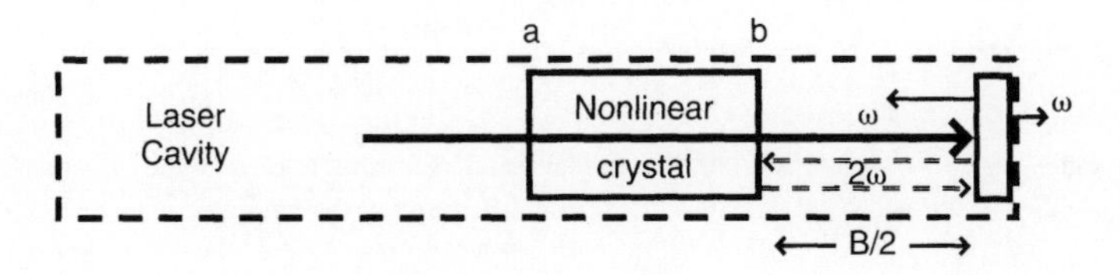

Figure 6.4 End cavity assembly constituting a nonlinear mirror. The end mirror is a total reflector for the SH and a partial reflector for the fundamental.

is combined with a dichroic mirror output coupler that totally reflects the SH beam and only partially reflects the fundamental. These two elements form a reflector, whose reflectivity at the fundamental wavelength can either increase or decrease, depending on the phases of the fundamental and SH radiation. These phase relations between the first and second harmonics can be adjusted inserting a dispersive element between the nonlinear crystal and the dichroic mirror. The dispersive element can be either air (the phase adjustable parameter is the distance between the end mirror and the crystal) or a phase plate (of which the angle can be adjusted).

At low intensity, the cavity loss is roughly equal to the transmission coefficient of the output coupler at the fundamental wavelength. At high intensities, more second harmonic is generated, reflected back and reconverted to the intracavity fundamental, resulting in an increase in the effective reflection coefficient of the crystal output coupler combination. The losses are thus decreasing with intensity, just as is the case with a saturable absorber. Figure 6.5 shows the variation of intensities of the fundamental and second harmonic in the first (left) and second (right) passage through the second harmonic generating crystal. Depletion of the

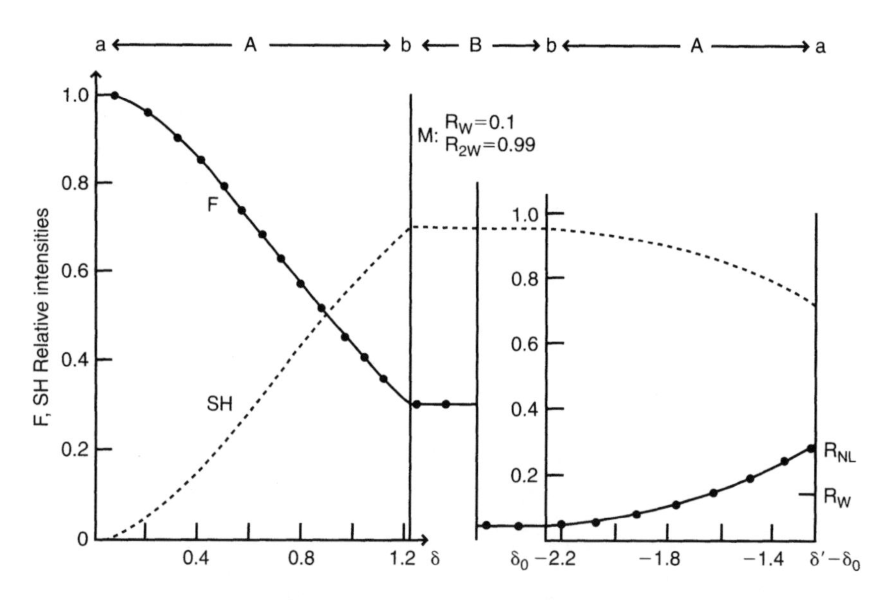

Figure 6.5 Variation of intensity of the fundamental (F, solid line) and the second harmonic (SH, dashed line) for two successive passages, A and B, through the nonlinear crystal. The entrance and exit surfaces a and b are labeled in Fig. 6.4. A fraction $R = 10\%$ of the fundamental intensity is reflected back into the crystal, together with the entire second harmonic. After propagation for a distance B in air, the phase of the second harmonic with respect to the fundamental has undergone a shift of π, resulting in a reconversion of second harmonic into fundamental at the second passage.

fundamental through SHG reduced the intensity to 30% of its initial value. Only 10% of that fundamental is reflected back through the second harmonic generating crystal. However, because the full SH signal that was generated in the first passage is reflected back, and because it has reversed phase with respect to the fundamental, 30% of the initial fundamental is recovered. At the first passage, the conversion to second harmonic should be sufficient to have a sizeable depletion of the fundamental. Therefore, this method works best for high-power lasers. The theoretical framework for the SHG has been set in Chapter 3 (Section 3.4.1) and can be applied for a theoretical analysis of this type of mode-locking. A frequency domain analysis of the mode-locking process using a nonlinear mirror can be found in Stankov [42]. Available software packages, such as—for example—*SNLO* software can be used to compute the transmission of fundamental and second harmonic at each passage [43].

The electronic nonlinearity for harmonic generation responds in less than a few femtoseconds. However, because of the need to use long crystals to obtain sufficient conversion, the shortest pulse durations that can be obtained by this method are limited to the picosecond range by the phase matching bandwidth. The method has been applied successfully to flashlamp pumped lasers [44] and diode pumped lasers [45–48]. A review can be found in Kubecek [49].

The same principle has also been applied in a technique of parametric mode-locking, which can be viewed as a laser hybridly mode-locked by a nonlinear process [50]. The third-order nonlinearity of a crystal applied to sum and difference frequency generation is used in the mode-locking process. The nonlinear mirror can also be used to provide negative feedback instead of positive feedback by adjusting the phase shift between fundamental and second harmonic by the dispersive element [51].

6.4.2. Polarization Rotation

Nonlinear polarization rotation because of the nonlinear index associated with elliptical polarization has been described in Section 5.4.2 as an example of a third-order nonlinear process. Again, a second-order nonlinearity can also be used for polarization rotation. As is the case when phase matched SHG is used, the minimum pulse duration is determined by the inverse of the phase matching bandwidth.

Under type II phase matching, the orientation of the fundamental field polarization (assumed to be linear) at the output of the nonlinear crystal is directly dependent on the relative intensity of the two orthogonal polarization components. The crystal cut and orientation is assumed to perfectly fulfill the phase matching conditions for SHG. If the linearly polarized incoming field is split into two orthogonal components with strongly unbalanced intensity, then the wave of smallest initial amplitude may be completely depleted because the SHG process diminishes each component by the same amount. If the nonlinear propagation continues beyond that point the SHG is replaced by difference frequency generation between the generated harmonic and the remaining fundamental component. The new fundamental field appears on the polarization axis where the fundamental had disappeared but the phase of the created field is now shifted by π with respect to the initial field. Difference frequency generation then goes on with propagation distance until the power of the second harmonic goes to zero. If we assume that the crystal behaves in the linear regime like a full-wave or half-wave plate then the output polarization state remains linear in the nonlinear regime, but the orientation of the output is intensity dependent. Two properly oriented polarizers placed on either side of the nonlinear crystal permit us to build a device with an intensity dependent transmission.

Details on the use of nonlinear polarization in a type II SHG for mode-locking of a cw lamp pumped Nd:YAG laser are given in Kubecek *et al.* [52].

6.5. NEGATIVE FEEDBACK

In this section we will describe a technique that limits the peak power of pulses circulating in the cavity. This can be accomplished by a combination of an element producing nonlinear defocusing and an aperture. Negative feedback has gained importance in Q-switched and mode-locked solid-state lasers because it tends to lengthen the pulse train by limiting the peak power and thereby reducing the gain depletion. Moreover, a longer time for pulse formation usually leads to shorter output pulses and more stable operation.[1]

We have seen that the pulse formation—in passively mode-locked lasers—is associated with a positive feedback element (Kerr lensing, saturable absorber) which enhances positive intensity fluctuations (generally through a decrease of losses with increasing intensity). Although a positive feedback leads to pulse formation, it is inherently an unstable process, because intensity fluctuations are amplified. Therefore, it is desirable, in particular in high-power lasers, to have a *negative* feedback element that sets in at higher intensities than the positive feedback element.

Pulses of 10, 5, and less than 1 ps have been generated with this technique with Nd:YAG, Nd:YAP, and Nd:glass lasers, respectively. More importantly for the fs field, the pulse-to-pulse reproducibility (better than 0.2% [53]) makes these lasers ideal pump sources for synchronous or hybrid mode-locking. The flashlamp pumped solid-state laser with negative feedback provides a much higher energy

[1] Note that in high-power solid-state lasers the typical Q-switched pulse is not much longer than a few cavity round-trips.

per pulse, at shorter pulse duration, than the cw mode-locked laser used conventionally as pump for fs systems. The use of negative feedback to effectively pump a fs dye laser was demonstrated by Angel *et al.* [10].

In semiconductor laser pumped solid-state lasers, negative feedback can be used to suppress Q-switched mode-locked operation, in favor of cw mode-locked operation [54]. The mechanism is the same as for the flashlamp pumped laser; the energy limiting prevents the total gain depletion that ultimately interrupts the pulse train.

Electronic Feedback

A typical flashlamp pumped, mode-locked Nd laser generates a train of only 5 to 10 pulses of all different intensities. In the first implementation of "negative feedback," an electronic feedback loop *increases* the cavity losses if the pulse energy exceeds a well-defined value. Martinez and Spinelli [55] proposed to use an electro-optic modulator to actively limit the intracavity energy in a passively mode-locked glass laser. They demonstrated that the pulse train could be extended. A fast high voltage electronics led to the generation of μs pulse trains in a passively mode-locked glass lasers [56] and in hybrid Nd:glass lasers [57].

Electronic Q-switching and negative feedback has the advantage that the timing of the pulses is electronically controlled. This is important in applications where several laser systems have to be synchronized. However, there is a minimum response time of one cavity round-trip before the feedback can react [57].

Passive Negative Feedback

A passive feedback system can provide immediate response—i.e., on the time scale of the pulse rather than on the time scale of the cavity round-trip. We will here restrict our description to the Nd laser using a semiconductor (GaAs) for passive negative feedback. The semiconductor used in a passive feedback system produces nonlinear lensing. The analysis of the beam focusing is identical to that of the Kerr lensing, except that the sign of the lensing is opposite. The nonlinear index change is initiated by electrons generated by two photon absorption into the conduction band. Various processes then contribute to the index change. The index change by free electrons, for example, can be estimated with the Drude model and is negative:

$$\Delta n_d(x, y, t) = -\frac{n_0 e^2}{2m^* \epsilon_0 \omega_\ell^2} N(x, y, t), \tag{6.7}$$

where N is the electron density, m^* is the electron's effective mass and n_0 is the linear index. We refer to the literature for additional contributions to Δn such as the interband contribution [58] and an additional electronic contribution [36,59].

Other implementations of passive negative feedback have used a SH crystal near phase matching ("cascaded nonlineary") to produce a large nonlinear index required for the energy limiter [60].

A typical laser using passive negative feedback generally includes a saturable absorber for Q-switching and mode-locking and an energy limiter. An energy limiter that can be used for passive negative feedback is illustrated in Figure 6.6. A two photon absorber (typically GaAs) is located near a cavity end mirror. After double passage through this sample, the beam is defocused by a self-induced lens originating mainly from the free carriers generated through two photon absorption. The defocused portion of the beam is truncated by an aperture. Self-defocusing in the semiconducting two photon absorber sets in at a power level that should be close to the saturation intensity of the saturable absorber used for Q-switching and mode-locking. Because the pulse intensity is close to the pulse saturation intensity, there is optimal pulse compression at the pulse leading edge by saturable absorption. Because of self-defocusing in GaAs, the pulse trailing edge is clipped off, resulting in further pulse compression and energy loss.

The stabilization and compression of the individual pulses result from a delicate balance of numerous physical mechanisms. Details of the experimental implementation and theoretical analysis can be found in the literature [61–63].

At the end of this section we will discuss an experiment that illustrates the saturation and focusing properties of a particular nonlinear element. Often the nonlinear element is just the substrate of a multiple quantum well [64]. In that case, one has combined in one element the function of saturable absorber (the MQW, excited by one photon absorption) and energy limiter (the substrate, excited by two photon absorption). The properties of the MQW on its substrate are well demonstrated by the measurement illustrated in Figure. 6.7 and 6.8. A diode pumped microchip YAG laser is used to focus pulses of 3 μJ energy and 1 ns duration at a repetition rate of 15.26 kHz into a sample consisting of 100 quantum

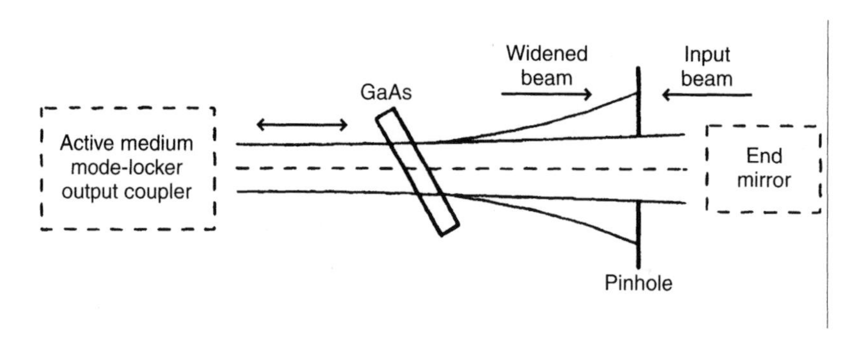

Figure 6.6 Passive negative feedback is typically achieved by inserting in the cavity an energy limiter, which can consist of a GaAs plate (acting as two photon absorber and subsequent defocusing element) and an aperture (pinhole).

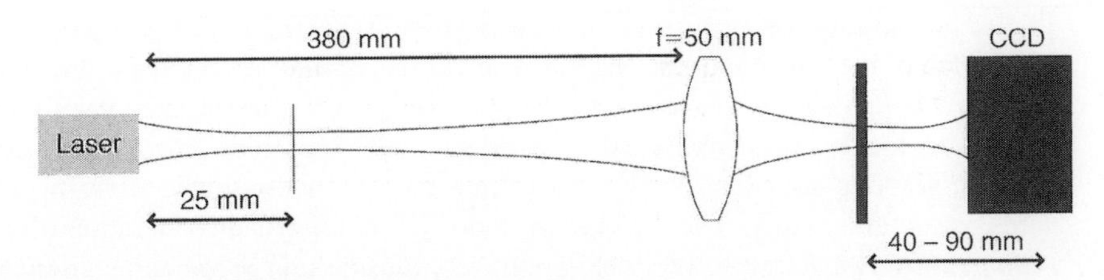

Figure 6.7 Experimental setup to observe the saturable absorption, two photon absorption and self-lensing in a sample of 100 quantum wells on a GaAs substrate located in front of a CCD camera (from Kubecek *et al.* [64]).

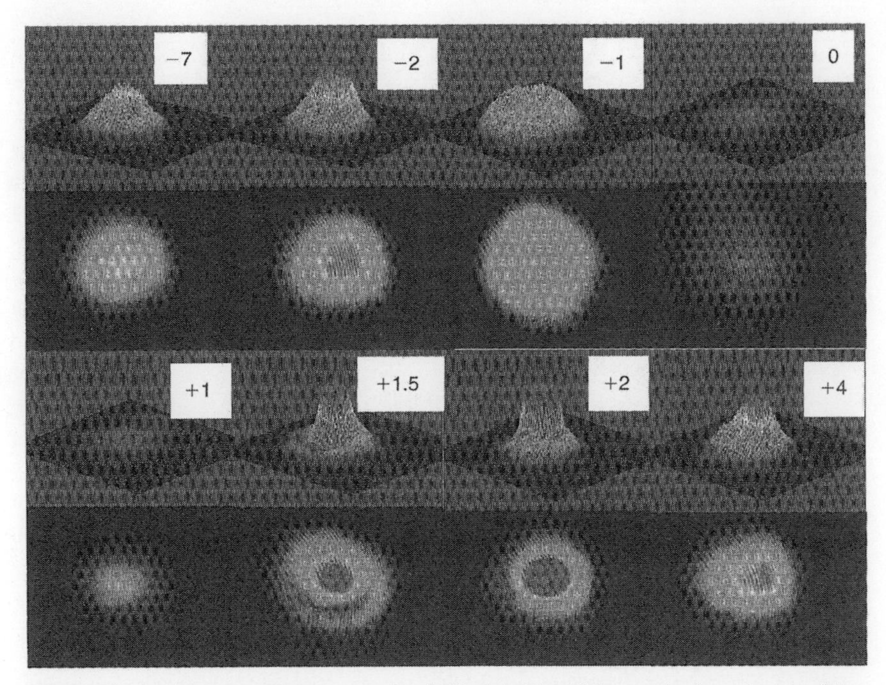

Figure 6.8 Spatial beam structure versus longitudinal position of the sample along the axis of the beam, after the lens. The distances from focus are indicated (in mm). The upper part of the figure corresponds to the positions left of the focus; the lower part right of the focus. (Adapted from Kubecek *et al.* [64].)

wells on a GaAs substrate. The lens has a focal distance of 50 mm. The output power from the laser was attenuated not to damage the MQW. The maximum power density in the focal point was 10 MW/cm^2. The spatial profile of the radiation transmitted through the sample was analyzed, using a CCD camera, as a function of the position of the sample. The various profiles are shown in Fig. 6.8. From this picture we can see that the initial low power transmission of 23% far from the focal point increases to 45% close to the focal point. The transmission of the GaAs plate alone is 52%, indicating that the nonsaturable losses in the MQW are about 10%. The increase in transmission reflects the saturation of the quantum wells. Close to the focal point, the transmission drops and significant defocusing is observed. This is a region of large two photon absorption, creating an electron plasma sufficiently dense to scatter the beam. Self-defocusing is observed with the sample positioned to the left of the focus, self-focusing to the right of the focus.

6.6. SEMICONDUCTOR-BASED SATURABLE ABSORBERS

Progress in the fabrication of semiconductors and semiconductor based structures, such as MQWs, has led to the development of compact and efficient saturable absorbers whose linear and nonlinear optical properties can be custom tailored. These devices are particularly suited for mode-locking solid-state lasers, fiber lasers and semiconductor lasers. They can conveniently be designed as laser mirrors, which makes them attractive for initiating and sustaining mode-locking in a variety of solid-state lasers and cavity configurations, for a review see Keller *et al.* [65].

In semiconductors a transition from the valence to the conduction band is mostly used. In MQWs an excitonic resonance near the band edge can be utilized [66], which leads to a lower saturation energy density [67].

As discussed in the previous chapter an important parameter is the relaxation time of the absorber. The recovery rate is the sum of the carrier relaxation rate $1/T_1$ and the rate of diffusion out of the excited volume $1/\tau_d$:

$$\frac{1}{\tau_A} = \frac{1}{T_1} + \frac{1}{\tau_d}. \tag{6.8}$$

For a beam waist w_0 at the absorber the characteristic diffusion time can be estimated by

$$\tau_d = \frac{w_0^2}{8D},$$

where D is the diffusion constant, which is related to the carrier mobility μ through the Einstein relation $D = k_B T \mu / e$. For a beam waist of 2 μm and $D = 10$ cm²/s for example, the diffusion time $\tau_d \approx 500$ ps.

Typical carrier lifetimes in pure semiconductors are ns and thus too long for most mode-locking applications, where the cavity round-trip time is of the order of a few ns. Several methods are available to reduce the effective absorption recovery rate of bulk semiconductors and MQWs:

1. tight focusing and
2. insertion of defects.

A commonly used technique to insert defects is proton bombardment with subsequent gentle annealing. For example, the bombardment of a MQW sample consisting of 80 pairs of 102 Å GaAs and 101 Å $Ga_{0.71}Al_{0.29}As$ wells, with 200-keV protons resulted in recovery times of 560 ps and 150 ps, respectively [67]. Structures with thinner wells (70 to 80 Å) separated by 100 Å barriers yield broader absorption bands [68], with the same recovery time of 150 ps after a 10^{13}/cm² proton bombardment and annealing.

Another technique to introduce defects is to grow the semiconductor at relatively low temperature. This can lead to a relatively large density of deep-level defects that can quickly trap excited carriers. As an example, Figure 6.9 shows

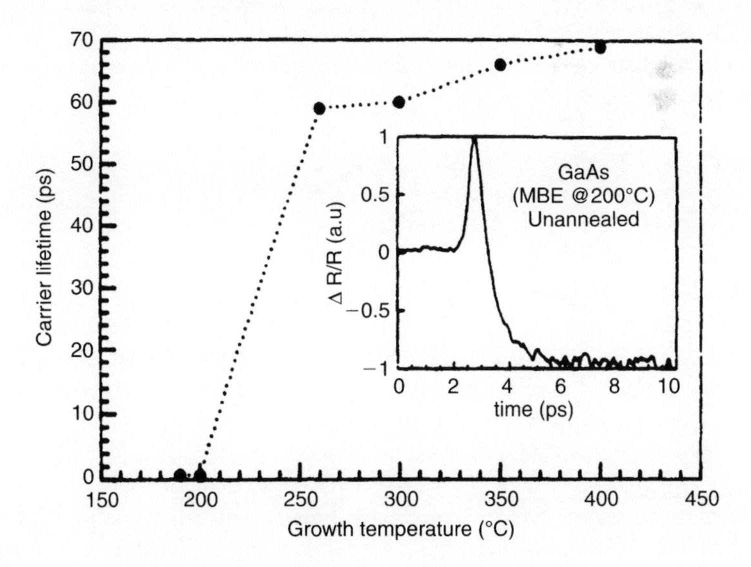

Figure 6.9 Carrier lifetime of GaAs versus MBE growth temperature. The inset shows the transient reflection measured in a pump probe experimental, for a 200°C grown unannealed sample. (Adapted from Gupta *et al.* [69].)

Table 6.1

Semiconducting materials with carrier lifetimes and mobilities. (Adapted from Gupta *et al.* [69].)

Material	Carrier lifetime T_1 (ps)	Mobility μ (cm^2/Vs)
Cr-doped GaAs	50–100	1000
Ion implanted InP	2–4	200
Ion-damaged Si-on-sapphire	0.6	30
Amorphous silicon	0.8–20	1
MOCVD CdTe	0.45	180
GaAs (MBE, 200°C)	0.3	150
$In_{0.42}Al_{0.48}As$ (MBE, 150°C)	0.4	5

a plot of the carrier lifetime versus MBE growth temperature. The measurement is performed by focusing a 100 fs pump pulse onto a 20–30 μm spot on the semiconductor. A 10 times attenuated (as compared to the pump) probe pulse is focused into a 10 μm island within the pumped region. Both pump and probe are at 620 nm. The reflectance of the probe is measured as a function of probe delay (inset in Fig. 6.9). The carrier lifetime is defined as the initial decay ($1/e$) of the reflectance versus delay.

Table 6.1 lists carrier lifetimes and mobilities of some representative semiconductor materials.

6.7. SOLID-STATE LASERS

6.7.1. Generalities

Most common solid-state lasers used for ultrashort pulse generation use materials with a long lifetime (compared to typical cavity round-trip times) as gain media. The laser efficiency can be high if pumped by other lasers, for example semiconductor lasers, tuned to the pump transition. This is especially the case for lasers such as Ytterbium YAG that have a small quantum defect.[2]

Because these solid-state lasers have small gain cross sections as compared to dye lasers and semiconductor lasers, gain modulation is ineffective

[2]The quantum defect is the difference in energy of the pump photon and the laser emitting photon.

for mode-locking. With an upper state lifetime many orders of magnitude longer than the round-trip time, synchronous pumping is seldom used.[3]

The relatively low gain calls for longer lasing media, of the order of several mm, as opposed to the typical 100 μm used with dye and semiconductor lasers. The long gain crystal in turn supports large SPM. Therefore, mode-locking will most often occur through Kerr lensing and chirping in the gain medium. Some exception where saturable absorbers are used are:

- Long pulse generation, tunable in wavelength.
- Mode-locking of LiCAF lasers, where the Kerr effect is small.
- Bidirectional mode-locking of ring lasers (Kerr lensing in the gain medium favors unidirectionality).

Also because of the longer gain medium, (as compared to dye and semiconductor lasers), the laser will be sensitive to any parameter that influences the index of refraction. These are:

- Laser pulse intensity—an effect generally used for passive mode-locking (Kerr lensing).
- Temperature dependence of the index of refraction, which leads to thermal lensing and birefringence.
- Change in index of refraction associated with the change in polarizability of optically pumped active ions.

The latter effect was investigated by Powell *et al.* [70] in Nd doped lasers, and found to be of the order of 50% of the thermal change in index.

Pumping of solid-state lasers is done either by another laser (for instance argon ion laser, or frequency doubled vanadate (YVO4) laser, for Ti:sapphire) or by a semiconductor laser (Cr:LiSAF, Nd:vanadate) or by flashlamps (Nd:YAG). Diode laser pumping is the most advantageous from the point-of-view of wall plug efficiency.

Mode-locked solid-state lasers tend to specialize according to the property that is desired. So far Ti:sapphire lasers have been the choice for shortest pulse generation and stabilized frequency combs. Diode pumped Cr:LiSAF lasers can reach pulse durations in the tens of fs and are the preferred laser when extremely low power consumption is desired. Nd:YAG lasers are most convenient for generating high-power Q-switched mode-locked ps pulse trains and are generally flashlamp pumped. Nd:vanadate is generally used as diode pumped Q-switched mode-locked source, although it is possible to achieve cw mode locked operation too. Both Nd:YAG and vanadate have a bandwidth that restricts their operation

---

[3]Synchronous pumping has been used with some Ti:sapphire lasers to provide the modulation necessary to start the Kerr lensing mode-locking, but not as a primary mode-locking mechanism.

to a shortest pulse of approximately 10 ps. The laser with the lowest quantum defect is sought for high power application where efficiency is an issue. Yb:YAG can be pumped with 940 nm diode lasers, to emit at 1.05 micron. An optical to optical conversion efficiency of 35% has been obtained [71].

6.7.2. Ti:sapphire Laser

The Ti:sapphire laser is the most popular source of fs pulses. The properties that make it one of the most attractive source of ultrashort pulses are listed in Table 6.2. Ti:sapphire is one of the materials with the largest gain bandwidth, excellent thermal and optical properties, and a reasonably large nonlinear index.

Table 6.2

Room temperature physical properties of Ti:sapphire. The gain cross-section increases with decreasing temperature, making it desirable to operate the laser rod at low temperatures. The values for the nonlinear index from Smolorz and Wise [73] take into account the conversion factor of Eq. (3.140). Some data are given for σ (perpendicular to the optical axis) and π (parallel to the optical axis) polarization.

Property	Value	Units	Reference
Index of refraction at 800 nm	1.76		[72]
Nonlinear index (electronic)	$10.5 \cdot 10^{-16}$	cm^2/W	[73]
Raman shift	419	cm^{-1}	[73]
Damping time T_R	6	ps	[73]
Raman contribution to $\bar{n}_2$	$1.7 \cdot 10^{-17}$	cm^2/W	[73]
Raman shift	647	cm^{-1}	[73]
Damping time T_R	6	ps	[73]
Raman contribution to $\bar{n}_2$	$0.8 \cdot 10^{-17}$	cm^2/W	[73]
Dispersion (k'') at 800 nm	612	fs^2/cm	
Peak absorption at	500	nm	
σ_π	$6.5 \cdot 10^{-20}$	cm^2	[74]
σ_σ	$2.5 \cdot 10^{-20}$	cm^2	[74]
Number density of Ti^{3+}	$3.3 \cdot 10^{19}$	cm^{-3}	
at a concentration of	0.1	wt.% Ti_2O_3	
Peak gain at	795	nm	
σ_π	$5 \cdot 10^{-20}$	cm^2	[74]
σ_σ	$1.7 \cdot 10^{-20}$	cm^2	[74]
Fluorescence lifetime τ_F	3.15	μs	[74]
$d\tau_F/dT$	−0.0265	$\mu s/K$	[74]

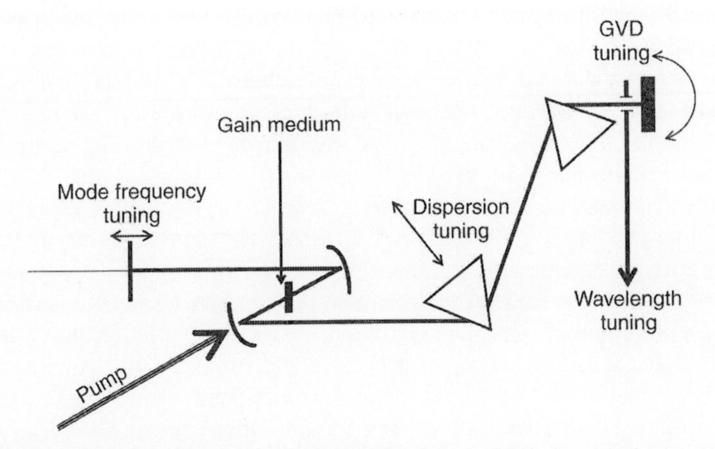

Figure 6.10 Typical Ti:sapphire laser cavity consisting (from the right) of an end mirror, an aperture, a prism pair, folding mirrors at both sides of the laser crystal, and an output coupler. The various controls that are possible on this laser are indicated.

A typical configuration is sketched in Figure 6.10. The pump laser is typically either a cw Ar ion laser or a frequency doubled Nd:vanadate laser. The operation of the Ti:sapphire laser is referred to as "self-mode-locked" [75]. The cavity configuration is usually linear, containing only the active element (the Ti:sapphire rod), mirrors and dispersive elements. The latter can be a pair of prisms (cf. Section 2.5.5), or negative dispersion mirrors (cf. Section 2.3.3), or other interferometric structures. Dispersion control by prisms [76] and by mirrors [77] led to the generation of pulses shorter than 12 fs in the early 90s. The output power typically can reach hundreds of mW at pump powers of less than 5 W. Sometimes, to start the pulse evolution and maintain a stable pulse regime, a saturable absorber, an acousto-optic modulator, a wobbling end mirror, or synchronous pumping is used.

The mode-locking mechanism most often used in the cavity of Fig. 6.10 is Kerr lens mode-locking. The cavity mode is adjusted in such a way that the lensing effect in the Ti:sapphire rod results in a better overlap with the pump beam, hence an increased gain for high peak power pulses (soft aperture). Another approach discussed in Section 5.4.3 and Appendix E is to insert an aperture in the cavity, at a location such that self-lensing results in reduced losses [increased transmission through the aperture (hard aperture)].

While Kerr lensing in conjunction with a soft or a hard aperture initiates the amplitude modulation essential to start the mode-locking, the succession of SPM and quadratic dispersion is responsible for pulse compression. The prism pair provides a convenient means to tune the dispersion to an optimal value that will compensate the SPM, by translating the prism into the path of the beam, as shown in Fig. 6.10.

The shortest pulse duration that can be achieved is ultimately determined by higher-order dispersion, which includes a contribution from the prism material, from the Ti:sapphire crystal, and the mirror coatings. To minimize the third-order dispersion from the gain medium, short crystal lengths (2 to 4 mm) with the maximum doping compatible with an acceptable optical quality of the Ti:sapphire crystal are generally used. If the shortest pulses are desired, quartz prisms are generally preferred because of their low third-order dispersion. However, because the second-order dispersion of quartz is also small, the shortest pulse is compromised against a long round-trip time, because the intra prism distance has to be large (>1 m) to achieve negative dispersion. Another choice of prism material is LaK16, which has a sufficient second-order dispersion to provide negative dispersion for distances of the order of 40 cm to 60 cm. Highly dispersive prisms such as SF10 or SF14 are used when a large number of dispersive intracavity elements has to be compensated with a large negative dispersion.

Several "control knobs" are indicated on the Ti:sapphire laser sketched in Fig. 6.10. After traversing the two prism sequence from left to right, the various wavelengths that constitute the pulse are displaced transversally before hitting the end mirror. An adjustable aperture located between the last prism and the end mirror can therefore be used either to narrow the pulse spectrum (hence elongate the pulse) or tune the central pulse wavelength. A small tilt of the end mirror—which can be performed with piezoelectric elements)—can by used to tune the group velocity (hence the cavity round-trip time, or the mode spacing) without affecting the optical cavity length at the average pulse frequency (no translation of the modes). The position of the modes—in particular the mode at the average pulse frequency—can be controlled by translation of the end mirror with piezoelectric transducers. Such a motion also affects the repetition rate of the cavity. Ideally, orthogonal control of the repetition rate and mode position requires two linear combinations of the piezo controls just mentioned.

Cavities with Chirped Mirrors

Instead of intracavity prisms, negative dispersion mirrors are the preferred solution for the shortest pulses, provided a short Ti:sapphire rod is available, and there is no other dispersive intracavity element. Continuous tuning of the dispersion is not possible as was the case with the intracavity prism pair. Discrete tuning however is possible, through the number of multiple reflection at the dispersive mirrors. The minimum increment of dispersion is the dispersion associated with a single reflection.

As we saw in Chapter 5, one of the applications of mode-locked lasers is to generate frequency combs for metrology. We will discuss such examples, and the

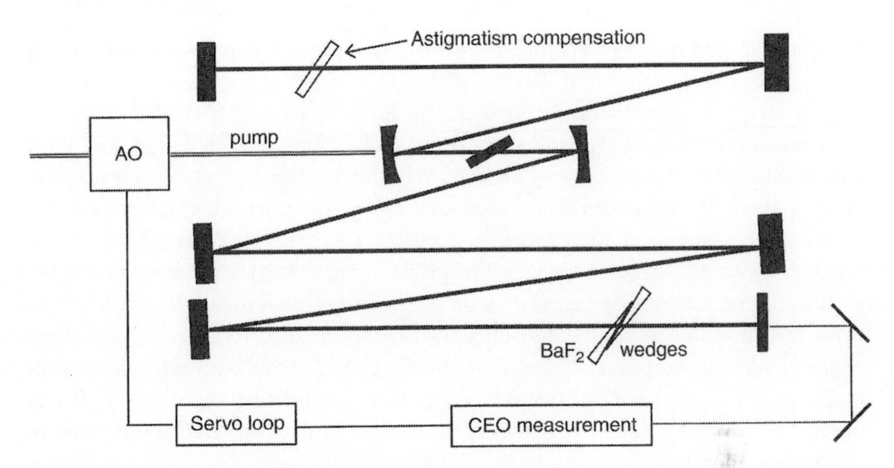

Figure 6.11 Ti:sapphire laser cavity with chirped mirrors for 5-fs pulse generation. Two wedges of BaF_2 are used for continuous dispersion control. The intracavity group velocity is tuned through the pump intensity. The servo loop takes the signal from the measurement of the CEO, and feeds it back to an acousto-optic modulator. Adapted from Ell *et al.* [1].

lasers applied in more detail in Chapter 13. For these applications it is desirable to have an octave spanning pulse spectrum, which implies pulses as short as 5 fs, or about two optical cycles [1]. This allows one to mix the second harmonic of the IR part of the mode comb with a mode from the short wavelength part of the fundamental spectrum—a technique to determine the carrier to envelope offset [78–81]. An example of such a 5-fs laser is sketched in Figure 6.11. Mirrors with a smooth negative dispersion over the whole spectrum have been developed (see Section 2.3.3) and double-chirped mirrors have been used for this laser [82]. Both the low and high index layers of these coatings are chirped. The spectral analysis of the reflectivity of these coatings still shows "phase ripples." To eliminate these ripples, the mirrors are used in pairs, manufactured in such a way that the ripples are 180 degrees out of phase.

Continuous dispersion tuning is achieved by the use of thin BaF_2 wedges. BaF_2 is the material with a low ratio of third- to second-order dispersion in the wavelength range from 600 to 1200 nm, and the slope of its dispersion is nearly identical to that of air. It is therefore possible to scale the cavity to, for instance, shorter dimensions, and maintain the same dispersion characteristics by adding the appropriate amount of BaF_2.

High Power from Oscillators

For some applications, for example laser micromachining, it is desirable to increase the pulse energy of the output of fs oscillators without amplification.

Because the pump power is limited an increase in pulse energy can only be at the expense of repetition rate. Several different techniques have been developed.

A cavity dumper can be inserted in the Kerr lens mode-locked Ti:sapphire laser resonator [83,84]. This allows the fs pulse to build up in a high Q cavity with essentially no outcoupling losses. When a certain energy is reached the outcoupler (typically based on an acousto-optic modulator) is turned on, and the pulse is coupled out of the cavity. Repetition rates typically range from a few 100 kHz to a few MHz. Pulse energies of up to the 100-nJ level are possible.

Another method tries to capitalize on the inherent trend in solid-state lasers to show relaxation oscillations and self Q switching. In such regimes the envelope of the mode-locked pulse train is modulated. The Q-switched and mode-locked output can be stabilized by (weakly) amplitude modulating the pump at a frequency of several hundred kHz that is derived from the Q-switched envelope in a feedback loop [85].

A third technique is based on long laser cavities (up to tens of meters) resulting in low repetition rates of a few MHz. Careful cavity and dispersion design are necessary to avoid the multiple pulse lasing and the instabilities that are usually associated with long cavities [86]. For example, 200 nJ, 30-fs pulses at a repetition rate of 11 MHz were obtained with a chirped mirror cavity and external pulse compression with prisms [87].

6.7.3. Cr:LiSAF, Cr:LiGAF, Cr:LiSGAF, and Alexandrite

The chromium ion has maintained its historical importance as a lasing medium. Ruby is produced by doping a sapphire host with Cr_2O_3. The ruby laser being a three-level system, requires high pump intensities to reach population inversion. It is a high gain, narrow bandwidth, laser, hence not suited for ultrashort pulse applications.

A broadband lasing medium is alexandrite, consisting of chromium doped chrysoberyl ($BeAl_2O_4{:}Cr^{3+}$). The alexandrite laser is generally flashlamp pumped (absorption bands from 380 to 630 nm), with a gain bandwidth ranging from 700 to 820 nm, and is therefore sometimes used as an amplifier (mostly regenerative amplifier) for pulses from Ti:sapphire lasers. It is one of rare laser media in which the gain cross section increases with temperature, from $7 \cdot 10^{-21}$ cm^2 at 300°K to $2 \cdot 10^{-20}$ cm^2 at 475°K [72].

Of importance for femtosecond pulse generation are the $Cr^{3+}{:}LiSrAlF_6$ or Cr:LiSAF, $Cr^{3+}{:}LiSrGaF_6$ or Cr:LiSGAF and $Cr^{3+}{:}LiCaAlF_6$ or Cr:LiCAF lasers. These crystals have similar properties as shown in Table 6.3. The gain cross section is relatively low compared with other diode pumped laser crystals (30× less than that of Nd:YAG for example). The thermal conductivity is 10 × smaller

Table 6.3

Room temperature physical properties of Cr:LiSAF, Cr:LiSGAF, and Cr:LiGAF. The second-order dispersion of LiSAF is indicated for two different Cr doping concentrations. *A*, *B*, *C*, and *D* are the parameters of the Sellmeir formula $n_i^2 = A_i + B_i/(\lambda_\ell^2 - C_i) - D_i\lambda_\ell^2$. with $i = o$ (ordinary) or *e* (extraordinary), and λ_ℓ expressed in μm.

Property	Cr:LiSAF	Cr:LiSGAF	Cr:LiCAF	Units	Ref.
Sellmeir coeff.					
A_o	1.95823	1.95733	1.91850		
A_e	1.95784	1.95503	1.91408		
B_o	0.00253	0.00205	0.00113	μm^2	
B_e	0.00378	0.00252	0.00155	μm^2	
C_o	0.02671	0.03836	0.04553	μm^2	
C_e	0.01825	0.03413	0.04132	μm^2	
D_o	0.05155	0.04765	0.02525	μm^{-2}	
D_e	0.02768	0.03822	0.01566	μm^{-2}	
n_o (850 nm)	1.38730	1.38776	1.37910		
Nonlinear index	$3.3\ 10^{-16}$	$3.3\ 10^{-16}$	$3.7\ 10^{-16}$	cm^2/W	[73]
Dispersion k'' (850 nm, 0.8%)	210	280		fs^2/cm	[91,92]
Dispersion k'' (850 nm, 2%)	250			fs^2/cm	[91]
Third-order dispersion k'''	1850	1540		fs^3/cm	[91,92]
Peak absorption	670	630		nm	
Peak gain at	850	835	763	nm	
cross section σ_π	$4.8\ 10^{-20}$	$3.3\ 10^{-20}$	$1.3\ 10^{-20}$	cm^2	[93]
Fluorescence τ_F (300°K)	67	88	170	μs	[93]
$T_{1/2}$	69	75	255	°C	[88]
Expansion coeff. along *c*-axis	−10	0	3.6	$10^{-6}/K$	[93]
along *a*-axis	25	12	22	$10^{-6}/K$	[93]
c-axis thermal conductivity	3.3	3.6	5.14	W/mK	[94]
Thermal index dependence dn/dT	−4.0		−4.6	$10^{-6}/K$	[94]

than for Ti:sapphire. Therefore, thin crystals are generally used for better cooling, which makes the mounting particularly delicate. The gain drops rapidly with temperature, because of increasing nonradiative decay. Stalder *et al.* [88] define a temperature $T_{1/2}$ at which the lifetime drops to half of the radiative decay time measured at low temperature. As shown in Table 6.3, this critical temperature is particularly low for Cr:LiSAF and Cr:LiSCAF (70°C) which, combined with their poor thermal conductivity, makes these crystals unsuitable for high power applications. Cr:LiCAF is preferred to the other two in applications such as regenerative amplifiers, because of its slightly larger saturation energy and better tolerance to a temperature increase.

The Cr^{3+}:$LiSrAlF_6$ is the most popular laser medium for low power, high efficiency operation. It is generally pumped by high brightness AlGaInP laser diodes. The emitting cross section of a typical laser diode is rectangular, with a thickness of only a few micron, and a width equal to that of the diode. A "high brightness" diode is one for which the width does not exceed 200 μm. The shorter the diode stripe, the higher the brightness, and the lower the threshold for laser operation. Pump threshold powers as low as 2 mW have been observed in diode pumped Cr^{3+}:$LiSrAlF_6$ lasers [89]. Mode-locked operation with 75-fs pulses was achieved with only 36 mW of pump power [90].

As can be seen from a comparison of Tables 6.2 and 6.3, the nonlinear index in LiSAF is significantly smaller than in Ti:sapphire. A careful design of the cavity including astigmatism compensation is required to have tighter focusing in the LiSAF crystal, leading to the same Kerr lensing than in a typical Ti:sapphire laser [90]. A pair of BK7 prisms (prism separation 360 mm) was found to be optimal for second- and third-order dispersion compensation, leading to pulses as short as 12 fs (200 MHz repetition rate) for a Cr:LiSAF laser, pumped by two diode lasers of 500 mW and 350 mW output power [95]. The average output power of the fs laser was 6 mW. Diode laser technology is the limiting factor in reaching high output powers. Indeed, 70 mW and 100 mW powers (14-fs pulse duration) are easily obtained by Kr-ion laser pumping of LiSAF and LiGAF, respectively [96]. One solution to alleviate the drawback of a reduced brightness for higher power pump diodes, is to pump with a diode laser master oscillator power amplifier system [97]. An output power of 50 mW was obtained with an absorbed pump power of 370 mW.

With chirped mirrors for dispersion compensation, the Cr:LiSAF laser should lend itself to compact structures at high repetition rate, although most lasers were operated at less than 100 MHz [90–92,96,97]. The 12 fs Cr:LiSAF laser operating with a BK7 prism pair however had the shortest cavity, with a repetition rate of 200 MHz [95].

Because of the small nonlinear index $\bar{n}_2$, it is often more convenient to use a single quantum well to initiate and maintain the mode-locking. Mode-locking with saturable absorber quantum wells was discussed in Section 6.6.

6.7.4. Cr:Forsterite and Cr:Cunyite Lasers

These two lasers use tetravalent chromium Cr^{4+} as a substitute for Si^{4+} in the host Mg_2SiO_4 (forsterite) [99,100] and as a substitute for Ge in the host

Table 6.4

Room temperature physical properties of Cr:Forsterite and Cr:Cunyite lasers.

Property	$Cr:Mg_2SiO_4$	$Cr:Ca_2GeO_4$	Units	Ref.
Nonlinear index	$2\ 10^{-16}$	$1.5\ 10^{-16}$	cm^2/W	[72,105]
Dispersion (k" at 1280 nm)	185		fs^2/cm	[106]
Peak absorption at	670		nm	
Peak gain (1240 nm)	14.4	80	$10^{-20}\ cm^2$	[72]
Fluorescence lifetime				[101]
τ_F	2.7	15	μs	[102,107]
Tuning range from	1167	1350	nm	[72]
to	1345	1500	nm	
Thermal conductivity		0.03	W/cm/K	

Ca_2GeO_4 (cunyite) [101,102]. The properties of these two laser materials are compared in Table 6.4. Forsterite-based lasers have become important because they operate in the 1.3 μm range (1167 to 1345 nm) and can be pumped with Nd:YAG lasers. Attempts have also been made at diode pumping [103]. By careful intracavity dispersion compensation with a pair of SF58 prisms complemented by double-chirped mirrors, a pulse duration of 14 fs was obtained [104]. This laser, pumped by a Nd:YAG laser, had a threshold of 800 mW for cw operation and 4 W for mode-locked operation. 100 mW output power could be achieved with a pump power of 6 W.

The forsterite laser produces pulses short enough to create an octave spanning spectral broadening in fibers as discussed in Section 13.4.1.[4] A prismless compact ring cavity was designed with combination of chirped mirrors (GDD of $-55\ fs^2$ from 1200 to 1415 nm) and Gires–Tournois interferometer mirrors (GDD of $-280\ fs^2$ from 1200 to 1325 nm) as sketched in Figure 6.12. This laser, pumped by a 10 W fiber laser, combined short pulse output (28 fs) with a high repetition rate of 420 MHz [98].

6.7.5. YAG Lasers

The crystal $Y_3Al_5O_{12}$ or YAG is transparent from 300 nm to beyond 4 μm, optically isotropic, with a cubic lattice structure characteristic of garnets. It is one of the preferred laser hosts because of its good optical quality and high thermal conductivity. Some of the physical–optical properties are listed in Table 6.5. The two most important lasers using YAG as a host are Nd:YAG and Yb:YAG.

[4]Germanium doped silica fiber with a small effective area of 14 μm^2 nonlinear coefficient of $8.5\ W^{-1}km^{-1}$, zero dispersion near 1550 nm.

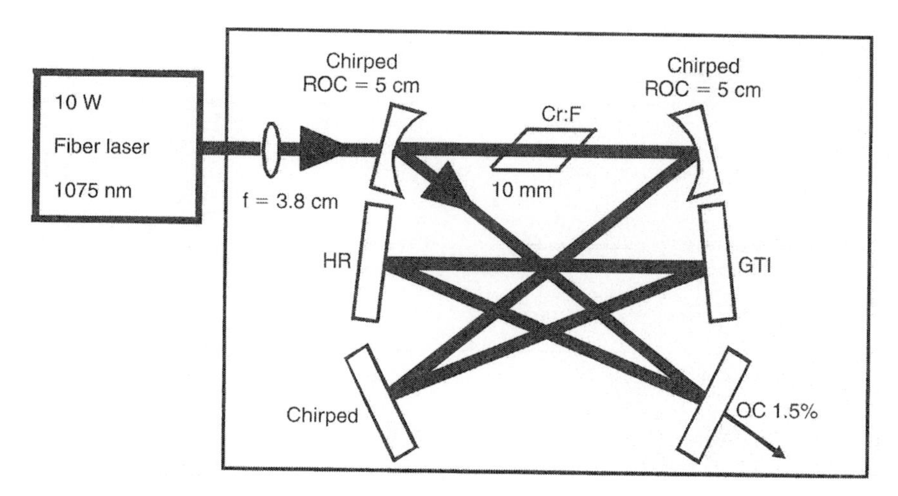

Figure 6.12 Compact ring cavity of a Cr:forsterite laser used in conjunction with HNLF fibers to generate an octave spanning continuum in the near IR. (Adapted from Thomann *et al.* [98].) The mirrors of 5 cm radius of curvature as well as the first folding mirror (HR) have chirped multilayer coatings. The second folding mirror is a Gires–Tournois Interferometer (GTI), the third one a standard high reflector, and the output coupler has a transmission of 1.5%.

Table 6.5

Room temperature physical properties of YAG. The second-order dispersion is calculated from the derivative of the Sellmeier equation: $n^2 = 1 + 2.2779\lambda_\ell^2/(\lambda_\ell^2 - 0.01142)$ with λ_ℓ in μm. The data are compiled from [70,72,104,108–110]

Property	YAG	Units
Index of refraction 1.064 μm	1.8169	
Index of refraction 1.030 μm	1.8173	
Dispersion (k'') at 1.064 μm	733	fs^2/cm
Dispersion (k'') at 1.030 μm	760	fs^2/cm
Nonlinear index	12.4	10^{-16} cm^2/W
Thermal expansion		
Ref. [100]	8.2	10^{-6} K^{-1}
Ref. [110]	7.7	10^{-6} K^{-1}
Ref. [111]	7.8	10^{-6} K^{-1}
Thermal conductivity	0.129	W cm^{-1} K^{-1}
dn/dT	8.9	10^{-6} K^{-1}

Nd:YAG

Typical doping concentrations of the Nd^{3+} ion (substitution of Y^{3+}) range from 0.2 to 1.4% (atomic). Larger doping degrades the optical quality of the crystal. Nd:YAG has been the workhorse industrial laser for several decades, because of its relatively high gain and broad absorption bands that makes it suitable for flashlamp pumping. It has a UV absorption band from 300 to 400 nm and absorption lines between 500 and 600 nm. It has also an absorption band at 808.6 nm which coincides with the emission of GaAlAs diode lasers. Being a four-level laser, Nd:YAG does not require as high a pump power to create an inversion as, for instance, the three level ruby laser or the Yb:YAG laser. The high gain is partly because of the narrow bandwidth of the fluorescence spectrum, limiting pulse durations to >10 ps. Despite this limitation, Nd:YAG has still a place as a source of intense femtosecond pulses. Intracavity pulse compression by passive negative feedback (Section 6.5) yields mJ pulses as short as 8 ps directly from the oscillator [63,64]. Efficient conversion to the femtosecond range has been achieved either by harmonic generation [111] or parametric oscillation [112,113]. The fundamentals of pulse compression associated with harmonic and parametric processes can be found in Sections 3.4.2 and 3.5.

Yb:YAG

Yb:YAG is a popular crystal for high average power, subpicosecond pulse generation. Up to 10 atomic percent of doping of the YAG crystal by Yb have been used. Table 6.6 compares some essential parameters of Nd:YAG and Yb:YAG. The main difference between the two crystals is that Yb:YAG is a quasi-three-level system, requiring large pump powers to reach an inversion. It does not have the broad absorption bands of Nd:YAG that would make it suitable for flashlamp pumping. The main advantage of Yb:YAG however is the small quantum defect,

Table 6.6

Comparison of Nd:YAG and Yb:YAG (data from [72,117]).

Property	Nd:YAG	Yb:YAG	Units
Lasing wavelength	1064.1	1030	nm
Doping density (1% at.)	1.38	1.38	10^{20} atoms/cm^3
Diode pump band	808.6	942	nm
Absorption bandwidth	2.5	18	nm
Emission cross section	28	2.1	10^{-20} cm^2
Emission bandwidth	0.45	≈ 8	nm
Fluorescent lifetime τ_F	230	951	μs

when pumped with InGaAs diode lasers at 942 nm. A small quantum defect implies that a minimum amount of energy is dissipated in the crystal in the form of heat.

The combination of diode pumping (high wall plug efficiency), broad bandwidth and small quantum defect has spurred the development of short pulse, high average power Yb:YAG sources. The main problem to be overcome in developing high average output power sources is the removal of the heat produced by pump intensities of the order of tens of kW/cm^2. Two solutions have been implemented, which led to pulse sources at 1.03 μm, subpicosecond pulse duration, and several tens of watts of average power:

1. A thin disk Yb:YAG laser [114] and
2. Laser rods with undoped endcaps.

The undoped endcaps allow for symmetrical heat extraction on either side of the beam waist. Typical average powers are between 20 and 30 W [71,115]. Quantum wells are generally used for mode-locking, with the exception of a 21 W, 124 MHz repetition laser using a variation of APML [71] (cf. Section 6.3).

In a thin-disk laser, the laser material has a thickness much smaller than the diameter of the pump and laser mode. One end face of the disk is coated for high reflectivity and put in direct contact with a heat sink. The resulting heat flow is longitudinal and nearly one-dimensional. Typical disks are 100 μm thick, for 10% doping with Yb. An average power of 60 W, for 810 fs pulses at a repetition rate of 34 MHz has been obtained [116].

6.7.6. Nd:YVO_4 and Nd:YLF

Both neodymium doped lithium yttrium fluoride (YLF) and vanadate (YVO_4) have gained importance as diode pumped lasers. The emission bandwidth is only slightly larger than that of Nd:YAG, hence the shortest pulse durations that are possible with these lasers are in the range of a few picoseconds (3 ps [118] to 5 ps [119] have been reported). The absorption bandwidth of Nd:vanadate is roughly 18 nm, as opposed to 2.5 nm for Nd:YAG, making it a preferred crystal for diode pumping.

Nd:YLF, like Alexandrite, is a long lifetime medium (twice as long as Nd:YAG), hence an ideal storage medium for regenerative amplifiers. Its natural birefringence overwhelms the thermal induced birefringence, eliminating the depolarization problems of optically isotropic hosts like YAG. For example, a 15 W cw diode array was used to pump a Nd:YLF regenerative amplifier, amplifying at 1 kHz 15 ps, 20 pJ pulses to 0.5 mJ [120].

The main parameters of Nd:YLF and Nd:YVO_4 are summarized in Table 6.7.

Table 6.7

Properties of $Nd:YVO_4$ and Nd:YLF (data from [72,117]).*

Property	$Nd:YVO_4$	Nd:YLF	Units
Lasing wavelength	1064.3	1053 (σ)	nm
		1047 (π)	nm
Index of refraction		1.4481 (n_o)	
		1.4704 (n_e)	
Absorption (1% doping)			
σ	9		cm^{-1}
at	809	806	nm
π	31	4.5	cm^{-1}
at	809	797	nm
Absorption bandwidth	15.7		nm
Emission cross section	15		
σ	21	12	$10^{-20}cm^2$
π	76	18	$10^{-20}cm^2$
Gain bandwidth	0.96	1.3	nm
Fluorescence lifetime τ_F	90	480	μs
Thermal conductivity	0.05	0.06	$W\ cm^{-1}\ K^{-1}$
Thermal expansion in σ	8.5	−2	$10^{-6}\ K^{-1}$
Thermal expansion in π	3	−4.3	$10^{-6}\ K^{-1}$

*Parameters are listed for the radiation polarized parallel (π) and orthogonal (σ) to the optical axis of the crystal

6.8. SEMICONDUCTOR AND DYE LASERS

One of the main advantages of semiconductor and dye lasers is that they can be engineered to cover various regions of the spectrum. As opposed to the solid-state lasers of the previous sections, the semiconductor and dye lasers are characterized by a high gain cross section, which implies also a short upper state lifetime, typically shorter than the cavity round-trip time. Consequently, mode-locking through gain modulation can be effective.

6.8.1. Dye Lasers

Over the past 15 years fs dye lasers have been replaced by solid-state and fiber lasers. It was, however, the dye laser that started the revolution of sub 100-fs laser science and technology. In 1981 Fork *et al.* [121] introduced the colliding pulse mode-locked (CPM) dye laser that produced sub 100-fs pulses.

In this dye laser, the ring configuration allows two counter-propagating trains of pulses to evolve in the cavity [121].[5] The gain medium is an organic dye in solution (for instance, Rh 6G in ethylene glycol), which, pumped through a nozzle, forms a thin ($\approx$100 μm) jet stream. Another flowing dye (for instance, diethyloxadicarbocyanine iodide, or DODCI, in ethylene glycol) acts as saturable absorber. The two counter-propagating pulses meet in the saturable absorber (this is the configuration of minimum losses).

A prism sequence (one, two, or four prisms) allows for the tuning of the resonator GVD. The pulse wavelength is determined by the spectral profiles of the gain and absorber dyes. Limited tuning is achieved by changing the dye concentration. Pulses shorter than 25 fs have been observed at output powers generally not exceeding 10 mW with cw pumping [122], and up to 60 mW with a pulsed (mode-locked argon laser) pump [123].

The palette of available organic dyes made it possible to cover practically all the visible to infrared with tunable and mode-locked sources. A table of gain absorber dye combinations used for passively mode-locked lasers can be found in Diels [124]. Hybrid mode-locking of dye lasers has extended the palette of wavelength hitherto available through passive mode-locking, making it possible to cover a broad spectral range spanning from covering the visible from the UV to the near infrared. A list of dye combinations for hybrid mode-locking is given in Table 6.8. Except when noted, the laser cavity is linear, with the absorber and the gain media at opposite ends. Another frequently used configuration is noted "antiresonant ring." The saturable absorber jet is located near the pulse crossing point of a small auxiliary cavity, in which the main pulse is split into two halves, which are recombined in a standing wave configuration in the absorber [16,125]. The ring laser appears only once in Table 6.8 [126], because of the difficulty of adjusting the cavity length independently of all other parameters.

Dye lasers have been particularly successful in the visible part of the spectrum, where virtually all wavelengths have been covered. The advantage of using an organic dye in a viscous solvent is that the flowing dye jet allows for extremely high pump power densities—in excess of 10 MW/cm^2—to be concentrated on the gain spot. The disadvantage of the dye laser lies also in the inconvenience associated with a circulating liquid system. One alternative for the liquid dye laser that conserves most of its characteristics is the dye doped, polymer nanoparticle gain medium. Significant progress has been made in developing a material with excellent optical quality [127,128]. These laser media have yet to be applied as a femtosecond source.

---

[5]The same ring configuration is sometimes used with a Ti:sapphire gain medium, when a bidirectional mode of operation is sought.

Table 6.8

Femtosecond pulse generation by hybrid mode-locking of dye lasers pumped by an argon ion laser, except as indicated. (Adapted from [124].) (ANR - antiresonant, p - pump laser)

Gain dye	Absorber[a]	λ_ℓ nm	Range	τ_{pmin} fs	at λ_ℓ nm	Remarks
Disodium fluorescein	RhB	535	575	450	545	
Rh 110	RhB	545	585	250	560	
Rh6G	DODCI	574	611	300	603	
Rh6G	DODCI			110	620	Ring laser
Rh6G	DODCI			60	620	ANR ring
Kiton red S	DQOCI			29	615	
Rh B	Oxazine 720	616	658	190	650	
SRh101	DQTCI	652	682	55	675	Doubled
	DCCI	652	694	240	650	Nd:YAG p
Pyridine 1[b]	DDI			103	695	
Rhodamine 700	DOTCI	710	718	470	713	
Pyridine 2	DDI, DOTCI			263	733	
Rhodamine 700	HITCI	770	781	550	776	
LDS-751	HITCI	790	810	100		
Styryl 8	HITCI		70	800		
Styryl 9[b]	IR 140[c]	840	880	65	865	Ring laser
Styryl 14	DaQTeC			228	974	

[a]See Appendix D for abbreviations.
[b]Solvent: propylene carbonate and ethylene glycol.
[c]In benzylalcohol.

Miniature Dye Lasers

The long (compared to the geometrical length of a fs pulse) cavity of most mode-locked lasers serves an essential purpose when a sequence of pulses—rather than a single pulse—is needed. Emission of a short pulse by the long resonator laser requires—as we have seen at the beginning of the previous chapter—a coherent superposition of the oscillating cavity modes with fixed phase relation. If, however, only a single pulse is needed, there is no need for more than one longitudinal mode within the gain profile. Ultrashort pulses are generated in small cavity lasers through resonator Q-switching and/or gain switching. Aside from gain bandwidth limitations, the pulse duration is set by the spectral width of the longitudinal mode, and hence the resonator lifetime. The latter in turn is limited by the resonator round-trip time $2L/c$. Ideally, the laser cavity should have a free spectral range c/2L exceeding the gain bandwidth.

Two methods of short pulse generation that use either ultrashort cavities (Fabry–Perot dye cells of thickness in the micron range) or no traditional cavity at all (distributed feedback lasers) have successfully been developed for (but are not limited to) dye lasers.

In distributed feedback lasers two pump beams create a spatially modulated excitation that acts as a Bragg grating. This grating serves as the feedback (resonator) of the laser and is destroyed during the pulse evolution. This short cavity lifetime together with the small spatial extend of the gain volume can produce subps pulses whose frequency can be tuned by varying the grating period [129,130]. The latter is determined by the overlap angle of the two pump beams.

In a typical "short cavity" laser, the wavelength is tuned by adjusting the thickness of the dye cell in a 3 to 5 μm range with a transducer bending slightly the back mirror of the cavity [131]. With a round-trip time of the order of only 10 fs, it is obvious that the pulse duration will not be longer than that of a ps pump pulse. As with the distributed feedback laser, the dynamics of pump depletion can result in pulses considerably shorter than the pump pulses. The basic operational principles of this laser can be found in Kurz *et al.* [132]. Technical details are given in Chin *et al.* [131]. For example, using an excimer laser, Szatmari and Schaefer [130] produced 500 fs pulses, tunable from 400 to 760 nm, in a cascade of distributed feedback and short cavity dye lasers. After SPM and recompression, pulses as short as 30 fs in a spectral range from 425 to 650 nm were obtained [133].

Another type of miniature laser is the integrated circuit semiconductor laser, which will be described in the next section.

6.8.2. Semiconductor Lasers

Generalities

Semiconductor lasers are obvious candidates for fs pulse generation, because of their large bandwidth. A lower limit estimate for the bandwidth of a diode laser is $k_B T$ (where k_B is the Boltzmann constant and T the temperature), which at room temperature is (1/40) eV, corresponding to a 15-nm bandwidth at 850 nm, or a minimum pulse duration of 50 fs. The main advantage of semiconductor lasers is that they can be directly electrically pumped. In the conventional diode laser, the gain medium is a narrow inverted region of a *p–n* junction. We refer to a publication of Vasil'ev [134] for a detailed tutorial review on short pulse generation with diode lasers. We will mainly concentrate here on problems associated with fs pulse generation in external and internal cavity (integrated) semiconductor lasers. The main technical challenges associated with laser diodes result from the small cross section of the active region (typically 1 μm by tens of μm), the large index of refraction of the material ($2.5 < n < 3.5$, typically) and the large nonlinearities of semiconductors.

The cleaved facets of a laser diode form a Fabry–Perot resonator with a mode spacing of the order of 1.5 THz. Two options are thus conceivable for the development of fs lasers: integrate the diode with a waveguide in the semiconductor, to construct fs lasers of THz repetition rates, or attempt to "neutralize" the Fabry–Perot effect of the chip, and couple the gain medium to an external cavity. We will consider first the latter approach.

External Cavity

Because of the high refractive index of the semiconductor, it is difficult to eliminate the Fabry–Perot resonances of the short resonator made by the cleaved facets of the crystal. Antireflection coatings have to be of exceptionally high quality. Even though reflectivities as low as 10^{-4} can be achieved, a good quality antireflection coating with a high optical damage threshold remains a technical challenge. A solution to this problem is the angled stripe semiconductor laser [135], which has the gain channel making an angle of typically 5° with the normal to the facets (Figure 6.13). Because of that angle, the Fabry–Perot resonance of the crystal can easily be decoupled from that of the external cavity. A standard antireflection coating applied to the semiconductor chip is sufficient to operate the laser with an external cavity.

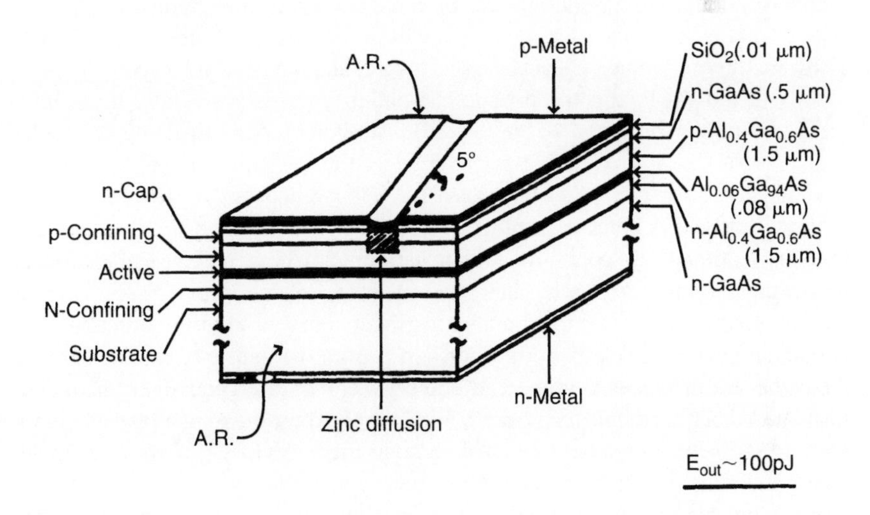

Figure 6.13 Structure of an angled stripe semiconductor laser. (Adapted from [135].)

Femtosecond pulse operation in a semiconductor laser with an external cavity is similar to that of a dye laser. The laser can be cw pumped, as in Delfyett *et al.* [136]. Best results so far were obtained in hybrid operation, using radio frequency current modulation for gain modulation (synchronous pumping), and a saturable absorber. The low intracavity power of the external cavity semiconductor laser—as compared to the dye laser—makes the use of conventional saturable absorbers (i.e., dyes, bulk semiconductors) impractical. It has been necessary to develop absorbing structures with a low saturation energy density. These are the MQW absorbers, which were analyzed in Section 6.6. The laser diode is modulated at the cavity round-trip frequency (0.5 W RF power applied via a bias tee [68]). Modulation of the index of refraction is associated with the gain depletion and the saturation of the MQW. Because the gain depletion results in an increase of the index, a negative dispersion line appears appropriate. Bandwidth-limited operation is difficult to achieve directly from a mode-locked semiconductor laser. An external dispersion line with gratings resulted in pulse durations of 200 fs [137].

The exact phase modulation mechanism of this laser is complex. The index of refraction of the diode is a function of temperature and free carrier density, which itself is a function of current, bias, light intensity, etc. As with other high gain solid-state lasers, changes in the pulse parameters can be as large as 50% from one element to the next [68].

Current Modulation To take full advantage of the fast lifetime of the gain in a semiconductor laser, one should have a circuit that drives ultrashort current pulses into the diode. As mentioned above, a feedback technique—generally referred to as regenerative feedback—can be used to produce a sine wave driving current exactly at the cavity repetition rate. The circuit consists essentially in a phase locked loop, synchronized by the signal of a photodiode monitoring the mode beat note of the laser, and a passive filter at the cavity round-trip frequency. A comb generator can be used to transform the sine wave in a train of short electrical pulses. A comb generator is a passive device which produces, in the frequency domain, a "comb" of higher harmonics which are integral multiples of the input frequency. As we had seen in the introduction of Chapter 5, to a regular frequency comb corresponds a periodic signal in the time domain. This periodic signal can correspond to ultrashort pulses, if—and only if—the teeth of the comb are in phase. Commercial comb generators are generally constructed to create higher harmonics, without being optimized for creating a phased comb. Therefore a selection should be made among these devices to find a generator with good temporal properties (shortest pulse generation).

To allow for the injection of a short current pulse into the laser diode, the latter should be designed with minimal capacitance. To this effect, the p and n

contacts of the angle striped diode of Fig. 6.13 should not cover the whole area of the strip, but be limited to a narrow stripe which follows the gain line.

Integrated Devices

Instead of trying to couple the semiconductor chip to a standard laser cavity, one can integrate the semiconductor into a waveguide cavity. Such devices ranging in length from 0.25 mm to 2 mm have been constructed and demonstrated for example by Chen and Wang [138]. The end mirrors of the cavity are—as in a conventional diode laser—the cleaved faces of the crystal (InP) used as substrate. Wave guiding is provided by graded index confining InGaAsP layers. Gain and saturable absorber media consist of MQWs of InGaAs. The amount of gain and saturable absorption is controlled by the current flowing through these parts of the device (reverse bias for the absorber). As shown in the sketch of Figure 6.14, the saturable absorber is located at the center of symmetry of the device, sandwiched between two gain regions. This configuration is analogous to that of the ring dye laser, in which the two counter-propagating pulses meet coherently in the absorber jet. In the case of this symmetric linear cavity, the laser operation of minimum losses will correspond to two circulating pulses overlapping as standing waves in the saturable absorber.

These devices are pumped continuously and are thus the solid-state equivalent of the passively mode-locked dye lasers. The laser parameters can, however, be significantly different. Although the average output power is only slightly inferior to that of a dye laser (1 mW), at the much higher repetition rate (up to 350 GHz), the pulse energy is only in the fW range! For these ultrashort cavity lengths, there are only a handful of modes sustained by the gain bandwidth.

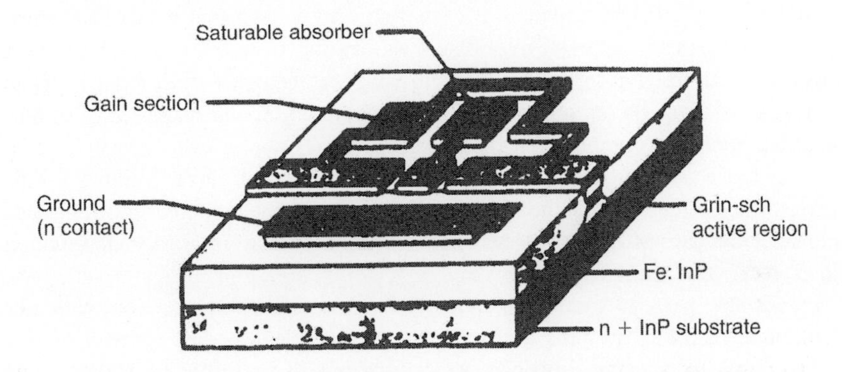

Figure 6.14 Layout of an integrated semiconductor fs laser. (Adapted from [138].)

Semi-Integrated Circuit Fs Lasers

Total integration as shown above results in a high duty cycle, at the expense of a lower energy per pulse. On can seek a compromise between the discrete elements semiconductor laser and the total integrated laser. For instance, the integration of the gain and saturable absorber of the integrated laser of Chen and Wang [138] can be maintained in a single element coupled to an external cavity. Such a design has been successfully tested by Lin and Tang [139]. The absorber consists of a 10 μm island in middle of the gain region, with an electrical contact, isolated from the gain structure by two 10-μm shallow etched regions (without any electrical contact). The absorption—as in the case of the totally integrated laser—can be controlled through the bias potential of the central contact. To prevent lasing action of the 330-μm long gain module, the end facets—after cleavage—are etched (chemically assisted ion beam etching) at 10° from the cleaved plane. The autocorrelation of the laser pulses from such a structure had a width of approximately 700 fs [139].

6.9. FIBER LASERS

6.9.1. Introduction

In most lasers discussed so far, the radiation is a free propagating wave in the gain or other elements of the cavity. The gain length is limited by the volume that can be pumped. The length of a nonlinear interaction is also limited by the Rayleigh range (ρ_0). By confining the wave in a wave guide, it is possible to have arbitrarily long gain media and nonlinear effects over arbitrarily long distances. A fiber is an ideal wave guide for this purpose. Its losses can be as small as a few dB/km. Yet the pulse confinement is such that substantial phase modulation can be achieved over distances ranging from cm to m. The fiber is particularly attractive in the wavelength range of negative dispersion (beyond 1.3 μm), because the combination of phase modulation and dispersion can lead to pulse (soliton) compression (see Chapter 8). The gain can be provided by Stimulated Raman Scattering (SRS) in the fiber material. Such "Raman soliton lasers" are reviewed in the next subsection. In the following subsection, we will consider the case of doped fibers, where the gain medium is of the same type as in conventional glass lasers.

Over the past 20 years ultrafast fiber lasers have matured dramatically. Compact, turn key systems are available commercially today and can deliver tens of mW of average power at pulse durations of the order of 100 fs. With amplification the micro Joule level is accessible. These lasers have applications as self-standing units or as compact seed sources for high-power fs amplifier systems.

6.9.2. Raman Soliton Fiber Lasers

SRS is associated with intense pulse propagation in optical fibers. A review of this topic can be found in Rudolph and Wilhelmi [140] for example. The broad Raman gain profile for the Stokes pulse extends up to the frequency of the pump pulse. An overlap region exists because of the broad pump pulse spectrum. The lower frequency components of the pulse can experience gain at the expense of attenuation of the higher frequency components. In addition, the amplification of spontaneously scattered light is possible. Either process leads to the formation of a Stokes pulse which separates from the pump pulse after the walk-off distance because of GVD. These processes can be utilized for femtosecond Raman soliton generation in fibers and fiber lasers [141–143]. An implementation of this idea is shown in Figure 6.15. The pulses from a cw mode-locked Nd:YAG laser

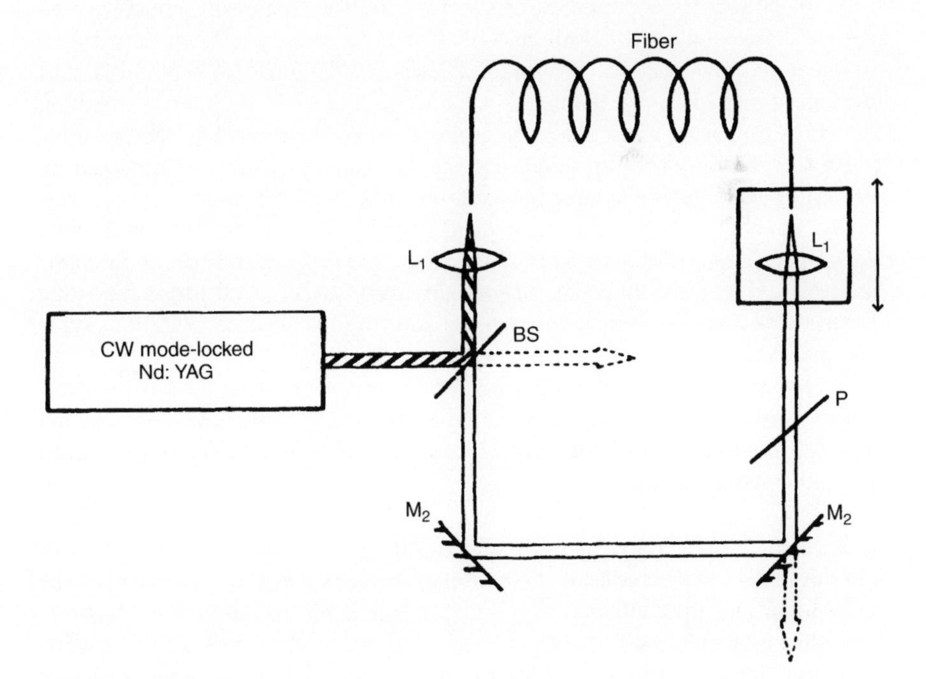

Figure 6.15 Experimental configuration of a synchronously pumped fiber ring Raman laser. (Adapted from Gouveia-Neto [143].)

(100 MHz, 100 ps, 1.32 μm) are coupled through a beam splitter BS into a ring laser containing an optical fiber. The fiber was tailored to have a negative dispersion for $\lambda_\ell > 1.46$ μm. While traveling through the fiber the pump pulses at 1.319 μm produce Stokes pulses at $\lambda_1 = 1.41$ μm. This first Stokes pulse in turn can act as pump source for the generation of a second Stokes pulse ($\lambda_2 = 1.495$ μm), which is in the dispersion region that enables soliton formation. Of course, for efficient synchronous pumping, the length of the ring laser had to be matched to the repetition rate of the pump. Second Stokes pulses as short as 200 fs were obtained.

6.9.3. Doped Fiber Lasers

Fibers can be doped with any of the rare earth ions used for glass lasers. Whether pumped through the fiber end, or transversely, these amplifying media can have an exceptionally large optical thickness ($a_g = \alpha_g d_g \gg 1$). An initial demonstration of this device was made by Duling [144,145]. Passively mode locked rare earth doped fiber lasers have since evolved into compact, convenient, and reliable sources of pulses shorter than 100 fs. The gain media generally used are Nd^{3+} operating at 1050 nm and Er^{3+} at 1550 nm. The erbium doped fiber is sometimes codoped with ytterbium, because of the broad absorption band of the latter centered at $\approx$ 980 nm and extending well beyond 1000 nm. Pump light at 1.06 μm can be absorbed by ytterbium, which then transfers the absorbed energy to the Er ions. High gain and signal powers can thus be obtained by using, for example, diode laser pumped miniature Nd:YAG lasers.

Because of the high gain in a typical fiber laser, it may include bulk optic components, e.g., mirror cavities, dispersion compensating prisms, or saturable absorbers. Obviously, the preferred configuration is that of an all-fiber laser, using a variety of pigtailed optical components and fused tapered couplers for output and pumping.

As compared to conventional solid-state lasers, fibers have the advantage of a large surface to volume ratio (hence efficient cooling is possible). The specific advantages of the single mode fiber geometry over bulk solid-state (rare earth) media for mode-locking are:

- Efficient conversion of the pump to the signal wavelength. Erbium, for example, is a three-level system and the tight mode confinement of the pump in a fiber allows for efficient depopulation of the ground state and thus high efficiency.
- Nonradiative ion–ion transitions that deplete the upper laser level are minimized. Such interactions are especially egregious in silica because of its high phonon energy, and because the trivalent dopants do not mix

well into the tetravalent silica matrix, tending instead to form strongly interacting clusters at the high concentrations necessary for practical bulk glass lasers [146]. The confinement of both the laser and pump modes allows the gain dopant to be distributed along greater lengths of fiber at lower concentration, obviating the need for high concentrations and so eliminating the interactions cited previously.
- Diode laser pumping is practicable (due in large part to the previous two points). Single mode laser diodes have been developed at 980 nm and 1480 nm for erbium fiber amplifiers in telecommunications applications. The four-level structure of neodymium allows for pumping even by multimode lasers, such as high-power laser diode arrays, by using fibers designed to guide the pump light in the cladding [147].
- Tight mode confinement and long propagation lengths maximize the SPM by the weak nonlinear index of silica ($\bar{n}_2 = 3\ 10^{-16}$ cm^2/W).
- The dispersion k" of fibers (including the sign) can be tailored to the application.

One drawback of the fiber laser is that the confinement limits the pulse energies that can be produced. In bulk-solid state lasers, the problem of material damage can be overcome by beam expansion.

A number of techniques have been developed to mode-lock fiber lasers. The most successful methods are:

1. nonlinear polarization rotation [148],
2. nonlinear loop mirrors [149],
3. mode-locking with semiconductor saturable absorbers [150].

Femtosecond pulse output with durations of 100 fs and below has been observed with a variety of gain media—Nd, Yb, Er, Er/Yb, Pr, and Tm. For a detailed overview on such lasers we refer the reader to a review paper by Fermann *et al.* [151].

6.9.4. Mode-Locking through Polarization Rotation

Because of its central importance in today's fs fiber lasers we will describe one of the mode-locking techniques—nonlinear polarization rotation—in more detail. As explained in Section 5.4.2 nonlinear polarization rotation in combination with polarizers can act as a fast saturable absorber, cf. Eq. (5.81). In a fiber laser using nonlinear polarization rotation, the differential accumulated phase yields an intensity-dependent state of polarization across the pulse. This polarization state is then converted into an intensity-dependent transmission by inserting a polarizer at the output of the birefringent element, oriented, for example, to transmit the high intensity central portion of the pulse and reject the wings. This approach is the fiber equivalent of the Kerr lens mode-locked Ti:sapphire laser. Pulses as short as 36 fs have been obtained from an Yb fiber laser that used nonlinear polarization rotation [152], to name just one example.

A standard single mode fiber serves as nonlinear element. Such a fiber has generally a weak birefringence. The degree of birefringence is defined by the parameter:

$$B = \frac{|k_x - k_y|}{2\pi/\lambda_\ell} = |n_x - n_y|, \tag{6.9}$$

where n_x and n_y are the effective refractive indices in the two orthogonal polarization states. For a given value of B, the power between the two modes (field components along $\hat{x}$ and $\hat{y}$) is exchanged periodically, with a period L_B called the "beat length" given by [153]:

$$L_B = \frac{\lambda_\ell}{B}. \tag{6.10}$$

The axis with the larger mode index is called the slow axis. In a typical single mode fiber, the beat length is around 2 to 10 m at 1.55 μm [154]. As shown by Winful [155], nonlinear polarization effects can be observed at reasonably low power in weakly birefringent fibers (as opposed to polarization preserving fibers).

In a typical fiber ring cavity a first polarization controller produces an elliptical polarization, whose major axis makes a small angle θ with the slow axis of the portion of fiber that follows. As shown in Section 5.4.2 the induced phase difference between two orthogonal polarization components depends on the propagation distance d and the pulse intensity. It can be adjusted such that after a distance d_m the polarization becomes linear. A polarizer can be used to maximize the loss for the lower intensities as compared to the higher intensities, as sketched in Figure 6.16.

We have derived in Section 5.4.2 the essential equations relating to nonlinear polarization rotation. To describe a fiber laser we need to track the evolution of two polarization components. This can conveniently be done using a column vector for the electric field at a certain point in the cavity

$$\begin{pmatrix} \tilde{\mathcal{E}}_x \\ \tilde{\mathcal{E}}_y \end{pmatrix}, \tag{6.11}$$

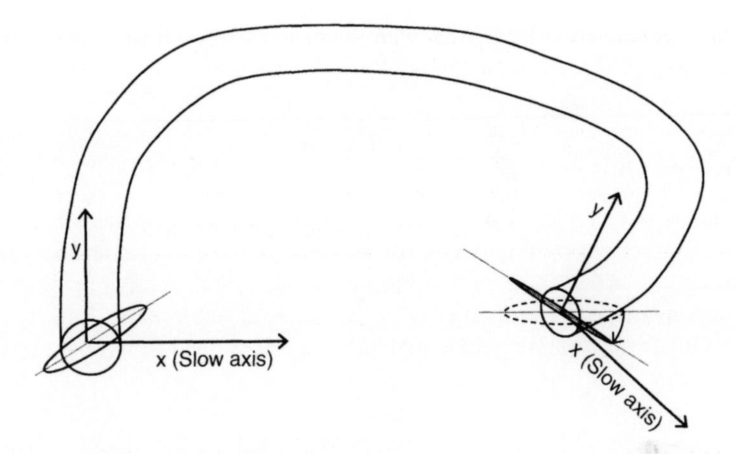

Figure 6.16 Sketch of the nonlinear polarization rotation in a fiber. The elliptically polarized input can be converted into linearly polarized light at the peak of the pulse for example.

and 2 × 2 matrices ($\mathcal{M}$) for the resonator elements [156,157]. The effect of the nonlinear birefringent fiber of length L is the combination of a linear propagation problem and nonlinear phase modulation. The resulting matrix is thus a product of two matrices, and the field vector is given by:

$$\begin{aligned}\begin{pmatrix}\tilde{\mathcal{E}}_x(L)\\ \tilde{\mathcal{E}}_y(L)\end{pmatrix} &= \begin{pmatrix}e^{-i\Phi_{NL,x}} & 0\\ 0 & e^{-i\Phi_{NL,y}}\end{pmatrix}\cdot\begin{pmatrix}e^{-ik_xL} & 0\\ 0 & e^{-ik_yL}\end{pmatrix}\cdot\begin{pmatrix}\tilde{\mathcal{E}}_x(0)\\ \tilde{\mathcal{E}}_y(0)\end{pmatrix}\\ &= \begin{pmatrix}e^{-i\Phi_x} & 0\\ 0 & e^{-i\Phi_y}\end{pmatrix}\cdot\begin{pmatrix}\tilde{\mathcal{E}}_x(0)\\ \tilde{\mathcal{E}}_y(0)\end{pmatrix},\end{aligned} \tag{6.12}$$

where

$$\Phi_{x,y} = \frac{2\pi n_2 L}{\lambda_\ell}\left[|\tilde{\mathcal{E}}_{x,y}|^2 + \frac{2}{3}|\tilde{\mathcal{E}}_{y,x}|^2\right] - \frac{2\pi n_{x,y} L}{\lambda_\ell}.$$

We have used here the same approximations for the nonlinear phase as in Section 5.4.2. The linear propagation constants $k_{x,y} = \omega_\ell n_{x,y}/c$. Matrices of common polarizing elements like wave plates and polarizers known from Jones calculus can easily be incorporated into this analysis.

Other components of the round-trip model like gain, saturable absorption, mirrors, etc., usually do not distinguish between the two polarization components. The transfer functions $\mathcal{T}$ are those introduced in Chapter 5. For implementing these elements in a way consistent with the matrix approach we define a transfer matrix

$$(\mathcal{M}) = \mathcal{T}\begin{pmatrix} 1 & 0 \\ 0 & 1 \end{pmatrix}. \tag{6.13}$$

Fiber lasers have typically high gain and losses. The laser operates in a regime of strong saturation, with pulses much shorter than the energy relaxation time of the lasing transition. The gain transition is generally sufficiently broad for phase modulation because of saturation to be negligible. Therefore the $\mathcal{T}$ factor in the transfer matrix describing gain is real and can be obtained from Eq. (3.55):

$$\mathcal{T}_g = \left[\frac{e^{W_0(t)/W_s}}{e^{-a} - 1 + e^{W_0(t)/W_s}}\right]^{1/2}. \tag{6.14}$$

An alternative approach is to consider the fiber laser as a continuous medium, which leads to a coupled system of differential equations for the components $\tilde{\mathcal{E}}_x$ and $\tilde{\mathcal{E}}_y$. This is essentially a two-field component extension of Eq. (3.190) without the transverse differential operators. We refer to the literature for a derivation of this system of equations and for their application to the modeling of a mode-locked fiber ring laser using nonlinear polarization rotation, Chang and Chi [157], Chi *et al.* [158], Agrawal [159], and Spaulding *et al.* [160].

6.9.5. Figure-Eight Laser

A widely studied fiber laser implementation of the nonlinear mirror is the figure-eight laser [144], so named for the schematic layout of its component fibers (Figure 6.17), with a nonlinear amplifying loop mirror [161]. In the example shown in Fig. 6.17, the laser consists of a nonlinear amplifying mirror (left loop) and an optical isolator with outcoupler (right loop). The two loops of the "figure-eight" are connected by a 50% beam splitter.

Let us follow a pulse that propagates counter clockwise in the right loop through the isolator (optical diode), through a polarization controller (to compensate for the natural birefringence of the fiber) and a 20% output coupler. The remaining part of the circulating pulse is equally split into the two directions of the left loop (nonlinear mirror). The counter-propagating pulses experience the same gain in the Er-doped fiber section of about 2 to 3 dB. The switching fiber introduces a phase shift through SPM. Being amplified before entering this fiber section, the counterclockwise circulating pulse experiences a larger phase shift than its replica propagating in the opposite direction. The two pulses arrive simultaneously at the beam splitter and recombine. The variation of the

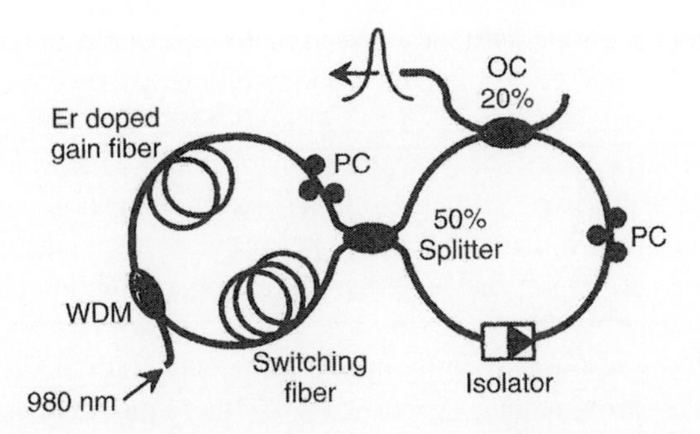

Figure 6.17 Schematic representation of the figure eight laser. The pump radiation at 980 nm is injected via the directional coupler WDM in the gain fiber (erbium doped). (Adapted from Duling [144].)

accumulated differential phase across the combined pulse will cause different parts of the pulse injected clockwise and counterclockwise into the left loop. From the point of view of the counterclockwise circulating pulse in the right loop, the left loop acts as a nonlinear mirror whose reflection varies sinusoidally as a function of intensity. Thus, the loop mirror behaves as a fast saturable absorber from low intensity to intensities corresponding to the first transmission maximum.

Fiber lasers operating on the 1050 nm transition of Nd^{3+} in silica require bulk optic elements (prism sequences) for compensating the substantial normal dispersion (30 ps/nm · km) of the gain fiber at the operating wavelength, and so are generally constructed as a bulk optic external cavity around the gain fiber. Passive mode-locking is obtained via nonlinear polarization rotation in the gain fiber, and the Brewster angled prisms serve as the polarizer. Pulses as short as 100 fs have been demonstrated [162].

Femtosecond fiber lasers operating in the 1530–1570 nm gain band of erbium are of obvious interest for their potential application in telecommunications. This wavelength range is in the low loss window of silica fibers, and such a source is obviously compatible with erbium fiber amplifiers. Of particular interest also is the anomalous dispersion exhibited by silica at this wavelength. The precise value of the dispersion may be tailored through the exact fiber design. This implies that a mode-locked laser with Er gain may be constructed entirely from fibers, with no need for dispersion compensating prisms as in the Nd fiber lasers or most other ultrafast sources. Indeed, subpicosecond erbium lasers have been demonstrated with all-fiber figure-eight, linear, and ring cavities, using both nonlinear mirrors and nonlinear polarization rotation. In addition, systems using semiconductor saturable absorbers have been demonstrated.

While soliton-like models have been used to describe a number of ultrafast laser systems as discussed in Chapter 5, the nonlinear dynamics of soliton propagation plays a more direct role in the erbium fiber laser than is seen in any other. The generated pulses are typically transform limited with a sech^2 intensity profile—the shape expected from the soliton solution of the nonlinear Schrödinger equation. The average intracavity energy per pulse corresponds reasonably well to the energy of a soliton of the same length propagating in fiber with dispersion equal to the average cavity dispersion.

It has been demonstrated that the minimum pulse length obtainable in erbium fiber lasers is approximately proportional to the total dispersion inside the cavity [163]. This is to some degree surprising: As the pulse propagation is soliton-like, the fiber dispersion is continuously balanced by the SPM of the fiber. In principle, solitons of any length will form as long as the amplitude of the input pulse exceeds the threshold value of Eq. (8.35) (cf. soliton description in Chapter 8). However, the coupling of energy into the dispersive wave increases exponentially as the pulse shortens, thus limiting the minimum obtainable pulse width [164]. This loss becomes important only when the cavity length is of the order of the characteristic soliton length defined in Eq. (8.37) . This is also why dispersive wave dynamics do not play an important role in other mode-locked lasers, that can be described by a soliton model. In such systems the soliton length corresponds to many cavity round-trips, much longer than the cavity lifetime of the dispersive wave. To obtain short pulses, then, it is necessary to minimize the total cavity dispersion, either by using dispersion shifted fiber components, or short cavities, or by including lengths of dispersion compensating fiber specially designed to have normal dispersion at 1550 nm. Pulse widths of less than 100 fs [165] have been achieved. With such short pulses, third-order dispersion plays an important role in limiting the pulse width and may impose a nonlinear chirp on the pulse [166].

BIBLIOGRAPHY

[1] R. Ell, U. Morgner, F. X. Kärtner, J. G. Fujimoto, E. P. Ippen, V. Scheuer, G. Angelow, T. Tschudi, M. J. Lederer, A. Boiko, and B. Luther-Davis. Generation of 5-fs pulses and octave-spanning spectra directly from a Ti:sapphire laser. *Optics Letters*, 26:373–375, 2001.

[2] D. E. Spence, J. M. Evans, W. E. Sleat, and W. Sibbett. Regeneratively initiated self-mode-locked Ti:sapphire laser. *Optics Letters*, 16:1762–1764, 1991.

[3] E. W. Van Stryland. The effect of pulse to pulse variation on ultrashort pulsewidth measurements. *Optics Communications*, 31:93–94, 1979.

[4] J.-C. Diels, J. J. Fontaine, I. C. McMichael, and F. Simoni. Control and measurement of ultrashort pulse shapes (in amplitude and phase) with femtosecond accuracy. *Applied Optics*, 24:1270–1282, 1985.

[5] G. H. C. New and J. M. Catherall. Advances in the theory of mode-locking by synchronous pumping. In G. R. Fleming and A. E. Siegman, Eds. *Ultrafast Phenomena V*, Springer-Verlag, Berlin, Germany, 1986, pp. 24–26.

[6] U. Stamm. Numerical analysis of pulse generation in synchronously mode locked cw dye lasers. *Applied Physics B*, B45:101–108, 1988.

[7] V. A. Nekhaenko, S. M. Pershin, and A. A. Podshivalov. Synchronously pumped tunable picosecond lasers (review). *Soviet Journal of Quantum Electronics*, 16:299–315, 1986.

[8] N. J. Frigo, H. Mahr, and T. Daly. A study of forced mode locked cw dye lasers. *IEEE Journal of Quantum Electronics*, QE-17:1134, 1977.

[9] A. M. Johnson and W. M. Simpson. Tunable femtosecond dye laser synchronously pumped by the compressed second harmonic of Nd:YAG. *Journal of the Optical Society of America B*, B4:619–625, 1985.

[10] G. Angel, R. Gagel, and A. Laubereau. Generation of femtosecond pulses by a pulsed laser system. *Optics Communications*, 63:259–263, 1987.

[11] L. Turi and F. Krausz. Amplitude modulation mode locking of lasers with regenerative feedback. *Applied Physics Letters*, 58:810, 1991.

[12] J. M. Catherall and G. H. C. New. Role of spontaneous emission in the dynamics of mode-locking by synchronous pumping. *IEEE Journal of Quantum Electronics*, QE-22:1593, 1986.

[13] D. S. Peter, P. Beaud, W. Hodel, and H. P. Weber. Passive stabilazation of a synchronously pumped mode locked dye laser with the use of a modified outcoupling mirror. *Optics Letters*, 16:405, 1991.

[14] J. P. Ryan, L. S. Goldberg, and D. J. Bradley. Comparison of synchronous pumping and passive mode-locking cw dye lasers for the generation of picosecond and subpicosecond light pulses. *Optics Communications*, 27:127–132, 1978.

[15] B. Couillaud, V. Fossati-Bellani, and G. Mitchel. Ultrashort pulse spectroscopy and applications. In *SPIE Proceedings*, volume 533, Springer-Verlag, Berlin, Germany, 1985, p. 46.

[16] N. Jamasbi, J.-C. Diels, and L. Sarger. Study of a linear femtosecond laser in passive and hybrid operation. *Journal of Modern Optics*, 35:1891–1906, 1988.

[17] V. Petrov, W. Rudolph, U. Stamm, and B. Wilhelmi. Limits of ultrashort pulse generation in cw mode-locked dye lasers. *Physical Review A*, 40:1474–1483, 1989.

[18] L. C. Foster, M. D. Ewy, and C. B. Crumly. Laser mode locking by an external Doppler cell. *Applied Physics Letters*, 6:6–8, 1965.

[19] R. W. Dunn, A. T. Hendow, J. G. Small, and E. Stijns. Gas laser mode-locking using an external acoustooptic modulator with a potential application to passive ring gyroscopes. *Applied Optics*, 21:3984–3986, 1982.

[20] K. J. Blow and D. Wood. Mode-locked lasers with nonlinear external cavities. *Journal of the Optical Society of America B*, 5:629–632, 1988.

[21] K. J. Blow and B. P. Nelson. Improved mode locking of an f-center laser with a nonlinear nonsoliton external cavity. *Optics Letters*, 13:1026–1028, 1988.

[22] P. N. Kean, X. Zhu, D. W. Crust, R. S. Grant, N. Langford, and W. Sibbett. Enhanced mode locking of color-center lasers by coupled-cavity feedback control. *Optics Letters*, 14:39, 1989.

[23] J. Mark, L. Y. Liu, K. L. Hall, H. A. Haus, and E. P. Ippen. Femtosecond laser pumped by a frequency-doubled diode-pumped Nd:YLF laser. *Optics Letters*, 14:48–50, 1989.

[24] C. P. Yakymyshyn, J. F. Pinto, and C. R. Pollock. Additive-pulse mode-locked NaCl:OH- laser. *Optics Letters*, 14:621–623, 1989.

[25] P. M. French, J. A. R. Williams, and R. Taylor. Femtosecond pulse generation from a titanium-doped sapphire laser using nonlinear external cavity feedback. *Optics Letters*, 14:686, 1989.

[26] J. Goodberlet, J. Wang, J. G. Fujimoto, and P. A. Schulz. Femtosecond passively mode-locked Ti:sapphire laser with a nonlinear external cavity. *Optics Letters*, 14:1125–1127, 1989.

[27] J. Goodberlet, J. Jacobson, J. G. Fujimoto, P. A. Schulz, and T. Y. Fan. Self-starting additive-pulse mode-locked diode-pumped Nd:YAG laser. *Optics Letters*, 15:504–507, 1990.

[28] L. Y. Liu, J. M. Huxley, E. P. Ippen, and H. A. Haus. Self-starting additive-pulse mode-locking of a Nd:YAG laser. *Optics Letters*, 15:553–555, 1990.

[29] J. M. Liu and J. K. Chee. Passive mode locking of a cw Nd:YLF laser with a nonlinear external coupled cavity. *Optics Letters*, 15:685, 1990.

[30] J. Goodberlet, J. Jackson, J. G. Fujimoto, and P. A. Schultz. Ultrashort pulse generation with additive pulse modelocking in solid state lasers: $Ti : Al_2O_3$, diode pumped Nd:YAG and Nd:YLF. In C. B. Harris, E. P. Ippen, G. A. Mourou, and A. H. Zewail, Eds., *Ultrafast Phenomena VII*, Springer-Verlag, Berlin, Germany, 1990, p. 11.

[31] F. Krausz, C. Spielmann, T. Brabec, E. Wintner, and A. J. Schmidt. Subpicosecond pulse generation from a Nd:glass laser using a nonlinear external cavity. *Optics Letters*, 15:737–739, 1990.

[32] W. Sibbett. Hybrid and passive mode-locking in coupled-cavity lasers. In C. B. Harris, E. P. Ippen, G. A. Mourou, and A. H. Zewail, Eds., *Ultrafast Phenomena VII*, Springer-Verlag, Berlin, Germany, 1990, p. 2–7.

[33] E. P. Ippen, H. A. Haus, and L. Y. Liu. Additive pulse mode locking. *Journal of the Optical Society of America B*, 6:1736, 1989.

[34] L. F. Mollenauer and R. H. Stolen. The soliton laser. *Optics Letters*, 9:13, 1984.

[35] U. Keller, G. W. 'tHooft, W. H. Knox, and J. E. Cuningham. Femtosecond pulses from a continously self-starting passively mode-locked Ti:sapphire laser. *Optics Letters*, 16:1022–1024, 1991.

[36] M. Sheik-Bahae, D. J. Hagan, and E. W. Van Stryland. Dispersion and band-gap scaling of the electronic Kerr effect in solids associated with two-photon absorption. *Physics Review Letters*, 65:96–99, 1990.

[37] M. Sheik-Bahae, D. C. Hutchings, D. J. Hagan, and E. W. Van Stryland. Dispersion of bound electronic nonlinear refraction in solids. *IEEE Journal of Quantum Electronics*, QE-27:1296–1309, 1990.

[38] G. Cerullo, M. B. Danailov, S. De Silvestri, P. Laporta, V. Magni, D. Segala, and S. Taccheo. Nonlinear mirror mode-locking of a cw nd:yag laser. *Applied Physics Letters*, 65:2392–2394, 1994.

[39] M. B. Danailov, G. Cerullo, V. Magni, D. Segala, and S. De Silvestri. Nonlinear mirror mode-locking of a cw nd:ylf laser. *Optics Letters*, 19:792–794, 1994.

[40] K. A. Stankov. A new mode-locking technique using a nonlinear mirror. *Optics Communications*, 66:41, 1988.

[41] K. A. Stankov. Pulse shortening by a nonlinear mirror mode locker. *Applied Optics*, 28:942–945, 1989.

[42] K. A. Stankov. Frequency domain analysis of the mode-locking process in a laser with second-harmonic nonlinear mirror. *Optics Letters*, 16:639–641, 1991.

[43] A. Smith. Free software Avaliable at: http://www.sandia.gov/imrl/X1118/xxtal.htm, 2003.

[44] K. A. Stankov. 25 ps pulses from a pulsed Nd:YAG laser mode locked by a frequency doubling bbo crystal. *Applied Physics Letters*, 58:2203–2204, 1991.

[45] V. Couderc, A. Barthelemy, V. Kubecek, and H. Jelinkova. Self-mode locked operation of a cw Nd:YAG laser using second-harmonic nonlinear mirror. In *International Symposium on Advanced Materials for Optics and Optoelectronics*, volume 2777, SPIE, Prague, 1995, pp. 216–218.

[46] A. Agnesi, C. Pennacchio, G. C. Reali, and V. Kubecek. High-power diode-pumped picosecond $Nd^{3+}:YVO_4$ laser. *Optics Letters*, 22:1645–1647, 1997.

[47] A. Agnesi, G. C. Reali, and V. Kubecek. Nonlinear mirror operation of a diode-pumped quasi-cw picosecond Nd:YAG laser. *Applied Physics B.*, 66:283–285, 1998.

[48] A. Agnesi, G. C. Reali, and V. Kubecek. Nonlinear mirror operation of a diode-pumped quasi-cw picosecond Nd:YAG laser. *Journal of the Optical Society of America B 16*, 66:1236–1239, 1999.

[49] V. Kubecek. Nonlinear mirror mode-locking of solid state lasers. volume 5259, SPIE, Bellingham, WA, 2003, pp. 403–410.

[50] X. M. Zhao and D. McGraw. Parametric mode locking. *IEEE Journal of Quantum Electronics*, QE-28:930, 1992.

[51] K. A. Stankov. Negative feedback by using a nonlinear mirror for generation of long train of short pulses. *Applied Physics B*, 52:158–162, 1991.

[52] V. Kubecek, V. Coudercand, B. Bourliaguetand, F. Loradour, and A. Barthelemy. 4-W and 23-ps pulses from a lamp-pumped Nd:YAG laser passively mode-locked by polarization switching in a KTP crystal. *Applied Physics B*, 69:99–101, 1999.

[53] A. V. Babushkin, N. S. Vorob'ev, A. M. Prokhorov, and M. Y. Shchelev. Stable picosecond $YAlO_3$:Nd crystal laser with hybrid mode-locking and passive intracavity feedback utilizing a GaAs crystal. *Soviet Journal of Quantum Electronics*, 19:1310–1311, 1989.

[54] T. T. Dang, A. Stintz, J.-C. Diels, and Y. Zhang. Active solid state short pulse laser gyroscope. Volume available on CD from ION, 1800 Diagonal Road, Suite 480, Alexandria, VA 22314, Albuquerque, New Mexico, 2001. ION 57th Annual Meeting–Session D4.

[55] O. E. Martinez and L. A. Spinelli. Deterministic passive mode locking of solid-state lasers. *Applied Physics Letters*, 39:875–877, 1981.

[56] K. Burneika, R. Grigonis, A. Piskarskas, G. Sinkyavichyus, and V. Sirutkaĭtis. Highly stable subpicosecond neodymium (Nd^{3+}) glass laser with passive mode-locking and negative feedback. *Kvantovaya Elektroniks*, 15:1658–1659, 1988.

[57] P. Heinz, W. Kriegleder, and A. Laubereau. Feedback control of an actively-passively mode-locked Nd:glass laser. *Applied Physics A*, 43:209–212, 1987.

[58] D. H. Auston, S. McAfee, C. V. Shank, E. P. Ippen, and O. Teschke. Picosecond spectroscopy of semiconductors. *Solid State Electronics*, 21:147, 1978.

[59] W. A. Schroeder, T. S. Stark, M. D. Dawson, T. F. Boggess, A. L. Smirl, and G. C. Valley. Picosecond separation and measurement of coexisting photo-refractive, bound-electronic, and free-carrier grating dynamics in GaAs. *Optics Letters*, 16:159–161, 1991.

[60] A. Agnesi, A. Guandalini, and G. Reali. Continuous passive mode-locked and passive stabilized low pump power Yb:YAG laser. In *Europhysics Conference Abstracts*, volume 28C, Lausanne, Switzerland, 2004.

[61] J. Schwartz, W. Weiler, and R. K. Chang. Laser-pulse shaping using inducible absorption with controlled q-switch time behavior. *IEEE Journal of Quantum Electronics*, QE-6:442–450, 1970.

[62] A. Hordvik. Pulse stretching utilizing two-photon-induced light absorption. *IEEE Journal of Quantum Electronics*, QE-6:199–203, 1970.

[63] A. Agnesi, A. Del Corno, J.-C. Diels, P. Di Trapani, M. Fogliani, V. Kubecek, G. C. Reali, C.-Y. Yeh, and X. M. Zhao. Generation of extended pulse trains of minimum duration by passive negative feedback applied to solid state q-switched lasers. *IEEE Journal of Quantum Electronics*, 28:710–719, 1992.

[64] V. Kubecek, A. Dombrovsky, J.-C. Diels, and A. Stintz. Mode-locked Nd:YAG laser with passive negative feedback using multiple quantum well saturable absorber. In *Proceedings of SPIE; GC:/HPL 2004*, SPIE, Prague, Czech Republic, 2004. pp. 81–85.

[65] U. Keller, K. J. Weingarten, F. X. Kartner, D. Kopf, B. Braun, I. D. Jung, R. Fluck, C. Honninger, N. Matuschek, and J. Aus der Au. Semiconductor saturable absorber mirrors (sesams) for femtosecond to nanosecond pulse generation in solid-state lasers. *IEEE Journal of Selected Topics Quantum Electronics*, 2:435–453, 1996.

[66] S. L. Chuang. *Physics of Optoelectronic Devices*. John Wiley & Sons, New York, 1995.

[67] P. W. Smith, Y. Silberberg, and D. Q. B. Miller. Mode-locking of semiconductor diode lasers using saturable excitonic nonlinearities. *Journal of the Optcial Soceity of America*, B-2:1228–1235, 1985.

[68] P. J. Delfyett, L. Flores, N. Stoffel, T. Gmitter, N. Andreadakis, G. Alphonse, and W. Ceislik. 200 femtosecond optical pulse generation and intracavity pulse evolution in a hybrid modelocked semiconductor diode laser-amplifier system. *Optics Letters*, 17:670–672, 1992.

[69] S. Gupta, J. F. Whitaker, and G. A. Mourou. Ultrafast carrier dynamics in III-V semiconductors grown by molecular-beam epitaxy at very low substrate temperatures. *IEEE Journal of Quantum Electronics*, 28:2464–2472, 1992.

[70] R. C. Powell, S. A. Payne, L. L. Chase, and G. D. Wilke. Index-of-refraction change in optically pumped solid-state laser materials. *Optics Letters*, 14:1204, 1989.

[71] M. Weitz, S. Reuter, R. Knappe, R. Wallenstein, and B. Henrich. In *CLEO, CTuCC1*, Optical Society of America, HC, San Francisco, 2004.

[72] W. Koechner and M. Bass. *Solid-State Lasers*. Springer-Verlag, Berlin, Heidelberg, New York, 2002.

[73] S. Smolorz and F. Wise. Time-resolved nonlinear refraction in femtosecond laser gain media. *Optics Letters*, 23:1381–1383, 1998.

[74] P. F. Moulton. Spectroscopic and laser characteristics of Ti:Al_2O_3. *Journal of the Optical Society of America B*, 3:125–133, 1986.

[75] D. E. Spence, P. N. Kean, and W. Sibbett. 60-fs pulse generation from a self-mode-locked Ti:sapphire laser. *Optics Letters*, 16:42–44, 1991.

[76] M. T. Asaki, C. P. Huang, D. Garvey, J. Zhou, H. Kapteyn, and M. M. Murnane. Generation of 11 fs pulses from a self-mode-locked Ti:sapphire laser. *Optics Letters*, 18:977–979, 1993.

[77] A. Stingl, Ch. Spielmann, F. Krausz, and R. Szipöcs. Generation of 11 fs pulses from a Ti:sapphire laser without the use of prisms. *Optics Letters*, 19:204–406, 1994.

[78] J. Reichert, R. Holzwarth, T. Udem, and T. W. Hänsch. Measuring the frequency of light with mode-locked lasers. *Optics Communications*, 172:59–68, 1999.

[79] H. R. Telle, G. Steinmeyer, A. E. Dunlop, S. Stenger, D. A. Sutter, and U. Keller. Carrier-envelope offset phase control: A novel concept for absolute optical frequency measurement and ultrashort pulse generation. *Applied Physics B*, 69:327, 1999.

[80] S. A. Diddams, D. J. Jones, J. Ye, S. T. Cundif, J. L. Hall, J. K. Ranka, R. S. Windeler, R. Holzwarth, T. Udem, and T. W. Hänsch. A direct link between microwave and optical frequencies with a 300 thz femtosecond laser comb. *Physical Review Letters*, 84:5102–5105, 2000.

[81] U. Morgner, R. Ell, G. Metzler, T. R. Schibli, F. X. Kärtner J. C. Fujimoto, H. A. Hauss, and E. P. Ippen. Nonlinear optics with phase-controlled pulses in the sub-two-cycle regime. *Physical Review Letters*, 86:5462–5465, 2001.

[82] F. X. Kärtner, N. Matuschek, T. Schibli, U. Keller, H. A. Haus, C. Heine, R. Morf, V. Scheuer, M. Tilsch, and T. Tschudi. Design and fabrication of double chirped mirrors. *Optics Letters*, 22:831–833, 1997.

[83] M. Ramaswamy, M. Ulman, J. Paye, and J. G. Fujimoto. Cavity dumped femtosecond Kerr-lens modelocked Ti:sapphire laser. *Optics Letters*, 18:1822–1884, 1993.

[84] M. S. Pshenichnikov, W. P. De Boeij, and D. A. Wiersma. Generation of 13 fs, 5-mw pulses from a cavity dumped Ti:sapphire laser. *Optics Letters*, 19:572–574, 1994.

[85] J. Jasapara, V. L. Kalashnikov, D. O. Krimer, G. Poloyko, W. Rudolph, and M. Lenzner. Automodulations in Kerr-lens modelocked solid-state lasers. *Journal of the Optical Society of America B*, 17:319–326, 2000.

[86] S. H. Cho, F. X. Kärtner, U. Morgner, E. P. Ippen, J. G. Fujimoto, J. E. Cunningham, and W. H. Knox. Generation of 90-nj pulses with a 4-mhz repetition rate Kerr-lens modelocked Ti:sapphire oscillator. *Optics Letters*, 26:560–562, 2001.

[87] A. Apolonski, A. Fernandez, T. Fuji, K. Krausz, A. Fürbach, and A. Stingl. Scaleable high-energy femtosecond Ti:sapphire oscillator. In *CLEO, CThA2*, San Francisco, 2004. Optical Society of America.

[88] M. Stalder, M. Bass, and B. H. T. Chai. Thermal quenching of fluorescence in chromium-doped fluoride laser crystals. *Journal of the Optical Society of America B*, 9:2271–2273, 1992.

[89] V. Kubecek, R. Quintero-Torres, and J.-C. Diels. Ultralow-pump-threshold laser diode pumped Cr:LiSAF laser. In V. Ya. Panchenko, G. Huber, Ivan A. Scherbakov, Eds., *Advanced Lasers and Systems*, volume 5137, SPIE, Bellingham, WA, 2003, pp. 43–47.

[90] G. J. Valentine, J. M. Hopkins, P. Loza-Alvarez, G. T. Kennedy, W. Sibbett, D. Burns, and A. Valster. Ultralow-pump-threshold, femtosecond Cr^{3+}:$LiSrAlF_6$. *Optics Letters*, 22:1639–1641, 1997.

[91] R. Szipocs and A. Köházi-Kis. Theory and design of chirped dielectric laser mirrors. *Applied Physics B*, 65:115–135, 1997.

[92] I. T. Sorokina, E. Sorokin, E. Wintner, A. Cassanho, H. P. Jenssen, and R. Szip ocs. Sub-20 fs pulse generation from the mirror dispersion controlled LiSGAF and LiSAF lasers. *Applied Physics B*, 65:245–253, 1997.

[93] L. K. Smith, S. A. Payne, W. L. Kway, L. L. Chase, and B. H. T. Chai. Investigation of the laser properties of Cr^{3+}:$LiSrGaF_6$. *IEEE Journal of Quantum Electronics*, 28:2612–2618, 1992.

[94] D. Kopf, K. J. Weingarten, G. Zhang, M. Moser, M. A. Emanuel, R. J. Beach, J. A. Skidmore, and U. Keller. High average power diode pumped femtosecond Cr:LiSAF lasers. *Applied Physics B*, 65:235–243, 1997.

[95] S. Uemura and K. Torizuka. Generation of 12-fs pulses from a diode-pumped Kerr-lens mode-locked CrLiSAF laser. *Optics Letters*, 24:780–782, 1999.

[96] I. T. Sorokina, E. Sorokin, E. Wintner, A. Cassanho, and H. P. Jenssen. 14 fs pulse generation in Kerr-lens mode-locked prismless LiSGAFand LiSAF lasers: Observation of pulse self-frequency shift. *Optics Letters*, 22:1716–1718, 1997.

[97] R. Robertson, R. Knappe, and R. Wallenstein. Kerr-lens mode-locked Cr:LiSAF laser pumped by the diffraction limited output of a 672 nm diode laser master-oscillator power-amplifier system. *Journal of the Optical Society of America B*, 14:672–675, 1997.

[98] I. Thomann, A. Bartels, K. L. Corwin, N. R. Newbury, L. Hollberg, S. A. Diddams, J. W. Nicholson, and M. F. Yan. 420 MHz Cr:forsterite femtosecond ring laser and continuum generation in the 1–2 μm range. *Optics Letters*, 28:1368–1370, 2003.

[99] V. Petricevic, S. K. Gayen, and R. Alfano. Laser action in chromium activated fosterite for near IR excitation, is Cr^{4+} the lasing ion? *Applied Physics Letters*, 53:2590, 1988.

[100] A. Seas, V. Petricevic, and R. R. Alfano. Self-mode-locked chromium doped forsterite laser generates 50 fs pulses. *Optics Letters*, 18:891–893, 1993.

[101] V. Petricevic, A. B. Bykov, J. M. Evans, and R. Alfano. Room temperature near IR tunable lasing operation of $Cr^{4+}Ca_2GeO_4$. *Optics Letters*, 21:1750–1752, 1996.

[102] B. Xu, J. M. Evans, V. Petricevic, S. P. Guo, O. Maksimov, M. C. Tamargo, and R. R. Alfano. Continuous wave and passively mode-locked operation of a cunyite laser. *Applied Optics*, 39(27):4975, 2000.

[103] L. Qian, X. Liu, and F. Wise. Cr:forsterite laser pumped by broad area laser diodes. *Optics Letters*, 22:1707–1709, 1997.

[104] Chudoba, J. G. Fujimoto, E. Ippen, H. A. Haus, U. Morgnere, F. X. Kärtner, V. Scheuer, G. Angelow, and T. Tschudi. All-solid-state Cr:forsterite laser generating 14 fs pulses at 1.3 μm. *Optics Letters*, 26:292–294, 2001.

[105] V. Yanovsky, Y. Pang, F. Wise, and B. I. Minkov. Generation of 25-fs pulses from a self-mode-locked Cr:forsterite laser with optimized group-delay dispersion. *Optics Letters*, 18:1541–1543, 1993.

[106] I. Thomann, L. Hollberg, S. A. Diddams, and R. Equal. Chromium-doped forsterite: Dispersion measurement with white-light interferometry. *Applied Optics*, 42:1661, 2003.

[107] J. M. Evans, V. Petricevic, A. B. Bykov, and R. R. Alfano. Direct diode-umped continuous wave near infrared tunable laser operation of Cr^{4+} forsterite and $Cr^{4+}Ca_2GeO_4$. *Optics Letters*, 22:1171–1173, 1997.

[108] R. Adair, L. L. Chase, and S. A. Payne. Nonlinear refractive index of optical materials. *Physical Review B*, 39:3337–3345, 1989.

[109] W. F. Krupke, M. D. Shinn, J. E.Marion, J. A. Caird, and S. E.Stokowski. Spectroscopic, optical, and thermomechanical properties of neodymium and chromium-doped gadolinium scandium gallium garnets. *Journal of the Optical Society of America B*, 3:102, 1986.

[110] T. Kushida, H. M. Marcos, and G. E. Geusic. Laser transition cross section and fluorescence branching ratio for nd^{3+} in yttrium aluminum garnet. *Physical Review*, 167:289, 1968.

[111] A. Umbrasas, J. C. Diels, G. Valiulis, J. Jacob, and A. Piskarskas. Generation of femtosecond pulses through second harmonic compression of the output of a Nd:YAG laser. *Optics Letters*, 20:2228–2230, 1995.

[112] A. Umbrasas, J. C. Diels, J. Jacob, and A. Piskarskas. Parametric oscillation and compression in KTP crystals. *Optics Letters*, 19:1753–1755, 1994.

[113] J. Biegert, V. Kubecek, and J. C. Diels. A new femtosecond UV source based on Nd:YAG. In *CLEO, 1999*, Optical Society of America, Baltimore, MD, 1999, p. 479.

[114] A. Giesen, H. Hügel, A. Voss, K. Wittig, U. Brauch, and H. Opower. Scalable concept for diode-pumped high-power solid-state lasers. *Applied Physics B*, 58:365, 1994.

[115] G. J. Spühler, T. Südmeyer, R. Paschotta, M. Moser, K. J. Weingarten, and U. Keller. Passively modelocked high-power Nd:YAG lasers with multiple laser heads. *Applied Physics B*, 71:19–25, 2000.

[116] E. Innerhofer, T. Südmeyer, F. Grunner, R. Häring, A. Aschwanden, and R. Paschotta. 60 w average power in 810 fs pulses from a thin disk Yb:YAG laser. *Optics Letters*, 28:367–369, 2003.

[117] A. Agnesi and G. C. Reali. Development of medium power, compact all-solid-state lasers. *La Rivista del Nuovo Cimento*, 21(4):1–31, 1998.

[118] J. R. Lincoln and A. I. Ferguson. All-solid-state self-mode-locking of a Nd:YLF laser. *Optics Letters*, 19:2119–2121, 1994.

[119] R. Fluck, G. Zhang, U. Keller, K. J. Weingarten, and M. Moser. Diode-pumped passively mode-locked 1.3 μm $Nd:YVO_4$ and Nd:YLF lasers by use of semiconductor saturable absorbers. *Optics Letters*, 21:1378–1380, 1996.

[120] L. Turi and J. Juhasz. High-power longitudinally end-diode-pumped Nd:YLF regenerative amplifier. *Optics Letters*, 20:154–156, 1995.

[121] R. L. Fork and C. V. Shank. Generation of optical pulses shorter than 0.1 ps by colliding pulse mode-locking. *Applied Physics Letters*, 38:671, 1981.

[122] A. Finch, G. Chen, W. Steat, and W. Sibbett. Pulse asymetry in the colliding-pulse mode-locked dye laser. *Journal of Modern Optics*, 35:345–349, 1988.

[123] H. Kubota, K. Kurokawa, and M. Nakazawa. 29 fs pulse generation from a linear cavity synchronously pumped dye laser. *Optics Letters*, 13:749–751, 1988.

[124] J. C. Diels. Femtosecond dye lasers. In F. Duarte and L. Hillman, Eds., *Dye Laser Principles: With Applications*, Academic Press, Boston, MA, 1990, pp. 41–132.

[125] A. E. Siegman. An antiresonant ring interferometer for coupled laser cavities, laser output coupling, mode-locking, and cavity dumping. *IEEE Journal of Quantum Electronics*, QE-9:247–250, 1973.

[126] M. C. Nuss, R. Leonhart, and W. Zinth. Stable operation of a synchronously pumped colliding-pulse mode-locked ring dye laser. *Optics Letters*, 10:16–18, 1985.

[127] F. J. Duarte and R. O. James. Tunable solid-state lasers incorporating dye-doped, polymer-nanoparticle gain media. *Optics Letters*, 21:2088–2090, 2003.

[128] F. J. Duarte and R. O. James. Spatial structure of dye-doped polymer nanoparticle laser media. *Applied Optics*, 43:4088–4090, 2004.

[129] Z. Bor and A. Muller. Picosecond distributed feedback dye lasers. *IEEE Journal of Quantum Electronics*, QE-22:1524–1533, 1986.

[130] S. Szatmari and F. P. Schaefer. Generation of high power UV femtosecond pulses. In T. Yajima, K. Yoshihara, C. B. Harris, and S. Shionoya, Eds., *Ultrafast Phenomena VI*, Springer-Verlag, Berlin, Germany, 1988, p. 82–86.

[131] P. H. Chin, P. Pex, L. Marshall, F. Wilson, and R. Aubert. A high energy, electromagnetically tuned single mode, short cavity picosecond dye laser system: Design and performances. *Optics and Lasers in Engineering*, 10:55–68, 1989.

[132] H. P. Kurz, A. J. Cox, G. W. Scott, D. M. Guthals, H. Nathel, S. W. Yeh, S. P. Webb, and J. H. Clark. Amplification of tunable, picosecond pulses from a single mode, short cavity dye laser. *IEEE Journal of Quantum Electronics*, QE-21:1795–1798, 1985.

[133] P. Simon, S. Szatmari, and F. P. Schafer. Generation of 30 fs pulses tunable over the visible spectrum. *Optics Letters*, 16:1569–1571, 1991.

[134] P. P. Vasil'ev. Ultrashort pulse generation in diode lasers. *Optical and Quantum Electronics*, 24:801–824, 1992.

[135] G. A. Alphonse, D. B. Gilbert, M. G. Harvey, and M. Ettenberg. High power superluminescent diodes. *IEEE Journal of Quantum Electronics*, 24:2454–2458, 1988.

[136] P. J. Delfyett, Y. Slberberg, G. A. Alphonse, and W. Ceislik. Hot-carrier thermalization induced self-phase modulation in semiconductor traveling wave amplifiers. *Applied Physics Letters*, 59:10, 1991.

[137] P. J. Delfyett, C. H. Lee, L. Flores, N. Stoffel, T. Gmitter, N. Andreadakis, G. Alphonse, and J. C. Connolly. Generation of subpicosecond high-power optical pulses from a hybrid mode-locked semiconductor laser. *Optics Letters*, 15:1371–1373, 1990.

[138] S. Chen and J. Wang. Self-starting issues of passive self-focusing mode-locking. *Optics Letters*, 16:1689–1691, 1991.

[139] C. F. Lin and C. L. Tang. Colliding pulse mode-locking of a semiconductor laser in an external ring cavity. *Applied Physics Letters*, 62:1053–1055, 1993.

[140] W. Rudolph and B. Wilhelmi. *Light Pulse Compression*. Harwood Academic, Chur, London, 1989.

[141] B. Zysset, P. Beaud, and W. Hodel. Generation of optical solitons in the wavelength region 1.37–1.43 μm. *Applied Physics Letters*, 50:1027–1029, 1987.

[142] J. D. Kafka and T. Baer. Fiber raman soliton laser pumped by a Nd:YAG laser. *Optics Letters*, 12:181–183, 1987.

[143] A. S. Gouveia-Neto, A. S. L. Gomes, and J. B. Taylor. Generation of 33 fs pulses at 1.32 μm through a high order soliton effect in a single mode optical fiber. *Optics Letters*, 12:395–397, 1987.

[144] I. N. Duling III. Subpicosecond all-fiber erbium laser. *Electronics Letters*, 27:544–545, 1991.

[145] I. N. Duling. All fiber ring soliton laser mode locked with a nonlinear mirror. *Optics Letters*, 16:539–541, 1991.

[146] S. P. Craig-Ryan and B. J. Ainslie. Glass structure and fabrication techniques. In P. W. France, Eds., *Optical Fiber Lasers and Amplifiers*, Blackies & Son, Ltd., Glasgow, Scotland, 1991, pp. 50–78.

[147] I. N. Duling, R. P. Moeller, W. K. Burns, C. A. Villaruel, L. Goldberg, E. Snitzer, and H. Po. Output characteristics of diode pumped fiber ase sources. *IEEE Journal of Quantum Electronics*, 27:995–1003, 1991.

[148] M. Hofer, M. E. Fermann, F. Haberl, M. H. Ober, and A. J. Schmidt. Mode locking with cross-phase and self-phase modulation. *Optics Letters*, 16:502–504, 1991.

[149] N. J. Doran and D. Wood. Nonlinear-optical loop mirror. *Optics Letters*, 13:56–58, 1988.

[150] M. Zirngibl, L. W. Stulz, J. Stone, D. DiGiovanni, and P. B. Hansen. 1.2 ps pulses from passively modelocked laser diode pumped Er-doped fiber ring laser. *Electronics Letters*, 27:1734–1735, 1991.

[151] M. E. Fermann, A. Galvanauskas, G. Sucha, and D. Harter. Fiber lasers for ultrafast optics. *Applied Physics B*, 65:259–275, 1997.

[152] F. Ö. Ilday, J. Buckley, L. Kuznetsova, and F. W. Wise. Generation of 36-fs pulses from an ytterbium fiber laser. *Optics Express*, 11:3550–3554, 2003.

[153] I. Kaminow. Polarization in optical fibers. *IEEE Journal of Quantum Electronics*, QE-17:15, 1981.

[154] K. Tamura, H. A. Haus, and E. P. Ippen. Self-starting additive pulse mode-locked erbium fiber ring laser. *Electronics Letters*, 28:2226–2228, 1992.

[155] H. G. Winful. Self-induced polarization changes in birefringent optical fibers. *Applied Physics Letters*, 47:213, 1985.

[156] C. W. Chang and S. Chi. Mode-locked erbium-doper fibre ring laser using nonlinear polarization rotation. *Journal of Modern Optics*, 45:355–362, 1998.

[157] C. W. Chang and S. Chi. Ultrashort pulse generation from mode-locked erbium-doper fibre ring lasers. *Journal of Modern Optics*, 46:1431–1442, 1999.

[158] S. Chi, C. W. Chang, and S. Wen. Ultrashort soliton pulse train propagation in erbium-doped fiber amplifiers. *Optics Communications*, 111:132, 1994.

[159] G. P. Agrawal. *Nonlinear Fiber Optics*. Academic Press, Boston, MA, 1995.

[160] L. M. Spaulding, D. H. Yong, A. D. Kimand, and J. N. Kutz. Nonlinear dynamics of mode-locking optical fiber ring lasers. *Journal of the Optical Society of America*, B-19:1045–1054, 2002.

[161] M. E. Fermann, F. Haberl, M. Hofer, and H. Hochreiter. Nonlinear amplifying loop mirror. *Optics Letters*, 15:752–754, 1990.

[162] M. H. Ober, F. Haberl, and M. E. Fermann. 100 fs pulse generation from an all-solid-state Nd:glass fiber laser oscillator. *Applied Physics Letters*, 60:2177–2179, 1992.

[163] M. L. Dennis and I. N. Duling. Role of dispersion in limiting pulse width in fiber lasers. *Applied Physics Letters*, 62:2911–2913, 1993.

[164] M. L. Dennis and I. N. Duling. Experimental study of sideband generation in femtosecond fiber lasers. *IEEE Journal of Quantum Electronics*, 30:1469–1477, 1993.

[165] M. Nakazawa, E. Yoshida, and Y. Kimura. Generation of 98 fs optical pulses directly from an erbium-doped fibre ring laser at 1.57 μm. *Electronincs Letters*, 29:63–65, 1993.

[166] M. L. Dennis and I. N. Duling. Third order dispersion effects in femtosecond fiber lasers. *Optics Letters*, 19:1750–1752, 1993.

7

Femtosecond Pulse Amplification

7.1. INTRODUCTION

As discussed in Chapters 5 and 6, femtosecond pulse oscillators typically generate pulse trains with repetition rates of about 100 MHz at mean output powers which range from several mW (passively mode-locked dye laser) to several hundred mW (Ti:sapphire laser). Corresponding pulse energies are between several tens of pJ and several nJ. Femtosecond pulses with larger energies are needed for a variety of practical applications. Therefore a number of different amplifier configurations have been developed (for a review see Heist *et al.* [1], Simon [2], and Knox [3]). These amplifiers differ in the repetition rate and energy gain factor that can be achieved, ranging from 0.1 Hz to several MHz, and from 10 to 10^{10}, respectively. Both parameters cannot be chosen independently of each other. Instead, in present amplifiers the product of repetition rate and pulse energy usually does not exceed several hundred mW, i.e., it remains in the order of magnitude of the mean output power of the oscillator. The power of a single amplified pulse, however, can be in the terawatt range [4, 5]. It is not necessary for many applications to reach this power level. The specific function of the fs pulse will dictate a compromise between single pulse energy and repetition rate. It should also be noted that it is mostly the *intensity* of the focused pulse that matters, rather than the pulse power. Therefore, a clean beam profile providing the possibility of diffraction limited focusing is desired, eventually at the expense of a reduction in pulse energy. For some applications in spectroscopy, it is desirable to generate a white light continuum in short bulk materials. Typical threshold intensities that have to be reached for this purpose are on the order of 10^{12} W/cm^2.

The basic design principles of amplifiers have already been established for ps and ns pulse amplification. The pulses to be increased in energy are sent through a medium which provides the required gain factor (Figure 7.1). However, on a femtosecond time scale, new design methods are required to (a) keep the pulse duration short and (b) prevent undesired nonlinear effects caused by the extremely high intensities of amplified fs pulses. A popular technique to circumvent the problems associated with high peak powers is to use dispersive elements to stretch the pulse duration to the ps scale, prior to amplification.

Femtosecond pulse amplification is a complex issue because of the interplay of linear and nonlinear optical processes. The basic physical phenomena relevant to fs amplification are discussed individually in the next sections.

7.2. FUNDAMENTALS

7.2.1. Gain Factor and Saturation

It is usually desired to optimize the amplifier to achieve the highest possible gain coefficient for a given pump energy. To simplify our discussion let us assume that the pump inverts uniformly the part of the gain medium (Fig. 7.1) that is traversed by the pulse to be amplified (single pulse). A longitudinal geometry is often used when pumping the gain medium with a laser of good beam quality. Transverse pumping is used for high gain amplifiers such as dyes, or when pumping with low coherence sources such as semiconductor laser bars. We discuss next the case of transverse pumping. To achieve uniform inversion with transverse optical pumping, we have to choose a certain concentration $\bar{N}$ of the (amplifying) particles which absorb the pump light and a certain focusing of the pump. The focusing not only determines the transverse dimensions of the

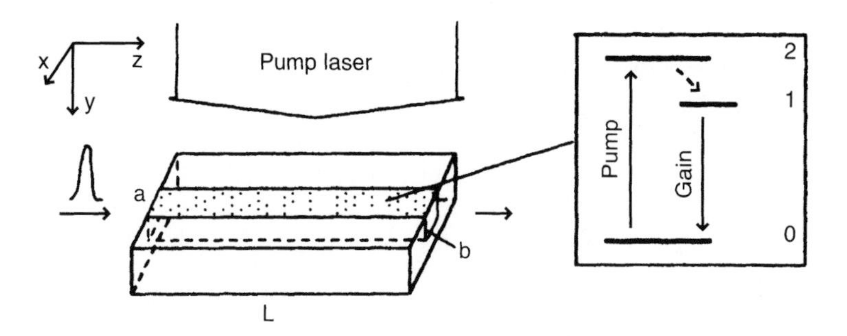

Figure 7.1 Light pulse amplification.

pumped volume, $\Delta x = a$, but also controls the saturation coefficient s_p and the depth $\Delta y = b$ of the inverted region. The saturation parameter $s_p = W_{p0}/W_s$ was defined as the ratio of incident pump pulse density and saturation energy density [see Eq. (3.59)].

In practice, the pump energy is set by equipment availability and other experimental considerations. Therefore, to change s_p, we have to change the focusing. Note that here the total number of excited particles corresponding to the number of absorbed pump photons remains constant. To illustrate the effect of the pump focusing for transverse pumping let us determine the depth distribution of the gain coefficient for various pump conditions. For simplicity, we assume a three-level system for the amplifier where $|0\rangle \rightarrow |2\rangle$ is the pump transition and $|1\rangle \rightarrow |0\rangle$ is the amplifying transition. The relaxation between $|2\rangle$ and $|1\rangle$ is to be much shorter than the pumping rate. Starting from the rate equations for a two-level system Eqs. (3.51) and (3.52), it can easily be shown that the system of rate equations for the photon flux density of the pump pulse F_p and the occupation number densities $N_i = \bar{N}\rho_{ii}$ $(i = 0, 1, 2)$ reads now:

$$\frac{\partial}{\partial t} N_0(y,t) = -\sigma_{02} N_0(y,t) F_p(y,t) \tag{7.1}$$

$$\frac{\partial}{\partial y} F_p(y,t) = -\sigma_{02} N_0(y,t) F_p(y,t) \tag{7.2}$$

and

$$N_1(y,t) = \bar{N} - N_0(y,t) \tag{7.3}$$

where σ_{02} is the interaction (absorption) cross-section of the transition $|0\rangle \rightarrow |2\rangle$. The coefficient of the small signal gain, a_g, is proportional to the occupation number difference of levels $|1\rangle$ and $|0\rangle$:

$$a_g = \sigma_{10}(N_1 - N_0)L = \sigma_{10}\Delta N_{10} L \tag{7.4}$$

where L is the amplifier length. With the initial conditions $N_0(y,0) = N_0^{(e)}(y) = \bar{N}$ (all particles are in the ground state) we find from Eqs. (7.1), (7.2), and (7.3) for the inversion density ΔN_{10} after interaction with the pump:

$$\Delta N_{10}(y) = \bar{N}\left\{1 - \frac{2}{1 - e^{a}\left(1 - e^{s_p}\right)}\right\} \tag{7.5}$$

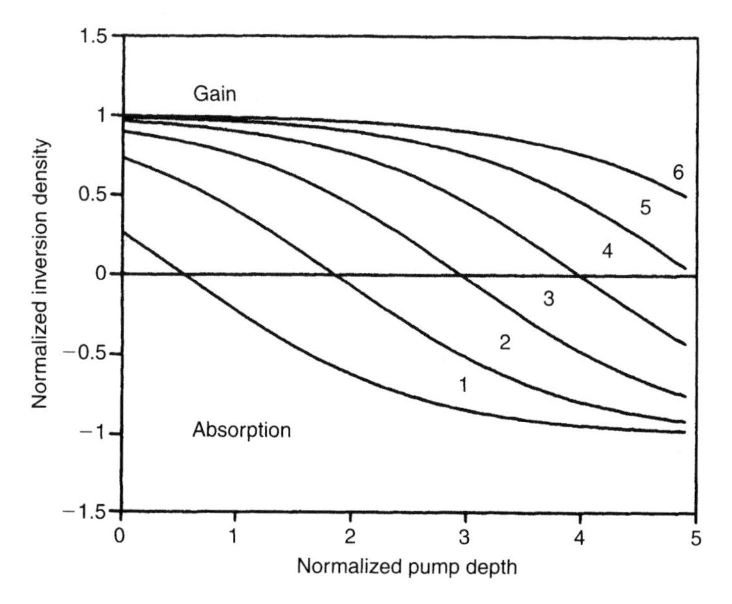

Figure 7.2 Inversion density ΔN_{10} as a function of normalized depth y/ℓ_a, where $\ell_a = |\sigma_{02}\bar{N}|^{-1}$, for different saturation parameters $s_p = W_{p0}/W_s$.

where $a = -\sigma_{02}\bar{N}y$ is the coefficient of the small signal absorption for the pump. Figure 7.2 shows some examples of the population inversion distribution for different intensities of the pump pulse.

In the limit of zero saturation the penetration depth is roughly given by the absorption length $\ell_a = |\sigma_{02}\bar{N}|^{-1}$ defined as the propagation length at which the pulse intensity drops to $1/e$ of its original value. If the pump density is large enough to saturate the transition $0 \rightarrow 2$ the penetration depth becomes larger and, moreover, a region of almost constant inversion (gain) is built.

Given a uniformly pumped volume, the system needs to be optimized for maximum energy amplification of the signal pulse. Using Eq. (3.57) the energy gain factor achieved at the end of the amplifier can be written as:[1]

$$G_e = \frac{W(L)}{W_0} = \frac{\hbar\omega_\ell}{2\sigma_{10}W_0} \ln\left[1 - e^{a_g}\left(1 - e^{2\sigma_{10}W_0}\right)\right]$$
$$= \frac{1}{s} \ln\left[1 - e^{a_g}\left(1 - e^{s}\right)\right]. \tag{7.6}$$

[1]Note that for the amplification, the relations found for the two-level system hold if we assume that during the amplification process no other transitions occur. This is justified in most practical situations.

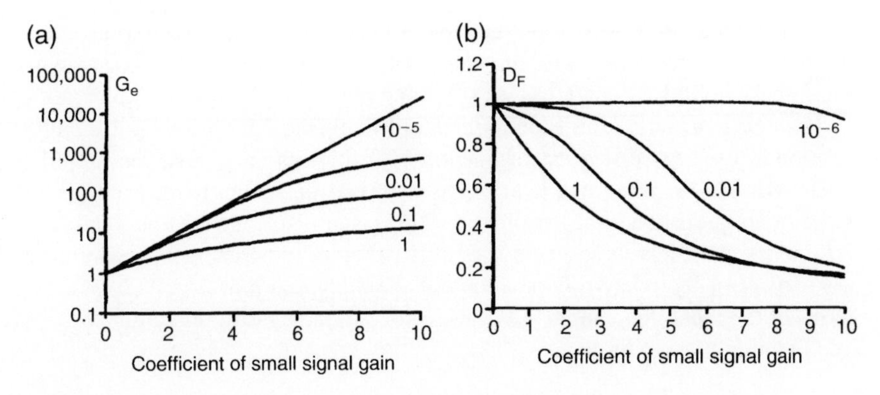

Figure 7.3 (a) Energy gain factor G_e and (b) damping of input fluctuations D_F as function of the coefficient of the small signal gain, a_g, for different saturation parameters $s' = 2s$.

Figure 7.3(a) shows this gain factor (on a logarithmic scale) as a function of the small signal gain a_g for different values of s. The saturation parameter s can be controlled by adjusting the beam cross section of the pulse to be amplified. As expected from Eq. (7.6), as long as saturation is negligible, the energy gain varies exponentially with a_g (linear slope for the logarithm of the gain G_e versus gain coefficient). The total gain is drastically reduced by saturation. It should be noted, however, that for the sake of high energy extraction from the system, the amplifier has to be operated near saturation. Active media with larger saturation energy densities (smaller gain cross sections) are therefore clearly favored if high pulse energies are to be reached. Table 7.1 shows some important parameters of

Table 7.1

Optical parameters of gain media.

Medium	λ_ℓ (μm)	$\Delta\lambda$ (nm)	σ_{10} (cm^2)	Lifetime (s)	Typical pump
Organic dyes	0.3...1	≥ 50	$\geq 10^{-16}$	$10^{-8} \ldots 10^{-12}$	Laser
Color centers	$1 \ldots 4$	≈ 200	$\geq 10^{-16}$	$\leq 10^{-6}$	Laser
XeCl	0.308	1.5	7×10^{-16}	$\approx 10^{-8}$	Discharge
XeF	0.351	≤ 2	3×10^{-16}	$\approx 10^{-8}$	Discharge
KrF	0.249	≈ 2	3×10^{-16}	$\leq 10^{-8}$	Discharge
ArF	0.193	≈ 2	3×10^{-16}	$\leq 10^{-8}$	Discharge
Alexandrite	≈ 0.75	≈ 100	7×10^{-21}	2.6×10^{-4}	Flashlamp
Cr:LiSAF	≈ 0.83	≈ 250	5×10^{-20}	6×10^{-5}	Diode laser
Ti:sapphire	≈ 0.78	≈ 400	3×10^{-19}	3×10^{-6}	Laser
Nd:glass	1.05	≈ 21	3×10^{-20}	3×10^{-4}	Flashlamp

gain media used in fs pulse amplification. The small σ_{10} (large saturation energy density W_s) in connection with the long energy storage time ($\sim$ fluorescence life time) make solid-state materials mostly attractive for high-energy amplification.

How close to saturation should an amplifier operate? If chirped pulse amplification is used (as discussed in Section 7.4), it is essential that the amplifier operates in the *linear* regime. In other cases, it is advantageous to have at least one stage of amplification totally saturated. The reason is that the saturated output of an amplifier is relatively insensitive to fluctuations of pulse energy. A quantitative assessment of the relative fluctuations of the amplified pulses $\Delta W(L)/W(L)$ in terms of the input fluctuations $\Delta W_0/W_0$ can be found by differentiating Eq. (7.6) and defining a damping factor:

$$D_F = \frac{\Delta W(L)/W(L)}{\Delta W_0/W_0} = \frac{e^{a_g} e^s}{[1 - e^{a_g}(1 - e^s)]} \frac{1}{G_e}. \tag{7.7}$$

For large saturation the output can be expected to be smoothed by a factor of $G_e{}^{-1}$ (Fig. 7.3b). This reduction in energy fluctuation is at the expense of a reduction of the amplification factor.

From the preceding discussion we may want to optimize the amplifier geometry as follows. From the given pump energy and the known absorption cross section we can estimate the focusing conditions for the pump pulse to achieve a uniformly pumped volume. For maximum amplification, the cross section of the signal beam has to be matched to this inverted region. If the saturation is too large or too small with respect to the overall design criteria, readjustment of either signal or pump focusing can correct the error. However, there are a number of additional effects that need to be considered in designing the amplification geometry, which are discussed in the following sections.

7.2.2. Shaping in Amplifiers

Saturation

Saturation has a direct and indirect pulse shaping influence. The direct impact of saturation arises from the time-dependent amplification. As the gain saturates, the dispersion associated with the amplifying transition changes, resulting in a phase modulation of the pulse. Although the phase modulation does not affect the pulse envelope directly, it does modify the propagation of the pulse through the dispersive components of the amplifier (glass, solvent, isolators).

The mathematical framework to deal with the effect of saturation on both the pulse shape and its phase was given in Chapters 3 and 4. We present here a few examples to illustrate the importance of these shaping mechanisms in specific configurations.

The change in the pulse intensity profile resulting from saturation—excluding dispersive effects—is found by evaluating Eq. (3.55), describing the intensity of a pulse at the output of an absorbing or amplifying medium as a function of the integrated intensity at the input $W_0(t) = \int_{\infty}^{t} I_0(t')dt'$:

$$I(z,t) = I_0(t)\frac{e^{W_0(t)/W_s}}{e^{-a_g} - 1 + e^{W_0(t)/W_s}}. \tag{7.8}$$

Figure 7.4 shows the normalized shape of the amplified pulse for different input pulse shapes and saturation. As expected, saturation in the amplification process favors the leading edge of the pulse. Thus the pulse center shifts toward earlier times whereby the actual change in pulse shape critically depends on the shape of the input pulse. In particular the wings of the amplified pulses are a sensitive function of the initial slope. To minimize pulse broadening or even to obtain shortening during amplification, it is recommended to have pulses with

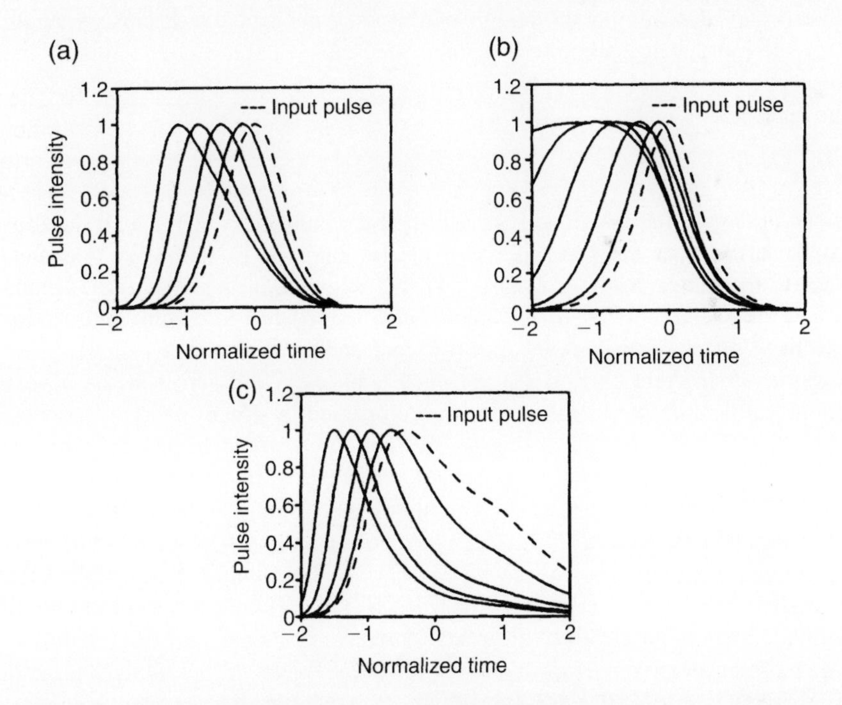

Figure 7.4 Behavior of the pulse shape in an amplifier for different shapes of the input pulse and different saturation parameters $s = W_0/W_s$ for a small signal gain $e^{a_g} = 10^4$. $2s$ varies from 10^{-4} to 1 (increment factor 10) in the order of increasing shift of the pulse maximum. The initial pulse shapes are a Gaussian $I(t) \propto \exp(-2t^2)$ (a) (t being the normalized time); a sech pulse $I(t) \propto \text{sech}^2(t)$ (b), and an asymmetric pulse, Gaussian in the wings $\exp(-t^2) + \frac{1}{2}\exp[-(t-1)^2]$ (c).

a steep leading edge. For reasons to be discussed later high power amplifiers usually consist of several stages isolated by saturable absorbers. These absorbers can also serve to steepen the pulses.

If the pulse is detuned from resonance, we have seen how saturation can also result in a chirp in Chapter 3. This effect will be discussed in Section 7.3.

Group Velocity Dispersion

While being amplified, the pulses travel through a certain length of material and are thus influenced by dispersion. In the case of *linear gain*, the pulse shaping is only because of GVD. Shortest amplified pulses will be obtained either by sending up- (down-)chirped pulses through the amplifier if its net GVD is positive (negative). Alternatively (the only appropriate procedure if the input pulses were bandwidth-limited) the broadened and chirped pulse at the end of the amplifier can be sent through a dispersive device, for example a prism or grating sequence, for recompression. For unchirped input pulses, the magnitude of the broadening that occurs depends on the length of the amplifier and the dispersive length (for the pulse being amplified) defined in Chapter 1 [cf. Eq. (1.128)]. Some dispersion parameters for typical materials relevant to fs amplification are shown in Table 7.2.

Gain Narrowing

In the preceding discussion we assumed the bandwidth of the gain medium to be larger than the spectral width of the pulse to be amplified. Depending on the active medium (Table 7.1) this assumption becomes questionable when the spectral width of the input pulse approaches a certain value. Now we have to take into account that different spectral components of the pulse experience different gain. Because typical gain curves of active media have a finite bandwidth, the amplification is accompanied by a narrowing of the pulse spectrum. Thus, in the linear regime (no saturation), an unchirped pulse broadens while being amplified. This behavior can easily be verified assuming an unchirped Gaussian pulse at the amplifier input, with a field spectrum varying as $\tilde{\mathcal{E}}_0(\Omega) = A_0 \exp[-(\Omega\tau_G/2)^2]$ [cf. Eq. (1.35)], and a small signal gain $G(\Omega) = e^{a_g(\Omega)}$ where $a_g = a_0 \exp[-(\Omega T_g)^2/2]$. For simplicity we expand the Gaussian distribution and use $a_g \simeq a_0[1-(\Omega T_g)^2/2]$. The spectral field amplitude behind the amplifier, neglecting saturation, is:

$$
\begin{aligned}
\tilde{\mathcal{E}}(\Omega) &= \tilde{\mathcal{E}}_0(\Omega) e^{a_g(\Omega)/2} \\
&\simeq A_0 e^{-(\Omega\tau_G)^2/4} e^{a_0/2[1-(\Omega T_g)^2/2]} \\
&= A_0 e^{a_0/2} e^{-\Omega^2(\tau_G^2 + a_0 T_g^2)/4}
\end{aligned}
\tag{7.9}
$$

Table 7.2

Optical parameters for typical materials used in fs pulse amplifiers. The dispersion length L_D is given for a pulse duration of 100 fs. $^1\lambda_\ell = 1.06$ μm, $^2\lambda_\ell = 1.06$ μm, and SF6.

Material	λ_ℓ (μm)	k''_ℓ (fs^2/cm)	L_D (cm)	$\bar{n}_2$ (cm^2/W)
water	0.6	480	21	0.67×10^{-16}
methanol	0.6	400	25	
benzene	0.6	1700	6	8.8×10^{-15}
ethylene glycol	0.6	840	12	3×10^{-16}
fused SiO_2	0.6	590	17	$^1 3 \times 10^{-16}$
fused SiO_2 [6]	1.06			4.7×10^{-16}
SF10	0.6	2530	4	$^2 1.3 \times 10^{-15}$
SF14	0.6	3900	2.5	
phosphate glass	1.06	330	30	1×10^{-15}
Ti:sapphire [6]	0.78	610	16	10.5×10^{-16}
diamond	0.6	1131	8.8	6.7×10^{-15}
diamond	0.25	3542	2.8	-8×10^{-15}
air (1 atm)	800	0.14	8.3×10^4	2.8×10^{-19}
air [7]	0.8	0.21	5.5×10^4	5.57×10^{-19}
air [8]	0.8			4×10^{-19}
air [9]	0.308			2.2×10^{-18}
air [10]	0.25	0.96	1.04×10^4	2.4×10^{-18}

where τ_G is a measure of the input pulse duration $\tau_p = \sqrt{2\ln 2}\tau_G \simeq 1.177\tau_G$ (Table 1.1) and $\Delta\omega_g \simeq 2.36/T_g$ is the FWHM of the gain curve. As can be seen from Eq. (7.9) the spectrum of the amplified pulse becomes narrower; the FWHM is given by $\simeq 2.36/\sqrt{\tau_G^2 + a_0 T_g^2}$. The corresponding pulse duration at the amplifier output is

$$\tau_p' \simeq \tau_p \sqrt{1 + a_0 (T_g/\tau_G)^2}. \tag{7.10}$$

If saturation occurs, the whole set of density matrix and Maxwell's equations has to be analyzed, as outlined in Chapters 3 and 4, to describe the behavior of the pulse on passing through the amplifier.

The situation is somewhat different in inhomogeneously broadened amplifiers if they are operated in the saturation regime. Roughly speaking, because field components that see the highest gain saturate the corresponding transitions first, those amplified independently by the wings of the gain curve can also reach the saturation level if the amplifier is sufficiently long. Therefore a net gain that is almost constant over a region exceeding the spectral FWHM of the small signal gain can be reached and thus correspondingly shorter pulses can be amplified. This was demonstrated by Glownia *et al.* [11] and Szatmari *et al.* [12] who succeeded in amplifying 150–200 fs pulses in XeCl.

7.2.3. Amplified Spontaneous Emission (ASE)

So far we have neglected one severe problem in (fs) pulse amplification, namely amplified spontaneous emission (ASE), which mainly results from the pump pulses being much longer than the fs pulses to be amplified. As a consequence of the medium being inverted before (and after) the actual amplification process, spontaneous emission traveling through the pumped volume can continuously be amplified and can therefore reach high energies. ASE reduces the available gain and decreases the ratio of signal (amplified fs pulse) to background (ASE), or even can cause lasing of the amplifier, preventing amplification of the seed pulse. For these reasons the evolution of ASE and its suppression has to be considered thoroughly in constructing fs pulse amplifiers. Here we shall illustrate the essential effects on basis of a simple model (illustrated in Fig. 7.5), and compare the small signal gain (for the signal pulse) with and without ASE [1]. For simplicity, let us assume that the ASE starts at $z = 0$ and propagates toward the exit of the amplifier while being amplified. The photon flux of the ASE is thus given by:

$$F_{ASE}(z,t) = F_{ASE}(0)\exp\left[\int_0^z \sigma_{ASE}\left(N_1(z',t) - N_0(z',t)\right)dz'\right] \tag{7.11}$$

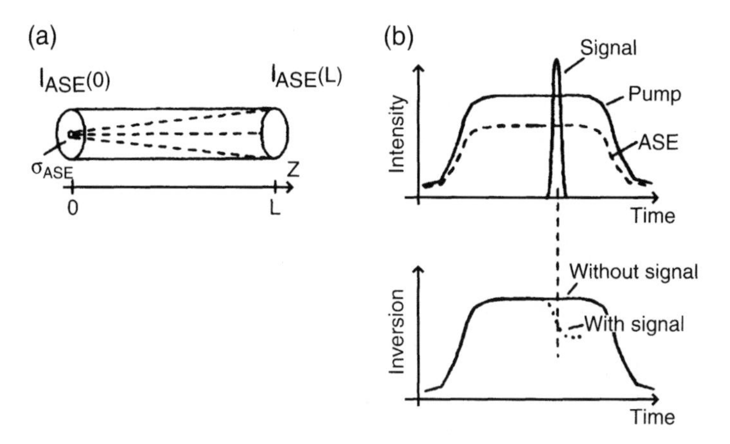

Figure 7.5 (a) Geometry of unidirectional ASE evolution. (b) Temporal behavior of pump pulse, ASE, signal pulse and inversion. (Adapted from [1].)

where σ_{ASE} is the emission cross section which, with reference to Fig. 7.1, describes the transition between levels $|1\rangle$ and $|0\rangle$. With strong pumping the ASE will follow the pump intensity almost instantaneously, and after a certain time a stationary state is reached in which the population numbers do not change. This means that additional pump photons are transferred exclusively to ASE while leaving the population inversion unchanged. Under these conditions the rate equations for the occupation numbers read:

$$\frac{dN_0(z,t)}{dt} = -\sigma_{02}N_0(z)F_p(t) + \sigma_{10}N_1(z)F_{ASE}(z,t) = 0 \tag{7.12}$$

and

$$N_0(z) + N_1(z) = \bar{N}. \tag{7.13}$$

Combination of Eq. (7.11) with Eqs. (7.12) and (7.13) yields an integral equation for the gain coefficient $a(z)$ for a signal pulse that has propagated a length z in the amplifier:

$$a(z) = \int_0^z \sigma_{10}\bar{N}\frac{F_p/F_{ASE}(0) - e^{a(z')}}{F_p/F_{ASE}(0) + e^{a(z')}}dz'. \tag{7.14}$$

In the absence of ASE the gain coefficient is:

$$a = \sigma_{10}\bar{N}z. \tag{7.15}$$

The actual gain in the presence of ASE is reduced to $G_a = \exp[a(z)]$ from the larger small signal gain in the ideal condition (without ASE) of $G_i = \exp(\sigma_{10}\bar{N}z)$. The ASE at $z = 0$ can be estimated from:

$$F_{ASE}(0) = \frac{\eta_F \Delta\Omega \hbar\omega_{ASE}}{4\sigma_{ASE}T_{10}} \tag{7.16}$$

where $\Delta\Omega = d^2/4L^2$ is the solid angle spanning the exit area of the amplifier from the entrance, T_{10} is the fluorescence life time, and η_F is the fluorescence quantum yield. Figure 7.6 shows the result of a numerical evaluation of Eq. (7.14). Note that a change in small signal gain at constant F_p can be achieved by changing either the amplifier length or the concentration $\bar{N}$. As can be seen at high small signal gain the ASE drastically reduces the gain available to the signal pulse. In this region a substantial part of the pump energy contributes to the build up of ASE.

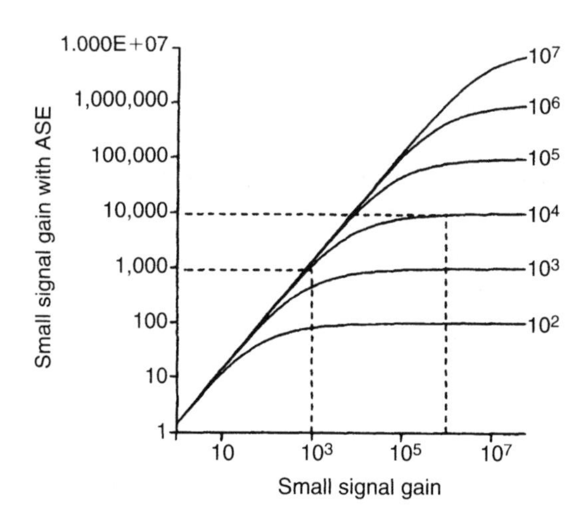

Figure 7.6 Comparison of small signal gain with and without ASE for different values of the normalized pump intensity $F_p/F_{ASE}(0)$.

One solution to the problem of gain reduction because of ASE is the segmentation of the amplifier in multiple stages. To understand the nature of this improvement, let us compare a single- and a two-stage amplifier. With a normalized pump power $F_p/F_{ASE}(0) = 10^4$ and $G_i = 10^6$, we expect a small signal amplification of about 10^4 in the single-stage amplifier (Fig. 7.6). In contrast, we obtain a gain of about 10^3 in one cell and thus 10^6 in the whole device when we pump two cells by the same intensity, and each has half the length of the original cell. Another advantage of multistage amplifiers is the possibility to place filters between the individual stages and thus to reduce further the influence of ASE. If saturable absorbers are used this may also lead to a favorable steepening of the leading pulse edge. Moreover, in multistage arrangements the beam size and pump power can be adjusted to control the saturation, taking into account the increasing pulse energy. For a more quantitative discussion of the interplay of ASE and signal pulse amplification as well as for the amplifier design, see, for example, Penzkofer and Falkenstein [13] and Hnilo and Martinez [14].

7.3. NONLINEAR REFRACTIVE INDEX EFFECTS

7.3.1. General

As discussed in previous chapters, the propagation and amplification of an intense pulse will induce changes of the index of refraction in the traversed medium. As a result, the optical pathlength through the amplifier varies along the beam and pulse profile, leading possibly to SPM and self-lensing. The origins of this pulse induced change in refractive index can be

(a) saturation in combination with off-resonant amplification (absorption) or
(b) nonresonant nonlinear refractive index effects in the host material.

The nonlinearity is somewhat more complex in semiconductor amplifiers, because it is related to the dependence of the index of refraction on the carrier density (which is a function of current, light intensity, and wavelength). The nonlinearities are nevertheless large and can contribute to significant spectral broadening in semiconductor amplifiers [15].

While SPM leads to changes in the pulse spectrum, self-lensing modifies the beam profile. Being caused by the same change in index, both effects occur simultaneously, unless the intensity of the input beam does not vary transversely to the propagation direction. Such a "flat" beam profile can be obtained by expanding the beam and filtering out the central part with an almost constant intensity.

It will often be desirable to exploit SPM in the amplifier chain for pulse compression, while self-focusing should be avoided. There are a number of successful attempts to achieve spectral broadening (to be exploited in subsequent pulse compression) through SPM in a dye amplifier [16, 17] and semiconductor amplifier [15]. We will elaborate on this technique toward the end of this chapter. At the same time, self-focusing should be avoided, because it leads to instabilities in the beam parameters such as filamentation or even to material damage.

We proceed with some estimates of the SPM that occurs in amplifiers. We derived in Chapter 3 an expression [Eq. (3.68)] for the change in instantaneous frequency with time because of gain depletion:

$$\delta\omega(t) = -\frac{(\omega_\ell - \omega_{10})T_2}{2} \frac{e^{-a} - 1}{e^{-a} - 1 + e^{W(t)/W_s}} \frac{I(t)}{W_s}. \tag{7.17}$$

The contribution from the nonlinear refractive index $\bar{n}_2$ of the host material is given by:

$$\delta\omega(t) = -k_\ell \bar{n}_2 \int_0^z \frac{\partial}{\partial t} I(z', t) dz' \tag{7.18}$$

A comparison of the functional behavior of the frequency modulation because of saturation [Eq. (7.17)] and because of the nonlinear index [Eq. (7.18)] is shown in Figure 7.7. The nonresonant refractive index change always results in up-chirp at the pulse center while the sign of the chirp because of gain saturation depends on the sign of the detuning $(\omega_\ell - \omega_{10})$. The corresponding refractive index

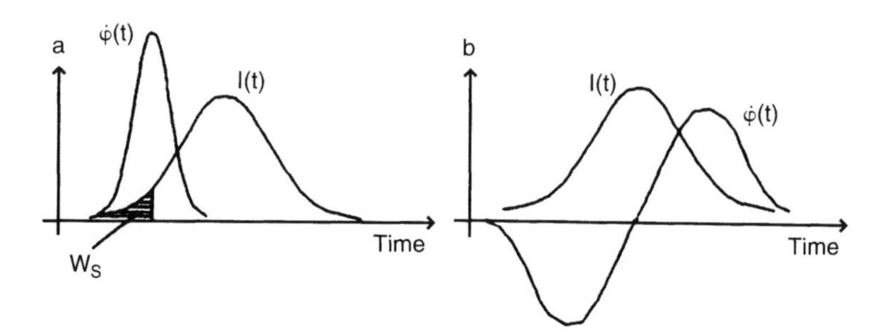

Figure 7.7 Comparison of the frequency modulation or chirp induced by saturation of an amplifier above resonance (a) and by Kerr effect (b). The frequency modulation (a) peaks at a time such that the integrated intensity equals the saturation energy density W_s. The Kerr effect induced phase modulation is proportional to minus the time derivative of the intensity.

variation transverse to the propagation direction can lead to self-focusing as well as to self-defocusing. Chirp because of saturation may play a role in dye as well as in solid-state amplifiers if they are highly saturated. Figure 7.8 illustrates the pulse shaping and chirping that arises for high values of the saturation parameter ($s \geq 0.0001$ at the amplifier input and $e^a = 10^4$).

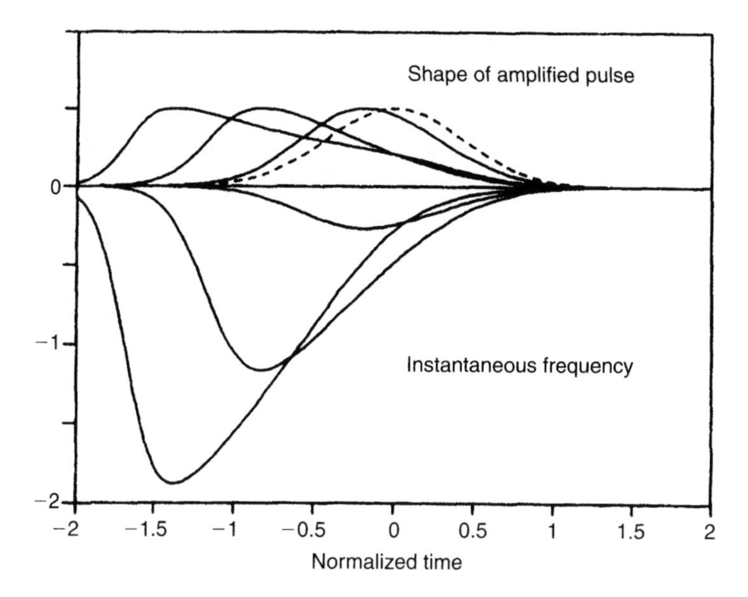

Figure 7.8 Chirp because of amplification of a Gaussian input pulse for different saturation parameters s ($s = 0.0001; 0.01; 1$ in the order of increasing chirp). The shape of the amplified pulse (solid lines) and incident pulse (dashed line) are also indicated. The other parameters are detuning $(\omega_\ell - \omega_{10}) = 0.5$ and small signal gain $e^a = 10^4$.

In amplifier chains operating up to high saturation parameters, saturable absorbers will generally be used to isolate stages of high gain, reducing the effect of ASE. In addition the saturable absorber can counteract the broadening effect of the amplifier because of:

- pulse shaping (steepening of the leading edge) by saturable absorption and thus elimination of subsequent broadening by the saturable gain (cf. Fig. 7.4); and
- pulse compression because of the combination of downchirping (if the pulse has a longer wavelength than that of the peak of the absorption band) by the absorber and propagation in a gain medium of normal (linear) dispersion.

7.3.2. Self-Focusing

The situation of pulse amplification is more complex if we consider self-focusing effects which were introduced in Chapter 3. In this section, to obtain some order of magnitude estimations, we will use the relations derived for cw Gaussian beams. If the instantaneous peak power of the amplified pulse exceeds the critical power for self-focusing defined in Eq. (3.171)

$$P_{cr} = \frac{(1.22\lambda_\ell)^2 \pi}{32\, n_0\, \bar{n}_2}, \tag{7.19}$$

particular attention has to be given to the beam profile. Even weak ripples in the transverse beam profile may get strongly amplified, and lead to breaking up of the beam in filaments. The critical transverse dimension of these beam fluctuations, w_{cr}, below which a beam of intensity I (and power $P > P_{cr}$) becomes unstable against transverse intensity irregularities can be estimated from:

$$P_{cr} = \frac{\pi w_{cr}^2}{2} I. \tag{7.20}$$

If a smooth transverse beam profile is used, the amplifier may be operated above the critical power, provided the optical path through the amplifier L does not exceed the *self-focusing length* L_{SF}:

$$L_{SF}(t) = \frac{\rho_0}{\sqrt{P/P_{cr} - 1}}, \tag{7.21}$$

where $P = P(t)$ refers to the instantaneous power on axis of the Gaussian beam, and the beam waist is at the sample input (cf. Eq. [3.189]).

Values of $\bar{n}_2$ are listed for various transparent materials in Table 7.2.[2] To a typical value of $\bar{n}_2 \approx 5 \times 10^{-16}$ cm^2/W at 0.6 μm corresponds a critical power of $P_{cr} \approx 700$ kW, or only 70 nJ for a 100 fs pulse. Much higher energies are readily obtained in fs amplifiers. To estimate the self-focusing length, let us consider a saturated amplifier. The pulse energy $W = P\tau_p$ is of the order of the saturation energy density $\hbar\omega_\ell/2\sigma_{01}$ times the beam area $S \approx 0.5\pi w_0^2$. For a dye amplifier operating around 600 nm ($\sigma_{01} \approx 10^{-16}$ cm^2), the saturation energy density is of the order of 3 mJ/cm^2; hence the peak power for a 100 fs pulse in a beam of 1 cm^2 cross section is 3×10^{10} W/cm^2. Inserting this peak power in Eq. (7.21) leads to a self-focusing length of about 2 m.

Self-focusing is therefore generally not a problem in dye amplifiers, because the gain medium saturates before L_{SF} is reduced to dimensions of the order of the amplifier. Solid-state media have a much lower cross section σ_{01}, hence a much higher saturation energy density. For instance, if a Ti:sapphire laser amplifier ($\sigma_{01} = 3 \times 10^{-19}$ cm^2, or a saturation energy density of 0.66 J/cm^2) were driven to full saturation as in the previous example, the peak power would be 0.66×10^{13} W/cm^2. At $\lambda = 1$ μm, the corresponding self-focusing length is only 4 cm.

The smaller the interaction cross section the shorter is L_{SF} and thus the more critical is self-focusing in a saturated amplifier. This problem has been a major obstacle in the construction of high power amplifier sources. The solution is to stretch the pulse *prior to amplification*, to reduce its peak power, and recompress it thereafter. This solution, known as chirped pulse amplification [5], is outlined in Section 7.4.

7.3.3. Thermal Noise

As the efficiency of an amplifier medium never approaches 100%, part of the pump energy is wasted in heat. In amplifiers as well as in lasers, thermally induced changes in index of refraction will result from a non-uniform heating. In most materials, all nonlinear lensing mechanisms are dwarfed by the thermal effects. A "z-scan" experiment can be performed to appreciate the size of this nonlinearity, for example [19].

Average power levels of a few mW are sufficient to detect a thermal nonlinear index. The problem of thermal lensing is much more severe in amplifiers than in lasers, because of the larger pump energies and larger volumes involved. For instance, the heat dissipated by the pump can easily be carried away by the transverse flow in a typical dye laser jet, with a spot size of the order of μm. The larger cross section of the amplifier calls generally for the use of cells.

---

[2]The values are often expressed in Gaussian units (such as for instance the values for air [9]), or have to be derived from values of the third order susceptibility. A detailed discussion of the conversion factors can be found in [18].

Non-uniform heating results in convection, turbulence, and a random noise in the beam profile and amplification. In the case of dye laser amplifiers, a simple but effective solution consists of using *aqueous dye solutions* cooled near 4°C, because $(dn/dT)|_{4^\circ C} \approx 0$ at that temperature (or 11.7°C for heavy water). Pulse-to-pulse fluctuations in the output of a Cu vapor laser pumped amplifier have been considerably reduced by this technique [20].

In the case of solid-state amplifiers, a careful design of a cylindrically symmetric pump (and cooling) geometry is required to prevent thermal lensing from causing beam distortion.

7.3.4. Combined Pulse Amplification and Chirping

The preceding section has established that self focusing sets a limit to the maximum power that can be extracted from an amplifier chain. Within that limit, SPM and subsequent compression can be combined with pulse amplification [15, 17]. In the implementation of the process by Heist *et al.* [17] the pulse is self-phase modulated through the nonlinear index of the solvent in the last stage of the amplifier, for the purpose of subsequent pulse compression (Figure 7.9). If the gain medium is not used at resonance, saturation can result in an even larger phase modulation, which can be calculated with Eq. (7.17).

With the constraints set above for the absence of self-focusing, the pulse propagation equation (1.194) is basically one-dimensional:

$$\frac{\partial}{\partial z}\tilde{\mathcal{E}} = \frac{1}{2} i k_\ell'' \frac{\partial^2}{\partial t^2}\tilde{\mathcal{E}} + \mathcal{B}_1 + \mathcal{B}_2; \tag{7.22}$$

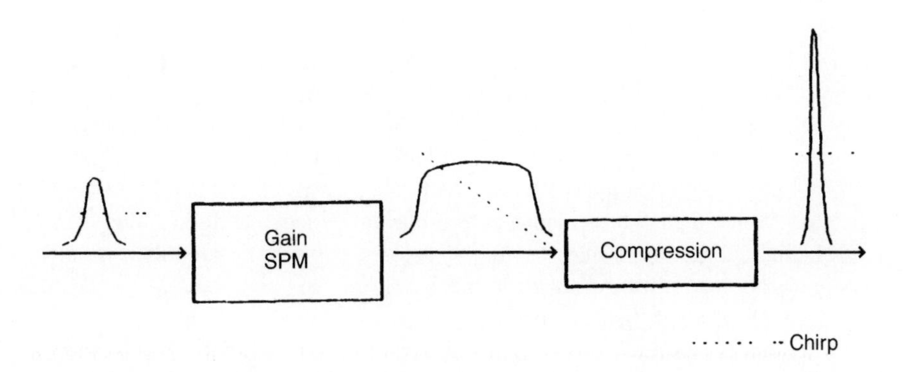

Figure 7.9 Block diagram showing the combination of pulse amplification, SPM, and compression.

where

$$\mathcal{B}_1 = -\frac{\omega_\ell^2 \mu_0}{2k_\ell} \mathcal{P}_{gain}(t, z) \tag{7.23}$$

and

$$\mathcal{B}_2 = -ik_\ell n_2 |\tilde{\mathcal{E}}|^2 \tilde{\mathcal{E}} \tag{7.24}$$

are nonlinear source terms (nonlinear polarization) responsible for the time-dependent gain and the nonlinear refractive index, respectively. The polarization for the time-dependent gain can be determined with Eq. (3.39), or with Eq. (3.43) if the rate equation approximation can be applied. We outlined in Chapter 1 the basic procedure to study the propagation of a given input pulse through such an amplifier. Saturated amplification and GVD will affect mainly the pulse temporal amplitude, while SPM will affect the pulse spectrum.

Figure 7.10(a) shows an example of pulse shape evolution, in amplitude and phase, through such an amplifier. The broadening and the development of a time-dependent frequency can be clearly seen. As for the case of optical fibers, the interplay of SPM and GVD leads to an almost linear chirp at the pulse center. The temporal broadening increases with the input pulse energy as a result of stronger saturation and larger SPM leading to a larger impact of GVD [Fig. 7.10(b)]. It is also evident that the chirped and amplified pulses can be compressed in a quadratic compressor following the amplifier. Detailed numerical and experimental studies show that an overall pulse compression by a factor of two is feasible for typical parameters of dye amplifiers [17].

7.4. CHIRPED PULSE AMPLIFICATION (CPA)

As mentioned earlier, the smaller the gain cross section σ_{10}, the larger the saturation energy density $\hbar\omega_\ell / 2\sigma_{10}$, which is a measure of the largest energy density that can be extracted from an amplifier. Because the maximum peak power is limited by self-focusing effects (the amplifier length has to be smaller than the self-focusing length), one solution is to limit the power by stretching the pulse in time. Dispersion lines with either positive or negative GVD can be made with combinations of gratings and lenses (cf. Chapter 2). This concept of chirped pulse amplification (CPA) was introduced by Strickland and Mourou [21] in 1985. Since then it has revolutionized ultrafast science and technology. CPA has facilitated the broad introduction of tabletop fs oscillator-amplifier systems, the generation of ultrafast pulses at the PW level, and continues to extend the frontiers of ultrahigh field science.

The idea is to stretch (and chirp) a fs pulse from an oscillator (up to 10,000 times) with a linear dispersion line, increase the energy by *linear amplification*,

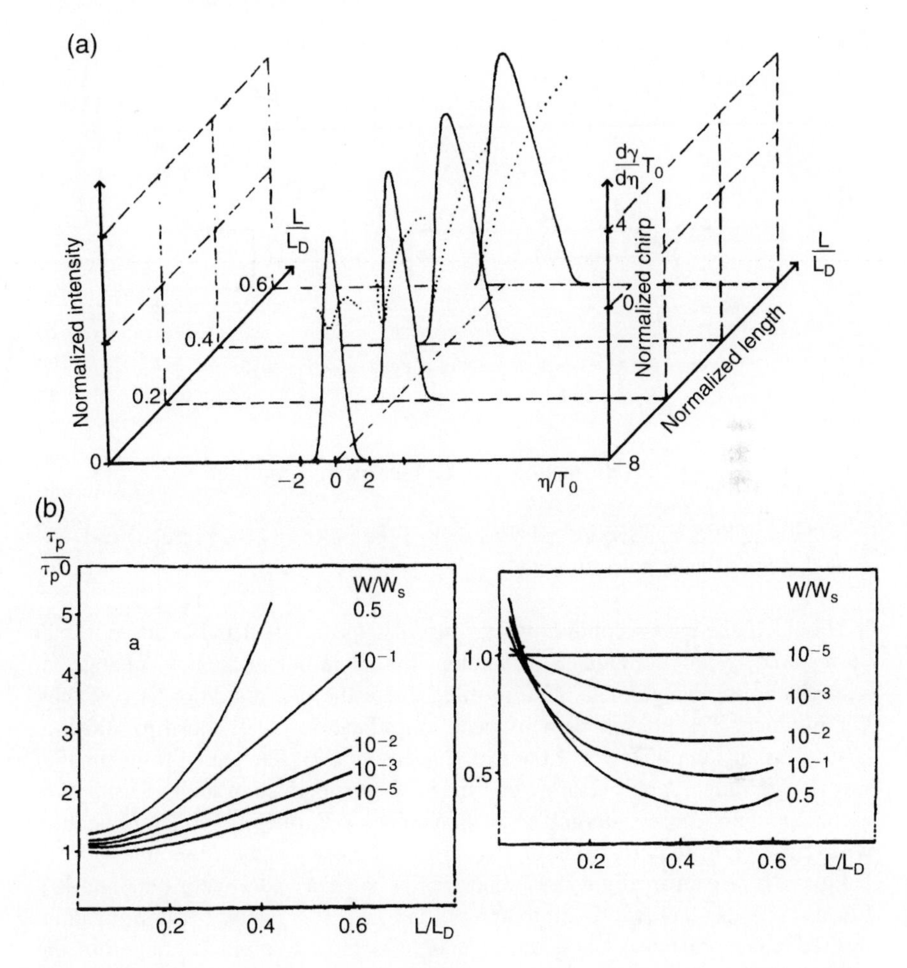

Figure 7.10 (a) Evolution of pulse shape and chirp in an amplifier with GVD and SPM, (b) broadening of pulses in an amplifier with GVD and SPM (left) and pulse duration normalized to that of the input pulse after an optimum quadratic compressor (right). The initial pulse shape is Gaussian, applied at resonance with the gain medium. Parameters: small signal gain 2×10^4, $\tau_{p0} = 100$ fs, $k_\ell'' = 6\times 10^{-26}\,\mathrm{s^2m^{-1}}$, $n_2 = 4\times 10^{-23}\,\mathrm{m^2V^{-2}}$, $\sigma_{10} = 10^{-16}\,\mathrm{cm^2}$. (Adapted from Heist *et al.* [17].)

and thereafter recompress the pulse to the original pulse duration and shape with the conjugate dispersion line (dispersion line with opposite GVD).

A block diagram illustrating the CPA concept is shown in Figure. 7.11. Stretching of a pulse up to 10,000 times can be achieved with a combination of gratings and a telescope, as discussed in Chapter 2. Such a combination of linear elements does not modify the original pulse spectrum. For the amplification to be truly

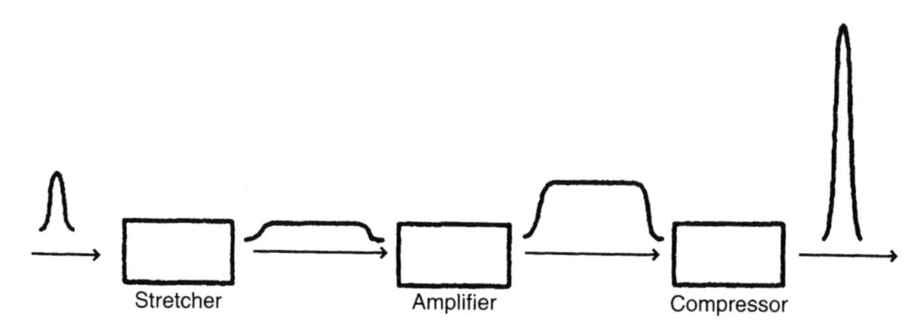

Figure 7.11 Block diagram of chirped pulse amplification.

linear, two essential conditions have to be met by the amplifier:

- the amplifier bandwidth exceeds that of the pulse to be amplified; and
- the amplifier is not saturated.

It is only if these two conditions are met that the original pulse duration can be restored by the conjugated dispersion line. It is not unusual to operate an amplifier in the wings of its gain profile, where the first condition is best met. For instance, Ti:sapphire, with its peak amplification factor close to 800 nm, is used as an amplifier for 1.06 μm, because of its flat gain profile in that wavelength range. A pulse energy of 1 mJ has been obtained in such a Ti:sapphire amplifier chain, corresponding to a gain of 10^7. Further *linear* amplification with Ti:sapphire requires rods of too large a diameter to be economical. With Nd:glass as a gain medium pulse energies as large as 20 J were obtained [5]. Because of the bandwidth limitation in the last 10^4 factor of amplification, the recompressed pulse has a duration of 400 fs, a fivefold stretch from the original 80 fs.

7.5. AMPLIFIER DESIGN

7.5.1. Gain Media and Pump Pulses

Parameters of gain media crucial for the amplification of fs pulses are:

- The interaction cross section. For a given amplifier volume (inverted volume) this parameter determines the small signal gain and the maximum possible energy per unit area that can be extracted from the system. The latter is limited by gain saturation.

- The energy storage time of the active medium. If there is no ASE, this time is determined by the lifetime of the upper laser level T_{10} and indicates (a) how long gain is available after pump pulse excitation for $\tau_{pump} < (<) T_g$ or (b) how fast a stationary gain is reached if $\tau_{pump} > (>) T_g$. The corresponding response time can be significantly shorter if ASE occurs, it is then roughly given by:

$$T_{ASE} = \frac{\hbar\omega_{ASE}}{I_{ASE}(L)\sigma_{ASE}}. \tag{7.25}$$

When short pump pulses are used, the latter quantity provides a measure of the maximum jitter allowable between pump and signal pulse without perturbing the reproducibility of amplification.

- The spectral width of the gain profile $\Delta\omega_g = (\omega_g/\lambda_g)\Delta\lambda_g$ and the nature of the line broadening. The minimum pulse duration (or maximum pulse bandwidth) that can be maintained by the amplification process is of the order of $2\pi/\Delta\omega_g$ (or $\Delta\omega_g$).

These parameters were given in Table 7.1 for typical materials used as gain media in fs amplifiers. The wavelength and bandwidth of the seed pulse dictates the choice of the gain medium. Presently, it is only in the near infrared that a selection can be made among various types of gain media, dyes, and solid-state materials for fs pulse amplification. At pulse durations of the order of 10^{-14} s gain narrowing effects of single dyes dominate [3] leading to pulse broadening. These difficulties can be overcome by using a mixture of several dyes with different transition frequencies providing optimum amplification for a broad input spectrum [22]. The achievable energies with dye amplifiers are on the order of 1 mJ. This value is determined by the saturation energy density and the dye volume that can be uniformly pumped with available pump lasers.

Shorter pulses and higher energies can, in principle, be extracted from certain solid-state amplifiers. With the additional advantage of compactness, such systems are attractive candidates for producing pulses in the TW and PW range. These systems are typically limited to the red and near infrared spectral range.

At certain wavelengths in the UV (Table 7.1) excimer gases can be used for fs pulse amplification (see for example, Glownia *et al.* [11], Szatmari *et al.* [12], Watanabe *et al.* [23], Taylor *et al.* [24], Heist *et al.* [25], and Mossavi *et al.* [26]). The interaction cross-section being similar to that of dyes, the saturation energy density of excimer gain media is also of the order of millijoules/cm^2.

Table 7.3

Typical parameters of pump lasers for fs pulse amplifiers.

Laser	Pulse energy (mJ)	Duration (ns)	Repetition rate (Hz)	λ (nm)
Ar^+(cavity dumped)	10^{-3}	15	3×10^6	514
Nd:YAG				
Q-switched	300	5	10	532
Regenerative amplified	2	0.07	10^3	532
Diode pumped	0.05	10	800	532
Nd:YLF (Q-switched)	10	400	10^4	523.5
Copper-vapor	2	15	5000	510, 578
Excimer	100	20	10	308

Much larger spulse energies however—ranging from millijoules to the Joule range—can be extracted, because the active volume that can be pumped is much larger than in dye amplifiers. As compared to solid-state or liquid materials, another advantage of excimer gases is the smaller susceptibilities associated with undesired nonlinear effects (such as self-focusing). Unfortunately, the relatively narrow gain bandwidth of excimers limits the shortest pulse duration that can be amplified and the tunability.

Essential pulse parameters, such as the achievable energy range and repetition rate, that can be reached are determined by the pump laser of the amplifier. Table 7.3 summarizes data on lasers that have successfully been used for pumping fs amplifiers. Usually these pump lasers have to be synchronized to the high repetition rate oscillators for reproducible amplification. On a nanosecond time scale this synchronization can be achieved electronically. With picosecond pump pulses, satisfactory synchronism requires generally that the pump pulses for the femtosecond oscillator and amplifier be derived from a single master oscillator.

7.5.2. Amplifier Configurations

Usually the amplifier is expected to satisfy certain requirements for the output radiation, which can be achieved by a suitable design and choice of the components. Table 7.4 shows some examples. Different applications of amplified pulses have different requirements, and subsequently various amplifier configurations have been developed. In particular, trade-off between pulse energy and repetition rate will call for a particular choice of amplifier design and pump source. A feature common to nearly all femtosecond amplifiers is that they are terminated by a linear optical element to recompress the pulses.

Table 7.4

Design requirements of a fs pulse amplifier.

Requirements	Realization
(a) Clean beam profile	Homogeneously inverted gain region, no self-focusing, proper (linear) optical design
(b) High peak power	Same as above, CPA
(c) High energy amplification	High pump power, amplification reaches saturation level
(d) Low background	ASE suppression through spatial and/or spectral filtering, filtering through saturable absorption
(e) Certain repetition rate	Repetition rate of pump, suitable gain medium
(f) No temporal broadening	GVD adjustment
(g) No spectral narrowing	Gain medium with broad band-width

Multistage Amplifiers

Low repetition rate systems (<500 Hz) used for high gain amplification consist mostly of several stages traversed in sequence by the signal pulse. A typical configuration is sketched in Figure 7.12. This concept, introduced by Fork *et al.* [27] for the amplification of fs pulses in a dye amplifier, has the following advantages. (a) Each stage can be adjusted separately for maximum gain, considering the particular signal pulse energy at that stage. The splitting of the pump energy among the various stages has to be optimized, as well as the pump focalization to match the volume to be pumped. Typically, only a few percent of the pump pulse is tightly focused into the first stage, resulting in a gain of several thousands. More than 50% of the pump energy is reserved for the last stage (to pump a much larger volume) resulting in a gain factor of about 10. (b) The unavoidable ASE can be suppressed with filters inserted between successive stages. Ideally, these filters are linear attenuators for the ASE, but are saturated by the signal pulse.

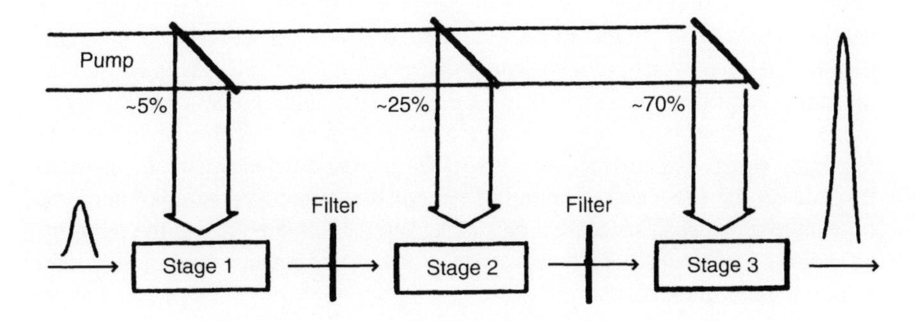

Figure 7.12 Sketch of a multistage amplifier.

Of course, the filter remains "open" after passage of the signal pulse for a time interval given by the energy relaxation time, and subsequently ASE within this temporal range cannot be suppressed. Edge filters, such as semiconductors and semiconductor doped glasses, can be used for ASE reduction whenever the ASE and the signal pulse are spectrally separated. Finally, because the beam characteristics of ASE and signal pulse are quite different, a spatial filter (for example a pinhole in the focal plane between two lenses) can enhance the signal-to-ASE ratio. Typical pump lasers for multistage dye amplifiers are Q-switched Nd:YAG lasers [27] and excimer lasers [28, 29] with pulse durations of about 5 ns and 20 ns, respectively. Typically the repetition rates are below 100 Hz, and pulse energies of a few hundred milliJoules have been reported. In the last decade many modifications of the setup shown in Figure 7.12 have been made. For example, multiple passages through one and the same stage to extract more energy or/and to use smaller pump lasers were implemented. To increase the homogeneity of the gain region longitudinal pumping is frequently used in the last amplifier stage(s).

An example for a fs multistage amplifier pumped by a XeCl excimer laser is shown in Figure 7.13. The excimer laser, consisting of two separate discharge channels, serves to pump the dye cells and to amplify the frequency doubled fs output at 308 nm. Another option is to generate a fs white light continuum and to amplify a certain spectral component. With the UV pump pulses and different dyes a wavelength range from the NIR to the UV can be covered.

7.5.3. Single-Stage, Multipass Amplifiers

For larger repetition rates one has to use pump lasers working at higher frequencies. Because the mean output power of tabletop pump lasers cannot be increased arbitrarily, higher repetition rates are achieved at the expense of energy per pulse. To obtain still reasonable gain factors one has to increase the efficiency of converting pump energy into signal energy as compared to the configurations described previously. In this respect the basic disadvantage of single-pass amplifiers is that only a fraction of the energy pumped into the gain media is used for the actual amplification. The main reason is that the pump process is often much longer than the recovery time of the gain medium; hence, a considerable part of the pump energy is converted into ASE. The overall efficiency can be enhanced by sending the pulse to be amplified several times through the amplifier. The time interval between successive passages should be of the order of the recovery time of the gain medium T_g. The number of passages should not exceed the ratio of pump pulse duration to T_g.

In a first attempt to amplify femtosecond pulses at high repetition rates, Downer *et al.* [30] used a cavity-dumped Ar^+-laser to pump a dye amplifier.

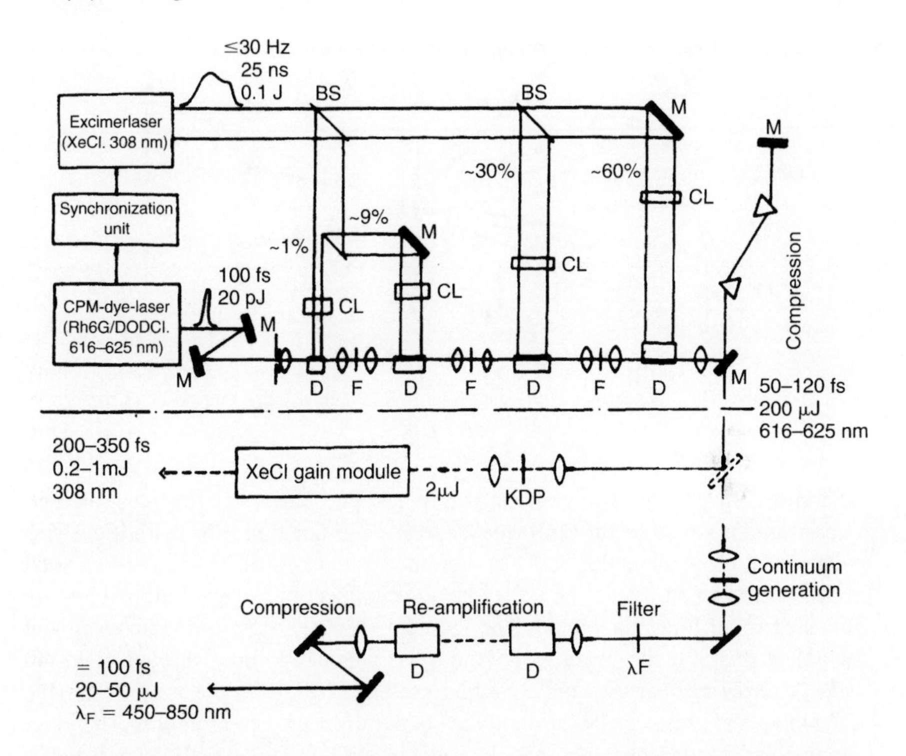

Figure 7.13 A fs dye amplifier pumped by an excimer laser. The amplification stages are decoupled by saturable absorbers (semiconductor doped glasses) or k-space filters. The prism sequence serves to compress the phase modulated, amplified pulses. To extend the wavelength range of available fs pulses, the frequency doubled output can be amplified in the second discharge channel of the excimer laser. Another option is to amplify the spectrally filtered white light continuum. (Adapted from Heist *et al.* [25].)

Despite the high repetition rate (3 MHz), this approach did not find broad application because of the relatively small net gain (~100) resulting from the low energy of the pump pulses. The use of copper-vapor lasers, working at repetition rates from 5 to 15 kHz, turned out to be a more practicable concept to pump dye amplifiers [31]. Knox *et al.* [32] used such a laser to pump a single dye jet and to amplify 100 fs pulses to microjoule energies. The dye jet was passed 6 times to match the pump pulse duration (~25 ns) where the reported small signal gain per pass was 5 to 8. The disadvantage of such a configuration, as sketched in Figure 7.14, is its complexity and large number of optical elements. Two saturable absorbers were implemented to suppress ASE. Higher output powers could be reached using a dye cell for the gain medium [33]. Other concepts distinguish themselves by a minimum number of optical components and simplicity

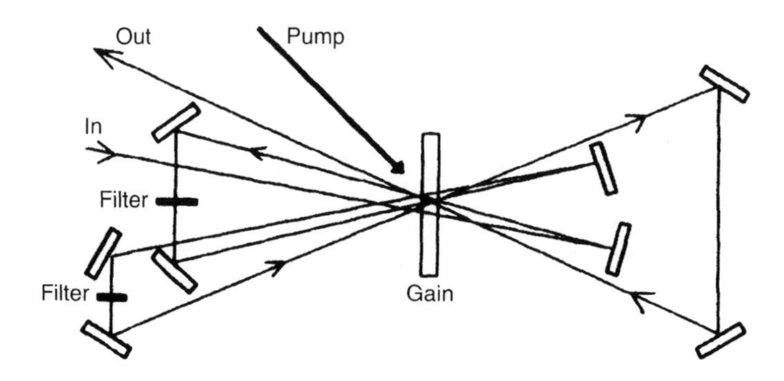

Figure 7.14 Sketch of a single-stage, multipass amplifier (after Knox *et al.* [32]).

of adjustment [20,34]. The gain medium is inserted in a resonator-like structure. A convenient structure for uniform transverse illumination of a cylindrical volume is the "Bethune cell" [35]. The volume to be pumped is inserted in a total reflection prism, at a location such that adjacent sections of the pump beam are reflected to all four quadrants of the cylinder (Figure 7.15). In the arrangement of Lai *et al.* [20], the beam to be amplified is sent 11 times through the gain cell. A series of apertures on a circular pattern (Fig. 7.15) are a guide for the alignment and prevent oscillation in the stable resonator configuration. The latter being close to concentric, the first 11 paths are focused to a small beam waist in the gain medium. This amplifier is intentionally operating at saturation for the

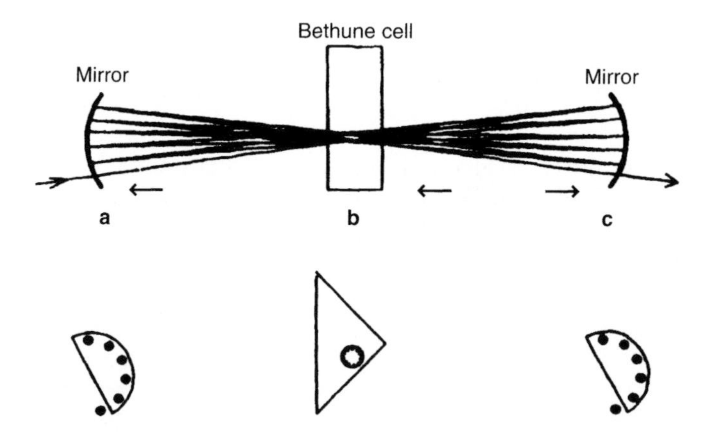

Figure 7.15 Schematic diagram of a multipass (copper-vapor laser pumped) amplifier and view on the beam geometry at the focusing mirrors and the Bethune cell. (Adapted from Lai *et al.* [20].)

last few passes, to reduce its sensitivity to fluctuations of the input. The beam is sent back for 2 more passes through the center of the 2 mm diameter amplifying cell, providing unsaturated amplification to 15 μJ. Copper-vapor laser pumped amplifiers have also been successfully applied to generate powerful femtosecond pulses in the near infrared [36]. Multipass amplifiers are also common for solid-state fs systems, for example based on Ti:sapphire, if moderate output energies (typically not exceeding several mJ) are desired.

7.5.4. Regenerative Amplifiers

As previously discussed, a broad gain bandwidth and high saturation energy density make some solid state materials (cf. Table 7.1) prime candidates for the generation of powerful femtosecond light pulses. A large energy storage time (10^{-6} s) is generally associated with the small gain cross section of these amplifying media. Regenerative amplification is the most efficient method to transfer efficiently energy to a fs pulse from an amplifier with a long storage time. The concept of regenerative amplification is illustrated in Figure 7.16.

The gain medium is placed in a resonator built by the high-reflecting mirrors M_1 and M_2. After the seed pulse is coupled into the resonator through a polarizer P_1, the Pockels cell PC1 is switched to rotate the polarization of the seed pulse and Q-switches the resonator. The pulse circulates in the resonator and is continuously amplified. After a certain number of round-trips (determined by the energy storage time and/or the time needed to reach saturation) a quarter-wave voltage is applied to the cavity-dumping Pockels cell PC2 and the amplified pulse is coupled out by reflection from the second polarizer P_2.

Regenerative amplifiers were originally developed to amplify the output of cw mode-locked solid-state (e.g., Nd:YAG, Nd:YLF) lasers at repetition rates up to 2 kHz and energies up to the milliJoule level (e.g., [37–39]), to obtain ps pulses at microjoule energies. These pulses served as pump for dye amplifiers [40]. Because the pump pulse duration is on the order of 100 ps the amplification

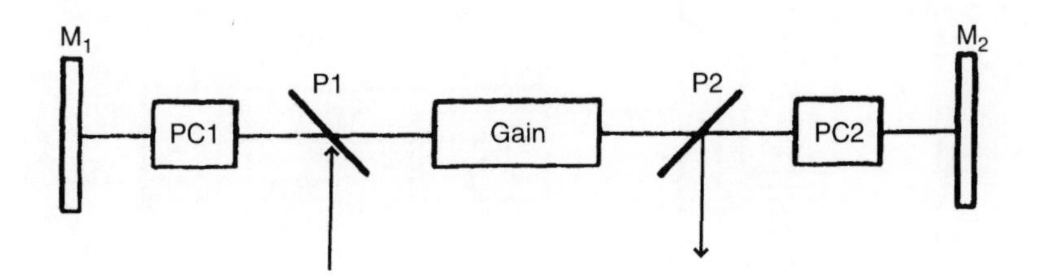

Figure 7.16 Principle of regenerative amplification.

process can be much more efficient than with ns pump lasers. Synchronization between pump and fs signal is achieved by pumping the dye laser synchronously with the same master oscillator used to provide the seed pulses for the regenerative amplifier. More recently regenerative solid-state amplifiers have been used to amplify fs light pulses directly in alexandrite and Ti:sapphire utilizing chirped pulse amplification [41,42]. Using a Q-switched Nd:YLF laser as pump for the Ti:sapphire crystal in the regenerative amplifier a repetition rate as high as 7 kHz could be reached [43].

A multiterawatt, 30 fs, Ti:sapphire laser system based on a combination of a regenerative amplifier and a multipass amplifier, operating at 10 Hz, was reported by Barty *et al.* [44]. A sketch of the system is shown in Figure 7.17. The 20-fs (5 nJ, 800 nm) pulses from a mode-locked Ti:sapphire laser are stretched to 300 ps. Amplification in a 14-pass regenerative amplifier (50 mJ pump pulse at 532 nm) yielded 9 mJ output pulses. A 4-pass amplifier (235 mJ pump pulse at 532 nm) increases the pulse energy to 125 mJ. Finally, after recompression, 30-fs pulses were obtained. To reduce the effect of gain narrowing the spectrum of the pulse prior to amplification was flattened. This was accomplished by spectral filtering (element M in Fig. 7.17, see also Chapter 8).

7.5.5. Traveling Wave Amplification

Amplified stimulated emission limits strongly the overall efficiency of the amplification process, in particular at shorter wavelengths where the ratio of spontaneous emission to stimulated emission is larger. One of the causes of a large

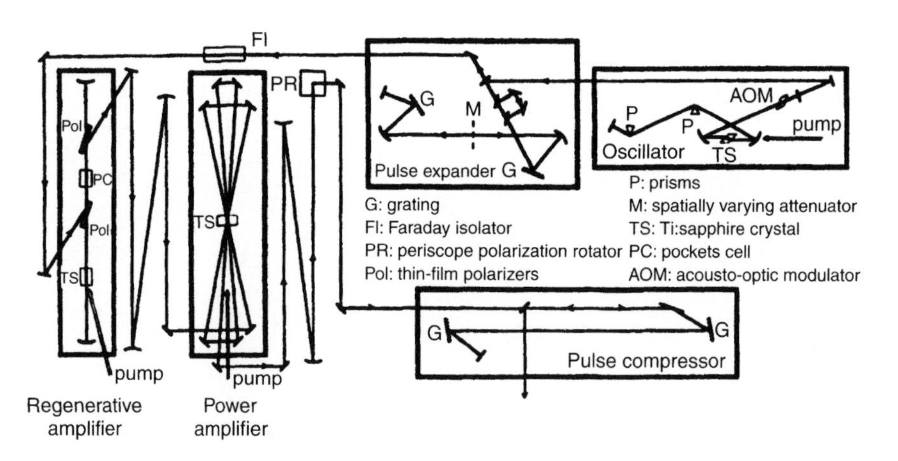

Figure 7.17 Multiterawatt fs laser system. (Adapted from Barty *et al.* [44].)

ASE-to-signal ratio is that the duration of the pump—and hence that of the gain—generally exceeds that of the pulse to be amplified by several orders of magnitude. Considerably higher conversion from pump energy into signal pulse energy is therefore expected when using femtosecond pump pulses. It may seem ludicrous to use a powerful fs light pulse, that is, a fs pulse already amplified, to amplify another fs pulse. However, such schemes offer the prospect of efficient frequency conversion with continuous tunability. A first implementation is the traveling wave amplifier (TWA) introduced by Polland *et al.* [45] and Klebniczki *et al.* [46] in a transverse pumping arrangement with ps light pulses. The TWA technique was successfully extended to the fs time scale by Hebling and Kuhl [47]. A theoretical analysis of TWA can be found in Klebniczki [46,48], Chernev and Petrov [48], and Werner *et al.* [49]. An example of implementation of TWA is sketched in Figure 7.18. The active medium can be a dye cell. The tilt of the pulse front of the pump with respect to the propagation direction of the signal pulse is chosen so as to invert the gain medium in synchronism with the propagating signal light. This TWA leaves practically no time for ASE in front of the signal pulse to develop.

It would seem that a tunable fs source—such as that provided by a spectrally filtered white light continuum—is required as seed pulse to produce fs light pulses at new frequencies by TWA. Such a sophisticated seed is fortunately not required, because the output pulse duration of the TWA amplifier does not depend on the

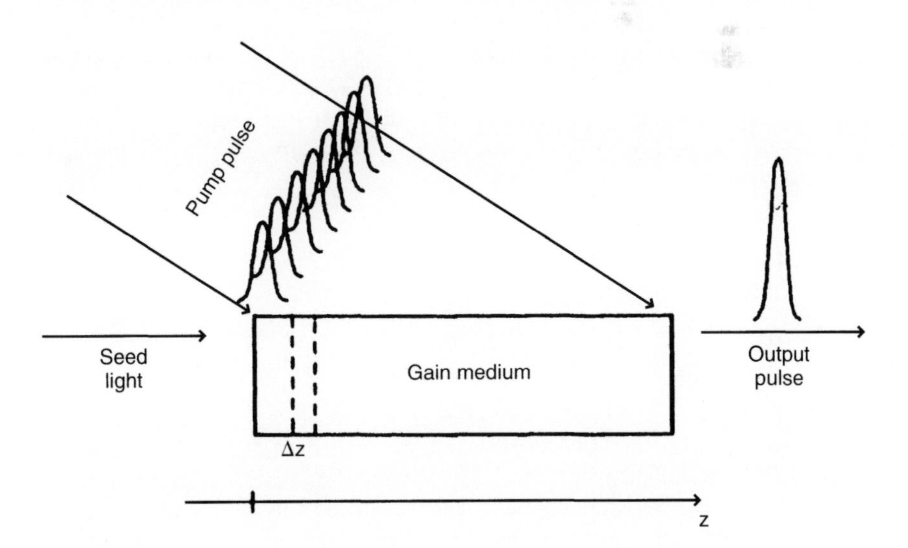

Figure 7.18 Transverse traveling wave amplification. A fs pulse, a spectrally filtered fs continuum or suitable (quasi-) cw radiation (for example, ASE) can serve as seed light.

input pulse duration. Hebling and Kuhl [50], for example, simply selected a certain spectral component from the ASE excited in the active medium as amplifier input. With three different dyes and pump pulses at 620 nm a spectral range between 660 nm and 785 nm could continuously be covered. In the absence of external seeding, the ASE excited in the first part of the amplifier serves as seed signal. The pulse evolution is similar to that in a synchronously pumped laser. The propagation from one TWA amplifier slice to the next compares to successive passages through the gain jet in a synchronously pumped laser. In view of the short pump pulse, the output will also be a fs light pulse with a mean wavelength given roughly by the maximum of the net gain profile of the amplifier. In a synchronously pumped laser the pulse evolution starts from noise (spontaneous emission) and a certain number of round trips are needed for the pulse to form (cf. Chapter 5). In the TWA the number of round-trips translates into a minimum number of "slices," which corresponds to a minimum amplification length. This analogy leads to conclude that the output pulse duration is shorter than the pump pulse and is a sensitive function of the tilt angle (the latter determines the timing mismatch of signal and pump pulse at any given location in the amplifier).

The general approach introduced in Chapter 1 can be applied to a quantitative study of TWA. The complex electric field amplitude of the signal pulse at position $z + \Delta z$, $\tilde{\mathcal{E}}_s(t, z + \Delta z)$ is related to the amplitude at z through:

$$\tilde{\mathcal{E}}_s(t, z + \Delta z) = \tilde{\mathcal{E}}_s(t, z) + \delta_g \tilde{\mathcal{E}}_s(t, z) + \delta_{nl} \tilde{\mathcal{E}}_s(t, z) + \delta_{k''} \tilde{\mathcal{E}}_s(t, z) \tag{7.26}$$

where Δz is the slice width, and $\delta_g \tilde{\mathcal{E}}_s$, $\delta_{nl} \tilde{\mathcal{E}}_s$, and $\delta_{k''} \tilde{\mathcal{E}}_s$ describe the amplitude change because of gain, nonlinear refractive index effects, and GVD, respectively. These changes can easily be calculated by means of Eqs. (7.8), (5.29), (3.146), and (1.198). Neglecting GVD, for the photon flux density and phase of the signal pulse we find:

$$F_s(t, z + \Delta z) = F_s(t, z) + |\tilde{L}(\omega_{10} - \omega_s)|^2 \sigma_{10}^{(0)} (N_1 - N_0) F_s(t, z) \Delta z \tag{7.27}$$

$$\phi_s(t, z + \Delta z) = \phi_s(t, z) - \frac{1}{2} \mathrm{Im}\left[\tilde{L}(\omega_{10} - \omega_s)\right] \sigma_{10}^{(0)} (N_1 - N_0) \Delta z \tag{7.28}$$

$$- i \left(\frac{2\omega_m}{c^2 n_0 \epsilon_0} \right) n_2 [F_s(t, z) + F_p(t, z)] \Delta z \tag{7.29}$$

where, for the sake of simplicity, we have introduced a mean frequency of signal and pump pulse, ω_m, and the linear refractive index n_0 at this frequency. F_p denotes the photon flux density of the pump pulse. As indicated in these equations, the phase modulation originates from (near) resonant (saturation) and nonresonant (host medium) contributions to the changes in index of refraction. The inversion density $\Delta N_{10}(t, z) = N_1 - N_0$ is obtained by solving a system of rate equations for the population numbers in each slice. Assuming a three-level system as shown in Fig. 7.1:

$$\frac{d}{dt}N_0(t,z) = -\sigma_{02}N_0F_p(t,z) + \sigma_{10}^{(0)}|\tilde{L}(\omega_{10}-\omega_s)|^2 N_1 F_s(t,z) \tag{7.30}$$

$$\frac{d}{dt}N_1(t,z) = -\sigma_{10}^{(0)}|\tilde{L}(\omega_{10}-\omega_s)|^2 N_1 F_s(t,z) + \frac{N_2}{T_{21}} \tag{7.31}$$

$$\bar{N} = N_0 + N_1 + N_2. \tag{7.32}$$

Because of the short pump pulse duration, we cannot neglect the population in level 2 and have to consider a nonzero relaxation time T_{21}. Starting either with the photon flux of a small seed pulse or with cw light at $z = 0$ the successive application of Eq. (7.26) yields the signal pulse for an amplifier length z. Figure 7.19 shows corresponding results for the evolution of pulse duration and pulse energy as a function of the propagation length [49]. Region I is characterized by a decrease of signal pulse duration and an almost exponential increase in energy. This is because of the fast rise of the gain (short pump pulse duration) resulting in a rapid buildup of a steep leading edge of the signal pulse, and the

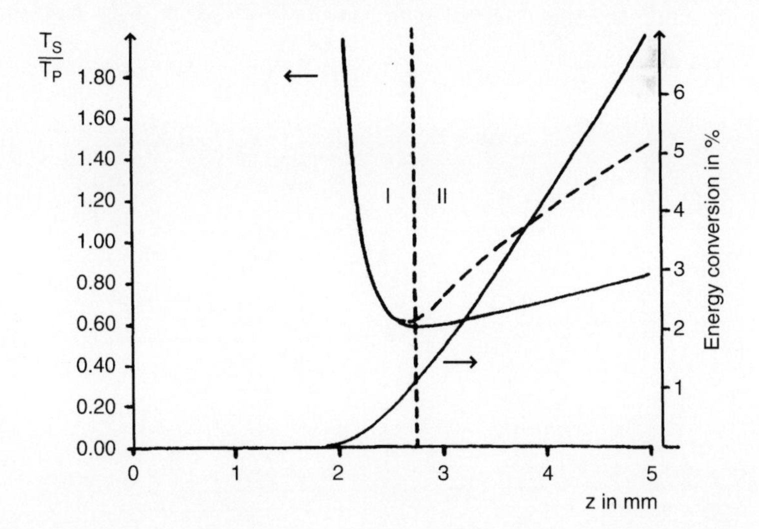

Figure 7.19 Ratio of signal pulse to pump pulse duration and energy conversion as function of the propagation length for typical parameters of a traveling wave dye amplifier. The dashed line describes the behavior of the pulse duration when GVD plays a role. (Adapted from Werner *et al.* [49].)

fact that the gain is essentially unsaturated until the pumping process stops. The latter is responsible for the shaping of the trailing edge of the signal pulse. With the onset of saturation (Region II) the pulse duration increases, a tendency which becomes more pronounced with significant GVD.

7.6. OPTICAL PARAMETRIC CHIRPED PULSE AMPLIFICATION (OPCPA)

The idea of optical parametric chirped pulse amplification (OPCPA) is to replace the laser gain media of a CPA system [21], (cf. Section 7.4), by a nonlinear crystal. Amplification by stimulated emission is substituted by parametric amplification of the signal pulse in the presence of a pump pulse. The apparatus is sketched in Figure 7.20. Since the introduction of the concept of chirped parametric amplification by Dubietis *et al.* [51] in 1992 and an analysis of the prospects for high-power amplification by Ross *et al.* [52] several amplifier systems have been developed producing sub-ps pulses in the TW range. A summary of various concepts and recent progress can be found in Ross *et al.* [52] and Butkus *et al.* [53].

The OPCPA concept relies on the fact that a chirped pulse can be parametrically amplified without significant distortion of the phase if the OPA bandwidth is large enough. Note that the bandwidth is determined by material parameters, the pump and signal wavelength, and the geometry favoring noncollinear

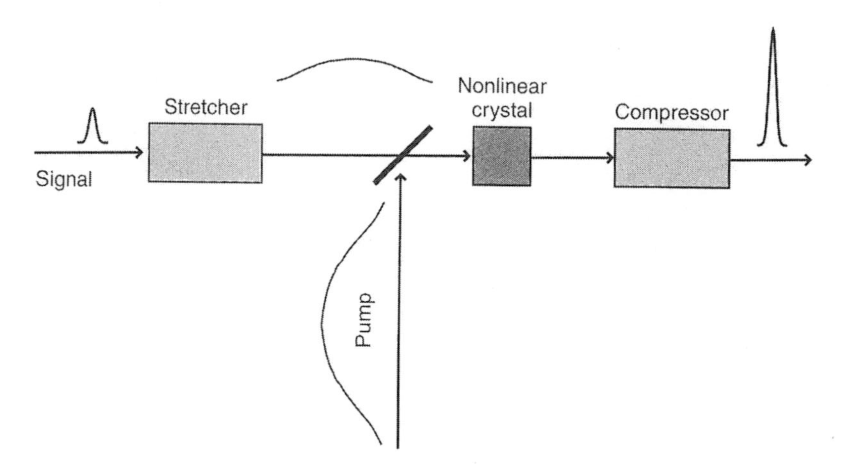

Figure 7.20 Schematic diagram of an OPCPA device. The pulse to be amplified is stretched and chirped and combined with a pump pulse in a nonlinear optical crystal. The parametrically amplified chirped output is subsequently compressed.

schemes (see for example DiTrapani *et al.* [54] and Wilhelm *et al.* [55]). Using a noncollinear geometry extremely large amplification bandwidths have been achieved, resulting in pulses as short as 4 fs after compression of the amplified spectrum (continuum) [56].

Because optical parametric amplification is a nonresonant nonlinear optical process the material does not absorb energy, and the heat load is considerably smaller than in ordinary laser gain media. High gain factors can be achieved at relatively small interaction lengths (single pass) which reduces the effect of dispersion and nonlinear phase modulation on the pulse and beam profile. Provided the gain bandwidth is large enough, gain narrowing, which limits the bandwidth of pulses from ordinary amplifiers, is absent. Another positive aspect of OPCPA is the greatly reduced background radiation, that usually results from ASE.

As discussed in Chapter 3, the parametric gain for the signal intensity is proportional to the pump intensity

$$I_s \propto I_p, \tag{7.33}$$

which is a manifestation of the instantaneous character of this nonlinear optical process. Therefore, for efficient energy conversion, the OPCPA requires a pump pulse whose duration matches that of the signal pulse. In addition, the fact that gain is present only during the pump pulse puts stringent requirements on the relative timing of signal and pump pulse. To achieve a large and homogeneous (over the beam profile) conversion bandwidth certain requirements on the beam profile and focusing of the pump exist.

The first OPCPA systems produced promising results. Pulse powers in the TW range have been reported [57, 58] and OPCPA and CPA were combined in a hybrid system to improve efficiency [59]. It has been experimentally verified that the (CEO) defined in Chapter 5, Section 5.1.3, is preserved throughout the stretching, amplification, and subsequent compression process [60]. To avoid phase-disturbing influences of diffraction gratings, the pump pulse was a regeneratively amplified picosecond pulse, and the stretching of the seed pulse (12 fs, 1 nJ) was achieved via an acousto-optic programmable dispersive filter or DAZZLER [61]. The output pulse was locked in CEO, with a duration of 17 fs, 85 μJ energy at a repetition rate of 1 kHz [60].

7.7. PROBLEMS

1. An important parameter of a gain medium is the energy storage time. Consider a one-stage amplifier of 5 mm length transversely pumped by a pulse of 20 ns duration and 5 mJ energy. A (fs) pulse with an energy of

100 pJ is to be amplified. The gain medium consists of a three-level system (cf. Fig. 7.1). The relaxation time from level 2 to level 1 is assumed to be extremely fast. Calculate and compare the energy amplification achievable in a single-pass configuration for a gain medium with a lifetime of the upper gain level, T_{10}, of (a) 100 ps, (b) 1 μs. For simplification you may assume a rectangular temporal and spatial profile for both pump and pulse to be amplified. Perform your calculation for a beam size of $50 \times 50\ \mu m^2$. Assume homogeneous gain and equal cross sections for absorption and amplification, $\sigma \approx 10^{-17}$ cm^2.

2. Explain the different effect of gain saturation on the shaping of the wings of a Gaussian and a sech pulse (cf. Fig. 7.4).
3. By means of Fig. 7.6 design a three-stage dye amplifier to amplify the output of a fs dye laser (100 fs, 100 pJ) to > 0.5 mJ. The second harmonic of a Nd:Yag (50 mJ, 10 ns) is to be used as pump. For the absorption and emission cross section use a value of 10^{-16} cm^2. Specify the split of the pump energy among the three stages and make an estimate for the focusing conditions.
4. Let us consider a 2 cm long cuvette filled with a solution of Rhodamine 6G. The dye solution is pumped longitudinally by a Nd:Yag laser beam of uniform intensity $I = 10$ MW/cm^2. The dye has an absorption coefficient of 5 cm^{-1} at the pump laser wavelength. The absorption cross section of the solution is $5\ 10^{-16}$ cm^2. Approximate the dye solution by a three-level system, with the upper level being common to the pump and lasing transition. The pump transition is from the ground state to the upper level. The lasing transition (cross section $\sigma_g = 5 \times 10^{-17}$ cm^2) is from the upper level (lifetime of 2.5 ns) to an intermediate level which relaxes to the ground state with a characteristic energy relaxation time of 1 ps.

   (a) Find the gain distribution $\alpha_g(z)$ along the propagation direction (z) of the pump beam.
   (b) A 50 fs pulse, with an energy density of 100 μJ/cm^2 is sent through the medium along the same z direction. Calculate the energy of the pulse exiting the gain cell. Solve the problem for a 100 μJ/cm^2 pulse sent in the direction opposite to the pump. How do the results differ?
   (c) Assume next that the pump beam diameter decreases linearly from 1 cm at the cell entrance down to 1 mm at z = 2 cm. The pump power at the cell entrance is 1 MW. The beam to be amplified has the same geometry, the initial pulse energy being 5 μJ. Find the amplified pulse energy for co- and counter-propagating pump pulse and pulse to be amplified.

BIBLIOGRAPHY

[1] P. Heist, W. Rudolph, and B. Wilhelmi. Amplification of femtosecond light pulses. In W. Koehler, Ed., *Laser Handbook* vol. 2, Vulkan Verlag, Essen, Germany, 1991, pp. 62–73.

[2] J. D. Simon. Ultrashort light pulses. *Review of Scientific Instruments*, 60:3597–3620, 1989.

[3] W. H. Knox. Femtosecond optical pulse amplification. *IEEE Journal of Quantum Electronics*, 24:388–397, 1988.

[4] S. Szatmari, F. P. Schaefer, E. Mueller-Horsche, and W. Mueckenheim. Hybrid dye-excimer laser system for the generation of 80 fs pulses at 248 nm. *Optics Communications*, 63:5–10, 1987.

[5] P. Maine, D. Strickland, P. Bado, M. Pessot, and G. Mourou. Generation of ultrahigh peak power pulses by chirped pulse amplification. *IEEE Journal of Quantum Electronics*, QE-24:398–403, 1988.

[6] S. Smolorz and F. Wise. Femtosecond two-beam coupling energy transfer from raman and electronic nonlinearities. *Journal of the Optical Society of America B*, 17:1636–1644, 2000.

[7] R. W. Hellwarth, D. M. Pennington, and M. A. Henesian. Indices governing optical self-focusing and self-induced changes in the state of polarization in N_2, O_2, H_2, and ar gases. *Physical Review A*, 41:2766, 1990.

[8] E. T. J. Nibbering, G. Grillon, M. A. Franco, B. S. Prade, and A. Mysyrowicz. Determination of the inertial contribution to the nonlinear refractive index of air, N_2, and O_2 by use of unfocused high-intensity femtosecond laser pulses. *Journal of the Optical Society of America*, B14:650–660, 1997.

[9] Y. Shimoji, A. T. Fay, R. S. F. Chang, and N. Djeu. Direct measurement of the nonlinear refractive index of air. *Journal of the Optical Society of America B*, 6(11):1994–1998, 1989.

[10] M. J. Shaw, C. J. Hooker, and D. C. Wilson. Measurement of the nonlineaer refractive index of air and other gases at 248 nm. *Optics Commnuications*, 103:153–160, 1993.

[11] J. H. Glownia, J. Misewich, and P. P. Sorokin. 160 fs XeCl excimer amplifier system. *Journal of the Optical Society of America B*, 4:1061, 1987.

[12] S. Szatmari, B. Racz, and F. P. Schaefer. Bandwidth limited amplification of 220 fs pulses in XeCl. *Optics Communications*, 62:271–276, 1987.

[13] A. Penzkofer and W. Falkenstein. Theoretical investigation of amplified spontaneous emission with picosecond light pulses in dye solutions. *Optics and Quantum Electronics*, 10:399–405, 1978.

[14] A. Hnilo and O. E. Martinez. On the design of pulsed dye laser amplifiers. *IEEE Journal of Quantum Electronics*, QE-23:593–599, 1987.

[15] G. P. Agrawal and N. A. Olsson. Self-phase modulation and spectral broadening of optical pulses in semiconducter laser amplifiers. *IEEE Journal of Quantum Electronics*, QE-25:2297–2306, 1989.

[16] W. Dietel, E. Doepel, V. Petrov, C. Rempel, W. Rudolph, B. Wilhelmi, G. Marowsky, and F. P. Schaefer. Self phase modulation in a femtosecond dye amplifier with subsequent compression. *Applied Physics B*, 46:183–185, 1988.

[17] P. Heist, W. Rudolph, and V. Petrov. Combined self-phase modulation and amplification of femtosecond light pulses. *Applied Physics B*, 49:113–119, 1989.

[18] D. C. Hutchings, M. Sheik-Bahae, D. J. Hagan, and E. W. Van Stryland. Kramers-kronig relations in nonlinear optics. *Optics and Quantum Electronics*, 24:1–30, 1992.

[19] M. Sheik-Bahae, A. A. Said, and E. W. Van Stryland. High sensitivity single beam n_2 measurement. *Optics Letters*, 14:955–957, 1989.

[20] M. Lai, J.-C. Diels, and C. Yan. A transversally pumped 11-pass amplifier for femtosecond optical pulses. *Applied Optics*, 30:4365–4367, 1991.

[21] D. Strickland and G. Mourou. Compression of amplified chirped optical pulses. *Optics Communications*, 56:219–221, 1985.

[22] G. Boyer, M. Franco, J. P. Chambaret, A. Migus, A. Antonetti, P. Georges, F. Salin, and A. Brun. Generation of 0.6 mj pulses of 16 fs duration through high-repetition rate amplification of self-phase modulated pulses. *Applied Physics Letters*, 53:823–825, 1988.

[23] S. Watanabe, A. Endoh, M. Watanabe, and N. Sakura. Terawatt XeCl discharge system. *Optics Letters*, 13:580–582, 1988.

[24] A. J. Taylor, T. R. Gosnell, and J. P. Roberts. Ultrashort pulse energy extraction measurements in XeCl amplifiers. *Optics Letters*, 15:39–41, 1990.

[25] P. Heist, W. Rudolph, and B. Wilhelmi. Amplification of femtosecond light pulses. *Experimental Technique of Physics*, 38:163–188, 1990.

[26] K. Mossavi, T. Hoffmann, and F. K. Tittel. Ultrahigh-brightness, femtosecond ArF excimer laser system. *Applied Physics Letters*, 62:1203–1205, 1993.

[27] R L. Fork, C. V. Shank, and R. Yen. Amplification of 70 fs optical pulse to gigawatt powers. *Applied Physics Letters*, 41:223–225, 1982.

[28] C. Rolland and P. B. Corkum. Amplification of 70 fs pulses in a high repetition rate XeCl pumped dye laser amplifier. *Optics Communications*, 59:64–67, 1986.

[29] T. Turner, M. Chatalet, D. S. Moore, and S. C. Schmidt. Large gain amplifier for subpicosecond optical pulses. *Optics Letters*, 11:357–359, 1986.

[30] M. C. Downer, R. L. Fork, and M. Islam. 3 MHz amplifier for femtosecond optical pulses. In D. H. Auston and K. B. Eisenthal, Eds., *Ultrafast Phenomena IV*, Springer-Verlag, Berlin Germany, 1984, pp. 27–29.

[31] D. B. McDonald and C. D. Jonah. Amplification of ps pulses using a copper vapor laser. *Review of Scientific Instruments*, 55:1166, 1984.

[32] W. H. Knox, M. C. Downer, R. L. Fork, and C. V. Shank. Amplifier femtosecond optical pulses and continuum generation at 5 kHz repetition rate. *Optics Letters*, 9:552–554, 1984.

[33] D. Nickel, D. Kuehlke, and D. von der Linde. Multipass dye cell amplifier for high repetition rate optical pulses. *Optics Letters*, 14:36–39, 1989.

[34] E. V. Koroshilov, I. V. Kryukov, P. G. Kryukov, and A. V. Sharkov. Amplification of fs pulses in a multi-pass dye amplifier. In T. Yajima, K. Yoshihara, C. B. Harris, and S. Shionoya, Eds., *Ultrafast Phenomena VI*, Springer-Verlag, Berlin, Germany, 1984, pp. 22–24.

[35] D. S. Bethune. Dye cell design for high-power low divergence excimer-pumped dye lasers. *Applied Optics*, 20:1897–1899, 1981.

[36] M. A. Kahlow, W. Jarzeba, and T. B. DuBruil. Ultrafast emission spectroscopy by time gated fluorescence. *Review of Scientific Instruments*, 59:1098–1109, 1988.

[37] P. Bado, M. Boovier, and J. S. Coe. Nd:YLF mode locked oscillator and regenerative amplifier. *Optics Letters*, 12:319–321, 1987.

[38] J. E. Murray and D. J. Kuizenga. Regenerative compression of laser pulses. *Applied Physics Letters*, 12:27–30, 1987.

[39] T. Sizer II, J. D. Kafka, A. Krisiloff, and G. Mourou. Generation and amplification of subps pulses using a frequency doubled Nd:Yag pumping source. *Optics Communications*, 39:259–262, 1981.

[40] I. N. Duling II, T. Norris, T. Sizer, P. Bado, and G. Mourou. Kilohertz synchronous amplification of 85 fs optical pulses. *Journal of the Optical Society of America B*, 2:616–621, 1985.

[41] M. Pessot, J. Squier, G. Mourou, and G. Vaillancourt. Chirped-pulse amplification of 100-fs pulses. *Optics Letters*, 14:797–799, 1989.

[42] J. Squier, F. Salin, J. S. Coe, P. Bado, and G. Mourou. Characteristics of an actively mode-locked 2-ps Ti:sapphire laser operating in the 1-μm wavelength regime. *Optics Letters*, 16:85–87, 1991.

[43] F. Salin, J. Squier, G. Mourou, and G. Vaillancourt. Millijoule femtosecond pulse amplification in Ti:Al_2O_3 at multi-khz repetition rates. In *Ultrafast Phenomena VIII*, Springer-Verlag, Berlin, Germany, 1993, pp. 203–205.

[44] C. P. J. Barty, C. L. Gordon III, and B. E. Lemoff. Multiterawatt 30-fs Ti:sapphire laser system. *Optics Letters*, 19:1442–1444, 1994.

[45] H. J. Polland, T. Elsaesser, W. Kaiser, M. Kussler, N. J. Marx, B. Sens, and K. Drexhage. Picosecond dye laser emission in the infrared. *Applied Physics*, B32:53–57, 1983.

[46] J. Klebniczki, Z. Bor, and G. Szabo. Theory of traveling wave amplified spontaneous emission. *Applied Physics B*, 46:151–155, 1988.

[47] J. Hebling and J. Kuhl. Generation of femtosecond pulses by amplified spontaneous emission. *Optics Letters*, 14:278–280, 1989.

[48] P. Chernev and V. Petrov. Numerical study of traveling wave amplified spontaneous emission. *Optics and Quantum Electronics*, 23:45–52, 1991.

[49] G. Werner, W. Rudolph, P. Heist, and P. Dorn. Simulation of femtosecond pulse traveling wave amplification. *Journal of the Optical Society of America B*, B9:1571–1578, 1992.

[50] J. Hebling and J. Kuhl. Generation of tunable fs pulses by traveling wave amplification. *Optics Communications*, 73:375–379, 1989.

[51] A. Dubietis, G. Jonusauskas, and A. Piskarskas. Powerful fs pulse generation by chirped and stretched pulse parametric amplification in bbo crystals. *Optics Communications*, 88:437–440, 1992.

[52] I. N. Ross, P. Matousek, M. Towrie, A. J. Langley, and J. L. Collier. The prospects for ultrashort pulse duration and ultrahigh intensity using optical parametric chirped pulse amplification. *Optics Communications*, 144:125–133, 1997.

[53] R. Butkus, R. Danielius, A. Dubietis, A. Piskarskas, and A. Stabinis. Progress in chirped pulse optical parametric amplifiers. *Applied Physics B*, 79:693–700, 2004.

[54] P. DiTrapani, A. Andreoni, C. Solcia, P. Foggi, R. Danielius, A. Dubietis, and A. Piskarskas. Matching of group velocities in three-wave parametric interaction with femtosecond pulses and application to traveling-wave generators. *Journal of the Optical Society of America B*, 12: 2237–2244, 1995.

[55] T. Wilhelm, J. Piel, and E. Riedle. Sub-20-fs pulses tunable across the visible from a blue-pumped single-pass noncollinear parametric converter. *Optics Letters*, 22:1494–1496, 1997.

[56] A. Baltuska, T. Fuji, and T. Kobayashi. Visible pulse compression to 4 fs by optical parametric amplification and programmable dispersion control. *Optics Letters*, 27:306–308, 2002.

[57] I. N. Ross, J. L. Collier, P. Matousek, C. N. Danson, D. Neely, R. M. Allot, D. A. Pepler, C. Hernandez-Gomez, and K. Osvay. Generation of terawatt pulses by use of optical parametric chirped pulse amplification. *Applied Optics*, 39:2422–2427, 2000.

[58] I. Jovanovic, B. J. Comaskey, C. A. Ebbers, R. A. Bonner, D. M. Pennington, and E. C. Morse. Optical parametric chirped-pulse amplifier as an alternative to ti:sapphire regenerative amplifiers. *Applied Optics*, 41:2923–2929, 2002.

[59] I. Jovanovic, C. A. Ebbers, and C. P. Barty. Hybrid chirped pulse amplification. *Optics Letters*, 27:1622–1624, 2002.

[60] C. P. Hauri, P. Schlup, G. Arisholm, J. Biegert, and U. Keller. Phase-preserving chirped pulse optical parametric amplification to 127.3 fs directly from a ti:sapphire oscillator. *Optics Letters*, 29:1–3, 2004.

[61] F. Verluise, V. Laude, Z. Cheng, C. Spielmann, and P. Tournois. Amplitude and phase control of ultrashort pulses by use of an acousto-optic programmable dispersive filter: pulse compression and shaping. *Optics Letters*, 25:575–577, 2000.

8

Pulse Shaping

On a fs time scale, many interactions depend on the particular temporal shape of the waveform being applied. For many applications it is desirable and necessary to modify the pulses from the source in a well-defined manner. A compression of the intensity profile leads to shorter pulses and higher peak powers. Closely spaced femtosecond pulses with controllable phase relations are needed for coherent multiphoton excitation and the selective excitation of, for example, certain molecular vibrations as detailed in Chapter 4.

The distortion of the complex pulse envelope caused by most linear and nonlinear optical processes was discussed in previous chapters. In this chapter we shall concentrate on techniques applied to compress or shape pulses in amplitude and phase. A comprehensive review on pulse compression can also be found in Rudolph and Wilhelmi [1]. While shaping of ns and ps pulses can be achieved by electronically driven pulse shapers, such as electro-optic modulators [2], all-optical techniques have to be applied for fs pulse shaping. Dispersion leads to pulse shortening or lengthening depending on the input chirp. Saturable absorption tends to steepen the leading edge of the pulse.

8.1. PULSE COMPRESSION

8.1.1. General

Optical pulse compression is the optical analogue of a well-established technique for the shaping of radar pulses [3]. Its implementation into optics was in the late 1960s for compression of ps pulses [4–7].

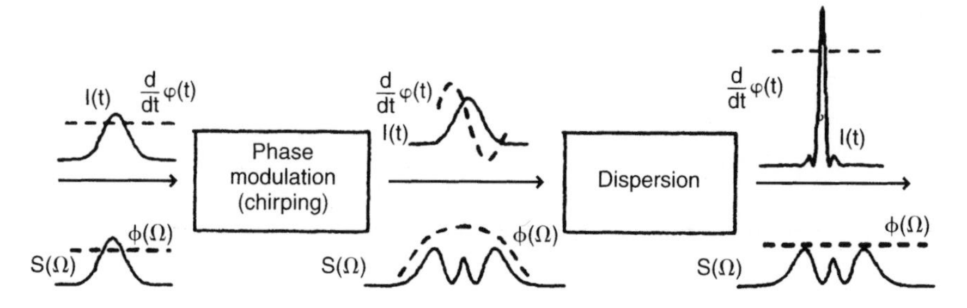

Figure 8.1 Sketch of optical pulse compression through SPM and chirp compensation.

Optical pulse compression is generally achieved by the two-step process sketched in Figure 8.1. Let us assume a bandwidth-limited input pulse. In the first step a phase modulation $\varphi(t)$ is impressed on the pulse, which for example can be obtained by SPM in a nonlinear refractive index material. The pulse is spectrally broadened, or, in the time domain, a chirped pulse. The temporal intensity profile $I(t)$ (or $|\tilde{\mathcal{E}}(t)|^2$) is generally unchanged by this first step, which modifies only the phase function of the pulse. As we saw in Chapter 3, for an instantaneous purely dispersive nonlinearity (n_2 real, no two photon absorption), SPM in an n_2 medium does not change the pulse shape. Only the spectral phase is modified by the nonlinear interaction.

The second step can be seen as the Fourier transform analog of the first one: The phase function $\phi(\Omega)$ of the pulse spectral field $\tilde{\mathcal{E}}(\Omega)$ is modified, without affecting the spectral intensity ($\propto |\tilde{\mathcal{E}}(\Omega)|^2$). The spectrally broadened, nonbandwidth-limited pulse is transformed into a bandwidth-limited pulse. This process is sometimes referred to as "chirp compensation." Because the spectrum does not change in the second step, the new pulse has to be shorter than the input pulse. The compression factor K_c is given roughly by the ratio of the spectral width before ($\Delta\omega_{in}$) and after ($\Delta\omega_{out}$) the nonlinear element (step 1):

$$K_c \sim \frac{\Delta\omega_{out}}{\Delta\omega_{in}}. \tag{8.1}$$

To analyze the second step, the amplitude at the output of the phase modulator is written in the frequency domain:

$$\tilde{\mathcal{E}}(\Omega) = |\tilde{\mathcal{E}}(\Omega)|e^{i\Phi(\Omega)}. \tag{8.2}$$

As discussed in Chapter 1, the action of a linear optical element is described by its optical transfer function $\tilde{H}(\Omega) = R(\Omega)e^{-i\Psi(\Omega)}$. The amplitude of the pulse transmitted by such an element is the inverse Fourier transform of $\tilde{H}(\Omega)\tilde{\mathcal{E}}(\Omega)$:

$$\tilde{\mathcal{E}}(t) = \frac{1}{2\pi} \int_{-\infty}^{\infty} R(\Omega)|\tilde{\mathcal{E}}(\Omega)|e^{i[\Phi(\Omega)-\Psi(\Omega)]}e^{-i\Omega t} d\Omega. \tag{8.3}$$

Let us consider a rather common situation where the amplitude response R is constant in the spectral range of interest. The peak amplitude will be highest if all spectral components add up in phase, which occurs if $\Phi(\Omega) = \Psi(\Omega)$. The linear element with the corresponding phase factor $\Psi(\Omega)$ is called an ideal compressor. Its phase response matches exactly the spectral phase of the chirped pulse. It was shown in Chapter 1 that the output pulse remains unchanged if the difference $(\Phi - \Psi)$ is a nonzero constant or a linear function of Ω. Possible techniques to synthesize the ideal compressor through spectral filtering are presented in Section 8.2 [8].

If a pulse is linearly chirped, its spectral phase varies quadratically with frequency. An ideal compressor is then simply an element with adjustable GVD, for example a prism or grating sequence, provided higher-order dispersion can be neglected. A piece of glass of suitable length can also compress a linearly chirped pulse, as discussed previously. If the spectral phase of the pulse deviates from a parabola by a term $b_3\Omega^3$, we can use two different linear elements in series to construct the corresponding phase response, provided their ratio of third- and second-order phase response is different. This can be a grating and a prism pair, which allows us to tune the overall GVD and third-order dispersion independently. To illustrate this procedure let us formally write the phase of the transfer function of a prism (P) and a grating (G) sequence as:

$$\Psi_{P,G} = L_{P,G}\left(b_2^{(P,G)}\Omega^2 + b_3^{(P,G)}\Omega^3\right) \tag{8.4}$$

where $L_{P,G}$ is the prism (grating) separation and $b_i^{(P,G)}$ are device constants (cf. Chapter 2). To fit a spectral phase

$$\Phi = a_2\Omega^2 + a_3\Omega^3 \tag{8.5}$$

we need to solve a simple system of algebraic equations to find the required prism and grating separation

$$a_2 = L_P b_2^{(P)} + L_G b_2^{(G)} \tag{8.6}$$

$$a_3 = L_P b_3^{(P)} + L_G b_3^{(G)} \tag{8.7}$$

A third element would be needed to compensate for the next term in the Taylor expansion of $\Phi(\Omega)$. Only one GVD element will be required if the phase behavior of the pulse to be compressed is sufficiently simple. However, even though the second-order dispersion dominates, higher-order dispersion terms are unavoidable in all linear elements. These terms become more important as broader pulse spectra have to be handled. If pure GVD is desired, two elements are needed to eliminate third-order dispersion.

As we saw in Chapter 3, focusing a pulse in a medium with a nonlinear index leads to a spectral broadening where the frequency is a complicated function of time. A comparison of the compression of such a pulse in an ideal and a quadratic compressor is made in Figure 8.2. Owing to the strong nonlinear behavior of the frequency modulation (upchirp in the center, downchirp in the wings), cf. Fig. 3.17, the spectral phase is far from being a parabola. Hence, a quadratic compressor cannot compensate the chirp well. We can, for example, adjust for compression in the pulse center but then encounter broadening of the wings. Much better results can be obtained using an ideal compressor. However, the compressed pulse exhibits also satellites in this "ideal" case, because of the particular shape imparted to the spectrum $|\tilde{\mathcal{E}}(\Omega)|^2$ by SPM. Because all frequency

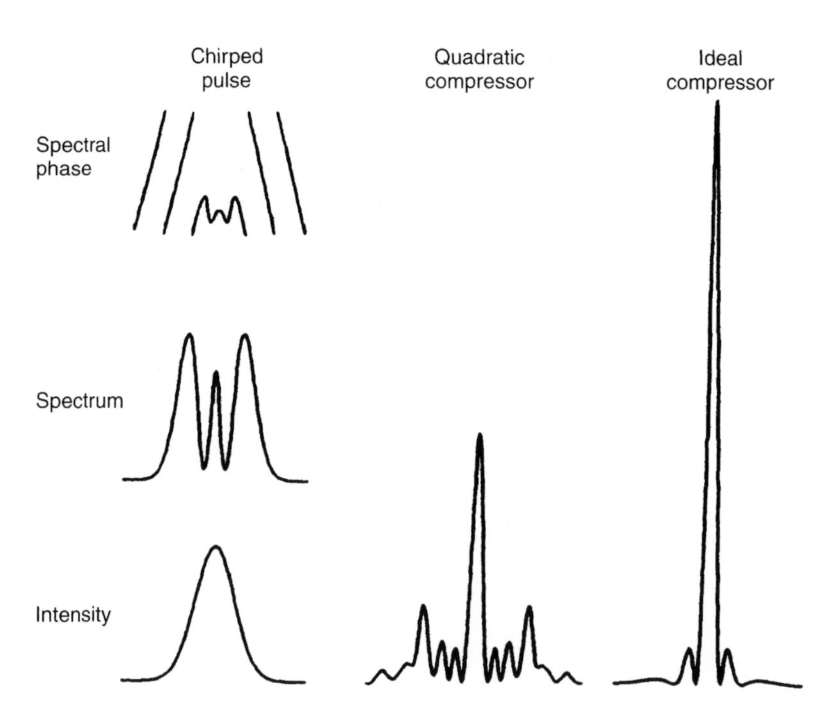

Figure 8.2 Gaussian pulse chirped in a nonlinear index medium and after compression. Note: The spectral phase is shown in the interval $(-\pi/2, +\pi/2)$.

components are in phase after passage through the ideal compressor, the temporal shape is that with the highest peak power, corresponding to the particular pulse spectrum.

For some applications, it may be more important to achieve a smooth, narrow temporal profile, rather than a maximum peak intensity. In this section we shall focus mostly on simple and practical devices, such as quadratic compressors.

The drawback of using bulk materials as a nonlinear medium is the strong nonlinear behavior of the frequency modulation and the poor pulse quality after compression. Moreover, as discussed in Chapters 3 and 5, SPM is associated with self-lensing if the transverse beam profile is not uniform. Therefore, the achievable spectral broadening is limited.

An approach other than Gaussian optics is needed to achieve larger frequency modulations. Of these, the most widely used for pulse compression is the optical single-mode fiber. As discussed next, it leads to the production of almost linearly chirped output pulses [9].

8.1.2. The Fiber Compressor

Pulse Propagation in Single Mode Fibers

An optical field can travel in single-mode fibers over long distances while remaining confined to a few microns in the transverse direction. This is a typical property of a guided wave; see, for example Marcuse [10]. Figure 8.3 summarizes some properties of single-mode fibers made from fused silica. To support only a single-mode, the core radius must satisfy the relation $R < \lambda_\ell/(\pi\sqrt{n^2 - n_c^2})$. Modes of higher order are generally undesired in guiding ultrashort light pulses because of mode dispersion. The core diameters are therefore not larger than a few microns in the VIS. This implies a limit for the pulse power to avoid

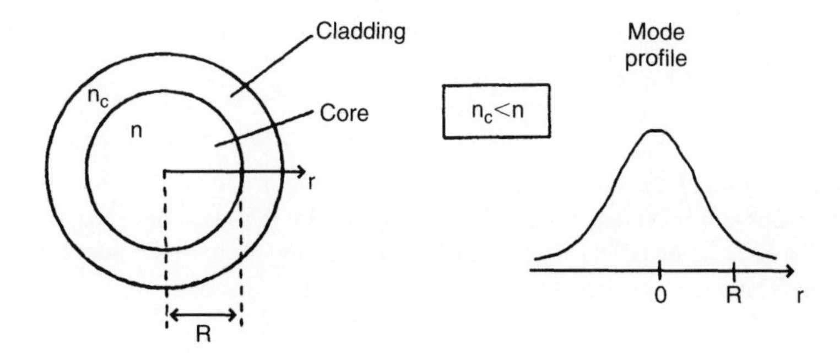

Figure 8.3 Cross section of a single-mode fiber and corresponding mode profile.

material damage. Typical damage intensities are in the order of $\mu J/\mu m^2$. With the focus on chirping for pulse compression, the only nonlinearity to be considered is the nonlinear refractive index effect. To describe fiber propagation under such conditions we have to use the three-dimensional wave equation supplemented by a nonlinear source term according to Eq. (3.143). A perturbative approach is used to study nonlinear pulse propagation in the fiber [11]. Assuming linear polarization the electric field in the fiber can be written:

$$E(x,y,z,t) = \frac{1}{2}\tilde{u}(x,y)\tilde{\mathcal{E}}(t,z)e^{i(\omega_\ell t - k_\ell z)} + c.c. \tag{8.8}$$

The quantity $\tilde{u}$ determines the mode profile which, to first order, is not affected by the nonlinear index change. Using the same procedure as in Chapters 1 and 3 to deal with the dispersion and the nonlinearity, respectively, the equation for the complex envelope is found to be:

$$\frac{\partial}{\partial z}\tilde{\mathcal{E}} - i\frac{k_\ell''}{2}\frac{\partial^2}{\partial t^2}\tilde{\mathcal{E}} = -i\gamma|\tilde{\mathcal{E}}|^2\tilde{\mathcal{E}}, \tag{8.9}$$

where $\gamma = n_2^{eff}\omega_\ell/c$. The differences as compared to the bulk medium can be roughly explained in terms of the refractive index, which we now write:

$$n_{NL}^{eff} = n_{eff} + n_2^{eff}|\tilde{\mathcal{E}}|^2. \tag{8.10}$$

The notation *effective* is to indicate that

(a) the refractive index which determines the dispersion is given not only by the material properties of the fiber core, but also by that of the cladding, as well as by the core shape and size, and
(b) the action of the nonlinearity must be averaged over the fiber cross section which means $n_2^{eff} = n_2/A_{eff}$ where the effective fiber area is given by $A_{eff} = \left(\int |u|^2 dA\right)^2 / \int |u|^4 dA$.

Also, the propagation constant k_ℓ'' differs slightly from its value in the bulk material, a small deviation that only becomes important in the vicinity of the zero dispersion wavelength λ_D where $k_\ell'' = 0$ (Figure 8.4). This zero dispersion wavelength can be shifted by suitable dopants and shaping of the core cross section. The effective area, which depends on the fiber geometry and the refractive indices n, n_c, can be somewhat smaller or larger than the core cross section. Its value ranges from 10 to 25 μm^2 in the visible and can be larger in the infrared because of the usually larger core radii.

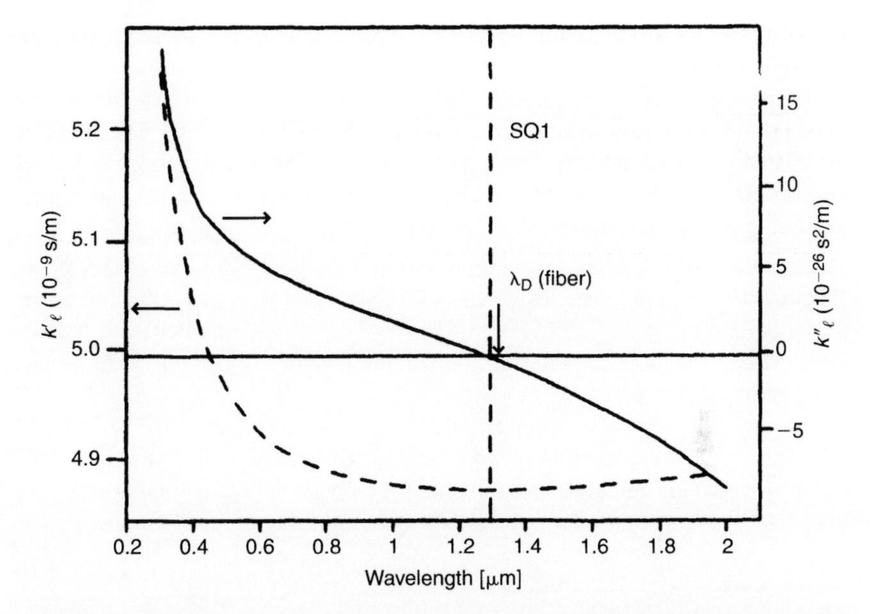

Figure 8.4 Dispersion parameters k'_ℓ and k''_ℓ of fused silica. For comparison, the zero dispersion wavelength λ_D of a typical single-mode fiber made from SQ1 is also shown. Frequently, the dispersion of fibers is expressed in terms of $D = dk'_\ell/d\lambda = -(2\pi c/\lambda^2)k''_\ell$ in units of ps/nm·km. It describes the group delay in ps per nm wavelength difference and per km propagation length.

It is sometimes useful to express Eq. (8.10) in terms of normalized quantities, as for example in Mollenauer *et al.* [12]. Dimensionless coordinates which are particularly convenient for the description of soliton propagation in the spectral region where $k''_\ell < 0$ (see the next section) are $s = t/t_c$ and $\xi = z/z_c$, the two normalization constants satisfying the relation:

$$k''_\ell = -t_c^2/z_c. \tag{8.11}$$

Using a normalized amplitude $\hat{u} = \sqrt{\gamma z_c}\tilde{\mathcal{E}}$, the propagation equation becomes:

$$\frac{\partial}{\partial \xi}\hat{u} + i\frac{1}{2}\frac{\partial^2}{\partial s^2}\hat{u} = -i\gamma z_c|\hat{u}|^2\hat{u}. \tag{8.12}$$

This equation governing the pulse propagation in fibers with GVD and a nonlinear refractive index resembles the Schrödinger equation known from quantum mechanics. This analogy becomes most obvious after associating the nonlinear term with a potential, and interchanging position and time coordinates in

Eq. (8.12). For this reason, this equation is often called the nonlinear Schrödinger equation (NLSE).

The propagation of ultrashort pulses in single-mode fibers is affected by dispersion and an n_2 (often referred to as Kerr type) nonlinearity. These effects were studied independently from each other in Chapters 1 and 3. For their characterization we introduced a dispersion length $L_D = \tau_{p0}^2/k_\ell''$ and a nonlinear interaction length $L_{NL} = (\gamma|\tilde{\mathcal{E}}_{0m}|^2)^{-1}$. Both quantities contain material parameters and properties of the input pulse. In terms of the two characteristic lengths, the limiting cases in which one effect dominates are valid for propagation lengths $L \approx L_D \ll L_{NL}$ and for $L \approx L_{NL} \ll L_D$, respectively. It is the intermediate situation characterized by the interplay of GVD and n_2 effect which shall be of interest now.

The behavior of pulses propagating through single-mode fibers is substantially different in the spectral range where $k_\ell'' > 0$ and $k_\ell'' < 0$. For wavelengths $\lambda_\ell < \lambda_D$ envelope and phase shaping appropriate for subsequent compression is achieved. For longer wavelengths $\lambda_\ell > \lambda_D$, soliton shaping may occur.

Compression of Pulses Chirped in the Normal Dispersion Regime ($k_\ell'' > 0$)

Grischkowsky and Balant [9] recognized the possibility of shaping optical pulses in single-mode fibers for subsequent pulse compression. To obtain the characteristics of a pulse as it travels through an optical single-mode fiber, we need to solve Eq. (8.10). The general case can only be solved numerically, for instance through the procedure outlined in Chapter 1. An example of such a calculation is shown in Figure 8.5.

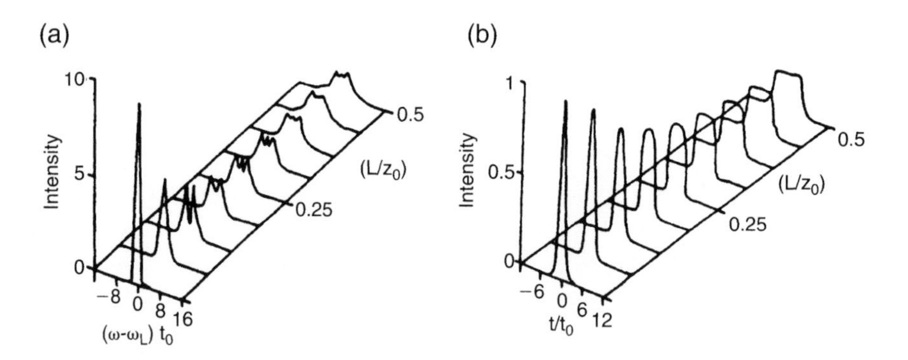

Figure 8.5 Development of spectrum and envelope of a pulse propagating through an optical fiber in the normal dispersion regime ($z_0 = 0.5L_D$). (a) Spectral intensity versus frequency. (b) Intensity versus normalized time. (Adapted from Tomlinson *et al.* [13].)

These results can easily be interpreted as follows. GVD broadens the initially unchirped pulse. Because $k''_\ell > 0$, longer wavelength components travel faster and accumulate along the leading edge of the pulse. SPM produces new frequency components; longer (shorter) wavelength components arise in the leading (trailing) edge. These new frequency components induce an even faster pulse broadening. Owing to GVD and the fact that new frequency components are preferably produced in the pulse edges, where the derivative of the intensity is large, the frequency develops an almost linear behavior over the main part of the pulse and the envelope approaches a rectangular shape. The broadening is associated with a decrease of the pulse intensity, and thus the SPM becomes less important for larger propagation lengths. Eventually a regime is reached where the spectrum remains almost unchanged, and the fiber acts like an element with linear GVD only.

Before discussing some quantitative results of the numerical evaluation of the nonlinear Schrödinger equation, we shall analyze the fiber propagation by means of a simple, heuristic model, describing temporal and spectral broadening under the simultaneous action of GVD and a nonlinearity. We will establish two ordinary differential equations. For simplicity and to exploit previously derived results, we make the approximation that the pulse is linearly chirped and Gaussian over the entire propagation length:

$$\tilde{\mathcal{E}}(z,t) = \mathcal{E}_m(z) e^{-[1+ia(z)][t/\tau_G(z)]^2} \tag{8.13}$$

with

$$\tau_G(z) = \sqrt{2 \ln 2}\ \tau_p(z) \tag{8.14}$$

and

$$\tau_p(z) \Delta\omega_p(z) = 4 \ln 2 \sqrt{1 + a^2(z)} \tag{8.15}$$

$$\mathcal{E}_m^2(z) \tau_p(z) = \mathcal{E}_m(0)^2 \tau_p(0) = \mathcal{E}_{0m}^2 \tau_{p0}. \tag{8.16}$$

The latter relations simply follow from the pulse duration-bandwidth product, cf. Eq. (1.39), and from the requirement of energy conservation. From Eqs. (1.123) and (8.15) we find for the change in pulse duration:

$$\frac{d}{dz}\tau_p(z) = \frac{4 \ln 2}{\tau_p(z)} k''_\ell \sqrt{\frac{\tau_p^2(z)\Delta\omega_p^2(z)}{(4 \ln 2)^2} - 1} + \frac{\Delta\omega_p^2(z) k''^2_\ell}{\tau_p(z)} z. \tag{8.17}$$

Next we need an equation for the change of the pulse spectrum because of SPM. Let us estimate the chirp coefficient as $a \approx 0.5\delta\omega_p\tau_p/(4\ln 2)$ with $\delta\omega_p\tau_p$ given by Eq. (3.152). We introduced the factor 0.5 here to account for the fact that the actual chirp from SPM is not monotonous over the entire pulse. Together with Eqs. (8.15) and (8.16) this chirp parameter a yields for the change of the spectral width

$$\frac{d}{dz}\Delta\omega_p(z) = \frac{\ln 2}{\tau_p^3(z)}\left(\frac{\tau_{p0}}{L_{NL}}\right)^2 \frac{z}{\sqrt{1+[\tau_{p0}/4\tau_p(z)]^2(z/L_{NL})^2}} \tag{8.18}$$

where $L_{NL} = (\gamma\tilde{\mathcal{E}}_{0m}^2)^{-1}$ as introduced earlier. In normalized quantities $\alpha = \tau_p(z)/\tau_{p0}$ for the temporal broadening, $\beta = \Delta\omega_p(z)/\Delta\omega_{p0}$ [where, from Eq. (8.15), $\Delta\omega_0 = 4\ln 2/\tau_{p0}$] for the spectral broadening, and $\xi = z/L_D$, the system of differential equations can be written as

$$\frac{d}{d\xi}\alpha = \frac{4\ln 2}{\alpha}\sqrt{\alpha^2\beta^2 - 1 + (4\ln 2)^2\frac{\beta^2}{\alpha}\xi} \overset{\xi\gg 1}{\rightarrow} 4\ln 2\beta\left(1+\frac{4\ln 2\beta}{\alpha}\xi\right), \tag{8.19}$$

$$\frac{d}{d\xi}\beta = \frac{1}{4\alpha^3}\xi\left(\frac{L_D}{L_{NL}}\right)\left[1+\frac{1}{4}\left(\frac{\xi}{\alpha}\right)^2\left(\frac{L_D}{L_{NL}}\right)^2\right]^{-\frac{1}{2}} \overset{\xi\gg 1}{\rightarrow} \frac{L_D}{L_{NL}}\frac{1}{\alpha^2}, \tag{8.20}$$

with the initial conditions $\alpha(\xi = 0) = 1$ and $\beta(\xi = 0) = 1$. It is interesting to note here that the parameters of the input pulse and the fiber enter this equation only as L_D/L_{NL} if we measure the propagation length in units of L_D. This set of ordinary differential equations can easily be integrated numerically. The results for $L_D/L_{NL} = 1600$ are depicted in Figure 8.6.

For fused silica ($k_\ell'' \approx 6 \times 10^{-26}$ s^2/m, $\bar{n}_2 \approx 3.2 \times 10^{-16}$ cm^2/W), an effective fiber area of 10 μm^2 and 500 fs input pulses at 600 nm, $L_D/L_{NL} = 1600$ corresponds to a peak power of 12 kW where $L_D \approx 4.2$ m and $L_{NL} \approx 2.6$ mm. The figure illustrates the properties discussed previously, in particular the saturation of the spectral broadening and the linear behavior of the temporal broadening for large propagation length. In our example the spectral broadening reaches a value of about 20, which sets a limit to the compression factor.

For the purpose of pulse compression, a large spectral broadening is desirable. Figure 8.6 suggests that long fibers are not essential, because most of the spectral broadening occurs within a finite length L_F (which, however, is still larger than L_{NL}). To obtain an approximate relationship between the maximum spectral broadening $\bar{\beta}$ and the fiber length L_F at which a certain percentage m of $\bar{\beta}$ is achieved, one can proceed as follows. Equations (8.19) and (8.20) are solved asymptotically in a perturbative approach. Substituting $\beta = \bar{\beta}$ into

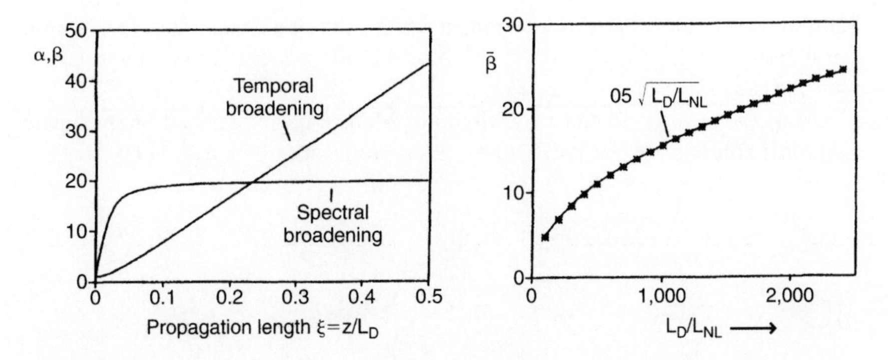

Figure 8.6 Temporal broadening α and spectral broadening β for pulse propagation in single-mode fibers as function of the normalized propagation length $\xi = z/L_D$ for $L_D/L_{NL} = 1600$ (left) and asymptotic spectral broadening, $\bar{\beta}$, versus L_D/L_{NL} (right).

Eq. (8.19) gives as solution for the temporal broadening $\alpha = 2(\sqrt{5}+1)\ln 2\bar{\beta}\xi$. This solution can be inserted into Eq. (8.20), which yields after integration from ξ_F to ∞:

$$\bar{\beta} - \beta(\xi_F) \approx \frac{L_D}{L_{NL}} \frac{1}{20\bar{\beta}} \frac{1}{\xi_F}. \tag{8.21}$$

If we chose $\xi_F = L_F/L_D$ so that at this length $m\%$ of the maximum spectral broadening occurs, we obtain:

$$\bar{\beta}^3 L_F \left(\frac{L_{NL}}{L_D^2}\right) \approx \frac{1}{20(1 - m/100)}. \tag{8.22}$$

The numerical evaluation of Eqs. (8.19) and (8.20) shown in Fig. 8.6 revealed that $\bar{\beta}$ varies as:

$$K_c \approx \bar{\beta} \approx 0.5\sqrt{\frac{L_D}{L_{NL}}}. \tag{8.23}$$

$\bar{\beta}$ is also a rough measure of the compression that can be achieved. To satisfy relation (8.22), the fiber length at which a certain spectral broadening can be expected must be proportional to:

$$L_F \propto \sqrt{L_D L_{NL}}. \tag{8.24}$$

In our example, the propagation length at which 95% of the maximum broadening occurs is $L_F \approx 2.9\sqrt{L_D L_{NL}} \approx 0.1z/L_D \approx 43$ cm, which is in good agreement with Fig. 8.6.

Using the inverse scattering technique, Meinel [15] found an approximate analytical solution for the pulse after a long propagation length L [14]:

$$\tilde{\mathcal{E}}(t) = \begin{cases} \mathcal{E}_m e^{ia(t/\tau_p)^2} & |t| \leq \tau_p/2 \\ 0 & |t| > \tau_p/2 \end{cases} \tag{8.25}$$

where

$$a \approx 0.7\frac{\tau_p}{\tau_{p0}}\sqrt{\frac{L_D}{L_{NL}}} \tag{8.26}$$

and

$$\tau_p \approx 2.9\frac{L}{\sqrt{L_D L_{NL}}}\tau_{p0}. \tag{8.27}$$

A linear element must be found for optimum pulse compression. For this particular pulse, to produce a chirp-free output, the b_2 parameter of a quadratic compressor should be chosen as [16]:

$$b_2 \approx \frac{\tau_p^2}{4a(1 + 22.5/\tau_p^2)}. \tag{8.28}$$

For the actual analysis of the compression step, one can favorably use the Poisson integral Eq. (1.108) to calculate the pulse behind the linear element and to determine its duration. The compression factor is found to be [15]:

$$K_c \approx 0.5\sqrt{\frac{L_D}{L_{NL}}}. \tag{8.29}$$

As mentioned previously the requirements for producing chirp-free output pulses of the shortest achievable duration and best quality cannot be satisfied simultaneously. Tomlinson *et al.* [13] solved the nonlinear Schrödinger equation numerically and varied b_2 in the compression step to obtain pulses with the highest peak intensity. They found this to be a reasonable compromise between pulse duration and pulse quality (small satellites). Figure 8.7 shows this optimum compression factor as a function of the fiber length for various values of $\sqrt{L_D/L_{NL}}$. Depending on the parameters of the input pulse, there is an optimum fiber length

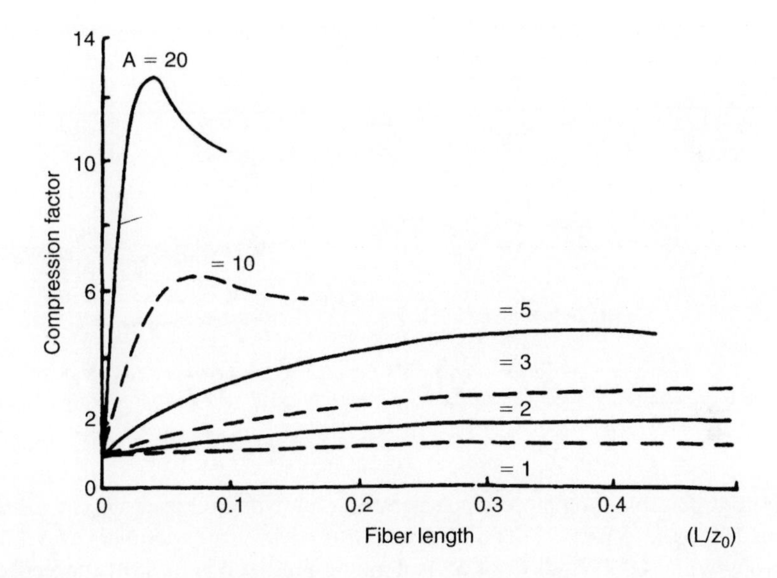

Figure 8.7 Optimum compression factor as a function of the fiber length. Notation: $A \approx 0.6\sqrt{L_D/L_{NL}}$, $z_0 = 0.5L_D$. (Adapted from Tomlinson *et al.* [13].)

at which pulse compression is most effective. Numerical simulations lead to an estimate for this optimum length:

$$L_{opt} \approx 1.4\sqrt{L_D L_{NL}}. \tag{8.30}$$

It was found that the corresponding compression factor, using such a fiber length, can be approximated by:

$$K_c \approx 0.37\sqrt{\frac{L_D}{L_{NL}}}. \tag{8.31}$$

Figure 8.8(a) shows a typical experimental setup for pulse compression. The input pulse is focused by a microscope objective into a single-mode optical fiber of suitable length. After recollimation, the chirped and temporally broadened pulse is sent through the linear element which, typically, is a grating or prism pair or a sequence of them. To achieve a substantial compression factor input pulses of certain power are necessary, as shown in Figs. 8.6 and 8.7. For compression of fs pulses, peak powers in the kW range are needed. Therefore, to apply the fiber compressor, pulses from most fs oscillators must be amplified first. For ps pulses L_D/L_{NL} takes on large values for peak powers even below 1 kW. It has been

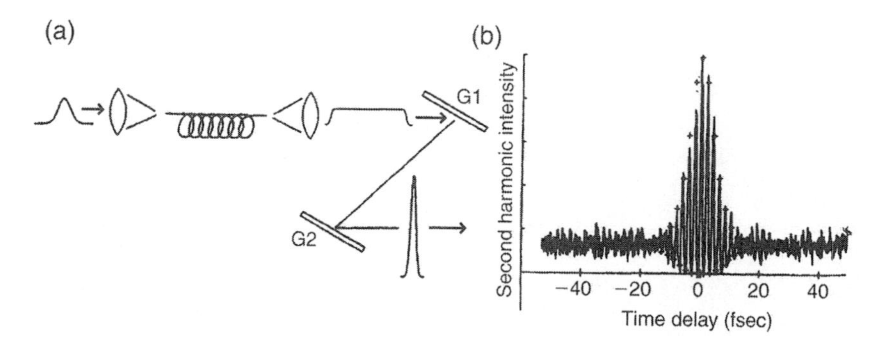

Figure 8.8 Sketch of a fiber-grating compressor (a) and interferometric autocorrelation of a compressed 6 fs pulse (b). For the compression a sequence of prism and gratings was used as linear element. (Adapted from Fork *et al.* [19].)

therefore possible to compress pulses from a cw Nd:YAG laser ($P_m \approx 100$ W) from 100 ps to 2 ps [17]. The fiber length used in this experiment was 2 km. Starting with 100 kW pulses of 65 fs duration Fujimoto *et al.* [18] succeeded in generating 16 fs pulses where the fiber length was only 8 mm.

Fork *et al.* [19] obtained 6 fs pulses with 65 fs, 300 kW pulses at the input of a 9-mm long fiber. At a wavelength of 620 nm this pulse duration corresponds to only three optical cycles. A corresponding interferometric autocorrelation function is shown in Fig. 8.8(b). As mentioned previously, the broader the pulse spectrum to be handled, the more important it becomes to match the third-order dispersion of the linear element. Therefore, in this experiment, the authors used a combination of a grating and a prism pair for the independent adjustment of GVD and third-order dispersion.

A practical limit for the achievable compression factor and pulse duration is of course given by the maximum power that can be propagated through the fiber without causing damage. However, before this limit is reached, other nonlinear and linear effects have to be considered. These include third-order material dispersion, Raman processes, the $|\tilde{\mathcal{E}}|^4$ dependence of the refractive index, and the effect of a shock term in the wave equation. A detailed discussion can be found, for example, in Rudolph and Wilhelmi [1], Vysloukh and Matveyeva [20], and Bourkhoff *et al.* [21].

The limitation on the fs pulse energy that can be handled by single mode fibers can be overcome by hollow fibers as demonstrated by Nisoli *et al.* [22]. Hollow cylindrical fibers were made from fused silica with bore diameters of the order of $2a \approx 100$ μm, filled with a noble gas. This provides the additional advantage of controlling the (n_2) nonlinearity by adjustment of the gas pressure. The EH_{11} hybrid mode with an intensity profile $I(r) \propto J_0^2(2.4r/a)$, where J_0 being the Bessel function of order zero, is the lowest order mode that can be supported by these fused silica hollow fibers. The dispersion includes a contribution from the waveguide and from the gas. Pulses as short as 5 fs with an energy of 70 μJ were obtained using Ar-filled hollow fibers (pressure of 3.3 bar) [23]. A prism sequence and a chirped mirror were used to compress the spectrally broadband pulse at the fiber output.

A modification of this technique is to operate in conditions of intensity and pulse duration such that self-wave guiding or filamentation of the pulse takes place in the gaseous nonlinear medium, hence eliminating the need for a capillary. This technique was successfully implemented by Hauri *et al.* [24], using 43 fs, 0.84 mJ pulses at 800 nm, sent successively through gas cells at pressures of 840 mbar and 700 mbar of argon. Pairs of chirped mirrors were used after each cell, resulting in an output pulse of 5.7 fs duration, 0.38 mJ at a repetition rate of 1 kHz. With the CEO introduced in Section 5.1.3 locked, the output pulses had also their CEO locked.

Soliton Compression in the Anomalous Dispersion Regime ($k_\ell'' < 0$)

In the spectral region where the nonlinearity (n_2) and dispersion (k_ℓ'') have opposite sign, the pulse propagation may have completely different properties than have been discussed so far. Theoretical studies of the NLSE for this case predicted the existence of pulses either with a constant, or with a periodically reproducing shape [25,26]. The existence of these solutions, designated as solitons, can be explained simply as follows. The nonlinearity ($n_2 > 0$) is responsible for spectral broadening and up-chirp. Because of the anomalous dispersion, $k_\ell'' < 0$, which in fused silica single-mode fibers occurs for $\lambda > 1.31$ μm, the lower frequency components produced in the trailing part travel faster than the long wavelength components of the pulse leading edge. Therefore, the tendency of pulse broadening owing to the exclusive action of GVD can be counterbalanced. Of course, the exact balance of the two effects is expected only for certain pulse and fiber parameters. For a rough estimate let us require that the chirp produced in the pulse center by the nonlinearity and the dispersion are of equal magnitude (but of opposite sign). Under this condition the pulse propagates through the fiber without developing a frequency modulation and spectral broadening. Let us use this requirement to estimate the parameters for form-stable pulse propagation. The effect of GVD is to create a pulse broadening and a down-chirp, with the change of the second derivative of the phase versus time equal to:

$$\Delta\left(\frac{\partial^2}{\partial t^2}\varphi(t)\right) = \frac{4k_\ell''}{\tau_{G0}^4}\Delta z \tag{8.32}$$

where we have used the characteristics of Gaussian pulse propagation (cf. Table 1.2). The chirp induced by SPM is [see Eq. (3.146)]:

$$\Delta\left(\frac{\partial^2}{\partial t^2}\varphi(t)\right) = -\frac{2\pi}{\lambda_\ell}\bar{n}_2\frac{\partial^2 I}{\partial t^2}\Delta z \tag{8.33}$$

Equating both relations for the chirp in the pulse center, we find as condition for the pulse parameters:

$$I_{om}\tau_{G0}^2 = \frac{\lambda_\ell k_\ell''}{\pi\bar{n}_2} \tag{8.34}$$

where I_{om} is the peak intensity.

The exact (analytical) solution of the nonlinear Schrödinger equation (8.11) shows that solitons occur if the pulse amplitude at the fiber input obeys the relation

$$|\tilde{\mathcal{E}}(s)| = \frac{N}{\sqrt{z_c\gamma}}\operatorname{sech}(s), \tag{8.35}$$

where N is an integer and refers to the soliton order, and $z_c \approx \tau_{p0}^2/(1.76k_\ell'')^2 = L_D/1.76^2$. For $N = 1$ the pulse propagates with a constant, stable shape through the fiber. The action of nonlinearity and GVD exactly compensate each other. The shape of the fundamental soliton is:

$$\tilde{\mathcal{E}}(\xi, s) = \frac{1}{\sqrt{z_c\gamma}}\operatorname{sech}(s)e^{-i\xi/2}. \tag{8.36}$$

The solution (8.36) can be verified by substitution of relation (8.36) into Eq. (8.11). Higher-order solitons ($N \geq 2$) periodically reproduce their shape after a distance given by:

$$L_p = \pi z_c/2. \tag{8.37}$$

Optical solitons in fibers were observed by Mollenauer *et al.* [27,28]. The potential application in digital pulse-coded communication, has spurred the interest in solitons, which offer the possibility of propagating ultrashort light pulses over thousands of km, while preserving their duration [12].

In relation to pulse compression, it is interesting to note that an arbitrary unchirped input pulse of sufficiently large energy will eventually develop into a soliton. The soliton order N depends on the power of the input pulse. The soliton formation will always lead first to a substantial narrowing of the central pulse peak at a certain propagation length L_S. This narrowing is independent of how complex the following behavior is, provided the amplitude of the input pulse corresponds to $N > 1$. The pulse narrowing was studied experimentally and theoretically in Mollenauer *et al.* [29]; see Fig. 8.9. It follows that pulses become shorter, the higher their input intensity, i.e., the higher the order of the soliton in which they finally develop. Narrowing factors up to 30 were measured. A disadvantage of this method as a compression technique is the relatively poor pulse quality, which manifests itself in broad wings and side lobes.

Soliton narrowing has successfully been applied in connection with a fiber grating setup in two-step compression experiments [17, 30]. By means of this technique, Gouveia-Neto *et al.* [30] succeeded in compressing 90 ps pulses from a Nd:YAG laser at $\lambda = 1.32$ μm to 33 fs. In the first stage the pulses were compressed to 1.5 ps by using a fiber-grating configuration. Because normal dispersion was required here, a fiber with a zero dispersion wavelength $\lambda_D = 1.5$ μm was chosen. Subsequent propagation of these pulses through 20 m of single-mode fiber with $\lambda_D \approx 1.27$ μm led to a pulse width of 33 fs through soliton narrowing where N was estimated to be 12. One drawback of fiber compressors and, in particular, of multistage configurations is the relatively high loss factor.

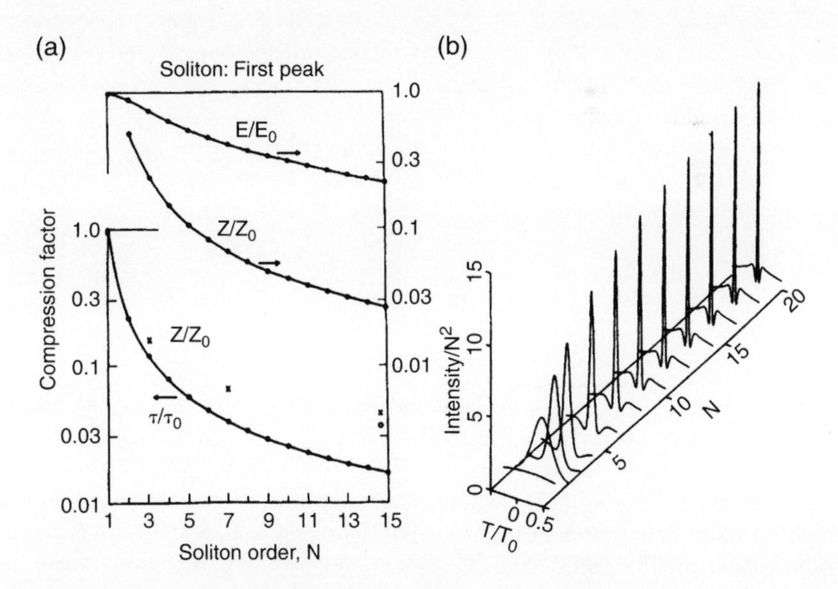

Figure 8.9 Pulse narrowing through soliton shaping. (a) Calculated parameters of the first optimum narrowing and related experimental data (∘, ×). (b) Pulse shape at optimal narrowing as a function of the soliton order. (Adapted from Mollenauer *et al.* [29].)

These losses are mainly associated with the coupling of the pulses into the fiber and diffraction at the gratings. The overall transmission in the experiment cited was about 7%.

8.1.3. Pulse Compression Using Bulk Materials

One drawback of fiber compressors is that only pulses of relatively small energy can be handled. With fs input pulses the possible energies do not exceed several tens of nJ. One possible solution is to spectrally broaden the pulse in the fiber before amplification and subsequent compression as demonstrated by Damm *et al.* [31] with pulses from a Nd:glass laser.

Previous attempts to use bulk materials for the chirping of high energy ps pulses resulted in a relatively poor pulse quality at the compressor output owing to the nonlinear chirp behavior. As discussed before, to obtain an almost linear chirp across the main part of the pulse, a certain ratio of L_D and L_{NL} is necessary. This could be achieved by utilizing pulse propagation in single-mode fibers. The fiber lengths needed become smaller with shorter durations of the input pulse and, for fs pulses, are of the order of several millimeters. From Eq. (8.30) it can easily be seen that $L_{opt} \propto \tau_{p0}$ if the peak intensity is kept constant. Over such propagation distances suitably focused Gaussian beams do not change their beam diameter much in bulk materials provided self-focusing can be neglected. The latter limits the possible propagation lengths to those shorter than the self-focusing length. The latter in turn can be adjusted by choosing an appropriate spot size of the focused beam, which sets an upper limit for the maximum pulse intensity. The maximum compression factor that can be achieved under such conditions can be estimated to be:

$$K_c \approx 0.3\sqrt{\frac{n_0 \bar{n}_2 P_0}{\lambda_\ell^2}} \tag{8.38}$$

where P_0 is the peak power of the input pulse (cf. [32]). Figure 8.10 shows results of a numerical evaluation of pulse compression using bulk SQ1 fused silica, and 60 fs input pulses of various energies.

Rolland and Corkum [33] demonstrated experimentally the compression of high-power fs light pulses in bulk materials. Starting from 500 μJ, 92 fs pulses from a dye amplifier, they obtained ≈20 fs compressed pulses at an energy of ≈100 μJ. The nonlinear sample was a 1.2 cm piece of quartz and the pulses were focused to a beam waist of 0.7 mm.

At high intensities a white light continuum pulse can be generated [34] as was discussed in Section 3.7. With fs pulses, SPM is expected to contribute

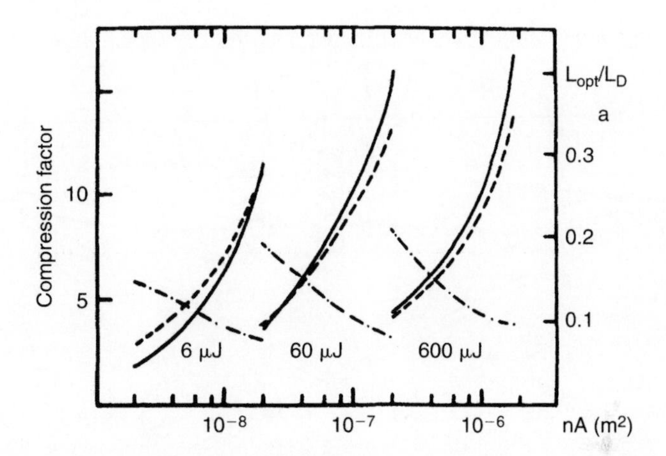

Figure 8.10 Plot of compression parameters after chirping in bulk SQ1 (fused silica) for different pulse energies and $\tau_{P0} = 60$ fs as a function of the beam cross section at the sample input (location of the beam waist). —·—·: Compression factor, — parameter of the optimum quadratic compressor ($a \approx 3b_2/\tau_{P0}^2$), ––– normalized optimum sample length ($L_D \approx 3$cm). (Adapted from Petrov *et al.* [32].)

substantially to the continuum generation. There have been successful experiments to compress continuum pulses produced by high-power fs pulses [35].

8.2. SHAPING THROUGH SPECTRAL FILTERING

On a ps and longer time scale optical pulses can be shaped directly by elements of which the transmission is controlled externally. An example is a Pockels cell placed between crossed polarizers and driven by an electrical pulse. The transients of this pulse determine the time scale on which the optical pulse can be shaped. The advantage of this technique is the possibility of producing a desired optical transmission by synthesizing a certain electrical pulse, as demonstrated in the picosecond scale by Haner and Warren [2]. The speed limitations of electronics have so far prevented the application of this technique to the fs scale.

A technique best suited for the shortest pulses consists of manipulating the pulse spectrum in amplitude and phase. This technique was originally introduced for ps light pulses [36–38] and later successfully applied and improved for fs optical pulses by Thurston *et al.* [39] and Weiner *et al.* [8]. The corresponding experimental arrangement is shown in Figure 8.11.

The pulse to be shaped is spectrally dispersed using a grating or a pair of prisms. The spectrum is propagated through a mask which spectrally filters

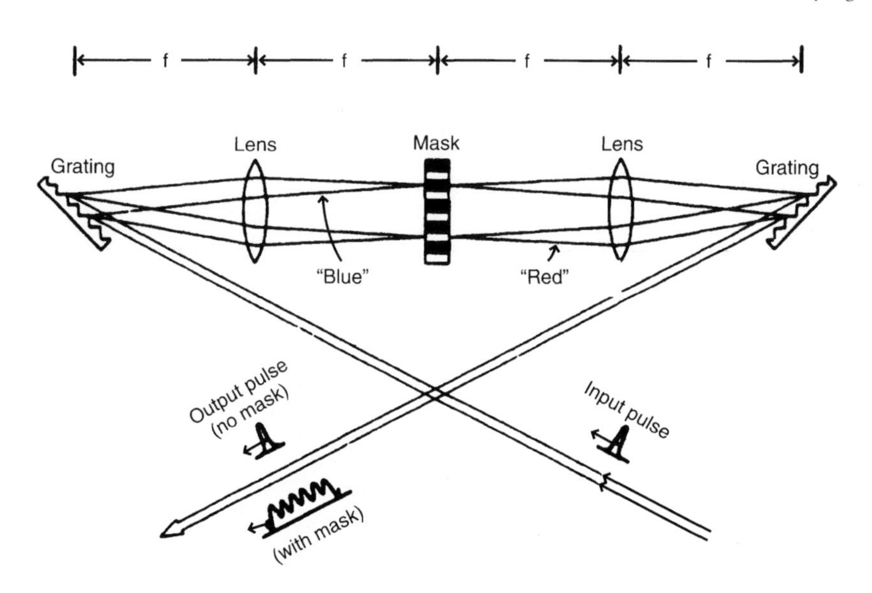

Figure 8.11 Spectral filtering in a dispersion-free grating-lens combination. (Adapted from Weiner *et al.* [8].)

the pulse. The spectral components are recollimated into a beam by a second grating or pair of prisms. In the arrangement of Fig. 8.11, the two-grating–two-lens combination has zero GVD, as can be verified easily by setting $z' = z = 0$ in Eq. (2.124). Each spectral component is focused at the position of the mask (the usual criterion for resolution applies). Because, to a good approximation, the frequency varies linearly in the focal plane of the lens, a variation of the complex transmission across the mask causes a transfer function of the form:

$$\tilde{H}(\Omega) = R(\Omega)e^{-i\Psi(\Omega)} \tag{8.39}$$

where $R(\Omega)$ represents the amplitude transmission and $\Psi(\Omega)$ the phase change experienced by a spectral component at frequency Ω. These masks can be produced by microlithographic techniques.[1] A pure phase filter, for example, could consist of a transparent material of variable thickness. If we neglect the effects caused by the finite resolution, the field at the device output is:

$$\tilde{E}_{out}(t) = \mathcal{F}^{-1}\{\tilde{E}_{out}(\Omega)\}, \tag{8.40}$$

[1]Another option is to use pixelated liquid-crystal arrays whose complex transmission can be controlled by applying voltages individually to each pixel.

where

$$\tilde{E}_{out}(\Omega) = \tilde{E}_{in}(\Omega)R(\Omega)e^{-i\Psi(\Omega)}, \tag{8.41}$$

and $\tilde{E}_{in}(\Omega)$ is the field spectrum of the input pulse. In principle, to achieve a certain output $\tilde{E}(t)$, we need to determine $\tilde{E}_{out}$ by a Fourier transform and divide it by the input spectrum $\tilde{E}_{in}$. This ratio gives the transmission function required for the mask. The mask can be a sequence of an amplitude only and a phase only filter to generate the desired $R(\Omega)\exp[-i\Psi(\Omega)]$.

A simple slit as mask acts as spectral window. Such spectral windowing can be used to enhance the pulse quality, in particular to reduce small satellites, in fiber grating compressors [40]. One aligns the window so that the wings of the spectrum, which are caused by the nonlinear chirp in the pulse edges, are blocked. The remaining linear chirp can then be compensated by a grating pair.

Many different output fields can be realized by using different mask functions. Of course, the shortest temporal structures that can be obtained are limited by the finite width of the input pulse spectrum. The narrowest spectral features that can be impressed on the spectrum are determined by the grating dispersion, mask structure, and size of the focal spot. Generally, amplitude masks introduce linear losses. For higher overall transmission pure phase masks are advantageous. Figure 8.12 shows some examples of intensity profiles obtained through spectral filtering of 75 fs pulses from a mode-locked dye laser.

Note that the action of any linear element can be interpreted as spectral filtering. If, for instance, the mirror separation d of a Fabry–Perot interferometer is larger than the original pulse width, the output will be a sequence of pulses, separated by $2d$. As explained in the chapter on coherent processes,

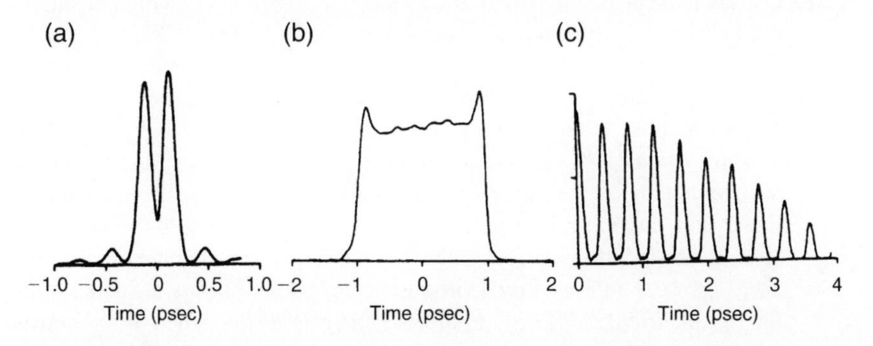

Figure 8.12 Cross-correlation measurements of intensity profiles generated by spectral filtering of 75 fs pulses. (a) Half of an originally symmetric spectrum was shifted by π resulting in a zero area pulse. (b) Square pulse produced by an amplitude mask resembling a sinc-function. (c) THz pulse train produced by a pure phase mask. (Adapted from Weiner *et al.* [8].)

pulse trains with well-defined phase relations are particularly interesting for coherent excitation in optical spectroscopy and coherent (quantum) control of photo induced processes in general.

8.3. PROBLEMS

1. A self-phase modulated pulse exhibits a spectrum that shows characteristic modulations, cf. Fig. 8.2. Assume a Gaussian pulse of duration τ_p and peak intensity I_0 which is propagated through a material of length L and nonlinear refractive index coefficient n_2. Explain why for large SPM the number of peaks in the spectrum is roughly given by ϕ_{max}/π. (ϕ_{max} is the maximum induced phase shift which occurs in the pulse center.)
2. SPM of Gaussian beams is associated with self-focusing in a bulk material with positive n_2. Estimate the achievable spectral broadening of a Gaussian pulse. Neglect dispersion and require that the beam diameter does not reduce to less than half of its original value.
3. Design a compressor for the pulses of a synchronously pumped and cavity-dumped dye laser ($\tau_p = 5$ ps, $W = 5$ nJ, $\lambda_\ell = 600$ nm). A fiber made from fused silica (core diameter 4 μm) and a pair of gratings (grating constant $= 2000\ \text{mm}^{-1}$) are to be used. Find a convenient fiber length, the required grating separation, and estimate the achievable compression factor.
4. Find an approximate expression that describes the pulse duration at the output of an n-stage compression unit. Each compression step introduces an energy loss of p. Use the pulse parameters from the previous problem and $p = 0.5$ to argue about the efficiency of such devices.
5. Compare the peak intensity of the fundamental soliton in an optical single-mode fiber as estimated in Eq. (8.34) with the exact value from Eq. (8.35). Explain the difference.
6. It is often desirable to generate a square-shaped pulse. One can think of a large number of methods to generate such a signal. The best solution depends often on the initial condition (pulse duration available) and the desired length of the square pulse. A square pulse can be obtained by linear and nonlinear optical processes. The process could be nonlinear absorption (for instance four photon absorption), self-defocusing through a $n_4 I^4$ process followed by spatial filtering, pulse splitting-sequencing by interferometric delay lines, or spectral filtering as described at the end of this chapter. Compare these methods in terms of energy loss and residual modulation of the "flat" part of the pulse. Assume you have initially a 1 mJ Gaussian pulse of 50 fs duration and want to approximate a square pulse of (a) 40 fs duration; (b) 150 fs duration; and (c) 900 fs duration.

7. How would one generate a train of identical pulses, each being 180° out of phase with the previous one?
8. Consider a bandwidth-limited pulse of duration τ_p. Is it possible to generate a temporal substructure with transients shorter than τ_p with spectral filtering? Explain your answer.

BIBLIOGRAPHY

[1] W. Rudolph and B. Wilhelmi. *Light Pulse Compression.* Harwood Academic, Chur, London, 1989.

[2] M. Haner and W. S. Warren. Generation of arbitrarily shaped picosecond optical pulses using an integrated electro-optic waveguide modulator. *Applied Optics*, 26:3687–3694, 1987.

[3] C. E. Cook. Pulse compression—key to more efficient radar transmission. *Proc. IRE*, 48:310–320, 1960.

[4] T. K. Gustafson, J. P. Taran, H. A. Haus, J. R. Lifsitz, and P. L. Kelley. Self-modulation, self-steepening, and spectral development of light in small-scale trapped filaments. *Physical Review*, 177:306–313, 1969.

[5] E. B. Treacy. Optical pulse compression with diffraction gratings. *IEEE Journal of Quantum Electronics*, QE-5:454–460, 1969.

[6] E. B. Traecy. Compression of picosecond light pulses. *Physics Letters*, 28A:34–38, 1969.

[7] A. Laubereau. External frequency modulation and compression of ps pulses. *Physics Letters*, 29A:539–540, 1969.

[8] A. M. Weiner, J. P. Heritage, and E. M. Kirschner. High resolution femtosecond pulse shaping. *Journal of the Optical Society of America B*, 5:1563–1572, 1988.

[9] D. Grischkowsky and A. C. Balant. Optical pulse compression based on enhanced frequency chirping. *Applied Physics Letters*, 41:1–3, 1981.

[10] D. Marcuse. *Light Transmission Optics.* Van Nostrand Reinhold, New York, 1982.

[11] G. P. Agrawal. *Nonlinear Fiber Optics.* Academic Press, Boston, MA, 1995.

[12] L. F. Mollenauer, J. P. Gordon, and N. N. Islam. Soliton propagation in long fibers with periodically compensated loss. *IEEE Journal of Quantum Electronics*, QE-22:157–173, 1986.

[13] W. J. Tomlinson, R. H. Stolen, and C. V. Shank. Compression of optical pulses chirped by self phase modulation in optical fibers. *Journal of the Optical Society of America*, B1:139–143, 1984.

[14] F. Calogero and A. Degasperis. *Spectral Transforms and Solitons.* North Holland, Amsterdam, 1982.

[15] R. Meinel. Generation of chirped pulses in optical fibers suitable for an effective pulse compression. *Optics Communications*, 47:343–346, 1983.

[16] B. Wilhelmi. Potentials and limits of femtosecond pulse compression [in German]. *Annals of Physics*, 43:355–368, 1986.

[17] K. Tai and A. Tomita. 1100 x optical fiber pulse compression using grating pair and soliton effect at 1319 nm. *Applied Physics Letters*, 48:1033–1035, 1986.

[18] J. G. Fujimoto, A. M. Weiner, and E. P. Ippen. Generation and measurement of optical pulses as short as 16 fs. *Applied Physics Letters*, 44:832–834, 1984.

[19] R. L. Fork, C. H. Cruz, P. C. Becker, and C. V. Shank. Compression of optical pulses to six femtoseconds by using cubic phase compensation. *Optics Letters*, 12:483–485, 1987.

[20] V. A. Vysloukh and T. A. Matveyeva. Feasibility of cascade compression of pulses in the near infrared. *Soviet Journal of Quantum Electronics*, 16:665–667, 1986.

[21] E. Bourkhoff, W. Zhao, R. I. Joseph, and D. N. Christodoulides. Evolution of fs pulses in single mode fibers having higher order nonlinearity and dispersion. *Optics Letters*, 12:272–274, 1987.

[22] M. Nisoli, S. De Sivestri, and O. Svelto. Generation of high-energy 10 fs pulses by a new pulse compression technique. *Applied Physics Letters*, 68:2793–2795, 1996.

[23] M. Nisoli, S. De Sivestri, O. Svelto, R. Szipocs, K. Ferencz, C. Spielmann, S. Sartania, and F. Krausz. Compression of high-energy laser pulses below 5 fs. *Optics Letters*, 22:522–524, 1997.

[24] C. P. Hauri, W. Kornelis, F. W. Helbing, A. Heinrich, A. Couairon, A. Mysirowicz, J. Biegert, and U. Keller. Generation of intense, carrier-envelope phase-locked few-cycle laser pulses through filamentation. *Applied Physics B*, 79:673–677, 2004.

[25] V. E. Zakharov and A. B. Shabat. Exact theory of two-dimensional self-focusing and one-dimensional self-modulation of waves in nonlinear media. *Translation of Zhurnal Eksperimental noi i Teoreticheskoi Fizik*, 34:62–69, 1972.

[26] A. Hasegawa and F. Tappert. Transmission of stationary nonlinear optical pulses in dispersive dielectric fibers. *Applied Physics Letters*, 23:142–144, 1973.

[27] L. F. Mollenauer, R. H. Stolen, and J. P. Gordon. Experimental observation of picosecond pulse narrowing and solitons in optical fibers. *Physical Review Letters*, 45:1095, 1980.

[28] L. F. Mollenauer and R. H. Stolen. The soliton laser. *Optics Letters*, 9:13, 1984.

[29] L. F. Mollenauer, R. H. Stolen, J. P. Gordon, and W. J. Tomlinson. Extreme picosecond pulse narrowing by means of soliton effect in single-mode fibers. *Optics Letters*, 8:289, 1983.

[30] A. S. Gouveia-Neto, A. S. L. Gomes, and J. B. Taylor. Generation of 33 fs pulses at 1.32 μm through a high order soliton effect in a single mode optical fiber. *Optics Letters*, 12:395–397, 1987.

[31] T. Damm, M. Kaschke, F. Noack, and B. Wilhelmi. Compression of ps pulses from a solid-state laser using self-phase modulation in graded index fibers. *Optics Letters*, 10:176–178, 1985.

[32] V. Petrov, W. Rudolph, and B. Wilhelmi. Compression of high energy femtosecond light pulses by self phase modulation in bulk media. *Journal of Modern Optics*, 36:587–595, 1989.

[33] C. Rolland and P. B. Corkum. Compression of high power optical pulses. *Journal of the Optical Society of America B*, 5:641–647, 1988.

[34] R. R. Alfano, Ed., *The Supercontinuum Laser Source*. Springer, New York, 1989.

[35] P. Simon, S. Szatmari, and F. P. Schafer. Generation of 30 fs pulses tunable over the visible spectrum. *Optics Letters*, 16:1569–1571, 1991.

[36] J. Desbois, F. Gires, and P. Tournois. A new approach to picosecond laser pulse analysis and shaping. *IEEE Journal of Quantum Electronics*, QE-9:213–218, 1973.

[37] J. Agostinello, G. Harvey, T. Stone, and C. Gabel. Optical pulse shaping with a grating pair. *Applied Optics*, 18:2500–2504, 1979.

[38] C. Froehly, B. Colombeau, and M. Vampouille. Shaping and analysis of picosecond light pulses. E. Wolf, Ed., In *Optics*. North Holland, Amsterdam, Vol. XX:115–125, 1981.

[39] R. N. Thurston, J. P. Heritage, A. M. Weiner, and W. J. Tomlinson. Analysis of picosecond pulse shape synthesis by spectral masking in a grating pulse compressor. *IEEE Journal of Quantum Electronics*, QE-22:682–696, 1986.

[40] A. M. Weiner, J. P. Heritage, and J. A. Saleh. Encoding and decoding of femtosecond pulses. *Optics Letters*, 13:300–302, 1988.

9

Diagnostic Techniques

The femtosecond time scale is beyond the reach of standard electronic display instruments. New methods have to be designed to freeze and time resolve events as short as a few optical cycles. Any measurement technique introduces some perturbation on the parameter to be measured. This problem is particularly acute in attempting to time resolve fs signals. As we have seen in the chapter on fs optics, reflection and transmission through most optical elements will modify the signal to be measured. In addition, most diagnostic schemes involve nonlinear elements, which may also have an influence on the amplitude and phase of the pulse to be measured. A careful analysis of the diagnostic instrument is required to find its exact transfer function. The inverse of this instrument transfer function should be applied to the result of the measurement, to obtain the parameters of the signal *prior* to entering the measuring device.

We will start this chapter with the description of simple, coarse methods that provide some estimate of the pulse duration and a description of some measurement techniques commonly used for recording pulse correlations (Sections 9.1–9.3). Many of these techniques were developed with the emergence of ns and ps laser technology; for a review see Shapiro [1]. In the second part of this chapter, Section 9.4, we will describe techniques that lead to a complete characterization of the pulse amplitude and phase.

9.1. INTENSITY CORRELATIONS

9.1.1. General Properties

The temporal profile $I_s(t)$ of an optical signal can be easily determined, if a shorter (reference) pulse of known shape $I_r(t)$ is available. The method is to measure the intensity cross correlation:

$$A_c(\tau) = \int_{-\infty}^{\infty} I_s(t) I_r(t-\tau)\, dt. \tag{9.1}$$

Let us define the Fourier transforms of the intensity profiles as:

$$\mathcal{I}_j(\Omega) = \int_{-\infty}^{\infty} I_j(t) e^{-i\Omega t} dt, \tag{9.2}$$

where the subscript j indicates either the reference (r) or signal (s) pulse. The Fourier transform of Eq. (9.2) should not be confused with the spectral intensity (proportional to $|\tilde{\mathcal{E}}(\Omega)|^2$). The Fourier transform of the correlation (9.1) is $A_c(\Omega)$, related to the Fourier transforms of the intensities by:

$$A_c(\Omega) = \mathcal{I}_r(\Omega)\mathcal{I}_s^*(\Omega). \tag{9.3}$$

The shape of the signal $I_s(t)$ can be determined by first taking the Fourier transform $A_c(\Omega)$ of the measured cross correlation and dividing by the Fourier transform $\mathcal{I}_r(\Omega)$ of the known reference pulse $I_r(t)$. The inverse Fourier transform of the complex conjugate of the ratio $A_c(\Omega)/\mathcal{I}_r(\Omega)$ is the temporal profile $I_s(t)$. In presence of noise, this operation leads to large errors unless the reference function is the (temporally) shorter of the two pulses being correlated (or the function with the broadest spectrum). The ideal limit is of course that of the reference being a delta function. In the frequency domain, we are dividing by a constant. In the time domain, the shape of the correlation $A_c(\tau)$ is identical to that of the signal $I_s(t)$. Even in that ideal case, the intensity cross-correlation has an important limitation: It does not provide any information on the phase content (frequency or phase modulation) of the pulse being analyzed.

9.1.2. The Intensity Autocorrelation

In most practical situations, a reference pulse much shorter than the signal cannot be generated. In the ideal cases where such a pulse is available, there is still a need for a technique to determine the shape of the reference signal.

It is therefore important to consider the limit where the signal itself has to be used as reference. The expression (9.1) with $I_s(t) = I_r(t) = I(t)$ is called an intensity autocorrelation. An autocorrelation is always a symmetric function—this property can be understood from a comparison of the overlap integral for positive and negative arguments τ. According to Eq. (9.3), the Fourier transform of the autocorrelation is a real function, consistent with a symmetric function in the time domain. As a result, the autocorrelation provides only little information on the pulse shape, because an infinity of symmetric and asymmetric pulse shapes can have similar autocorrelations. Nevertheless, the intensity autocorrelation is a widely used diagnostic technique, because it can be easily implemented, and is the first tool used to determine whether a laser is producing short pulses rather than intensity fluctuations of a continuous background. Typical examples are given in Section 9.3. The intensity autocorrelation is also used to quote a "pulse duration." The most widely used procedure is to *assume* a pulse shape (generally a sech^2 or a Gaussian shape), and to "determine" the pulse duration from the known ratio between the FWHM of the autocorrelation and that of the pulse. The parameters pertaining to the various shapes are listed in Table 9.1 in Section 9.4.

9.1.3. Intensity Correlations of Higher Order

Let us look at an intensity correlation of a higher order, defined as:

$$A_n(\tau) = \int_{-\infty}^{\infty} I(t) I^n(t - \tau) dt. \tag{9.4}$$

For $n > 1$, the function defined by Eq. (9.4) has the same symmetry as the pulse. In fact, for a reasonably peaked function $I(t)$, $\lim_{n\to\infty} I^n(t) \propto \delta(t)$, and the shape of the correlation $A_n(\tau)$ approaches the pulse shape $I(t)$. Such higher-order correlations are convenient and powerful tools to determine intensity profiles.

9.2. INTERFEROMETRIC CORRELATIONS

9.2.1. General Expression

We have analyzed in Chapter 2 the Michelson interferometer and defined the field[1] correlation measured by that instrument as:

$$G_1(\tau) = \tilde{A}_{12}^{+}(\tau) + c.c = \frac{1}{4} \int_{-\infty}^{\infty} \tilde{\mathcal{E}}_1(t) \tilde{\mathcal{E}}_2^*(t - \tau) e^{i\omega_\ell \tau} dt + c.c. \tag{9.5}$$

[1]The field correlation is often referred to as a first-order correlation.

We have seen also that the Fourier transform of $A_{12}^{+}(\tau)$ is equal to $\tilde{E}_1^*(\Omega)\tilde{E}_2(\Omega)$ [Eq. (2.6)]. Hence, the Fourier transform of the autocorrelation (identical fields) is proportional to the spectral intensity of the pulse. Therefore, a first-order field autocorrelation does not carry any other information than that provided by a spectrometer.

In a Michelson interferometer, let us add to the detector a second harmonic generating crystal (type I) and a filter to eliminate the fundamental. Instead of the expression (9.5), the detected signal is a second-order interferometric correlation, proportional to the function:

$$G_2(\tau) = \int_{-\infty}^{\infty} \left\langle \left| [E_1(t-\tau) + E_2(t)]^2 \right|^2 \right\rangle dt. \tag{9.6}$$

Here $\langle\rangle$ denotes averaging over the fast oscillations of the electric field and the integral stands for integration over the pulse envelope. A Mach–Zehnder interferometer can also be substituted for a Michelson interferometer for such a measurement [2]. Replacing the fields by the usual envelope and phase functions, $E_{1,2} = (\mathcal{E}_{1,2}e^{i\omega_\ell t}e^{i\varphi_{1,2}} + c.c.)/2$ and performing the $\langle\rangle$ average yields for the correlation apart from a constant factor:

$$G_2(\tau) = A(\tau) = A_0(\tau) + \mathrm{Re}\left[A_1(\tau)e^{-i\omega_\ell\tau}\right] + \mathrm{Re}\left[A_2(\tau)e^{-2i\omega_\ell\tau}\right], \tag{9.7}$$

where

$$A_0(\tau) = \int_{-\infty}^{\infty} \left[\mathcal{E}_1^4(t-\tau) + \mathcal{E}_2^4(t) + 4\mathcal{E}_1^2(t-\tau)\mathcal{E}_2^2(t)\right] dt \tag{9.8}$$

$$A_1(\tau) = 4\int_{-\infty}^{\infty} \mathcal{E}_1(t-\tau)\mathcal{E}_2(t)\left[\mathcal{E}_1^2(t-\tau) + \mathcal{E}_2^2(t)\right] e^{i[\varphi_1(t-\tau)-\varphi_2(t)]}dt \tag{9.9}$$

$$A_2(\tau) = 2\int_{-\infty}^{\infty} \mathcal{E}_1^2(t-\tau)\mathcal{E}_2^2(t)e^{2i[\varphi_1(t-\tau)-\varphi_2(t)]}dt. \tag{9.10}$$

The purpose of the decomposition (9.7) is to show that the correlation has three frequency components[2] centered respectively around zero frequency, around ω_ℓ and $2\omega_\ell$. Most often, the detection system of the correlator will act as a low pass filter, eliminating all but the first term of the expansion. The interferometric correlation reduces then to $A_0(\tau)$—the sum of a background term and the

[2]Here "frequency" refers to the variation of the function $G_2(\tau)$ as a function of its argument τ. The latter argument τ is the delay parameter, which is continuously tuned in the correlation measurement.

(background-free) intensity correlation [labeled $A_c(\tau)$ in Eq. (9.1)]. The terms A_0, A_1, and A_2 of the expansion (9.7) can be extracted from a measurement by taking the Fourier transform of the data, identifying the cluster of data near the three characteristic frequencies, and recovering them by successive inverse Fourier transforms. Fast data acquisition and processing can also perform this task in real time when working with fs oscillators [3]. The components A_1 and A_2 contain phase terms $[\varphi_1(t-\tau) - \varphi_2(t)]$, and thus carry information about pulse chirp.

Similarly the third-order interferometric correlation

$$G_3(\tau) = B(\tau) = \int \left\langle \left| [E_1(t-\tau) + E_2(t)]^3 \right|^2 \right\rangle dt \tag{9.11}$$

has four frequency components. In terms of pulse envelopes and phases it can be written as

$$B(\tau) = B_0(\tau) + \mathrm{Re}\left[B_1(\tau)e^{-i\omega_\ell \tau}\right] + \mathrm{Re}\left[B_2(\tau)e^{-2i\omega_\ell \tau}\right] + \mathrm{Re}\left[B_3(\tau)e^{-3i\omega_\ell \tau}\right], \tag{9.12}$$

where

$$B_0(\tau) = \int \left\{ \mathcal{E}_1^6(t-\tau) + \mathcal{E}_2^6(t) + 9\mathcal{E}_1^2(t-\tau)\mathcal{E}_2^2(t)\left[\mathcal{E}_1^2(t-\tau) + \mathcal{E}_2^2(t)\right] \right\} dt \tag{9.13}$$

$$B_1(\tau) = 6 \int \left[\mathcal{E}_1^4(t-\tau) + \mathcal{E}_2^4(t) + 3\mathcal{E}_1^2(t-\tau)\mathcal{E}_2^2(t)\right] \times \mathcal{E}_1(t-\tau)\mathcal{E}_2(t)\, e^{i[\varphi_1(t-\tau) - \varphi_2(t)]} dt \tag{9.14}$$

$$B_2(\tau) = 6 \int \left[\mathcal{E}_1^2(t-\tau) + \mathcal{E}_2^2(t)\right] \mathcal{E}_1^2(t-\tau)\mathcal{E}_2^2(t)\, e^{2i[\varphi_1(t-\tau) - \varphi_2(t)]} dt \tag{9.15}$$

$$B_3(\tau) = 2 \int \mathcal{E}_1^3(t-\tau)\mathcal{E}_2^3(t)\, e^{3i[\varphi_1(t-\tau) - \varphi_2(t)]} dt. \tag{9.16}$$

Again we have omitted a constant factor. The zero frequency component of this interferometric correlation, $B_0(\tau)$, corresponds to the third-order intensity correlation with background.

9.2.2. Interferometric Autocorrelation

9.2.2.1. General Properties

Let us consider in more detail the particular case of the previous expressions for the cross correlation (9.6) through (9.10) where the two fields $E_1 = E_2 = E$. At $\tau = 0$, the peak value of the function $A(\tau = 0) = 16 \int \mathcal{E}^4(t)dt$. For large delays compared to the pulse duration, cross products containing terms with $\mathcal{E}(t - \tau)\mathcal{E}(t)$ vanish, leaving a background of $A(\tau = \infty) = 2 \int \mathcal{E}^4(t)dt$. The peak to background ratio for the interferometric autocorrelation is thus 8 to 1. The "d.c." term of the interferometric autocorrelation, $A_0(\tau)$,—which is in fact an intensity autocorrelation—has a peak to background of 3 to 1. The measurement leading to $A_0(\tau)$ is generally referred to as the *intensity autocorrelation with background*, as opposed to the *background free autocorrelation* leading to the expression $A_c(\tau)$ (9.1).

The fourth term of the expansion of Eq. (9.7) can be regarded as a correlation of the second harmonic fields. In the absence of phase modulation—i.e., for bandwidth-limited pulses—this function is identical to the intensity autocorrelation. This property has been exploited to determine if a pulse is phase modulated or not [4].

As any autocorrelation, the interferometric autocorrelation is a symmetric function. However, as opposed to the intensity autocorrelation, it contains phase information. The shape and phase sensitivity of the interferometric autocorrelation can be exploited to:

1. *qualitatively* test the absence or presence of phase modulation and eventually determine the type of modulation;
2. quantitatively measure a *linear* chirp; and
3. determine, in combination with the pulse spectrum and linear filtering, the pulse shape and phase by fitting procedures (see Section 9.4).

9.2.2.2. Linearly Chirped Pulses

The sensitivity of the interferometric autocorrelation to chirp is well illustrated by the experimental recordings made with the beam from a Ti:sapphire laser in Spence *et al.* [5]. The lower and upper envelopes of the interference pattern split evenly from the background level in Figure 9.1 pertaining to an unchirped 45 fs pulse. In the case of a phase modulated pulse as in Figure 9.2, the interference pattern is much narrower than the pulse intensity autocorrelation. The wings of the interferometric autocorrelation are identical to those of the intensity autocorrelation. The level at which the interference pattern starts relative to the peak (2.8/8 in the case of Fig. 9.2) is a measure of the chirp, as explained later.

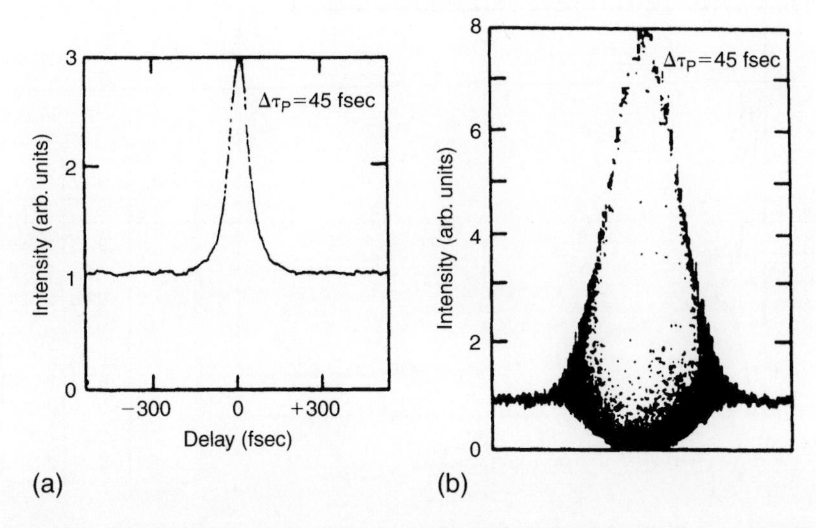

Figure 9.1 Intensity (with background) (a) and interferometric (b) autocorrelation traces of a mode-locked Ti:sapphire laser pulse after extracavity pulse compression. Note the peak to background ratios of 3/1 and 8/1 for the intensity and interferometric autocorrelations, respectively (from Spence *et al.* [5]).

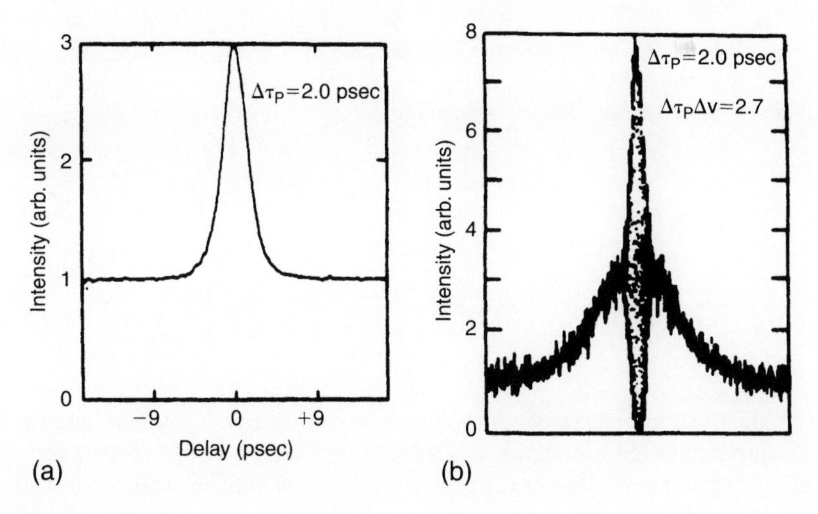

Figure 9.2 Intensity (a) and interferometric (b) autocorrelation traces of a mode-locked phase modulated Ti:sapphire laser pulse (from Spence *et al.* [5]).

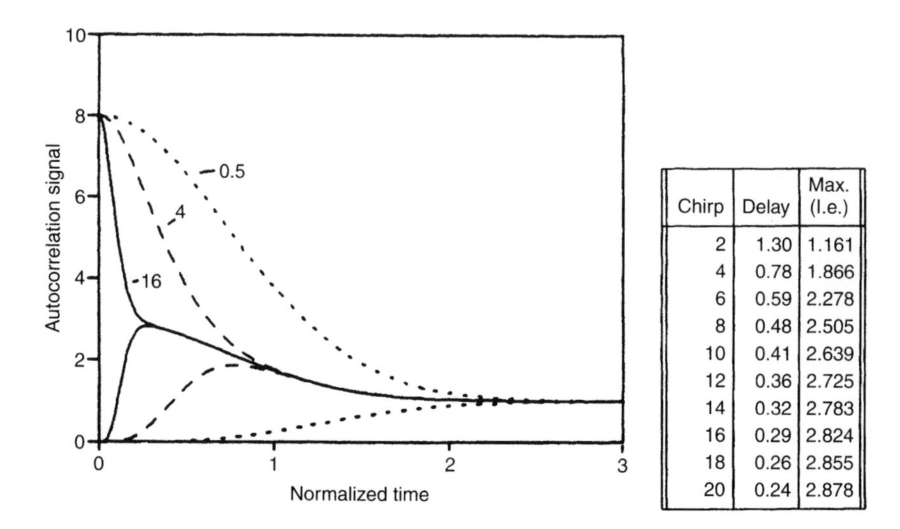

Chirp	Delay	Max. (l.e.)
2	1.30	1.161
4	0.78	1.866
6	0.59	2.278
8	0.48	2.505
10	0.41	2.639
12	0.36	2.725
14	0.32	2.783
16	0.29	2.824
18	0.26	2.855
20	0.24	2.878

Figure 9.3 Interferometric autocorrelations of Gaussian pulses $\tilde{\mathcal{E}}(t) = \exp\{-(1+ia)(t/\tau_G)^2\}$ for various values of the linear chirp parameter a. The upper and lower envelopes of the autocorrelations are plotted for three values of the chirp parameter a. The upper and lower envelopes merge with the intensity autocorrelation. The table on the right shows the position (delay and value) of the maxima of the lower envelope (l.e.) of the interferometric autocorrelation as a function of chirp parameter a.

A simple tabulation of the chirp can be made by considering a linearly chirped Gaussian pulse ($\tilde{\mathcal{E}}E(t) = \exp[-(1+ia)(t/\tau_G)^2]$), for which the interferometric autocorrelation can be determined analytically:

$$G_2(\tau) = \left\{ 1 + 2\exp\left[-\left(\frac{\tau}{\tau_G}\right)^2\right] + 4\exp\left[-\frac{a^2+3}{4}\left(\frac{\tau}{\tau_G}\right)^2\right]\cos\left[\frac{a}{2}\left(\frac{\tau}{\tau_G}\right)^2\right] \right.$$
$$\left. \times \cos(\omega_\ell \tau) + \exp\left[-(1+a^2)\left(\frac{\tau}{\tau_G}\right)^2\right]\cos 2\omega_\ell \tau \right\}. \tag{9.17}$$

A graphical representation of the upper and lower envelopes as a function of the chirp parameter a is shown in Figure 9.3. Comparison of Figs. 9.2 and 9.3 indicate a chirp parameter of roughly $a = 20$ for the experimental pulse. This is of course only an approximation, but it gives a good estimate of the magnitude of the frequency modulation near the pulse center.

9.2.2.3. Averaging over the Pulse Train

Most often, measurements are an average over a pulse train. It has been demonstrated by Van Stryland [6] that the intensity autocorrelation of pulses with a statistical distribution of pulse durations is shaped like a double-sided exponential. This is precisely the shape of the autocorrelation of a single-sided exponential pulse. The autocorrelation of the output of synchronously pumped lasers have typically such a double-sided exponential shape. It would be incorrect to conclude that the pulses generated by these lasers have a single sided exponential shape. Theoretical simulations have indeed confirmed that fluctuations of the pulse duration along the train are at the origin of the observed autocorrelation [7, 8].

A similar ambiguity exists in the case of the interferometric autocorrelation of a train of pulses. The measurement can be either interpreted as the interferometric autocorrelation of identical chirped pulses, or as the average of interferometric autocorrelations of bandwidth-limited pulses of different frequencies. To illustrate this point, let us consider unchirped Gaussian pulses with a Gaussian distribution of frequencies $F(\Delta\Omega)$ centered at ω_ℓ ($\Omega = \omega_\ell + \Delta\Omega$):

$$F(\Delta\Omega) = \frac{\tau_G}{b\sqrt{\pi}} e^{-(\Delta\Omega\ \tau_G/b)^2}. \tag{9.18}$$

The total autocorrelation (averaged over many pulses at each delay) is the statistical average of the autocorrelations at each frequency Ω:

$$G_2(\tau) = \int_{-\infty}^{\infty} \left\{ 1 + 2\exp\left[-\left(\frac{\tau}{\tau_G}\right)^2\right] + 4\exp\left[-\frac{3}{4}\left(\frac{\tau}{\tau_G}\right)^2\right] \cos\Omega\tau \right.$$
$$\left. + \exp\left[-\left(\frac{\tau}{\tau_G}\right)^2\right] \cos 2\Omega\tau \right\} F(\Delta\Omega) d\Omega. \tag{9.19}$$

The integration over frequency can easily be performed. For small chirps, we can make the approximation in Eq. (9.17) that $\cos[\frac{a}{2}(\frac{\tau}{\tau_G})^2] \approx 1$. The equation obtained after substitution of (9.18) into Eq. (9.19) and subsequent integration is undistinguishable from Eq. (9.17) for $b = a$. Therefore, it is important to verify that the pulse train is constituted of *identical* pulses, in amplitude (energy), duration and frequency. It is relatively easy to check whether the pulses have constant energy and duration by displaying simultaneously the fundamental and the second harmonic of the pulse train. If both show no fluctuation, it can be said with reasonable certainty that the intensity autocorrelation represents pulses having the same energy and duration. It can be verified that there are no pulse to pulse variation in frequency by displaying the pulse train on an oscilloscope, after transmission through a spectrometer or reflection off a thin ($\ll$100 μm) etalon.

9.3. MEASUREMENT TECHNIQUES

9.3.1. Nonlinear Optical Processes for Measuring Femtosecond Pulse Correlations

In ultrafast optics, SHG is the most widely used technique for recording second-order correlations. Because it is a nonresonant process of electronic origin, the nonlinearity is fast enough to measure pulses down to $10^{-14}s$ duration. While applicable through the IR and visible spectrum, the method is limited at short wavelength ($\lambda < 380$ nm) by the UV absorption edge of optical crystals. Techniques that have been used successfully for shorter wavelengths include multiphoton ionization [9] surface SHG [10], and two-photon luminescence [11]. Third-order processes such as the optical Kerr effect have also been applied to the diagnostic of fs UV pulses [12–14]. These third-order correlations, as discussed in Section 9.1.3, have the additional advantage of being sensitive to pulse asymmetry. Photodetectors excited by a multiphoton absorption are also convenient tools for correlation measurements as they produce a signal (current) that can be processed directly unlike most other nonlinear optical techniques [15–18].

9.3.2. Recurrent Signals

It is assumed here that we have an ensemble of identical fs pulses, so that the correlations can be constructed from a large number of measurements taken for different delay parameters τ.

An example of a simple second-order correlator is sketched in Figure 9.4(a). The beams to be correlated are cross-polarized, and combined with a polarizing beam splitter. An optical delay is used to adjust the delay of the reference signal $I_r(t - \tau)$. The cross-polarized beams are sent orthogonally polarized into a nonlinear crystal phase matched for type II SHG. If the conditions outlined below are satisfied, the second harmonic signal is proportional to the function $A_c(\tau)$ defined in Eq. (9.1). This measurement—or function—is generally referred to as the background-free correlation, as opposed to the correlation with background. The latter is obtained by frequency doubling the output of a Michelson interferometer (parallel polarization) in a crystal phase matched for type I SHG. An alternate technique to generate $A_c(\tau)$ is to use beams with parallel polarization intersecting in a nonlinear crystal. The background free signal is the second

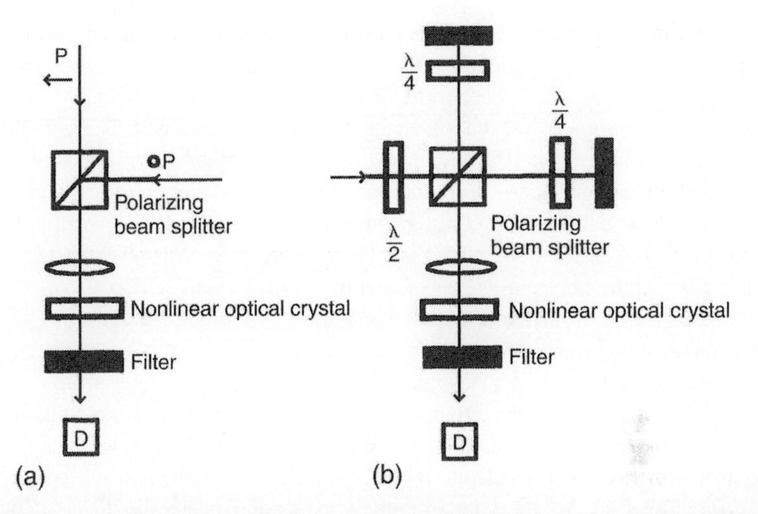

Figure 9.4 (a) Basic intensity cross-correlator, using second harmonic type II detection. A polarizing beam splitter cube combines the beams to be correlated. The SH signal generated by the crystal is proportional to the product of the fundamental intensities along the two orthogonal directions of polarization. The sketch on the right side, (b), shows a simple autocorrelator using the same type of detection. The same "recombining" polarizing beam splitter cube can be used to split the beams into the two arms of the interferometer. Two quarter wave plates are used to rotate the polarization of the beam reflected by the mirrors of the two delay arms by 90°.

harmonic generated with wave vector k_2 along the bisector of the two wave vectors k_s and k_r. The crystal orientation has to satisfy the phase matching condition $k_2 = k_s + k_r$.

With either background-free techniques, the SH field is proportional to the product of the fundamental fields:

$$\mathcal{E}_{SHG}(\Omega) = \eta(\Omega)\mathcal{E}_s(\Omega)\mathcal{E}_r(\Omega), \tag{9.20}$$

or, for the spectral intensities:

$$I_{SHG}(\Omega) = \frac{2\mu_0 c}{n}|\eta(\Omega)|^2 I_s(\Omega) I_r(\Omega). \tag{9.21}$$

In order for the instrument sketched in Fig. 9.4 to measure the true cross-correlation or autocorrelation, it is essential that the SH conversion efficiency $\eta(\Omega)$ be a constant over the frequency range of the combined pulses. Another way to express the same condition is to state that the effective crystal length should be shorter than the coherence length of harmonic generation over the pulse bandwidth (cf. Chapter 3). The "effective length" can be either the physical crystal thickness, Rayleigh range ρ_0 of the focused fundamental beams, or the overlapping region of the fundamental pulses in a noncollinear geometry. The shorter the crystal, the broader the bandwidth over which phase matched harmonic conversion is obtained, but the lower the conversion efficiency. There is clearly a compromise to be reached between bandwidth and sensitivity. In Eq. (9.21), the bandwidth efficiency factor $\eta(\Omega)$ includes only the frequency dependence of the phase matching condition, and not a finite response time for the harmonic generation process. It is assumed here that the response time of the second harmonic process is much shorter than the pulses to be measured, which is a reasonable assumption since the second-order nonlinearity of wide bandgap crystals is a nonresonant electronic process.

Provided the pulses can be approximated by a Gaussian, a simple test can be performed to determine whether the proper focusing and crystal thickness has been chosen. In the case of the autocorrelation [Fig. 9.4(b)], a standard spectrometer (a 25-cm spectrometer is generally sufficient) is used to record the spectral intensities of the fundamental and second harmonic [2]. In the case of perfect phase matching and zero dispersion, and for a conversion efficiency independent of frequency, the ratio of second harmonic to the *square* of the fundamental spectral intensity will be a constant in the case of Gaussian pulses, according to Eq. (3.106). The spectrum of the second harmonic will be narrower than the squared fundamental spectrum if the effective crystal length is too long. As a consequence of the SH conversion efficiency being frequency dependent, the measured correlation width will be longer than the exact correlation length.

The background-free autocorrelation function A_c [Eq. (9.1)] for a fluctuating cw signal consists of a symmetric "bump" riding on an infinite background (Figure 9.5). The width of the bump is a measure of the temporal width of the fluctuations, and the contrast ratio (peak-to-background ratio of A_c) is a measure of the modulation depth. A 100% modulation depth results in a peak-to-background ratio of 2 to 1 [19]. Any background-free signal of finite duration results in a function A_c of finite width [Fig. 9.5(c)]. If that signal has some fine structure (amplitude modulation), a narrow spike will appear in the middle of the correlation function [Fig. 9.5(d)]. This is the coherence spike, typical of a signal consisting of a burst of *amplitude* noise [20]. These considerations do not apply to the phase content or phase coherence of the pulse: The intensity correlations are the same whether the pulse is at a fixed carrier frequency or has a random or deterministic frequency modulation.

9.3.3. Single Shot Measurements

Not all lasers provide a train of identical pulses and/or work at high repetition rate. Pulse to pulse fluctuation can be particularly severe in oscillator amplifier systems. Single shot autocorrelators are therefore highly desirable. In this

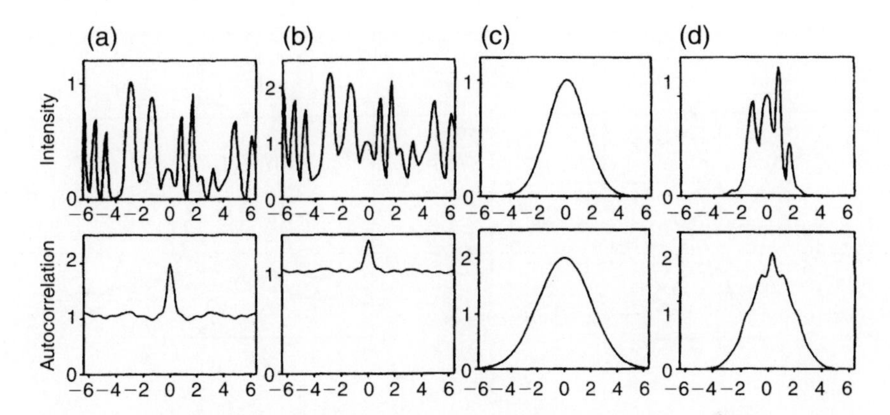

Figure 9.5 Some typical waveforms (intensity versus time) (top) and corresponding intensity autocorrelation $A_C(\tau)$ (bottom). From left to right: (a) continuous signal with 100% amplitude modulation; (b) noisy cw signal; (c) pulse; and (d) noisy pulse.

section we will discuss the simplest single shot autocorrelators. More sophisticated instruments for single shot amplitude and phase retrieval will be described in Section 9.4.

9.3.3.1. Intensity Autocorrelators

One of the first intensity autocorrelators for mode-locked lasers was a single-shot instrument [21]. The beam to be measured is split in two beams, which are thereafter sent with opposite propagation vector into a nonlinear medium. The first autocorrelator was based on two photon excitation rather than SHG: The medium (for instance a dye solution) was selected for its large two photon absorption and subsequent fluorescence. Because of the larger optical field in the region where the two counter-propagating pulses collide, the observed pattern of two-photon fluorescence essentially displays the intensity autocorrelation (with background). Because of the higher conversion efficiency of SHG, two-photon fluorescence is not widely used in the fs time scale, except in the UV, where no transparent nonlinear crystals can be found. To circumvent the difficulty of spatially resolving the μm size of the two-photon fluorescence trace, the beams are made to intersect at a small angle, thereby magnifying the fluorescence trace [11, 22].

Single-shot autocorrelators using SHG have also been designed. In an arrangement developed for ps pulses by Jansky *et al.* [23] and Gyuzalian *et al.* [25] the autocorrelation in time is transformed into a spatial intensity distribution. This method has been applied by numerous investigators to the fs scale [24, 26, 27]. The instrument is a typical noncolinear SH autocorrelator. The nonlinear crystal is oriented for phase matched type I SHG for two beams intersecting at an

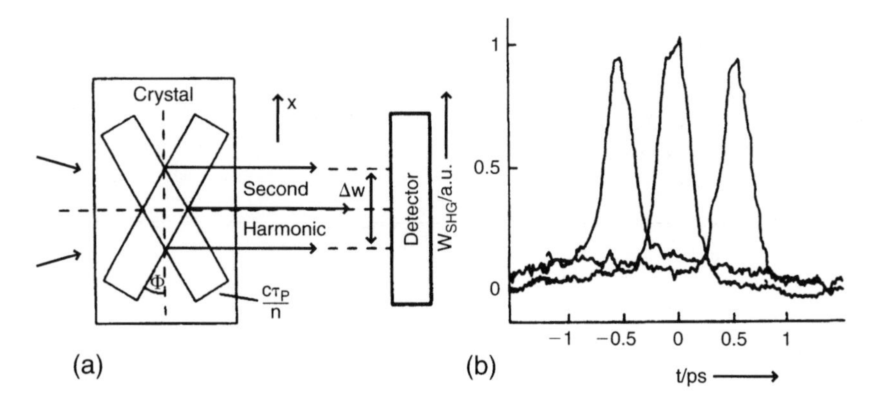

Figure 9.6 (a) Basic principle of the single-shot autocorrelator of Jansky *et al.* [23]. The nonlinear crystal is oriented for phase matched type I SHG for two beams intersecting at an angle 2Φ. Arrows indicate the propagation direction of the fundamental and the second harmonic (z direction). (b) Typical recording for a 250-fs pulse. Temporal calibration is made by recording the shift of the trace resulting from insertion of a 0.36 mm glass plate in either arm of the autocorrelator (Adapted from Rempel [24]).

angle 2Φ [Figure 9.6(a)]. Provided the beam waist in the overlap region w_0 is much larger than the pulse length $\nu_g \tau_p$, the intensity distribution across the SH beam corresponds to the intensity autocorrelation function. Cylindrical focusing and a linear array detector can be used to capture the intensity profile of the autocorrelation. A typical measurement is shown in Fig. 9.6(b). The width of the autocorrelation function $\Delta\tau_a$ is proportional to the diameter Δw of the SH beam:

$$\Delta w = \frac{\Delta\tau_a \nu_g}{\sin \Phi}. \tag{9.22}$$

Let us consider the projection of the two intersecting beams on axis z (along the direction of propagation) and x (the orthogonal direction). The components of the pulse envelope are copropagating along the z direction, counter-propagating along the x axis. For a pulse with a temporal profile $\tilde{\mathcal{E}}(t)$, the component of the two envelopes propagating along the x direction are $\tilde{\mathcal{E}}_x = \tilde{\mathcal{E}}(t - x/L_x)$ and $\tilde{\mathcal{E}}_{-x} = \tilde{\mathcal{E}}(t + x/L_x)$ with $L_x = \nu_g/\sin \Phi$. The SH intensity along the x direction is proportional to the product $\tilde{\mathcal{E}}_x \tilde{\mathcal{E}}_{-x}$ [23]. If we consider for instance a linearly chirped Gaussian pulse, the second harmonic just behind the crystal is also a Gaussian:

$$\tilde{\mathcal{E}}_{SHG}(t) \propto \exp\left[-2\left(\frac{t^2}{\tau_G^2} + \frac{x^2 \sin^2 \Phi}{\nu_g^2 \tau_G^2}\right)\left(1 + i\frac{a}{2}\right)\right]. \tag{9.23}$$

The spatial dependence of the SH intensity is indeed an intensity autocorrelation, expanded transversely by the factor $1/\sin\Phi$. As noted by Jansky *et al.* [23] and further investigated by Danielius *et al.* [28] the spatial phase dependence of the SH field indicates that the wavefront is no longer a plane wave, but has a (cylindrical) curvature proportional to the chirp parameter a. Referring to Gaussian beam propagation Eq. (1.180), the curvature is approximately $R = k_\ell v_G^2 \tau_G^2/(2a\sin^2\Phi)$.

An analysis of the beam propagation after the crystal can be made by spatial Fourier transforms, starting with the wave equation in the retarded frame Eq. (1.177):

$$\frac{\partial}{\partial z}\tilde{\mathcal{E}}_{SHG} = -\frac{i}{2k_\ell}\frac{\partial^2}{\partial x^2}\tilde{\mathcal{E}}_{SHG}. \tag{9.24}$$

Taking the Fourier transform along the transverse spatial coordinate x and integrating along the propagation direction z:

$$\tilde{\mathcal{E}}_{SHG}(k_x, z, t) = \tilde{\mathcal{E}}_{SHG}(k_x, z = 0, t)e^{\frac{i}{2k_\ell}k_x^2 z}. \tag{9.25}$$

Taking the inverse Fourier transform leads to the field distribution after a propagation distance z:

$$\tilde{\mathcal{E}}_{SHG}(x, z, t) = \int_{-\infty}^{\infty} \tilde{\mathcal{E}}_{SHG}(k_x, z = 0, t)e^{-ik_x x}e^{\frac{i}{2k_\ell}k_x^2 z}dk_x. \tag{9.26}$$

Assuming that the overlapping region is short enough to produce an undistorted second harmonic, as discussed in Chapter 3, the SH field at the end of the crystal is:

$$\tilde{\mathcal{E}}_{SHG}(x, z = 0, t) = \eta\tilde{\mathcal{E}}\left(t - \frac{x}{L_x}\right)\tilde{\mathcal{E}}\left(t + \frac{x}{L_x}\right), \tag{9.27}$$

where η is the proportionality factor of Eq. (9.20) assumed to be constant over the frequency range of interest. Taking the Fourier transform gives us the function $\tilde{\mathcal{E}}_{SHG}(k_x, z = 0, t)$ to be inserted in Eq. (9.26):

$$\tilde{\mathcal{E}}_{SHG}(k_x, z = 0, t) = \int \eta e^{ik_x x}\tilde{\mathcal{E}}\left(t - \frac{x}{L_x}\right)\tilde{\mathcal{E}}\left(t + \frac{x}{L_x}\right)dx. \tag{9.28}$$

The spatial distribution given by Eq. (9.26) focuses after a distance z_F for which the phase factor $(i/2k_\ell)k_x^2 z_F$ compensates the phase factor produced by chirp in Eq. (9.23). It is possible to reconstruct the chirp by matching the intensity distributions measured at $z = 0$ and $z = z_F$ to the distributions calculated with help of Eqs. (9.24) through (9.28).

9.3.3.2. Interferometric Autocorrelator

By recording the spatial profile of the second harmonic with interferometric accuracy, Salin *et al.* [29] showed that it was actually possible to record an interferometric autocorrelation. Because of diffraction effects, as explained earlier, the method is difficult to implement, and usually does not provide a recording with an 8 to 1 peak to background contrast. Simpler methods are available, making use of the tilt in energy front introduced by a dispersive element. We have seen in Chapter 2 that the energy front is tilted with respect to the wavefront by a prism (or any other element introducing angular dispersion). As shown by Szabo *et al.* [30] this property can be exploited to provide a variable delay along a transverse coordinate of the beam. A glass wedge is inserted in one or both arm(s) of a Michelson type autocorrelator (Figure 9.7). As in the previous method, the spatial (transverse) distribution of second harmonic is proportional to the pulse

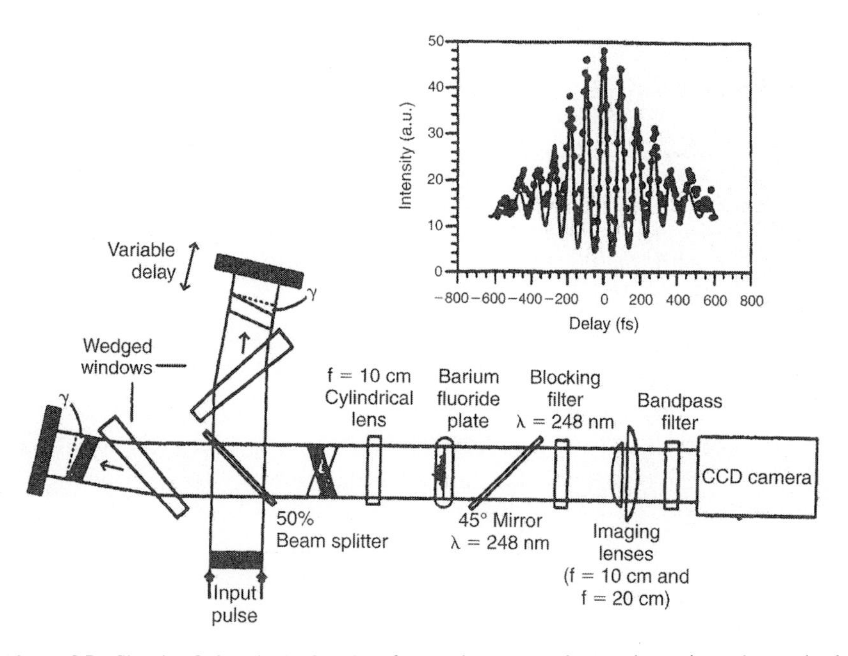

Figure 9.7 Sketch of the single-shot interferometric autocorrelator using prisms (or wedged windows) to transfer the delay to the transverse coordinate of the beam, and typical recording of a 250-fs UV pulse. To record fs pulses, the angle of the prisms should not exceed a few degrees. The fringe spacing is adjusted by the tilt of a mirror in one arm of the autocorrelator (Adapted from LeBlanc *et al.* [22]).

intensity autocorrelation. Obtaining such an intensity autocorrelation, however, assumes that the beams coming from both arms of the correlator add constructively toward the detector, destructively toward the source. Such a condition is difficult to implement, because it requires subwavelength stability and accuracy in controlling either arm. For this particular correlator, it is more convenient to introduce a small tilt of either end mirror of the Michelson interferometer. Such a tilt produces a pattern of parallel fringes at the output. Before the frequency doubling crystal, we have thus generated a first-order correlation. The second harmonic of such a first-order correlation is an interferometric correlation [22]. This arrangement has the advantage that one has complete control over the spacing of the fringes, which can be adjusted to accommodate the spatial resolution of the array detector used in this measurement. An example of an interferometric autocorrelation obtained with a fs UV pulse is shown in Fig. 9.7. For this particular case, the nonlinearity is two-photon fluorescence in BaF_2.

9.4. PULSE AMPLITUDE AND PHASE RECONSTRUCTION

9.4.1. Introduction

Because the second-order autocorrelations are symmetric and do not provide any information about the pulse asymmetry, either an additional measurement or a new technique is required to determine the signal shape. We will start with simple methods that complement the information of the autocorrelations, and proceed with an overview of various methods that have been introduced to provide amplitude and phase information on fs signals. The ideal diagnostic instrument is obviously one that would give a real time display of all pulse parameters. Because of the ambiguity associated with an average over a large number of pulses, a single-shot method is also desirable. The challenge in fs pulse characterization is that a temporal resolution is needed that is faster than the pulse itself. The solution, as sketched in Figure 9.8, is to apply a transfer function to expand the signal. From the knowledge of the transfer function and the expanded signal the shape of the original object is recovered.

As introduced in Chapter 1 a light pulse in the time domain is characterized by its electric field

$$E(t) = \frac{1}{2}\mathcal{E}(t)e^{i\varphi_0}e^{i\varphi(t)}e^{i\omega_\ell t} + c.c. \tag{9.29}$$

In this section we are concerned with the retrieval of the pulse envelope $\mathcal{E}(t)$ and the time-dependent phase $\varphi(t)$ only. The measurement and control of the absolute phase φ_0 are described in Chapter 13.

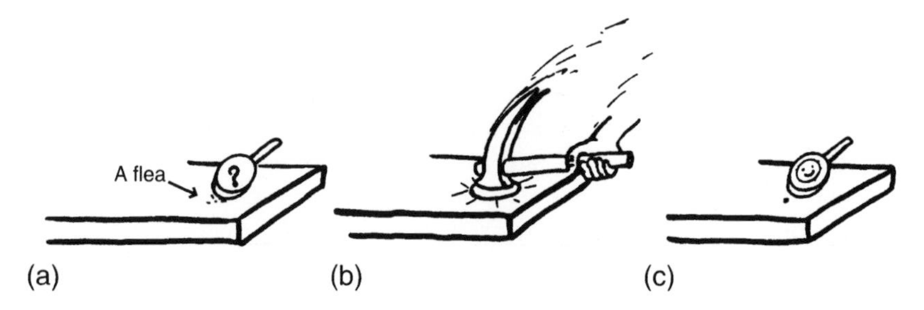

Figure 9.8 Applying a transfer function (b) to magnify the object (a) to be observed. The original object can be reconstructed by applying the inverse transfer function to the observed figure (c).

More than 30 years ago Treacy [31] measured a "sonogram" of ps pulses. It took more than a decade before its importance for the full-field characterization of fs pulses was recognized [32]. Phase and amplitude of a pulse stretched in an optical fiber was measured using a cross-correlation with a short sample pulse (compressed output pulse) using an interferometric technique [33].

Early on methods based on interferometric autocorrelations and the pulse spectrum were developed to retrieve the complex field [2, 4, 34]. Efforts to use the pulse spectrum in conjunction with some kind of (nonlinear) correlation have been pursued [35–37].

With the development of more reliable and powerful fs lasers in the 1990s a variety of other pulse characterization schemes have been discussed and demonstrated [38–46]. While some rely on matching the measurements with a complex pulse amplitude iteratively, others permit a direct reconstruction. In the past decade, two techniques have emerged as most successful and versatile for a variety of different application scenarios—frequency resolved optical gating (FROG) [14] and spectral phase interferometry for direct electric field reconstruction (SPIDER) [47].

In the next section we will discuss the general requirements on an experimental apparatus to retrieve amplitude and phase of ultrashort light pulses. In Section 9.4.3 we will review techniques based on the measurement of pulse correlations and spectrum with subsequent fitting. FROG and SPIDER will be introduced in Sections 9.4.4 and 9.4.5, respectively.

9.4.2. Methods for Full-Field Characterization of Ultrashort Light Pulses

Walmsley and colleagues [48–50] used an elegant approach to discuss the general requirements on measurement techniques that permit the retrieval of both amplitude and phase of short light pulses. A necessary and sufficient condition is that the instrument contains at least one-time nonstationary and one-time stationary element. Four interferometric and four noninterferometric schemes that consist of a minimal number of filters exist that satisfy these requirements [49, 50]. The detector is assumed to be time integrating, that is, it has zero bandwidth.

Time stationary filters whose output do not depend on the arrival time of the pulse act on the pulse field according to

$$E_{out}(t) = \int S(t-t')\, E_{in}(t')\, dt'. \tag{9.30}$$

Examples are passive devices such as mirrors, gratings, spectrometers, and dispersive delay lines. Time nonstationary (or frequency stationary) filters produce an output that does not change with arbitrary spectral shifts of the input

$$E_{out}(\Omega) = \int N(\Omega-\Omega')E_{in}(\Omega')d\Omega'. \tag{9.31}$$

Examples are shutters, which may be controlled externally or by the light pulse itself.

Each of the two filter classes can be further divided into phase-only (P) and amplitude-only (A) filters. Examples of corresponding filter functions are:

$$N_A(t,\tau) = e^{-\Gamma^2(t-\tau)^2} \tag{9.32}$$

$$N_P(t,a) = e^{iat^2} \tag{9.33}$$

$$S_A(\Omega,\omega_c) = e^{-(\Omega-\omega_c)^2/\gamma^2} \tag{9.34}$$

$$S_P(\Omega,b) = e^{-ib\Omega^2}, \tag{9.35}$$

which represent a time gate, a phase modulator, a spectral filter, and a dispersive delay line, respectively.

In Chapter 1 we introduced the Wigner function $\mathcal{W}$ as a convenient tool to completely characterize the field of an ultrashort pulse. It is related to the nonstationary two-time correlation function

$$C(t,t') = \langle E_{in}(t)E_{in}^*(t')\rangle$$

by

$$\mathcal{W}(\Omega,t) = \int dt' C\,(t+s/2, t-s/2)\, e^{i\Omega s}. \tag{9.36}$$

Note, that the correlation function C is what one could theoretically measure with a quadratic detector after a suitable set of filters. Any apparatus for complete pulse characterization must produce a function of two independent variables to carry the information about the pulse contained in $C(t, t')$ or $W(\Omega, t)$. The required two independently adjustable parameters have to be provided by the set of filters used. The detector then produces an output signal

$$D(p_i) = \int d\Omega \int dt \mathcal{W}(\Omega, t) F(\Omega, t, p_i). \tag{9.37}$$

Here F is a window function determined by the filter parameters p_i (type and sequence) of the instrument. If the window function of the apparatus is known the Wigner function (and from that the pulse parameters) can be retrieved from the measurement D. If the apparatus consists of only time stationary (frequency–stationary filters) the window function becomes independent of time (frequency). The measured signal in these cases is the overlap of one of the marginals of the Wigner function, cf. Eqs. (1.41) and (1.42), and therefore contains no phase information. A necessary requirement for full-field reconstruction is thus the presence of at least one of either filters.

The general layout of a noninterferometric and an interferometric system is sketched in Figure 9.9(a) and (b) consisting of a minimum set of filters. Because the detector responds only to intensities (square in the field) only amplitude filters are meaningful elements just preceding the detector. A noninterferometric system consists of filters in sequence. In an interferometric device the pulse is split, each replica filtered separately before the combined and filtered output is detected. While FROG belongs to the first group of techniques, SPIDER is an example of an interferometric technique.

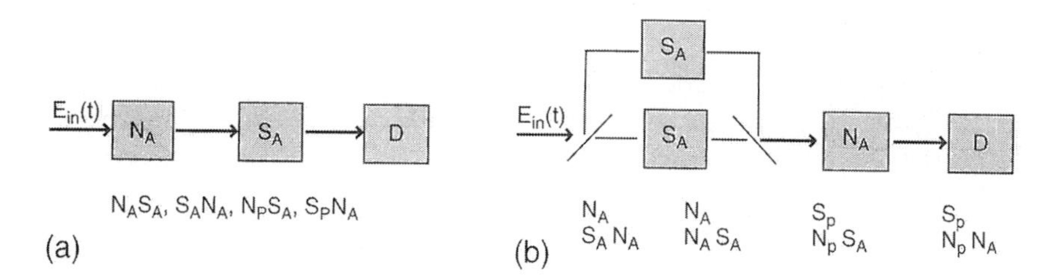

Figure 9.9 (a) Noninterferometric and (b) interferometric techniques for the measurement of amplitude and phase of short light pulses. Four possible combinations of filters (labeled by their transfer functions) are shown for each concept. Adapted from Iaconis *et al.* [47] and Walmsley and Wong [49].

9.4.3. Retrieval from Correlation and Spectrum

While being always symmetric, the shape of an intensity and interferometric autocorrelation is (somewhat) sensitive to the pulse shape. It is conceivable that the pulse spectrum can complement the information provided by the symmetric second-order autocorrelations, to determine the signal shape. As an illustration of this, Table 9.1 shows analytical expressions [2] for the pulse spectrum, the intensity correlation, and the envelope of the interferometric correlation for various pulse shapes. For some typical shapes of the temporal intensity profile given in the first column, the spectral intensity (column 2) is used to compute the duration–bandwidth product $\tau_p \Delta\nu$ listed in column 3. The unit for the time t is such that the functional dependence takes the simplest form in column 1. The inverse of that time unit is used as unit of frequency Ω. The most often quoted parameter is the ratio of the FWHM τ_{ac} of the intensity autocorrelation (column 4) to the pulse duration τ_p, and is given in column 5. Finally, the upper and lower envelopes of the interferometric autocorrelation $G_2(\tau)$ can be reconstructed from the expressions given in column 6.

Table 9.1

Typical pulse shapes, spectra, intensity, and interferometric autocorrelations. To condense the notation, x has been substituted for $\frac{2}{3}\tau$, y for $\frac{4}{7}\tau$, ch for cosh, sh for sinh. τ_{ac} is the FWHM of the intensity autocorrelation. τ_p is the FWHM of the pulse intensity given in column 1. In the last column, $Q = \pm 4[\tau \mathrm{ch}2\tau - \frac{3}{2}\mathrm{ch}^2 x\,\mathrm{sh}x(2 - \mathrm{ch}2x)]/[\mathrm{sh}^3 2x]$.

$\mathcal{E}^2(t)$	$\lvert\mathcal{E}(\Omega)\rvert^2$	$\tau_p\Delta\nu$	$A_c(\tau)$	τ_{ac}/τ_p	$G_2(\tau) - [1 + 3A_c(\tau)]$
e^{-t^2}	$e^{-\Omega^2}$	0.441	$e^{-\tau^2/2}$	1.414	$\pm e^{-(3/8)\tau^2}$
$\mathrm{sech}^2(t)$	$\mathrm{sech}^2\left(\frac{\pi\Omega}{2}\right)$	0.315	$\frac{3\tau(\mathrm{ch}\tau - \mathrm{sh}\tau)}{\mathrm{sh}^3\tau}$	1.543	$\pm\frac{3(\mathrm{sh}2\tau - 2\tau)}{\mathrm{sh}^3\tau}$
$[e^{t/(t-A)} + e^{-t/(t+A)}]^{-1}$					
$A = \frac{1}{4}$	$\frac{1 + 1/\sqrt{2}}{\mathrm{ch}\frac{15\pi}{16}\Omega + 1/\sqrt{2}}$	0.306	$\frac{1}{\mathrm{ch}^3\frac{8}{15}\tau}$	1.544	$\pm 4\left(\frac{\mathrm{ch}\frac{4}{15}\tau}{\mathrm{ch}\frac{8}{15}\tau}\right)^3$
$A = \frac{1}{2}$	$\mathrm{sech}\frac{2\pi}{4}\Omega$	0.278	$\frac{3\mathrm{sh}4x - 8\tau}{4\mathrm{sh}^3\frac{4}{3}\tau}$	1.549	$\pm Q$
$A = \frac{3}{4}$	$\frac{1 - 1/\sqrt{2}}{\mathrm{ch}\frac{7\pi}{16}\Omega - 1/\sqrt{2}}$	0.221	$\frac{2\mathrm{ch}4y + 3}{5\mathrm{ch}^3 2y}$	1.570	$\pm 4\frac{\mathrm{ch}^3 y(6\mathrm{ch}2y - 1)}{5\mathrm{ch}^3 2y}$

As we have seen in the previous section, the interferometric autocorrelation also carries information about the pulse chirp. At the same time these correlation functions are one of the data sets that require relatively little experimental effort. It is therefore tempting to explore the feasibility of obtaining amplitude and phase of the optical pulse from such measurements. Indeed some of the earliest successful retrievals of the complex field of fs pulses were based on the simultaneous fitting of spectrum and interferometric autocorrelation [2, 4]. The reconstruction was facilitated after replacing the autocorrelation by a cross-correlation of pulses of different duration. The result of that cross-correlation approximated the longest of the two pulses. Figure 9.10 shows an example of a Michelson interferometer unbalanced with a phase-only filter (block of glass), and the recorded intensity and interferometric correlation functions.

In most cases the problem reduces to the task of measuring the pulse spectrum and suitable correlations and finding an amplitude and time-dependent phase that fits the data best. Care has to be taken to guarantee a unique retrieval, which is to avoid ambiguities hidden in the data sets (see, for example, Problem 3 at the end of this chapter). For this reason unbalanced correlators where a

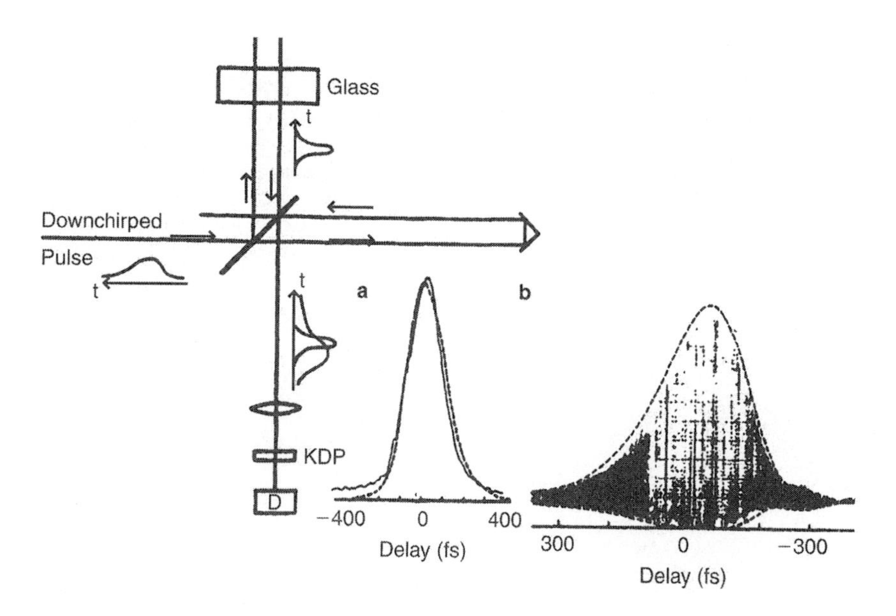

Figure 9.10 Sketch of an asymmetric correlator to record second-order correlation functions G_2 through SHG. An example of intensity (a) and interferometric (b) cross-correlation measured with this setup is also shown. The input was an asymmetric downchirped pulse, which is compressed in the arm containing a 5-cm BK7 block. Adapted from Diels *et al.* [2].

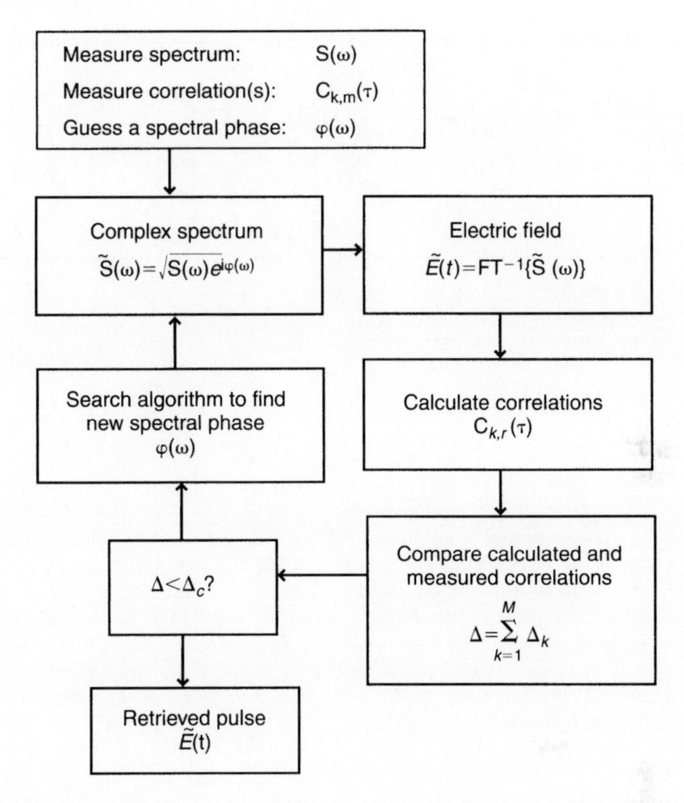

Figure 9.11 Schematic diagram of the retrieval of phase and intensity from correlation and spectrum only (PICASO) [36,51].

linear optical element of known transfer function is inserted into one arm of the correlator have been implemented [34,36].

A possible retrieval algorithm is sketched in Figure 9.11. The pulse spectrum, $S(\Omega) = |\mathcal{E}(\Omega)|^2$, and a pulse correlation of type k or an ensemble of M correlations, $C_{k,m}$ are measured. The retrieval starts by guessing a spectral phase, $\varphi(\Omega)$, which combined with the measured spectrum results in an initial pulse $\sqrt{S(\Omega)}e^{i\varphi(\Omega)}$. This pulse is used to calculate the correlation(s) $C_{k,r}(\tau_i)$ that are recorded in the measurement. A root mean square deviation of measured and calculated correlation can be defined by

$$\Delta_k = \sum_{k=1}^{M} \sqrt{\frac{1}{N} \sum_{i=1}^{N} \left[C_{k,r}(\tau_i) - C_{k,m}(\tau_i)\right]^2}, \tag{9.38}$$

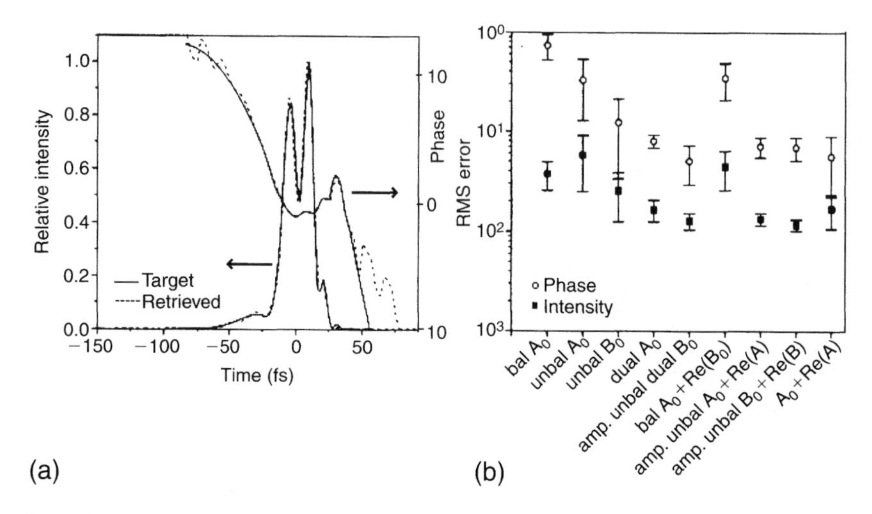

Figure 9.12 (a) Target and retrieved pulse amplitude and phase using a third-order correlation, which was unbalanced with an amplitude-only filter. Field noise (3% additive and multiplicative) was considered. (b) rms errors for different retrieval scenarios for the pulse shown in the left part of the figure. The labels refer to the components of the correlations used in the retrieval (amp., unbal., dual B_0, for example, means that two intensity correlations of third order, where used, measured with an amplitude unbalanced Michelson interferometer. In one of the measurements an additional filter was placed in front of the detector). The labels were defined in Eqs. (9.7) and (9.12). Adapted from Nicholson and Rudolph [37].

which serves as the figure of merit to be minimized during the search. The quality of the retrieval and its robustness against experimental noise depends on the data sets used for the correlation functions.

If the correlation is chosen properly, the pulse can be retrieved reliably together with the spectrum. This is illustrated in Figure 9.12. Part (a) shows a test pulse and its retrieved replica using a third-order correlation, which can be based on third harmonic generation for example. Figure 9.12(b) depicts the rms error of the retrieval results for different measurement scenarios, that is, different types of correlation in addition to the spectrum.

9.4.4. Frequency Resolved Optical Gating (FROG)

Frequency resolved optical gating (FROG) was introduced by Kane and Trebino [13, 14] in 1993, and since then has developed into a field of its own. Numerous versions of the original scheme have been introduced to increase accuracy, sensitivity, versatility, and practicality. Details of these developments can be found in Trebino's [52] book devoted entirely to this subject.

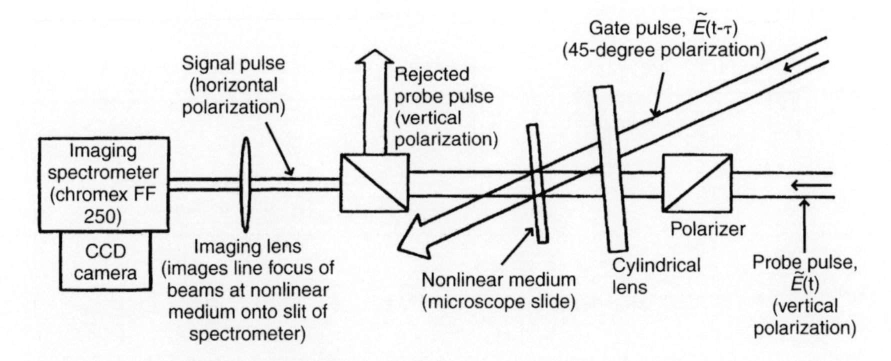

Figure 9.13 Experimental arrangement for frequency resolved optical gating. The pulse to be analyzed $\tilde{E}(t)$ is gated by its delayed replica $\tilde{E}(t-\tau)$ in a Kerr shutter assumed to respond instantaneously (Adapted from Trebino and Kane [13]).

A FROG measurement records a two-dimensional trace of the form

$$S_E(\Omega, \tau) = \left| \int_{-\infty}^{\infty} dt\ \tilde{\mathcal{E}}(t) g(t-\tau) e^{-i\Omega t} \right|^2, \tag{9.39}$$

where g is a gate function of variable delay τ, and $\tilde{\mathcal{E}}(t)$ is the complex pulse amplitude to be determined. The gate function is usually provided by a nonlinear optical process. Assuming instantaneous response, $g(t-\tau) \approx |\tilde{\mathcal{E}}'(t-\tau)|^2$ for a Kerr nonlinearity and $g(t-\tau) \approx \tilde{\mathcal{E}}'^*(t-\tau)$ for sum frequency generation. $\tilde{\mathcal{E}}'$ can be the original pulse in which case an autocorrelation is measured or a reference field for a cross correlation (XFROG). The Fourier transform in Eq. (9.39) is realized using an optical spectrometer in front of the detector.

The original FROG apparatus involved a Kerr shutter and is sketched in Figure 9.13. The signal transmitted by the Kerr gate, the sequence polarizer—Kerr medium—polarizer, is a pulse of electric field (complex) amplitude[3]:

$$\tilde{\mathcal{E}}_s(t, \tau) = \tilde{\mathcal{E}}(t) g(t-\tau), \tag{9.40}$$

A CCD camera at the output of the spectrometer records the spectrogram $S_E(\Omega, \tau)$ of $\tilde{\mathcal{E}}_s(t, \tau)$. The delay τ varies parallel to the entrance slit of the spectrometer because gate and signal pulse intersect at an angle in the Kerr medium.

The function represented by Eq. (9.39) is well-known in acoustics and used to display acoustic waves [53]. The spectrogram of a strongly chirped pulse shown on the left side of Figure 9.14 seems identical to the writing of many successive

[3]Apart from constants that do not affect the shape and that we will omit here.

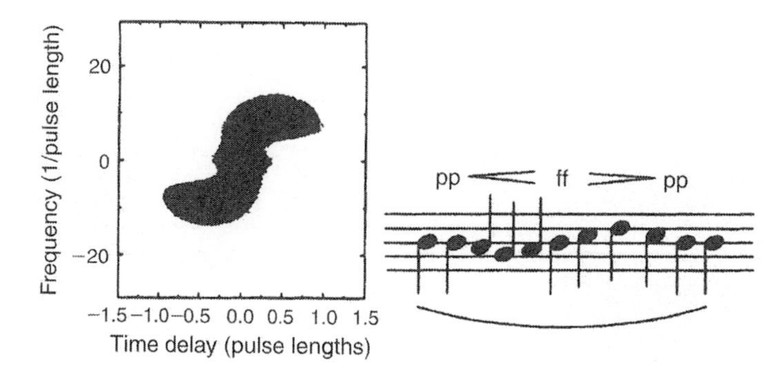

Figure 9.14 Spectrogram of a pulse with strong SPM. The frequency sweeps from −12 to +12 inverse pulse lengths during the pulse (Adapted from Trebino and Kane [13]). The acoustic analog (see also the reconstruction in Fig. 9.16).

notes (right side of Fig. 9.14), which can be considered a temporal sequence of spectral components. The difference is that only, in the case of music, the carrier frequency is in the KHz rather than PHz range, and the time delays are seconds rather than fs. The problem of reconstruction reduces essentially to extracting the function $\tilde{\mathcal{E}}_s(t,\tau)$ from the spectrogram. Indeed, we note that the integral of Eq. (9.40) over the delay τ is simply proportional to the pulse itself[4]

$$\tilde{\mathcal{E}}(t) \propto \int_{-\infty}^{\infty} \tilde{\mathcal{E}}_s(t,\tau)d\tau. \tag{9.41}$$

The FROG trace is related to the Fourier transform of $\tilde{\mathcal{E}}_s(t,\tau)$, $\tilde{\mathcal{E}}_s(t,\Omega_\tau)$, by

$$S_E(\Omega,\tau) = \left|\int_{-\infty}^{\infty} dt \int_{-\infty}^{\infty} d\Omega_\tau \tilde{\mathcal{E}}_s(t,\Omega_\tau)e^{-i\Omega t + i\Omega_\tau \tau}\right|^2. \tag{9.42}$$

Extraction of the unknown signal $\tilde{\mathcal{E}}_s(t,\Omega_\tau)$ from the spectrogram (9.42) is a two-dimensional (the two dimensions are t and τ) phase retrieval problem. This (phase retrieval) problem is known to have a unique solution for two and higher dimensions [54, 55]. The flowchart of the original FROG algorithm, which can be found in Trebino and Kane [13, 14], is shown in Figure 9.15 for illustration.

Because we can write Eq. (9.39) as

$$S_E(\Omega,\tau) = |\tilde{\mathcal{E}}_s(\Omega,\tau)|^2,$$

[4] The integration over the variable τ is equivalent to opening the gate function for all times in Eq. (9.40).

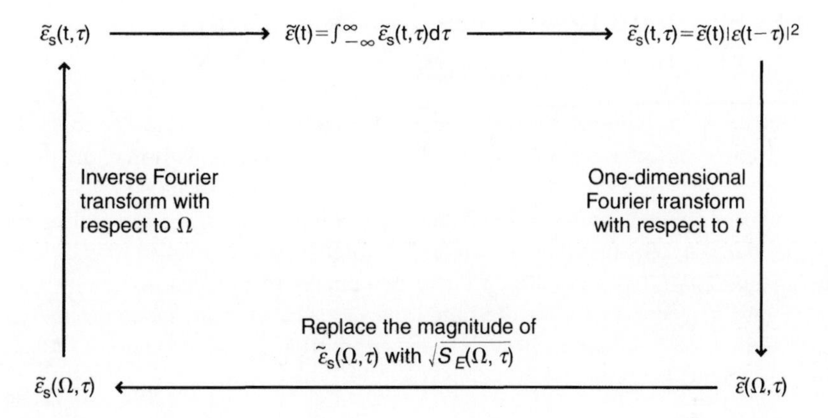

Figure 9.15 Flowchart of an iterative reconstruction algorithm. Adapted from Trebino and Kane [13].

by replacing the magnitude of $\tilde{\mathcal{E}}_s(\Omega,\tau)$ by $\sqrt{S_E(\Omega,\tau)}$ during each iteration cycle, Eq. (9.39) is always satisfied. Improved algorithms, based for example on generalized projections known from phase retrieval of images, have been developed over the years [52, 56]. Here the integration step to obtain $\tilde{\mathcal{E}}(t)$ is replaced by a search to minimize the figure of merit

$$Z = \sum_{i,j} \left| \tilde{\mathcal{E}}_s(t_i,\tau_j) - \tilde{\mathcal{E}}(t_i)|\tilde{\mathcal{E}}(t_i-\tau_j)|^2 \right| .$$

The reconstruction of the pulse defined by the spectrogram on the left side of Fig. 9.14 leads to the instantaneous frequency versus time and spectrum of Figure 9.16.

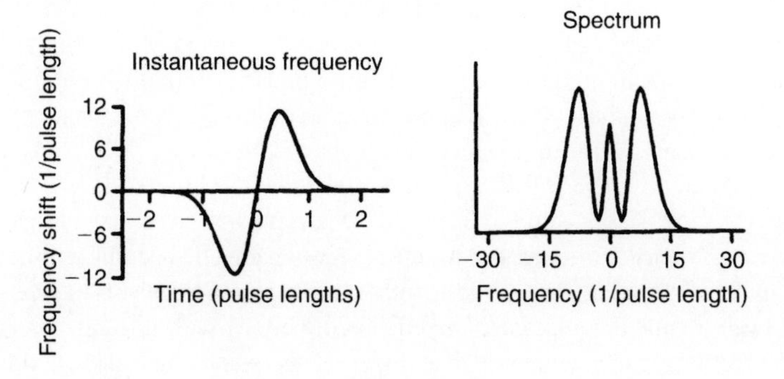

Figure 9.16 Instantaneous frequency versus time and spectrum reconstructed from the spectrogram on the left Fig. 9.14 (Adapted from Kane and Trebino [14]).

9.4.5. Spectral Phase Interferometry for Direct Electric Field Reconstruction (SPIDER)

Unlike the techniques discussed before SPIDER allows one to determine the pulse shape and phase noniteratively by a set of linear transformations of the measured data. The technique developed by Iaconis and Walmsley [47] is an example of spectral shearing interferometry. Suppose we have two pulses that are identical except that they are shifted in frequency with respect to each other by an amount ω_s (spectral shear). These two pulses are delayed in time by τ and send into a spectrometer. At the output of the spectrometer, using a quadratic integrating detector, we measure a signal proportional to the square of the sum of the spectral fields $\tilde{E}(\Omega + \omega_s)$ and $\tilde{E}(\Omega)e^{-i\Omega\tau}$.

$$S(\Omega) = |\tilde{E}(\Omega + \omega_s)|^2 + |\tilde{E}(\Omega)|^2 + 2|\tilde{E}(\Omega + \omega_s)||\tilde{E}(\Omega)| \cos(\Delta\Phi) \tag{9.43}$$

where

$$\Delta\Phi = [\Phi(\Omega + \omega_s) - \Phi(\Omega) + \Omega\tau]. \tag{9.44}$$

Because of the cosine term the spectrogram is modulated with a period of about τ^{-1}. The data set $S(\Omega)$ can be processed noniteratively using a retrieval procedure known from spectral interferometry to obtain the spectral phase difference $\Phi(\Omega + \omega_s) - \Phi(\Omega)$ and from this the spectral phase $\Phi(\Omega)$ [57,58]. Note that the spectral phase $\Phi(\Omega)$ and amplitude $|E(\Omega)|$, which can be obtained from the square root of the spectral envelope, determine the pulse amplitude and phase unambiguously.

For example, the spectral interferogram is Fourier transformed using a computer. The resulting spectrum (in time) has components centered at $t = 0$ (carrying information about the spectral envelope) and at $t = \pm\tau$. The $t = 0, -\tau$ components are removed by filtering and the result is inverse Fourier transformed. After removing the component $\Omega\tau$ the spectral phase difference is obtained, from which the spectral phase $\Phi(\Omega)$ can be calculated through concatenation.

The question is how to produce two pulse replicas that differ only in their center frequencies? A SPIDER apparatus is sketched in Figure 9.17. The pulse to be characterized is split into two replicas. One replica is stretched and chirped in a dispersive device, for example a grating sequence. The second replica is split again into two time-delayed pulses, for example in a Michelson interferometer. These two (identical) pulses are mixed (upconverted) with different parts of the stretched pulse, each centered at a different frequency because of the chirp. The result is a pair of (upconverted) pulses that are identical except for a spectral shear.

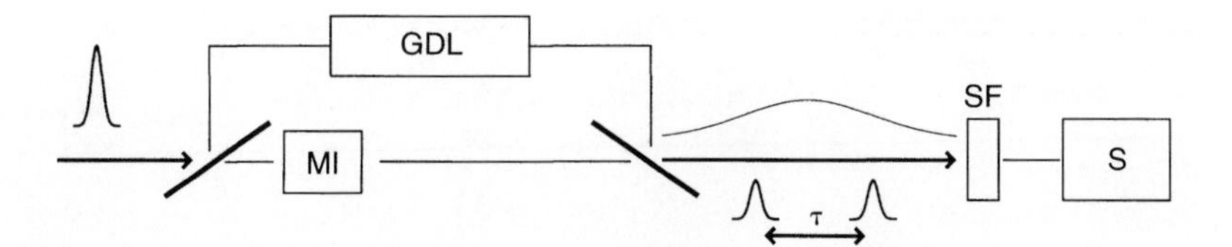

Figure 9.17 Schematic diagram of a SPIDER apparatus. One replica of the pulse to be characterized is stretched and chirped in an element with group delay dispersion (GDL), the other replica is split into identical time-delayed pulses in a Michelson interferometer (MI). The sum frequency is produced in a nonlinear crystal (SF) and recorded in a spectrometer (S). Adapted from Iaconis *et al.* [47].

The stretcher has to be designed so that the frequency of the stretched pulse does not change (much) during the time of the original pulse duration. Pulses consisting of few optical cycles have been successfully characterized using the SPIDER technique [60]. By combining it with homodyne detection sensitivity and versatility are improved [61]. True single shot implementation of SPIDER has been demonstrated at a repetition rate of 1 KHz [62].

9.5. PROBLEMS

1. Show that the statistical average of the autocorrelation of Gaussian pulses distributed in frequency [distribution given by Eq. (9.18)] is approximately identical to the autocorrelation of a simple Gaussian pulse [perform the integration of Eq. (9.19), and compare to Eq. (9.17)].
2. Consider a Gaussian pulse of 50-fs duration at 800 nm, with an upchirp corresponding to $a = 1.5$. Determine analytically the result of a measurement using the cross-correlator in which a block of BK7 glass has been inserted in one arm. Calculate the amount of glass required for pulse broadening by a factor 5, after double passage through the glass. Calculate the envelopes $A_1(\tau)$, $A_2(\tau)$, and $A_3(\tau)$ that will be obtained by measuring a second-order cross-correlation, cf. Eq. (9.7). Write the expressions corresponding to the various steps of the procedure leading to the reconstruction of the original pulse following Section 9.2.2.2.
3. Derive Eq. (9.12), the expression for the third-order interferometric correlation. Determine the peak to background ratio of the fringe resolved autocorrelation and of the intensity autocorrelation. Assume equal pulses, $E_1 = E_2$.
4. Let us assume that the spectral phase of a pulse is given by $\Phi(\Omega) = \phi_2(\Omega - \omega_\ell)^2 + \phi_3(\Omega - \omega_\ell)^3$ and the complex field is to be retrieved from the measurement of the pulse spectrum and (a) an amplitude unbalanced

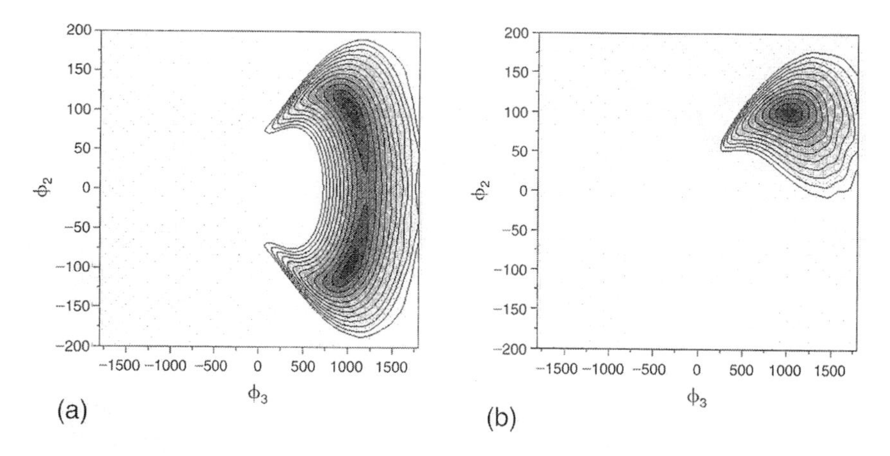

Figure 9.18 Figure of merit, cf. Eq. (9.38), of pulse retrieval based on a spectral measurement and (a) an amplitude unbalanced third-order correlation and (b) an amplitude and phase unbalanced third-order correlation.

third-order correlation, and (b) an amplitude and phase unbalanced third-order correlation using the PICASO scheme, cf. Fig. 9.11. The search space $\Delta(\phi_2, \phi_3)$ is depicted in Figure 9.18, with dark zones representing minima of the root-mean square deviation, cf. Eq. (9.38). Explain the reason for the ambiguity in case (a). How do the two possible pulses differ in the time domain? Why does adding a phase filter resolve the ambiguity?

BIBLIOGRAPHY

[1] S. L. Shapiro, Ed. *Ultrashort Light Pulses*. Springer-Verlag, Berlin, Germany, 1977.

[2] J.-C. Diels, J. J. Fontaine, I. C. McMichael, and F. Simoni. Control and measurement of ultrashort pulse shapes (in amplitude and phase) with femtosecond accuracy. *Applied Optics*, 24:1270–1282, 1985.

[3] M. Sheik-Bahae, and T. Hirayama. Real-time chirp diagnostic for ultrashort laser pulses. *Opt. Lett.*, 27(10):860–862, 2002.

[4] K. Naganuma, K. Mogi, and H. Yamada. General method for ultrashort light pulse chirp measurement. *IEEE J. of Quantum Electronics*, QE-25:1225–1233, 1989.

[5] D. E. Spence, P. N. Kean, and W. Sibbett. 60-fs pulse generation from a self-mode-locked Ti:sapphire laser. *Optics Lett.*, 16:42–44, 1991.

[6] E. W. Van Stryland. The effect of pulse to pulse variation on ultrashort pulsewidth measurements. *Opt. Comm.*, 31:93–94, 1979.

[7] G. H. C. New and J. M. Catherall. Advances in the theory of mode-locking by synchronous pumping. In *Ultrafast Phenomena V*, Springer-Verlag, Berlin, Germany, 1986, pages 24–26.

[8] J. M. Catherall and G. H. C. New. Role of spontaneous emission in the dynamics of mode-locking by synchronous pumping. *IEEE Journal of Quantum Electronics*, QE-22:1593, 1986.

[9] S. Szatmari and F. P. Schaefer. Generation of high power UV femtosecond pulses. In T. Yajima, K. Yoshihara, C. B. Harris, and S. Shionoya, Eds., *Ultrafast Phenomena VI*, Springer-Verlag, Berlin, Germany, 1988, pp. 82–86.

[10] E. S. Kintzel and C. Rempel. Near surface second harmonic generation for autocorrelation measurements in the UV. *Appl. Phys.*, B28:91–95, 1987.

[11] R. Deich, F. Noack, W. Rudolph, and V. Postovalos. Two-photon luminescence in CsI(Na) under UV femtosecond pulse excitation. *Solid state communications*, 74:269–273, 1990.

[12] H. S. Albrecht, P. Heist, J. Kleinschmidt, D. van Lap, and T. Schroeder. Single-shot measurement of femtosecond pulses using the optical kerr effect. *Meas. Sci. Technol.*, 4:1–7, 1993.

[13] R. Trebino and D. J. Kane. Using phase retrieval to measure the intensity and phase of ultrashort pulses: frequency-resolved optical gating. *Journal Opt. Soc. Am. A.*, 10:1101–1111, 1993.

[14] D. J. Kane and R. Trebino. Single-shot measurement of the intensity and phase of an arbitrary ultrashort pulse using frequency-resolved gating. *Optics Lett.*, 18:823–825, 1993.

[15] D. T. Reid, M. Padgett, C. McGowan, W. E. Sleat, and W. Sibbett. Light-emitting diodes as measurement devices for femtosecond laser pulses. *Optics Lett.*, 22:233–235, 1997.

[16] W. Rudolph, M. Sheik-Bahae, A. Bernstein, and L. F. Lester. Femtosecond autocorrelation measurements based on 2-photon photoconductivity in ZnSe. *Opt. Lett.*, 22:313–315, 1997.

[17] A. M. Streltsov, K. D. Moll, A. L. Gaeta, P. Kung, D. Walker, and M. Razeghi. Pulse autocorrelation measurements based on two- and three-photon conductivity in a GaN photodiode. *Appl. Phys. Lett.*, 75:3778–3780, 1999.

[18] P. Langlois and E. P. Ippen. Measurement of pulse asymmetry by three-photon-absorption autocorrelation in a GaAsP photodiode. *Opt. Lett.*, 24(24):1868–1870, 1999.

[19] H. P. Weber and H. G. Danielmeyer. Multimode effects in intensity correlations measurements. *Phys. Rev. A*, 2:2074–2078, 1970.

[20] G. H. C. New. Pulse evolution in mode-locked quasi-continuous lasers. *IEEE J. of Quantum Electron.*, QE-10:115–124, 1974.

[21] J. A. Giordmaine, P. M. Rentzepis, S. L. Shapiro, and K. W. Wecht. Two-photon excitation of fluorescence by picosecond light pulses. *Applied Physics Letters*, 11:216–218, 1967.

[22] S. P. LeBlanc, G. Szabo, and R. Sauerbrey. Femtosecond single-shot phase-sensitive autocorrelator for the ultraviolet. *Opt. Lett.*, 16:1508–1510, 1991.

[23] J. Jansky, G. Corradi, and R. N. Gyuzalian. On a possibility of analyzing the temporal characteristics of short light pulses. *Optics Comm.*, 23:293–298, 1977.

[24] C. Rempel and W. Rudolph. Single shot autocorrelator for femtosecond pulses. *Experimentell Technik der Physik*, 37:381–385, 1989.

[25] R. N. Gyuzalian, S. B. Sogomonian, and Z. G. Horvath. Background-free measurement of time behavior of an individual picosecond laser pulse. *Optics Comm.*, 29:239, 1979.

[26] F. Krausz, T. Juhasz, J. S. Bakos, and C. Kuti. Microprocessor based system for measurement of the characteristics of ultrashort laser pulses. *J. Phys. E: Sci. Instrum.*, 19:1027–1029, 1986.

[27] F. Salin, P. Georges, G. Roger, and A. Brun. Single-shot measurement of a 52-fs pulse. *Applied Optics*, 26:4528–4531, 1987.

[28] R. Danielius, V. Sirutkaitis, G. Valikulis, A. Stabinis, and A. Yankauskas. Characterization of phase modulated ultrashort pulses using single-shot autocorrelator. *Optics Comm.*, 105:67–71, 1994.

[29] F. Salin, P. Georges, G. Le Saux, G. Roger, and A. Brun. Autocorrelation interferometrique monocoup d'impulsions femtosecondes. *Revue de Physique Appliquée*, 22:1613–1618, 1987.

[30] G. Szabo, Z. Bor, and A. Mueller. Phase-sensitive single pulse autocorrelator for ultrashort laser pulses. *Opt. Lett.*, 13:746–748, 1988.

[31] E. B. Treacy. Measurement and interpretation of dynamic spectrograms of picosecond light pulses. *J. Appl. Phys.*, 42:3848–3858, 1971.

[32] J. L. A. Chilla and O. E. Martinez. Analysis of a method of phase measurement of ultrashort pulses in the frequency domain. *IEEE J. of Quant. Electr.*, 27:1228–1235, 1991.

[33] J. E. Rothenberg and D. Grischkowsky. Measurement of the optical phase with sub ps resolution. *Opt. Lett.*, 12:99–101, 1987.

[34] C. Yan and J.-C. Diels. Amplitude and phase recording of ultrashort pulses. *J. of the Opt. Soc. Am. B*, 8:1259–1263, 1991.

[35] A. Baltuska, A. Pugzlys, M. Pshenichnikov, and D. A. Wiersma. Rapid amplitude-phase reconstruction of femtosecond pulses from intensity autocorrelation and spectrum. In *Conference on Lasers and Electro-Optics*, Optical Society of America, Washington, D.C., 1999, pp. 264–265.

[36] J. W. Nicholson, J. Jasapara, W. Rudolph, F. G. Omenetto, and A. J. Taylor. Full-field characterization of femtosecond pulses by spectrum and cross-correlation measurements. *Opt. Lett.*, 24(23):1774–1776, 1999.

[37] J. W. Nicholson and W. Rudolph. Noise sensitivity and accuracy of femtosecond pulse retrieval using PICASO. *J. Opt. Soc. of Am. B*, 19(2):330–339, 2002.

[38] J.-P. Foing, J.-P. Likforman, M. Joffe, and A. Migus. Femtoseond pulse phase measurement by spectrally resolved up-conversion. *IEEE J. Quant. Electron.*, 28:2285–2290, 1992.

[39] J. Paye, M. Ramaswamy, J. G. Fujimoto, and E. P. Ippen. Measurement of the amplitude and phase of ultrashort light pulses from spectrally resolved autocorrelation. *Opt. Lett.*, 18:1946–1948, 1993.

[40] M. Beck, M. G. Raymer, I. A. Walmsley, and V. Wong. Chronocyclic tomography for measuring the amplitude and phase structure of optical pulses. *Opt. Lett.*, 18(23):2041–2043, 1993.

[41] B. S. Prade, J. M. Schins, E. T. J. Nimbering, M. A. Franco, and A. Mysyrowicz. A simple method for the determination of the intensity and phase of ultrashort optical pulses. *Optics Comm.*, 113:79–84, 1994.

[42] S. P. LeBlanc and R. Sauerbrey. Ultrashort pulse characterization using plasma-induced cross-phase modulation. *Opt. Commun.*, 111:297–302, 1994.

[43] K. C. Chu, J. P. Heritage, R. S. Grand, K. X. Lie, A. Dienes, E. E. White, and A. Sullivan. Direct measurement of the spectral phase of femtosecond pulses. *Opt. Lett.*, 20:904–906, 1995.

[44] S. Prein, S. Diddams, and J.-C. Diels. Complete characterization of femtosecond pulses using an all-electronic detector. *Optics Comm.*, 123:567–573, 1996.

[45] B. Haase. Determination of weak optical pulses in amplitude and phase by measurement of the transient polarization state. *Opt. Lett.*, 24(8):543–545, 1999.

[46] N. Nakajima and Y. Tomita. New approach to determine the intensity and phase of ultrashort pulses by use of time-to-space conversion and a noniterative phase-retrieval algorithm. *J. Opt. Soc. Am. A*, 16(6):1268–1276, 1999.

[47] C. Iaconis, V. Wong, and I. A. Walmsley. Direct interferometric techniques for characterizing ultrashort optical pulses. *IEEE Selected Topics in Quant. Electron.*, 4(2):285–294, 1998.

[48] V. Wong and I. A. Walmsley. Analysis of ultrashort pulse-shape measurement using linear interferometers. *Optics Lett.*, 19:287–289, 1994.

[49] I. A. Walmsley and V. Wong. Characterization of the electric field of ultrashort optical pulses. *J. Opt. Soc. B*, 13:2453–2463, 1996.

[50] C. Iaconis, V. Wong, and I. A. Walmsley. Direct interferometric techniques for characterizing ultrashort optical pulses. *IEEE Selected Topics in Quant. Electron.*, 4(2):285–294, 1998.

[51] J. W. Nicholson, M. Mero, J. Jasapara, and W. Rudolph. Unbalanced third order correlations for full characterization of femtosecond pulses. *Opt. Lett.*, 25(24):1801–1803, 2000.

[52] R. Trebino. *Frequency-Resolved Optical Gating, The Measurement of Ultrashort Laser Pulses.* Kluwer, New York, 2002.

[53] W. Koenig, H. K. Dunn, and L. Y. Lacy. The sound spectrograph. *J. Acoust. Soc. Amer.*, 18:19–49, 1946.

[54] D. Israelevitz and J. S. Lim. A new direct algorithm for image reconstruction from fourier transform magnitude. *IEEE Trans. Acoust. Speech Signal Process.*, ASSP-35:511–519, 1987.

[55] J. R. Fienup. Reconstruction of a complex-valued object from the modulus of its fourier transform using a support constraint. *J. Opt. Soc. Amer. A.*, 4:118–123, 1987.

[56] K. W. DeLong, D. N. Fittinghoff, R. Trebino, B. Kohler, and K. Wilson. Pulse retrieval in frequency-resolved optical gating based on the method of generalized projections. *Opt. Lett.*, 19(24):2152–2154, 1994.

[57] M. Takeda, H. Ina, and S. Kobayashi. Fourier-transform method of fringe-pattern analysis for computer-based topography and interferometry. *J. Opt. Soc. Am.*, 72:156, 1982.

[58] L. Lepetit, G. Cheriaux, and M. Joffre. Linear techninques of phase measurement by fs spectral interferometry for applications in spectroscopy. *J. Opt. Sco. Am. B*, 21:2467, 1995.

[59] L. Gallmann, D. H. Sutter, N. Matuschek, G. Steinmeyer, U. Keller, C. Iaconis, and I. A. Walmsley. Characterization of sub-6-fs optical pulses with spectral phase interferometry for direct electric-field reconstruction. *Opt. Lett.*, 24(18):1314–1316, 1999.

[60] B. Schenkeland, J. Biegert, U. Keller, C. Vozzi, M. Nisoli, G. Sansone, S. Stagira, S. De Silvestri, and O. Svelto. Generation of 3.8-fs pulses from adaptive compression of a cascaded hollow fiber supercontinuum. *Opt. Lett.*, 28:1987–1989, 2003.

[61] C. Dorrer, P. Londero, and I. A. Walmsley. Homodyne detection in spectral phase interferometry for direct electric-field reconstruction. *Opt. Lett.*, 26(19):1510–1512, 2001.

[62] W. Kornelis, J. Biegert, J. Tisch, M. Nisoli, G. Sansone, C. Vozzi, S. De Silvestri, and U. Keller. Single-shot kilohertz characterization of ultrashort pulses by spectral phase interferometry for direct electric-field reconstruction. *Opt. Lett.*, 28:281–283, 2003.

10

Measurement Techniques of Femtosecond Spectroscopy

10.1. INTRODUCTION

Femtosecond pulses are an ideal tool to investigate ultrafast processes of various origins. There is usually more than one parameter that varies with time in any particular experiment. One of these parameters will often be the position. As an example of the types of time dependence that have to be distinguished, let us consider the example of an apple falling from a tree (Figure 10.1). A photograph taken with a sufficiently short exposure time can freeze the motion of the falling apple as it reaches the position x. We can compare this picture with one of the apple still on the tree, which provides some information about the aging process. To establish either the law of motion $x(t)$ or the temporal behavior of aging, we need to know exactly the time elapsed from the moment the apple was shaken loose from the tree to the moment the photograph was taken. The standard experimental technique is to *trigger* the fall (for instance, ignite a small explosion) and simultaneously start a clock that synchronizes the shutter of the camera. By triggering the event, we do not have to wait days for the apple to fall down.

A pump–probe femtosecond experiment has analogies as well as fundamental differences with the falling apple measurement. The basic analogy is that a powerful light pulse—usually labeled the "pump pulse" or "excitation pulse"—interacts with the sample and excites it into a nonequilibrium state (Figure 10.2). The sample thereafter relaxes toward a new equilibrium state. This process can be mapped by sending a second (much weaker) pulse, called a probe or test

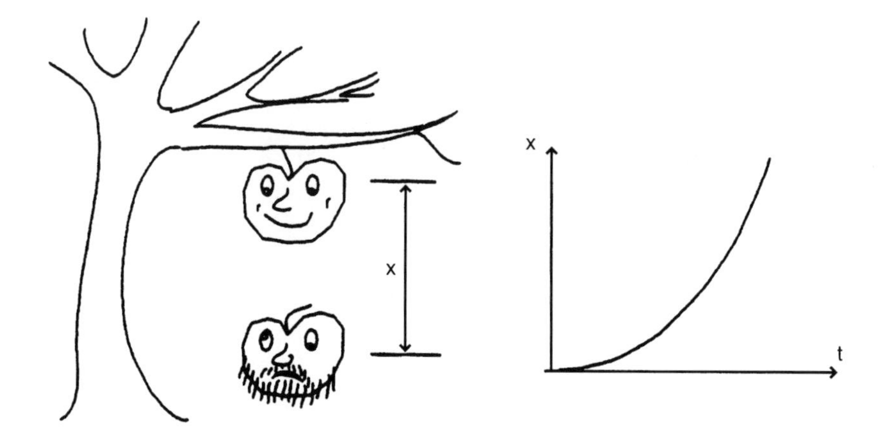

Figure 10.1 The falling apple and the aging apple.

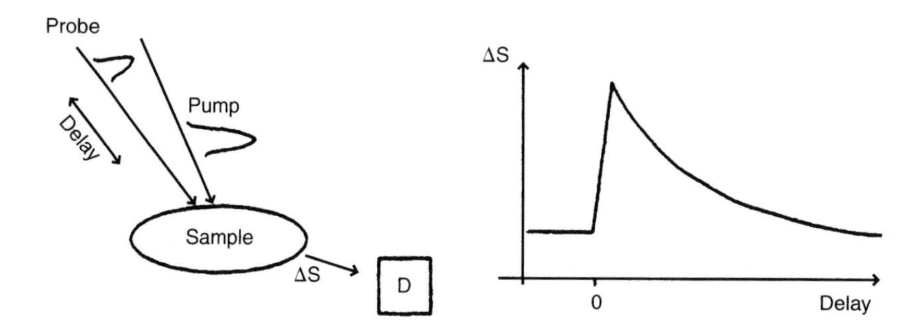

Figure 10.2 Schematic representation of a pump–probe experiment in ultrafast spectroscopy. The pump and probe pulses can be obtained from a single source and delayed with respect to each other in a Michelson or a Mach–Zehnder interferometer, for example.

(pulse), onto the sample. The probe is the analog of the snapshot photograph, aimed at detecting a change of optical properties without disturbing the object under investigation. The difference with the falling apple is that the speed of light is infinite compared to the velocity of the apple, whereas the propagation time of the probe radiation through the sample can be long compared with the event to be observed. Therefore, the geometry of interaction, the angle between probe and pump, the interaction length, and the group velocities in the sample are essential parameters in femtosecond pump–probe experiments.

In a typical pump–probe experiment, the delayed weak pulse probes the change of an optical property ΔS, such as transmission or reflection, induced by the pump. Repeating the experiments for various delays τ_d provides the function $\Delta S(\tau_d)$. Depending on the actual light–matter interaction, $\Delta S(\tau_d)$ is related to material parameters such as occupation numbers, carrier density, and molecular orientation. In this chapter we discuss a selection of fs experimental techniques, many of them originally developed for ps and ns spectroscopy.

The obvious temporal limitation of the pump–probe technique is the duration of the pump and probe. The medium preparation should be completed before the material can be probed. If the physics of the interaction is well understood, a theoretical modeling (deconvolution) can provide some interpretation of data corresponding to partial temporal overlap of pump and probe.

A compromise often must be sought between spectral and temporal resolution. Either the probe or the pump pulse has to *select* a specific spectral feature. To excite the desired transition, rather than an adjacent one, the excitation spectrum (i.e., the pulse spectral width $\Delta\omega_p$, augmented by the Rabi frequency $\kappa\mathcal{E}$ or power broadening of the transition, if needed) should not exceed the separation between lines Δ. The spectral resolution imposes therefore a limit to the temporal resolution, because the pulse duration should not be less than $\tau_p \approx 1/(\Delta + \kappa\mathcal{E})$.

10.2. DATA DECONVOLUTIONS

In most fs time-resolved experiments, a signal $S(\tau_d)$ is measured as a function of position or delay τ_d of a reference probe pulse of intensity $I_{\text{ref}}(t)$. We will consider the large class of measurements where the measured quantity is proportional to the product of a gating function I_g times the physical quantity $f(t)$ to be analyzed. The gate $I_g(t)$ is a direct function of the reference intensity $I_{\text{ref}}(t)$. Because—as pointed out in the previous chapter—the detection electronics has no fs resolution, the measured signal will be the time integral:

$$S(\tau_d) = \int_{-\infty}^{\infty} I_g(t - \tau_d) f(t) dt. \tag{10.1}$$

Deconvolution procedures should thus be applied to retrieve the physical quantity $f(t)$ from the measurement $S(\tau_d)$. A typical example is a measurement of time resolved fluorescence by upconversion. As detailed in Section 10.7, the detected upconversion radiation results from mixing the signal (fluorescence) and the reference pulse in a nonlinear crystal. Therefore, in that particular case, the gate function is the reference pulse itself $I_g(t) = I_{\text{ref}}(t)$. It is often assumed that the gating function in the correlation product [Eq. (10.1)] is much shorter than the fastest transient of the signal and thus can be approximated by a δ function. With that simplifying assumption, the signal is directly proportional to the physical parameter to be measured: $S(\tau_d) \propto f(\tau_d)$. There are, however, fast events—such as the rise of fluorescence—for which this simplifying assumption is not valid. The exact temporal dependence $f(\tau_d)$ can be extracted from the data if the gating function $I_g(t)$ is known. Indeed, if $\mathcal{I}(\Omega)$ is the Fourier transform of the gate function $I_g(t)$, and $\mathcal{S}(\Omega)$ is the Fourier transform of the measured signal $S(\tau_d)$, the Fourier transform $f(\Omega)$ of the physical quantity $f(t)$ is just the ratio:

$$f(\Omega) = \frac{\mathcal{S}(\Omega)}{\mathcal{I}(\Omega)}. \tag{10.2}$$

The physical quantity $f(t)$ can be calculated by taking the inverse Fourier transform of Eq. (10.2). This deconvolution technique can be applied in numerous cases where the gate function $I_g(t)$ does not depend on the phase of the interaction.[1]

10.3. BEAM GEOMETRY AND TEMPORAL RESOLUTION

To obtain a better quantitative understanding of the influence of the beam geometry on the temporal resolution, let us analyze a pump–probe experiment as sketched in Figure 10.3. The pump pulse creates a small change of the transmission coefficient, $\Delta a(x, y, z, t)$, which is sampled by the time-delayed test

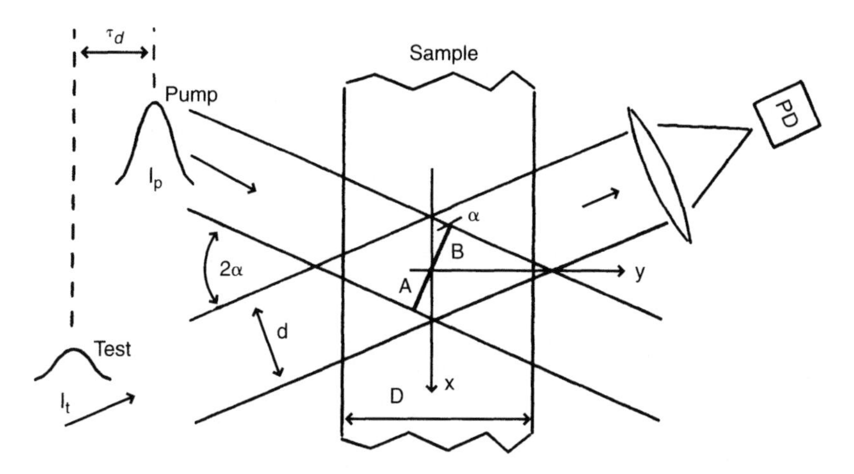

Figure 10.3 Schematic diagram of a pump–probe transmission experiment in noncollinear geometry. The line AB shows the position of the pump pulse maximum at $t = 0$. Refraction at the sample interfaces has been neglected.

---

[1]The gate depends on the phase of the interaction in the case of coherent interaction treated in Chapter 4. In that case the measured signal cannot be described by the simple expression (10.1).

pulse of intensity $I_t(t)$. The signal measured by the detector PD as function of the delay τ_d can be written as the sum of the transmitted test pulse energy $\mathcal{W}_{t0}$ in the absence of a pump and a pump-induced change $\Delta\mathcal{W}_t(\tau_d)$:

$$\mathcal{W}_t(\tau_d) = \mathcal{W}_{t0} + \Delta\mathcal{W}_t(\tau_d)$$

$$\propto \int_{-\infty}^{\infty} dt \int\int\int dx\, dy\, dz\ [1 + a_o + \Delta a(x, y, z, t)]\, I_t(x, y, z, t - \tau_d) \tag{10.3}$$

where a_o is the transmission coefficient in the absence of the pump and $|a_0| \ll 1$ has been assumed: $e^a = 1 + a_0 + \Delta a$. In the overlapping volume of the two beams, a complex mixing of spatial and temporal effects occurs. We want to derive conditions under which the excitation geometry does not affect substantially the outcome of the experiment. For simplicity, the beam profiles of pump and test pulse are assumed to be uniform and of rectangular shape, the temporal profiles are Gaussian of equal FWHM τ_p, and the overlapping region is symmetric with respect to the sample center. The time axis is chosen so that the pump pulse maximum reaches the origin of the coordinate system at $t = 0$, and the test pulse reaches the origin at $t = \tau_d$. The sample response is assumed to follow the pump pulse instantaneously, $\Delta a \propto I_p(t)$, and we expect a signal $\Delta W_t(\tau_d)$ resembling the pulse autocorrelation in the absence of geometrical effects. An increase of the correlation FWHM is then a measure of the loss in temporal resolution of any pump–probe experiment because of geometrical effects.

In the following considerations we will omit constants for the sake of brevity. The delay-dependent part of the measured signal is

$$\Delta\mathcal{W}_t(\tau_d) \propto \int_{-\infty}^{\infty} dt \int dx \int dy I_t(x, y, t - \tau_d) I_p(x, y, t) \tag{10.4}$$

where we have already carried out the z-integration yielding a constant. The pulses propagate through the sample with the group velocity ν_g. Lines of constant intensity (parallel to AB in Fig. 10.3) obey the equation

$$(x - \nu_g t \sin\alpha) = -(y - \nu_g t \cos\alpha)\frac{\cos\alpha}{\sin\alpha} \tag{10.5}$$

for the pump pulse and

$$\left[x + \nu_g(t - \tau_d)\sin\alpha\right] = \left[y - \nu_g(t - \tau_d)\cos\alpha\right]\frac{\cos\alpha}{\sin\alpha} \tag{10.6}$$

for the test pulse. Hence, the corresponding pulse intensities which are needed in the integral (10.4) can be written as

$$I_p = I_{p0} \exp\left\{-\frac{4 \ln 2}{(\nu_g \tau_p)^2}[y \cos\alpha + x \sin\alpha - \nu_g t]^2\right\} \tag{10.7}$$

$$I_t = I_{t0} \exp\left\{-\frac{4 \ln 2}{(\nu_g \tau_p)^2}\left[y \cos\alpha - x \sin\alpha - \nu_g(t - \tau_d)\right]^2\right\}. \tag{10.8}$$

Inserting these expressions into Eq. (10.4) and carrying out the time integration yields after some algebra

$$\Delta W_t(\tau_d) \propto e^{-2 \ln 2(\tau_d/\tau_p)^2} \int_{-y_m}^{y_m} dy \int_{l(y)}^{u(y)} dx \exp\left\{-\frac{8 \ln 2}{(\tau_p \nu_g)^2}\left[x^2 \sin^2\alpha - \nu_g x \tau_d \sin\alpha\right]\right\} \tag{10.9}$$

where

$$y_m = \min\left(\frac{d}{2 \sin\alpha}, \frac{D}{2}\right) \tag{10.10}$$

$$l(y) = -\frac{d}{2 \cos\alpha} + y \tan\alpha \tag{10.11}$$

$$u(y) = \frac{d}{2 \cos\alpha} - y \tan\alpha. \tag{10.12}$$

The value of y_m depends on whether or not the overlapping area of the beams is completely inside the sample. The upper and lower limit of the x-integration form the diamond-shaped boundary of the overlapping region as sketched in Fig. 10.3.

The exponential function in front of the integrals is the autocorrelation function of a Gaussian pulse and represents the result of an ideal measurement where the geometrical effects do not play a part, i.e., the spatial integration yields a constant which does not depend on τ_d. This is obviously the case for a collinear beam geometry ($\alpha = 0°$). For all other cases one can evaluate Eq. (10.9) numerically. Figure 10.4 shows the FWHM of $\Delta W_t(\tau_d)$ normalized to its value at $\alpha = 0°$ as a function of α and for different values of the parameter $K = \nu_g \tau_p / d$. The latter describes the ratio of the geometrical pulse length and the beam width. As can be seen, the shorter the pulses at a given beam width, the more critical becomes the beam geometry in a noncollinear experiment. The temporal broadening of ΔW_t

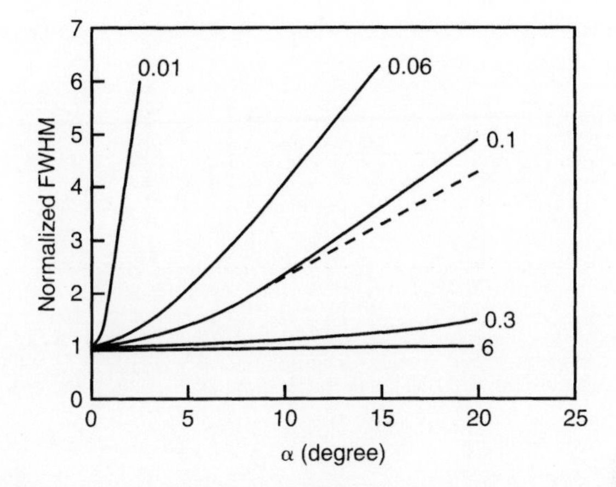

Figure 10.4 FWHM of $\Delta\mathcal{W}_t(\tau_d)$ according to Eq. (10.9), normalized to its value at $\alpha = 0°$ and shown as a function of the half angle α between pump and probe pulse. The curves are depicted for different values of the ratio of the geometrical pulse length and beam width, $K = \nu_g \tau_p / d$. The sample thickness was chosen to be $D = 3\nu_g \tau_p$. For $K = 0.1$, a second curve for $D = 10\nu_g \tau_p$ is also shown for comparison (from Krueger [1]).

can be substantial, causing a loss in time resolution of a pump–probe experiment. The effect of the crystal thickness, on the other hand, is small at moderate values of the angle α. In the following sections we will always assume an experimental geometry that justifies neglecting these geometric effects.

10.4. TRANSIENT ABSORPTION SPECTROSCOPY

Transient absorption spectroscopy is a widely used form of a pump–probe technique. As a simple example to illustrate the method, we consider an ensemble of two-level systems at resonance with a fs pulse source. With all the particles in the ground state at thermal equilibrium, the sample acts as a saturable absorber. The physical quantity to be determined is the energy relaxation time T_1 of the excited state. This parameter is to be extracted from the measurement of attenuation of the probe versus delay.

A typical experimental arrangement is sketched in Figure 10.5. In order for the probe to be much weaker than the pump, the reflectivity of the beam splitter (called BS in Fig. 10.5) should be larger than 0.5. Because only the transmission of the *probe* is measured, there is a need to devise a means to shield the pump pulse from the detector. Because pump and probe have the same wavelength,

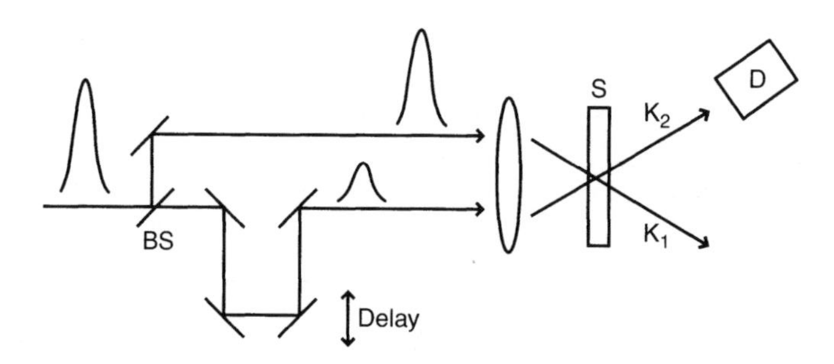

Figure 10.5 Typical geometry for measuring transient absorption. The two relatively delayed pulses are spatially separated before being sent onto the focusing optics, to provide for **k** vector separation in the sample S.

one is left with the following choices:

- separation by polarization,
- separation by wave vector, or
- temporal separation in combination with a gated detector.

In the example sketched in Fig. 10.5, it is the wave vector separation that is used.

In a typical experiment, the first (pump) pulse will saturate the absorber, and the delayed probe pulse will sample the absorption coefficient. The interpretation of the data is straightforward if the transition can be considered as homogeneously broadened. For delays longer than the phase relaxation time T_2 of the transition, the probe pulse samples the absorption coefficient α:

$$\alpha = \sigma \Delta N \tag{10.13}$$

where σ is the absorption cross section and ΔN is the population difference (density) between the upper and lower level of the transition. According to the rate equation (3.51), after excitation, the absorption coefficient relaxes exponentially with time:

$$\alpha(\tau_d) = \alpha_0 + \Delta\alpha e^{-\tau_d/T_1}, \tag{10.14}$$

where $\Delta\alpha$ is the change in absorption produced by the pump pulse, τ_d the delay of the probe relative to the pump, and T_1 is the energy relaxation time of the absorbing transition.

Extraction of T_1 from the measurement can be made under a variety of experimental conditions. In the considerations that follow, we will not attempt to select the experimental conditions for best signal-to-noise ratio, but the ones that lead to the simplest analytical expression relating the measurement to T_1, without the need for numerical modeling. We assume a uniform beam profile. In addition to being "optically thin," the sample thickness d is assumed to be negligible compared with the overlap length of pump and probe beams. Finally, the pump and probe pulse duration is assumed to be much shorter than the relaxation time to be measured ($\tau_p \ll T_1$), to avoid the need of deconvolution procedures. The completion of the pumping process is taken as time origin. The measured signal $S(\tau_d)$ is the energy of the transmitted probe versus delay. For a probe signal of energy density $W = \int I dt$ sent through a sample of thickness d:

$$S(\tau_d) = A \int_{-\infty}^{\infty} I(t - \tau_d) e^{\alpha(t)d} dt$$

$$\approx AWe^{\alpha(\tau_d)d} \approx AW[1 + \alpha(\tau_d)d], \tag{10.15}$$

where A is the beam cross section. Inserting Eq. (10.14) into Eq. (10.15) yields:

$$S(\tau_d) = AW\left[1 + \alpha_0 d + \Delta\alpha d e^{-\tau_d/T_1}\right] \tag{10.16}$$

$$\approx S_{-\infty} + AW\Delta\alpha d e^{-\tau_d/T_1}, \tag{10.17}$$

where $S_{-\infty}$ is the probe transmission in the absence of the pump pulse. The energy relaxation time T_1 can be obtained directly from a logarithmic plot of $S(\tau_d) - S_{-\infty}$ versus τ_d. The crucial approximation in Eq. (10.15) is that the temporal variations of the absorption coefficient be slow compared to the duration of the probe pulse. A numerical deconvolution of the data can be made if this latter condition is not satisfied.

The expression (10.14) is only valid for delays sufficiently large such that the excitation of the pump has dephased before the arrival of the leading edge of the probe ($\tau_d \gg T_2 + \tau_p$). For short delays that do not satisfy the latter condition, the probe coherently interacts with the polarization created by the pump. For pump and probe collinear and having the same polarization, the induced dipoles created by the pump will be in or out-of-phase with the probe field, depending on whether the delay is an even or odd number of half wavelengths. The transmitted energy versus delay will have an interference-like pattern similar to that observed in a zero area pulse experiment (see Chapter 4 and the end of this chapter). This pattern is often referred to as the "coherent spike" of a pump–probe experiment. In the case of noncollinear pump–probe experiments, a "transient grating" is created by the spatial–temporal superposition of the probe and the polarization created by the pump. In the latter geometry, the coherent spike can be

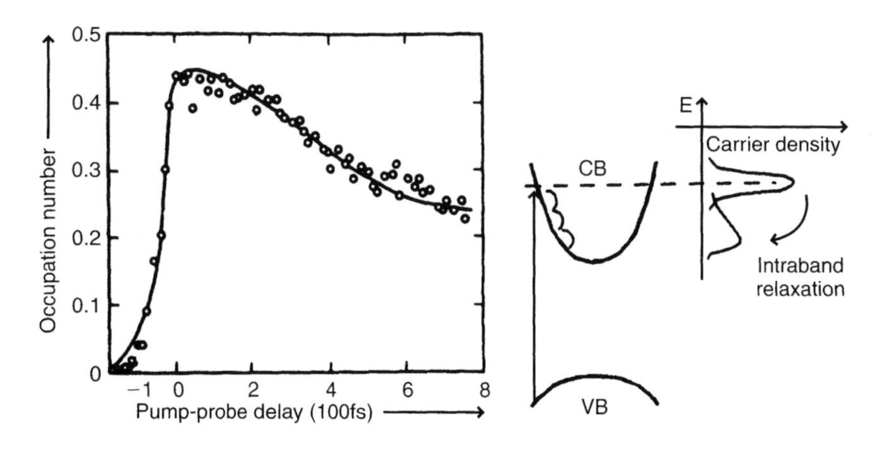

Figure 10.6 Transient transmission in CdS_xSe_{1-x}. The population numbers in the states excited by the pump can be determined from the measured changes in transmission of the 100 fs probe pulses at 618 nm. The change in occupation numbers is a measure of intraband relaxation (Adapted from Rudolph *et al.* [2]).

explained as a result of partial diffraction of the pump pulse into the direction of the probe.

As a typical example of transient absorption, Figure 10.6 shows the absorption recovery of a mixed crystal, CdS_xSe_{1-x}, after excitation with a 618 nm pulse. The pump pulse creates free carriers, i.e., electrons in the conduction band and holes in the valence band, which occupy states and subsequently increase the transmission of a test pulse at the corresponding wavelength. As the carriers relax toward the bottom of the band, the transmission decreases. The decay time is a measure of the intraband relaxation.

Much more information can be gained by using a fs white light continuum instead of a probe at the excitation frequency. In the previous example, the test pulse monitored the change in carrier density only for specific states above the band gap. A continuum fs pulse can probe simultaneously all states in a broad energy range, providing detailed information on the time-dependent carrier density distribution. An example of a pump–probe transmission experiment using a white light continuum is discussed in the next chapter.

10.5. TRANSIENT POLARIZATION ROTATION

A linearly polarized pump pulse can induce anisotropy in a sample, which can be probed subsequently with a delayed pulse. Anisotropy means here that the transmission depends on the polarization of the probing radiation. The decay in anisotropy can often be related to orientational relaxation of the dipoles excited by the pump. Such measurements have been applied successfully on a fs time scale to the determination of momentum (**k**-space) relaxation of photo-excited electrons in condensed matter (for instance, GaAs [3]).

A polarization rotation can also be induced in transparent media. The pump pulse acts through the optical Kerr effect causing birefringence. This polarization anisotropy can be seen as a polarization direction dependence of the refractive index experienced by the probe.

A standard experimental arrangement to measure the temporal change in pump induced polarization anisotropy is shown in Figure 10.7(a). Let us assume an absorbing sample. Subsequent to the excitation by the pump pulse, a probe pulse—of the same wavelength—is sent with its electric field vector oriented at an angle β with respect to that of the pump. The pump induces a polarization change (rotation) for the probe pulse, $\Delta\beta$.

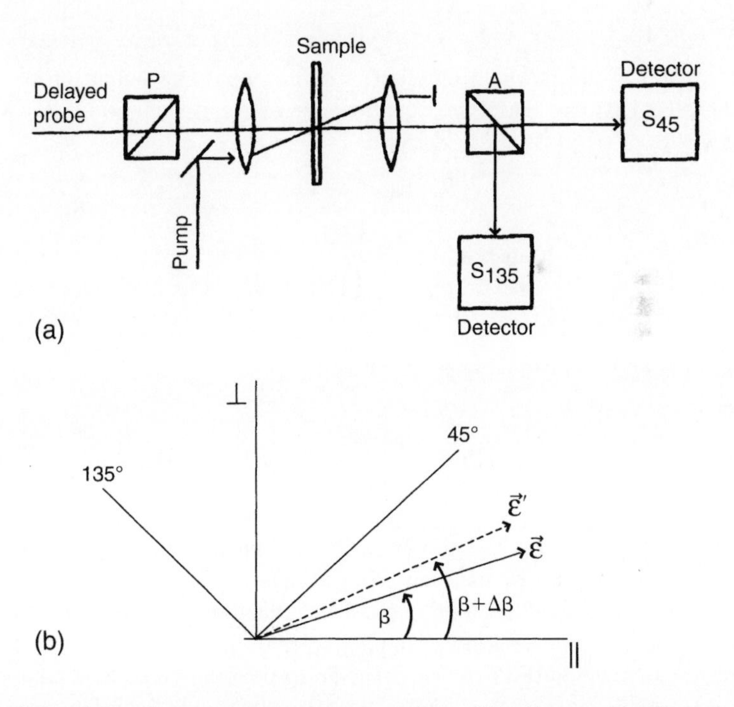

Figure 10.7 (a) Experimental setup to measure pump-induced birefringence. The Glan polarizer P sets the polarization of the probe at an angle β with respect to that of the pump. The analyzer A extracts the components of the transmitted probe at an angle of 45° and 135° with respect to the pump polarization. (b) Sketch showing the relation between the rotation angle $\Delta\beta$ of the probe $\mathcal{E}$ and the induced anisotropy in absorption. In and out denote the probe polarization before and after the sample.

We will first derive a relationship between the polarization rotation $\Delta\beta$ of the probe as a function of the anisotropy produced by the pump and then relate $\Delta\beta$ to experimental parameters that are easily accessible. We make the assumptions of optically thin samples so that the absorption $e^a \approx 1 + a$ causing small rotations $\Delta\beta$ for which $\sin\Delta\beta = \tan\Delta\beta \approx \Delta\beta$ and $\cos\Delta\beta \approx 1$. Let us assume that the pump pulse causes the sample to have an absorption coefficient $a_\parallel$ and $a_\perp$ for the components of the probe field, $\mathcal{E}_\parallel$ and $\mathcal{E}_\perp$, that are polarized parallel and perpendicular to the pump, respectively. The two field components after the absorber are:

$$\mathcal{E}'_\parallel \approx \left(1 + \frac{1}{2}a_\parallel\right)\mathcal{E}_\parallel = \left(1 + \frac{1}{2}a_\parallel\right)\mathcal{E}\cos\beta = \mathcal{E}'\cos(\beta + \Delta\beta) \tag{10.18}$$

$$\mathcal{E}'_\perp \approx \left(1 + \frac{1}{2}a_\perp\right)\mathcal{E}_\perp = \left(1 + \frac{1}{2}a_\parallel\right)\mathcal{E}\sin\beta = \mathcal{E}'\sin(\beta + \Delta\beta) \tag{10.19}$$

where $\mathcal{E}$ and $\mathcal{E}'$ are the field amplitudes of the input and output probe fields, respectively. Dividing these two equations, $\mathcal{E}'_\perp/\mathcal{E}'_\parallel$, and using only the two last terms yield:

$$\tan(\beta + \Delta\beta) = \frac{\left(1 + \frac{1}{2}a_\perp\right)}{\left(1 + \frac{1}{2}a_\parallel\right)}\tan\beta. \tag{10.20}$$

For $|a|, \Delta\beta \ll 1$ this can be approximated by

$$\Delta\beta \approx \frac{\sin(2\beta)}{4}\left(a_\perp - a_\parallel\right). \tag{10.21}$$

Clearly, the polarization rotation is caused by an induced anisotropy of the optical thickness of the sample by the pump, $\left(a_\perp - a_\parallel\right)$.

To measure the small rotation angle it is advantageous to monitor the transmitted probe components at polarization angles of 45° and 135° with respect to the pump polarization [Fig. 10.7(b)]. The corresponding probe field components after the sample are:

$$\mathcal{E}'_{45} = \mathcal{E}'\cos\left(\frac{\pi}{4} - \beta - \Delta\beta\right) \approx \frac{1}{2}\sqrt{2}\mathcal{E}'\left[\cos\beta + \sin\beta + \Delta\beta(\cos\beta - \sin\beta)\right] \tag{10.22}$$

and

$$\mathcal{E}'_{135} = \mathcal{E}' \cos\left(\frac{\pi}{4} + \beta + \Delta\beta\right) \approx \frac{1}{2}\sqrt{2}\mathcal{E}' \left[\cos\beta - \sin\beta - \Delta\beta(\cos\beta + \sin\beta)\right]. \tag{10.23}$$

What is recorded in such a pump probe experiment is the ratio of the intensities (pulse energies):

$$R' = \frac{(\mathcal{E}'_{45})^2}{(\mathcal{E}'_{135})^2}. \tag{10.24}$$

Eqs. (10.21) and (10.23) can be inserted into the equation for R'. After some basic algebra and making use of $\Delta\beta \ll 1$ again we find

$$R' \approx R\left(1 + \frac{4\Delta\beta}{\cos(2\beta)}\right), \tag{10.25}$$

where $R = [1 + \sin(2\beta)]/[1 - \sin(2\beta)]$ is the ratio of the two probe components in the absence of the pump. From relation (10.25) the rotation angle $\Delta\beta$ can be determined from the measurement of R and R' and the known angle β.

The scattered pump intensity adds a noise component to the signals S_{135} and S_{45}. The angle β has to be sufficiently large, such that the scattered noise from the pump be negligible compared with the measured probe component $\mathcal{E}^2 \sin^2\beta$. An angle of $\beta = 15°$ is chosen in most applications [3,4]. The measured anisotropy (rotation in probe beam polarization of the order of one degree) decays with probe delay and is a measure of relaxation processes following the excitation. An example of a polarization rotation measurement is shown in Figure 10.8.

10.6. TRANSIENT GRATING TECHNIQUES

10.6.1. General Technique

There are numerous variations of transient grating techniques, providing a wide array of information on sample properties. A general review of these techniques is given by Eichler *et al.* [6]. The basic experimental setup is sketched in Figure 10.9. Two pump pulses of different propagation direction overlap in the sample. If their relative delay (τ_1) is less than the phase relaxation time of the interaction, they produce an interference pattern of the sample excitation. The modulation of the sample excitation can manifest itself in a periodically changing

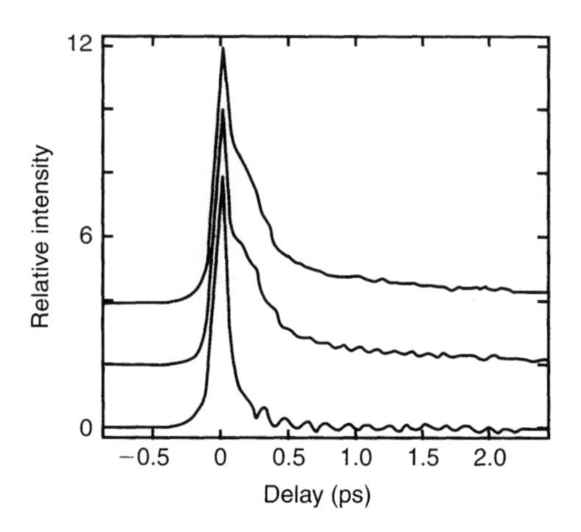

Figure 10.8 Optical Kerr signals for the liquids CH_2Cl_2, $CHCl_3$ and CCl_4 (top to bottom). Different time constants can be identified for each sample. They represent a complex interplay of intramolecular processes as well as interaction with the local environment (Adapted from [5]).

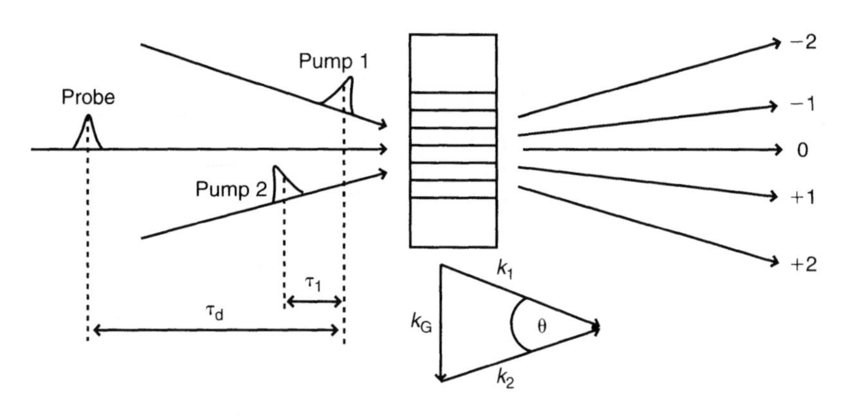

Figure 10.9 Schematic representation of a transient grating experiment.

transmission (amplitude grating) and/or a refractive index (phase grating). The grating vector is:

$$\mathbf{k}_G = \mathbf{k}_1 - \mathbf{k}_2, \tag{10.26}$$

where $\mathbf{k}_{1,2}$ are the propagation vectors of the pump pulses in the sample. The existence and dynamics of the grating can be probed by the diffraction that a delayed (weak) probe pulse experiences. A series of detectors can probe simultaneously the behavior in several diffraction orders. If the excitation is weak, the absorption and refractive index modulation are small [$|\Delta\alpha/\alpha_0| \ll 1$, $|\Delta n/n| \ll 1$], and the relative diffracted probe intensity (first order) is given by [6]:

$$\frac{\Delta I_{\text{diff}}}{I_p} \propto (\Delta n)^2 + \left(\frac{\lambda_p}{2\pi}\right)^2 (\Delta\alpha)^2. \tag{10.27}$$

One has generally to distinguish between two mechanisms for the decay of diffraction efficiency with delay:

1. The pump-induced changes in the sample relax locally. For example, if $\Delta\alpha$ and Δn are the result of free carrier generation in a semiconductor, carrier relaxation toward the original equilibrium state will lower the modulation depth. The diffraction as a function of delay provides information on the carrier relaxation time.
2. The sample excitation diffuses spatially (nonlocal mechanism). In the example of free carrier generation in a semiconductor, the pump modulates the carrier density, and thus triggers diffusion of carriers into the low excitation regions (minima of the induced grating). The result is a gradual wash-out of the modulation and decline of the diffraction signal. In many cases, the diffusion process can be described by a diffusion equation with a characteristic diffusion constant. From the characteristic decay time of the diffraction efficiency and the grating period one can determine the diffusion constant.

Both processes (1) and (2) have to be taken into account in the data evaluation. To distinguish between the local and nonlocal relaxation mechanisms (in particular when they occur on a comparable time scale), a series of measurements can be performed at various angles θ between the two pump beams producing the grating. Because the grating constant is modified by changing the angle θ [Eq. (10.26)], the decay component resulting from diffusion will also be modified. The local relaxation component to the decay should not depend on the angle θ. Another possibility to distinguish between local and nonlocal contributions to the decay is to compare transmitted (zero diffraction order) and diffracted signals.

Grating techniques also provide the possibility of measuring coherent effects by varying the delay τ_1 between the two pump pulses, at constant probe delay τ_d. The first arriving pump pulse generates a polarization oscillation in the sample which decays with the characteristic transverse relaxation time T_2. The second pump interferes with this polarization, which results in a modulation of the excitation (e.g., occupation numbers). The modulation depth and thus the diffraction efficiency experienced by the probe pulse and measured as function of τ_1 contain information on T_2.

The actual data evaluation in a transient grating experiment can be complex and requires a detailed model of the processes involved. An example of determination of phase relaxation times using collinear counter-propagating pump pulses is detailed in the next subsection.

10.6.2. Degenerate Four Wave Mixing (DFWM)

In this particular variation of transient grating experiment, the two pump pulses are two strong counter-propagating waves $\tilde{\mathcal{E}}_{p1}(t)\exp[i(\omega_\ell t - k_p z)]$ and $\tilde{\mathcal{E}}_{p2}(t)\exp[i(\omega_\ell t + k_p z)]$. The probe wave is sent along an intersecting direction x and has as electric field $\tilde{\mathcal{E}}_3(t-\tau_d)\exp[i(\omega_\ell t - k_x x)]$. The nonlinear interaction results in the generation of a signal $\tilde{\mathcal{E}}_4(t)\exp[i(\omega_\ell t + k_x x)]$, which, for momentum conservation, is counter-propagating to the probe direction (Figure 10.10). In the case of continuous waves, and, for instance, a quadratic nonlinearity, it can be shown that the wavefront of the generated signal wave $\tilde{\mathcal{E}}_4$ is the reverse of the wavefront of the probe $\tilde{\mathcal{E}}_3$ [7]. This property of spatial phase conjugation does not transpose directly in the time domain. Temporal phase conjugation is chirp reversal, which can be shown to occur only when the following conditions are met [8]:

- instantaneous nonlinearity,
- medium thickness $\ll$ than the pulse length, and
- weak interaction ($|\tilde{\mathcal{E}}_4| \ll |\tilde{\mathcal{E}}_3|$).

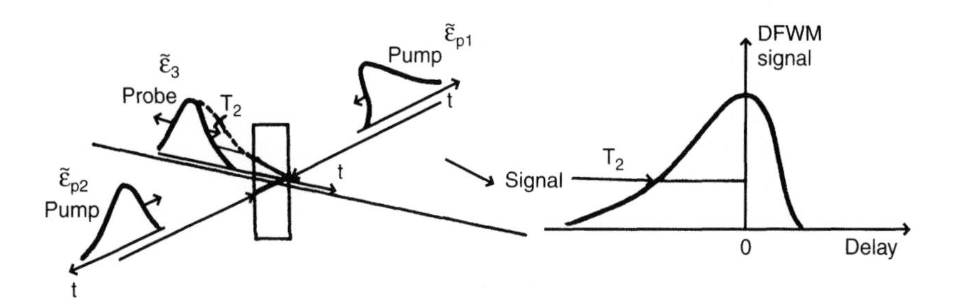

Figure 10.10 Coherent single-photon resonant DFWM. The probe pulse is trailed by a polarization wave, that forms a population grating with one of the pump pulses that follows. The other pump pulse scatters off that grating into the direction from which the probe originates. The *rise* of the signal energy versus delay is thus a measure of the phase relaxation time of the single-photon resonance.

It can easily be seen that if all but the second condition are met, each depth of the medium will generate a DFWM signal, resulting in a square pulse $\tilde{\mathcal{E}}_4$ with a length equal to twice the sample thickness [8].

We have so far assumed that all three waves meet simultaneously in the nonlinear medium. Interesting information on the dynamics of the interaction can be gathered from the study of the DFWM signal when all three waves are applied in a particular time sequence.

We assume in the following discussion that the nonlinear medium is shorter than the optical pulses and is either at single- or at two-photon resonance with the radiation. Let us first consider the case of a single-photon resonant absorber being excited first by a weak probe, followed by two simultaneous strong counter-propagating pump pulses (Fig. 10.10). As we saw in Chapter 4, the short pulse creates a pseudo-polarization $\tilde{Q}_3 = w_0 \sin\theta_0 \exp[-ik_x x]$ that decays with a characteristic time T_2. If a strong pump pulse enters the interaction region within that characteristic time, it will form a population grating [as seen from the Bloch equation (4.7)] corresponding to the interferences between waves of vector $\mathbf{k}_x$ and $\mathbf{k}_z$. If the second pump pulse impinges on this grating, it will be diffracted along the opposite direction as the signal (wave vector $-\mathbf{k}_x$) according to the Bloch equation (4.6). The longer T_2 is, the more the probe can be launched in advance of the two pump pulses, and still produce a signal. As illustrated in Fig. 10.10, the rise time of the signal versus delay is a measure of the phase relaxation time of a single photon transition.

The same experiment performed on a two-photon resonant transition, as sketched in Figure 10.11(a), leads to different results and interpretation. Because the interaction is a two-photon process, the weak probe alone cannot have any significant effect on the system, and there will be no signal if the probe is ahead of the pump pulses. We saw in Chapter 4 that for a two-photon transition, Bloch's equations (4.6), (4.7) apply, except that the driving term is proportional to the square of the field. The two counter-propagating pump pulses can produce a two-photon excitation oscillating at $2\omega_\ell$ [see Eq. (4.95)], which will decay with the phase relaxation time $T_{2(2ph)}$ of the two-photon transition. One component of this two-photon excitation, ϱ_{12}, with no spatial modulation (zero spatial frequency), will interact with a probe to generate a counter-propagating signal by two-photon stimulated emission. A probe pulse sent through the interaction region with a subsequent delay τ_d will induce a signal by two-photon stimulated emission, $\tilde{\mathcal{E}}_4 \propto \varrho_{12}(\tau_d)\tilde{\mathcal{E}}_3^*$. Because the probe field corresponds to a phase factor $\omega_\ell + k_x x$ and the two-photon excitation to a phase factor $2\omega_\ell$, the signal E_4 has a phase factor $2\omega_\ell - \omega_\ell - k_x x = \omega_\ell - k_x x$, which describes a wave propagating in the direction opposite to the probe. Because the two-photon excitation ϱ_{12} is the amplitude of an off-diagonal matrix element decaying with a two-photon phase relaxation time $T_{2(2ph)}$, the two-photon stimulated emission being proportional to ϱ_{12} will only exist within $T_{2(2ph)}$ of the pump excitation. In the case

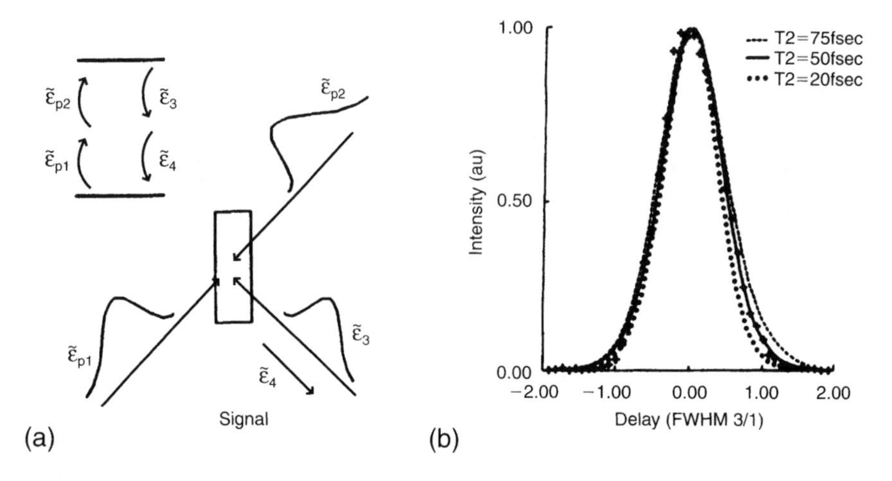

Figure 10.11 Coherent two-photon resonant DFWM. (a) The counter-propagating pump pulses create a two-photon excitation of frequency $2\omega_\ell$, which decays with the two-photon phase relaxation time $T_{2(2ph)}$. The signal is two-photon stimulated emission induced by subsequent passage of the probe pulse. It is thus the *fall* of the signal energy versus delay that is a measure of the phase relaxation time of the two-photon resonance. (b) Intensity of the DFWM signal versus probe delay. The sample is the saturable absorber jet of a mode-locked dye laser. The crosses indicate the experimental data points. Theoretical curves for the two-photon resonant interaction are plotted for three values of the phase relaxation time: 20 fs (dotted), 50 fs (solid), and 75 fs (dashed).

of two-photon resonance, it is thus the trailing edge of the signal energy versus delay that is a measure of the phase relaxation time $T_{2(2ph)}$.

An example of determination of phase relaxation times through DFWM is given in Fig. 10.11(b). In this particular case, the sample is the saturable absorber jet (dye DODCI) of a mode-locked dye laser [9]. The two pump pulses are the counter-propagating pulses circulating inside the dye laser cavity. The probe is taken from one of the outputs of the dye laser, and focused with a 25-mm focal distance lens into the interaction region. Figure 10.11(b) shows the average intensity of the signal retro-reflected into the probe direction, as a function of the delay of the probe. The leading edge of the signal matches exactly the instantaneous response, given the pulse shape $\tilde{\mathcal{E}}(t) = \exp[-0.15ix^2]/\{\exp[-1.33x] + \exp[0.8x]\}$ (where $x = t/\tau_s$, and the pulse FWHM is $1.72\tau_s = 76$ fs). The instantaneous response, for the single-photon transition model, is calculated by taking the steady-state solution of Bloch's equations (4.10) and (4.11) for the field consisting of the sum of the probe and pump fields. The trailing edge of the DFWM signal versus delay shows clearly the effect of a two-photon coherence. Following the procedure outlined above, the DFWM signal can be calculated as a function of delay for the two-photon excitation [9]. The result of the calculation (for the particular pulse shape mentioned above) is plotted in Fig. 10.11(b) for three values of the two-photon phase relaxation time $T_{2(2ph)} = 20$ fs (dotted line), $T_{2(2ph)} = 50$ fs (solid line), and $T_{2(2ph)} = 75$ fs (dashed line).

The experiment thus indicates a two-photon resonant DFWM and a phase relaxation time of 50 fs (decay of the DFWM signal versus delay) for the two-photon transition. There is no resolvable effect of a single-photon resonant DFWM (rise time of DFWM signal versus delay). The dominance of the two-photon enhancement of DFWM in DODCI at 620 nm is confirmed by theory. Simple numerical estimates indicate that indeed, the contribution of the two-photon resonance dominates the DFWM signal [9].

10.7. FEMTOSECOND RESOLVED FLUORESCENCE

If an excitation is followed by fluorescence (luminescence), the time resolved measurement of the transients of this radiation provides useful information on the evolution of occupation numbers and relaxation channels. Streak cameras are often used to measure fluorescence decay. The temporal resolution of this instrument is limited to approximately one-half of a picosecond. As noted previously, all-optical techniques are needed to obtain even better time resolution. The general method of correlation introduced in Chapter 9 applies also to fluorescence measurements.

A pump pulse provides the time-dependent excitation to be analyzed. The radiation to be measured as a function of time is correlated with a delayed replica of the pump (reference pulse). This cross correlation is achieved by upconverting (sum frequency generation) the fluorescence with the fs reference pulse. This technique, pioneered by Mahr and Hirsch [10] with ps pulses, was first applied to the fs range to measure the rise time of fluorescence in organic dyes [11].

The basic experimental setup as sketched in Figure 10.12 includes a polarizing beam splitter, two quarter wave plates, and a nonlinear crystal for type II sum frequency generation. Type II sum frequency generation is essential to provide an optimum signal-to-background ratio. In the first experiment, an unamplified dye laser at 620 nm was used. After the calcite polarizing beam splitter, one of the polarization components of the main pulse is focused into the sample, e.g., a concentrated solution of oxazine dye in ethylene glycol. The backscattered fluorescence radiation (at ω_f) is collected by the focusing lens and recollimated toward the calcite prism and the nonlinear detection. In this reflective geometry, the temporal resolution is limited by the optical depth of the sample or the confocal parameter of the focused beam, whichever is shorter. With the concentrated solution of oxazine dyes used, the optical depth of the sample was approximately 2 μm, limiting the temporal resolution to 6 fs. A half wave plate can be introduced before the calcite polarizer to control the fraction of radiation sent to the sample.

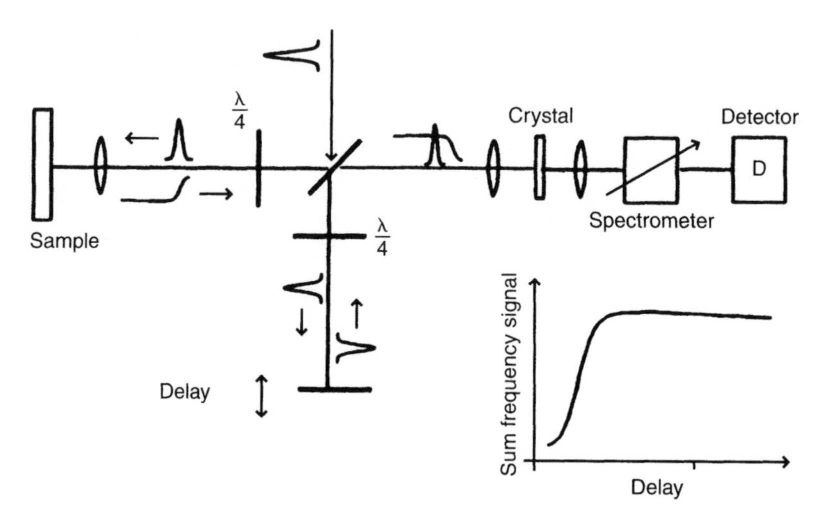

Figure 10.12 Setup for femtosecond resolved detection of fluorescence.

The nonlinear crystal generates the sum frequency (of intensity I_{sum}) of the radiation components polarized along two orthogonal axes. If $I_{ref}(t - \tau_d)$ is the reference signal delayed by an amount τ_d and polarized along $\hat{x}$, and $I_s(t)$ is the fluorescence signal polarized along the orthogonal direction $\hat{y}$:

$$I_{sum}(\tau_d) \propto \int_{-\infty}^{\infty} I_{ref}(t - \tau_d) I_s(t) dt. \tag{10.28}$$

To reach the ultimate resolution, the bandwidth of the conversion process should be larger than the bandwidth of the reference pulse. Both the imperfection of the crystal and the nonperfect rejection factor of the polarizing beam splitter contribute to a fraction ϵ_y of the reference beam polarized along the axis y, and hence a (τ_d independent) background signal:

$$I_b \propto \int_{-\infty}^{\infty} \epsilon_y I_{ref}^2(t - \tau_d) dt. \tag{10.29}$$

The optical quality of the nonlinear crystal is essential in this experiment, because it helps discriminate between the signal and a background caused by SHG of the more intense reference beam. Additional background rejection can be obtained by spectrally separating the gated fluorescence (at $\omega_\ell + \omega_f$) from the second harmonic of the reference signal (at $2\omega_\ell$). The bandwidth of this filter should correspond to the pulse spectral bandwidth $\Delta\omega_p$ to ensure a temporal resolution given by the pulse duration. The focusing lens can be replaced by a parabolic mirror which collects fluorescence from a larger solid angle [12].

The number of upconverted photons per excitation pulse can formally be written as

$$N_{up} \approx V_1 V_2 Q \left(\frac{\nu_F \mathcal{W}_0}{\hbar\omega_F} \right) \frac{\Delta\omega_p}{\Delta\omega_F} \frac{\tau_p}{T_F}, \tag{10.30}$$

where V_1 is the linear loss of the experimental setup, V_2 is the fractional solid angle (i.e., solid angle divided by 4π) from which the focusing optics gather the fluorescence, the term in parentheses describes the total number of fluorescence photons excited, $\Delta\omega_p/\Delta\omega_F$ is the fraction of the fluorescence spectrum which is upconverted and reaches the detector, τ_p/T_F with T_F as fluorescence lifetime is the fraction of fluorescence within the time window set by the pulse, and Q is the conversion efficiency of the sum frequency generation. For an upconversion experiment using a passively mode-locked dye laser to resolve the fluorescence dynamics of an organic dye and urea as nonlinear crystal, the following parameters are typical: $V_1 = 10^{-1}$, $V_2 = 2 \times 10^{-3}$, $\mathcal{W}_0 = 150$ pJ, $\tau_p = 100$ fs, $T_F = 1$ ns, $Q = 5 \times 10^{-4}$, $\nu_F = 1$. This yields $N_1 = 1.5 \times 10^{-4}$ upconverted photons per pump pulse photon. This weak signal is detectable because the repetition rate of the source is $\approx 10^8$ Hz, resulting in a photon flux of 1.5×10^4 s^{-1}. Figure 10.13 shows as an example the onset of fluorescence for the dye oxazine 720.

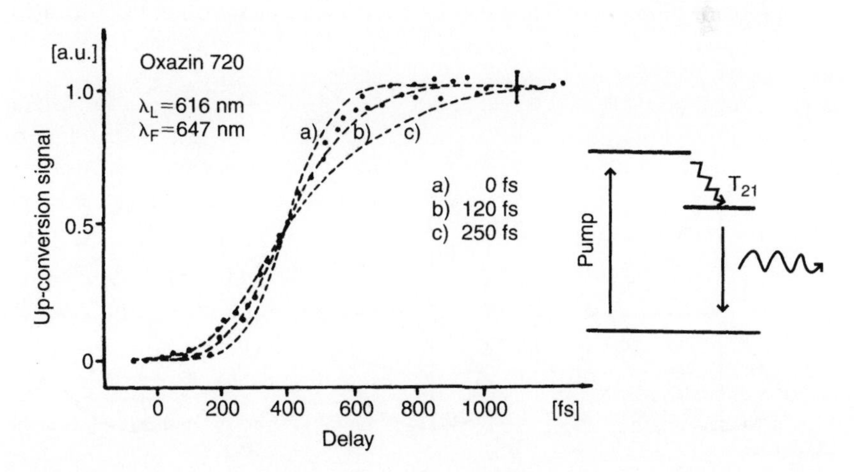

Figure 10.13 Onset of fluorescence of an organic dye (Oxazin 720) measured by up-conversion. The number of photon counts is plotted versus delay. Theoretical curves corresponding to different relaxation times T_{21} of a simple three-level model system are shown.

10.8. PHOTON ECHOES

Photon echo is the standard method—directly derived from spin echoes—to determine the phase relaxation time T_2 of a transition [13,14]. In the basic photon echo experiment, a sequence of two pulses is sent through the sample. Ideally, the first pulse will be a $\pi/2$ pulse, and the second a π pulse.

In an inhomogeneously broadened medium, the $\pi/2$ pulse excites the electric dipoles to oscillate with their characteristic frequency ω_0. Immediately after excitation all dipoles are in phase and the macroscopic polarization is maximum. As time progresses, because of their different eigenfrequencies, the dipoles dephase relative to each other. The macroscopic polarization is damped with a time constant given by the inverse of the width of the inhomogeneous line profile $g_{inh}(\omega_0 - \omega_{ih})$. The individual dipole groups still oscillate, with an amplitude damped with the phase relaxation time T_2 corresponding to the homogeneous line profile. The π pulse at delay τ_d adds a phase of π to each individual oscillator, which causes them to add again in phase after a time τ_d ($2\tau_d$ after the $\pi/2$ pulse). The associated macroscopic polarization results in a collective radiation effect called an echo. The explanation of the echo in the Bloch vector model is as follows.

After the first pulse ($\pi/2$ pulse), the pseudo-polarization vector is aligned along the ν axis, as shown in Figure 10.14(a). Each component of the line $g_{inh}(\omega_0 - \omega_{ih})$ will precess around the w axis—thus in the u-ν plane—at an angular velocity $(\omega_0 - \omega_\ell)$, for a time equal to the delay τ_d between the $\pi/2$ and π pulses. The effect of the π pulse, however, is to create the mirror image of the component of the pseudo-polarization vector, with respect to the u axis, as shown in Fig. 10.14(b). Each component of the line $g_{inh}(\omega_0 - \omega_{ih})$ is subsequently precessing at the

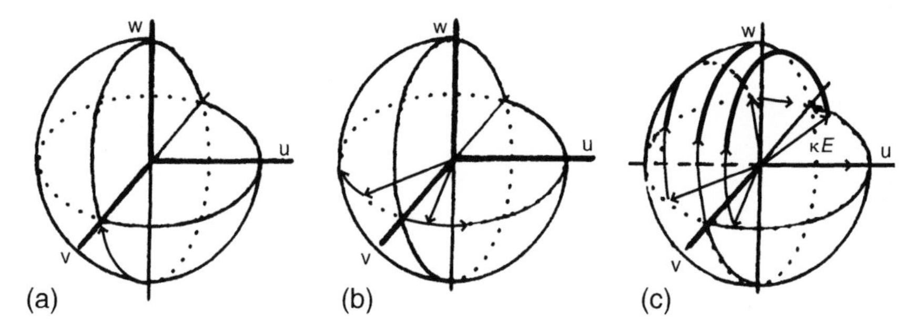

Figure 10.14 Photon echoes: A $\pi/2$ pulse (a) creates a macroscopic polarization (pseudo-polarization vector aligned along the ν axis). Following the excitation, the components of the pseudo-polarization vector precess (b). After a π pulse is applied, the spreading process is reversed (c).

velocity $(\omega_\ell - \omega_0)$, hence reversing the course of the previous "spreading." After a time τ_d following the π pulse, the components of the pseudo-polarization vector will be lined up again, resulting in a macroscopic echo signal [Fig. 10.14(c)]. The only decay mechanism for the echo is the nonreversible homogeneous decay. The pseudo-polarization vector of initial amplitude $\mathcal{P}_0$, after the delay $2\tau_d$, has been reduced exponentially to $\mathcal{P}_0 \exp(-2\tau_d/T_2)$; hence the echo intensity decays as $\exp(-4\tau_d/T_2)$. It should be noted that the $\pi/2$ and π pulse areas need not be reached to observe an echo. The amplitude, however, is maximum for this particular choice.

The dephasing in condensed matter at room temperature is extremely fast. The challenge in photon echo measurement is to resolve the fast decaying echo from scattering from the tail of the preceding π pulse. The various possibilities to separate the signal are:

1. temporal gating of the echo,
2. **k** vector separation,
3. separation by polarization, or
4. separation by focalization.

The time resolution necessary for the first technique could be achieved by upconverting the echo using type II SHG with a delayed excitation pulse (of polarization orthogonal to that of the echo). The other three techniques are commonly used. If $\mathbf{k_1}$ is the wave vector of the first $(\pi/2)$ pulse, and $\mathbf{k_2}$ the wave vector of the second one, it can be shown that the angle of emission of the third pulse (the echo) is twice the angle between the direction of the two first pulses—hence $\mathbf{k_e} = 2\mathbf{k_2} - \mathbf{k_1}$ (cf. Figure 10.15). This property can be easily understood from momentum conservation considerations. Indeed, a photon echo is a particular case of degenerate four wave mixing experiment, in which the first two waves form a grating. The latter waves do not need to be simultaneous; their interval only needs to be shorter than the phase relaxation time, to form a grating. The grating vector is $\mathbf{k_2} - \mathbf{k_1}$. The second pulse with wave vector $\mathbf{k_2}$ scatters off that grating, to generate a first-order diffracted wave in the direction $\mathbf{k_e} = \mathbf{k_2} - (\mathbf{k_1} - \mathbf{k_2})$. The latter property is directly related to the focusing properties of the echo. If the radius of curvature of the first $\pi/2$ pulse is R_1, and that of the π pulse R_2, the echo has the wavefront curvature given by [15]:

$$\frac{1}{R_e} = \frac{2}{R_2} - \frac{1}{R_1}. \tag{10.31}$$

Polarization can also be used to distinguish the echo from the intense excitation pulses. The polarization dependence of the echo has been investigated by

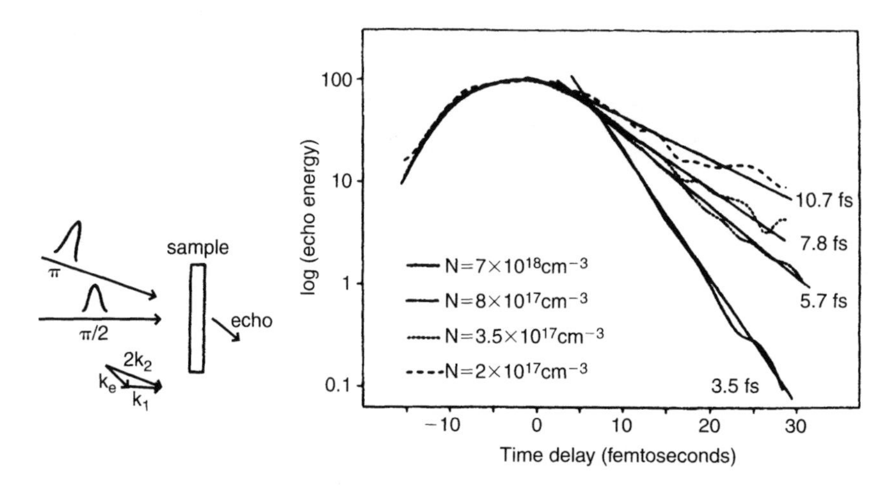

Figure 10.15 Photon echo applied to GaAs. Left: wave vector diagram. Right: echo amplitude versus delay for different carrier densities (adapted from Becker *et al.* [17]).

Alekseef and Evseev [16] and shown to depend on the total angular momentum number J of each of the two levels involved in the transition. For a $J = 1/2 \rightarrow J = 1/2$ transition, the polarization of the echo makes an angle 2ψ with that of the first (linearly polarized) pulse (ψ being the angle between the polarization of the first and second pulse). In the latter case also, a linearly polarized pulse following a circularly polarized pulse, produces a photon echo with circular polarization. For transitions $J = 0 \leftrightarrow J = 1$ and $J = 1 \rightarrow J = 1$, the echo has the polarization of the second pulse, with an amplitude proportional to $\cos\psi$.

None of the echo separation techniques totally eliminates the background provided by the first pulses. The duration of the exciting pulses should be shorter than the phase relaxation T_2 to be measured.

The fs photon echo technique has been applied to the study of dephasing of band-to-band transitions in the direct gap semiconductor GaAs [18]. Dephasing in this system is because of momentum relaxation of the carriers, as verified by an independent method that specifically probes momentum relaxation (see previous sections). The data (Fig. 10.15) indicate a carrier concentration dependent phase relaxation time ranging from 14 to 44 fs, [17, 18] fitting the power law $T_2 = 6.2 \times 10^6 \times N^{-0.3}$ (T_2 in fs, concentration of excited carriers N in cm^{-3}). This power law is characteristic of a three-dimensional screening. A similar experiment performed on two-dimensional MQWs gave a density law $T_2 = 6.8 \times 10^7 \times N^{-0.55}$, reflecting a two-dimensional screening of carriers [19].

Recent advances in fs photon echo spectroscopy of molecules and solids are summarized in papers by Ashbury *et al.* [20] and Dao *et al.* [21].

In summary, the photon echo method is quite powerful and useful for the determination of relaxation times longer than the pulse duration. It has been one of the most commonly used.

10.9. ZERO AREA PULSE PROPAGATION

The photon echo experiment is based on a sequence of two nonoverlapping pulses whose relative phase is unimportant. An essential feature of coherent excitation is that the excitation depends on the phase of the applied signal. We saw in Chapter 4 that a sequence of two pulses 180° out-of-phase applied at resonance to a two-level system, will return that system to the ground state. There will be no energy loss for this particular pulse sequence, while there will be maximum absorption if the pulses are in phase. The contrast in absorption for the in phase pulse sequence—as opposed to the sequence of pulses out-of-phase—can be used as a measure of coherent interaction, and to determine T_2. The experimental setup consists essentially of a Michelson or a Mach–Zehnder interferometer (Figure 10.16) to produce a zero area pulse.

The measurement is particularly simple and clear in the case of a single homogeneously broadened line. A linear (i.e., with a small area pulse) measurement provides all the information needed in that case. The zero area pulse sequence has a zero spectral Fourier component at the average pulse frequency. The *linear* absorption for that pulse sequence—when applied at resonance with the line—is proportional to the spectral overlap of the line and the pulse spectrum. For $T_2 = \infty$, the infinitely narrow line coincides with the node of the spectrum of the zero area pulse, and there is no absorption. The smaller T_2, the broader the line and its overlap with the pulse spectrum. With decreasing T_2, the ratio of

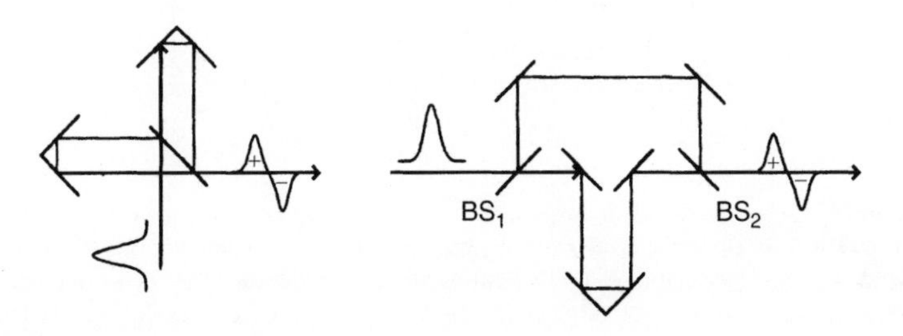

Figure 10.16 Michelson or Mach–Zehnder interferometers for the generation of zero area pulses. The beam splitters BS_1 and BS_2 should be identical, to produce a zero area pulse. The field envelopes of the pulses are shown.

absorption for an out-of-phase (zero area) pulse sequence to the absorption for an in phase pulse sequence will also decrease. An illustration of such a measurement in Li vapor is shown in Figure 10.17. The energy of the second harmonic of the transmitted pulse sequence is plotted as function of the delay between the two components of the pulse.

In the time domain the experiment can be explained as follows. The first signal emerging out of the interferometer of Fig. 10.16 excites the resonant transition in lithium vapor. The induced dipoles reradiate a field which opposes the applied field, and therefore cause absorption. The energy stored in the medium will be restituted to the second signal emerging out of the interferometer if the latter is 180° out-of-phase with the first pulse (the reradiated field adds in phase with the applied electromagnetic signal). Maximum absorption occurs for in phase pulse sequences. The signal versus delay should therefore show an interference pattern with a periodicity in delay equal to the light period.

The constructive–destructive interferences that extend beyond the region of pulse overlap decay with the collision time of the resonant sodium atoms with a buffer gas (Ar, 1000 torr pressure). SH detection was used in that particular example [22]. By using SH detection, the transition between the region corresponding to pulse interferences, and coherent interaction effects, can easily be identified.

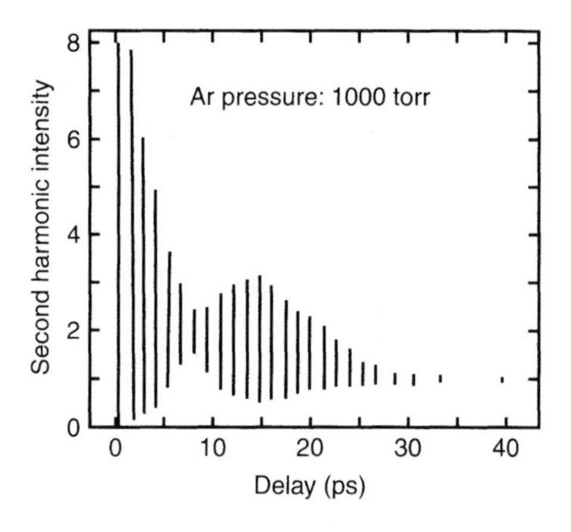

Figure 10.17 SH detection of the transmission of a zero area pulse sequence consisting of two delayed pulses through lithium vapor in the presence of argon as buffer gas. The vertical lines indicate the contrast between in-phase and out-of-phase transmission. The second harmonic of the transmitted zero area pulse sequence versus delay is recorded. The advantage of the SH detection is that the first portion of the curve is approximately the interferometric autocorrelation of the pulse. The transmission corresponding to out-of-phase pulse sequence is the lower envelope near zero delay (weaker pulse because of destructive interference) and becomes the upper envelope for larger delays (larger transmission on resonance for out-of-phase pulse sequences).

For delays smaller than the pulse duration, the pulse interference pattern is an interferometric autocorrelation (see Chapter 9).

In the case of an inhomogeneously broadened line, the phase dependence of the interaction disappears in the weak pulse limit. We have seen in Chapter 4 that the weak pulse absorption is proportional to the spectral overlap of the line and pulse. As shown by Eq. (4.35), in the case of purely inhomogeneous broadening and no saturation:

$$\frac{dW}{dz} \propto \int_{-\infty}^{\infty} g_{inh} |\mathcal{E}(\Omega)|^2 d\Omega = -\alpha_0 W, \tag{10.32}$$

and the absorbed energy is independent of the phase content of the pulse. There is, however, a difference in *nonlinear* transmission [for which the approximation of Eq. (10.32) does not apply] of in-phase and out-of-phase pulse sequences, even in the case of inhomogeneous broadening. Let us consider a sequence consisting of two pulses. If each half of the pulse sequence has an area between 0 and π, the zero area pulse sequence will be absorbed more strongly than the in-phase pulse sequence. The physical reason can be explained simply by considering the originally uniform absorption spectrum [Figure 10.18(a)] in which the first pulse burns a hole, which is seen by the second pulse as an inverted homogeneously broadened line [Fig. 10.18(b)]. For the in-phase pulse sequence, there is less absorption because of the reduced absorption of spectral components at the pulse average frequency [Fig. 10.18(c)]. In contrast, the zero area pulse sequence does not have spectral components overlapping with the center of the hole [Fig. 10.18(d)]. However, if each half of the pulse sequence is a π pulse, the system will be returned to ground state independently of the relative phase of the pulses.

The phase relaxation time T_2 can be extracted by measuring the ratio of the energy transmission factor $\Delta\mathcal{W}/\mathcal{W}$ for a sequence of pulses 180° out-of-phase to the same transmission factor for the in-phase pulse sequence as a function of total energy $\mathcal{W}$ in the pulse sequence [23]. The corresponding values of α_π/α_0 are plotted in Figure 10.19 for three values of the phase relaxation time T_2.

The main advantages of zero area pulse excitation as applied to the determination of phase relaxation times are:

- the pulses of the sequence can overlap,
- phase relaxation times shorter than the pulse duration can be measured,
- the experimental technique is particularly simple, and
- a 180° pulse sequence has zero area for transitions of different degeneracy and dipole moment.

The last property results in an easier interpretation of the data when the measurement covers more than one type of transition. The extension of this method

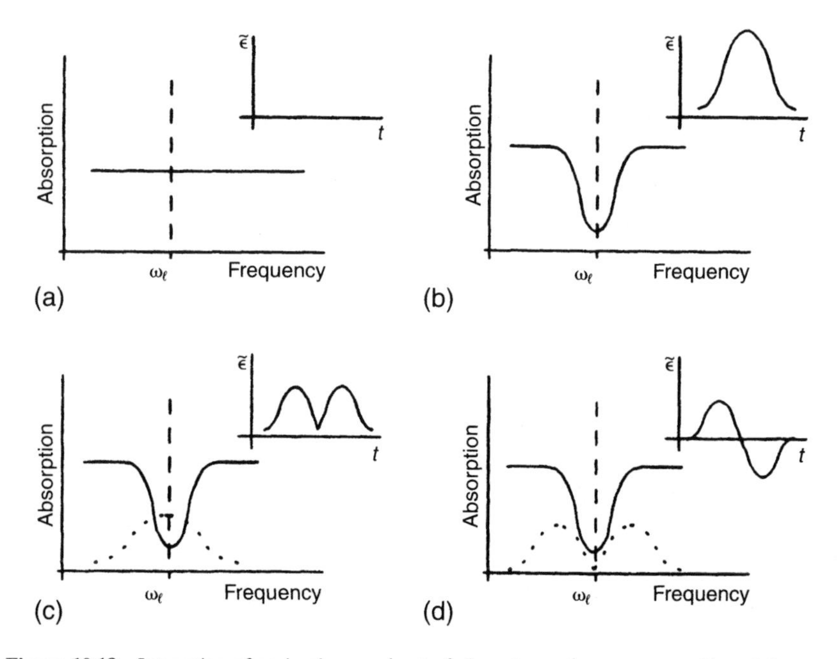

Figure 10.18 Interaction of an in-phase and out-of-phase two-pulse sequence with an inhomogeneously broadened absorption transition. Initially (a), the line profile is uniform. At $t = t_1$, the first of a two-pulse sequence burns a hole in the uniformly inhomogeneously broadened absorption line (b). The second pulse no longer sees a uniform line, but an inverted homogeneously broadened line. The absorption will be smaller for the in-phase pulse sequence (c) which has more spectral components (dotted line) overlapping with the center of the hole, than with the out-of-phase pulse sequence (d).

to molecular multiphoton transitions has been discussed in Diels *et al.* [24] and Besnainou [25]. In addition to the measurement of a dephasing time for a multiphoton transition, the pattern of absorption versus relative phase of the pulse sequence can be used to identify the type of resonance [25].

10.10. IMPULSIVE STIMULATED RAMAN SCATTERING

10.10.1. General Description

Some molecular vibrations—for instance the stretching mode of a symmetric diatomic molecule such as N_2—cannot be directly excited by a resonant electromagnetic field. However, such dipole forbidden transitions between states of equal parity and angular momentum can be accessed by a transition involving

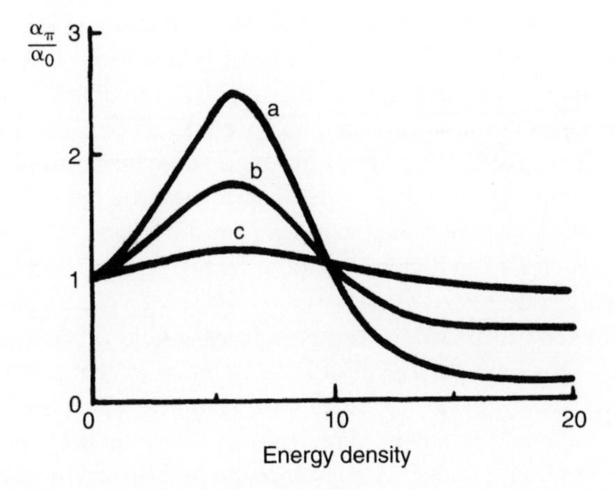

Figure 10.19 Ratio of the relative energy transmission for the 180° out-of-phase α_π to the in-phase α_0 pulse sequence, as a function of the normalized pulse energy, for various values of the phase relaxation time T_2, for an absorption line with infinite inhomogeneous broadening. The pulses are Gaussian, with a temporal separation equal to twice their duration (FWHM). The phase relaxation times are 100× (a), 6× (b), and 1× (c) the pulse duration.

two photons. In resonant Raman scattering, the difference between the optical frequencies of the two photons involved in the transition is equal to the frequency of the mechanical vibration being excited. Let us consider, for instance, a molecular vibration of frequency ω_{21} between two states $|2\rangle$ and $|1\rangle$ of identical parity [Figure 10.20(a)]. The molecule can be brought in the vibrationally excited

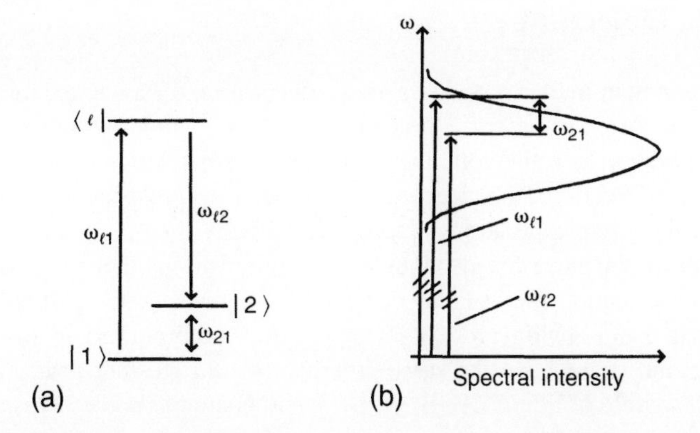

Figure 10.20 (a) Excitation of a Raman transition $|1\rangle \rightarrow |2\rangle$ via an electronically excited state $\langle \ell |$. (b) For a fs pulse of average carrier frequency ω_ℓ, the two frequency components of the Raman transition are contained within the pulse spectrum. The medium itself selects the frequency pairs suitable to drive the Raman transition.

state $|2\rangle$ by a succession of optical excitations via the dipole allowed transitions $|1\rangle \rightarrow \langle\ell|$ and $\langle\ell| \rightarrow |2\rangle$, involving photons of frequencies $\omega_{\ell 1}$ and $\omega_{\ell 2}$.

Because of the broad bandwidth of the fs pulse, stimulated Raman scattering can occur through the mixing of various spectral components of the ultrashort optical pulse [Fig. 10.20(b)]. For example, a molecular vibration is initiated by the sudden impulse exerted by the electric field of the pulse. The sample selects a pair of frequency components whose difference is in resonance with the eigenfrequencies of a Raman transition. This type of Raman scattering is called "impulsive stimulated Raman scattering" [26]. The fs excitation makes it possible to excite in phase a macroscopic ensemble of vibrating molecules. In solids, it is a coherent excitation of lattice vibrations that is achieved.

It is generally not possible to achieve a complete excitation of the Raman transition with a single fs pulse. Many Raman active modes can sometimes be accessed by the same fs pulse. However, if the process of impulsive stimulated Raman scattering is repeated at each cycle $2\pi/\omega_\nu$ of a Raman transition, the excitation will be enhanced. The selectivity of the process is also increased by the periodic excitation [27].

Impulsive stimulated Raman scattering can be used to analyze vibrational motions—for instance to determine their decay through pump–probe techniques. Synchronous excitation by a train of pulses can lead to substantially larger amplitudes of motion. This excitation process can generate high frequency vibration. A train of pulses spaced by a picosecond can generate THz LO phonons in semiconductors, which have a wavelength in the 100 Å range, and can therefore be used for high resolution imaging in solids.

10.10.2. Detection

The change in matter properties associated with the Raman excitation can be probed in a variety of ways. One can, for instance, probe an induced birefringence, in which case the rotation of the probe polarization will be measured, as detailed in Section 10.5. In parallel polarization (probe polarization parallel to that of the pump), the attenuation of the probe will be modulated with delay, because the probe pulse can also induce Raman transitions. The phase of the oscillations of attenuation versus delay of the probe depends on the particular spectral component that is being probed. In the example reproduced in Figure 10.21, the pump and probe have the same polarization and are sent nearly collinearly through a sample of liquid CH_2Br_2 [26]. The transmitted probe is dispersed by a monochromator. Two frequency components (609 nm and 620 nm) are displayed as a function of delay in Fig. 10.21.

Both spectral components are seen to oscillate with delay at the molecular vibration frequency, but with opposite phase. A simple explanation is that at a

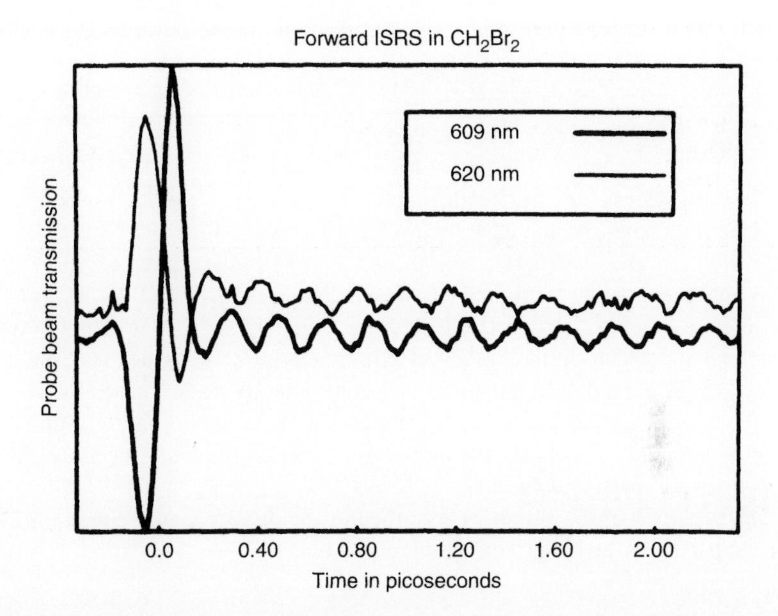

Figure 10.21 Transmission versus delay for two spectral components of the probe signal, for a sample of CH_2Br_2 pumped by a 65 fs pulse of a few μJ energy at 615 nm. The excitation and probe pulses are focused to a 200-μm spot size, at an angle of 5°, into a 2-mm sample cuvette (Adapted from Ruhman *et al.* [26]).

particular delay, the position of the vibrating coordinates is such that the 609 nm radiation is absorbed, and the 620 nm reinforced by the Raman transition. For a delay corresponding to half a vibration cycle later, the 609 nm transition will be reinforced and the 620 nm attenuated.

So far we have assumed a single pump pulse to induce the Raman signal. A standing wave pattern can also be generated for the impulsive stimulated Raman signal, either through a periodic configuration of the sample, or through the use of two intersecting pump pulses.

An example of sample periodicity is an MQW structure, of which the spacing between wells is made to match the wavelength of the phonon to be generated. The phonon can be generated by a train of fs pulses spaced by the phonon period, tuned to the intraband absorption in the quantum wells. The periodic spatial structure that is excited is responsible for the spatial coherence of the phonon [28]. The excitation by a periodic pulse sequence ensures temporal coherence of the created phonons.

It is also possible to create a standing wave Raman excitation with two intersecting pump pulses of the same frequency [26]. The temporal evolution of the vibration is easily analyzed through diffraction of a probe pulse by the standing wave pattern.

10.10.3. Theoretical Framework

The same density matrix formalism as in Chapters 3 and 4 can be used to describe impulsive stimulated Raman scattering. As in Fig. 10.20, we will consider Raman transitions between a ground state $|1\rangle$ and a first excited state $|2\rangle$ of a vibrational mode with frequency ω_{21}. The states $|1\rangle$ and $|2\rangle$ are infrared inactive, i.e., there is no dipole allowed transition $|1\rangle \rightarrow |2\rangle$. Coupling between these two states can occur via any electronic state $\langle \ell|$. All states $|\ell\rangle$ connected to $|1\rangle$ and $|2\rangle$ via a dipole transition will contribute to the Raman transition. We assume the optical field E to be off-resonant with all single-photon transitions. For this assumption to hold, the detuning of the intermediate levels $\langle \ell|$ has to exceed several pulse bandwidths.

The evolution of the system is described by the density matrix equations (4.1). For the particular level system being considered:

$$\frac{\partial \rho_{12}}{\partial t} - i\omega_{21}\rho_{12} = -\frac{i\tilde{E}^+}{\hbar}\sum_{\ell}(\rho_{1\ell}p_{\ell 2} - p_{1\ell}\rho_{\ell 2})$$

$$\frac{\partial \rho_{22}}{\partial t} = -\frac{i\tilde{E}^+}{\hbar}\sum_{\ell}(\rho_{2\ell}p_{\ell 2} - p_{2\ell}\rho_{\ell 2})$$

$$\frac{\partial \rho_{1\ell}}{\partial t} - i\omega_{\ell 1}\rho_{1\ell} = -\frac{i\tilde{E}^+}{\hbar}\sum_{j}(\rho_{1j}p_{j\ell} - p_{1j}\rho_{j\ell}), \tag{10.33}$$

where the sum over j applies to any level connected to $\langle \ell|$ by a dipole transition, including levels 1 and 2. A similar equation applies for $\rho_{2\ell}$.

As we have seen in Chapter 4, it is more convenient to decompose the off-diagonal matrix elements $\rho_{1\ell}$ and $\rho_{2\ell}$ into an envelope and fast varying phase term. For instance:

$$\rho_{1\ell} = \varrho_{1\ell}e^{i\omega_{\ell}t} \tag{10.34}$$

and a similar equation for $\rho_{2\ell}$. Substituting Eq. (10.34) into the third equation (10.33), and keeping only the levels 1 and 2 as levels that are dipole connected to levels ℓ:

$$\frac{\partial \varrho_{1\ell}}{\partial t} + i(\omega_{\ell} - \omega_{\ell 1})\varrho_{1\ell} = -\frac{i\tilde{\mathcal{E}}}{2\hbar}(\rho_{11}p_{1\ell} + \rho_{12}p_{2\ell}). \tag{10.35}$$

The assumption of the intermediate levels $\langle \ell |$ being off-resonance enables us to use the adiabatic approximation. This is a standard approximation used routinely in the context of deriving interaction equations in condition of two- (and more) photon resonance [29]. A detailed analysis of the use of the adiabatic approximation in the context of two-photon transitions can be found in [29]. Essentially, the second term in the left-hand side of Eq. (10.35) dominates, and we can approximate $\varrho_{1\ell}$ by its steady-state value:

$$\varrho_{1\ell} = \frac{-\tilde{\mathcal{E}}(\rho_{11}p_{1\ell} + \rho_{12}p_{2\ell})}{2\hbar(\omega_\ell - \omega_{\ell 1})}, \tag{10.36}$$

and a similar equation for $\varrho_{2\ell}$. Substituting into the first equation (10.33), we find the evolution equation for the coherent Raman excitation:

$$\frac{\partial \rho_{12}}{\partial t} - i\omega_{21}\rho_{12} = -\frac{i}{4\hbar^2}\tilde{\mathcal{E}}\tilde{\mathcal{E}}^* \sum_\ell \left(\rho_{11}\frac{p_{1\ell}p_{\ell 2}}{\omega_\ell - \omega_{\ell 1}} - \frac{p_{1\ell}p_{\ell 2}}{\omega_\ell - \omega_{\ell 2}}\rho_{22} \right). \tag{10.37}$$

Of particular interest is the amplitude of the off-diagonal element ρ_{12}. Let us define a (complex) amplitude ϱ_{12} similarly as in Eq. (10.34): $\rho_{12} = \varrho_{12}\exp(i\omega_{12}t)$. In addition, to simplify the discussion, let us assume, that there is only one level $\langle \ell |$ that dominates the interaction. We note that $\omega_\ell - \omega_{\ell 2} = (\omega_\ell - \omega_{\ell 1})\,[1 + \omega_{21}/(\omega_\ell - \omega_{\ell 1})]$. Substituting in Eq. (10.37) yields:

$$\frac{\partial \varrho_{12}}{\partial t} e^{i\omega_{21}t} = \frac{i p_{1\ell}p_{\ell 2}}{4\hbar^2(\omega_\ell - \omega_{\ell 1})}\tilde{\mathcal{E}}\tilde{\mathcal{E}}^*\,(\varrho_{22} - \rho_{11}) \tag{10.38}$$

where $\varrho_{22} = \rho_{22}/[1 + \omega_{12}(\omega_\ell - \omega_{\ell 1})]$. We recognize in Eq. (10.38) a Rabi frequency similar to the two-photon Rabi frequency discussed in Chapter 4:

$$\frac{p_{1\ell}p_{\ell 2}}{4\hbar^2(\omega_\ell - \omega_{\ell 1})}\tilde{\mathcal{E}}(t)\tilde{\mathcal{E}}^*(t) = \frac{r_{12}}{\hbar^2}\tilde{\mathcal{E}}(t)\tilde{\mathcal{E}}^*(t). \tag{10.39}$$

The evolution equations for the density matrix components can be rewritten:

$$\frac{\partial \varrho_{12}}{\partial t} = i\frac{r_{12}}{\hbar^2}\tilde{\mathcal{E}}\tilde{\mathcal{E}}^* e^{-i\omega_{21}t}\,[\varrho_{22} - \rho_{11}]$$

$$\frac{\partial \rho_{22}}{\partial t} = -2\mathrm{Im}\left[\frac{r_{12}}{\hbar^2}\tilde{\mathcal{E}}\tilde{\mathcal{E}}^* \varrho_{12}e^{i\omega_{21}t}\right]. \tag{10.40}$$

The form of the set of equations (10.40) is similar to Bloch's equations (4.6) and (4.7). In the weak pulse approximation ($\rho_{11} \approx 1$), after passage of the fs excitation, the off-diagonal matrix element oscillates at the Raman frequency:

$$\begin{aligned}\rho_{12} &\approx -ie^{i\omega_{21}t}\left[\int_{-\infty}^{\infty} \frac{r_{12}}{\hbar^2}\tilde{\mathcal{E}}(t')\tilde{\mathcal{E}}^*(t')e^{-i\omega_{21}t'}dt'\right] \\ &= -ie^{i\omega_{21}t}\left[\int_{-\infty}^{\infty} \frac{r_{12}}{\hbar^2}\tilde{\mathcal{E}}(\Omega)\tilde{\mathcal{E}}^*(\Omega-\omega_{21})d\Omega\right]. \end{aligned} \tag{10.41}$$

Equation (10.41) is obtained by integrating the first differential equation (10.40) with $\rho_{11} \approx 1$ and $\rho_{22} \approx 0$. It can be seen immediately from the convolution product in Eq. (10.41) that, for efficient Raman excitation, the pulse spectrum should be broad compared with the Raman frequency ω_{21}. Indeed, for $\omega_{21} \gg \tau_p^{-1}$, there is no overlap between $\tilde{\mathcal{E}}(\Omega)$ and $\tilde{\mathcal{E}}(\Omega-\omega_{21})$. The dimensionless quantity

$$\theta_R = \int_{-\infty}^{\infty} \frac{r_{12}}{\hbar^2}\tilde{\mathcal{E}}(\Omega)\tilde{\mathcal{E}}^*(\Omega-\omega_{21})d\Omega \tag{10.42}$$

is the analog of the tipping angle of the polarization in the Bloch vector model. We recognize from the analogy between Eqs. (10.40) and Bloch's equations (4.6) and (4.7), and the description of the vector model in Chapter 4, that an angle θ_R on the order of unity will be required to bring the ground state population to a vibrational excited state of energy $\hbar\omega_{21}$. It is left as a problem at the end of this chapter to demonstrate that the convolution in θ_R can be maximized by using, instead of a single pulse, a train of pulses spaced in time by $\tau_d = 2n\pi/\omega_{21}$ (n integer). The increase in selectivity can be inferred from the form of θ_R in the frequency domain [Eq. (10.42)]: θ_R vanishes for a pulse spacing $\Delta t \neq 2\pi/\omega_{21}$, in the case of a large number of pulses and undamped oscillations. The technique of using a synchronized pulse train can also lead to much larger amplitudes of motion than a single pulse [27]. Methods of generating such pulse trains have been presented in Chapter 8.

10.10.4. Single Pulse Shaping Versus Mode-Locked Train

The expression (10.42) for θ_R can be maximized by a pulse train whose repetition rate is any submultiple of the frequency $\omega_{21}/2\pi$. With ω_{21} in the

THz range, one technique is to shape a fs pulse into a sequence of pulses. Another possibility is to tune the mode-locked period of the laser to $T = 2n\pi/\omega_{21}$ (n integer). Such a technique is reminiscent of high-resolution coherence spectroscopy, where the repetition rate of mode-locked trains is tuned to a submultiple of an atomic resonance, leading, for instance, to enhanced quantum beats [30]. With advances in semiconductor lasers, repetition rates in the GHz to THz range are accessible with fs pulses. The repetition rate of passively mode-locked lasers can also be tuned continuously by adjusting the cavity length [31].

A question that arises is: what is the lowest repetition rate that can be used to excite a particular resonance ω_{21}? That question can be simply answered by modeling the resonant system by a classical oscillator, driven by an infinite series of δ function forces separated by a time T. Each successive pulse excites the particular oscillation corresponding to the resonance. This oscillation is represented in the classical model by the displacement x of an oscillator of mass m, restoring force–Kx, and damping constant b. The oscillation is not completely damped before the time of arrival of the next pulse, which, if T is a multiple of $2\pi/\omega_{21}$, will reinforce the motion. After an infinite number of driving pulses, the damped oscillation between two successive driving pulses will be stationary (see Figure 10.22). Assuming $(2N + 1)$ pulses in the train, the periodic driving force is represented by a series of δ functions: $F = F_0 \sum_{j=-N}^{N} \delta(t - jT)$. The equation of motion for the classical oscillator is:

$$m\ddot{x} + b\dot{x} + Kx = F_0 \sum_{j=-N}^{N} \delta(t - jT). \tag{10.43}$$

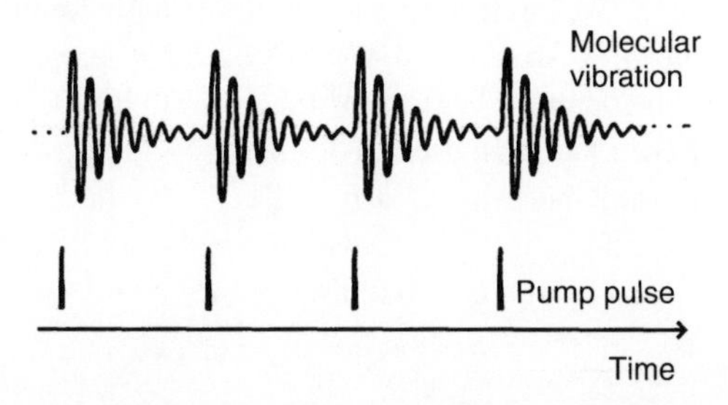

Figure 10.22 Damped molecular vibration, following impulsive stimulated Raman excitation by a train of ultrashort pulses. The amplitude of the oscillation will be maximum for a pulse separation equal to a multiple of the period of the molecular vibration.

Simple Fourier transformation of this equation leads to a solution for the amplitude $x(\omega)$. Taking the inverse Fourier transform of that solution yields $x(t)$:

$$x(t) = \frac{F_0}{2\pi} \sum_j I_j \tag{10.44}$$

with

$$I_j = \int_{-\infty}^{\infty} \frac{e^{i\omega(t-jT)}}{(K - m\omega^2) + ib\omega} d\omega. \tag{10.45}$$

The integrand I_j in Eq. (10.45) has two poles at $\omega = i\Gamma \pm \omega_{21}$, and $\omega_{21}^2 = K/m - \Gamma^2$ and $\Gamma = b/2m$. The stationary solution for the oscillator is found by contour integration and summation over j of the geometric series:

$$x(t) = \frac{1}{2} A(T) e^{-\Gamma t + i\omega_{21} t} + c.c. \tag{10.46}$$

with

$$\begin{aligned} A(T) &= \frac{iF_0}{\omega_{21}} \frac{1}{1 - e^{\Gamma T + i\omega_{21} T}} \\ &= \frac{iF_0}{\omega_{21}} \frac{1 - e^{\Gamma T - i\omega_{21} T}}{1 + e^{2\Gamma T} - 2e^{\Gamma T} \cos \omega_{21} T}. \end{aligned} \tag{10.47}$$

$A(T)$ is essentially the amplitude of the first cycle of oscillation. Its value is maximum and equal to $iF_0/[\omega_{21}(1 - e^{\Gamma T})]$ when $\omega_{21} T = 2n\pi$, and minimum, equal to $iF_0/[\omega_{21}(1 + e^{\Gamma T})]$ for $\omega_{21} T = 2(n+1)\pi$. The modulation depth $(1 - e^{\Gamma T})/(1 + e^{\Gamma T})$ is thus determined solely by the damping rate and the period of the driving force T. When driving a system at a subharmonic of the resonant frequency, the term $\omega_{21} T$ in Eq. (10.47) can be large ($\omega_{21} T = 2n\pi$, with n a large integer). The resonances (values of the periodicity T that satisfy the resonance condition) are closely spaced. The damping factor Γ determines which subharmonic N can still be used to drive effectively the resonance ω_{21}. Each δ function force sets off an oscillation, which should not be completely damped before being reinforced by the next exciting pulse.

10.11. SELF-ACTION EXPERIMENTS

Pump–probe experiments are intended to provide information on linear and nonlinear properties of matter. As noted previously, there is a fundamental

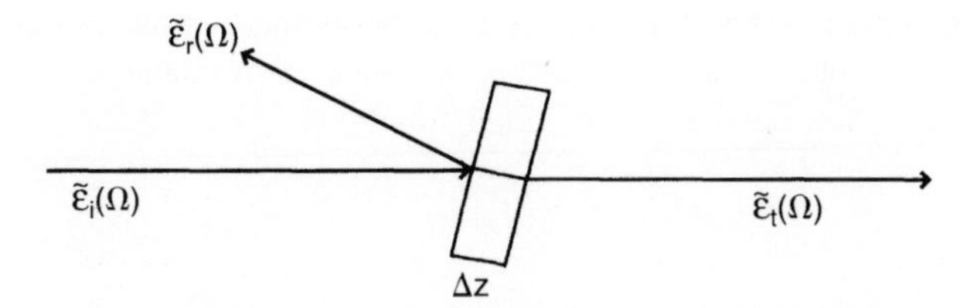

Figure 10.23 For linear systems, and some simple nonlinear systems, the complex susceptibility can be completely determined from single pulse transmission (reflection) measurements, provided the amplitude and phase of the incident, transmitted, and reflected signals can be completely determined.

temporal limitation: For the measurement interpretation the pump or excitation process should be completed before the medium is probed. One could try to obtain information on the properties of matter by measuring the time resolved fields of a single pulse incident, reflected and/or transmitted by a thin sample (Figure 10.23), using some of the techniques outlined in Chapter 9.

In the case of a linear interaction with the medium, the problem is analogous to the analysis of a linear circuit. For instance, referring to Chapter 1 [Eqs. (1.73) through (1.79)], the complex dielectric constant $\tilde{\epsilon}(\Omega) = \epsilon_0[1 + \tilde{\chi}(\Omega)]$ can be extracted by taking the Fourier transform $\tilde{E}(\Omega)$ of the incident (i) and transmitted (t) fields:

$$1 + \tilde{\chi}(\Omega) = -\frac{c^2}{z^2\Omega^2} \ln^2 \left[\frac{\tilde{E}_t(\Omega)}{\tilde{E}_i(\Omega)} \right] \tag{10.48}$$

where z is the sample thickness.

There is no simple algorithm that can solve the general problem of retrieval of a nonlinear susceptibility $\chi^{(n)}(\Omega)$ from a series of measurements of incident, transmitted, and reflected fields. Some assumptions have to be made—for instance, that all nonlinear susceptibilities except the third order, $\chi^{(3)}$, can be neglected. Within this approximation, measurement of the third harmonic transmitted field $\tilde{\mathcal{E}}_{3\omega}$ leads to a determination of the third-order susceptibility:

$$\chi^{(3)}(\Omega) = \frac{2c\Delta z}{\omega_\ell} \frac{\tilde{\mathcal{E}}_{3\omega}(\Omega)}{\tilde{\mathcal{E}}^2(\Omega)}. \tag{10.49}$$

The transmission measurements provide information on the bulk properties of the sample. Properties at the surface can be analyzed by measuring the reflected field. For instance, at normal incidence, the reflection coefficient is approximately $\chi_s(\Omega)/[4 + \chi_s(\Omega)]$, where $\chi_s(\Omega)$ is the complex susceptibility at the surface (assumed to be $\ll 1$). In the presence of resonances, these complex susceptibilities may have a complicated functional dependence on the optical field.

10.12. PROBLEMS

1. Referring to Section 10.3, derive in detail Eqs. (10.5) through (10.9). Find the effect of the beam geometry on the temporal resolution for a square temporal profile and square spatial profile in x and y.
2. A transient grating experiment is performed with a semiconductor. Let us assume that we have an amplitude grating only and that the carrier density $n(x, t)$ obeys the equation for ambipolar diffusion (one-dimensional model):

$$\frac{\partial n}{\partial t} - D\frac{\partial^2 n}{\partial x^2} = 0. \tag{10.50}$$

Derive a formula that relates the diffraction signal versus τ_d to the diffusion parameter D to be determined. From the diffusion parameter one can then obtain the carrier mobility $\mu = eD/(k_B T)$, where e is the electron charge, k_B Boltzmann's constant, and T the temperature.
3. Prove that in a transient grating experiment the diffraction of a probe pulse measured as function of the delay between the two pump pulses contains the information on the transverse relaxation time T_2. Assume an ensemble of homogeneously broadened two-level systems, weak excitation, thin samples.
4. The purpose of this problem is to compare impulsive stimulated Raman scattering excited by a single pulse and a train of identical pulses. The period of the Raman oscillation to be excited is 1 ps, and its damping time is 500 ps. The molecular system has a resonant absorption at 750 nm. The laser system delivers a Gaussian pulse of 50-fs duration, 1 nJ energy, at 770 nm, focused into the sample with a beam waist of $w_0 = 200$ μm. The dipole moment of the transitions $p_{1\ell} = p_{\ell 2} = 6\ 10^{-29}$ Cm. Calculate the off-diagonal matrix element ρ_{12} resulting from the excitation by the Gaussian pulses. Assume next each fs pulse is replaced by a train of 10 Gaussian pulses of 50-fs duration, but of 0.1 nJ energy each. Calculate the off-diagonal matrix element as a function of the period of this pulse train (in the range 1–10 ps).

BIBLIOGRAPHY

[1] J. Krueger. Masters thesis. Friedrich-Schiller-University Jena, 1988.

[2] W. Rudolph, J. Puls, F. Henneberger, and D. Lap. Femtosecond studies of transient nonlinearities in wide-gap II-VI semiconductor compounds. *Phys. Stat. sol. (b)*, 159:49–53, 1990.

[3] M. T. Portella, J.-Y. Rigot, R. W. Schoenlein, C. V. Shank, and J. E. Cunningham. k-space carrier dynamics in GaAs. In C. B. Harris, E. P. Ippen, G. A. Mourou, and A. H. Zewail, Eds., *Ultrafast Phenomena VII*, Springer-Verlag, Berlin, Germany, 1990, pp. 285–287.

[4] L. Oudar, A. Migus, D. Hulin, G. Grillon, J. Etchepare, and A. Antonetti. Femtosecond orientational relaxation of photoexcited carriers in GaAs. *Phys. Rev. Lett.*, 53:384–387, 1984.

[5] D. McMorrow, W. T. Lotshaw, and G. A. Kenney-Wallace. Femtosecond optical studies on the origin of the nonlinear responses in simple liquids. *IEEE J. of Quantum Electronics*, QE-24:443–454, 1988.

[6] H. J. Eichler, P. Guenther, and D. W. Pohl. *Laser-Induced Dynamic Gratings*. Springer, Berlin, 1986.

[7] Y. R. Shen. *The Principles of Nonlinear Optics*. John Wiley & Sons, New York, 1984.

[8] J.-C. Diels, W.-C. Wang, and H. Winful. Dynamics of the nonlinear four-wave mixing interaction. *Applied Physics*, B26:105–110, 1981.

[9] J.-C. Diels and I. C. McMichael. Degenerate four-wave mixing of femtosecond pulses in an absorbing dye jet. *J. Opt. Soc. Am. B*, 3:535–543, 1986.

[10] H. Mahr and M. D. Hirsch. An optical upconversion light gate with picosecond resolution. *Optics Commun.*, 13:96–99, 1975.

[11] W. Rudolph and J.-C. Diels. Femtosecond time resolved fluorescence. In A. E. Siegman, Ed., *Picosecond Phenomena V*, Springer-Verlag, Berlin, Germany, 1986, pp. 71–74.

[12] J. Shah. Ultrafast luminescence spectroscopy using sum frequency generation. *IEEE J. of Quantum Electron.*, QE-24:276–288, 1988.

[13] I. D. Abella, N. Q. Kurnit, and S. R. Hartmann. Photon echoes. *Phys. Rev.*, 141:391–406, 1966.

[14] E. L. Hahn. Spin echoes. *Phys. Rev.*, 80:580–594, 1950.

[15] C. V. Heer. Focusing of carr-purcell photon echoes, and collisional effects. *Phys. Rev. A*, 13:1908–1920, 1976.

[16] A. I. Alekseef and I. V. Evseev. Photon echo polarization in a gas medium. *Soviet Physics JETP*, 29:1139–1143, 1969.

[17] P. C. Becker, H. C. Fragnito, C. H. Brito-Cruz, R. L. Fork, J. E. Cunningham, J. E. Henry, and C. V. Shank. Femtosecond photon echoes from band-to-band transitions in GaAs. *Phys. Rev. Lett.*, 61:647–649, 1988.

[18] C. V. Shank, P. C. Becker, H. L. Fragnito, and R. L. Fork. Femtosecond photon echoes. In T. Yajima, K Yoshihara, C. B. Harris, and S. Shionoya, Eds., *Ultrafast Phenomena VI*, Springer-Verlag, Berlin, Germany, 1988, pp. 344–388.

[19] J.-Y. Rigot, M. T. Portella, R. W. Schoenlein, C. V. Shank, and J. E. Cunningham. Two-dimensional carrier-c1arrier screening studied with femtosecond photon echoes. In C. B. Harris, E. P. Ippen, G. A. Mourou, and A. H. Zewail, Eds., *Ultrafast Phenomena VII*, Springer-Verlag, Berlin, Germany, 1990, pp. 239–243.

[20] J. B. Ashbury, T. Steinel, and M. D. Fayer. In P. Hannaford, Ed., *Femtosecond Laser Spectroscopy*, Springer-Verlag, Berlin, Germany, 2005, pp. 167–196.

[21] L. V. Dao, C. Lincoln, M. Lowe, and P. Hannaford. In P. Hannaford, Ed., *Femtosecond Laser Spectroscopy*, Springer-Verlag, Berlin, Germany, 2005, pp. 197–224.

[22] J.-C. Diels, W. C. Wang, and R. K. Jain. Experimental demonstration of a new technique to measure ultrashort dephasing times. In K. B. Eisenthal, R. M. Hochstrasser, W. Kaiser, and A. Laubereau, Eds., *Ultrafast Phenomena III*, Springer-Verlag, Berlin, Germany, 1982, pp. 120–122.

[23] J.-C. Diels. Feasibility of measuring phase relaxation time with subpicosecond pulses. *IEEE Journal of Quantum Electron.*, QE-16:1020–1021, 1980.

[24] J.-C. Diels, J. Stone, S. Besnainou, M. Goodman, and E. Thiele. Probing the phase coherence time of multiphoton excited molecules. *Optics Communications*, 37:11–14, 1981.

[25] S. Besnainou, J.-C. Diels, and J. Stone. Molecular multiphoton excitation of phase coherent pulse pairs. *J. of Chem. and Phys.*, 81:143–149, 1984.

[26] S. Ruhman, A. Joly, and K. Nelson. Coherent molecular vibrational motion observed in the time domain through impulsive stimulated raman scattering. *IEEE Journal of Quantum Electron.*, 34:460–468, 1988.

[27] A. Weiner, D. Leaird, G. Wiederrecht, M. Banet, and K. Nelson. Spectroscopy with shaped femtosecond pulses: Styles for the 1990s. In K. A. Nelson, editor, *SPIE Proceedings 1209*, Los Angeles, CA, 1990, pp. 185–195.

[28] H. T. Grahn, H. J. Maris, and J. Tauc. Picosecond ultrasonics. *IEEE J. of Quantum Electron.*, 25:2562–2568, 1989.

[29] D. Grischkowsky, M. M. T. Loy, and P. F. Liao. Adiabatic following model for two-photon transitions: Nonlinear mixing and pulse propagation. *Phys. Rev. A*, 12:2514–2533, 1975.

[30] J. Mlynek, W. Lange, H. Harde, and H. Burggraf. High-resolution coherence spectroscopy using pulse trains. *Phys. Rev. A*, 24:1099–1101, 1981.

[31] N. Jamasbi, J.-C. Diels, and L. Sarger. Study of a linear femtosecond laser in passive and hybrid operation. *Journal of Modern Optics*, 35:1891–1906, 1988.

11

Examples of Ultrafast Processes in Matter

11.1. INTRODUCTION

A microscopic analysis of many fundamental processes in matter starts at the ps or fs time scale. Primary events associated with macroscopic transformations that appear relatively slow, such as chemical reactions, photosynthesis, phase changes, and human vision, evolve on a fs time scale. A mere listing of all processes in biology, chemistry, and physics that are being actively investigated is already beyond the scope of this book. A detailed introduction of these topics can be found for example in the books by Kaiser and Auston [1], De Schryver *et al.* [2], Shah [3], and Mukamel [4]. A periodic update of these topics is published in the proceedings of the biannual conferences on Ultrafast Phenomena [5].

Rather than to attempt an extensive review, this chapter will focus on a few examples of ultrafast events in matter and their measurement. We will proceed by order of material systems of increasing complexity. The simplest system is the single atom, in which wave packets representing the motion of the electron in a Rydberg orbit can be analyzed with ultrafast techniques. Next, we proceed from the single atom to simple molecules to dissociating molecules—a step toward chemical reactions. The next form of arrangement of atoms is condensed matter, in which fs techniques are particularly powerful in analyzing changes of phase. Finally, biological systems offer the ultimate in molecular complexity. Femtosecond techniques are an essential tool in unraveling, for example, the primary processes of vision and photosynthesis.

11.2. ULTRAFAST TRANSIENTS IN ATOMS

11.2.1. The Classical Limit of the Quantum Mechanical Atom

Bohr's model of the hydrogen atom was based on the concept of the electron describing a classical trajectory in the attractive potential of the nucleus. The quantization relation introduced empirically by Bohr (see, for instance Cohen-Tannoudji *et al.* [6]) states that the angular momentum of the orbits is quantized:

$$\frac{m_e m_p}{m_e + m_p} \nu r = pr = \hbar n \tag{11.1}$$

where v is the radial velocity of the electron (p its linear momentum) along the orbit of radius r, m_e, and m_p are the masses of the electron and proton, and n is the quantum number. The classical picture of the orbiting electron violates the uncertainty principle for small quantum numbers n. To be able to describe the electron motion by a classical trajectory, the uncertainty in position (Δr) and momentum (Δp) should be smaller than r and p, respectively, or:

$$\frac{\Delta r}{r} \frac{\Delta p}{p} \ll 1 \tag{11.2}$$

$$\gg \frac{\hbar}{n\hbar}. \tag{11.3}$$

The last inequality (11.3) is simply obtained by substituting the uncertainty principle and the quantization condition (11.1) into the classical representation condition (11.2). The two conditions (11.2) and (11.3) are only compatible for large values of the quantum number n, or large orbits. States characterized by a high principal quantum number are called Rydberg states. The classical orbit becomes a reasonable approximation for these states with large quantum number n.

11.2.2. The Radial Wave Packet

A fs pulse cannot be used to excite an atom from its ground state to a single Rydberg state, because Rydberg states are closely spaced as compared to the bandwidth of ultrashort pulses. Instead, a fs pulse will excite a superposition of many Rydberg states. This superposition is a wave packet localized in the *radial* coordinate. The period of oscillation corresponds to the period of the Kepler orbit of a classical particle with the energy corresponding to that of the average Rydberg state excited.

The experimental technique to observe the radial motion of the electrons along these Kepler orbits is a pump–probe experiment. The pump pulse excites the atoms to a superposition of Rydberg states. The number of ions (or free electrons) produced by a subsequent probe is recorded as a function of delay. As explained below, the number of ions can be related to the position (velocity) of the electron along its Kepler orbit [7–9].

There is no localization in the angular coordinates. If the ground state is an S state, the states forming the wave packet are P states with various principal quantum numbers. Each of these states has as an angular dependence proportional to the square of a single spherical harmonic, a dependence in $\sin^2\theta$ in the case of the $l = m = 1$ state (where l and m are the usual eigenvalues of the orbital angular momentum and its projection along a z axis). The classical description of a Kepler orbit applies: Rather than a single orbiting electron, we should visualize an ensemble of noninteracting particles orbiting the nucleus, with their principal axes distributed according to the $\sin^2\theta$ distribution (Figure 11.1). This "radial Rydberg wave packet" will move in the effective atomic potential between the

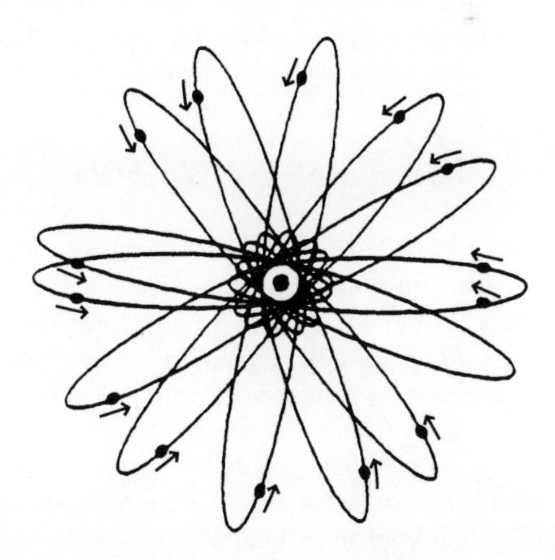

Figure 11.1 An ensemble of classical Kepler orbits make up the radially localized wave packet created by fs excitation of the ground state atom. The major axes of the ellipses are randomly distributed over all directions with a $\sin^2\theta$ distribution, but each electron has the same angular coordinate in various ellipses. (From Stroud [7].)

two classical turning points. It is a radial wave packet, because only a few angular momentum eigenstates can be excited (selection rule $l \rightarrow l \pm 1$), and the angular coordinates of the Rydberg electron are delocalized in a quantum mechanical sense [10]. Each of these orbits correspond to approximately the same energy, hence the same classical period. Therefore, with all particles moving in phase along the various elliptical orbits, they arrive at the same time close to the nucleus, as illustrated in Fig. 11.1. To the motion of the charged particle is associated an electric current $\mathbf{J} = e\mathbf{v}$ proportional to its velocity $\mathbf{v}$. The Rydberg wave packet is excited by a pump pulse. The energy absorbed by a delayed probe pulse of electric field $\mathbf{E}_p$ is proportional to $\mathbf{J} \cdot \mathbf{E}_p$. The absorbed energy is large if the delay is such that the pulse reaches the atoms with the electron near the nucleus (maximum velocity), and substantial ionization will result. At the other turning point far away from the core, the electron is nearly a free particle (which will not absorb radiation).

The photoionization versus probe delay is shown in Figure 11.2. The signal oscillates at the classical orbital frequency. However, because the Rydberg states are not equally spaced in frequency, the states get out-of-phase, and the wave packet decays. Owing to the finite number of states excited, the observed decay of the wave packet shown in Fig. 11.2 is not an irreversible process: After a large number of cycles, the components of the wave packet come back in phase [11], a process called "revival" of the wave packet. One can also observe "fractional revivals" [9]. For instance, during the one-half fractional revival, every other state in the superposition comes into phase, leading to the formation of two wave packets. Experimental evidence of the formation of two wave packets is the change in oscillation frequency to twice the orbital frequency in Fig. 11.2.

11.2.3. The Angularly Localized Wave Packet

Radial localization was obtained by creating a superposition of states corresponding to a large radial quantum number n, spanning a group of values Δn. Similarly, angular localization will require the superposition of excitations to a large angular momentum l, spanning a group of values Δl. Because a single photon carries only one unit of angular momentum, many photons are required to reach the high angular momentum states from the ground state. The technique devised by Yeazell and Stroud [12] is to excite sodium atoms from the ground state to the $n \approx 50$ manifold of states via a 2-photon transition, using circularly polarized light at 483.7 nm from an excimer pumped dye laser. A radio frequency (rf) field is used to create the high angular momentum wave packet through 30-photon excitation from the $50d$ state to states grouped around $l = 32$ ($n = 50$, $29 < l < 37$, $m = l$). The orientation of the wave packet lies along the direction

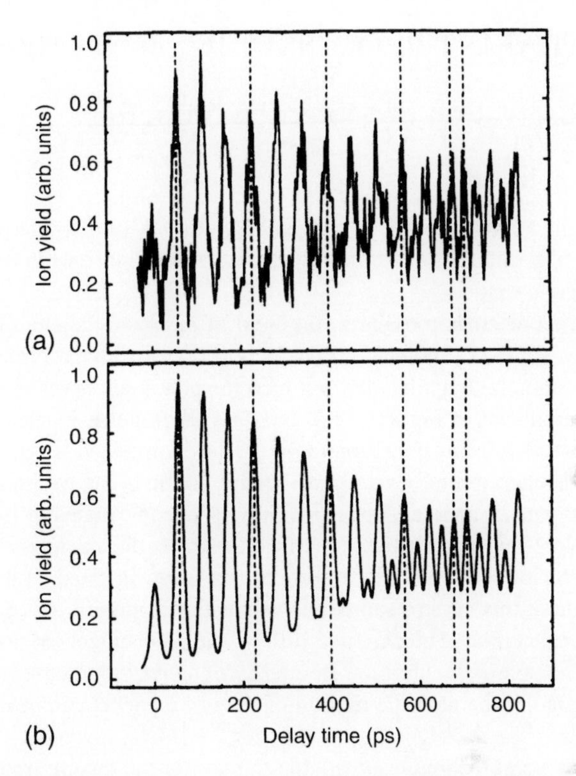

Figure 11.2 Photoionization signal as a function of probe delay. (a) Experimental recording. (b) Theoretical simulation. The Rydberg wave packet spans $\Delta n = 5$ and is centered on the Rydberg state $n = 72$ of atomic potassium (wavelength of pump pulse 285.6 nm). The probe is at 571.2 nm. (From Yeazell and Stroud [9].)

of the rf field vector at the time of the optical excitation. Because the precession and rate of dispersion of the wave packet are slow (order of ms), detection can be made through ionization with a pulsed electric field [12]. The wave packet is localized in the angular direction but not in the radial direction. The classical description is that of an ensemble of elliptical orbits, all with their axes aligned along the direction of the rf field. However, the phases of the motion along the ellipses are not determined, resulting in an elliptical distribution in space that is approximately stationary in time.

Techniques involving ultrashort optical and electrical pulses have been proposed by Gaeta *et al.* [13] to localize wave packets in the radial and angular coordinates. This would produce an atomic electron in a classical orbit.

11.3. ULTRAFAST PROCESSES IN MOLECULES

11.3.1. Observation of Molecular Vibrations

11.3.1.1. Simple Molecules

When single atoms combine to form molecules, additional internal degrees of freedom, such as rotations and vibrations, arise, with transients in the picosecond and femtosecond range.

Instead of an electron moving in the field of an atom, we shall now consider the case of an atom in a molecule. For the purpose of illustration let us consider as specific system the I_2 molecule for which the potential curves of the electronic states are reproduced in Figures 11.3 and 11.4. A fs pulse is used to excite the $\mathbf{X}(\nu''s = 0) \rightarrow \mathbf{B}(\nu' = n)$ transition in the 500–600 nm wavelength range (where ν characterizes the vibronic excitation). Owing to the broad excitation spectrum, the fs pulse creates a coherent superposition of vibronic states of mean quantum number n. Note that during the short fs interaction the nuclear motion can be neglected, which corresponds to a vertical transition in Figs. 11.4 or 11.3. The time evolution of this system can be viewed as the motion of this wave packet in the molecular potential. The classical limit is the mechanical (harmonic) oscillation with a characteristic vibration frequency ω_{vib}. As in the case of the electron in a Rydberg atom, the periodic motion of the wave packet can be observed with fs techniques.

There are several techniques available to monitor the quantum state of excited molecules. They are based on the fact that the interaction strength with a second light pulse depends on the instantaneous location and shape of the wave packet. If the experiment is carried out in a molecular beam, a delayed fs pulse can be used to excite the molecule from state **B** to a dissociative state. The fragments can be monitored with a mass spectrometer. If the measurement is carried out in a cell, the population of the **B** state can be observed simply through the fluorescence from the **B** state to the ground state. To probe the dynamics of the vibration, the molecule can be irradiated by probe pulses identical to those that created the excitation (except for the timing and phase). Because the excited state still represents a stable molecule, return to ground state stimulated by the second pulse will be possible at periodic intervals corresponding to the vibrational period of the electronically excited I_2 molecule (Fig. 11.3).

The experimental technique is essentially that of the zero area pulse experiment described in the previous chapter. In the case of I_2, the excitation wavelength should be in the range of 608 to 613 nm. The zero area pulses are generated in a Mach–Zehnder interferometer and sent through a 5 cm long room temperature I_2 cell at 0.25 torr [14]. The fluorescence is detected at a right angle. The return

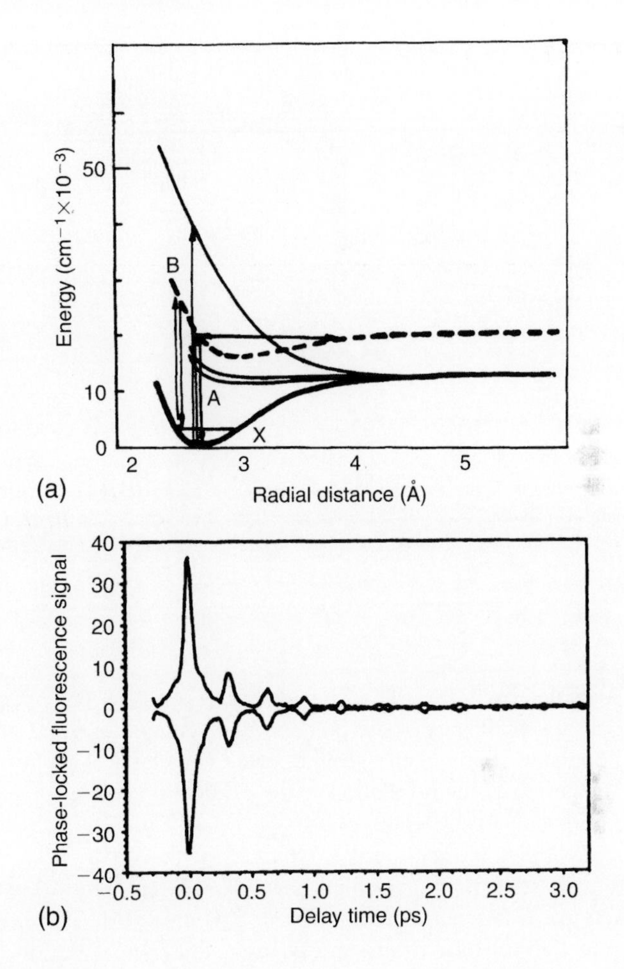

Figure 11.3 Study of the potential surface of I_2, using a pair of identical pulses with adjustable relative phase and delay. (a) Sketch of the potential surfaces. (b) Envelope of the fluorescence signal (only the contribution because of the two-pulse excitation) from the **B** state, corresponding to an in phase sequence (upper envelope) and an out-of-phase sequence (lower envelope). (Adapted from Scherer *et al.* [14].)

to ground state will occur if the delayed probe of the same wavelength as the pump is π out-of-phase with the exciting pulse. If instead the probe is in phase with the first pulse, the excitation will be reinforced. As in the case of the radial Rydberg wave packet, the classical picture for an oscillating particle fully applies. The envelope of the fluorescence pattern for in phase and out-of-phase pulse sequences is shown in Fig. 11.3(b) (from Scherer *et al.* [14]). The successive

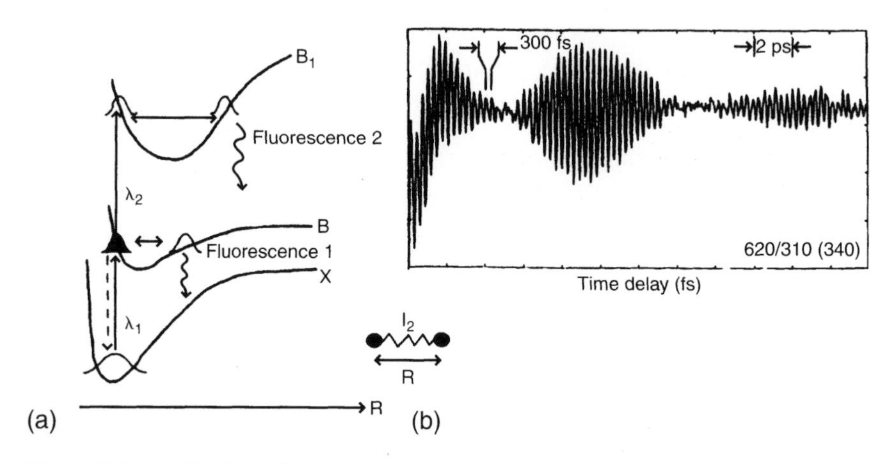

Figure 11.4 (a) Sketch of the bound potential energy surfaces relevant to the study of iodine through excited state fluorescence. (b) Fluorescence from the excited state $\mathbf{B}_1$ of I_2 as function of the delay between the two excitation pulses of wavelength λ_1 and λ_2. (Adapted from Bowman *et al.* [15].)

peaks are separated by 278 fs. This spacing corresponds to the superposition of the vibrational levels of the **B** state pumped by excitation at 611.2 nm from the thermally populated levels of the ground state. The classical picture is that we are seeing the period of the oscillation of the excited molecule, corresponding to a vibration frequency of 3.6 THz.

Another possibility to measure the dynamics of the wave packet is to probe the excitation of state **B** into a bound state $\mathbf{B}_1$ [Figure 11.4(a)] with a time-delayed pulse of different frequency ($\lambda_2 = 310$ nm) [15]. This was done by measuring the fluorescence from state $\mathbf{B}_1$ as a function of the delay between the excitation pulses, as shown in Fig. 11.4(b). The short time oscillation has the period of vibration of the molecule or period of wave packet motion in the **B** state. The periodical behavior of the oscillation period results from a revival of the wave packet (see, for instance, Yeazell *et al.* [11]). The wave packet consists of a finite number of *nearly* equally spaced energy states (anharmonic potential). This causes the wave packet to spread as time progresses so that it is no longer localized. As a result, the periodic behavior of the fluorescence disappears. However, because only a finite number of states is excited by the fs pulse and forms the wave packet, a rephasing of the states occurs after a certain time period. As in the case of the Rydberg states, the wave packet again becomes localized, which manifests itself in an increased modulation amplitude of the fluorescence.

11.3.1.2. Complex Molecules

Vibrations and Other Motions

The interpretation of the vibrational studies is particularly simple for isolated diatomic molecules. Molecular vibration and vibrational relaxation, of course, occur in more complex systems, too. As an example, let us consider organic dye molecules in solution. As outlined in previous chapters these systems have gained importance as laser dyes and saturable absorbers, and have therefore been extensively studied. Because of the large number of internal degrees of freedom and the strong interaction with the solvent, the damping of coherently excited wave packets and the vibrational relaxation often proceed on a subpicosecond time scale. Wise *et al.* [16] observed a damped sinusoidal decay in a pump–probe absorption experiment. For the dye Nile Blue, for example, they could identify eight different oscillation frequencies, documenting the large manifold of molecular eigenmodes of this complex system. Femtosecond techniques have also been successfully applied to the spectroscopic characterization of clusters, see, for example, Baumert *et al.* [17].

An absorption spectrum of a dye solution taken with an ordinary spectrophotometer typically exhibits a resonance corresponding to the $S_0 \rightarrow S_1$ transition with a spectral width of several tens of nanometers. This broad absorption profile results from a large number of rotational and vibrational states within one electronic state. An interesting question is whether the transition is homogeneously or inhomogeneously broadened. As explained in Chapter 3 (Fig. 3.2), the answer depends on the time scale on which the experiment is performed. A convenient experimental technique is time resolved hole burning.

Hole Burning

Hole burning or saturation spectroscopy is the standard technique to determine the homogeneous linewidth (T_2^{-1}) in gases and vapors. In the case of condensed matter, a fs variation of that technique can be used. First, an intense fs pump pulse is applied to saturate a particular transition. Let us consider as a specific example a hole burning experiment performed on cresyl violet [18]. The 60-fs wide pump pulse (centered at 618 nm) excites the $S_0(\nu = 0) \rightarrow S_1(\nu = 0)$ transition of the molecule (Figure 11.5). Because the occupation numbers of the $\nu = 0$ transition in the S_0 and S_1 electronic state are modified, an absorption change in the $0 \rightarrow 1$ and $1 \rightarrow 0$ transition is also observed immediately after excitation. A fraction of the pump pulse is chirped and compressed—and hence spectrally broadened—to probe the modified absorption spectrum of the sample. Twelve millimeters of fiber and a pair of gratings compress that fraction of the pump down to 10 fs, using the technique outlined in Chapter 8. For every delay increment, the difference spectrum (with and without pump) is recorded.

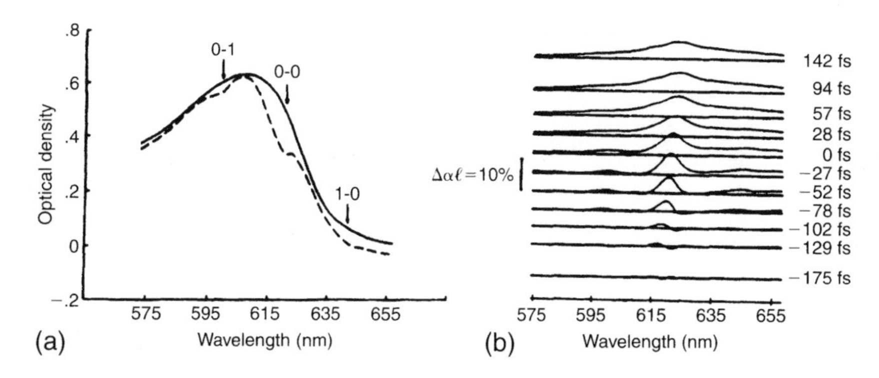

Figure 11.5 (a) Absorption spectrum of cresyl violet near the region of pumping. The dashed line illustrated qualitatively the spectral modification immediately after the 60-fs pump pulse. (b) Differential absorption spectra for successive delay increments after excitation of cresyl violet with a 60-fs pump pulse at 618 nm (Adapted from Brito-Cruz *et al.* [18]).

The resulting plot reproduced in Fig. 11.5 (b) shows clearly that three successive holes are burned in the absorption profile. The inverse of the linewidth of the hole indicates a phase relaxation time of $T_2 = 75$ fs. A plot of the differential absorption versus time also shows a fast transient. The decay of the hole structure with time is a measure of the cross relaxation. The red shift of the peak in the differential absorption indicates vibrational relaxation. The spectral feature gives in this case a more positive identification of the homogeneous broadening than the more complex temporal transient.

11.3.2. Chemical Reactions

One of the great frontiers in chemistry is detailed experimental investigations of chemical reactions in progress from reactants through a transition state to products. Previously, understanding of the evolution of the transition state relied almost exclusively on theoretical treatments. For a three-atom system with a small number of electrons, calculations may provide potential energy surfaces on which to compute classical trajectories to simulate chemical reactivity. To adequately reflect observable chemical phenomena, the accuracy of these energy potential surfaces needs to be on the order of 1 kcal/mol, an appalling figure for spectroscopists, because it corresponds to a spectral uncertainty of 350 cm^{-1} (or 10^{13} s^{-1})! The uncertainty is even worse for more complex molecular systems with many more internal degrees of freedom, hence the need for an experimental technique that will directly measure the potential energy surfaces of the transition states.

Femtosecond pulses offer the possibility of separating the electronic and nuclear parts of the wave function and therefore work directly within the framework of the Born–Oppenheimer approximation (see, for instance, Cohen-Tannoudji *et al.* [6]). One of the new methods discussed in the beginning of this section is to coherently excite or de-excite a transition from a ground to a higher energy potential surface. The advantage of the fs pulse excitation is that no substantial change in nuclear coordinates can take place during the interaction with light.

For the study of chemical reactions, the experimental difficulty is that measurements cannot be made on a single isolated molecule, and it seems difficult at best to synchronize (for instance) pairs of molecules involved in bimolecular reactions. In the case of unimolecular reactions, however, it is possible to use a femtosecond pulse to initiate synchronously the dissociation of a group of molecules (excitation to a repulsive potential surface V_1), and to monitor subsequently their evolution to products with delayed probe pulses as sketched in Figure 11.6 [19].

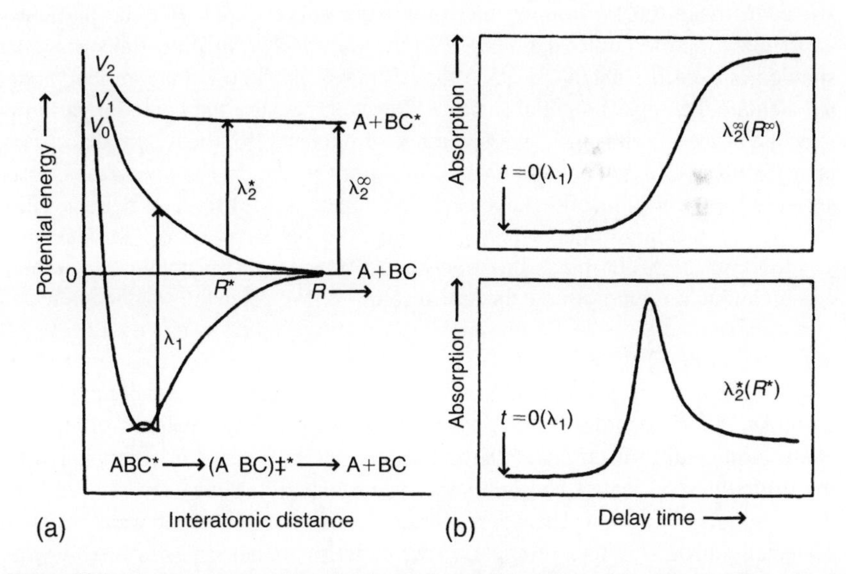

Figure 11.6 Pump–probe experiment to observe the transition region of a reaction. (a) The molecule is first excited by a pump pulse at λ_1 from the potential energy curves for the bound molecule V_0, to the dissociative state V_1. After a delay τ_d during which the fragment evolves along the repulsive potential V_1, a probe pulse at λ_2^* excites the complex to the dissociative (eventually ionizing) potential surface V_2. As the fragments recoil, the pulse at λ_2^* (or λ_2^∞) probes the transition region (or the free fragments). (b) The expected fs transients, signal versus delay τ_d at λ_2^* and λ_2^∞ (Adapted from Zewail *et al.* [20].)

Potential curves have been extracted from such measurements [21]. The probe pulse induces transitions from the repulsive potential surface V_1 (along which the dissociating molecule is moving, following excitation by the pump), to another potential surface V_2. For a given delay τ_d, the probe absorption versus probe wavelength approximates a step function, where the wavelength at the step is a measure of the energy difference between potential energy surfaces V_2 and V_1 at the particular delay τ_d. It is usually not practical to measure the absorption of the probe with low density molecules. Instead, one can determine the number of molecules excited to the V_2 potential surface by laser induced fluorescence. Another possibility is to use a photoionizing probe (in which case the number of transitions is directly measured by an ion count). This technique has been applied to a detailed study of the unimolecular dissociation [19]:

$$\mathrm{ICN}^* \rightarrow \mathrm{I} + \mathrm{CN}. \tag{11.4}$$

The value of the resonance energy versus delay is a measure of the *difference* between the potential energy surfaces $V_2 - V_1$. To obtain an absolute measurement of an energy potential curve, it is necessary to know the shape of the upper curve V_2, or to make the assumption that this upper curve is flat. For the particular experiment reported in Rosker *et al.* [19] the variation of the potential surface V_2 should not exceed 100 cm^{-1} [19]. Another approach is to use a theoretical model to calculate the upper potential surface. However, because the potential variation is on the order of 6000 cm^{-1} over the range of interest [19], the required accuracy is on the order of a few percent. Procedures to invert the data to obtain the energy surfaces for the ICN reaction have been developed by Bernstein and Zewail [21]. Because of the large energy changes along the potential energy surface in a short delay, the probe pulse duration has to be selected to obtain the optimal combination of temporal and spectral resolution. We refer the reader interested in a general overview of fs probing of dissociative reaction to Zewail [20, 22] and Cong *et al.* [23].

The technique of probing chemical reactions has been successfully applied to unimolecular dissociations. The possibility of using a femtosecond technique to study bimolecular reactions at the individual collision level is complicated by the difficulties of spatial and temporal synchronization. One way to overcome this problem is through the use of van der Waals complexes of weakly bound molecular clusters. In these complexes the moieties are held in a reasonably well-defined geometry, so that the prospective bimolecular reagents or their precursors may be frozen into a convenient geometry in preparation for reaction initiation. There are well-established techniques to produce clusters of heterodimers [24, 25]. Once frozen collision complexes have been prepared by expansion of molecular beams after a supersonic nozzle, a bimolecular chemical reaction can be initiated by a fs photodissociation pulse producing a pair of reagents. Such a technique has been applied to the reaction [26]

$$H + OCO \rightarrow [HOCO]^{\ddagger} \rightarrow OH + CO. \tag{11.5}$$

The van der Waals "precursor molecule" was $[IH \cdots OCO]$ formed in a free-jet expansion of a mixture of HI and CO_2 in an excess of helium carrier gas. To clock the reaction, an ultrashort laser pulse photodissociates HI, ejecting an H atom toward the O atom of the CO_2. The delayed probe detects the formation of OH. Such an experiment establishes clearly that the reaction proceeds via an intermediate state, as shown in Eq. (11.5) and gives values for the lifetime of the intermediate complex $[HOCO]^{\ddagger}$.

Femtosecond techniques are not limited to the *observation* of chemical reactions, but can even be exploited to influence the course of the reaction (see, for instance, Potter *et al.* [27]). This can open new relaxation channels or increase the yield of certain reaction products.

11.3.3. Molecules in Solution

Considerable progress has been made toward the microscopic understanding of molecular vibration and chemical reactions in solution. For instance, we have shown at the beginning of this section techniques to study the wave packet dynamics of the nuclear motions of iodine in the **B** state, in the collision-free limit. These techniques can be applied to solutions of different densities and liquids. The primary effects of the solvent on the fs wave packet are dephasing, energy relaxation, caging, and recombination [28]. Except for collision-induced rapid nonradiative transitions in the liquid state, which cause the main fluorescence emission to originate from a lower transition, the experimental techniques are similar to the one used in the gaseous phase.

Femtosecond techniques have also been applied to more complex chemical problems, such as the study of photodissociation. The influence of the solvent on the dynamics of photodissociation of ICN can be dramatic [29]. The knowledge gained of how the solvent influences the decay of photofragment translation and rotation is useful in understanding the dynamics of thermally activated chemical reactions [29]. Theoretical simulations have indeed shown that the fluctuations in reactant and product translational and rotational motions of thermally activated reactions proceed on the fs scale [30].

11.4. ULTRAFAST PROCESSES IN SOLID-STATE MATERIALS

11.4.1. Excitation Across the Band Gap

Femtosecond techniques made it possible to resolve fundamental interaction mechanisms and times in solids at room temperature. These processes are of tremendous importance. For instance, they determine physical limits for speed and miniaturization in semiconductor devices. Figure 11.7 illustrates essential processes in semiconductors, following optical excitation above the band gap. For a comprehensive review of ultrafast processes in semiconductors probed by laser pulses, see Shah [3].

An ultrashort light pulse of frequency ω_ℓ creates electron–hole pairs in states above the band gap. Their mean excess energy is $\Delta E = \hbar(\omega_\ell - \omega_{gap})$, and their initial energy distribution resembles the excitation spectrum. With large excited carrier densities, mainly carrier–carrier scattering leads to a thermalization within the Γ-valley without changing the mean carrier energy. This means that some carriers scatter out of their initial states, so that the distribution of occupied states becomes broader. Such processes are generally associated with momentum relaxation and are responsible for the dephasing of the polarization. Corresponding T_2 times can be measured by means of photon echo experiments as described in the previous chapter. The temperature that can be attributed to the thermalized

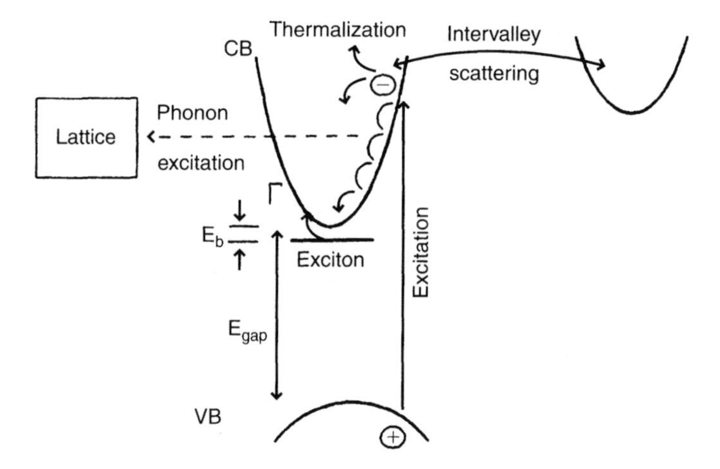

Figure 11.7 Simplified diagram of ultrafast processes occurring after above band gap excitation in semiconductors.

electronic system can exceed the lattice temperature by far. Depending on the band structure and photon energy, intervalley scattering can occur.

Energy is transferred to the lattice (heating) by inelastic electron–phonon collisions, and the carriers relax into states at the bottom of the band. The Fermi distribution which is finally reached, can be characterized by a temperature which is equivalent to the lattice temperature. If the excitation density is sufficiently high, a local change in the lattice temperature can readily be observed. Extremely high excitation can even result in melting. While the initial carrier scattering proceeds on a time scale of tens of fs or less, the intraband energy relaxation times can amount to a few ps.

11.4.2. Excitons

Another interesting feature of the excitation spectrum of solids is the exciton resonance. Excitons can be viewed as an electron–hole pair bound together through the Coulomb attraction, with properties similar to a hydrogen atom. Because of the positive Coulomb interaction, the corresponding energy levels are below the band gap (cf. Fig. 11.7). If the energy of the exciton is raised by an amount larger than the binding energy (E_b), the bound systems decay into a free electron and hole (exciton ionization). Such a process can be induced, for example, by longitudinal optical (LO) phonon scattering and typically proceeds on a time scale of about 100 fs in bulk materials at room temperature.

Owing to the strong excitonic oscillator strength and nonlinear susceptibilities, transient properties of excitons have attracted much attention. In particular in MQW structures, the exciton resonances can be clearly distinguished from the bulk absorption at room temperature. Figure 11.8 displays the absorption spectrum of a CdZnTe–ZnTe MQW and the results of a pump–probe experiment [31]. The pump spectrum was chosen to excite predominantly excitons. The differential transmission at the exciton resonance shows a fast increase and a partial recovery. Its dynamics can be explained by exciton excitation, exciton ionization because of LO-phonon scattering, and the presence of a coherent artifact. The increase of the transmission at photon energies, which probe the occupation of states at the bottom of the bands ($\lambda = 610$ nm), is a direct indication of the exciton ionization into free carriers. The characteristic ionization time was determined to be about 110 fs [31].

11.4.3. Intraband Relaxation

Intraband relaxation processes can conveniently be observed using pump–probe absorption techniques. A pump pulse of certain energy creates carriers at

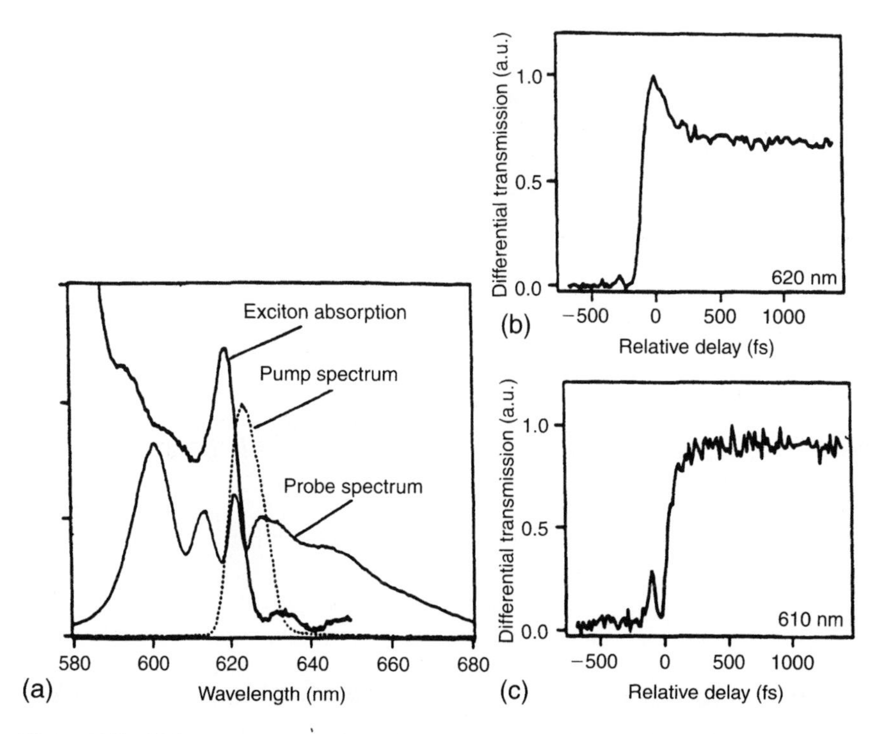

Figure 11.8 (a) Room temperature absorption spectrum of a CdZnTe–ZnTe MQW and the spectra of the 80-fs pump pulse and 14-fs probe pulse. The latter is a self-phase modulated and compressed part of the pump pulse. (b) Differential transmission at 620 nm and (c) 610 nm for a pump excitation level of 2×10^{11} carriers/cm^2. The wavelength filtering was done after the sample with a filter with a bandwidth of $\approx$8 nm (from Becker *et al.* [31]).

corresponding states above the band gap. Temporally delayed probe pulses of various frequencies test the occupation of states at different energies above the gap. The results of such an experiment for $Al_{0.2}Ga_{0.3}As$ are shown in Figure 11.9 [32]. A quantitative evaluation of the data is rather complicated, in view of the complexity of the processes involved in highly excited semiconductors. The interested reader is referred to the book by Haug and Koch [33]. Qualitatively, however, the time resolved transmission data follow a pattern consistent with the basic properties of the band model.

A rapid transmission change occurs not only at the excitation energy, but over a broader spectral range, indicating a thermalization within a time range significantly shorter than 100 fs. At 1.88 eV and 1.94 eV, a reduced change in transmission can be attributed to the cooling of the electronic system through energy transfer to the phonon system (lattice). This cooling results in a relaxation of carriers toward the bottom of the band, thus emptying higher energy states.

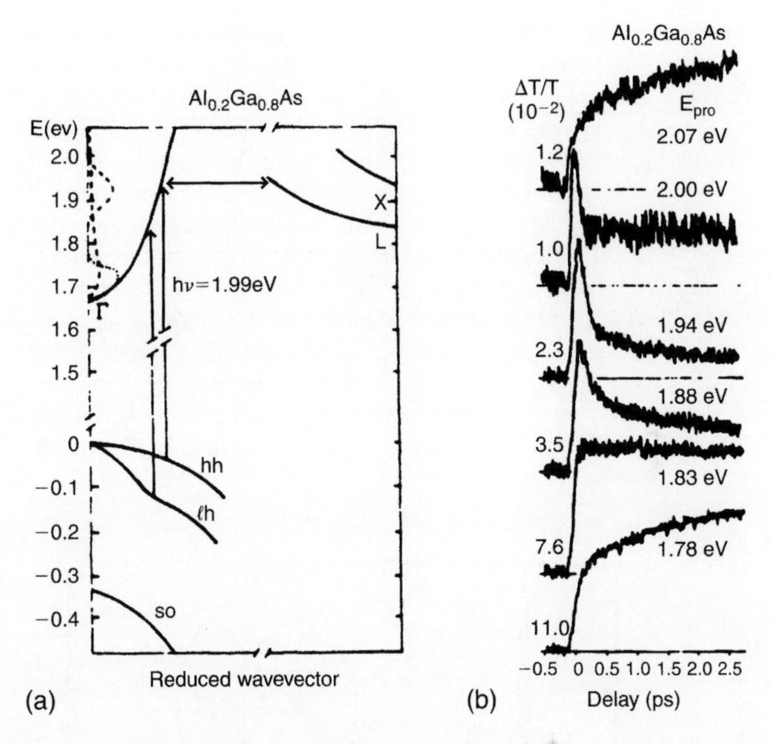

Figure 11.9 (a) Band structure for $Al_{0.2}Ga_{0.8}As$. The excitation photon energy being 1.99 eV, absorption from the light hole (ℓh) and heavy hole (hh) subbands into the Γ band are allowed. The carrier distribution is sketched for different instants (upper left corner). (b) Relative transmission change as seen by the test pulse for different photon energies and as function of the delay with respect to the pump pulse. (Adapted from Lin *et al.* [32].)

The increase of transmission at 1.78 eV accounts for the increase of occupied states at the band edge, with a characteristic time constant of 1 to 2 ps. The transmission features observed at 2.07 eV are explained by intervalley scattering ($\Gamma \leftrightarrow L$) and confirmed by additional probing of the split-off transition [32].

11.4.4. Phonon Dynamics

Phonons represent lattice vibrations. Just as in the case of molecular vibrations, they can be probed either by Raman techniques (frequency domain spectroscopy) or by ultrafast probing (time domain spectroscopy). The latter has the additional advantage of being able to retrieve not only the amplitude but also the phase of the vibration. A simple model for coherent excitation of phonons was introduced in Section 10.10. The phonon vibration of frequency ω_{phonon} is excited by pairs of spectral components of the pulse spectrum ω_1 and ω_2 such that $\omega_{phonon} = \omega_2 - \omega_1$. The measurement of such collective atomic motion in crystals can be performed in reflection as well as in transmission (see, for example, Kütt *et al.* [34] and references therein). The vibrations are observable through optical probing because the atomic displacements directly affect the band structure, and consequently the dielectric function through the deformation potential and electro-optic coupling. In addition, in polar crystals, direct excitation of phonons is possible by an electric field containing suitable frequency components.

Transient reflectivity measurements performed on GaAs are presented in Figure 11.10. A 50-fs pump pulse at 2 eV [34] is followed by an orthogonally polarized probe. The [010] crystal axis is oriented at an angle $\vartheta = 45°$ and 135° with respect to the probe polarization. After an initial peak, the reflectivity versus delay shows an oscillatory behavior, with a characteristic frequency of 8.8 THz that matches the frequency of the longitudinal (LO) phonons in GaAs.

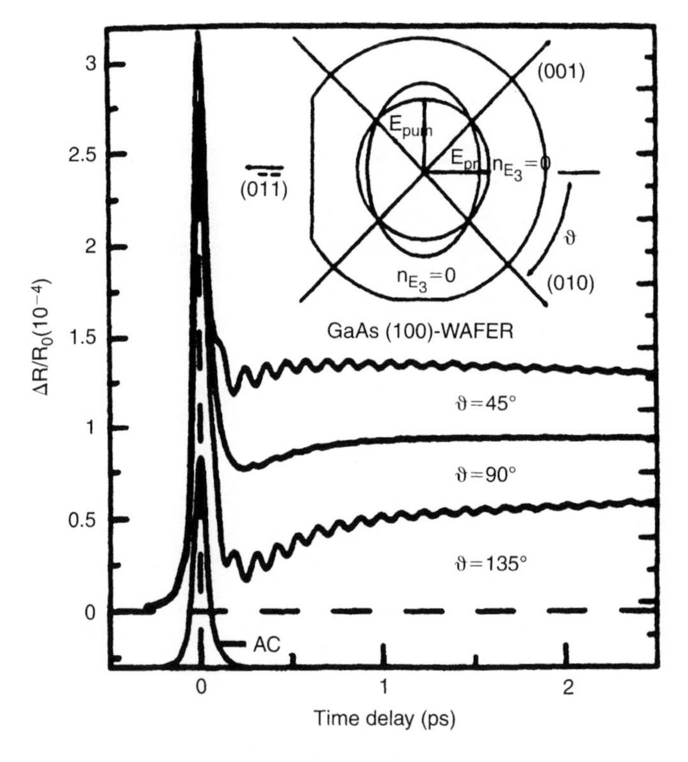

Figure 11.10 Relative reflection change of a [100] GaAs crystal excited by a 50-fs pulse at 2 eV. The generated carrier density is 10^{18} cm^{-3}. ϑ is the angle between the probe polarization and the [010] crystal axis. AC denotes the autocorrelation of the pump (probe) pulse (from [34]).

The dependence of the modulation amplitude on ϑ results from the electro-optic effect, which here is responsible for the phonon-induced reflectivity change.

11.4.5. Laser-Induced Surface Disordering

In a strongly absorbing material, the energy deposited in a small surface layer by a short light pulse can locally raise the temperature beyond the melting point. What is the response of matter to a δ-function impulse of energy, sufficient to cause melting? "How fast does melting occur?" is a question of fundamental interest, which involves changes in order and structure. In addition to the possibility of observing melting through "femtosecond photography" [35], nonlinear techniques sensitive to the material symmetry can be applied [36]. Results of an experiment to monitor changes in symmetry in GaAs during melting are shown in Figure 11.11. In GaAs, melting can be considered as a transition from a noncentrosymmetric material to an isotropic liquid. The second-order nonlinear susceptibility $\chi^{(2)}$ is therefore expected to change from a relatively large value (for the crystal) to (almost) zero (for the liquid) during the phase transition. This second-order susceptibility can be monitored by measuring the second harmonic in reflection generated by a delayed probe [38–40]. The reflectivity for the fundamental increases from the solid reflectance value to that of liquid GaAs with a characteristic time of about 1 ps. On the other hand, the SH signal drops substantially on a time scale of about 100 fs. These data suggest an intermediate state between the non-centrosymmetric crystal structure and the molten material. Note that a transition to a centrosymmetric crystal would only require a small

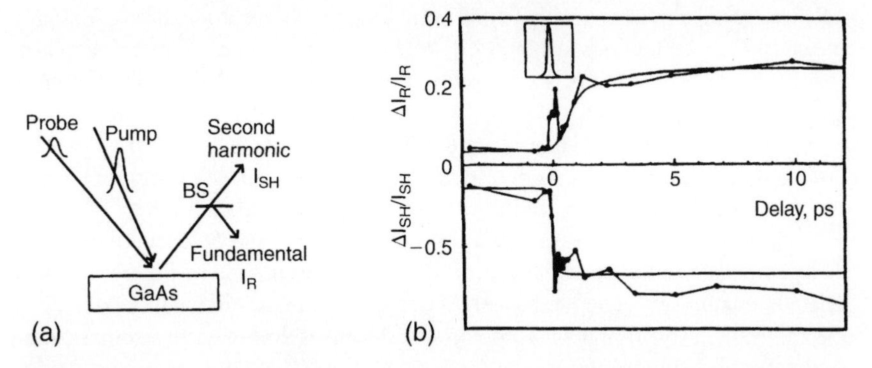

Figure 11.11 (a) Experimental setup to monitor ultrafast phase changes on a GaAs surface. A strong pump pulse induces the phase change. (b) The upper curve shows the reflectivity of a delayed probe. The lower curve is a plot of the second harmonic (in reflection) of the probe signal, which is a measure of the change in symmetry associated with the phase transition. (Adapted from Govorkov *et al.* [37].)

displacement of the atoms and substantially less energy than required for the actual bond breaking that occurs with melting.

With the availability of ultrafast X-ray sources, X-ray diffraction became a powerful new spectroscopic tool for time-resolved spectroscopy [41]. Because X-ray diffraction is sensitive to the crystal symmetry laser-induced phase changes can be probed directly by this technique [42].

11.5. PRIMARY STEPS IN PHOTO–BIOLOGICAL REACTIONS

In the progression of increasingly complex systems, we have come to the role of fs tools in analyzing the most complex biological systems. Two important biological problems connected to fs spectroscopy are photosynthesis and vision. In both cases, light energy is converted to biochemical energy, either for the purpose of energy storage–transfer or for the purpose of detection. The primary processes in the complex chain of reactions following light absorption, in vision, or photosynthesis, takes place on a fs time scale. The quantum yield of these ultrafast transformations is remarkably high—typically between 50 and 100%.

11.5.1. Femtosecond Isomerization of Rhodopsin

The primary process of vision takes place in rhodopsin, a pigment embedded in the membranes of specialized photoreceptor cells, the rod and cone cells of the retina. The role of the pigment is light absorption followed by a molecular conformational change, which leads eventually to a change in membrane potential. This change in electrical potential across the photoreceptor cells is eventually transmitted to the nervous system [43]. We are interested here in the primary process of vision, which is the isomerization of the pigment following absorption of a photon.

The pigment is a complex molecule called rhodopsin consisting of an "opsin" protein bound to the 11-cis form of retinal chromophore. The absorption band of rhodopsin peaks at 500 nm, which corresponds to the peak sensitivity of vision. This main absorption corresponds to a transition from the S_0 ground state to a S_1 excited state in the potential energy surface representation of Figure 11.12(a). The potential energy is plotted as a function of an angular torsional coordinate of the molecule. Absorption of a photon at 500 nm is followed by isomerization to the red-absorbing trans-isomer bathorhodopsin. The classical representation of the transformation is a "twist" of the chain [Fig. 11.12(b)]. The potential surfaces as a function of the corresponding coordinate angle have a minimum corresponding to the cis- and trans-configurations.

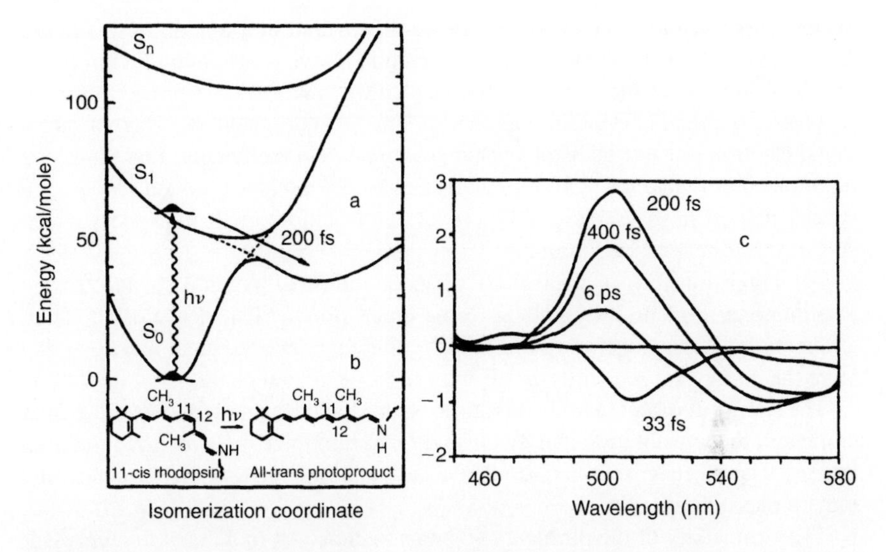

Figure 11.12 Schematic representation of the photo-isomerization reaction of rhodopsin. (a) Ground and excited state potential energy curves as a function of the torsional coordinate. The spectrum is red-shifted after absorption of a photon at 500 nm. The classical sketch of the cis-trans transformation is shown in (b). (c) Difference spectra measurements of 11-cis-rhodopsin at various delays following a 35-fs pump pulse at 500 nm ($\approx$10 fs probe) (from Schoenlein *et al.* [44]).

The quantum efficiency of this reaction is exceptionally high (0.67). The radiation lifetime of the excited state of rhodopsin is 10^{-8} s. The extinction coefficient is $6.4\ 10^4\ \mathrm{M}^{-1}\mathrm{cm}^{-1}$, a typical value for a strongly absorbing dye.

The reaction of photo-isomerization was studied through transient transmission spectroscopy through a jet of rhodopsin [45]. Adequate spectral selectivity was achieved with a pump pulse of 35-fs at 500 nm. A 10-fs probe pulse with a spectrum in the range of 450 nm to 570 nm was used. The differential transmission spectrum versus delay shown in Fig. 11.12 indicate disappearance of the 500 nm peak, and increased absorption at 530 nm, in the first 150 fs following excitation. The speed of that isomerization calls for a better classical representation of the cis- versus trans-configuration than Fig. 11.12(b). It is doubtful that the large motion of nuclei implied by the sketch could take place in a time as short as 100 fs.

11.5.2. Photosynthesis

Photosynthesis is the process by which plants convert solar energy into chemical energy. Its importance is obvious, because it is at the origin of life on our planet. This topic is too vast to be adequately covered in a section of this book. A general overview of the topic can be found in a review article by Fleming and van Grondelle [46], and in topical books [1,43].

There are pigment–protein complexes called reaction centers, where a directional electron transfer takes place across a biological membrane. Light harvesting molecules ("antenna" chlorophylls) transfer electronic excitation energy to a special pair (P in the sketch of Figure 11.13) of chlorophyll molecules, which acts as the primary electron donor.

The latter transfers an electron to a pheophytin (H_A) within 3 ps, and from it to a quinone (Q_A) in 200 ps, then to the other quinone Q_B, hence establishing a potential difference across a biological membrane. Biochemical reactions that store the energy subsequently occur with these separated charges.

The energy dissipation in the first processes should be small (about 0.25 eV) as compared to the excitation energy (1.38 eV), to minimize the waste of excitation energy. The electron transfer should be fast to compete with fluorescence and radiationless decay.

The complexity of the problem can be appreciated by looking at the representation of the molecular structure of a bacterium's photosynthetic reaction center, which was determined to atomic resolution by Deisenhofer and Michel [46,48]. A block diagram of the electron-carrying pigments in the reaction center is shown in Fig. 11.13.

Recent transient absorption experiments have concentrated on the fast initial electron transfers [49]. In a model proposed by Zinth *et al.* [47,49], the bacteriochlorophyll anion B_A^- is created in the first 3-ps reaction. The subsequent

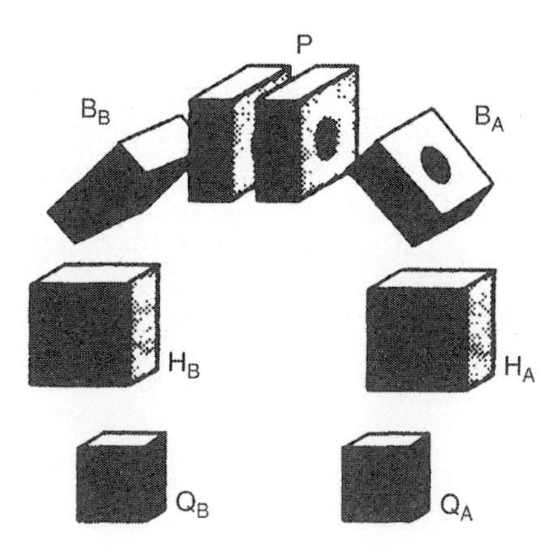

Figure 11.13 Sketch of the molecular arrangement of the four bacteriochlorophylls (P, B_A, B_B), the two bacteriopheophytins (H_A, H_B), and the two quinones (Q_A, Q_B) in reaction centers (from Zinth *et al.* [47]).

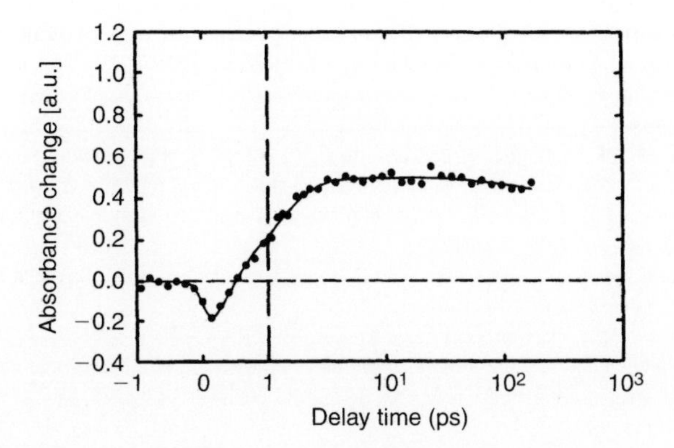

Figure 11.14 Transient absorption changes for native reaction centers at 1.02 μm associated with the photosynthesis. The large initial absorption increase has a time constant of 0.9 ps. It is related to the absorption change due to the formation of a bacteriochlorophyll anion (from Zinth *et al.* [49]).

electron transfer to the bacteriopheophytin H_A is faster, taking only 0.9 ps. In the experiment, after the main absorption band of the pigment is pumped, a probe is sent in the near IR, where the bacteriochlorophyll anions ($P^+B_A^-$) have a strong absorption. The transient absorption change at 1.02 μm is shown in Figure 11.14. The 0.9 ps time constant would correspond to the electron transfer from the bacteriochlorophyll to the bacteriopheophytin ($P^+H_A^-$).

BIBLIOGRAPHY

[1] W. Kaiser and D. H. Auston, Eds. *Ultrashort laser pulses*, volume 60, Springer-Verlag, Berlin, Germany, 1988.

[2] F. C. De Schryver, S. De Feyter, and G. Schweitzer (Eds.). *Femtochemistry: With the Nobel Lecture of A. Zewail*. John Wiley & Sons, New York, 2001.

[3] J. Shah. *Ultrafast Spectroscopy of Semiconductors and Semiconductor Nanostructures*. Springer, Berlin, Germany, 1999.

[4] S. Mukamel. *Principles of Nonlinear Optical Spectroscopy*. Oxford University Press, New York, 1995.

[5] *Ultrafast Phenomena*. Biannual conference, proceedings, Springer .

[6] C. Cohen-Tannoudji, B. Diu, and F. Laloe. *Quantum Mechanics*. John Wiley & Sons, New York, 1977.

[7] C. R. Stroud. The classical limit of an atom. In W. T. Grandy and P. W. Milonni, Eds. *Physics and Probability: A symposium in honor of E. T. Jaynes*, Cambridge University Press, Cambridge, MA, 1993, pp. 1–9.

[8] A. ten Wolde, L. D. Noordam, H. G. Muller, A. Lagendijk, H. B. van Linden, and van den Heuvel. Observation of radially localized atomic wave packets. *Phys. Rev. Lett.*, 61:2099–2102, 1988.

[9] J. A. Yeazell and C. R. Stroud Jr. Observation of fractional revivals in the evolution of a Rydberg atomic wave packet. *Phys. Rev. A*, 4:5153–5156, 1991.
[10] G. Alber and P. Zoller. Laser-induced excitation of electronic Rydberg wave packets. *Contemporary Physics*, 32:185–189, 1991.
[11] J. A. Yeazell, M. Mallalieu, J. Parker, and C. R. Stroud Jr. Observation of the collapse and revival of a Rydberg electronic wave packet. *Phys. Rev. Lett.*, 64:2007–2010, 1990.
[12] J. A. Yeazell and C. R. Stroud Jr. Observation of spatially localized atomic electron wave packets. *Phys. Rev. Lett.*, 60:1494–1497, 1988.
[13] Z. D. Gaeta, M. Noel, and C. R. Stroud. Excitation of the classical-limit state of an atom. *Phys. Rev. Lett.*, 73:636–639, 1994.
[14] N. F. Scherer, R. J. Carlson, A. Matro, M. Du, A. J. Ruggiero, V. Romero-Rochin, J. A. Cina, G. F. Fleming, and S. A. Rice. Fluorescence-detected wave packet interferometry: Time resolved molecular spectroscopy with sequences of femtosecond phase-locked pulses. *J. Chem. Phys.*, 95:1487–1511, 1991.
[15] R. M. Bowman, M. Dantus, and A. H. Zewail. Femtosecond transition state spectroscopy of iodine: From strongly bound to repulsive surface dynamics. *Chem. Phys. Lett.*, 161:297–302, 1989.
[16] F. W. Wise, M. J. Rosker, and C. L. Tang. Oscillatory femtosecond relaxation of photoexcited organic molecules. *J. Chem. Phys.*, 86:2827–2832, 1987.
[17] T. Baumert, R. Thalweiser, and G. Gerber. Femtosecond two-photon ionization spectroscopy of the B state of Na_3 clusters. *Chem. Phys. Lett.*, 209:29–32, 1993.
[18] C. H. Brito-Cruz, R. L. Fork, W. H. Knox, and C. V. Shank. Spectral hole burning in large molecules probed with 10 fs optical pulses. *Chem. Phys. Lett.*, 132:341–344, 1986.
[19] M. J. Rosker, M. Dantus, and A. H. Zewail. Femtosecond real-time probing of reactions, I. The technique. *J. Chem. Phys.*, 89:6113–6127, 1988.
[20] A. H. Zewail. Laser femtochemistry. *Science*, 242:1645–1653, 1988.
[21] R. B. Bernstein and A. H. Zewail. Femtosecond real-time probing of reactions, III. Inversion to the potential from femtosecond transition-state spectroscopy experiments. *J. Chem. Phys.*, 90:829–842, 1989.
[22] A. H. Zewail. The birth of molecules. *Scientific American*, 262:76–82, 1990.
[23] P. Cong, A. Mokhtari, and A. H. Zewail. Femtosecond probing of persistent wave packet motion in dissociative reactions: Up to 40 ps. *Chemical Phys. Lett.*, 172:109–113, 1990.
[24] J. R. Grover and E. A. Walters. Optimization of weak neutral dimers in nozzle beams. *J. Phys. Chem.*, 90:6201–6206, 1986.
[25] J. R. Grover, E. A. Walters, D. L. Arneberg, and C. J. Santandrea. Competitive production of weakly bound heterodimers in free jet expansions. *Chem. Phys. Lett.*, 146:305–307, 1988.
[26] N. F. Scherer, L. R. Khundkar, R. B. Bernstein, and A. H. Zewail. Real-time picosecond clocking of the collision complex in a bimolecular reaction: The birth of OH from $H + CO_2$. *J. Chem. Phys.*, 87:1451–1453, 1987.
[27] E. D. Potter, J. L. Herek, S. Petersen, Q. Liu, and A. H. Zewail. Femtosecond laser control of a chemical reaction. *Nature*, 355:66–68, 1992.
[28] Y. Yan, R. M. Whitnell, K. R. Wilson, and A. H. Zewail. Femtosecond chemical dynamics in solution: Wavepacket evolution and caging of I_2. *Chem. Phys. Lett.*, 193:402–412, 1992.
[29] I. Benjamin and K. R. Wilson. Proposed experimental probes of chemical reaction molecular dynamics in solution: ICN photodissociation. *J. Chem. Phys.*, 90:4176–4197, 1989.
[30] L. L. Lee, Y. S. Li, and K. R. Wilson. Reaction dynamics from liquid structure. *J. Chem. Phys.*, 95:2458–2464, 1991.
[31] P. C. Becker, D. Lee, M. R. Xavier de Barros, A. M. Johnson, A. G. Prosser, R. D. Feldman, R. F. Austin, and R. E. Behringer. Femtosecond dynamic exciton bleaching in room temperature II-VI quantum wells. *IEEE J. of Quantum Electron.*, 28:2535–2542, 1992.

[32] W. Z. Lin, R. W. Schoenlein, J. G. Fujimoto, and E. P. Ippen. Femtosecond absorption saturation studies of hot carriers in GaAs and AlGaAs. *IEEE J. of Quantum Electron.*, 24:267–275, 1988.

[33] H. Haug and S. W. Koch. *Quantum theory of the optical and electronic properties of semiconductors*. World Scientific, Singapore, 1990.

[34] W. Kütt, W. Albrecht, and H. Kurz. Generation of coherent phonons in condensed media. *IEEE J. of Quantum Electron.*, 28:2434–2444, 1992.

[35] M. C. Downer, R. L. Fork, and C. V. Shank. Femtosecond imaging of melting and evaporation of a photoexcited silicon surface. *J. of Optical Soc. Am. B*, 2:595–599, 1985.

[36] N. Bloembergen, A. M. Malvezzi, and J. M. Lin. Second harmonic generation in reflection from crystalline GaAs under intense picosecond laser irradiation. *Appl. Phys. Lett.*, 45:1019–1021, 1984.

[37] S. V. Govorkov, I. L. Shumay, W. Rudolph, and T. Schroeder. Time-resolved second-harmonic study of femtosecond laser-induced disordering of GaAs surfaces. *Optics Lett.*, 16:1013–1015, 1991.

[38] P. Saeta, J. K. Wang, Y. Siegal, and N. Bloembergen. Ultrafast electronic disordering during femtosecond laser melting of GaAs. *Phys. Rev. Lett.*, 67:1023–1025, 1991.

[39] K. Sokolowski-Tinten, H. Schultz, J. Bialkowski, and D. von der Linde. Two distinct transitions in ultrafast solid-liquid phase transformations of GaAs. *Appl. Phys.*, A53:227–234, 1991.

[40] T. Schroeder, W. Rudolph, S. V. Govorkov, and I. L. Shumay. Femtosecond laser-induced melting of GaAs probed by optical second-harmonic generation. *Appl. Phys.*, A51:49–51, 1990.

[41] C. Rischel, A. Rousse, I. Uschmann, P. A. Albouy, J. P. Geindre, P. Audebert, J. C. Gauthier, E. Forster, J. L. Martin, and A. Antonetti. Femtosecond time-resolved x-ray diffraction from laser-heated organic films. *Nature*, 390:490–492, 1997.

[42] S. H. Chin, R. W. Schoenlein, T. E. Glover, P. Balling, W. P. Leemans, and C. V. Shank. Ultrafast structural dynamics in InSb probed by time-resolved x-ray diffraction. *Phys. Rev. Lett.*, 83:336–339, 1999.

[43] R. R. Alfano, Ed. *Biological Events Probed by Ultrafast Laser Spectroscopy*. Academic Press, New York, 1982, and M. M. Martin and J. T. Hynes, eds. *Femtochemistry and Femtobiology*. Elsevier, Amsterdam, 2004.

[44] R. W. Schoenlein, L. A. Peteanu, R. A. Mathies, and C. V. Shank. The first step in vision: femtosecond isomerization of rhodopsin. *Science*, 254:412–415, 1991.

[45] R. W. Schoenlein, L.-A. Poteanis, R. A. Mathias, and C. V. Shank. Femtosecond isomerization of rhodopsin. *Ultrafast Phenomena in Spectroscopy*, UPS'91, October 1991. Bayreuth, Germany.

[46] G. R. Fleming and R. van Grondelle. The primary steps of photosynthesis. *Physics Today*, 47:48–55, 1994.

[47] W. Zinth, C. Lauterwasser, U. Finkele, P. Hamm, S. Schmidt, and W. Kaiser. Molecular processes in the primary reaction of photosynthetic reaction centers. In J.-L. Martin, A. Migus, G. A. Mourou, and A. H. Zewail, Eds., *Ultrafast Phenomena VIII*, Springer-Verlag, Berlin, Germany, 1994, pp. 535–538.

[48] J. Deisenhofer and H. Michel. The photosynthetic reaction center from the purple bakterium rhodopseudomonas viridis. *EMBO Journal*, 8:2419–2170, 1989.

[49] W. Zinth, S. Schmidt, T. Arlt, J. Wachtveitl, H. Huber, T. Naegele, M. Meyer, and H. Scheer. Direct observation of the accessory bacteriochlorophyll in the primary electron transfer in bacterial reaction centers. In G. A. Mourou and A. H. Zewail, Eds., *Ultrafast Phenomena IX*, Springer-Verlag, Berlin, Germany, 1994, pp. 437–438.

12

Generation of Extreme Wavelengths

Presently, the generation of fs light pulses in lasers covers a spectral range from the UV to the NIR. Figure 12.1 gives an overview on laser materials matched to specific wavelength ranges. Nonlinear optical processes such as sum and difference frequency generation and parametric generation are typically used to extend the spectral range covered by laser materials.

Beyond the traditional nonlinear optics, new techniques have emerged leading to electromagnetic pulses of extremely short (X-ray) and extremely long wavelengths (far infrared or FIR). These novel sources in turn opened completely new application fields, for example, time-resolved X-ray spectroscopy, short pulse (single-cycle) radar, and FIR coherent spectroscopy. Femtosecond light pulses can generate the shortest electrical pulses, which in turn serves to characterize the fastest electronic components where purely electronic means must fail. Moreover, with fs optical pulses short acoustic pulses can be produced and launched into materials for diagnostic purposes. The geometrical length of these pulses is only a few nm. In this chapter we will discuss the basic principles leading to the formation of pulses at these extreme carrier frequencies. The relevant processes are a fascinating example of the complexity of the interaction of ultrashort light pulses with matter. From a naive view point all four types of pulses arise simply when a fs pulse is focused onto a solid surface. Another area where fs light pulses help to push the limits of our present knowledge is the physics of extremely intense electromagnetic fields (larger than the atomic field strengths), which can be reached at the focus of high power pulses.

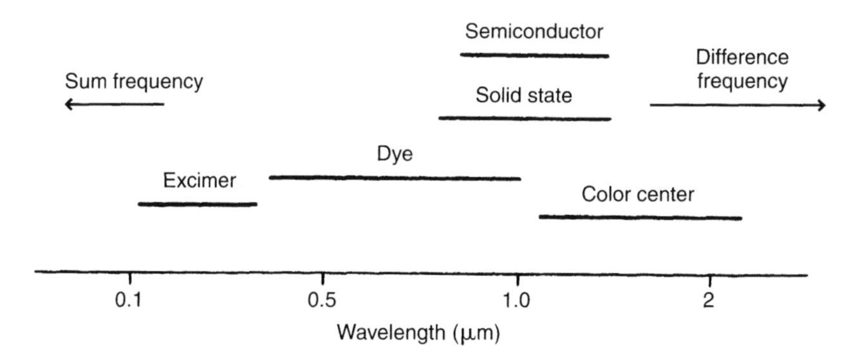

Figure 12.1 Femtosecond pulse generation in different wavelength regions.

12.1. GENERATION OF TERAHERTZ (THz) RADIATION

Two mechanisms can lead to terahertz (THz) radiation through focalization of an ultrashort light pulse onto a sample: (a) optical rectification and (b) a radiative current transient. In the original experiment on optical rectification using femtosecond light pulses Auston *et al.* [1] demonstrated the generation of THz waves through Cherenkov radiation. A fs pulse with an energy of about 100 pJ was focused into a $LiTaO_3$ crystal (Figure 12.2). The optical pulse propagating through the crystal produces a polarization pulse via optical rectification. The latter is a second-order nonlinear optical process that occurs simultaneously with SHG. Indeed, an instantaneous polarization quadratic in the electric field is:

$$P^{(2)} = \epsilon_0 \chi^{(2)} \mathcal{E}^2(t) \cos^2[\omega_\ell t + \varphi(t)]$$

$$= \frac{1}{2} \epsilon_0 \chi^{(2)} \mathcal{E}^2(t) \{\cos 2[\omega_\ell t + \varphi(t)] + 1\}, \tag{12.1}$$

which includes a SH term centered at $2\omega_\ell$ and a dc field of the same amplitude centered at zero frequency. Optical rectification can also be understood as difference frequency generation. A frequency domain form of Eq. (12.1) is

$$P^{(2)}(\omega_d = \omega_1 - \omega_2) = \epsilon_0 \chi^{(2)} E(\omega_1) E(\omega_2) \tag{12.2}$$

where ω_1 and ω_2 can be any two frequencies from the pulse spectrum. Thus, the difference frequency ω_d can cover a spectral range from zero to several THz, which corresponds to far infrared (FIR) radiation. For a Gaussian pulse with a

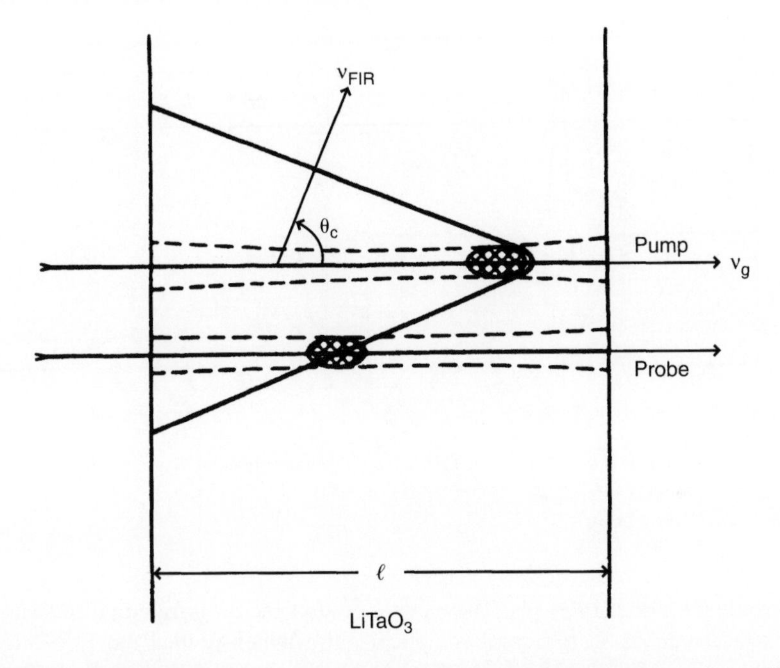

Figure 12.2 Schematic diagram for the generation of terahertz radiation through optical rectification. (Adapted from Auston *et al.* [1].) A time-delayed optical pulse sampled the FIR field by probing the induced birefringence.

temporal intensity profile $I(t) \propto \exp[-4\ln 2(t/\tau_p)^2]$, the THz radiation can be described as a single-cycle infrared radiation of frequency $\sqrt{\ln 2}(2\tau_p)^{-1}$ [2].

The optical rectification pulse is basically an electric dipole field that moves with the group velocity of the optical pulse. At a given position along its path in the crystal, this dipole generates a field that travels with the group velocity ν_{FIR} associated with its low carrier frequency (of the order of the inverse pulse duration τ_p^{-1}). For $LiTaO_3$, $\nu_{FIR} \approx 0.153c$, which is rather low because (quasi-resonant) lattice vibrations contribute substantially to the dispersion behavior in the FIR spectral range. Because the group velocity of the optical pulse is $\nu_g \approx 0.433c$, we have an interesting situation where the source velocity is larger than the velocity of the emitted wave. This condition leads to the formation of an electromagnetic shock wave (Cherenkov) radiation propagating on a conical surface in the crystal. The characteristic angle between the surface normal of the cone and its symmetry axis (propagation direction of the optical pulse) is

$$\cos\theta_c = \frac{\nu_{FIR}}{\nu_g} \tag{12.3}$$

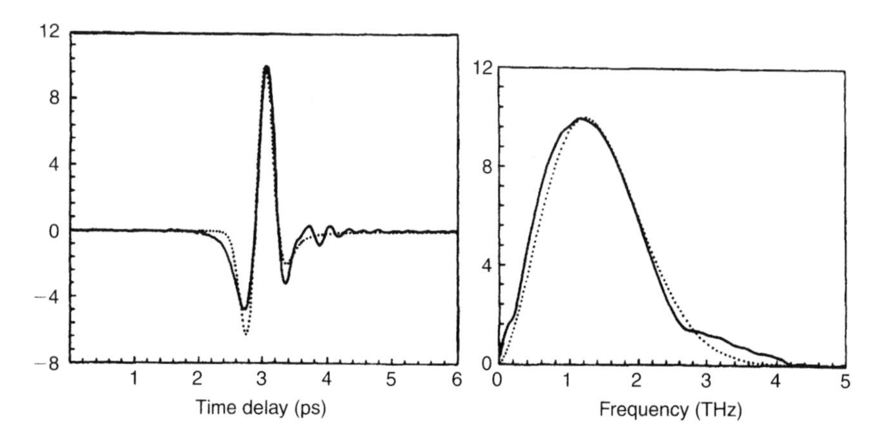

Figure 12.3 Temporal behavior of the THz field and the corresponding Fourier spectrum obtained from optical rectification (Adapted from Auston *et al.* [1]).

which is 69° for $LiTaO_3$ [1]. The electric field of the far infrared pulse transient was measured using a second fs optical pulse which probed the FIR-induced birefringence. By varying the time delay between pump and probe pulse the FIR field can be sampled. This detection gives a complete recording of the FIR waveform—hence, the FIR radiation is completely determined in amplitude and phase. Figure 12.3 displays the temporal behavior of the FIR field as well as its Fourier spectrum. If the (external) angle of incidence of the optical pulse is chosen to be $\alpha = 51°$, a portion of the Cherenkov cone propagates normally to the crystal surface and thus can propagate into free space (air) [3]. In this manner the crystal acts like an emitter for THz radiation.

Instead of a moving dipole in a dielectric medium, an ultrashort electrical pulse on a coplanar transmission line can also serve as source for THz radiation, as reported by Fattinger and Grischkowsky [4]. The physical situation is sketched in Figure 12.4. The transmission line consisted of two 5-μm wide, 5-μm thick aluminum lines separated by 10 μm. The substrate was heavily implanted silicon on sapphire to ensure a short carrier lifetime of about 600 fs [5]. The latter is crucial for generating the short electrical transients and for a short response time of the detection switch.

More recently even simpler techniques turned out to be effective means to generate THz radiation. These include biased metal–semiconductor interfaces [5] and semiconductor surfaces between biased metal electrodes [6], which is similar to a technique already used by Mourou *et al.* [7] to produce ps microwave pulses. Their common operational principle rests on the production of a fast carrier transient in an external bias field, following the optical excitation.

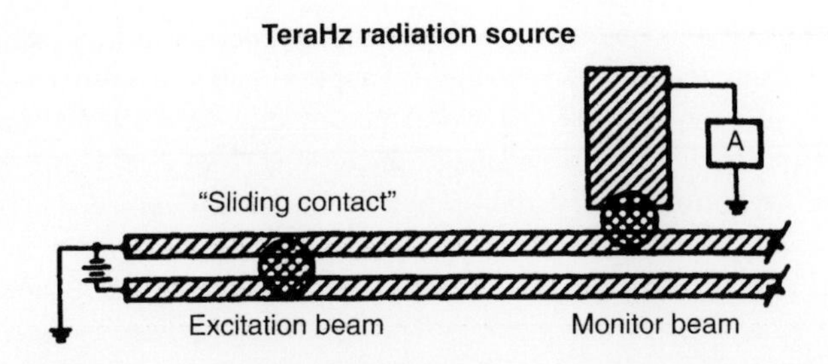

Figure 12.4 THz radiation through ultrafast switching of a charged coplanar transmission line. A fs pulse shortens the transmission line creating a subps electrical pulse. Its transients after propagation over a certain length L can be measured by a time-delayed second pulse that excites a fast photoconductive switch (Adapted from Fattinger and Grischkowsky [4]).

The rise time is roughly given by the carrier generation process, i.e., by the optical pulse duration. The fall time is limited by the finite transit time of the carriers across the region of the electric field and by the carrier lifetime, whichever is shorter.

Surprisingly, semiconductor surfaces excited above the band gap and without external bias voltage were found to be emitters of THz radiation, too [8]. A static built-in field normal to the semiconductor air surface acts as bias and may drive the carrier current. The origin of this field is a downward bending of the conduction and valence band creating a charge depletion region near the semiconductor surface. The generation of photocarriers in this layer then initiates an electron and hole current in opposite directions.

A coherent contribution to this FIR generation process on a fs time scale was recently suggested and analyzed [9]. The effect relies on the formation of an instantaneous nonlinear polarization because of excited electron–hole pairs and the coherent evolution of this system. These coherent effects—as is the case for any type of coherent light–matter interactions, (cf. Chapter 4)—play an important role if the Rabi frequency becomes comparable or larger than the dephasing rate T_2^{-1}. Therefore, the coherent processes are expected to produce a significant contribution to the THz emission process when short and/or powerful optical pulses are used [9]. The FIR radiation originating from optical rectification, as mentioned, is another example of a coherent generation process.

The sources of FIR radiation described so far are mainly point sources. An extended source (area of linear dimensions that are large compared with the THz wavelength) has been demonstrated by sending fs pulses of large energy onto a large area biased semiconductor [10]. Such an extended source has interesting far- and near-field properties. The radiation is coherently emitted from an area of cm dimensions and has diffraction properties similar to a Gaussian beam of cm beam waist at a wavelength of approximately 0.3 mm (Raleigh range of the order of 1 meter). The inconvenience of this source is the large optical power (distributed over the large area of the emitter) required. However, because of the large excitation energy and area, the emitted FIR power is considerably larger than with the point source devices [9, 10]. The THz pulses can therefore be detected directly with bolometric structures. Pyroelectric detectors can even be used for nJ and larger energies [9]. The bandwidth limitation associated with the photoconducting receiver structures is eliminated by recording THz pulse interferograms directly. To this aim, the FIR radiation is sent into a Michelson interferometer, and the output signal is measured as a function of length detuning. A corresponding interferogram of a THz pulse with a field FWHM as short as 85 fs is shown in Figure 12.5.

The large majority of the THz emitters can be considered as point sources (dimensions $\ll$ wavelength). For most applications, it is desirable to collimate the emitted waves with a system of lenses or mirrors, propagate the beam over long distances, and finally detect it. A simple means of collimating the beam is to attach a lens, for example a silicon lens, directly to the radiation source [3,11,12]. Collimation of the point source radiation was achieved by a combination of silicon lenses and parabolic mirrors, [11, 13] as sketched in Figure 12.6. The transmitted radiation is focused by a combination of parabolic mirror and lens

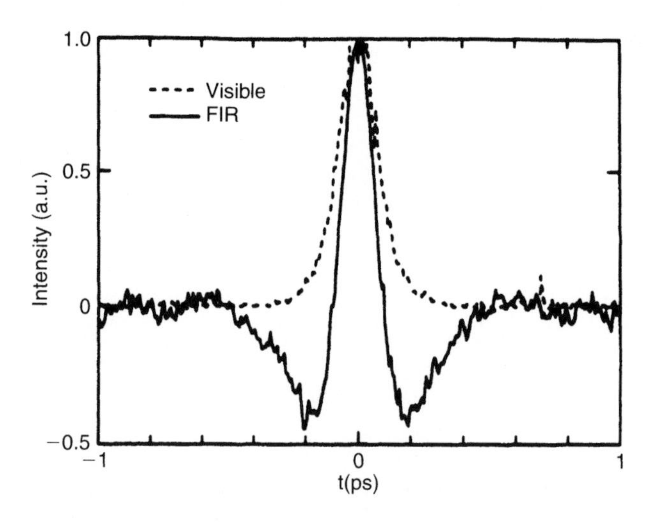

Figure 12.5 Second-order autocorrelation of the visible excitation pulse (broken line) and interferogram of the FIR pulse (solid line). The THz pulse was produced from an unbiased, n-type InP (111) wafer doped at 10^{17} cm^{-3} (Adapted from Green *et al.* [9]).

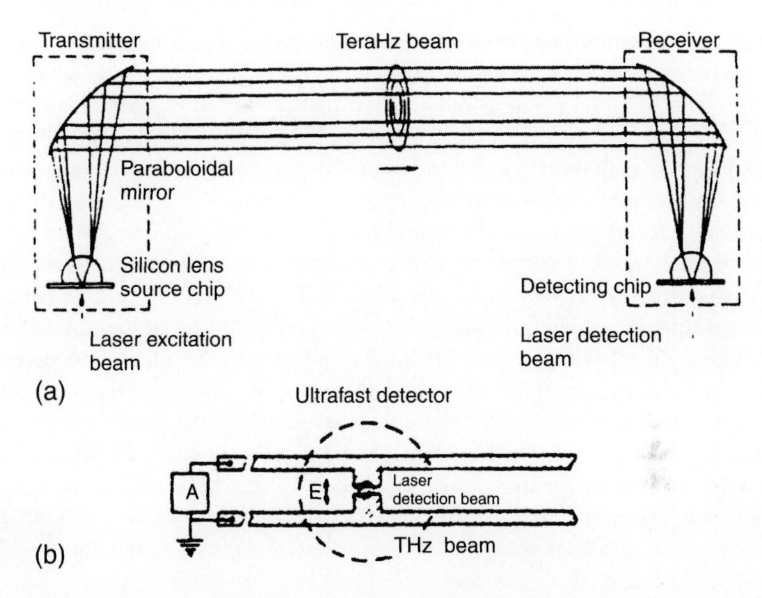

Figure 12.6 (a) Setup for generating and transmitting ultrashort THz radiation. (b) Detector for the time resolved measurement of the THz field (Adapted from van Exter and Grischkowsky [11]).

onto a photoconducting antenna, which produces a transient bias voltage across the 5-μm gap [Fig. 12.6(b)]. The transient voltage is gated by a photoconductive switch driven by a time-delayed optical pulse. The recorded average current versus delay is a convolution among the FIR field, the optical pulse, and the response of the receiver. It is the last that limits the temporal resolution of the detection system.

An interesting application of the THz source is coherent rotational spectroscopy [14]. The THz radiation is propagated through an 88-cm long sample cell. The time domain spectrometer has a transmitter and a receiver part which are replicas of each other. The rotational spectrum of N_2O between 0 and 1 THz consists essentially of regularly spaced narrow lines [14]. The rotational frequencies are given by:

$$\nu_R = 2B(J+1) - 4D(J+1)^2 \tag{12.4}$$

where J is the rotational quantum number, B the rotational constant, and D the centrifugal stretching constant. As many as 70 transitions can be excited simultaneously by the THz pulse. Bloch's equations (cf. Chapter 4) apply to this system. They can be simplified assuming small pulse area and a long T_2 time.

Indeed, at a pressure of 600 torr (which corresponds to an optical thickness of the sample $\alpha d = 1.2$), the dephasing time because of collisions is 65 ps, thus much longer than the exciting pulse. Each of the excited molecules, for a particular J value, acts as a vibrating dipole. These dipoles—nearly equally spaced in frequency—radiate in phase following the pulsed excitation (free induction decay), resulting in a train of ultrashort pulses (Figure 12.7). The directly transmitted pulse is followed by a series of THz pulses at a repetition rate of 25.1 GHz, equal to the frequency separation between adjacent lines [$2B$ in Eq. (12.4) if the anharmonicity is negligibly small], as shown in Fig. 12.7(a). Because the incident and transmitted fields are measured with high precision (the signal-to-noise ratio in this experiment is better than 20,000), in amplitude and phase (the detection measures the field magnitude rather than the radiation intensity), accurate fit with the theory is possible [Fig. 12.7(b) as compared to Fig. 12.7(a)]. In addition to a verification of the rotational constant B, the data lead to an improved determination of the centrifugal stretching constant $D \approx 5.28$ kHz, and the interatomic distances in the molecule (N $\leftrightarrow$ N $\approx$ 1.125 Å; and N $\leftrightarrow$ O $\approx$ 1.191 Å) [14]. An example of measurement and fitting of successive pulses in the train is shown in Fig. 12.7(c) and (d).

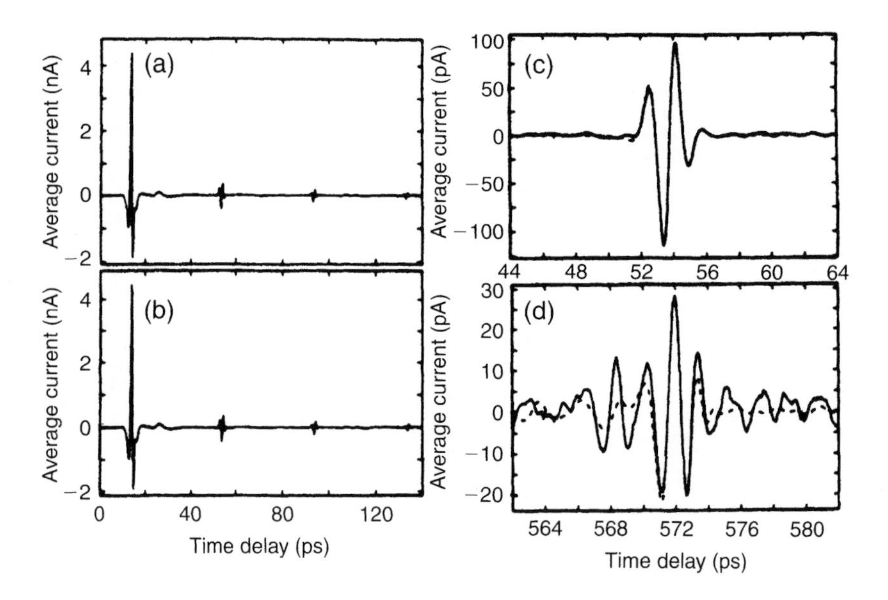

Figure 12.7 Coherent THz rotational spectroscopy of N_2O. THz radiation transmitted (a) and calculated (b) through N_2O. Measurement (solid line) and calculation (dashed lines) of (c) the first, and (d) the 14th radiated coherent pulse (Adapted from Harde and Grischkowsky [14]).

12.2. GENERATION OF ULTRAFAST X-RAY PULSES

12.2.1. Incoherent Bursts of X-Rays

When a powerful light pulse is focused onto a solid, a high-temperature plasma is produced, which can emit X-rays. Stimulated by laser fusion research, the physics of laser induced X-ray generation has been extensively investigated. The availability of high energy (milliJoule) tabletop lasers has made laser-produced plasmas a unique and practical source of ultrashort X-ray pulses [15–21]. A diagram of optical pulse induced X-ray production is sketched in Figure 12.8(a). In the absorption process, which takes place in a thin surface layer, the optical energy is initially transferred to the electronic system. Depending on the target and the laser pulse intensity multiphoton processes ionize the material, and the plasma is subsequently heated. The cooling of the laser-produced plasma is accompanied by the emission of a burst of X-rays. Figure 12.8(b) shows the time integrated spectral distribution from a Si target as was obtained by Murnane *et al.* [22].

Several processes contribute to the X-ray emission spectrum:

- the emission of distinct lines resulting from transitions between inner subshells of the ions,
- broadband emission from the recombination of unbound (free) electrons and ions, and
- broadband emission consisting of bremsstrahlung radiation.

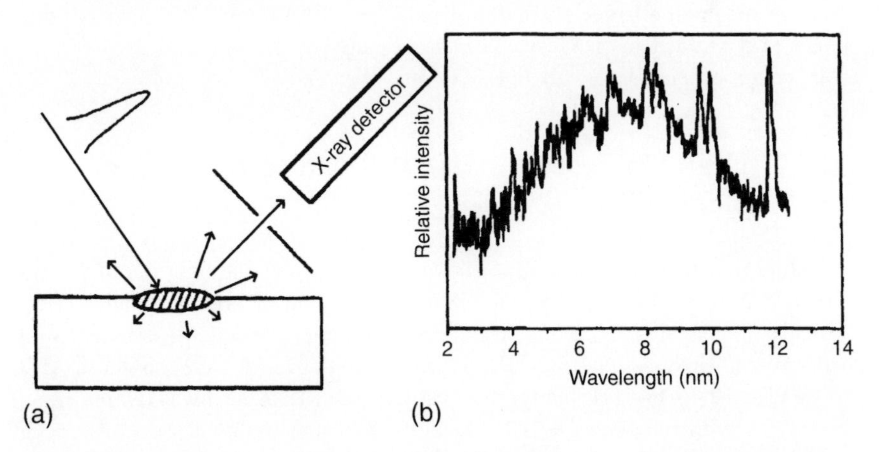

Figure 12.8 (a) Sketch of laser-generated X-ray plasmas. (b) Time integrated soft X-ray emission. Pulses of 160-fs duration and 5 mJ of energy were focused onto an Si target at an intensity of 10^{16} W/cm^2 (Adapted from Murnane *et al.* [22]).

There are substantial differences between the plasma generation with pulses on the order of ps or longer and fs pulses. The physical difference is related to the time scale of the plasma expansion for a distance equal to the absorption length, into the surrounding vacuum. Expansion velocities are comparable with the sound velocity, typically about 0.1 nm/fs, while the absorption length of metals is of the order of 10 nm. The energy will thus be deposited in a time shorter than the expansion time with a 100-fs pulse. Consequently a fs pulse interacts with a solid or near-solid density plasma. With ps and longer pulses, the leading part generates an expanding plasma that creates a density gradient above the target surface. The later parts of the pulse are absorbed in this low density plasma up to a density at which the electron plasma frequency equals the laser frequency.

The duration of the emitted X-ray burst follows roughly the duration of the optical excitation pulse if it is of ps or longer duration. To obtain sub-ps X-ray pulses, fs excitation pulses are needed to produce a fast enough rise of the electron temperature, and the plasma must cool rapidly to terminate the emission. While the former is responsible for the rise time of the X-ray pulse, it is the latter that will determine its decay characteristics. There are a number of processes which allow a rapid cooling of the plasma if excited to high densities [22]. Among them are high thermal and pressure gradients which drive the hot electrons into the bulk and expand the plasma into the vacuum. Moreover, because the electron temperature exceeds the temperature of the atoms and ions by far, inelastic collisions between them and the electrons are an efficient cooling mechanism. Using 160-fs excitation pulses, X-ray pulses as short as ~ 1.1 ps could be measured [22], where this value represents the detector limit of the X-ray streak camera. While the corresponding spectrum was in the soft X-ray range, a two order of magnitude larger excitation density ($\sim 10^{18}$ W/cm^2) and a heavy metal target (Ta) yielded hard X-rays in a range between 20 keV and 2 MeV [18]. Conversion efficiencies of about 0.3% were observed.

12.2.2. High Harmonics (HH) and Attosecond Pulse Generation

While the techniques described above lead to incoherent emission of X-rays, coherent X-rays can be generated via the production of high harmonics (HH). Already in the 1970s, using CO_2 laser radiation at intensities exceeding 10^{14} W/cm^2 and Al targets, HH up to order $n = 11$ were observed [23]. In the late 1980s the 17th harmonic of a KrF laser (248 nm) was demonstrated in Ne gas with intensities $> 10^{15}$ W/cm^2 [24], and harmonics up to $n = 33$ were observed in Ne with fundamental laser radiation at 1.06 μm [25]. These experiments demonstrated the potential of HH generation to enter the soft X-ray spectral range. Real breakthroughs of this technique were made in the 1990s when

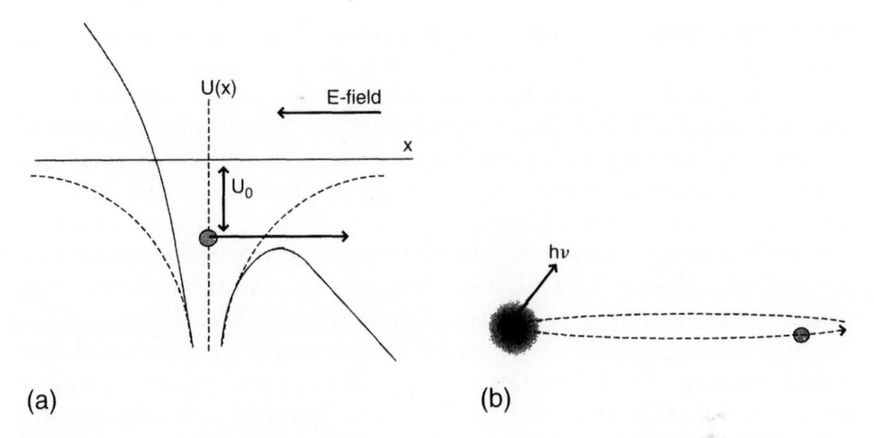

Figure 12.9 (a) Above threshold ionization of an atom. The atomic Coulomb potential (dashed line), $U(x)$, is bent (solid line) because of the electric field of the pulse, leading to ionization. (b) The electron trajectory in the vicinity of the ion caused by the electric field of the pulse. Recombination with the ion leads to the emission of HH.

sub-ps pulses from various sources were focused on jets of noble gases. Summaries of HH generation and their application including its history can be found in reviews by Brabec and Krausz [26], Eden [27], and Kapteyn *et al.* [28].

A schematic diagram of HH generation using an atomic gas jet and its interpretation are shown in Figure 12.9. According to a semiclassical model [29–31] the atom is ionized by the incident laser pulse. The electron driven by the electric field of the laser can remain in the vicinity of the ion over several light oscillations, while the oscillation amplitude can amount to several Angstroms. In the nonrelativistic regime an oscillating free electron does not emit radiation. The electron, however, can recollide with the ionized atom a fraction of an optical period after the ionization event. When this occurs the electron loses an energy

$$\mathcal{W} = U_0 + \kappa U_p, \tag{12.5}$$

that is emitted as radiation. Here U_0 is the ionization potential energy of the atom and

$$U_p = \frac{e^2 \langle E^2 \rangle}{4m\omega_\ell^2} \tag{12.6}$$

is the kinetic energy (averaged over on optical cycle) of the electron (mass m) wiggling in the laser field. The quantity U_p is also called ponderomotive potential. The energy $\mathcal{W}$ depends on the time within one optical cycle at which the atom is ionized and the electron recollides. This defines a maximum possible energy the electron can acquire before recombination, for which $\kappa \approx 3.2$. As a result, the emission spectrum has a sharp cutoff at the short wavelength side, $\lambda_{\text{min}} \approx hc/\sqrt{U_0 + 3.2U_p}$. Recollisions can occur with certain probability in each half-cycle of the laser field, producing bursts of (attosecond) emissions coherent to each other. In the frequency domain this corresponds to a comb of frequencies representing the odd harmonics of ω_ℓ, the frequency of the driving field.[1] HH with order n as large as 221 were observed [32]. With few cycle fundamental pulses, only few emission bursts occur and the discrete (comb) structure of the emission spectrum disappears. Extending the coherent continuum to the X-ray water window (2.3 nm–4.4 nm) was accomplished with sub-10-fs pump pulses [33].

Single attosecond pulses were predicted theoretically for few-cycle fundamental pump pulses and demonstrated experimentally [34, 35]. This opened up exciting new applications of high field physics and attosecond spectroscopy.

12.3. GENERATION OF ULTRASHORT ACOUSTIC PULSES

An acoustic wave is a strain or shear wave that can be produced by piezoelectric transducers in mechanical contact with the material. The transducers are usually driven by rf voltages. Such techniques have gained importance for the design of acousto-optical modulators for actively mode-locked lasers. Moreover, acoustic pulse propagation, scattering, and reflection can conveniently be used for material characterization.

The generation of acoustic waves with (fs) optical pulses is another example that illustrates the complexity of processes associated with the interaction of light pulses with matter. Let us assume that a fs pulse is incident on a solid surface of a highly absorbing material and that its energy density is below the threshold for plasma generation and other irreversible processes. Two effects are responsible for launching an ultrasonic wave (pulse) into the material. Their relative importance depends on the material properties and the parameters of the (optical) excitation pulse (see, for example, Grahn *et al.* [36] and references therein).

(a) If the pulse is absorbed in a semiconductor, a high-density distribution of electron–hole pairs is created in a thin layer at the material surface. This electron–hole plasma changes locally the effective potential that determines the arrangement of the atoms in the lattice. The resulting stress is given by $\sigma = N(dV_g/d\eta)$ where N is the excited carrier density and $dV_g/d\eta$ is the deformation potential.

[1] For symmetry reasons (inversion symmetry) only odd harmonics are possible.

(b) The carriers excited to states above the band gap relax (cool) mainly because of interaction with the lattice through electron–phonon collisions. Consequently, the temperature of the excited volume rises, producing an elastic stress. The stress amplitude is directly related to the thermal expansion coefficient and the absorbed energy density. For a more detailed analysis one has to account for a possible diffusion of the excited carriers during the electron–phonon interaction. This may increase the effective excited region.

Figure 12.10 illustrates a simple model of acoustic pulse formation. At $t = 0$ elastic stress is generated with a depth profile that follows approximately the absorbed energy density. This stress distribution acts as source for a strain wave (pulse) propagating into the material and toward the interface. On reflection at the interface, the latter experiences a phase change of π. This gives rise to the final shape of the strain pulse as shown in Fig. 12.10 for $t = d/\nu_{ac}$. According to our model the width of the stress pulse is determined by the absorption length z_a for the optical pulse if the propagation length of the strain wave during the optical excitation is smaller than z_a. If ν_{ac} is the sound velocity this condition can be expressed as

$$\tau_p < \frac{z_a}{\nu_{ac}}. \tag{12.7}$$

For an order of magnitude estimate let us assume $z_a \approx 10^{-8}$ m and $\nu_{ac} \approx 10^4$ m/s, which yields $\tau_p < 1$ ps. For optical pulses shorter than 1 ps, the duration of the acoustic pulse depends mainly on material parameters rather than on the particular

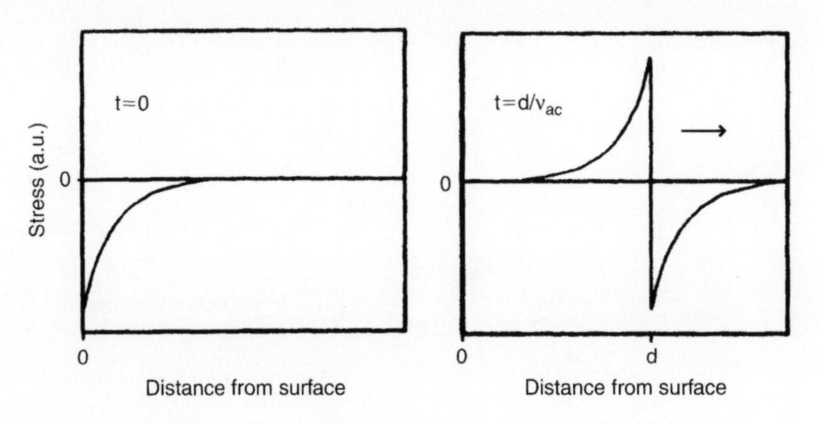

Figure 12.10 Propagation of a stress pulse generated at $t = 0$ through absorption of an optical pulse. (Adapted from Grahn *et al.* [36].)

shape of the exciting pulse. In reality, of course, the formation of the stress is not instantaneous. This concerns, for example, the finite thermalization time of excited carriers (~0.1 . . . 2 ps) and the rise of the temperature, respectively. Such effects, in addition, limit the achievable duration of the acoustic pulse.

It is worth mentioning that the geometrical length of the acoustic pulse amounts only to few nm, i.e., a fraction of an optical wavelength, which enables a variety of attractive applications. One example is shown in Figure 12.11. A pump pulse creates an acoustic pulse that subsequently bounces back and forth in a thin material layer while being damped. This was monitored by testing the reflectivity of the surface with a time-delayed probe pulse, see Fig. 12.11(b). The temporal separation between the peaks in the reflectivity change corresponds to the transit time of the acoustic pulse in the film. From the decrease in the peak height, information on the damping of acoustic waves can be obtained.

Periodic structures such as MQWs can be used to generate acoustic waves with some degree of spatial coherence [37]. For instance, the energy can be deposited in an array of MQWs spaced by the wavelength of a longitudinal optical (LO) phonon generated in the process of thermalization of the excited carriers. Such a structure is the acoustical analogue of the distributed feedback laser [38–40]. The LO phonons add coherently along the normal to the MQW planes. The frequency of the phonon can be in the hundreds of GHz, with a corresponding wavelength of the order of 100 Å. The phonon generation can be understood as a particular case of impulsive Raman scattering. As mentioned in Chapter 10, coherent addition of the phonons at the frequency ν_{ph} will be achieved if a train of pulses separated by the period $1/\nu_{ph}$ is used instead of a single pulse. The pulse shaping techniques mentioned in Chapter 8 can generate such

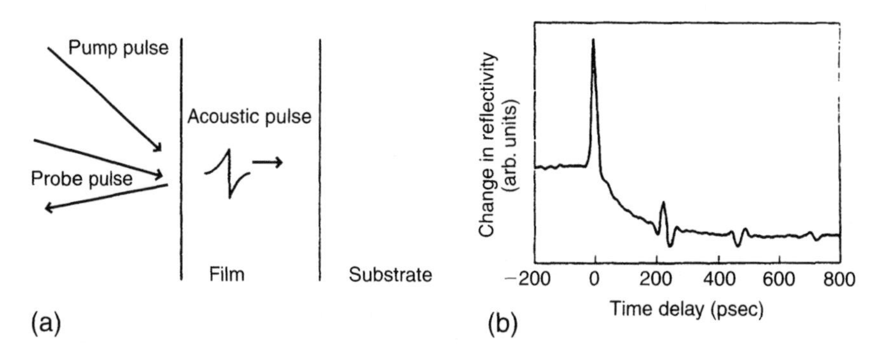

Figure 12.11 (a) Diagram for the generation and detection of acoustic pulses. (b) Time dependence of the reflectivity change of a 220 nm As_2Te_3 film as seen by the optical probe pulse. The echoes corresponding to subsequent reflections of the acoustic pulse at the film–air interface appear clearly as reflection changes (Adapted from Grahn *et al.* [36]).

pulse trains. Femtosecond technology applied to MQW structures can thus lead to the generation of temporally and spatially coherent acoustic waves of short ($\approx$ 100Å) wavelength.

12.4. GENERATION OF ULTRAFAST ELECTRIC PULSES

The rapid progress in microelectronics not only makes circuits smaller and more powerful but also faster. The fastest all-electronically produced transients ($\approx$0.5 ps) are still a few orders of magnitude slower than what can be obtained by all-optical techniques. Of course, optics and electronics cover different applications and compete directly only in a limited application field. In many cases a combination of optical and electronic means can be regarded as the optimum. In light of this, to generate ultrashort electrical transients, one can advantageously use fs light pulses to trigger photoconductive switches. Such switches made from semiconductors and driven by ps optical pulses were first demonstrated by Auston [41] and Lawton and Scavannec [42]. The development of fs pulse sources and progress in material fabrication have since made possible the production of sub-ps electrical pulses [43]. These pulses are employed to test ultrafast electronic circuits and components and measure their temporal response; see Frankel *et al.* [44] for example.

The basic operational principle of a photoconductive switch can be explained by means of Figure 12.12. A metallic microstrip line on top of a semiconductor is interrupted by a narrow gap. The implementation of the switch in a high speed transmission line such as a strip line is necessary to propagate ultrafast electric pulses while limiting the broadening effect of dispersion. The photoconductor

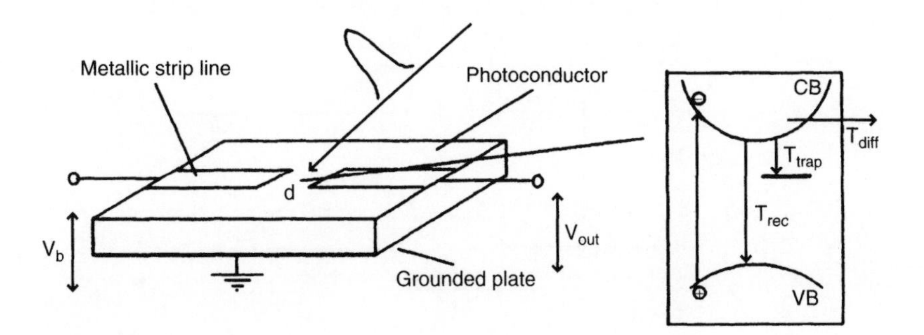

Figure 12.12 Sketch of a photoconductive switch. The inset shows the excitation and relaxation processes of the semiconductor.

can also be a thin film deposited on an insulator. The bottom of this substrate is metal-coated and grounded. The strip line is connected to a bias voltage V_b. Because of the low (ideally zero) dark conductivity of the gap, the output voltage is zero. When a light pulse of suitable frequency is focused into the gap free carriers are generated which increase the conductivity. Subsequently, a current develops which can be measured as a certain output voltage V_{out}. In the first demonstration of a ps-pulse triggered switch [41], the voltage was again set to zero by a second light pulse of certain delay but of longer wavelength (1.06 μm) which produced a short circuit between the excited surface layer and the grounded plate. This was possible since the absorption length of the 0.53 μm excitation pulse in Si was much smaller than that of the "turn-off" pulse. Without the second pulse, the drop of the output voltage is given by the decrease of free carriers in the gap. This in turn is determined by local processes such as carrier recombination and carrier trapping and by nonlocal processes such as carrier diffusion.

An exact analysis of the switching behavior in the fs regime is a rather difficult task. It requires not only the solution of Maxwell's equations with corresponding initial and boundary conditions but also an accurate modeling of the matter response to the fs excitation pulse. To explain the basic operational principle, however, a strongly simplified approach is possible and provides excellent results. We will briefly explain this model for a photoconductive switch in a transmission line and follow the discussion of Auston [45,46]. The main idea is to model the switch by an equivalent circuit consisting of a capacitor of capacitance C and a parallel (time varying) resistor of conductance $G(t)$; see Figure 12.13. The conductance can be written as

$$G(t) = G_0 + g(t) \tag{12.8}$$

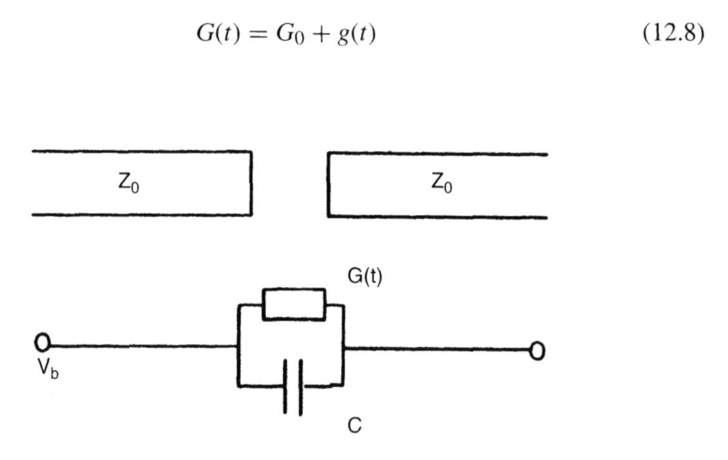

Figure 12.13 Equivalent circuit for a photoconductive switch in a transmission line. (Adapted from Auston [45].)

where G_0 is the dark conductivity and $g(t)$ is the pulse induced conductance. The latter can be expressed in terms of the free carrier concentration N and the electron (hole) mobility μ_n (μ_p), as

$$g(t) = \frac{1}{V_b^2} \int d^3x (Ne\mu_n + Ne\mu_p) |\tilde{E}|^2 \tag{12.9}$$

where V_b is the voltage across the gap, E is the local field strength in the gap and d^3x denotes integration over the excited volume layer. The local field is a complicated function of the gap geometry and the carrier concentration and distribution. To get an explicit expression for the photoconductance let us assume that the perturbation by the pulse is small so that $g \ll G$ holds.[2] The electric field in the gap can then be approximated by $E = V_b/d$ where d is the gap length. Assuming homogeneous gap illumination and complete absorption in a thin surface layer, the total number of excited carriers is given by $(1 - r)\mathcal{W}/(\hbar\omega_\ell)$ where r is the reflection coefficient and $\mathcal{W}/(\hbar\omega_\ell)$ is the number of incident photons. Inserting the values for the number of carriers and the electric field thus obtained in Eq. (12.9) yields for the photoconductance after pulse absorption at $t = 0$

$$g(t > 0) = \frac{1 - r^2}{d^2} e(\mu_n + \mu_p) \frac{\mathcal{W}}{\hbar\omega_\ell}. \tag{12.10}$$

This expression shows the importance of high carrier mobilities for the switch sensitivity.

Let us next investigate the fundamental switching properties by idealizing the process of carrier generation and using the equivalent circuit model. For a step function conductance change

$$G(t) = \begin{cases} 0 & t < 0 \\ G_1 & t \geq 0 \end{cases} \tag{12.11}$$

and a dc bias voltage V_b, the transmitted voltage signal for a photoconductor in a transmission line of impedance Z_0 is [45]:

$$V(t) = \frac{V_b}{2} \frac{2Z_0G_1}{1 + 2Z_0G_1} \left\{ 1 - \exp\left[-(1 + Z_0G_1)\frac{t}{CZ_0} \right] \right\} \tag{12.12}$$

where Z_0 is the impedance of the strip line. Obviously the transmitted signal increases with G_1 and saturates at $V_b/2$ for $Z_0G_1 \gg 1$. For small excitation,

---

[2]This simplifying assumption is for the purpose of analytical evaluation only: It is neither desired for high switching efficiencies nor typical for ultrashort pulse excitation.

i.e., $Z_0 G_1 \ll 1$, the signal's rise time is limited by the capacitance and is expected to decrease with increasing G_1.

As mentioned previously the drop in the transmitted signal can be controlled with such material parameters as carrier recombination, trapping, and diffusion. These parameters depend sensitively on the material and the fabrication process. Diffusion and recombination times typically range well above several 10 ps and thus would not lead to a sub-ps signal switch-off. Additional relaxation channels can be opened by the introduction of local defects which act as trapping centers. These defects are implanted by doping with impurities or through radiation damage. Effective carrier lifetimes as short as 600 fs were measured for radiation damaged silicon on sapphire [5]. Other examples are CdTe [47] and GaAs [48] where recombination centers were introduced during the material growth process. With the implementation of such techniques there is usually a trade-off to be made between the decrease of the free carrier lifetime and a decrease of the mobility. These materials, however, enabled the generation of sub-ps electrical pulses [43,47]. An example is shown in Figure 12.14.

An important question is how to transmit and measure sub-ps electrical transients with THz bandwidths. Strip lines like those depicted in Fig. 12.12 have too large a dispersion to be ideal transmitters for sub-ps electrical pulses. Mode dispersion is the most severe problem. For only one single mode to exist, the distance between the metal strip and the grounded plate has to be small (a few microns), which is difficult to achieve (mechanical tolerances). Electrical pulses of less than 10 ps double their length after propagation distances shorter than 1 mm [49].

Better results have been obtained with coplanar strip lines deposited on top of the photoconductor as depicted in Fig. 12.14. Their geometrical separation can

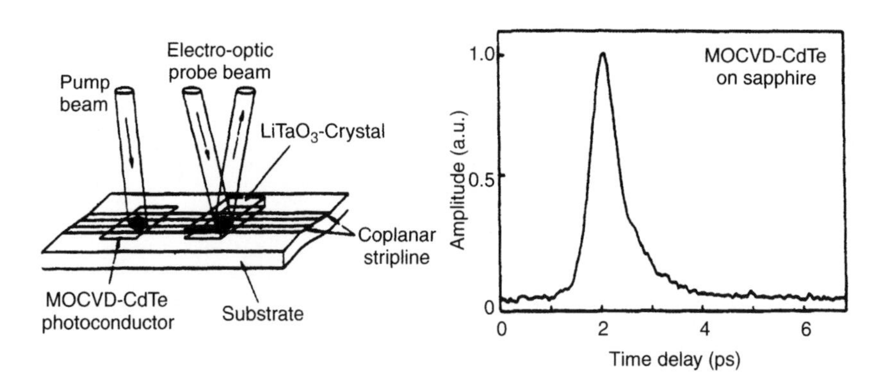

Figure 12.14 Sub-ps electrical pulse generated with a CdTe coplanar strip line and 100 fs excitation pulses. Left: experimental setup. Right: correlation signal obtained through electro-optic sampling (Adapted from Nuss *et al.* [47]).

easily be controlled during the fabrication process, which is essential for single mode propagation. Using "sliding contact" excitation the effective capacitance was found to be zero to first order [50], which is most desirable for short electric transients, cf. Eq. (12.12). But even with coplanar strip lines, sub-ps electrical pulses (or slopes) cannot be propagated over more than several mm [51]. As mentioned at the beginning of this chapter, they can be sent through considerable distances in air or a dielectric bulk medium if suitable antenna–receiver structures are used. There are speculations about utilizing soliton mechanisms to transmit sub-ps electrical pulses over long distances in strip lines [52]. In analogy to solitons in optical fibers, this requires an interplay between nonlinear and linear processes.

Because of the dispersion problem, the measurement of sub-ps electrical pulses has to be performed close to the location of their generation. This can be done by implementing a second photoelectronic switch in the transmission line. The voltage pulse generated in the first gap, propagates to the second gap which is illuminated by a time-delayed second pulse [41]. The voltage signal at the output of such a device measured as a function of the delay time is thus an autocorrelation of the electrical pulse. Electro-optic techniques have been applied to measure cross-correlations between the electrical and optical pulses. If the substrate has a large enough electro-optic coefficient, the electrical pulse induced birefringence can be probed by a time-delayed optical pulse of suitable wavelength [53]. What is measured then is the polarization rotation experienced by the test pulse. Another approach is to bring an electro-optic material in the vicinity of the location to be probed [53]. The experiment depicted in Fig. 12.14 utilizes this method. For good spatial resolution, the probe can be a needle produced from a suitable material such as $LiTaO_3$. The advantage of this technique is that the behavior of the electrical pulse can be sampled along the transmission line and possibly in following electronic components.

BIBLIOGRAPHY

[1] D. H. Auston, K. P. Cheung, J. A. Valdmanis, and D. A. Kleinman. Cherenkov radiation from femtosecond optical pulses in electrooptic media. *Phys. Rev. Lett.*, 35:1555–1558, 1984.

[2] D. A. Kleinman and D. H. Auston. Theory of electrooptic shock radiation in nonlinear optical media. *IEEE J. Quantum Electron.*, QE-20:964–970, 1984.

[3] B. B. Hu, X. C. Zhang, D. H. Auston, and R. R. Smith. Free-space radiation from electrooptic crystals. *Appl. Phys. Lett.*, 56:506–508, 1990.

[4] C. Fattinger and D. Grischkowsky. Point source terahertz optics. *Appl. Phys. Lett.*, 53: 1480–1482, 1988.

[5] F. E. Doany, D. Grischkowsky, and C. C. Chi. Carrier life time versus ion-implantation dose in silicon on sapphire. *Appl. Phys. Lett.*, 50:460–462, 1987.

[6] X. C. Zhang and D. H. Auston. Optoelectronic measurements of semiconductor surfaces with femtosecond optics. *J. Appl. Phys.*, 71:326–338, 1992.

[7] G. Mourou, C. V. Stancampiano, A. Antonetti, and A. Orszag. Picosecond microwave pulses generated with subpicosecond laser driven switch. *Appl. Phys. Lett.*, 39:295–296, 1981.

[8] X. C. Zhang, B. B. Hu, J. T. Darrow, and D. H. Auston. Generation of FIR electromagnetic pulse from semiconductor surfaces. *Appl. Phys. Lett.*, 56:1011–1013, 1990.

[9] B. I. Green, P. N. Saeta, D. R. Dykaar, S. Schmitt-Rink, and S. L. Chuang. Far-infrared light generation at semiconductor surfaces and its spectroscopic applications. *IEEE J. of Quantum Electron.*, QE-28:2302–2312, 1992.

[10] P. K. Benicewitz, J. P. Roberts, and A. J. Taylor. Scaling of terahertz radiation from large aperture biased photoconductors. *J. of the Optical Soc. Am.*, 11:2533–2546, 1994.

[11] M. van Exter and D. Grischkowsky. Characterization of an optoelectronic terahertz beam system. *IEEE J. of Quantum Electron.*, QE-28:1684–1691, 1990.

[12] C. Fattinger and D. Grischkowsky. A Cherenkov source for freely propagating THz beams. *IEEE J. of Quantum Electron.*, QE-25:2608–2610, 1988.

[13] D. Grischkowsky, S. Keiding, M. van Exter, and C. Fattinger. Far infrared time-domain spectroscopy with terahertz beams of dielectrics and semiconductors. *J. Opt. Soc. B*, 7:2006–2015, 1990.

[14] H. Harde and D. Grischkowsky. Coherent transients excited by subpicosecond pulses of terahertz radiation. *J. Opt. Soc. Am. B*, 8:1642–1651, 1991.

[15] R. W. Falcone and M. M. Murnane. Proposal for a fs x-ray light source. In D. T. Atwood and J. Bokor, Eds., *Short Wavelength Coherent Radiation: Generation and Applications*, American Institute of Physics, New York, vol. 147, 1986, pp. 81–85.

[16] D. Kuehlke, U. Herpers, and D. von der Linde. Soft x-ray emision from subps produced laser-induced plasmas. *Appl. Phys. Lett.*, 50:1785–1787, 1987.

[17] G. Kuehnle, F. P. Schaefer, S. Szatmari, and G. D. Tsakiris. X-ray production by irradiation of solid targets with subps excimer laser pulses. *Appl. Phys.*, B47:361–366, 1988.

[18] J. D. Kmetec. Ultrafast laser generation of hard x-rays. *IEEE J. of Quantum Electron.*, QE-28:2382–2387, 1992.

[19] J. A. Cobble, G. A. Kyrala, A. A. Hauser, A. J. Taylor, C. C. Gomez, N. D. Delamater, and G. T. Schappert. Kilovolt x-ray spectroscopy of subps-laser-excited source. *Phys. Rev. A*, 39: 454–457, 1989.

[20] H. M. Milchberg, I. Lyubomirsky, and C. G. Durfee III. Factors controlling the x-ray pulse emission from an intense laser heated solid. *Phys. Rev. Lett.*, 67:2654–2657, 1991.

[21] O. R. Wood, W. T. Silfvast, H. W. K. Tom, W. H. Fork, C. H. Brito-Cruz, M. C. Downer, and P. J. Moloney. Effect of laser pulse duration on short wavelength emission from femtosecond and picosecond laser induced Ta plasmas. *Appl. Phys. Lett.*, 53:654–656, 1988.

[22] M. M. Murnane, H. C. Kapteyn, M. D. Rosen, and R. Falcone. Ultrafast x-ray pulses from laser produced plasmas. *Science*, 25:531, 1991.

[23] N. H. Burnett, H. A. Baldis, M. C. Richardson, and G. D. Enright. Harmonic generation in CO_2 laser target interaction. *Appl. Phys. Lett.*, 31:172–175, 1977.

[24] A. McPherson, G. Gibson, H. Jara, U. Johann, T. S. Luk, I. A. McIntyre, K. Boyer, and C. K. Rhodes. Studies of multiphoton production of vacuum-ultraviolet radiation in the rare gases. *J. Opt. Soc. B*, 4:595–601, 1987.

[25] X. F. Li, A. L'Huillier, M. Ferray, L. A. Lompre, and G. Mainfray. Multiple-harmonic generation in rare gases at high laser intensity. *Phys. Rev. A*, 39:5751–5761, 1989.

[26] T. Brabec and F. Krausz. Intense few-cycle laser fields: Frontiers of nonlinear optics. *Rev. Mod. Phys.*, 72:545–591, 2000.

[27] J. G. Eden. High-order harmonic generation and other intense optical field-matter interactions: Review of recent experimental and theoretical advances. *Progr. in Quant. Electron.*, 28:197–246, 2004.

[28] H. C. Kapteyn, M. M. Murnane, and I. P. Christov. Extreme nonlinear optics: Coherent x-rays from lasers. *Physics Today*, March:39–44, 2005.

[29] P. Corkum. Plasma perspective on strong-field multiphoton ionization. *Phys. Rev. Lett.*, 71:1994, 1993.

[30] K. C. Kulander, K. J. Schafer, and J. L. Krause. Super-intense laser-atom physics. In B. Piraux, A. L'Huillier, and K. Rzazewski, Eds., *Proceedings of the NATO Advanced Research Workshop, SILAP III, Han-sur-Lesse, Belgium*, volume 316, Plenum Press, New York, 1993, pp. 95.

[31] M. Lewenstein, P. Balcou, M. Y. Ivanov, A. L'Huiliier, and P. B. Corkum. Theory of high-harmonic generation by low-frequency laser fields. *Phys. Rev. A*, 49:2117–2132, 1994.

[32] Z. Chang, A. Rundquist, H. Wang, M. M. Murnane, and H. C. Kapteyn. Generation of coherent soft x-rays at 2.7 nm using high harmonics. *Phys. Rev. Lett.*, 79:2967–2970, 1997.

[33] C. Spielmann, N. H. Burnett, S. Sartania, R. Koppitsch, M. Schnuerer, C. Kan, M. Lenzner, P. Wobrauschek, and F. Krausz. Generation of coherent x-rays in the water window using 5-femtosecond laser pulses. *Science*, 278:661–664, 1997.

[34] I. P. Christov, M. M. Murnane, and H. C. Kapteyn. High-harmonic generation of attosecond pulses in the "single-cycle" regime. *Phys. Rev. Lett.*, 78:1251–1254, 1997.

[35] E. Goulielmakis, M. Uiberacker, R. Kienberger, A. Baltuska, V. Yakovlev, A. Scrinzi, Th. Westerwalbesloh, U. Kleineberg, U. Heinzmann, M. Drescher, and F. Krausz. Direct measurement of light waves. *Science*, 305:1267–1269, 2004.

[36] H. T. Grahn, H. J. Maris, and J. Tauc. Picosecond ultrasonics. *IEEE J. Quantum Electron.*, QE-25:2562–2569, 1989.

[37] V. F. Sapega, V. I. Belitsky, and A. J. Shields. Resonant one-acoustic-phonon raman scattering in multiple quantum wells. *Solid State Commun.*, 84:1039, 1992.

[38] Z. Bor. A novel pumping arrangement for tunable single picosecond pulse generation with a N_2 laser pumped distributed feedback dye laser. *Optics Comm.*, 29:103–108, 1979.

[39] F. P. Schaefer. New methods for the generation of ultrashort laser pulses. In H. Walther and K. W. Rothe, Eds., *Laser Spectroscopy IV, Proceedings of the fourth Int. Conf. Rottach-Egern*, Springer-Verlag, Berlin, Germany, 1979, pp. 590–596.

[40] Z. Bor and A. Muller. Picosecond distributed feedback dye lasers. *IEEE Journal of Quantum Electronics*, QE-22:1524–1533, 1986.

[41] D. H. Auston. Picosecond optoelectronic switching and gating in silicon. *Appl. Phys. Lett.*, 26:101–103, 1975.

[42] R. A. Lawton and A. Scavannec. Pulsed laser application to sampling oscilloscope. *Electron. Lett.*, 11:138, 1975.

[43] M. B. Ketchen, D. Grischkowsky, T. C. Chen, C. C. Li, I. N. Duling III, N. J. Halas, J. M. Halbout, J. A. Kash, and C. P. Li. Generation of subpicosecond electrical pulses on coplanar transmission lines. *Appl. Phys. Lett.*, 48:751–753, 1986.

[44] M. Y. Frankel, J. F. Whitaker, and G. A. Mourou. Optoelectronic transient characteristic of ultrafast devices. *IEEE J. Quantum Electron.*, 28:2313–2324, 1992.

[45] D. H. Auston. Impulse response of photoconductors in transmission lines. *IEEE J. Quantum Electron.*, QE-19:639–648, 1983.

[46] D. H. Auston. Ultrafast optoelectronics. In W. Kaiser, Ed., *Ultrashort Laser Pulses and Application*, Springer, Berlin, Germany, 1988, pp. 183–233.

[47] M. C. Nuss, D. W. Kisker, P. R. Smith, and T. E. Harvey. Efficient generation of 480 fs electrical pulses on transmission lines by photoconductive switching in metalorganic chemical vapor deposited cdte. *Appl. Phys. Lett.*, 54:57–59, 1989.

[48] S. Gupta, M. Y. Frankel, J. A. Valdmanis, J. Whitaker, G. Mourou, F. W. Smith, and A. R. Calawa. Subpicosecond carrier lifetime in GaAs grown by molecular beam epitaxy at very low temperatures. *Appl. Phys. Lett.*, 59:3276–3278, 1991.

[49] G. A. Mourou. High-speed electronics. In B. Kallback and H. Beneking, Eds., *Springer series Electronics and Photonics*. Springer, Berlin, Germany, 1986.

[50] D. R. Grisckkowsky, M. B. Ketchen, C. C. Chi, I. N. Duling III, N. J. Halas, J. M. Halbout, and P. G. May. Capacitance-free generation and detection of subpicosecond electrical pulses on coplanar transmission lines. *IEEE J. Quantum Electron.*, QE-24:221–225, 1988.

[51] M. Y. Frankel, S. Gupta, J. A. Valdmanis, and G. A. Mourou. Terahertz attenuation and dispersion characteristics of coplanar transmission lines. *IEEE Trans. Microwave Theory Tech.*, 39: 910–915, 1991.

[52] P. Paulus, B. Wedding, A. Gaseh, and D. Jaeger. Bistability and solitons observed in nonlinear ring resonators. *Phys. Lett.*, 102A:89–92, 1984.

[53] J. A. Valdmanis and G. A. Mourou. Subpicosecond electrooptic sampling: Principles and applications. *IEEE Journal of Quantum Electron.*, QE-22:69–78, 1986.

13

Selected Applications

In previous chapters we have emphasized the role of femtosecond pulses in basic research. Ultrashort pulses are not limited to esoteric research on ultrafast events. We want to emphasize here more down-to-earth applications, for which the femtosecond source can have practical advantages. The topics covered are short pulse imaging, solitons, fs lasers as sensors, and stabilized fs lasers for applications in metrology.

13.1. IMAGING

13.1.1. Introduction

It does not come as a surprise that ultrashort pulses contribute to the most fundamental function of light: imaging. The intensity information of light is sufficient to record two-dimensional images. Additional information provided by the phase of the optical field makes it possible to record an image along all three space coordinates. This technique, combining phase and amplitude retrieval of the light scattered by objects, is called holography, and is the most accurate of all macroscopic imaging methods. It can easily measure deformations much smaller than one wavelength. The price to pay for the high accuracy of holography is that an excessive amount of data has to be recorded. Because all the three-dimensional information of the object has to be stored in a single recording, high optical energy densities are used, with the possibility of laser damage if the material absorbs light.

"Range gating" is another method to obtain depth information, by measuring the transit time of the radiation from the source to the object and thereafter to the detector. If—as is often the case—the source and detector are colocated, the distance z from source to object is simply $c\times$ (roundtrip time)/2. This technique has been used since World War II for localizing and tracking moving objects. The resolution has shifted from meter (radar) to centimeter (lidar). Femtosecond pulses offer the possibility of a depth resolution of a few microns. Another function of ultrashort range gating is to discriminate against scattering, as will be shown later in this chapter.

13.1.2. Range Gating with Ultrashort Pulses

A basic sketch of principle for range gating 3D images with fs pulses is shown in Figure 13.1. The source fs beam is split into a reference and an object beam. The reference beam, after an appropriate delay line, triggers the optical gate at a delay time τ. The light backscattered from various depths z of the object reaches the optical gate at time intervals spaced out by $2z/c$. A particular depth is selected by the delay time at which the shutter is opened. The signal S received by the detector, as a function of delay τ, for a particular position (x, y) in the transverse plane of the beam, is the correlation of the gating function $g(t)$ and the intensity from the object $I_s(t)$:

$$S(\tau) = \int_{-\infty}^{\infty} I_s(t)g(t-\tau)dt. \tag{13.1}$$

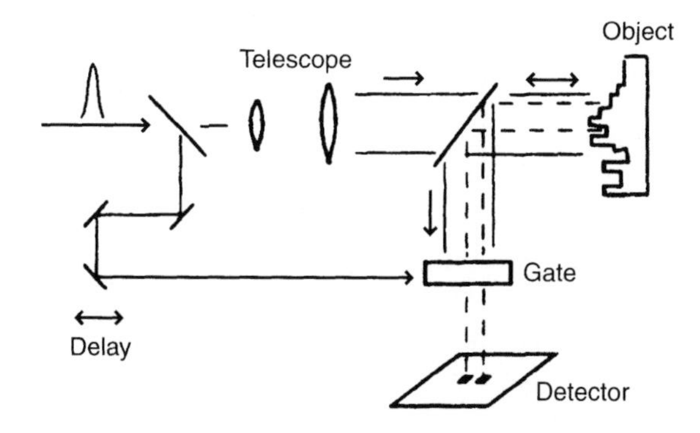

Figure 13.1 Recording of a three-dimensional object through range gating with ultrashort pulses. The gate is "opened" by a fs reference pulse derived from the illuminating source and appropriately delayed. An ultrafast gating function can be achieved, for example, with a Kerr gate or a nonlinear crystal for sum frequency generation. In the latter case, transverse resolution is generally obtained by scanning a narrow beam across the object.

If the gating function can be assimilated to a δ function with respect to the variations of the signal $I_s(t)$, the measured transmitted signal is simply $S(\tau) \approx I_s(\tau) = I_s(2z/c)$, where the depth of observation z is determined by the position of the reference mirror (delay). In general, the higher the order of the gating process is, the better is the depth resolution. For instance, if three-photon interaction ($\omega_d = 2\omega_r + \omega_s$, where ω_d, ω_r, and ω_s are respectively the frequencies of the detected, reference, and signal photons) is used to gate the signal light, because $g(t) \propto I_r(t)^2$, the gating signal is approximately $\sqrt{2}\times$ shorter than the signal pulse. A compromise has to be reached: The higher the order of the gating process, the greater the required reference intensity. Given the power limitation of the source, the intensity requirement for the reference pulse limits the beam cross section that can be utilized. Lateral scanning is therefore often used to obtain the transverse information on the object.

Bruckner [1] proposed using a fast Kerr shutter to gate the radiation reflected by index discontinuities in eyes. A Kerr shutter consists of a Kerr liquid between crossed polarizers. The shutter is "opened" by an intense ultrashort pulse inducing birefringence in the liquid. The gating time is either the pulse duration or the response time of the liquid, whichever is longer. Kerr gates have been applied to picosecond gating [2,3]. Femtosecond temporal resolution can be achieved, for example, by gating through second harmonic or sum frequency generation [4,5]. The technique consists of generating a second harmonic signal $I_2(t)$ proportional to the product of a reference pulse $I_r(t)$, derived directly from the source, and the signal $I_s(t)$. The SH energy $S_{2\omega}(\tau)$ recorded as a function of reference delay τ is simply the correlation function defined in Chapter 9. This technique has been applied in one dimension to fibers, to locate defects in fibers and connectors with a resolution of the order of a few microns [6]. The method has been extended to three dimensions (three-dimensional imaging of the eyes) by scanning the beam transversely [7]. The transverse resolution, limited by the size of the beam, can be improved by illuminating each point of the 3D object and using tomographic reconstruction algorithms [4].

Linear correlation techniques, such as heterodyning, can also be applied. Here the reference pulse is frequency shifted. Gating is achieved by interfering reference and object pulse and detection at the heterodyne frequency. Because no nonlinear optical processes are involved, these linear techniques are sensitive even at low illumination power. The backscattered signal is mixed with the reference signal, which is continuously scanned. The mixing signal is at the Doppler frequency, thus measuring the speed of the scanner. The mixing beat note is observable only in the regions where the reference and signal are coherent with each other. In this particular application, either ultrashort pulses, or light with a short coherence length are used. The technique was initially applied to fibers and integrated optics structures [8].

Huang *et al.* [9] introduced a greatly improved method as optical coherence tomography (OCT) in the early 1990s. Since then OCT has developed into a field of its own with impressive applications initially in the imaging of eyes [10], and later in biomedical noninvasive imaging in general. We refer the reader to several books devoted to OCT [11, 12].

Speed is an essential element in 3D imaging of in vivo biological objects. There is a compromise between speed and sensitivity: Sensitive detectors require a longer integration time. One possibility to reduce the time needed for data acquisition is to reduce the multidimensional scanning to only one dimension (the depth). It is possible in the case of nonlinear gating (second harmonic or parametric generation) to record a single-shot transverse picture for each depth increment. Direct recording of 2D images in "depth slices" has been demonstrated with high contrast objects [5, 13]. The optical arrangement is sketched in Figure 13.2. The laser beam is expanded to the size of the object after being split by a calcite prism (polarizing beam splitter) between a reference and probing beam. The amount of beam splitting is controlled by a half wave plate, to have the maximum probing intensity that the sample can accommodate. A quarter wave plate in both the reference and object arms ensures that the returning beams

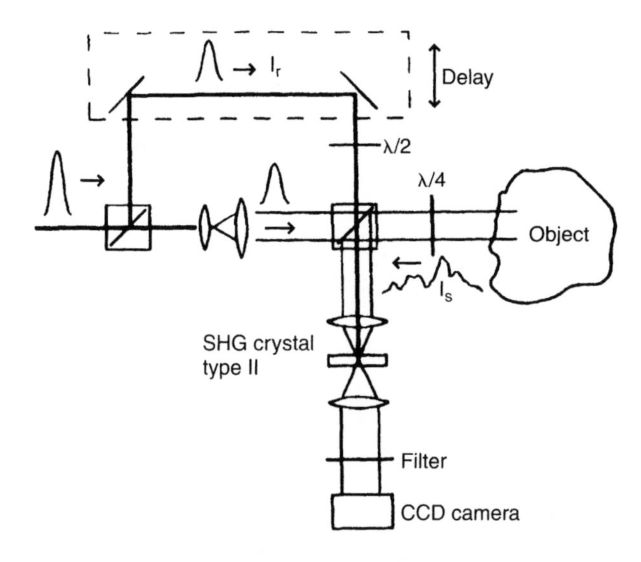

Figure 13.2 Setup for recording successive two-dimensional "slices" of a transparent 3D object based on SHG. The backscattered radiation I_s from the object is stretched out in time, corresponding to the time of arrival from various depths. In the nonlinear crystal, the second harmonic, being proportional to the product of the reference I_r and signal I_s, selects a portion of the signal corresponding to a certain depth (set by the reference delay).

are redirected toward the detector. The backscattered signal beam and the orthogonally polarized retro-reflected reference are sent into a nonlinear crystal cut for type II phase matched SHG. Assuming the reference has a uniform transverse intensity profile, the second harmonic contains the image information contained in the fundamental beam. The time of arrival of the reference ultrashort pulse determines the depth d at which a cross section through the object is imaged into the CCD, as illustrated in Fig. 13.2.

The ultrashort pulse source used for the preliminary tests was a fs dye laser operating at 620 nm [5]. Urea crystals were chosen for these tests as being the only phase-matchable type II crystals at that wavelength. Unfortunately, good quality urea crystals are not readily available. Despite these limitations, a spatial transverse resolution of the order of 100 μm has been achieved in experiments with a configuration in which the reference and object beams have the same size [5, 13]. The titanium sapphire femtosecond lasers appear promising for this particular application because of the better transmission of biological tissues in its wavelength range of 750 nm to 850 nm and the possibility of using KDP crystals (type II) for the gating.

There is a subtle interplay among sensitivity, depth, and transverse resolution in the setup shown in Fig. 13.2. One cannot have the three parameters simultaneously optimized. For instance, an optimum conversion efficiency of one single SH photon for one signal photon can be achieved, provided the crystal length and the reference power density are sufficient. A simple estimate given as a problem at the end of this chapter illustrates this problem. A minimum crystal length is required to achieve single photon upconversion (i.e., one second harmonic photon for each signal photon). But to this minimum crystal length corresponds a phase matching bandwidth, hence a limitation to the temporal resolution of the up-conversion. In addition, the waist of the reference beam acts as a spatial filter, limiting the transverse resolution of the imaging system.

13.1.3. Imaging through Scatterers

One of the main medical motivations for this type of research is early detection of tumors. The photon energy of visible and infrared light is too small for direct ionization of most tissues. Hence, an optical method seems to be an attractive and safe alternative to X-rays. Large differences in absorption have been reported between in vivo normal tissue and some types of tumor [14].

In medical and biological imaging, the biggest challenge is generally posed by scattering. To illustrate the effect of scattering on short light pulses, let us consider a femtosecond pulse being incident on a slab of thickness L made of isotropic scatterers as shown in Figure 13.3. If we time resolve the transmission, we observe a peak on the leading edge of the transmitted pulse, which corresponds

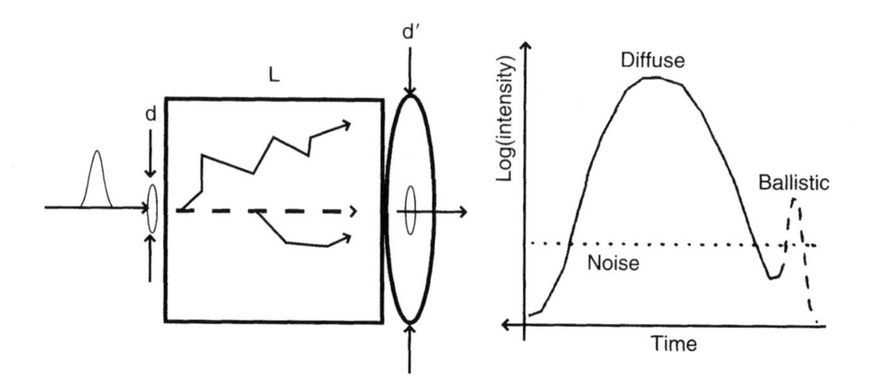

Figure 13.3 Sketch of pulse propagation through a scattering medium.

to the unscattered light ("ballistic" component) followed by a broad distribution of scattered light. Diffraction-limited resolution in an imaging process can only be obtained with the ballistic light. The latter can be separated from the diffuse light by appropriate time gating. The ballistic component of the transmitted light is attenuated exponentially:

$$I_{\text{ball}} = I_0 e^{-\mu_s L}, \tag{13.2}$$

where $\mu_s = l_t^{-1}$ is the scattering coefficient, and l_t is the scattering mean free pathlength. There have been numerous attempts to visualize objects embedded in dense scatterers by range gating the backscattered or transmitted ballistic radiation [3,13,15]. High sensitivity is required to compensate for the large attenuation of the return signal. In the case of nonlinear detection, the second harmonic that is recorded is proportional to the product of a reference intensity by the weak backscattered radiation. Therefore, high peak powers (of the reference signal) are required to obtain a good conversion efficiency at the detection.

Problems arise if the scatterer is dense, and I_{ball} approaches the noise level of the detection system. Here techniques are needed which provide not only the time gating but also an optical amplification. Nonlinear techniques, such as Raman amplification, and linear methods, such as heterodyning, have been applied successfully [16–19] leading to micrometer resolution through dense scatterers. The ultimate limit to the resolution is the quantum noise (photon shot noise), which essentially implies that at least one photon should be detected per element or pixel of the image. In the case of biological and medical samples that can only withstand average powers of irradiation of a few mW, these noise considerations limit the applicability of diffraction limited imaging to scattering densities $\mu_s L \leq 35$.

The multiply scattered photon path can be described by a diffusion model [20]. For $L \gg l_t$, it can be shown that a collimated input beam of diameter $d < l_t$ at the sample input broadens to a diameter d', which is approximately given by [21]:

$$d' = 0.2L, \tag{13.3}$$

if we refer to the early-arriving scattered light that exceeds the detection noise and to illumination intensities below the critical values for biomedical samples. The quantity d' gives a reasonable measure of the resolution that can be achieved in imaging an object buried in a dense scatterer. With a sample thickness of several mm to several cm, the achievable resolution cannot be better than a few mm.

Several approaches are being attempted to utilize the large diffuse light component for imaging through dense scatterers, such as several cm of tissue. One direction that promises to improve the resolution is to use the earlier portion of the scattered light, which may or may not follow a diffusion-like path [17,22] for imaging.

An overview of activities in the field of imaging with short light pulses can be found for example in the annual meetings of SPIE and OSA [23] and Kempe and Rudolph [24].

13.1.4. Prospects for Four-Dimensional Imaging

Ultrashort pulses can be used as a substitute for holographic techniques to record three-dimensional images, provided the object does not move on the time scale of the ultrashort pulses. Holography with ultrashort pulses should be used to record the temporal evolution of ultrafast 3D events. A method called "light-in-flight holography" (LIFH) has been proposed by N. Abramson [25,26] to convert the rapid time information obtained with ultrashort illumination holography into space information that is stored. The basic principle is that the holographic fringes can only be recorded if the object and reference beams arrive simultaneously at the recording medium. An ultrashort reference beam sent at oblique incidence to the recording medium sets a time axis (Figure 13.4). Point A is illuminated first and point B last after a time interval $\Delta t_{AB} = D \tan\theta / c$. At any point on the line AB, a hologram of the object beam corresponding to a particular instant within the time interval Δt_{AB} is recorded.

This technique has been used to detect the first arriving light through scattering media—the ballistic component cited previously. The holographic data can be recorded on a CCD camera (provided the reference and object beam make a

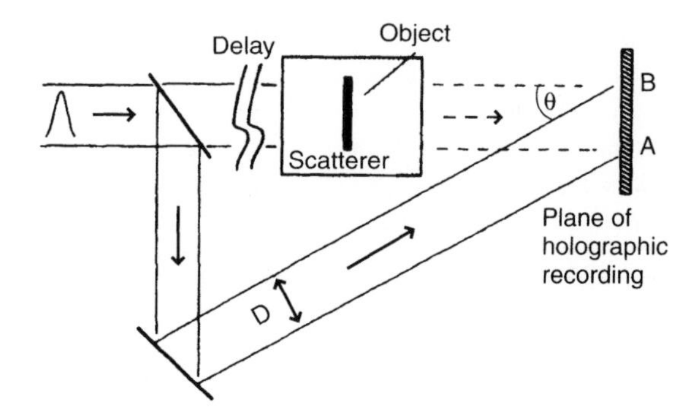

Figure 13.4 Light-in-flight holography illustrated for a simple one-dimensional transparent object (uniform in the transverse dimension of the object).

sufficiently small angle for the fringes to be resolved by the camera), and reconstructed through numerical fast Fourier transform [27–29]. The advantage of the electronic recording and numerical processing is the possibility to integrate a large number of successive (reconstructed) images. The speckle pattern is averaged out if the time interval between exposures is greater than the correlation time of the speckle.

13.1.5. Microscopy

Laser-scanning microscopy (see, for example, Wilson and Sheppard [30]) is ideally suited to combine microscopic imaging with fs pulse illumination. There are several attractive application fields of femtosecond microscopy—(a) nonlinear microscopy, (b) microscopy with simultaneous space and time resolution, and (c) microscopy of structures immersed in a scattering environment. An early review can be found in Kempe and Rudolph [19]. In nonlinear microscopy the image signal is generated by a nonlinear optical process, such as surface SHG and two-photon excited fluorescence. An image is a map of the distribution of the corresponding nonlinear susceptibility. Multiphoton fluorescence is particularly attractive for microscopy of biological cells because of the depth selectivity of the excitation process. Since its invention in 1990 by Denk *et al.* [31] the two-photon fluorescence microscope has greatly improved the microscopic imaging capabilities, in particular in the life sciences. Femtosecond pulses are needed because of their great peak power at comparatively small pulse energy (i.e., small heat consumption in the specimen). An overview of ultrafast optics for biological imaging can be found in a review by Squier [32].

Simultaneous μm spatial and temporal resolution is of great desire for the inspection of ultrafast opto-electronic circuits. Another direction is to combine techniques of ultrafast spectroscopy with spatial resolution—to monitor, for example, diffusion and relaxation of excited carriers in semiconductors. In fluorescence microscopy additional information can be gained by measuring the lifetime of the fluorescence. Because this relaxation depends sensitively on the interaction of the fluorescing dye with the environment, the lifetime image can describe local field and ion concentrations in cells [33].

Another example where scanning can be complemented by temporal correlation with a reference pulse is confocal imaging of objects buried under scattering layers. Confocal microscopy distinguishes itself by its depth selectivity [see Figure 13.5]. Enhanced depth selectivity and optical amplification of the image signal are desirable for imaging through scattering layers that strongly attenuate the ballistic light. Both aspects can be addressed with scanning microscopy based on a sensitive correlation technique, such as heterodyning [34]. A realization of such a microscope is shown in Figure 13.6. It consists of a Michelson interferometer which contains a scanning microscope in one arm and a piezoelectric transducer for Doppler shifting the reference pulse in the other arm. The role of the pinhole is played by the coherent overlap of the plane reference wave and the image light. A maximum heterodyne signal is obtained if the wave front from the object is plane (parallel to the reference wave front), that is, if the object is in focus.

Assuming a layer of thickness L with scattering coefficient μ_s on top of an object with reflectivity R, the image signal of the correlation microscope can be

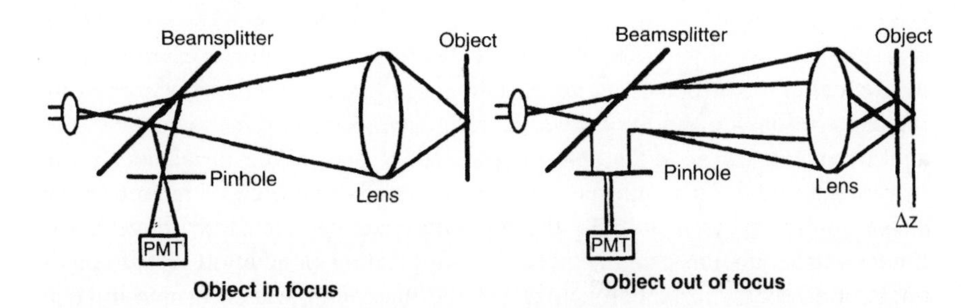

Figure 13.5 Schematic diagram of (reflection) confocal microscopy. Only light from layers that are in focus can pass through the pinhole and reach the detector. The beam (or object) is scanned in transverse direction to obtain a two-dimensional image of the layer that can be displayed by a computer.

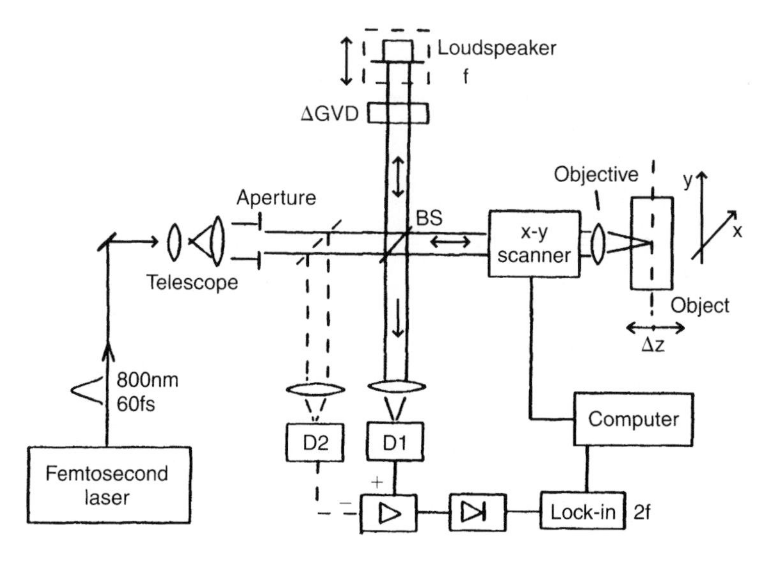

Figure 13.6 Schematic diagram of a correlation microscope based on heterodyne detection (from Kempe and Rudolph [35]).

written as:

$$S_d \propto \left(\mathcal{E}_{r0}\mathcal{E}_{s0}e^{-2\mu_s L}\right)\left(R \otimes \tilde{h}^2\right)\left[\frac{\langle \mathcal{E}(t)\mathcal{E}(t-\tau)\rangle}{\mathcal{E}_{r0}\mathcal{E}_{s0}}\right]. \tag{13.4}$$

$\mathcal{E}_{r0}$, $\mathcal{E}_{s0}$ are the amplitudes of the reference and the object pulse, respectively, $\tilde{h}$ is the amplitude point spread function (APF) of the objective, $\otimes$ describes convolution and $\langle\rangle$ denotes correlation. As can be seen from the first term in Eq. (13.4), the attenuation of the ballistic light can be compensated by a large enough amplitude of the reference wave (optical amplification). The second term is the convolution of the object response with the APF. This term is essentially the square root of the response of a confocal microscope and describes the transverse and depth resolution. Additional depth selectivity and discrimination of scattered light from layers close to the object is possible because of the correlation (third) term in Eq. (13.4). It is nonzero only if the length mismatch of reference and image arm is smaller than the pulse duration (coherence length). Monte–Carlo simulations of photon paths showed that time gating in addition to the confocal (spatial) gate substantially increases the maximum scattering density $\mu_s L$ through which nearly diffraction limited imaging is possible, see for example Magnor *et al.* [36].

The depth resolution of a microscope is usually measured by scanning a reflecting object through the focus. The improved depth resolution of the correlation as

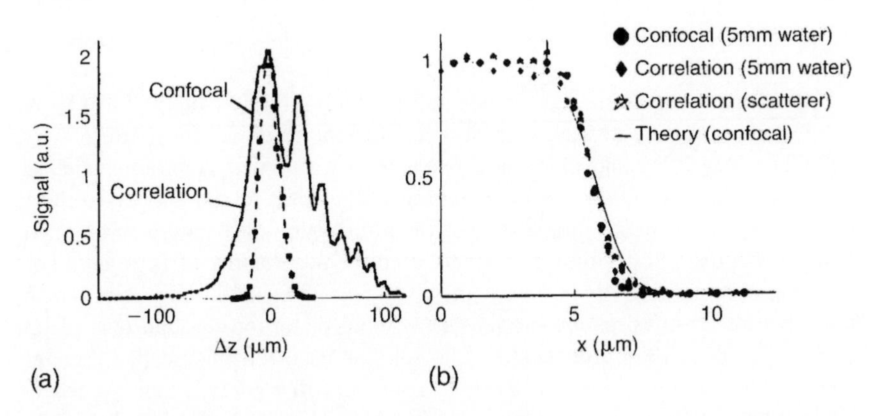

Figure 13.7 (a) Depth scan through a 10-mm glass layer that introduces spherical aberration and decreases the depth resolution of the confocal microscope (from Kempe and Rudloph [35]). (b) Scan over a straight edge. The scatter density was $2\mu_S L \approx 20$.

compared to the confocal microscope is shown in Figure 13.7(a). Figure 13.7(b) illustrates the transverse resolution as measured by scanning the beam focus over a straight edge buried under 5 mm of scattering material (96 nm latex spheres dissolved in water).

If the system is able to detect ballistic light, there is no loss in resolution. Microscopic techniques through scatterers have great potentials for noninvasive imaging of biological and medical samples, see, for example Schmitt *et al.* [37]. Figure 13.8 shows the images from a confocal and a correlation microscope of a cell layer of a leaf. The depth position of the layer was 80 μm from the lower epidermis.

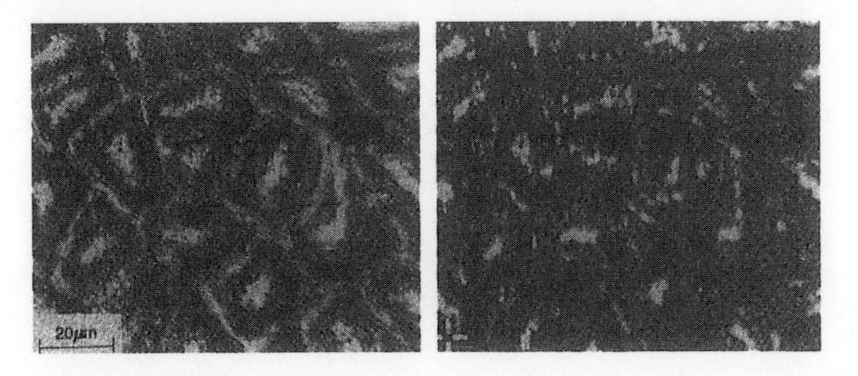

Figure 13.8 Correlation (left) and confocal (right) image of a cell layer 80 μm buried under the lower epidermis (from Kempe and Rudolph [38]).

13.2. SOLITONS

The concept of solitons has already appeared in various chapters of this book. For instance, we saw in Chapter 5 how an elementary model for a laser cavity, including only SPM and dispersion, leads to the nonlinear Schrödinger equation, which has steady-state soliton solutions. The same model applied to fibers finds stable pulse shapes propagating without distortion over long distances. We saw in Chapter 8 how solitons could be used for the shaping of fs pulses. This application transcends the fs time domain: In communication, pulse durations in the range of 20 to 80 ps are propagated without distortion through tens of km of fibers or over 10^6 pulse lengths. The solitons used for pulse-coded communication, which are *solitons in the time domain*, will be briefly reviewed later. The nonlinear Schrödinger equation was first derived and solved by Zacharov and Shabat [39] in the context of self-focusing and self-filamentation. The solutions of this equation describing stable filaments are *solitons in the space domain*. The high intensities of fs pulses can also lead to spatial solitons, such as the observed filamentation in air of fs pulses [40–42]. A more complex problem is that of solitons both in the temporal and spatial domain, which we will discuss at the end of this section.

13.2.1. Temporal Solitons

We saw in Chapter 8 that ultrashort pulses of sufficient intensity launched in a single-mode glass fiber above the zero dispersion wavelength evolve into a soliton. As mentioned previously, because the soliton maintains its characteristics over long distances, it is an ideal signal for pulse-coded long-distance communication. Because any wavelength above the zero dispersion point can be used, wavelength multiplexing is possible. The following problems must be overcome for long-distance propagation:

1. decrease of the soliton pulse energy over long distances because of linear fiber losses;
2. variations of the soliton propagation velocity resulting in timing jitter, hence loss of information; and
3. sliding of the soliton frequency, resulting in a mixture of adjacent frequency channels.

The first problem is solved with erbium-doped fibers (cf. Chapter 6) used as optical amplifiers. For example, in a test of a trans-Pacific soliton link [43], an Er-doped amplifier was located every 26 km to restore the original soliton pulse energy, as sketched in Figure 13.9.

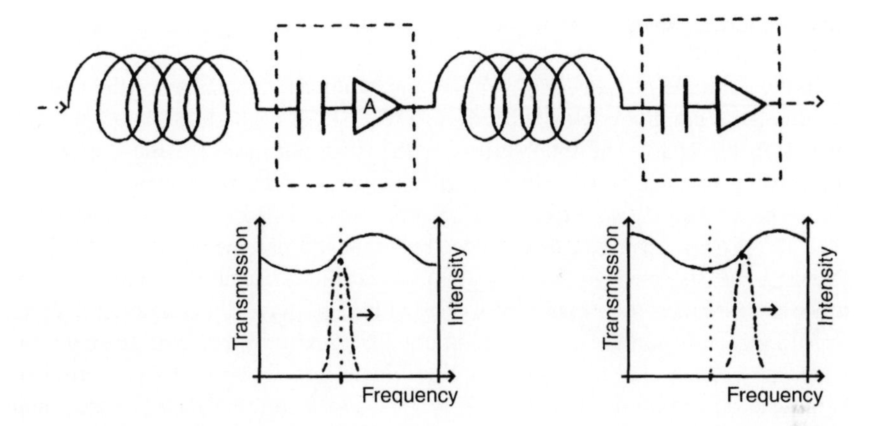

Figure 13.9 Optical soliton transmission line, showing schematically two units of fiber repeater. Each unit consists of a single-mode, low-loss communication fiber, a Fabry–Perot filter, and an erbium-doped amplifier. The soliton spectrum makes its way through the successive filters with sliding transmission peaks, while the noise sees the overall attenuation of the overlapping filters.

Gordon and Haus [44] showed that the timing jitter is related to a jitter in the pulses' central frequency. They found that a certain component of noise added to the soliton will instantly shift its optical frequency and consequently its velocity, hence its time of arrival. A consequence of the Gordon–Haus theory is that a soliton has the property to shift its average frequency toward the frequency of maximum transmission of a filter. Let us consider, for instance, a soliton whose frequency is slightly shifted from the filter peak transmission. The differential loss across the soliton spectrum, *in conjunction* with the ability of the nonlinear effect to generate new frequency components, provides a force to push the soliton back toward the filter peak. A frequency filter can therefore solve the time jitter problem by preventing the soliton frequency from drifting around.

In an actual system, to accommodate frequency multiplexing, a Fabry–Perot filter is used at the same location as the amplifier. Each transmission peak of the Fabry–Perot interferometer defines a particular communication channel. One disadvantage of the optical amplifiers is that the noise is also amplified. Because of the property of the soliton to shift its frequency in the presence of differential noise, filters of slightly different frequency can be put at the successive amplifier location (cf. Fig. 13.9). The soliton makes its way through the different filters located at each successive amplifier, while the noise, being linear, sees the attenuation provided by the overlapping filters centered at different frequencies. Typical etalons used as filters are Fabry–Perot interferometers of 1.5 mm length and 9% reflectivity [43]. The fiber can support communication channels at frequency intervals of 100 GHz, which is the free spectral range of such an etalon.

In the example of trans-Pacific communication, the frequency of the etalons at successive amplifiers was shifted by 0.18 GHz at each successive amplifier spaced 26 km apart. The total shift over the 9,000 km trans-Pacific distance is still smaller than the 100 GHz frequency spacing between channels [43]. The soliton pulse duration was 16 ps at 1557 nm. The fiber had an average dispersion $D = 0.5$ ps/(nm km). Recall (cf. Fig. 8.4) that the parameter $D = dk'_\ell/d\lambda = -(2\pi c_0/\lambda^2)k''_\ell$ is generally used to characterize fibers, because it relates directly to the group delay (in ps) per nm of bandwidth and per km propagation length.

Although this particular example is not specifically in the femtosecond time scale, the concept and implementation originate directly from the properties of fs pulse propagation discussed in Chapters 1 and 2, and the pulse compression techniques explained in Chapter 8. Stable soliton propagation requires a balance of positive (negative) SPM and negative (positive) GVD and negligible losses. The diffraction losses are eliminated in fibers by confining the high intensity pulse in a wave guide. In a bulk material, a mechanism for transverse confinement of the beam is required to compensate for diffraction losses. Such a mechanism is provided by self-focusing and self-filamentation, a problem addressed in the next subsection.

13.2.2. Spatial Solitons and Filaments

In this section, spatial solitons relate to the confinement of pulses in a self-guided wave guide. Because of their high peak power, femtosecond pulses are a primary source to observe this phenomenon.

Chiao *et al.* [45] showed that the propagation equations for a time-independent field, in the presence of a self-focusing nonlinearity, reduce to the nonlinear Schrödinger equation. As we have seen in Chapters 5 and 8, the nonlinear Schrödinger equation has stationary solutions. These solutions were precisely investigated in the context of self-focusing [39]. Steady-state solutions, however, are not a proof of the existence of stable filaments, in particular for pulsed radiation.

As detailed in Chapter 3, Akhmanov *et al.* [46] showed that a nonlinearity of order larger than n_2 and of opposite sign can result in the formation of stable filaments. As the beam collapses because of self-focusing, the intensity on axis increases until the self-defocusing (for instance, from a negative term in $\bar{n}_4 I^2$) balances the self-focusing produced by the positive $\bar{n}_2 I$ term. The filament stabilizes at a diameter w such that the defocusing and focusing are in equilibrium. Such filaments have been observed with continuous radiation in materials of large, slow nonlinearities, such as suspensions of latex spheres, aerosols, and

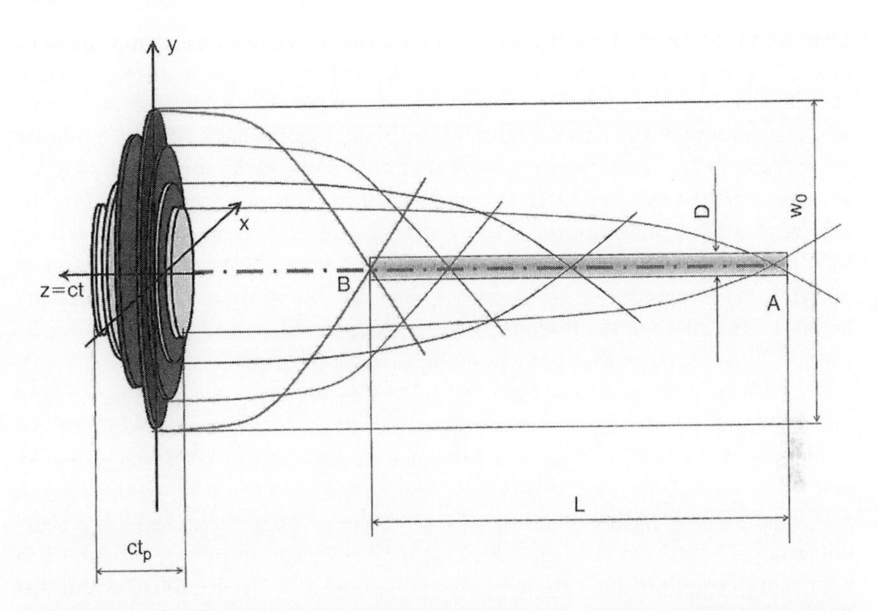

Figure 13.10 Illustration of the moving focus phenomena, for a pulse of duration t_p. Each disk represents a slice of the input pulse of certain power (darker shades correspond to higher powers). Self-focusing leads to focusing at different locations depending on the power of a particular slice.

microemulsions [47,48]. In the case of pulsed radiation, the existence of filaments has been questioned. Instead, it has been postulated that a "moving focus," which in solids leaves a filament-like trace, can be generated with pulses [49, 50].

Figure 13.10 illustrates the mechanism by which a self-focused pulse is expected to create a moving focus. On the left of the figure the initial pulse is shown at the beam waist characterized by w_0. The pulse is divided in successive slices of equal energy approximating the pulse profile. In the sketch of Fig. 13.10 the first and last slice start at the pulse FWHM. This first slice is assumed to be sufficiently above the critical power to focus at point A at a distance z_{SF} given by Eq. (3.173); the central slice focuses at point B. The length $AB = L$ is equal to $z_F(P_{max/2}) - z_F(P_{max})$. It is interesting to note that the pulse intensity of the input pulse is actually higher than the intensity in any plane within the focal region L for fs pulses. For ns pulses, the opposite is true. While this can be shown numerically, a simple model shall suffice to bring this point home.

Each slice of the original pulse gets focused at various (cylindrical) focal volumes distributed along the line AB. The diameter of the cylinders (focus spot) D is finite for the reasons discussed in Section 3.9.1; from experiments $D \approx 100\ \mu$m [51, 52]. Because each of the focal volumes contains the same number of photons as the original slice, the total energy in the cylindrical volume made up of all the focused slices is equal to the energy of the pancake shaped original pulse. We neglect the contribution from the out-of-focus light in a particular slice because the intensity is mainly determined by the in-focus components. The total energy of the pulse is initially distributed in a volume of approximately $V_i = c\tau_p \times \pi w_0^2/2$. This energy is thereafter distributed along the "focal volume" of approximately $V_f = \pi D^2 L/4$, which controls the pulse intensity. For the comparison of fs and ns pulse focusing let us assume that in both cases $L = 1$ m. In the case of a 10 ns pulses of $w_0 = 1$ cm, V_i is of the order of $0.5 \cdot 10^5$ mm^3, while V_f is of the order of 10^{-3} mm^3. A considerable increase in energy density is thus taking place in the filament region, hence the filament like damage tracks observed after high power nanosecond irradiation of solids.

In the case of fs pulses as sketched in Fig. 13.10, the pancake shaped initial pulse volume is only $V_i \leq 0.5$ mm^3, while the intense pulse sweeps a focal volume of $V_f \geq 15$ mm^3. In the case of a fs pulse, there is thus *less energy density* in the focal region than in the original pulse. Therefore, in the case of self-trapping of fs pulses in air, the strong nonlinear phenomena such as conical emission [41, 42] and multiphoton ionization [40] can only be explained if a significant portion of the original pulse is trapped as a "light bullet" within the filament as opposed to a moving focus.

We discussed in Section 3.9 possible mechanisms of beam trapping. In the simple theory of steady-state self-focusing, the filament remains confined through a balance of the self-focusing (term $n_2\mathcal{E}^2$) and a self-defocusing [term $n_3\mathcal{E}^3$ in Eq. (3.179)] of opposite sign. The simple interpretation most commonly cited is that the stabilizing higher order index produced by an electron plasma leads to defocusing. In the case of air, the electron plasma is created by (multiphoton) ionization. A second contribution to the negative index change stems from the shift of the absorption edge toward shorter wavelengths (because of the replacement of neutral molecules by ions).

The physical reality is more complex, because numerous effects other than the negative lensing of an electron plasma contribute to compensate self-focusing. Some of these effects are illustrated in a model of filamentation without ionization [53]. Any nonlinear phenomenon (of order higher than 3) that limits the pulse intensity will have a stabilizing influence on the filament. One example is third harmonic generation, that has been shown to be quasi-phase matched and to play a role in sustaining the propagation of filaments produced by IR pulses [54–56]. At the opposite end of the spectrum, optical rectification has been observed and—because of the short duration of the propagating optical pulse—has resulted in the generation of THz radiation. The experimental observation of a THz pulse emitted by a filament was explained as being the result of a longitudinal plasma oscillation created by the Lorenz force [57, 58]. Both of these nonlinear effects reduce the $n_2\mathcal{E}^2$ term by drawing power from the beam, hence acting similarly as a saturation of the focusing term.

A similar saturation effect can occur as a result of pulse splitting [59]. Multiple pulse splitting results from SPM and dispersion only, far below the power required for plasma formation, as demonstrated by Bernstein *et al.* [60, 61] in a measurement of the pulse temporal and spatial profiles for a self-focusing beam. Because the beam, with an initially Gaussian profile (4 mm diameter FWHM) focused down to not less than 1 mm over 23 m, the intensity never reached a level at which conical emission, harmonic generation, or ionization become significant. Therefore, the multiple pulse splitting that was observed resulted purely from phase modulation and dispersion. The phase modulation leads to a lower frequency of the pulse leading edge and to a higher frequency at the pulse trailing edge. Because of the normal dispersion of air, the leading edge travels faster than the trailing edge, resulting in pulse splitting. As the pulse splits, the peak intensity is reduced, resulting again in an apparent saturation of the self-focusing effect.

13.2.2.1. Application to Remote Sensing

Associated with the filaments is an intense white light or conical emission, which can be used to probe remotely the atmosphere [62, 63]. Two physical phenomena giving rise to that white light emission are illustrated in Figure 13.11. Figure 13.11(a) is the pulse intensity profile. SPM results in a similar phase modulation profile [Fig. 13.11(b)], leading to a frequency sweep [Fig. 13.11(c)]. The latter frequency excursion adds frequency components to the pulse, thereby broadening its spectrum. In combination with normal dispersion, the SPM results in pulse splitting. The low frequency components generated in the leading edge of the pulse propagate faster than the high frequency components generated in the tail of the pulse. This mechanism accounts for spectral broadening in the formation stage of filaments. As filamentation sets in, further spectral broadening is mainly because of amplitude modulation rather than phase modulation [64, 65]. This results from the self-steepening effect because of the first-order correction to the SVEA ($\partial \mathcal{P}_{NL}/\partial t$), as detailed in Section 3.6.3 and illustrated in Figs. 13.11(d) through (f). The responsible nonlinear polarization is sketched in Fig. 13.11(d). Its impact on the propagating pulse can be illustrated by subtracting a field distribution of shape similar to curve (e) from the original pulse, leading to Fig. 13.11(f).

The white light emission of filaments has been studied extensively both theoretically and experimentally (see for instance Aközbek *et al.* [65] and Kasparian *et al.* [66]). Launched vertically, the white light emission has been used to irradiate the atmosphere up to 13 km altitude [62, 63]. Spectral analysis of the time-gated return provides a means to study the composition of the atmosphere [67]. Measurements of the angular distribution of the supercontinuum emission show that it is peaked toward the backward direction, an effect that could be attributed to an inversion created in multiphoton excited nitrogen [68, 69].

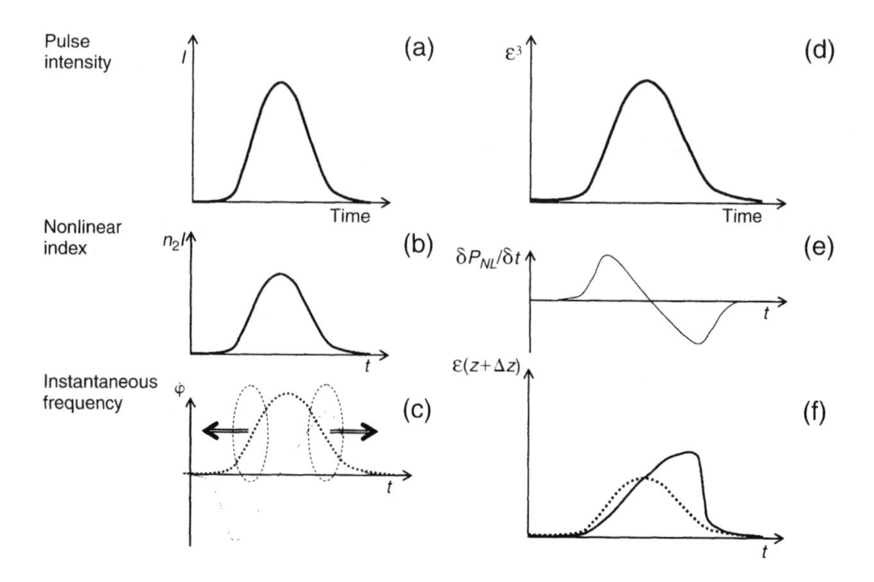

Figure 13.11 Left: illustration of the influence on pulse propagation of the Kerr effect nonlinearity. The pulse intensity represented in (a) gives rise to a nonlinear index (b), hence the frequency modulation shown in (c). Right: influence of the shock term, proportional to the derivative (e) of the nonlinear polarization (d), leading to the asymmetric shape (f) for the field.

13.2.2.2. Application to Laser-Induced Discharges

The ionizing properties of a filament, combined with the relatively long spatial extension, should make the filament an ideal tool for laser-induced discharges. The process by which seeding of a low density of charges in a uniform electric field results in a discharge has been investigated theoretically and experimentally [40]. High voltage electrical discharges were triggered and guided with UV pulses (248 nm) of only a few mJ energy over a gap of the order of 0.5 meter [70, 71]. The gap or reduction in breakdown voltage do not scale with the pulse energy. Even with 400 times higher pulse energies, the length over which a discharge could be guided reliably was only 2 or 3 times longer [72]. Plasma interferometry measurements have shown that the ionization created in air by a filament exists for only 200 ps [73]. A typical electron density of 10^{17} cm^{-3} in air causes a decrease in the index of refraction, which lasts for about 200 ps because of electron–ion recombination and the attachment of electrons to oxygen. There is a subsequent decrease in the index of refraction because of expansion of the air heated by the laser excitation. The rarefied air can provide a preferential path for the discharge, but does not efficiently reduce the minimum field required to produce a discharge over a given gap. This explains the difficulties in achieving breakdown reduction and discharge guiding over large gaps. A solution to the problem of triggering and guiding discharges over long distances is to maintain the ionized channel created by the UV or IR filament through inverse Bremstrahlung (plasma heating) and photodetachment of oxygen. Pulse intensities ranging from 1 to 10 MW/cm^2 have to be maintained in the channel, for the time duration required to trigger the discharge [71].

The ability to trigger a discharge depends also on the initial electron density deposited by the fs pulse in the filament path. Various evaluations of the electron density have been based on conductivity measurements. The values reported vary between 10^{12} cm^{-3} [74], 10^{14} cm^{-3} [51], and 10^{16} cm^{-3} [75]. Measurements performed with the same setups and 1-ps UV pulses (250 nm) and 100-fs IR pulses (800 nm) [51] indicate a 20 $\times$ larger conductivity induced by the UV filament than by an IR filament, the latter produced by a 10 $\times$ more energetic pulse. The larger conductivity in the UV filament is important for laser discharge applications and is attributed to the fact that the nonlinear ionization is only a three-photon process compared to a 9- to 10-photon process for the IR filaments. In both cases a diameter of the filaments of 100 μm was obtained, implying an intensity of 1 TW/cm^2 in the UV filament versus 100 TW/cm^2 in the IR filament. Using these intensities and typical cross sections for the multiphoton absorption one estimates electron densities that are consistent with the conductivity measurements.[1]

13.2.3. Spatial and Temporal Solitons

The GVD parameter k''_ℓ of air is approximately 0.15 fs^2/cm at 800 nm [78,79]. For a 50-fs pulse that is often used to produce filaments, this dispersion corresponds to a characteristics distance [as defined by Eq. (1.128)] of $L_D \approx 160$ m. All dispersion effects of the atmosphere are thus negligible at that wavelength.

If we consider instead the dispersion of air at 248 nm, $k''_\ell \approx 1$ fs^2/cm and for a 50-fs pulse $L_D \approx 25$ m. Thus pulse broadening should occur over distances of the order of 10 m with UV fs pulses. Therefore, the existence of filaments over tens of meters requires that the pulse be trapped in space *and* in time.

A similar situation arises with filaments created at 800 nm with pulses of less than 10 fs duration: the dispersion length in air is now only $L_D \approx 6$ m. Using gases other than air (such as Ar) at higher pressure the characteristic length can be made of the order of tens of cm, and spatial-temporal soliton formation is possible. A recent application of filaments involves compressing intense fs pulses

---

[1]Couairon and Bergé [76] using the Keldysh formula [77] for the evaluation of the three-photon ionization of oxygen, infer from their calculations a beam diameter of 40 μm for the UV filament and 200 μm for the IR, leading to the same intensity in UV and IR filaments.

(i.e., <20 fs, ≈1 mJ) down to a few fs (5.7 fs to 5.1 fs reported in Hauri *et al.* [80, 81]). This mechanism of (soliton) compression is based on phase modulation and dispersion, as explained in Chapter 8. Numerical simulations relating to this compression mechanism have been published by Couairon [82].

13.3. SENSORS BASED ON FS LASERS

13.3.1. Description of the Operation

This section is dedicated to some applications of femtosecond lasers as sensors. Rather than to use the beam radiated by the laser to perform measurements, this type of metrology uses the laser as a differential interferometer. Two or more pulses that are circulating in the laser resonator are made to interfere and a beat note is measured. A form of amplitude coupling is required to ensure that the pulses cross at the same two points during each roundtrip, which also ensures that the two output pulse trains corresponding to each of these pulses have the same repetition rate. Any phase coupling (for instance backscattering from one pulse into the other) at a crossing point should be avoided, because it leads to frequency locking of the two pulses, which washes out the differential measurement. Two possible ways to achieve the desired two pulse per cavity roundtrip operation are:

(1) inserting of a saturable absorber *flowing dye jet* in the laser resonator and

(2) use of an intracavity pumped Optical Parametric Oscillator (OPO).

The first method is relatively straightforward but limited to laboratory applications [83, 84]. The pulses meet at the saturable absorber, because this is the configuration of minimum loss, because—as discussed in Section 5.3.2—standing wave saturation is more effective than traveling wave saturation. To prevent locking of the two pulse frequencies to each other, liquid jet saturable absorbers are used, because the phase of the backscattered pulse is averaged out [83]. In the case of a ring laser (Figure 13.12) the two pulses circulate in opposite directions in the cavity. The pulse crossing point is "imaged" on a detector via a "detection interferometer" (the optical path for the two pulses from the crossing point to the detector is the same).

The second method consists essentially in having a fs pump laser, in the cavity of which an OPO crystal is inserted (Figure 13.13). That same crystal is also part of the signal cavity [85]. At each passage of the pump pulse through the OPO crystal, a signal pulse is generated. There are therefore two signal pulses per cavity roundtrip time. The repetition rate is the same for both pulse trains, because it is uniquely determined by the length of the pump cavity ($\tau_{RT} = L_p/\nu_{gp}$ where L_p is the pump cavity length, and ν_{gp} the group velocity averaged over

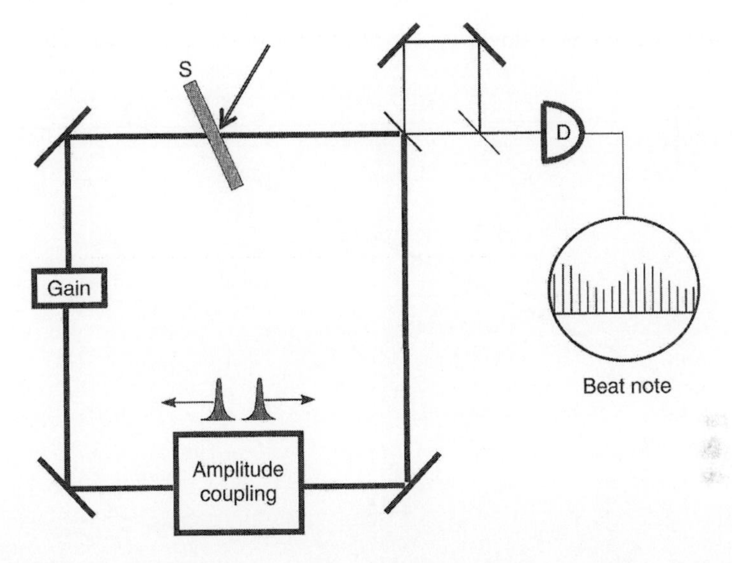

Figure 13.12 General configuration of sensors based on interfering the outputs of a laser with two intracavity pulses per roundtrip, in a ring configuration. A saturable absorber dye jet sets the intracavity pulse crossing point. An extracavity detector is located at equal optical path from the pulse crossing point in the cavity. The gain medium is located at 1/4 cavity perimeter from the pulse crossing point to ensure equal gain recovery after each pulse passage. *S* is a possible sample of which certain properties are to be measured (sensed). For this an excitation can be applied synchronized to the pulse roundtrip if needed.

one round-trip) in the pump cavity. In the case of the intracavity pumped OPO, it is the fixed repetition rate for the two pulses that results in a fixed crossing point.

Whether in a linear or a ring cavity, two circulating pulse trains (labeled by the index "1" and "2" below) of identical repetition frequency Δ are generated, which, in the frequency domain, correspond to mode combs of frequency

$$\nu_{m,1} = f_{0,1} + m\Delta$$

and

$$\nu_{m,2} = f_{0,2} + m\Delta,$$

as discussed in detail in Chapter 5 [cf. Eq. (5.11)]. Mixing these frequencies in a quadratic detector (photodiode) produces a beat note $\Delta\nu$ at the difference frequency of the two carrier to envelope offsets:

$$\Delta\nu = |f_{0,1} - f_{0,2}|. \tag{13.5}$$

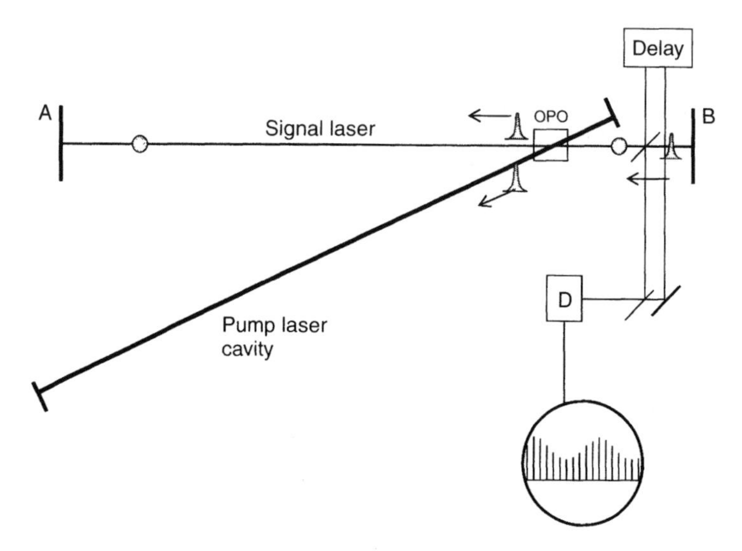

Figure 13.13 Linear cavity configuration of an intracavity pumped OPO. An OPO crystal is common to the pump and signal cavities. The two pulses generated in the signal cavity cross at the two points marked by a dot. As in Fig. 13.12, the two outputs are made to interfere after an appropriate delay line at the detector *D*.

Let us now assume that by some process (to be defined later) the dispersive properties of sample *S* (Fig. 13.12) are different for each circulating pulse, which changes the effective index of the cavity $n(\nu)$ defined in Eq. (5.1). As a result the mode combs become

$$\nu'_{m,j} = f'_{0,j} + m\Delta'$$

resulting in a new beat note of

$$\Delta\nu = |f'_{0,1} - f'_{0,2}|, \tag{13.6}$$

because the repetition rate Δ' is locked to the same value for either train of pulses. Experiments have shown that the difference $|f'_{0,1} - f'_{0,2}|$ is proportional to the phase difference $\Delta\varphi$ between the two pulses experienced at each roundtrip:

$$\Delta\nu \propto \frac{\Delta\varphi}{2\pi\tau_{RT}}. \tag{13.7}$$

In many cases the constant of proportionality is close to one, some examples will be discussed below. Thus, the sensor described in this section is basically a phase detector ($\Delta\varphi$). In contrast to standard techniques that convert a phase difference into an amplitude difference in an interferometer, this method converts the phase difference into a beat frequency. As we saw in Fig. 5.6, even with an unstabilized laser it is possible to record a change in $\Delta\nu$ of 1 Hz, which, according to Eq. (13.7) corresponds to a phase sensitivity of 10^{-7} in a cavity of $\tau_{RT} = 16$ ns. If—as is often the case—the phase difference is because of a differential change in cavity length or perimeter ΔP: $\Delta\varphi = k_\ell \Delta P$, then Eq. (13.7) can be written:

$$\frac{\Delta\nu}{\nu} = \frac{\Delta P}{P}. \tag{13.8}$$

With $\nu \approx 3 \cdot 10^{14}$ Hz, $\Delta\nu \approx 1$ Hz and a cavity perimeter P (corresponding to $2L$ for a linear cavity), of the order of 1 m, this sensor should be sensitive to changes of cavity length of the order of 10^{-15} m.

The different types of sensors distinguish themselves by the particular process of converting a physical quantity into a phase shift. Some examples will be given next, pertaining to two categories of detectors:

1. Detector of nonreciprocal effects. The "sample" can be the laser itself (rotation sensing), a flowing fluid (motion measurement by Fresnel drag), a material with a high Verdet constant (magnetic field measurement), or a resonant atomic vapor (intracavity phase spectroscopy). This type of response exists also with cw lasers, but, with mode-locked ring lasers, the sensitivity is not limited by a dead band.
2. Detector of changes of the optical cavity length externally synchronized by processes such as the electro-optic effect or the displacement of reflecting surfaces by phonons or the change in cavity length due to nonlinear indices.

The latter type of measurement is unique to mode-locked lasers, because it exploits the property that the two intracavity pulses occupy different positions in the cavity at different times and can thus be distinguished. This is not the case if the cavity is filled uniformly by a cw beam, as is the case in a He–Ne ring laser for instance.

13.3.2. Inertial Measurements (Rotation and Acceleration)

13.3.2.1. Rotation

The mechanical gyroscope (gyro) is an instrument based on the conservation of angular momentum of a spinning wheel. The fixed orientation of the angular momentum provides information on the motion of a moving frame of reference. An optical gyro based on a fs laser is essentially the instrument sketched in Fig. 13.12, without any intracavity addition.

There are three possible descriptions of the operational principle of this instrument that were initially introduced for cw lasers. It turns out that, with some caution, one can also apply the same arguments to explain the behavior of mode-locked laser gyros. The key is that an intracavity element always ensures that the pulse roundtrip time τ_{RT} remains the same for both pulse trains. Let us assume a ring laser of diameter R and rotating with angular velocity Ω. The first model considers the interference pattern created by the two "counter-rotating" beams. The two beams having the same frequency, this standing wave pattern (of period $\lambda_\ell/2$) is fixed in an absolute (i.e., nonaccelerating) frame of reference. A detector in the rotating laboratory frame will produce a sinusoidal signal from these interference fringes passing by at a rate $2R\Omega/\lambda$.

A second approach is to consider that, for the observer in the laboratory frame rotating at the angular velocity Ω, the two circulating beams will be Doppler shifted up and down by $\nu_\ell R\Omega/c$, resulting in a total shift (or beat note) of $\Delta\nu = 4A\Omega/(P\lambda_\ell)$ (A and P being the area and perimeter of the ring, respectively). The factor $\mathcal{R} = 4A/(\lambda P)$ is called the "scale factor" of a ring laser and is valid for cavities of arbitrary shape [86, 87].

A third point-of-view, now with the laboratory frame at rest and the laser rotating, is that the two counter-rotating beams are resonating in a cavity that is lengthened in the sense of rotation (cavity perimeter P_2), shortened in the other direction (cavity perimeter P_1). Hence, the corresponding mode combs [cf. Eq. (13.6)] will be shifted in frequency by the amount $\Delta\nu = |f'_{0,1} - f_{0,2}| = \nu_\ell(P_2 - P_1)/c = 4A\Omega/(P\lambda_\ell)$.

Compact single-mode He–Ne ring lasers are extensively used as navigation gyroscopes in commercial aircrafts. CW laser gyros are plagued by a phenomenon called lock-in: the response of CW laser gyros is zero for a range of small rotation rates. This dead band is because of the scattering of one circulating beam of the ring laser into the other direction. This weak coupling may "injection lock" the counter-propagating modes, i.e., force each of them to operate at the same frequency as radiation injected from the other mode.[2]

The "lock-in" problem can be avoided with ultrashort pulse lasers, where the two counter-propagating pulses meet in only two places. If the phase coupling at these meeting points is avoided, (as is the case in the examples presented so far), the dead band is eliminated. There is no phase coupling in the case of an intracavity pumped OPO such as sketched in Fig. 13.13. Such is also the case

[2]Hence the label mode-locking, which is sometimes given to this effect, because the scattering of one circulating mode locks the frequency of the other. This is not to be confused with the mode-locking creating ultrashort pulse trains.

for a phase conjugated interaction through degenerate four wave mixing, such as occurs in a saturable absorber dye jet as in Fig. 13.12 [88]. This lock-in problem will be dealt with in more detail in Section 13.3.3.1 to follow.

13.3.2.2. Acceleration

We have seen in the previous section that the ring configuration of Fig. 13.12 leads naturally to a form of inertial sensing of rotation. Similarly, the linear configuration of Fig. 13.13 has an inertial response as accelerometer. Let us consider indeed that the whole laser is accelerating along the direction BA (the laser cavity is rigid, and the distance from A to B is L). Let us consider a pair of pulses issued at the crossing point at a distance ℓ from mirror B. One of the two intracavity pulses traveling to the right hits mirror B, receives a Doppler shift $\nu_\ell v/c$, before proceeding to the left and sending an output to the delay line. Meanwhile, the other pulse travels to the right, receives a Doppler shift $\nu_\ell[v + a(L - 2\ell)/c]/c$, before proceeding to the right and sending an output to the detector, to interfere with the other output. The measured beat note is thus $\Delta\nu = a(L - 2\ell)\nu_\ell/c^2$. If the detection delay arm is increased by the amount $N\tau_{RT}$, where N is a large integer, the beat note is

$$\Delta\nu = a[N\tau_{RT} + (L - 2\ell)/c]\nu_\ell/c. \tag{13.9}$$

13.3.3. Measurement of Changes in Index

In this subsection we will describe how a change in refractive index synchronized to the cavity repetition rate can be sensed. A straight forward measurement is that of the nonlinear index of a sample inserted in the cavity of Fig. 13.13. The sample can be the lithium niobate crystal of the OPO itself [89]. If the two pulses circulating in the OPO cavity have an intensity I_1 and I_2, a beat note appears because of the different phase shift introduced by the nonlinear index of $LiNbO_3$. The beat note frequency

$$\Delta\nu \propto \frac{2\pi}{\lambda_\ell}\bar{n}_2\langle I_2 - I_1\rangle \tag{13.10}$$

is proportional to the nonlinear index of the crystal. $\langle\rangle$ denotes averaging over the pulse and beam profile.

In Fig. 13.12, an arrow at the sample S indicates a possibility to change some property of a cavity element by an external signal that has the same periodicity as the pulse rate. An example is a voltage applied to a Pockels cell. The synchronization can be ensured by using the voltage from a photodetector monitoring the pulse train. To demonstrate the concept, a Pockels cell oriented as a

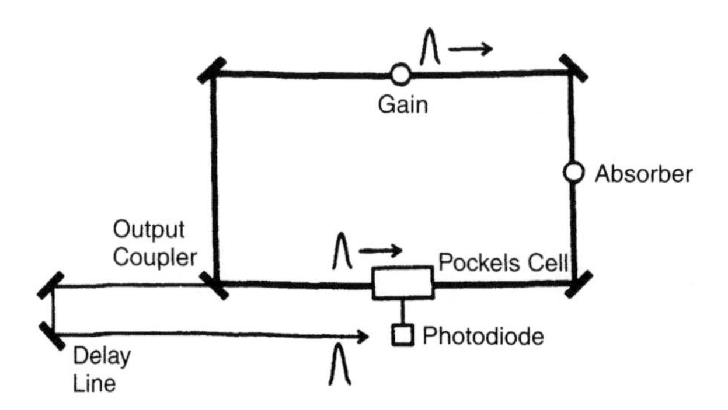

Figure 13.14 The output of the clockwise pulse of the ring cavity reaches the avalanche photodiode shortly before the time of arrival of the counterclockwise pulse in the Pockels cell. The electrical pulse from the photodiode is applied to the Pockels cell, resulting in a change in index, and therefore also in cavity length for the counter-clockwise pulse.

phase modulator is inserted in the cavity as shown in Figure 13.14. An avalanche photodiode detects the pulse train from the laser and applies an electrical pulse to the Pockels cell. Appropriate optical delay ensures temporal coincidence of the electrical pulses and *one* of the cavity pulses at the electro-optic crystal. Because the other cavity pulse always reaches the Pockel's cell between electrical pulses, the two counter-propagating pulses experience different indices in the cell. The optical length of the cavity is therefore different for the two senses of circulation of the intracavity pulses. Therefore, interfering the two outputs on a detector will result in a beat frequency between the two outputs that is equal to the difference in the CEO frequencies. The voltage $V_0 \cos(2\pi t/\tau_{RT})$ applied across the thickness e of the electro-optic crystal results in a change of index $\Delta n = (d_{\text{eff}} V_0/e) \cos 2\pi t/\tau_{RT}$ over a crystal length ℓ. The resulting beat note is:

$$\Delta \nu = K \frac{\Delta n \ell}{\lambda_\ell \tau_{RT}} = K \frac{V_0 d_{\text{eff}} \ell}{e \lambda_\ell \tau_{RT}}, \tag{13.11}$$

where K is a constant of proportionality shown to be close to one in this experiment and d_{eff} is the electrooptic constant.

Figure 13.15(a) shows a plot of the beat frequency versus the amplitude of the electric pulse applied to the modulator. By varying the optical delay, one can record the temporal response of the detector–electro-optic crystal combination as shown in Fig. 13.15(b). The temporal resolution is that of the detector–crystal combination. The particular measurement reproduced here was performed with an avalanche photodiode and 1 m cable delay to the crystal [83]. Figure 13.15(b) shows a resolution of about 300 ps. The intrinsic resolution of the method,

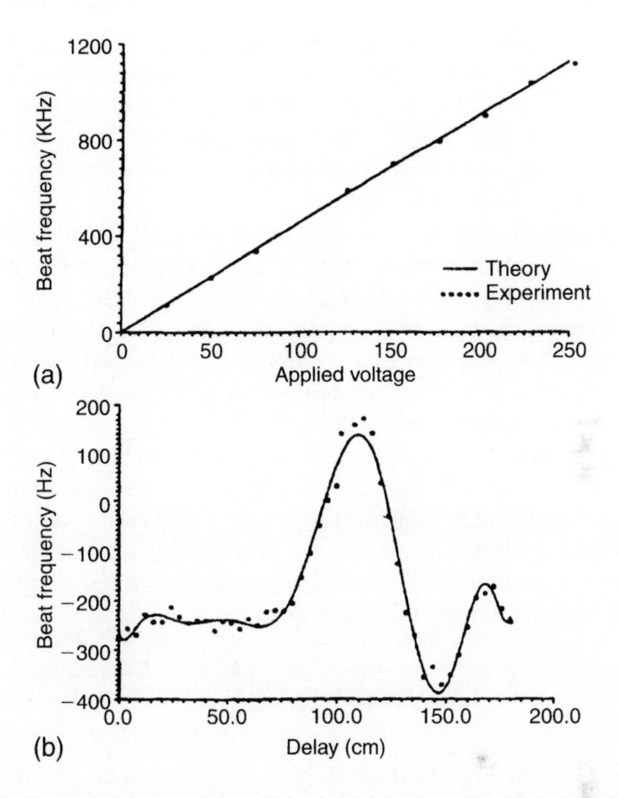

Figure 13.15 Beat note versus amplitude of the signal from the avalanche photodiode (a), all other parameters being kept constant. (b) Change in beat note versus the optical delay of the pulse impinging on the photodiode.

however, is in the fs regime, limited only by the pulse duration. This arrangement is a fast and sensitive tool for studying the intrinsic response of photodetectors (by measuring directly the change of index because of the generated carriers) or photodetector–modulator combinations. The best sensitivity in these measurements can be achieved when all other contributions to the beat note (such as rotation [90] or air currents [91]) can be eliminated. The linear laser is therefore to be preferred for this class of measurements. There is no "bias beat note" in the linear laser, because the intracavity circulating pulses travel through the same optical elements in the same order.

13.3.3.1. Dead Band and Measurement of Low-Level Scattering

If there is an optical element at one of the crossing points of the two intracavity pulses, the fields may couple into each other. This injection lock-in modifies the otherwise linear relationship between beat note and mode spacing between the two pulse trains generated by the laser, leading to what is known as the "dead band" in cw laser gyros. This problem is similar to frequency pulling and locking though injection seeding. Theoretical treatments for the cw case can be found in Siegman [92] for example. With small modifications they can also be applied to mode-locked lasers as described below.

The phenomenon of coupling in the case of the mode-locked laser with two intracavity pulses can be modeled best by considering the time evolution of the spectral modes of the pulse trains. We note that this problem involves only one time scale t: the time scale of the evolution of the phase and amplitude of the pulses from the two trains over a large number of roundtrips. Obviously this time scale is much larger than the pulse duration. It is because of this that the main characteristics of the cw models apply.

We consider each of the pulses of index $i = 1, 2$ to be represented by their spectral field envelope $\tilde{\mathcal{E}}_i(\Omega, t)\exp[i\phi_i(\Omega, t)]$. If $\tilde{r}$ is the complex scattering coefficient for one field into the counter-propagating field, at each roundtrip, a fraction $\tilde{r}\tilde{\mathcal{E}}_1$ of the field $\tilde{\mathcal{E}}_1$ of pulse 1 is injected into the field $\tilde{\mathcal{E}}_2$, and vice versa. We make the approximation that the pulses are unchirped, and preserve their shape. The pulse belonging to train i can be written as:

$$\tilde{\mathcal{E}}_i(\Omega, t)\, e^{i[\phi_i(\Omega,t)]} \approx \tilde{\mathcal{E}}(\Omega)\left[\mathcal{E}_{0,i}(t)\, e^{[i\phi_i(t)]}\right]. \tag{13.12}$$

The equations of (slow) motion for each frequency mode of each pulse are:

$$\begin{aligned}
\frac{d\mathcal{E}_{0,1}(t)e^{i[\phi_1(t)]}}{dt/\tau_{RT}} &= \frac{\alpha_1}{2}\mathcal{E}_{0,1}(t)e^{i[\phi_1(t)]} + re^{i\theta}\mathcal{E}_{0,2}(t)e^{i[\phi_2(t)]} \\
\frac{d\mathcal{E}_{0,2}(t)e^{i[\phi_2(t)]}}{dt/\tau_{RT}} &= \frac{\alpha_2}{2}\mathcal{E}_{0,2}(t)e^{i[\phi_2(t)]} + re^{i\theta}\mathcal{E}_{0,1}(t)e^{i[\phi_1(t)]}
\end{aligned} \tag{13.13}$$

where $\tilde{r} = r\exp(i\theta)$ is the (complex) backscattering coefficient coupling the two pulses at their meeting point. The net gain coefficient α_i at each roundtrip is defined as the difference of the saturated gain and the loss–roundtrip. Separating real and imaginary parts leads to the system of equations:

$$\frac{d\left(\mathcal{E}_{0,1}/\mathcal{E}_{0,2}\right)}{dt/\tau_{RT}} = r\cos(\theta - \psi) - r\frac{\mathcal{E}_{0,1}^2}{\mathcal{E}_{0,2}^2}\cos(\theta + \psi) \tag{13.14}$$

$$\frac{d\psi}{dt} = \Delta\omega + \frac{r}{\tau_{RT}}\left[\frac{\mathcal{E}_{0,2}}{\mathcal{E}_{0,1}}\sin(\theta - \psi) - \frac{\mathcal{E}_{0,1}}{\mathcal{E}_{0,2}}\sin(\theta + \psi)\right], \tag{13.15}$$

where $\psi = \phi_1 - \phi_2$, and we made the approximation that $\alpha_1 \approx \alpha_2$, consistent with a laser with a high Q cavity. The term $\Delta\omega$ on the right side of Eq. (13.1[illegible]) is the externally imposed relative shift between the two mode combs. The instantaneous angular beat note frequency is thus $\dot{\psi} = 2\pi\Delta\nu_b$ as defined by Eq. (13.15). Instead of a linear relation between the differential CEO and the beat note of Eq. (13.6), the coupling through the scattering results in a smaller beat note signal characteristic for frequency pulling:

$$\Delta\nu_b = \frac{\Delta\omega}{2\pi} + \frac{r}{2\pi\tau_{RT}}\left\{\sqrt{\frac{I_{0,2}}{I_{0,1}}}\sin(\theta - \psi) - \sqrt{\frac{I_{0,1}}{I_{0,2}}}\sin(\theta + \psi)\right\}, \tag{13.16}$$

where we have substituted the intensities for the fields. Note that these intensities have the meaning of average pulse intensities changing on a time scale of several roundtrips. If the scattering occurs through randomly moving scatterers (such as in a dye jet), expression (13.16) has to be averaged over r and θ. For a given r, the largest coupling occurs for $\theta = 0$. For this case and steady-state where the time derivatives in Eqs. (13.14) and (13.15) are zero, we find a particular solution:

$$\psi = 0 \quad \text{and} \quad \frac{I_{0,1}}{I_{0,2}} = 1 \quad \text{for} \quad \frac{|\Delta\omega|}{2\pi} \leq \frac{r}{\tau_{RT}}. \tag{13.17}$$

This steady-state with its zero beatnote frequency relates the width $\Delta\omega$ of the dead band to the magnitude of the backscattering coefficient.

Measurements such as those shown in Figure 13.16 yield the largest value of $\Delta\omega = \Delta\omega_{max}$ for which the beat note response $\Delta\nu_b = 0$. This value corresponds to the equal sign in the last expression of Eq. (13.17), and thus leads directly to the measurement of the scattering coefficient r [93]. As shown in the example of Fig. 5.6 the beat note resolution corresponds to 1 Hz in a laser with a roundtrip time of 10 ns. Therefore one can resolve an intensity backscattering coefficient of $r^2 \approx 10^{-16}$. Figure 13.16 shows the change in linear response of the beat note, when a dielectric mirror is inserted at a pulse crossing point. The observed dead band corresponds to an intensity backscattering coefficient of that mirror of about $2 \cdot 10^{-11}$, obtained when the mirror was used at an angle of incidence of 20°.

13.3.3.2. Reduction of the Beat Note Bandwidth through Stabilization

The 1-Hz linewidth of the beat note noticed in the previous example originates mainly from vibrations of the mirrors. Let us consider a particular mirror of the cavity, vibrating with an amplitude $a \approx 0.1\ \mu$m, at a mechanical resonance frequency corresponding to a period $T_b \approx 10$ ms. The vibration of the mirrors

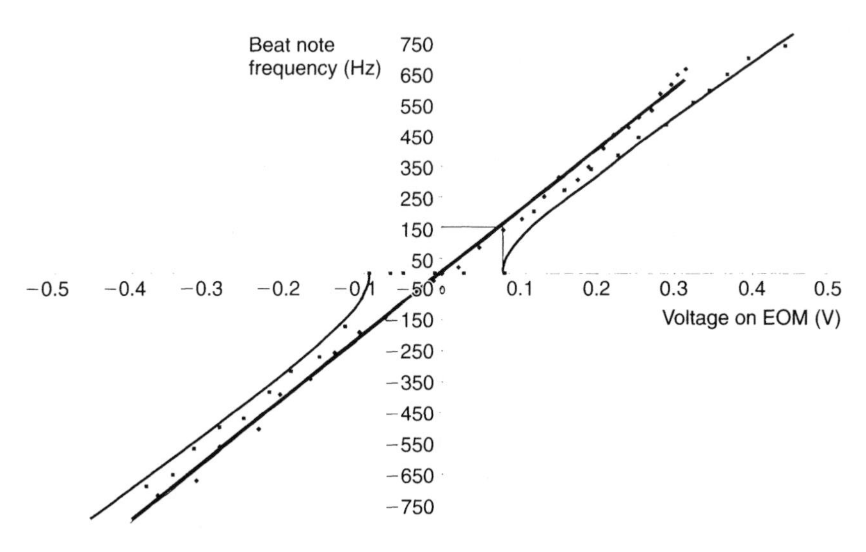

Figure 13.16 Beat note versus amplitude of the signal applied to the Pockel's cell (artificial rotation). The straight line is the empty cavity response. With a mirror inserted at 20° at the pulse crossing point, the beat note characteristics shows a dead band due to a small amount of backscattering from the mirror.

contribute to the bandwidth $\Delta\nu_{\text{laser}}$ of the individual modes of the two frequency combs emitted by the laser. The maximum possible change in phase per roundtrip because of vibration is $8\pi a/\lambda$; for a mean value averaged over many roundtrips and vibration periods let us use a/λ. The broadening of a laser mode can now be estimated:

$$\Delta\nu_{\text{laser}} = \frac{a}{\lambda\tau_{RT}} = \frac{\nu_\ell a}{P}. \tag{13.18}$$

Obviously, this bandwidth depends on the pulse repetition rate or the perimeter ($P = 2L$ for a linear cavity) of the cavity.

During the short time of a pulse roundtrip, the two pulses hit the vibrating mirror at a different location. Hence a slightly different cavity perimeter ΔP_b seen by the two pulses. If δt is the difference in time of arrival at the mirror, a corresponding cavity length difference of $\Delta P_b \approx a \times 2\pi\delta t/T_b$ results. For instance, to $\delta t = 1$ ns there corresponds a mirror displacement between the two pulses of $2\pi \cdot 10^{-14}$ m. The cumulative effect of interfering the two pulse trains on the detector results in a beat note bandwidth $\Delta\nu_b$:

$$\Delta\nu_b \approx \frac{\Delta P_b \nu_\ell}{P} = \frac{a\tau_{RT}}{T_b}\frac{\nu_\ell}{P} = \left(\frac{\tau_{RT}}{T_b}\right)\Delta\nu_{\text{laser}} \tag{13.19}$$

where we have made use of Eq. (13.18). The beat note bandwidth is thus typically three orders of magnitude smaller than that of an isolated mode of the laser. Because the beat note is *proportional* to the laser bandwidth, a stabilization that reduces the laser mode bandwidth by three orders of magnitude will also reduce that of the beat note by the same amount, hence a sensitivity to rotation of the order of 10^{-4} degrees per hour can be expected. The stabilization should be applied to both pulses circulating in the cavity simultaneously. This requirement puts some constraints on the geometry of the cavity and the location of the control elements. For example, in the OPO configuration of Fig. 13.13, the beat note bandwidth of the signal can be reduced by mode stabilization, provided the correction to the cavity length is applied symmetrically with respect to the pulse crossing point (in such a configuration, both pulses traveling in opposite direction receive the same modification at various points of the cavity).

13.4. STABILIZED MODE-LOCKED LASERS FOR METROLOGY

In Chapter 5 we explained the ability of a stabilized fs laser to act as an extremely sensitive frequency ruler and most accurate clock at the same time. Recall that the frequency comb describing the laser output in the spectral domain

$$\nu_m = f_0 + m\Delta \tag{13.20}$$

relates optical frequencies ν_m to radio frequencies $\Delta = 1/\tau_{RT}$. As is obvious from this relation, two parameters are required to define the femtosecond frequency comb corresponding to a mode-locked pulse train. Likewise, to stabilize the comb, two parameters must be controlled (and stabilized) independently. These two parameters can for instance be the pulse repetition rate, and the exact frequency of a particular optical mode. A second option is to make an exact measurement of the frequency of two modes, which requires two optical standards within the bandwidth of the pulse. A third option is to measure one optical mode and the carrier to envelope offset (CEO). Only one calibrated measurement is required, if the carrier to envelope offset can be measured, controlled and set to a constant value. We will next present the technique to measure the CEO. As will be shown in the next two sections, this technique is limited to pulses shorter than 100 fs. For picosecond pulses, the CEO can be extracted from a precise measurement of the repetition rate and one optical mode. At the end of this section we will then describe how a fs laser can be stabilized to external cavities.

13.4.1. Measurement of the Carrier to Envelope Offset (CEO)

We will first describe the most commonly applied techniques to measure the CEO based on "f to 2f interferometry." We will then introduce a technique to produce the required octave-spanning spectrum using continuum generation in special fibers.

13.4.1.1. f to 2f Interferometry

This is a self-referencing method, in which a mode from the high frequency part of the spectrum (mode number m_h) is made to interfere with a frequency doubled mode from the low frequency portion of the spectrum (mode number m_l) [94]. The lowest component of the beat note spectrum

$$\Delta \nu = |2(f_0 + m_l \Delta) - (f_0 + m_h \Delta)| = |f_0 + (2m_l - m_h)\Delta| \tag{13.21}$$

is the CEO f_0.

An experimental setup of this self-referencing technique is shown in Figure 13.17. A two-prism spectrometer is used to separate the red and blue parts of the spectrum. An aperture A selects the desired blue portion of the spectrum, which is retro-reflected through the two prisms before being sent by a beam splitter to the f/2f interferometer. In one arm of the interferometer, a nonlinear crystal is inserted to frequency double the infrared pulse (shown as dotted line in Fig. 13.17). With a frequency doubling crystal type I, (BBO crystal phase matched for the infrared end of the pulse spectrum) the second harmonic is polarized orthogonally to the fundamental and can thus be combined with a polarizing beam splitter with the blue portion of the fundamental pulse. A polarizer oriented to project equal components of the two orthogonally polarized signals along its axis ensures maximum contrast of the beat signal on the photodetector APD.

There is no need for a wavelength selective aperture in the infrared portion of the spectrum, because the phase matching condition of the SHG crystal will provide the required spectral selection. Ideally, the two prisms should be configured for zero GVD, according to the calculations of Section 2.5.5, to prevent pulse broadening and phase modulation because of dispersion. The amount of intracavity glass and the prism separation can be selected for zero GVD according to Eq. (2.101).

13.4.1.2. Creating Pulse Spectra Spanning an Octave

There are two main approaches to produce pulses whose spectra span a full octave—(a) build lasers that emit extremely short (5 fs in the NIR) pulses, and

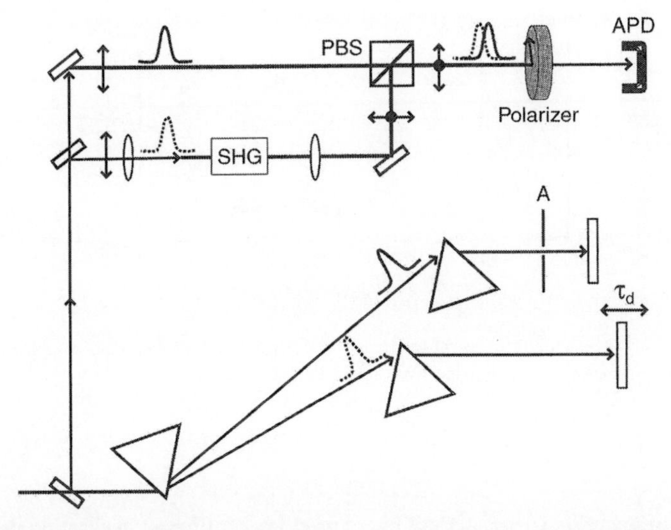

Figure 13.17 Measurement of the CEO frequency of a fs comb by f to 2f interferometry. The relevant high and low frequency parts of the spectrum are selected by a prism spectrometer with zero GVD. A high frequency part of the spectrum is selected by an aperture *A*. The pulses from the blue and IR part of the spectrum are sent to a Mach–Zehnder interferometer. The beams are reflected after the prism pairs at a lower level, so that they can be picked up by a mirror diverting them toward the Mach–Zehnder interferometer. A frequency doubling crystal (phase matched for SHG type I) is inserted in one arm of a Mach–Zehnder interferometer. The orthogonally polarized blue part of the pulse spectrum and the second harmonic of the infrared are combined with a polarizing beam splitter (PBS). An adjustable delay τ_d ensures that these two orthogonally polarized pulses meet on the beam splitter. A polarizer selects an equal component of both pulses to record their beat note with an avalanche photodetector (APD).

(b) broaden the spectrum of longer pulses outside the laser oscillator without destroying the mode structure and coherence. Because the first approach was described in Section 6.7.2 we will concentrate now on the spectral broadening.

Techniques based on optical fibers to generate a broad spectral continuum, while preserving the coherence of the comb, have been developed, cf. Section 3.7. Microstructured fibers [95], tapered fibers [96], and highly nonlinear dispersion shifted (HNLF) fibers [97–99] have been used to demonstrate octave-spanning continua. Figure 13.18 compares the dispersion of microstructered fibers and HNLFs. In all these fibers the crucial issue is low dispersion at the wavelength of the input pulse allowing for long interaction lengths. This together with the small confined beam diameter (guided mode) can produce large nonlinear effects, like continuum generation, with the relatively low pulse energies available from laser oscillators. Some characteristic parameters of fibers for continuum generation with low power lasers are listed in Table 13.1. The HNLF are ideally suited for

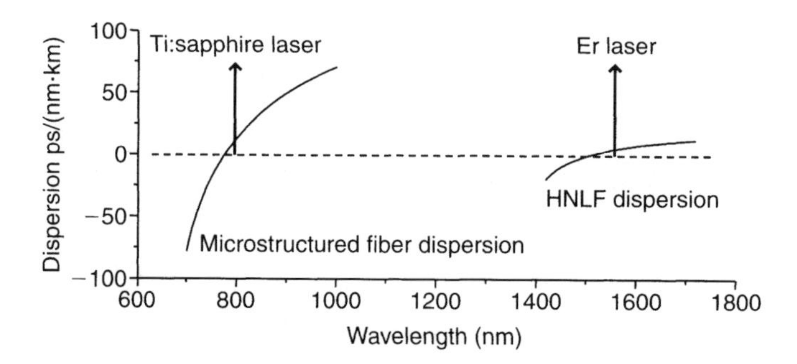

Figure 13.18 Dispersion of a typical small-core microstructured fiber in relation to the Ti:sapphire wavelength, compared to HNLF dispersion and an Er fiber laser (from Nicholson *et al.* [99]).

Table 13.1

Some characteristic parameters of fibers used for continuum generation compared to standard single-mode silica fibers (SMFs). The last column lists the Kerr nonlinearity in terms of parameters introduced in Chapter 3. The parameter γ is related to the nonlinear phase shift Φ_{NL} that is induced by a laser power P over a propagation distance L by $\gamma = \Phi_{NL}/(PL) = \omega\bar{n}_2/(cA_{\text{eff}})$, where A_{eff} is the effective fiber cross section, cf. Chapter 8.

Type	Wavelength (nm)	GVD ps/(nm km)	Nonlinearity
Standard SMF	800	−110	$\bar{n}_2 = 310^{-16}$ cm^2/W
Microstructure	770	0	
	780	10	$\gamma = 0.07$ W^{-1}m^{-1}
	900	70	
Tapered	850	122	$L_{NL} = 0.6$ mm
HNLF	1550	2.2	$\gamma = 9$ W^{-1}m^{-1}

the continuum generation with Er-based fiber lasers operating at pulse durations in the range of 50 fs to 200 fs.

The nonlinear Schrödinger equation, (3.190), has been used successfully to simulate the pulse propagation and broadening in microstructure fibers [100, 101]. It is observed that the broadest continuum is generated when the laser pulse is in the anomalous-dispersion regime of the fiber, cf. Fig. 13.18.

The method of f to $2f$ interferometry requires that some phase correlation between the pulse, and the train is maintained in the continuum generation. One method to study the coherence properties of the continuum is to interfere independently generated continua. Two possible setups to perform such experiments are sketched in Figure 13.19. A continuum is generated in each branch of a Michelson interferometer [Fig. 13.19(a)]. The length of one arm can be varied to introduce a delay τ_d between the spectrally broadened pulses entering the spectrometer. This delay is observed as fringes across the combined spectrum of the pulses (spectral interferometry). The visibility of fringes recorded as the spectrum is being scanned is measured. The modulation depth of these fringes is a measure of the coherence. Bellini and Hänsch [102] performed a coherence experiment with continua generated in microstructure fibers by a Ti:sapphire laser using Young's double slit instead of the spectrometer. There, the visibility of the spatial fringe pattern was analyzed.

The configuration shown in Fig. 13.19(b) with an integrated fiber interferometer was used to study the coherence of continua produced by fiber lasers operating at 1550 nm [103]. The result of the measurements of Bellini and Hänsch [102] and Nicholson and Yan [103] is that the continuum can be highly coherent (fringe visibility ≈ 1) when generated by ultrashort pulses (<150 fs). The fringe structure disappeared in a broad continuum generated by 1-ps pulses.

Experimental investigation of pulses of 190-fs duration in microstructure fibers shows that the pulse initially begins to self-Raman shift to longer wavelengths [104]. The interplay of anomalous dispersion and Kerr nonlinearity

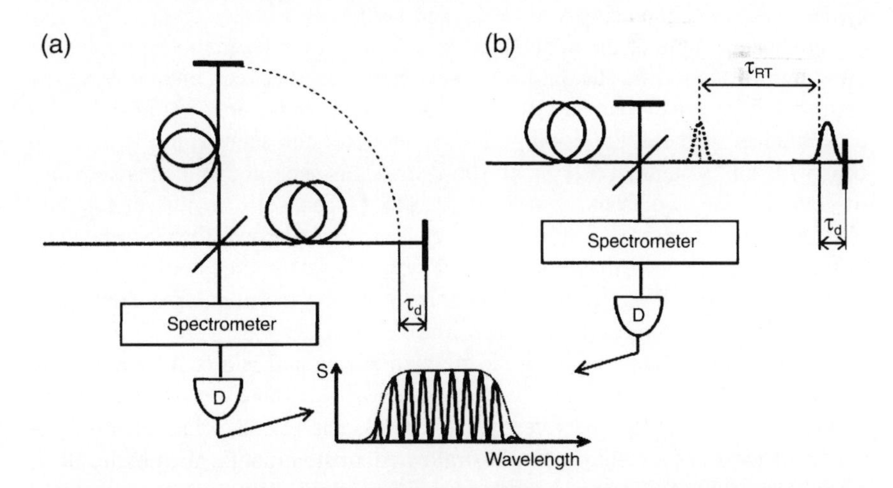

Figure 13.19 Spectral interferometry applied to the study of the coherence of the continuum. (a) The continuum is generated in each arm of the interferometer. (b) The continuum is generated outside the interferometer, but the relative delay of the two arms is close to the delay between successive pulses of the train. Therefore, it is the coherence between the continuum generated by two successive pulses that is being analyzed.

enables the formation and propagation of solitons propagating at different velocities. As these higher-order solitons break up, parametric four wave mixing generates frequencies at wavelengths shorter than the zero dispersion wavelength, eventually leading to an intense blue radiation coexisting with a broad infrared supercontinuum [104]. One conclusion was that special fibers can only be used to extend the frequency comb of sub-150-fs pulse trains.

13.4.2. Locking of fs Lasers to Stable Reference Cavities

Earlier stabilization experiments with mode-locked lasers often used short Fabry–Perot etalons to stabilize the *average position* of the comb, directly employing the techniques developed for single-mode lasers [105, 106]. More recent experiments have used external Fabry–Perot cavities as "mode filters," transmitting every 20th comb component, such that any individual mode of the comb can be unambiguously identified with a wavemeter [107, 108]. Ultimate stability is reached by stabilizing one mode to an atomic resonance while at the same time keeping the CEO f_0 constant. The use of the frequency comb as a "ruler" has been introduced by the group of Hänsch as a powerful means to compare frequency standards [108–110]. It was first applied to a measurement of the cesium D_1 line using a mode-locked laser [107].

Another example of the application of a frequency ruler involves linking the ytterbium to the iodine standard. The beat frequency between one mode of the comb and a frequency standard such as an Yb^+ stabilized laser at 871 nm provides a calibration of the entire comb. Another mode of the same comb is made to beat with an I_2 stabilized laser at 1064 nm. This measurement thus provides the ratio of the two frequency standards [111]. The noise in this direct ratio measurement is the same as that of a direct comparison between two independent I_2 standards, indicating that this ratio measurement was limited only by the noise of the I_2 stabilized laser. It is estimated that the excess fractional frequency noise introduced by the comb can be as low as 10^{-19} [112].

A building block of this time and frequency standard is a fs laser stabilized to an external Fabry–Perot cavity. A close-up of such an experiment is shown in Figure 13.20 [113]. The reference cavity is a long (62.5 cm) Fabry–Perot resonator made of a solid block of ultralow expansion quartz, with high reflectivity mirrors of the same substrate material optically contacted on both ends. The cavity was placed in a vacuum chamber to isolate it from thermal and acoustic noise and, if needed, to control the ambient pressure. The vacuum chamber usually kept the pressure inside the reference cavity below 15 mTorr. The laser cavity length is twice the length of the reference cavity, so that every other fs

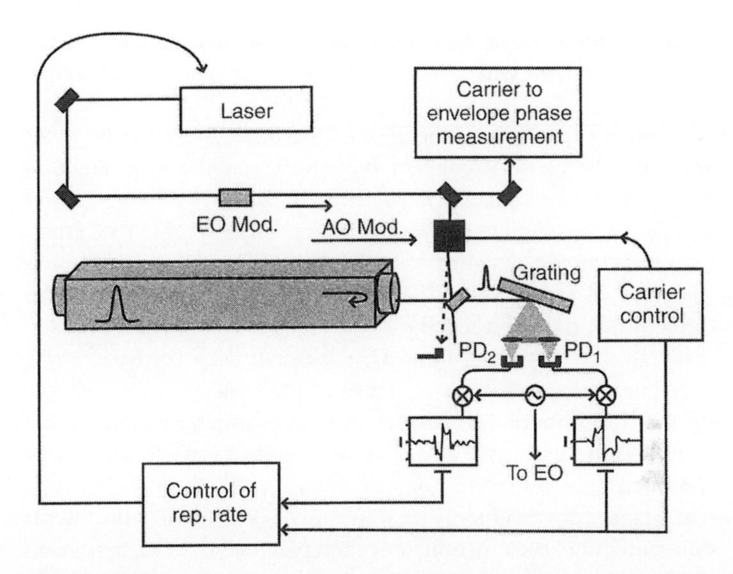

Figure 13.20 Stabilization of a mode-locked laser to a reference cavity. A phase modulator (EO) produces sidebands of opposite sign for each mode of the laser. The beam is mode-matched to the reference cavity. The beam reflected off the reference cavity is dispersed by a grating. The spectral component picked up by detector PD_1 is mixed with the modulation signal (which was applied to the phase modulator) to produce an error signal which is amplified and added as a correction frequency to the laser via an acousto-optic modulator. The difference between the signals from PD_2 and PD_1 provides an error signal for the repetition rate. The group velocity correction is either a tilt of an end mirror (following a prism sequence in the laser cavity) or an intensity adjustment of the pump laser of the fs laser.

laser mode can be transmitted.[3] The Kerr lens mode-locked laser produced pulses of $\approx$ 40-fs duration.

The standard technique to stabilize a single-mode laser is the Pound-Drever-Hall method [114]. The basic principle is to create sidebands (of opposite phase) of the laser with a phase modulator (EO Mod. in Fig. 13.20). The phase modulated signal is detected after reflection by the reference cavity, and the detector output is mixed with the original modulation signal (not shown). At exact resonance (between laser and reference cavity mode), the reflected signal has still equal and opposite sidebands, and the result of the mixing is a null signal. The balance between the two sidebands is lost outside of resonance, resulting in a negative or positive mixing signal below or above resonance, which is the error signal to

[3]A shorter cavity is not desirable: if N is the ratio of the laser to reference cavity lengths, the effective cavity finesse is reduced since the intensity of the pulse in the reference cavity is reduced by a factor $(1 - R)^N$ between incident laser pulses due to reflections at the mirrors.

apply as correction to some controlling element within the laser. In the case of a single-mode laser, this will generally be a piezo-element (PZT) controlling the cavity length.

As we saw in Chapter 5, in the case of the mode-locked laser, both the repetition rate and the carrier frequency have to be stabilized. A single mode can be stabilized using the same Pound-Drever-Hall [114] technique as for the cw laser, by selecting a single mode with a grating.[4] The single mode error signal is obtained by mixing the signal from the detector PD_1 with the phase modulation (at $f_{EO} = 10.7$ MHz, for example). Ideally this error signal should be applied to an element that only corrects the mode frequency, without affecting the repetition rate. Such a correcting intracavity element does not exist. A PZT on an end-cavity mirror does offer good control of the mode position, but not without affecting the repetition rate somewhat. An acousto-optic modulator outside the laser cavity can be used as a fast frequency shifter, which does not affect the repetition rate.

Two techniques are commonly used to apply a correction to the laser repetition rate, with minimum perturbation on other parameters. The first one is to tilt (using PZT) the end mirror that follows the prism sequence [108]. This technique modifies the group velocity through the prism sequence and affect the cavity length (mode spacing) only to second order. A major disadvantage for some applications is that it also affects the overall spectrum of the pulse train. A second technique is to act on the pump beam intensity [115].

Once a mode of the laser comb has been "locked" to a mode of the reference cavity, the mode comb is still free to "breathe" about that central mode. Fluctuations in the laser repetition rate result in the largest mode frequency excursion between the extreme ends of the spectrum. A convenient error signal is provided by mixing the *difference* between the signals of PD_1 and PD_2 with the original modulation at frequency f_{EO}. The error signal obtained in this way is the composite signal from *all* longitudinal modes detected within the spectral region spanned by the detectors.

13.5. PROBLEM

Consider a range gating setup like that in Fig. 13.1, with SHG as gating mechanism. Both the reference pulse and the backscattered radiation are focused, with crossed polarization, into a SH crystal phase matched for SHG type II at 620 nm. The crystal is urea, with a nonlinear coefficient of $d = 1.04 \cdot 10^{-23}$ C/V^2, and an index of refraction of 1.48. Given that both the reference beam and the signal beam are collimated with a diameter of 0.7 mm and focused into the

---

[4]In practice it will be a group of modes that will be selected by the grating.

crystal with a 2.5 cm focal distance lens, find the peak power of the reference beam required to achieve single photon up-conversion. What should the crystal length be? Given that crystal length, is there a limitation to the depth resolution of this 3D imaging system? Is there a limitation to the transverse resolution? Estimate these limits.

BIBLIOGRAPHY

[1] A. P. Bruckner. Some applications of picosecond optical range gating. *SPIE*, 94, 1976.

[2] J. L. Martin, A. Antonetti, A. Astier, J. Etchepare, G. Grillon, and A. Migus. Subpicosecond spectroscopic techniques in biological materials. *SPIE*, 211:128–132, 1979.

[3] J. L. Martin, Y. Lecarpentier, A. Antonetti, and G. Grillon. Picosecond laser stereometry light scattering measurements on biological material. *Medical & Biological Engineering & Computing*, 18:250–252, 1980.

[4] J.-C. Diels and J. J. Fontaine. Imaging with short optical pulses. *Optics and Lasers in Engineering*, 4:145–162, 1983.

[5] J.-C. Diels, J. J. Fontaine, and W. Rudolph. Ultrafast diagnostics. *Revue Phys. Appl.*, 22:1605–1611, 1987.

[6] J. J. Fontaine, J.-C. Diels, C. Y. Wang, and H. Sallaba. Subpicosecond time domain reflectometry. *Optics Lett.*, 6:495–498, 1981.

[7] J. G. Fujimoto, S. De Silvestri, E. P. Ippen, C. A. Puliafito, R. Margolis, and A. Oscroff. Femtosecond optical ranging in biological systems. *Opt. Lett.*, 11:150–152, 1986.

[8] R. C. Younquist, S. Carr, and D. E. Davies. Optical coherence-domain reflectometry: A new optical evaluation technique. *Optics Lett.*, 12:158–160, 1987.

[9] D. Huang, E. A. Swanson, C. P. Lin, J. S. Schuman, W. G. Stinson, W. Chang, H. R. Hee, T. Flotte, K. Gregory, C. A. Puliafito, and J. G. Fujimoto. Optical coherence tomography. *Science*, 254:1178–1181, 1991.

[10] E. A. Swanson, J. A. Izatt, M. R. Hee, D. Huang, C. P. Liu, J. S. Schuman, G. A. Puliafito, and J. G. Fujimoto. In vivo retinal imaging by optical coherence tomography. *Optics Lett.*, 18:1864–1866, 1993.

[11] B. E. Bouma and G. J. Tearney, Eds. *Handbook of Optical Coherence Tomography*. Marcell Dekker, New York, 2001.

[12] J. S. Schuman, C. Puliafito, and J. G. Fujimoto, Eds. *Optical Coherence Tomography of Ocular Diseases*. Slack, 2004.

[13] C. Yan and J.-C. Diels. 3D imaging with ultrashort pulses. *Appl. Opt.*, 34:6869–6873, 1991.

[14] S. Ertefai and A. E. Profio. Spectral transmittance and contrast in breast diaphanography. *Med. Phys.*, 12:393–400, 1985.

[15] K. M. Yoo, Q. Xing, and R. R. Alfano. Imaging objects hidden in highly scattering media using femtosecond second-harmonic-generation cross-correlation time gating. *Optics Letters*, 16:1019–1021, 1991.

[16] M. D. Duncan, R. Mahon, L. L. Tankersley, and J. Reintjes. Time-gated imaging through scattering media using stimulated Raman amplification. *Optics Lett.*, 16:1868–1870, 1991.

[17] M. Bashkansky, C. L. Adler, and J. Reintjes. Coherently amplified Raman polarization gate for imaging through scattering media. *Optics Lett.*, 19:350–352, 1994.

[18] J. Izatt, M. R. Hee, and G. M. Owen. Optical coherence microscopy in scattering media. *Optics Lett.*, 19:590–592, 1994.

[19] M. Kempe and W. Rudolph. Microscopy with ultrashort light pulses. *Nonlinear Optics*, 7:129–151, 1994.

[20] A. Ishimaru. *Wave Propagation and Scattering in Random Media.* Academic Press, New York, 1978.

[21] J. A. Moon, R. Mahon, M. D. Duncan, and J. Reintjes. Resolution limits for imaging through turbid media with diffuse light. *Optics Lett.*, 18:1591–1593, 1993.

[22] J. C. Hebden and D. T. Delpy. Enhanced time resolved imaging with a diffusion model of photon transport. *Optics Lett.*, 19:311–313, 1994.

[23] *Proceedings of SPIE and OSA on optical biomedical imaging*, annual meetings.

[24] M. Kempe and W. Rudolph. Topical review: Trends in optical biomedical imaging. *J. Mod. Optics*, 44:1617–1642, 1997.

[25] N. Abramson. Light-in-flight recording: High speed holographic motion pictures of ultrafast phenomena. *Appl. Opt.*, 22:215–232, 1983.

[26] N. Abramson. Light-in-flight recording 2: Compensation for the limited speed of the light used for observation. *Appl. Opt.*, 23:1481–1492, 1984.

[27] H. J. Gerritsen. *Proc. Soc. Photo-Opt. Instrum. Eng.*, 519:250, 1984.

[28] H. Chen, Y. Chen, D. Dilworth, E. Leith, J. Lopez, and J. Valdmanis. Two-dimensional imaging through diffusing media using 150-fs gated electronic holography techniques. *Opt. Lett.*, 16:487–489, 1991.

[29] E. Leith, C. Chen, Y. Chen, D. Dilworth, J. Lopez, J. Rudd, P. C. Sun, J. Valdmanis, and G. Vossler. Imaging through scattering media with holography. *J. Opt. Soc. Am. A*, 9:1148–1153, 1992.

[30] T. Wilson and C. J. R. Sheppard. *Theory and Practice of Scanning Optical Microscopy.* Academic Press, London, 1984.

[31] W. Denk, J. M. Strickler, and W. W. Webb. Two-photon laser scanning fluorescence microscopy. *Science*, 248:73–76, 1990.

[32] J. Squier. Ultrafast optics: Opening new windows in biology. *Optics and Photonics News*, April:42–45, 2002.

[33] J. R. Lakowicz. Fluorescence lifetime sensing. *Laser Focus World*, 28-5:60–80, 1992.

[34] T. Sawatari. Optical heterodyne scanning microscope. *Applied Optics*, 12:2768–2772, 1973.

[35] M. Kempe and W. Rudolph. Scanning microscopy through thick layers based on linear correlation. *Optics Lett.*, 19:1919–1921, 1994.

[36] M. Magnor, P. Dorn, and W. Rudolph. Simulation of confocal microscopy though scattering media with and without time gating. *J. Opt. Soc. A*, 18:1695–1700, 2001.

[37] J. M. Schmitt, M. Yadloswky, and R. F. Bonner. Subsurface imaging of living skin with optical coherence microscopy. *Dermatology*, 191:93–98, 1995.

[38] M. Kempe and W. Rudolph. Analysis of heterodyne and confocal microscopy for illumination with broad-bandwidth light. *J. Mod. Optics*, 43:2189–2204, 1997.

[39] V. E. Zakharov and A. B. Shabat. Exact theory of two-dimensional self-focusing and one-dimensional self-modulation of waves in nonlinear media. *Sov. Phys. JETP*, 34:62–69, 1972.

[40] X. M. Zhao, J.-C. Diels, A. Braun, X. Liu, D. Du, G. Korn, G. Mourou, and J. Elizondo. Use of self-trapped filaments in air to trigger lightning. In P. F. Barbara, W. H. Knox, G. A. Mourou, and A. H. Zewail, Eds., *Ultrafast Phenomena IX*, Springer-Verlag, Berlin, Germany, 1994, pp. 233–235.

[41] A. Braun, C. Y. Chien, S. Coe, and G. Mourou. Long range, high resolution laser radar. *Opt. Comm.*, 105:63–66, 1993.

[42] A. Braun, G. Korn, X. Liu, D. Du, J. Squier, and G. Mourou. Self-channeling of high-peak-power fs laser pulses in air. *Opt. Lett.*, 20:73–75, 1994.

[43] L. F. Mollenauer. Soliton transmission speeds greatly multiplied by sliding-frequency guiding filters. *Optics and Photonics News*, 1994, pp. 15–19.

[44] J. P. Gordon and H. A. Haus. Random walk of coherently amplified solitons in optical fiber. *Opt. Lett.*, 11:665–667, 1986.

[45] R. Y. Chiao, E. Garmire, and C. H. Townes. Self-trapping of optical beams. *Physics Review Letter*, 13:479–482, 1964.

[46] S. A. Akhmanov, A. P. Sukhorukov, and R. V. Khokhlov. Self focusing and self trapping of intense light beams in a nonlinear medium. *Sov. Phys JETP*, 23:1025–1033, 1966.

[47] M. Giglio and A. Vendramini. Soret-type motion of macromolecules in solution. *Phys. Rev. Lett.*, 38:26–29, 1977.

[48] E. Freysz, M. Afifi, A. Ducasse, B. Pouligny, and J. R. Lalanne. Critical microemulsions as optically nonlinear media. *Journal of the Optical Society of America B*, 1:433, 1984.

[49] Y. R. Shen and M. M. Loy. Theoretical interpretation of small-scale filaments of light originating from moving focal spots. *Physical Review A*, 3:2099–2105, 1971.

[50] G. K. L. Wong and R. Y. Shen. Transient self-focusing in a nematic liquid crystal in the isotropic phase. *Physical Review Letter*, 32:527–530, 1974.

[51] J. Schwarz, P. K. Rambo, J.-C. Diels, M. Kolesik, E. Wright, and J. V. Moloney. UV filamentation in air. *Optics Comm.*, 180:383–390, 2000.

[52] A. Braun, G. Korn, X. Liu, D. Du, J. Squier, and G. Mourou. Self-channeling of high-peak-power femtosecond laser pulses in air. *Optics Lett.*, 20:73–75, 1995.

[53] G. Mechain, A. Couairon, Y. B. Andre, C. D'Amico, M. Franco, B. Prade, S. Tzortzakis, Q. Mysyrowicz, and R. Sauerbrey. Long-range self-channeling of infrared laser pulses in air: A new propagation regime without ionization. *Applied Physics B*, 79:379–382, 2004.

[54] L. Bergé, S. Skupin, C. Méjean, J. Kasparian, J. Yu, S. Frey, E. Salmon, and J. P. Wolf. Supercontinuum emission and enhanced self-guiding of infrared femtosecond filaments sustained by third harmonic generation in air. *Phys. Rev. E*, 71:016602-1–016602-13, 2005.

[55] N. Aközbek, M. Scalora, C. M. Bowden, and S. L. Chin. Third-harmonic generation and self-channeling in air using high-power femtosecond laser pulses. *Phys. Rev. Lett.*, 89:143901, 2002.

[56] H. Yang, J. Zhang, L. Z. Zhao, Y. J. Li, H. Teng, Y. T. Li, Z. Wang, Z. Chen, Z. Wei, J. Ma, W. Yu, and Z. M. Sheng. Third-order harmonic generation by self-guided femtosecond pulses in air. *Phys. Rev. E*, 015401-1–015401-4(R), 2003.

[57] A. Proulx, A. Talebpour, S. Petit, and S. L. Chin. Fast pulsed electric field created from the self-generated filament of a femtosecond Ti:sapphire laser pulse in air. *Optics Comm.*, 174:305–309, 2000.

[58] C.-C. Cheng, E. M. Wright, and J. V. Moloney. Generation of electromagnetic pulses from plasma channels induced by femtosecond light strings. *Phys. Rev. Lett.*, 87:213001-1-213001-4, 2001.

[59] J. E. Rothenberg. Pulse splitting during self-focusing in normally dispersive media. *Optics Lett.*, 17:583–586, 1992.

[60] A. C. Bernstein, T. S. Luk, T. R. Nelson, A. McPherson, J.-C. Diels, and S. M. Cameron. Asymmetric ultra-short pulse splitting measured in air using FROG. *Applied Physics B (Lasers and Optics)*, B75(1):119–122, July 2002.

[61] A. C. Bernstein, T. R. Nelson, T. S. Luk, A. McPherson, S. M. Cameron, and J.-C. Diels. Time-resolved measurements of self-focusing pulses in air. *Optics Lett.*, 28:2354–2355, 2003.

[62] P. Rairoux, H. Schillinger, S. Niedermeier, M. Rodriguez, F. Ronneberger, R. Sauerbrey, B. Stein, D. Waite, C. Wedeking, H. Wille, L. Woeste, and C. Ziener. Remote sensing of the atmosphere using ultrashort laser pulses. *Appl. Phys. B*, 71:573–580, 2000.

[63] L. Woeste, S. Wedeking, J. Wille, P. Rairouis, B. Stein, S. Nikolov, C. Werner, S. Niedermeier, F. Ronneberger, H. Schillinger, and R. Sauerbrey. Femtosecond atmospheric lamp. *Laser und Optoelektronic*, 29:51–53, 1997.

[64] A. L. Gaeta. Catastrophic collapse of ultrashort pulses. *Phys. Rev. Lett.*, 84:3582–3585, 2000.

[65] N. Aközbek, M. Scalora, C. M. Bowden, and S. L. Chin. White-light continuum generation and filamentation during the propagation of ultra-short laser pulses in air. *Optics Comm.*, 191:353–362, 2001.

[66] J. Kasparian, R. Sauerbrey, D. Mondelain, S. Niedermeier, J. Yu, J. P. Wolf, Y.-B. André, M. Franco, B. Prade, S. Tzortzakis, A. Mysyrowicz, M. Rodriguez, H. Wille, and L. Wöste. Infrared extension of the supercontinuum generated by femtosecond terawatt laser pulses propagating in the atmosphere. *Opt. Lett.*, 25:1397–1399, 2000.

[67] J. Kasparian, M. Rodriguez, J. Méjean, J. Yu, E. Salmon, H. Wille, R. Bourayou, S. Frey, Y.-B. André, A. Mysyrowicz, R. Sauerbrey, J. P. Wolf, and L. Wöste. White-light filaments for atmospheric analysis. *Science*, 301:61–64, 2003.

[68] J. Yu, D. Mondelain, G. Ange, R. Volk, S. Niedermeier, J. P. Wolf, J. Kasparian, and R. Sauerbrey. Backward supercontinuum emission from a filament generated by ultrashort laser pulses in air. *Optics Lett.*, 26:533–535, 2001.

[69] Q. Luo, W. Liu, and S. L. Chin. Lasing action in air induced by ultra-fast laser filamentation. *Appl. Phys. B*, 76:337–340, 2003.

[70] J.-C. Diels, R. Bernstein, K. Stahlkopf, and X. M. Zhao. Lightning control with lasers. *Scientific American*, 277:50–55, 1997.

[71] P. K. Rambo, J. Schwarz, and J.-C. Diels. High voltage electrical discharges induced by an ultrashort pulse UV laser system. *Journal of Optics A*, 3:146–158, 2001.

[72] M. Rodriguez, R. Sauerbrey, H. Wille, L. Woeste, T. Fujii, Y. B. Andre, A. Mysyrowicz, L. Klingbeil, K. Rethmeier, W. Kalkner, et al. Triggering and guiding megavolt discharges by use of laser-induced ionized filaments. *Optics Lett.*, 27:772–774, 2002.

[73] P. K. Rambo, J. Schwarz, and J.-C. Diels. Interferometry with 2-dimensional spatial and high temporal resolution. *Opt. Comm*, 197:145–159, 2001.

[74] H. Schillinger and R. Sauerbrey. Electrical conductivity of long plasma channels in air generated by self-guided femtosecond laser pulses. *Applied Physics B*, 68:753–756, 1999.

[75] S. Tzortzakis, M. A. Franco, Y. B. André, A. Chiron, B. Lamouroux, B. S. Prade, and A. Mysyrowicz. Formation of a conducting channel in air by self-guided femtosecond laser pulses. *Phys. Rev. E*, 60:3505–3507, 1999.

[76] A. Couairon and L. Bergé. Light filaments in air for ultraviolet and infrared wavelengths. *Phys. Rev. Lett*, 88:13503-1–13503-4, 2002.

[77] L. V. Keldysh. Ionization in the field of a strong electromagnetic wave. *Soviet Physics JET*, 20(5):1307–1314, 1965.

[78] S. Diddams, A. Van Engen, and T. S. Clement. Measurement of air dispersion by white light interferometry. In *CLEO*, Optical Society of America, Baltimore, MD, 1999.

[79] J.-C. Diels, J. Jones, and L. Arissian. Applications to sensors of extreme sensitivity. In J. Ye and S. Cundiff, Eds., *Femtosecond Optical Frequency Comb: Principle, Operation and Applications*, Springer, New York, NY, 2005, pp. 333–354.

[80] C. P. Hauri, W. Kornelis, F. W. Helbing, A. Heinrich, A. Couairon, A. Mysirowicz, J. Biegert, and U. Keller. Generation of intense, carrier-envelope phase-locked few-cycle laser pulses through filamentation. *Applied Physics B*, 79:673–677, 2004.

[81] C. P. Hauri, A. Guandalini, P. Eckle, W. Kornelis, J. Biegert, and U. Keller. Generation of intense few-cycle laser pulses through filamentation parameter dependence. *Optics Express*, 13:7541, 2005.

[82] A. Couairon, J. Biegert, C. P. Hauri, U. Keller, and A. Mysyrowicz. Self-generation of near-single-cycle pulses through filamentation. *Journal of Modern Optics*, 53:75–85, 2005.

[83] S. Diddams, B. Atherton, and J.-C. Diels. Frequency locking and unlocking in a femtosecond ring laser with the application to intracavity phase measurements. *Applied Physics B*, 63:473–480, 1996.

[84] M. J. Bohn, J.-C. Diels, and R. K. Jain. Measuring intracavity phase changes using double pulses in a linear cavity. *Optics Lett.*, 22:642–644, 1997.

[85] X. Meng, R. Quintero, and J.-C. Diels. Intracavity pumped optical parametric bidirectional ring laser as a differential interferometer. *Opt. Comm*, 233:167–172, 2004.

[86] E. J. Post. Sagnac effect. *Rev. Mod. Phys.*, 39:475–493, 1967.

[87] E. O. Schulz-Dubois. Alternative interpretation of rotation rate sensing by ring laser. *IEEE J. Quant. Electr.*, QE-2:299–305, 1966.

[88] J.-C. Diels and I. C. McMichael. Degenerate four-wave mixing of femtosecond pulses in an absorbing dye jet. *J. Opt. Soc. Am. B*, 3:535–543, 1986.

[89] X. Meng, J.-C. Diels, D. Kuehlke, R. Batchko, and R. Byer. Bidirectional, synchronously pumped, ring optical parametric oscillator. *Opt. Letters*, 26:265–267, 2001.

[90] M. Lai, J.-C. Diels, and M. Dennis. Nonreciprocal measurements in fs ring lasers. *Optics Letters*, 17:1535–1537, 1992.

[91] M. L. Dennis, J.-C. Diels, and M. Lai. The femtosecond ring dye laser: A potential new laser gyro. *Optics Letters*, 16:529–531, 1991.

[92] A. Siegman. *Lasers*. University Science Books, Mill Valley, CA, 1986.

[93] R. Quintero-Torres, M. Ackerman, M. Navarro, and J.-C. Diels. Scatterometer using a bidirectional ring laser. *Opt. Comm.*, 240:179–183, 2004.

[94] J. Reichert, R. Holzwarth, T. Udem, and T. W. Hänsch. Measuring the frequency of light with mode-locked lasers. *Optics Comm.*, 172:59–68, 1999.

[95] J. K. Ranka, R. S. Windeler, and A. J. Stentz. Visible continuum generation in air-silica microstructure optical fibers with anomalous dispersion at 800 nm. *Optics Lett.*, 25:25–27, 2000.

[96] T. A. Birks, W. J. Wadsworth, and P. S. Russell. Supercontinuum generation in tapered fibers. *Optics Lett.*, 25:1415–1417, 2000.

[97] T. Okuno, M. Nishi, and M. Nishimura. Silica-based functional fibers with enhanced nonlinearity and their applications. *IEEE J. Sel. Top. Quantum Electron.*, 5:1385–1391, 1999.

[98] N. Nishizawa and T. Goto. Widely broadened super continuum generation using highly nonlinear dispersion shifted fibers and femtosecond fiber laser. *Jpn. J. Appl. Phys.*, 40:L365–L367, 2001.

[99] J. W. Nicholson, M. F. Yan, P. Wisk, J. Fleming, F. DiMarcello, E. Monberg, A. Yablon, C. Jorgensen, and T. Veng. All-fiber, octave-spanning supercontinuum. *Optics Lett.*, 28:643–645, 2003.

[100] S. Coen, A. H. L. Chau, R. Leonhardt, J. D. Harvey, J. C. Knight, W. J. Wadsworth, and P. St. J. Russell. Supercontinuum generation by stimulated Raman scattering and parametric four-wave mixing in photonic crystal fibers. *J. Opt.Soc. Am. B*, 19:753–764, 2002.

[101] B. R. Washburn, S. E. Ralph, and R. S. Windeler. Ultrashort pulse propagation in air-silica microstructure fiber. *Optics Express*, 10:575–580, 2002.

[102] M. Bellini and T. W. Hänsch. Phase-locked white-light continuum pulses: Towards a universal optical frequency-comb synthesizer. *Optics Lett.*, 25:1049–1051, 2000.

[103] J. W. Nicholson and M. F. Yan. Cross-coherence measurements of supercontinua generated in highly-nonlinear, dispersion shifted fiber at 1550 nm. *Optics Express*, 12:679–688, 2004.

[104] L. Tartara, I. Cristiani, and V. Degiorgio. Blue light and infrared continuum generation by soliton fission in a microstructured fiber. *Applied Physics B: Lasers and Optics*, 77:307–311, 2003.

[105] A. I. Ferguson and R. A. Taylor. Active mode stabilization of a synchronously pumped mode-locked dye laser. *Optics Comm.*, 41:271–276, 1982.

[106] E. Krüger. Frequency stabilization and control of a mode-locked dye laser. *Rev. Sci. Instrum.*, 66:4806–4812, 1995.

[107] T. Udem, J. Reichert, R. Holzwarth, and T. W. Hänsch. Absolute optical frequency measurement of the cesium D_1 line with a mode-locked laser. *Phys. Rev. Lett.*, 82:3568–3571, 1999.

[108] J. Reichert, R. Holzwarth, T. Udem, and T. W. Hänsch. Measuring the frequency of light with mode-locked lasers. *Optics Comm.*, 172:59–68, 1999.

[109] T. Udem, J. Reichert, R. Holzwarth, and T. W. Hänsch. Accurate measurement of large optical frequency differences with a mode-locked laser. *Opt. Lett.*, 24:881–883, 1999.

[110] J. Reichert, M. Neiring, R. Holzwarth, M. Weitz, T. Udem, and T. W. Hänsch. Phase coherent vacuum-ultraviolet to radio frequency comparison with a mode-locked laser. *Phys. Rev. Lett.*, 84:3232–3235, 2000.

[111] H. Schnatz, J. Stenger, B. Lipphardt, N. Haverkamp, and C.-O. Weiss. Optical frequency measurement using frequency multiplication and frequency combs. In J. Ye and S. Cundiff, Eds., *Femtosecond Optical Frequency Comb: Principle, Operation and Applications*, Springer, New York, NY, 2005, pp. 198–224.

[112] S. Diddams, J. Ye, and L. Hollberg. Femtosecond lasers for optical clocks and low noise frequency synthesis. In J. Ye and S. Cundiff, Eds., *Femtosecond Optical Frequency Comb: Principle, Operation and Applications*, Springer, New York, NY, 2005, pp. 225–262.

[113] R. J. Jones, J.-C. Diels, J. Jasapara, and W. Rudolph. Stabilization of the frequency, phase, and repetition rate of an ultra-short pulse train to a Fabry-Perot reference cavity. *Optics Comm.*, 175:409–418, 2000.

[114] K. H. Drever, J. L. Hall, F. V. Kowalski, J. Hough, G. M. Ford, A. J. Munley, and H. W. Ward. Laser phase and frequency stabilization using an optical resonator. *Appl. Phys. B*, 31:97–105, 1983.

[115] J. Ye and S. Cundiff. *Femtosecond Optical Frequency Comb: Principle, Operation and Applications*. Springer, New York, NY, 2005.

Appendix A

The Uncertainty Principle

This demonstration of the uncertainty principle is an example of application of various Fourier transform properties. We will give a simple derivation of the uncertainty relation (1.57) that we wrote between the conjugated variables time t and frequency Ω. This uncertainty relation and its derivation also apply in all generality between any pair of conjugated variables, for instance between the transverse beam dimension x and the corresponding wave vector k. To derive the uncertainty relation (1.57), we will use a family of functions defined by the relation:

$$g(t) = tf(t) + \mu \frac{d}{dt} f(t) \tag{A.1}$$

where t and μ are real variables. The total "energy" associated with that distribution is proportional to:

$$\int_{-\infty}^{\infty} |g(t)|^2 \, dt = \int t^2 \, |f|^2 \, dt + \mu \int \left[tf \frac{df^*}{dt} + tf^* \frac{df}{dt} \right] dt + \mu^2 \int \left| \frac{df}{dt} \right|^2 dt$$
$$\geq 0. \tag{A.2}$$

While the first term of the inequality defines the second order moment $\langle t^2 \rangle$,

$$\int t^2 |f(t)|^2 dt = \langle t^2 \rangle \int |f(t)|^2 dt. \tag{A.3}$$

we can apply Parseval's theorem to the last term to obtain:

$$\int_{-\infty}^{\infty} \left|\frac{df}{dt}\right|^2 dt = \frac{1}{2\pi} \int \left|\mathcal{F}\left(\frac{df}{dt}\right)\right|^2 d\Omega$$
$$= \frac{1}{2\pi} \int \Omega^2 |f(\Omega)|^2 d\Omega = \langle\Omega^2\rangle \int |f(t)|^2\, dt. \qquad \text{(A.4)}$$

If the function $|f|$ has finite boundaries so that $\lim_{t\to\pm\infty}\left(t|f|^2\right) = 0$, as is true for the modulus of the electric field of a laser pulse for example, one can write:

$$\int_{-\infty}^{\infty} \frac{d}{dt}\left[tf(t)f^*(t)\right] dt = \int_{-\infty}^{\infty} |f(t)|^2\, dt + \int_{-\infty}^{\infty} tf^*(t)\frac{df}{dt} dt$$
$$+ \int_{-\infty}^{\infty} tf(t)\frac{df^*}{dt} dt = 0. \qquad \text{(A.5)}$$

Substituting the terms of Eqs. (A.4), (A.3), and (A.5) into the inequality (A.2) leads, after division by $\int_{-\infty}^{\infty} |f(t)|^2\, dt$, to:

$$\langle\Omega^2\rangle\mu^2 - \mu + \langle t^2\rangle \geq 0. \qquad \text{(A.6)}$$

The left-hand side of the inequality is a quadratic polynomial in μ. Because $\langle\Omega^2\rangle > 0$, the polynomial is nonnegative if the ordinate of the vertex of the parabola is ≥ 0. This is the case if

$$1 - 4\langle t^2\rangle\langle\Omega^2\rangle \leq 0, \qquad \text{(A.7)}$$

which is equivalent to the uncertainty relation (1.57).

Appendix B

Phase Shifts on Transmission and Reflection

B.1. THE SYMMETRICAL INTERFACE

Let us consider first the simple situation sketched in Figure B.1. The interface can be a mirror with a reflecting coating on the front face and an antireflection coating on the back face. We are only interested in fields propagating *outside* the mirror. The energy conservation relation between the reflected (field reflection coefficient $\tilde{r}$) and transmitted (field transmission coefficient $\tilde{t}$) waves implies:

$$|\tilde{r}|^2 + |\tilde{t}|^2 = 1, \tag{B.1}$$

where we assumed a unity field amplitude.

Another relation can be found by adding another incident field of amplitude 1 (beam 2 in the figure), and taking advantage of the symmetry. Summing the intensities:

$$|\tilde{r} + \tilde{t}|^2 + |\tilde{r} + \tilde{t}|^2 = 2. \tag{B.2}$$

Combination of Eqs. (B.1) and (B.2) leads to

$$2[\tilde{r}\tilde{t}^* + \tilde{r}^*\tilde{t}] = 0, \tag{B.3}$$

which implies that the phase shifts on transmission and reflection are complementary:

$$\varphi_r - \varphi_t = \frac{\pi}{2}. \tag{B.4}$$

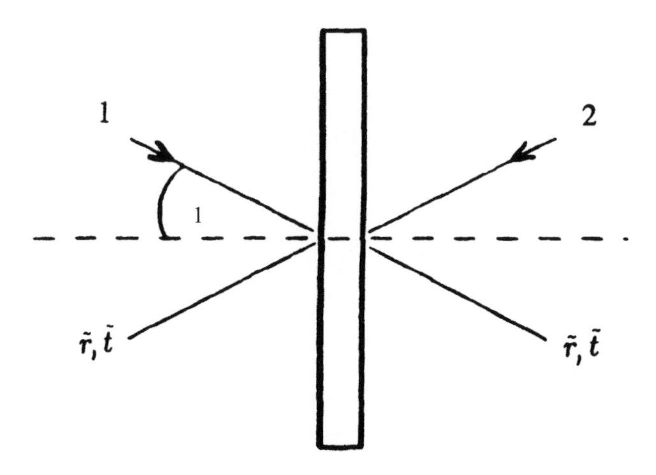

Figure B.1 Reflection and transmission by a plane mirror between two identical media.

It is because of the latter phase relation that the antiresonant ring reflects back all the incident radiation and has zero losses if $|\tilde{r}|^2 = |\tilde{t}|^2 = 0.5$.

B.2. COATED INTERFACE BETWEEN TWO DIFFERENT DIELECTRICS

Let us consider—as in Figure B.2—a partially reflecting coating at an interface between air (index 1) and a medium of index n. A light beam of amplitude $\mathcal{E}_1 = 1/\sqrt{\cos\,\theta_1}$ is incident from the air, at an angle of incidence θ_1. The transmitted beam is refracted at the angle θ_2 and has an amplitude $\tilde{t}_1/\sqrt{\cos\,\theta_1}$. The reflected beam has an amplitude $\tilde{r}_1/\sqrt{\cos\,\theta_1}$. Energy conservation leads to the relation:

$$|\tilde{r}_1|^2 + |\tilde{t}_1|^2 \frac{n\cos\theta_2}{\cos\theta_1} = 1, \tag{B.5}$$

where we took into account the change in beam cross section on refraction. We have a similar energy conservation equation for a beam of amplitude $\mathcal{E}_2 = 1/\sqrt{n\cos\theta_2}$ incident at an angle θ_2 on the dielectric–air interface:

$$|\tilde{r}_2|^2 + |\tilde{t}_2|^2 \frac{\cos\theta_1}{n\cos\theta_2} = 1. \tag{B.6}$$

The amplitude of the reflection coefficient is equal on both sides of the interface. For the phase, the only sign relation consistent with energy conservation in a

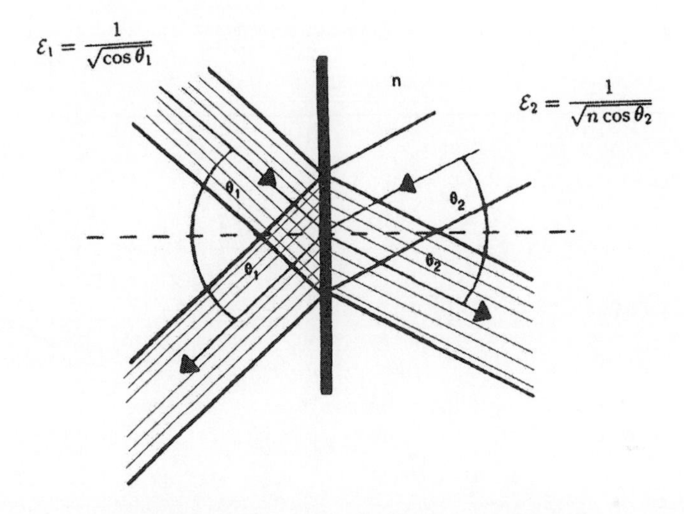

Figure B.2 Reflection and transmission at an interface.

Gires–Tournois interferometer and with the known phase shift on pure dielectric interfaces, is:

$$\tilde{r}_1 = -\tilde{r}_2^*, \tag{B.7}$$

or, $r_1 = r_2$, with the relation between phase angles:

$$\varphi_{r,1} = -\varphi_{r,2} - \pi. \tag{B.8}$$

To find a relation between the phase shift on transmission and reflection, we consider the energy conservation for light incident from the upper half of the figure (the axis of symmetry being the dashed normal to the interface):

$$1 + 1 = \cos\theta_1 \left| \frac{\tilde{r}_1}{\sqrt{\cos\theta_1}} + \frac{\tilde{t}_2}{\sqrt{n\cos\theta_2}} \right|^2 + n\cos\theta_2 \left| \frac{\tilde{r}_2}{\sqrt{n\cos\theta_2}} + \frac{\tilde{t}_1}{\sqrt{\cos\theta_1}} \right|^2. \tag{B.9}$$

Taking into account the energy conservation relations (B.5) and (B.6) leads to the following trigonometric relations between phase shifts on transmission and reflection:

$$\frac{\cos(\varphi_{r,1} - \varphi_{t,2})}{\cos(\varphi_{r,2} - \varphi_{t,1})} = -1, \tag{B.10}$$

which leads to the relation between phase angles:

$$\varphi_{t,2} - \varphi_{r,2} = \varphi_{r,2} - \varphi_{t,1} + (2n+1)\pi. \tag{B.11}$$

A direct consequence of this phase relation is:

$$\tilde{t}_1\tilde{t}_2 - \tilde{r}_1\tilde{r}_2 = 1. \tag{B.12}$$

Appendix C

Slowly Evolving Wave Approximation

The derivation here essentially follows Brabec and Krausz [1]. We start with the wave equation in the frequency domain for a scalar electric field propagating in the z direction, including a nonlinear polarization and the diffraction term

$$\left[\frac{\partial^2}{\partial z^2} + \nabla_\perp^2 + \tilde{k}^2(\Omega)\right] E(x, y, z, \Omega) = \mu_0 \mathcal{F}\left\{\frac{\partial^2}{\partial t^2} P^{NL}(t, x, y, z)\right\} \tag{C.1}$$

where $\nabla_\perp^2 = \frac{\partial^2}{\partial x^2} + \frac{\partial^2}{\partial x^2}$. Note that the effect of the linear polarization is included in $\tilde{k}^2(\Omega) = \Omega^2 \epsilon(\Omega)\mu_0 = \Omega^2 n^2/c^2$. We start with the following ansatz for the electric field and the nonlinear polarization:

$$E(\Omega) = \frac{1}{2}\tilde{\mathcal{E}}(\Omega - \omega_\ell, x, y, z)e^{-ik_\ell z} + c.c. \tag{C.2}$$

$$P(t) = \frac{1}{2}\tilde{\mathcal{P}}(t, x, y, z)e^{i(w_\ell t - k_\ell z)} + c.c. \tag{C.3}$$

and neglect processes leading to backscattering, that is, coupling of opposite propagation directions. This ansatz inserted in Eq. (C.1) yields

$$\left[\left(\frac{\partial}{\partial z} - ik_\ell\right)^2 + \nabla_\perp^2 + \tilde{k}^2(\Omega)\right]\tilde{\mathcal{E}}(\Omega - \omega_\ell) = \mu_0 \mathcal{F}\left\{e^{i\omega_\ell t}\left(\frac{\partial}{\partial t} - i\omega_\ell\right)^2 \tilde{\mathcal{P}}(t)\right\}. \tag{C.4}$$

We now expand the complex $\tilde{k}(\Omega) = k + i\alpha/2$ into a Taylor series about ω_ℓ

$$\tilde{k}(\Omega) = k_\ell + i\frac{\alpha_0}{2} + k_1(\Omega - \omega_\ell) + \tilde{D}(\Omega), \tag{C.5}$$

where

$$\tilde{D}(\Omega) = i\frac{\alpha_1}{2}(\Omega - \omega_\ell) + \sum_{m=2}^{\infty} \frac{k_m + i\alpha_m/2}{m!}(\Omega - \omega_\ell)^m \tag{C.6}$$

with

$$k_m = \left. \frac{\partial^m}{\partial \Omega^m} \mathrm{Re}(\tilde{k}) \right|_{\omega_\ell} \tag{C.7}$$

$$\alpha_m = \left. \frac{\partial^m}{\partial \Omega^m} \mathrm{Im}(\tilde{k}) \right|_{\omega_\ell} . \tag{C.8}$$

The quantities α_m are related to linear loss coefficients for the field intensity. The next step is to insert Eq. (C.6) into Eq. (C.4) and inverse Fourier transform the resulting expression into the time domain. For this step we use the fact that an expression of the kind $(\Omega - \omega_\ell)^m A(\Omega - \omega_\ell)$ transforms into $\left(-i\frac{\partial}{\partial t}\right)^m A(t)$ where $A(t) = \mathcal{F}^{-1}\{A(\Omega - \omega_\ell)\}$. The resulting wave equation now reads

$$\left\{ \left(\frac{\partial}{\partial z} - ik_\ell \right)^2 + \nabla_\perp^2 + \left[k_\ell - ik_1 \frac{\partial}{\partial t} + i\frac{\alpha_0}{2} + \hat{D}(t) \right]^2 \right\} \tilde{\mathcal{E}}(t)$$
$$= \mu_0 \left(\frac{\partial}{\partial t} - i\omega_\ell \right)^2 \tilde{\mathcal{P}}(t), \tag{C.9}$$

where

$$\hat{D}(t) = \frac{\alpha_1}{2}\frac{\partial}{\partial t} + \sum_{m=2}^{\infty} \frac{k_m + i\alpha_m/2}{m!} \left(-i\frac{\partial}{\partial t} \right)^m . \tag{C.10}$$

The terms in Eq. (C.9) can be regrouped, factoring out the operator $(1 - \frac{i}{\omega_\ell}\frac{\partial}{\partial t})$. To this end, the third term in Eq. (C.9) can be written as:

$$
\begin{aligned}
&\left[k_\ell - ik_1\frac{\partial}{\partial t} + i\frac{\alpha_0}{2} + \hat{D}(t)\right]^2 \\
&= k_\ell^2 - 2ik_\ell k_1\frac{\partial}{\partial t} + 2ik_\ell\left(1 - i\frac{k_1}{k_\ell}\frac{\partial}{\partial t}\right)\left(\frac{\alpha_0}{2} - i\hat{D}\right) \\
&\quad + i\alpha_0\hat{D} - k_1^2\frac{\partial}{\partial t} - \frac{\alpha_0^2}{4} + \hat{D}^2 \\
&= k_\ell^2 - 2ik_\ell k_1\frac{\partial}{\partial t} + 2ik_\ell\left(1 - \frac{i}{\omega_\ell}\frac{\partial}{\partial t}\right)\left(\frac{\alpha_0}{2} - i\hat{D}\right) \\
&\quad + 2ik_\ell\left(\frac{i}{\omega_\ell}\right)\left(1 - \frac{c}{n v_g}\right)\frac{\partial}{\partial t}\left(\frac{\alpha_0}{2} - i\hat{D}\right) + i\alpha_0\hat{D} - k_1^2\frac{\partial}{\partial t} - \frac{\alpha_0^2}{4} + \hat{D}^2.
\end{aligned}
\tag{C.11}
$$

We use a retarded frame of reference, i.e., the transformation $\xi = z$ and $\eta = t - z/v_g$ as in Chapter 1, Eqs. (1.86) and (1.87). We note that the first squared expression in Eq. (C.9) is:

$$
\begin{aligned}
\left(\partial\xi - \frac{1}{v_g}\partial\eta - ik_\ell\right)^2 &= -2ik_\ell\left(1 - \frac{i}{k_\ell v_g}\partial_\eta\right)\partial_\xi - k_\ell^2 + 2ik_\ell\frac{1}{v_g}\partial_\eta \\
&= -2ik_\ell\left(1 - \frac{i}{\omega_\ell}\partial_\eta\right)\partial_\xi - k_\ell^2 + 2ik_\ell\frac{1}{v_g}\partial_\eta \\
&\quad + 2\left(\frac{n}{c} - \frac{1}{v_g}\right)\partial_\xi\partial_\eta.
\end{aligned}
\tag{C.12}
$$

Substituting Eq. (C.11) and Eq. (C.12) into Eq. (C.10) yields:

$$
\begin{aligned}
&\left(1-\frac{i}{\omega_\ell}\frac{\partial}{\partial\eta}\right)\left[\left(\frac{\partial}{\partial\xi}-\frac{\alpha_0}{2}+i\hat{D}\right)\tilde{\mathcal{E}}+i\frac{\omega_\ell c\mu_0}{2n_0}\left(1-\frac{i}{\omega_\ell}\frac{\partial}{\partial\eta}\right)\tilde{\mathcal{P}}\right]-\frac{1}{2ik_\ell}\nabla_\perp^2\tilde{\mathcal{E}} \\
&=\left(1-\frac{c}{n v_g}\right)\frac{i}{\omega_\ell}\frac{\partial}{\partial\eta}\left(\frac{\partial}{\partial\xi}-\frac{\alpha_0}{2}+i\hat{D}\right)\tilde{\mathcal{E}}-\frac{1}{2ik_\ell}\left(\frac{\partial^2}{\partial\xi^2}+\hat{D}^2-\frac{\alpha_0^2}{4}+i\alpha_0\hat{D}\right)\tilde{\mathcal{E}}.
\end{aligned}
\tag{C.13}
$$

So far the propagation equation is still exact, and no approximations have been made. The terms on the right-hand side are small compared to those on the left-hand side and can be neglected if Eqs. (3.94)–(3.96) are satisfied. In this case we obtain Eq. (3.97). One should exercise caution in the application of Eq. (3.97) up to an arbitrary order in the expansion of $\hat{D}$. Depending on the particular pulse duration, the higher order terms in the expansion of $\hat{D}$ on the right-hand side of Eq. (C.13) may not be negligible as compared to the highest-order term of the left-hand side of that expression.

Note, in the main text we frequently used t, z as local coordinates in a frame moving with the group velocity.

BIBLIOGRAPHY

[1] T. Brabec and F. Krausz. Nonlinear optical pulse propagation in the single-cycle regime. *Phys. Rev. Lett.*, 78:3282–3284, 1997.

Appendix D

Four-Photon Coherent Interaction

The procedure to calculate the coherent four-photon resonant coherent interaction is detailed in Mukherjee *et al.* [1] and Mukherjee [2]. The density matrix of the multilevel atomic system is reduced to a two-by-two matrix associated with the resonant levels. The temporal evolution of the density matrix elements associated with all the off-resonant levels is calculated by an adiabatic expansion to fourth order [1,2]. We refer to the literature for the "straightforward but tedious" derivation of the time evolution of the diagonal (ρ_{44}, ρ_{00}) and off-diagonal [$\rho_{04} = \varrho_{04} \exp(4i\omega_\ell t)$] elements. We will instead concentrate on the physical meaning of the various terms entering these equations. In analogy with the two-photon coherent interaction equations, we introduce an amplitude function for the four-photon excitation of the atom $\tilde{Q}_4 = 2i\varrho_{04}$. In addition to the fundamental frequency ω_ℓ, fields at the third harmonic ($\omega_3 = 3\omega_\ell$) and the fifth harmonic ($\omega_5 = 5\omega_\ell$) frequency are also generated.

One of the main driving terms of the electronic excitation ρ_{04} oscillating at $4\omega_\ell$ is a term proportional to the fourth power of the applied field. The time evolution of the amplitude $\tilde{Q}_4$ of the harmonic electronic excitation of the atom is given by the system of interaction equations:

$$\left\{\frac{\partial}{\partial t} + i(\omega_{04} - 4\omega_\ell + \delta\omega_2) + \frac{\gamma_0 + \gamma_4}{2} + \frac{1}{T_2}\right\} \tilde{Q}_4 = (\rho_{44} - \rho_{00}) \left\{ \tilde{V}_1^4 + (\xi_{40} + a_3 |\tilde{V}_1|^2) \tilde{V}_1 \tilde{V}_3 + (\zeta_{40}^* + c_5^* |\tilde{V}_1|^2) \tilde{V}_1^* \tilde{V}_5 + d_5^* \tilde{V}_1^2 \tilde{V}_5 \tilde{V}_3^* \right\} \tag{D.1}$$

$$\frac{\partial \rho_{44}}{\partial t} + \gamma_4 \rho_{44} + \frac{\rho_{44}}{T_1}$$

$$= \mathrm{Re}\left\{ Q_4^* [\tilde{V}_1^4 + (\xi_{40} + a_3 |\tilde{V}_1|^2)\tilde{V}_1 \tilde{V}_3 + (\zeta_{40}^* + c_5^* |\tilde{V}_1|^2)\tilde{V}_1^* \tilde{V}_5 \right.$$

$$\left. + d_5^* \tilde{V}_1^2 \tilde{V}_5 \tilde{V}_3^*] \right\} \tag{D.2}$$

$$\frac{\partial(\rho_{44} + \rho_{00})}{\partial t} = -\gamma_4 \rho_{44} - \gamma_0 \rho_{00}, \tag{D.3}$$

where ξ_{40} is the ratio of the sum frequency $(\omega_\ell + \omega_3)$ to four-photon excitation coefficients, a_3 being the intensity-dependent part of this ratio; ζ_{40} and b_3 are the ratio—and the intensity-dependent part thereoff—of the difference frequency $(\omega_5 - \omega_\ell)$ to four-photon excitation coefficients; d_5 is the ratio of the excitation process via coupling of frequencies $(2\omega_\ell + \omega_5 - \omega_3)$ to the process $(4\omega_\ell)$.

In the equations (D.1) through (D.3), the fields $\tilde{V}_i$ are in units of the Rabi cycle for the four-photon transition:

$$\tilde{V}_i^4 = \frac{2r_{04}}{(2\hbar)^4}\tilde{\mathcal{E}}_i^4 = \frac{p_{0f}p_{fk}p_{kj}p_{j4}}{(\omega_{j1} - 3\omega_\ell)(\omega_{k0} - 2\omega_\ell)(\omega_{f0} - \omega_\ell)} \frac{\tilde{\mathcal{E}}_i^4}{8\hbar^4}, \tag{D.4}$$

where $\tilde{\mathcal{E}}_i$ is the amplitude of the field at $\omega_{i\ell} = i\omega_\ell$, and the summation convention over the repeated indices is assumed. Equation D.4 defines a complex parameter r_{04}. All the fields are expressed in frequency units through the conversion factor $(r_{04})^{1/4}/\hbar$.

As in the case of the two-photon resonant two-level system, the detuning is time-dependent through the Stark shift $\delta\omega_2$:

$$\delta\omega_2 = [\alpha_0' - \alpha_4'] |\tilde{V}_1|^2 \tag{D.5}$$

where the "prime" indicates the real part of the *normalized* susceptibility α_i given by:

$$\alpha_i = \frac{1}{\sqrt{r_{04}}} \sum_k \left[\frac{|p_{ki}|^2}{(\omega_{ki} - \omega_\ell)} + \frac{|p_{ki}|^2}{(\omega_{ki} + \omega_\ell)} \right]. \tag{D.6}$$

The coherence is expressed by the function $\tilde{Q}_4 = 2i\rho_{04}e^{-4\omega_\ell t}$, which will be the main driving term of Maxwell's wave equations. For instance, for the fundamental:

$$\frac{\partial \tilde{V}_1}{\partial z} = \frac{\mu\omega_\ell cN}{2}\frac{2\sqrt{2r_{04}}}{\hbar}[\tilde{V}_1^{*3}\tilde{Q}_4$$
$$+(\xi_{40}+a_3|\tilde{V}_1|^2)\tilde{Q}_4\tilde{V}_3^* + (\zeta_{40}^* + c_5^*|\tilde{V}_1|^2)\tilde{Q}_4^*\tilde{V}_5 + 2d_5\tilde{V}_1^*\tilde{V}_3\tilde{V}_5^*\tilde{Q}_4$$
$$+\frac{3}{4}\frac{\alpha_3}{r_{04}}\tilde{V}_3\tilde{V}_1^{*2}\rho_{11}] + \frac{\mu\omega_\ell cN}{2}[\alpha_0(\omega_\ell)\rho_{00} + \alpha_4(\omega_\ell)\rho_{44}]\tilde{V}_1, \tag{D.7}$$

where α_0 and α_4 are the linear susceptibilities of levels 0 and 4, respectively.

In Eq. (D.7), α_3 is the (complex) polarizability, responsible for *nonresonant* third harmonic generation:

$$\alpha_3 = \frac{p_{0f}p_{fk}p_{kj}p_{j0}}{(\omega_{f0}-\omega_\ell)(\omega_{k0}-2\omega_\ell)(\omega_{j0}-3\omega_\ell)}. \tag{D.8}$$

As source terms opposing or enhancing the fundamental field we recognize in Eq. (D.7) all the combinations of amplitude terms for which the corresponding phase factor is at the frequency ω_ℓ: $\tilde{Q}_4\tilde{V}_1^{*3}$ $(4\omega_\ell - 3\omega_\ell)$, $\tilde{Q}_4\tilde{V}_3^*$ $(4\omega_\ell - \omega_{3\ell})$, $\tilde{Q}_4\tilde{V}_1^*\tilde{V}_3\tilde{V}_5^*$ $(4\omega_\ell + \omega_{3\ell} - \omega_{5\ell} - \omega_\ell)$, and $\tilde{V}_3\tilde{V}_1^{*2}$ $(\omega_{3\ell} - 2\omega_\ell)$. The latter term, in contrast to the formers, does not involve the resonant four-photon process, but is a second-order Raman term associated with nonresonant third harmonic generation. The nonresonant third harmonic generation term can interfere constructively or destructively with the resonant generation process, as can be seen in Maxwell's wave equations for the field at $\omega_{3\ell}$:

$$\frac{\partial \tilde{V}_3}{\partial z} = \frac{\mu\omega_\ell cN}{2}\frac{2\sqrt{2r_{04}}}{\hbar}[(\xi_{40}+a_3|\tilde{V}_1|^2)\tilde{Q}_4\tilde{V}_1$$
$$+2d_5^*\tilde{V}_1^2\tilde{V}_5\tilde{Q}_4$$
$$+\frac{3}{4}\frac{\alpha_3}{r_{04}}\tilde{V}_1^3\rho_{11}] + \frac{\mu\omega_\ell cN}{2}[\alpha_0(\omega_{3\ell})\rho_{00} + \alpha_4(\omega_{3\ell})\rho_{44}]\tilde{V}_3. \tag{D.9}$$

The objective of this display of equations is not to confuse the reader, but point to interesting transient phenomena and intriguing possibilities of applications for femtosecond pulses.

BIBLIOGRAPHY

[1] N. Mukherjee, A. Mukherjee, and J.-C. Diels. Four-photon coherent resonant propagation and transient wave mixing: Application to the mercury atom. *Phys. Rev. A*, A38:1990–2004, 1988.

[2] N. Mukherjee. *Coherent resonant interaction and harmonic generation in atomic vapors.* Ph.D. thesis, University of North Texas, Denton, 1987.

Appendix E

Kerr Lensing in a Cavity

We will consider first the sequence of a nonlinear lens and an aperture and discuss the transmission of such an element based on simple ray optics. A more accurate description based on Gaussian beam propagation was given in Section 5.4.3. We will then apply Gaussian optics to study the effect of a nonlinear lens in a specific cavity. This is an example of the procedure outlined in Section 5.5.3.

E.1. ELEMENTARY KERR LENSING MODEL

The main ideas of an intensity dependent transmission based on a Kerr lens and subsequent aperture can be understood with simple arguments based on paraxial ray optics. We are assuming in the following uniform beam profiles to analyze the transmission through an aperture. Because a flat top beam does not produce self-lensing the nonlinear lens must be introduced somewhat artificially. To illustrate the effect of Kerr lensing, we consider here only the simplest approximation of paraxial geometric optics, applied to a nonlinear lens located in a collimated beam of radius r_0, as in Figure E.1. The transmission of the beam of radius r through the aperture or radius r_a is simply:

$$T=\left(\frac{r_a}{r_0}\cdot\frac{r_0}{r}\right)^2=T_0\left(\frac{r_0}{r}\right)^2=T_0\left(\frac{1}{1-\frac{L}{f_{\text{nl}}}}\right)^2\approx T_0\left[\frac{1}{1+CI_0(t)}\right]^2, \qquad \text{(E.1)}$$

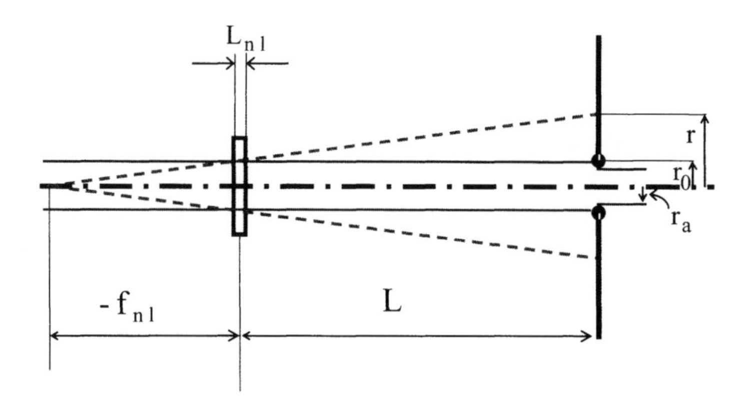

Figure E.1 A simplified power limiter, including a nonlinear lens acting on a collimated beam of radius r_0, and an aperture of radius r_a. At high power, the nonlinear lens acquires a focal length of f_{nl}. The example shown refers to a negative nonlinearity leading to defocusing.

where $T_0 = (r_a/r_0)^2$ is the transmission without the nonlinear lens, $r = r_0 - Lr_0/f_{\mathrm{nl}}$, and C is a positive constant (for defocusing). Clearly the transmission is maximum for low intensities and negligible self-lensing. If $I_0(t)$ represents a pulse envelope the transmission decreases in the center, which may lead to pulse broadening and limits the overall energy transmission.

E.2. EXAMPLE OF A NONLINEAR CAVITY AND GAUSSIAN BEAM ANALYSIS

We proceed next to a numerical example of nonlinear lensing in a cavity using the Gaussian beam analysis of Section 5.5.3. Figure E.2 shows a ring cavity and the equivalent unit cell of the unfolded (linear) cavity. The cavity contains two identical focusing elements, a nonlinear lensing element and an aperture. We choose the distance between L_1 and L_2, $2d_1 = 52.632$ mm, and $f = 25$ mm, and the wavelength of the radiation $\lambda = 1$ μm. The length of the cavity segment between L_1 and L_2 that contains the aperture has a length L that we write as

$$L = \ell - 2\frac{d_1 f}{d_1 - f}. \tag{E.2}$$

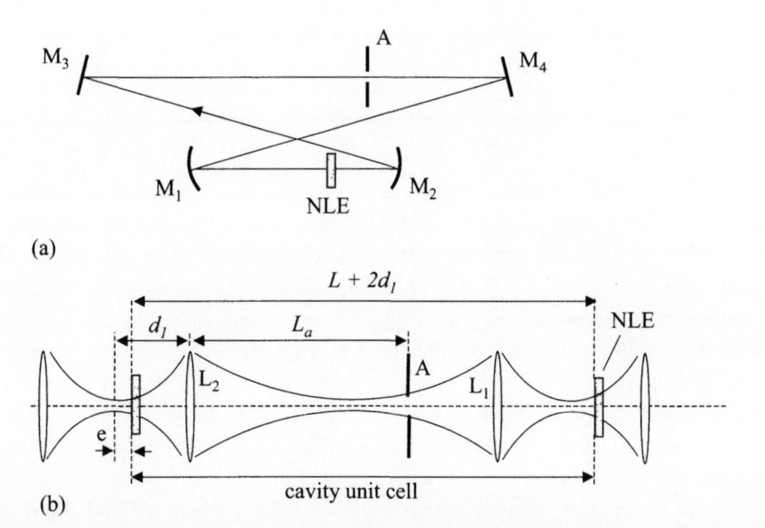

Figure E.2 (a) Ring cavity with a nonlinear lensing element NLE, two identical focusing mirrors (M_1 and M_2) with focal length f, two flat mirrors (M_3 and M_4), and an aperture A. (b) Unit cell of the equivalent unfolded (linear) cavity. For symmetry reasons two beam waists are formed, on either side, halfway in between lenses (mirrors) L_1 and L_2, that is, a distance d_1 and $L/2$ from lens L_1.

We first determine the system matrix of this unit cell, $\mathcal{M}_1$, starting from the position of the NLE:

$$\mathcal{M}_1 = \begin{pmatrix} A_1 & B_1 \\ C_1 & D_1 \end{pmatrix} \tag{E.3}$$

where the elements of the matrix are:

$$A_1 = 1 + \frac{2d_1}{f} + \frac{2d_1 e}{f(d-f)} + \frac{\ell}{f} - \frac{(d_1+e)(2f+\ell)}{f^2} = 2.324$$

$$B_1 = \left\{ \left[\frac{2d_1 f}{d_1 - f} \right] - \ell \right\} \left[\left(1 - \frac{d_1}{f} \right)^2 - \frac{e^2}{f^2} \right] + 2d_1 - 2\frac{d_1^2 - e^2}{f} = -1.58\ \text{mm}$$

$$C_1 = -\frac{2}{f} + \frac{2d_1}{f(d_1-f)} - \frac{\ell}{f^2} = 1.44\ \text{mm}^{-1}$$

$$D_1 = 1 + \frac{2e+\ell}{f} - \frac{2d_1 e}{f(d_1-f)} - \frac{(d_1-e)\ell}{f^2} = -0.545. \tag{E.4}$$

The numerical values correspond to $e = 1$ mm and $\ell = 50$ mm.

The stability criterium of the cavity takes the simple form:

$$\left|\frac{A_1+D_1}{2}\right| = 1 - \frac{(d_1-f)\ell}{f^2} < 1. \tag{E.5}$$

The limit $\ell = 0$ marks a stability limit of the cavity, corresponding to a concentric type cavity with a beam waist $w_0 \to 0$. The parameter d_1 determines the length of the cavity [perimeter equal to $2d_1^2/(d_1-f)$ at the position of the NLE]. To obtain a stable cavity we choose a cavity perimeter shorter by the amount ℓ (50 mm in our numerical example).

The eigenvalue of the system matrix is $\tilde{s}_1 = (0.909–0.2895\text{i})\ \text{mm}^{-1}$, to which corresponds a spot size of $w_1 = 33\ \mu\text{m}$, and a radius of curvature of $R_1 = 1.1$ mm at the position of the NLE, as given by the Eq. (5.118). We note that the nonlinear crystal (NLE) is outside the Rayleigh range of the beam waist, because propagation by -1 mm shows a beam waist of about 7 μm.

The matrix for translating from the crystal to the aperture located at a distance L_a from L_2 is:

$$\begin{pmatrix} A_m & B_m \\ C_m & D_m \end{pmatrix} = \begin{pmatrix} 1 & L_a \\ 0 & 1 \end{pmatrix} \begin{pmatrix} 1 & 0 \\ -\frac{1}{f} & 1 \end{pmatrix} \begin{pmatrix} 1 & d_1-e \\ 0 & 1 \end{pmatrix}$$

$$= \begin{pmatrix} 1-\frac{L_a}{f} & (d_1-e)-\frac{L_a}{f}(d_1-f-e) \\ -\frac{1}{f} & -\frac{d_1-f-e}{f} \end{pmatrix} \tag{E.6}$$

If we choose as distance to the aperture from L_2 a length $L_a = 300$ mm, we find for the matrix:

$$\begin{pmatrix} A_m & B_m \\ C_m & D_m \end{pmatrix} = \begin{pmatrix} -11 & 21.52\ \text{mm} \\ -0.04\ \text{mm}^{-1} & -0.01264 \end{pmatrix} \tag{E.7}$$

The complex beam parameter at the location of the aperture, in the absence of nonlinear lensing, is thus:

$$\tilde{s}_m = \frac{C_m + D_m\tilde{s}_1}{A_m + B_m\tilde{s}_1} = (-0.0336 - 0.00258)\ \text{mm}^{-1}, \tag{E.8}$$

which corresponds to a beam parameter of $w_m = 350\ \mu\text{m}$. Let us first look at the change in beam size induced by the nonlinear lensing at the location of the NLE, as given by Eq. (5.122). We find that

$$\delta\tilde{s} = \frac{1}{f_{\text{nl}}}(0.4521 + 0.6458i). \tag{E.9}$$

We assume nonlinear lensing that produces a lens of focal length $f_{nl} = 500$ mm, which gives for the change in the complex s parameter $\delta s = 0.0009 + 0.00129i$. From that, the relative change in beam waist $\delta w_1/w_1 \sim 0.5 \times 0.00129/0.289$ is about 0.2%.

Application of Eq. (5.127) yields the change in complex beam parameter at the location of the aperture:

$$\delta\tilde{s}_m = \frac{1}{f_{nl}}\left[\frac{(0.4521 + 0.6458i)}{(A_m + B_m\tilde{s}_1)^2}\right] = \frac{1}{f_{nl}}[-0.0042 - 0.0056i]. \tag{E.10}$$

The relative change in beam waist at the aperture, for $f_{\text{nl}} = 500$, is $\delta w_m/w_m = 0.5 \times 0.0056/(500 \times 0.00258)$ and is about 0.2%.

The location of the aperture should be away from a beam waist. If we chose for instance $L_a = [2d_1 f/(d_1 - f) - \ell]/2 = 475$ mm, which brings us close to the second beam waist of the cavity, we find $\tilde{s}_m = 0.001 - 0.16i$, corresponding to a beam waist of 45 μm, and $\delta\tilde{s}_m = 0.017(1 + i)/f_{\text{nl}}$. The relative beam waist change $\delta w_m/(w_m)$ is only 0.01%.

Appendix F

Abbreviations for Dyes

Abbreviation	**Full name**
RhB	Rhodamine B or Rhodamine 610
Rh 110	Rhodamine 110 or Rhodamine 560
Rh6G	Rhodamine 6G or Rhodamine 590
DODCI	3,3'-diethyloxadicarbocyanine iodide
DQOCI	1,3'-diethyl-4,2'-quinolyoxacarbocyanine iodide
SRh101	Sulforhodamine 101 or sulforhodamine 640
DOTCI	3,3'-diethyloxatricarbocyanine iodide
DCCI	or DCI, 1,1'-diethyl-2,4'-carbocyanine iodide
DQTCI	1,3'-diethyl-4,2'-quinolthiacarbocyanine iodide
DDI	= DDCI 1,1'-diethyl-2,2'-dicarbocyanine iodide
HITCI	1,1',3,3,3',3'-hexamethylindotricarbocyanine iodide
DaQTeC	2-(p-dimethylaminostyryl)-benzothiazolylethyl iodide

List of Symbols

Frequently used symbols are listed in the table below. The list is neither complete nor absolute: some duplication of notation takes place because of the large amount of symbols required. The duplicate symbols are generally used in different context and different chapters. To simplify notations we will often use $f(x')$ as the Fourier transform of $f(x)$, where x' and x are the conjugate Fourier variables.

Physical parameter or constant	Symbol	Unit
Instantaneous real electric field	$E(t)$	V/m
Instantaneous complex electric field	$\tilde{E}^{+}(t)$ or $\tilde{E}(t)$	V/m
Complex electric field amplitude	$\tilde{\mathcal{E}}(z,t)$	V/m
Electric field envelope (real)	$\mathcal{E}(z,t)$	V/m
Intensity	$I(t)$	W/cm^2
Pulse energy density	$W = \int_{-\infty}^{\infty} I(t)dt$	J/cm^2
Accumulated energy density	$W(t) = \int_{-\infty}^{t} I(t')dt'$	J/cm^2
Pulse energy	$\mathcal{W} = \int_S W dS$	J
Phase of the electric field	$\varphi(z,t)$	radian
Average carrier frequency	ω_ℓ	radian/s
Instantaneous frequency	$\omega(t) = \omega_\ell + \dot{\varphi}$	radian/s
Resonance frequency	ω_0	radian/s
Carrier to envelope offset (CEO)	f_0	s^{-1}
Pulse duration (FWHM of intensity)	τ_p	s
Pulse duration (parameter of Gaussian)	τ_G	s
Detuning	$\Delta\omega = \omega_0 - \omega_\ell$	radian/s
Pulse bandwidth	$\Delta\omega_p$	radian/s
Frequency	Ω	radian/s
Spectral complex electric field	$\tilde{E}(\Omega)$	Vs/m
Spectral complex field amplitude	$\tilde{\mathcal{E}}(\Omega)$	Vs/m
Spectral electric field envelope (real)	$\mathcal{E}(\Omega)$	Vs/m
Spectral phase of the field	$\phi(\Omega)$	radian
k -vector (magnitude)	$k(\Omega)$ or $\tilde{k}(\Omega)$	m^{-1}
k -vector (at ω_ℓ)	k_ℓ	m^{-1}
Permittivity of free space	ϵ_0	F/m
Index of refraction	n or $n(\Omega)$	dimensionless

(Continued)

Physical parameter or constant	Symbol	Unit
Dielectric constant (complex)	$\tilde{\epsilon}$ or $\tilde{\epsilon}(\Omega) = \epsilon_0 n^2$	F/m
Dielectric constant, Real part	ϵ_r	F/m
Dielectric constant, Imaginary part	ϵ_i	F/m
Linear susceptibility	$\chi^{(1)}$ or χ	dimensionless
Linear polarization	$P = \epsilon_0 \chi^{(1)} E$	Cm
Nonlinear index $n_2 \mathcal{E}^2$	n_2	cm^2/V^2
Nonlinear index $\bar{n}_2 I$	$\bar{n}_2$	cm^2/W
Phase velocity	$v_p = \Omega/k(\Omega)$	m/s
Group velocity	$v_g = d\Omega/dk$	m/s
Parameter of group velocity dispersion	$k''_\ell = d^2k/d\Omega^2\|_{\omega_\ell}$	s^2/m
Dispersion length	$L_D = \tau_{p0}^2/\|k''_\ell\|$	m
Rayleigh range	ρ_0	m
Phase relaxation time	T_2	s
Energy relaxation time	T_1	s

Index

Printed and bound by CPI Group (UK) Ltd, Croydon, CR0 4YY

10/05/2025

01866533-0001